Дорогому Валерию Николаевичу
с глубоким уважением
от авторов,

Z. Shabarova, A. Bogdanov

Advanced Organic Chemistry of Nucleic Acids

Distribution:

VCH, P.O. Box 10 1161, D-69451 Weinheim (Federal Republic of Germany)

Switzerland: VCH, P.O. Box, CH-4020 Basel (Switzerland)

United Kingdom and Ireland: VCH (UK) Ltd., 8 Wellington Court,
Cambridge CB1 1HZ (England)

USA and Canada: VCH, 220 East 23rd Street, New York, NY 10010–4606 (USA)

Japan: VCH, Eikow Building, 10-9 Hongo 1-chome, Bunkyo-ku, Tokyo 113 (Japan)

ISBN 3-527-29021-4 (VCH, Weinheim)

Z. Shabarova, A. Bogdanov

Advanced Organic Chemistry of Nucleic Acids

Weinheim · New York
Basel · Cambridge · Tokyo

Zoe A. Shabarova
Moscow State University
Korpus M, apt. 139
Moscow 117234
Russia

Alexey A. Bogdanov
Moscow State University
Korpus K, apt. 102
Moscow 117234
Russia

Published jointly by
VCH Verlagsgesellschaft mbH, Weinheim (Federal Republic of Germany)
VCH Publishers, Inc., New York, NY (USA)

Translator: Vladimir Vopyan
Editorial Director: Dr. Thomas Kellersohn
Production Manager: Dipl.-Ing. (FH) Hans Jörg Maier

Library of Congress Card No. applied for.

British Library Cataloguing-in-Publication Data: A catalogue record for this book is available from the British Library.

Deutsche Bibliothek Cataloguing-in-Publication Data:
Šabarova, Zoe A.:
Advanced organic chemistry of nucleic acids / Z. Shabarova ; A. Bogdanov. [Transl.: Vladimir Vopyan]. - Weinheim ; New York ; Basel ; Cambridge ; Tokyo : VCH, 1994
ISBN 3-527-29021-4
NE: Bogdanov, Aleksej A. :

Printed on acid-free and chlorine-free paper.

Composition: Fa. Hagedornsatz GmbH, D-68519 Viernheim. Printing: betz-druck gmbh, D-64291 Darmstadt. Bookbinding: IVB, D-64646 Heppenheim. Cover design: Graphik & Text Studio Zettlmeier-Kammerer, D-93164 Laaber-Waldetzenberg.
Printed in the Federal Republic of Germany.

Dedicated to

Professor Mikhail A. Prokofiev,

the Teacher

Preface

In the early sixties, we started teaching an advanced course of chemistry of nucleic acids to Moscow University chemistry majors already with a solid organic and physical chemistry background. To teach this particular subject was most exciting at the time when virtually every year was marked by stunning discoveries in the field of nucleic acids. We still derive a great deal of pleasure and satisfaction from teaching the course.

Our main difficulty throughout the years has been the absence of a suitable textbook on nucleic acid chemistry. Of course, publication of Michelson's, Khorana's and other scientists' books, especially the monograph written by Kochetkov and coworkers, did help, but these books had been written with professionals, rather than students, in mind. Therefore, putting doubts and wavering aside, we decided to write a textbook ourselves. In making this decision we were wholeheartedly encouraged and supported by Professor M. A. Prokofiev, founder of nucleic acid chemistry in our country, to whom this book is dedicated.

The Russian edition of the textbook was published in 1978 by "Khimiya" (Moscow) under the title "The Chemistry of Nucleic Acids and Their Components". To the best of our knowledge, for many years it had been the only (and still is one of the few) textbook on the subject worldwide. However, access to the book has been limited outside our country. This is why we accepted with enthusiasm VCH's proposal to prepare the English edition.

Naturally, the English edition of a textbook to be published almost 15 years after the Russian predecessor could by no means be a mere translation of the latter – too many important events have taken place in DNA and RNA science not to leave an imprint on the chemistry of nucleic acids. Our greatest efforts have been spent in revising the chapters concerned with determination of the primary structure of nucleic acids, their synthesis, macromolecular structure, and chemical modification. Many chapters, including all of Chapter 10 dealing with ribozymes, have entirely been written anew. However, at the same time (much to our surprise), many fundamentals of nucleic acid chemistry are still as valid as 15 years ago, which is yet another proof of the great strides made in the field already in the seventies.

We are convinced that the chemistry of nucleic acids has always formed and still provides the basis for elaboration of methods as well as key concepts of molecular and cell biology. It should be remembered that, having emerged from organic and physical chemistry, the chemistry of nucleic acids is now exerting major influence on the two fields of knowledge. This gives us every hope that this book will be of use to all those who wish to become versed in both sciences.

Chapters 1 through 6, most of Chapter 9, and Chapter 11 have been written by Z. A. Shabarova, whereas A. A. Bogdanov is author of Chapters 7, 8, part of Chapter 9, and Chapter 10.

We should like to express our gratitude to Professor D. Soll (of Yale University) for giving us the impetus to embark on the English edition. We are thankful to our colleagues and co-workers from the Department of Chemistry of Natural Compounds, Moscow State University, for their assistance and support. In particular, we gratefully thank D. Chernov, I. Kozlov and N. Naryshkin for preparation of all formulas and some figures. Special thanks should be expressed to N. Naryshkin and I. Kozlov for their help with correction of the manuscript and proof-reading. And we also thank V. G. Vopian, our translator, for his cooperation and long patience.

Z. Shabarova
A. Bogdanov

Contents

1 Structure of Nucleosides

1.1 Introduction

The term "nucleoside" was introduced in 1909 by Levene and Jacobs to denote carbohydrate derivatives of the purine bases isolated from yeast nucleic acid hydrolysates. Later, the term was expanded to additionally cover compounds containing pyrimidines and other heterocyclic bases. Today, it is applied to a broad class of naturally occurring and synthetic compounds that are essentially *N*- and *C*-glycosides – derivatives of various carbohydrates and heterocyclic compounds.

Nucleic acids contain nucleosides of two types: derivatives of D-ribose, known as ribonucleosides, and those of 2-deoxy-D-ribose, known as deoxyribonucleosides or, sometimes, deoxynucleosides.

$HOCH_2$ B O HO OH

ribonucleoside

$HOCH_2$ B O HO H

deoxyribonucleoside

Such classification of nucleosides stems from the nature of their constituent sugar. According to the structure of the heterocyclic base (B), which is the other component of nucleosides, the latter are also divided into pyrimidine and purine bases (see Tables 1-1 through 1-4)[1].

[1] If X and Y = OH, then the base exists in the oxo form.

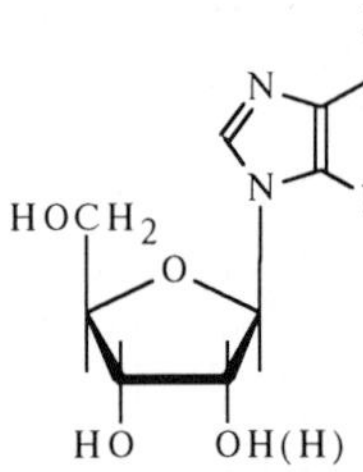

purine nucleoside

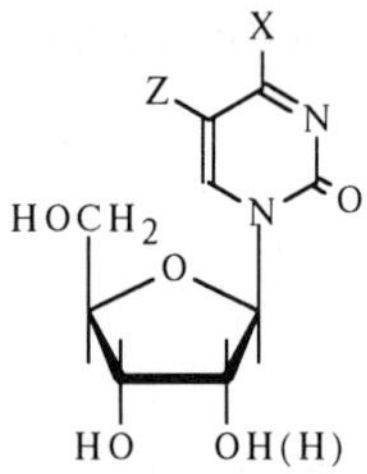

pyrimidine nucleoside

X = OH or NH_2; Y = H,OH or NH_2; Z = H or CH_3

Nucleosides can be isolated from nucleic acids (NA) after chemical or enzymatic hydrolysis. Ribonucleic acids (RNA) are hydrolysed to nucleosides when boiled with aqueous pyridine, with a diluted ammonia solution or in an ammonium formiate buffer at pH 4.0. Nucleotides are yielded as intermediates:

$$\text{RNA} \rightarrow \text{Ribonucleotides} \rightarrow \text{Ribonucleosides}$$

Deoxynucleosides are produced by the enzymatic hydrolysis of deoxyribonucleic acids (chemical hydrolysis being accompanied by some side processes). To this end, use is normally made of snake venom containing the enzymes phosphodi- and phosphomonoesterase.

Some nucleosides occur in a free state and can be isolated by direct extraction.

The isolation and identification of nucleosides usually involve ion-exchange, thin-layer and gas–liquid chromatography as well as UV spectroscopy and mass spectrometry.

Acid hydrolysis of the nucleosides present in nucleic acids yields a heterocyclic (pyrimidine or purine) base and pentose. Nucleosides do not react at the aldehyde group, the implication being that the linkage between the sugar and bases is via a glycosidic bond. To establish the structure of each nucleoside present in a nucleic acid one must determine: (1) the structure of the constituent base; (2) the structure of the constituent sugar; (3) the type of the bond linking the two (the site at which the sugar is attached to the base); (4) the size of the oxide ring in the carbohydrate moiety; and (5) the configuration of the glycosyl (anomeric) moiety.

1.2 Pyrimidine and Purine Bases

Pyrimidines and purines, first isolated from hydrolysates of nucleic acids (1874–1900), were identified using classical methods of organic chemistry (see Table 1-1). An important contribution was made by Emil Fischer who must be credited with the earliest synthesis of purines (1897). Today, pyrimidines and purines can be identified rapidly and reliably by chromatographic and spectrophotometric techniques.

1.2.1 Pyrimidines

Pyrimidines, also known as *meta*-diazines, are structurally akin to benzene and pyridine. Pinner (1885), who had noticed this analogy, coined a new term from the words "pyridine" and "amidine", emphasizing thereby that apart from aromaticity pyrimidines also exhibit properties inherent in amidines. Pyrimidines are numbered according to the Chemical Abstracts Service registry.

Given below are the formulas of the pyrimidines whose compounds are nucleic acid constituents, such as uracil, thymine, and cytosine (see Table 1-1).

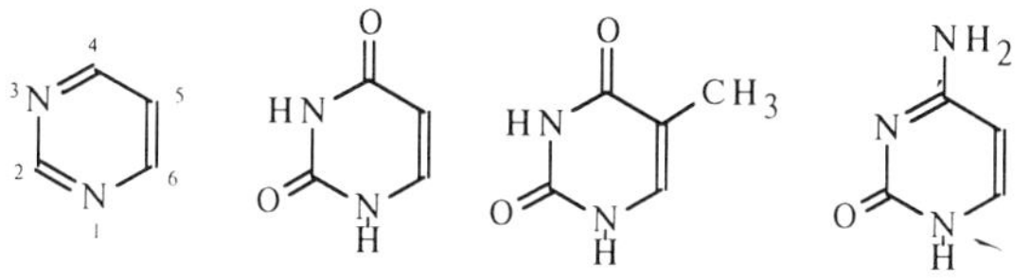

pyrimidine (*m*-diazine) uracil thymine cytosine

Thymines and cytosines are usually present in DNA, while those of uracil and cytosine are found in RNA. The DNAs of some phages are exceptional (see Table 1-4) in that they contain 5-hydroxymethylcytosine or its glycosides associated with the 5-hydroxymethyl group instead of cytosine (even-numbered T phages) and 5-hydroxymethyluracil (phage SP8) or uracil (phage PBS1) instead of thymine.

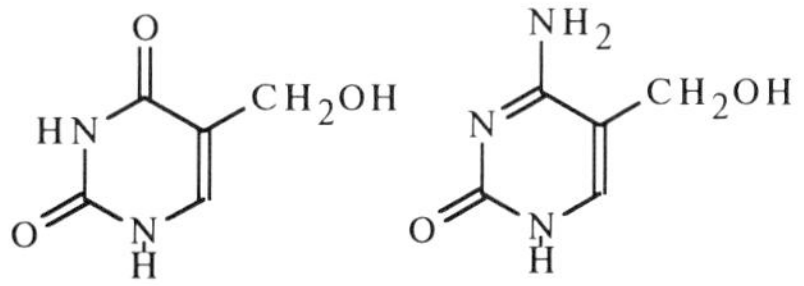

5-hydroxymethyluracil 5-hydroxymethylcytosine

Uracil, thymine, and cytosine are ubiquitously present in nucleic acids in the form of the corresponding nucleosides in significant amounts (each accounting for at least 5 % of the total bases in nucleic acids).

These compounds are usually referred to as the major pyrimidine bases of nucleic acids. In certain DNA and RNA species, some uracil and cytosine derivatives (usually *N*-alkyl ones) have been found. Such pyrimidine bases are referred to as rare or minor. Their structural formulas (as part of the corresponding nucleosides) are given in Tables 1-5 and 1-6. Minor pyrimidine bases do not occur in all nucleic acids, and the content of each base is usually below one or two per cent.

1.2.2 Purines

Purines are heterocyclic systems consisting of a pyrimidine and an imidazole condensed at the 4–5 bond. The term "purine" (from "purum" and "uricum") was introduced in 1898 by Emil Fischer. Purines are also numbered according to the Chemical Abstracts Service registry.

Both DNA and RNA contain two major purine substituents – adenine and guanine:

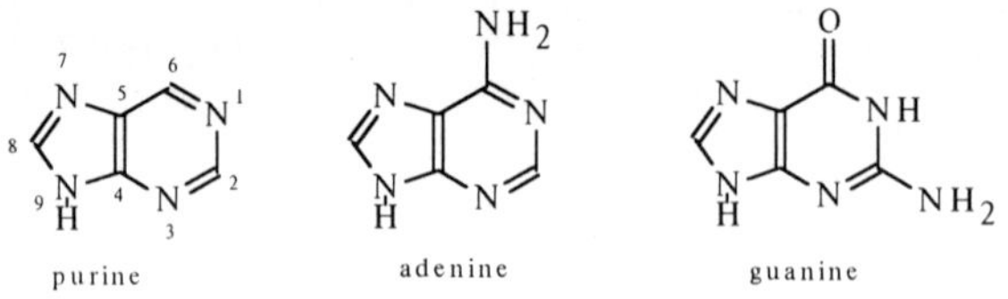

purine adenine guanine

Minor purines differing from adenine or guanine by the presence of alkyl (more commonly, methyl), acyl and other groups have also been isolated from some nucleic acids. The structural formulas of the minor purines present in nucleosides are presented in Tables 1-5 and 1-6.

1.2.3 Nomenclature of Pyrimidines and Purines

In the case of natural pyrimidines and purines, the above-mentioned common names are widely used. As regards synthetic bases and various analogues or modifications of natural pyrimidines and purines, specialists resort to the nomenclature usually applicable to heterocyclic bases, with appropriate numbering of atoms in the pyrimidine or purine ring. For example, thymine is called 2,4-dioxo-5-methyltetrahydropyrimidine or 2,4-dihydroxy-5-methylpyrimidine, whereas guanine is called 2-amino-6-oxodihydropurine or 2-amino-6-hydroxypurine.

Since the plane of symmetry in unsubstituted pyrimidine passes through C^2 and C^5, positions 4 and 6 are equivalent.

1.2.4 Abbreviations

Bases are very often designated by abbreviations (symbols) when writing structural formulas of nucleosides and their derivatives. Table 1-1 lists the names and symbols of the major pyrimidine and purine bases constituting RNA and DNA.

Table 1-1. Names and Abbreviated Symbols of Major Pyrimidines and Purines Constituting Nucleic Acids.

Name	Symbols one-letter	three-letter
Uracil (2,4-dioxotetrahydropyrimidine)	U	Ura
Thymine (2,4-dioxo-5-methyltetrahydropyrimidine)	T	Thy
Cytosine (2-oxo-4-aminodihydropyrimidine)	C	Cyt
Adenine (6-aminopurine)	G	Gua
Guanine (2-amino-6-oxodihydropurine)	A	Ade
Pyrimidine	–	Py*
Purine	–	Pu*

* Use is made of two-letter symbols, although three-letter ones (Pyr, Pur) can also be used.

The symbolic notation of the formulas in this and other chapters follows the recommendations of the International Union of Pure and Applied Chemistry (IUPAC) and the International Union of Biochemistry (IUB) and has been approved by the 3rd All-Union Working Session on Nucleotide Chemistry.

1.3 Carbohydrate Moieties of Nucleosides

The nuclesides isolated from nucleic acids contain only two simple sugars, D-ribose and 2-deoxy-D-ribose, which, as has already been mentioned, determine the nucleic acid type (RNA and DNA).

As far back as 1891, Albrecht Kossel pointed out that hydrolysis of RNA (this name was non-existent at that time) yielded a carbohydrate which was isolated in a crystalline state 18 years later and identified as a yet unknown sugar D-ribose. When oxidized under certain conditions, this simple sugar underwent conversion first into D-ribonic acid and then into optically inactive trihydroxyglutaric acid.

CHO / –OH / –OH / –OH / CH_2OH (D-ribose) → COOH / –OH / –OH / –OH / CH_2OH (D-ribonic acid) → COOH / –OH / –OH / –OH / COOH (trihydroxyglutaric acid)

The structure of D-ribose was confirmed by its synthesis. Synthetic D-ribose turned out to be identical with the crystalline simple sugar isolated from RNA nucleosides.

Serious difficulties had to be overcome in establishing the nature of the sugar contained in the nucleosides that are the building blocks of DNA (i.e. deoxyribonucleosides). It was found that the monosaccharide was more labile than D-ribose. Another finding was that it formed a hydrazone readily soluble in water, but no osazone, which rendered isolation of its hydrolysates difficult. It was long believed that this sugar was a hexose because hydrolysis of nucleosides isolated from DNA yielded levulinic acid. In 1929, however, it became possible to isolate, through mild hydrolysis, a crystalline carbohydrate from the guanine nucleoside of DNA, which turned out to be deoxypentose. Its structure was established beyond any doubt after the synthesis of 2-deoxy-L-ribose. Both preparations exhibited similar porperties as well as identical absolute, but opposite in sign, values of optical rotation. The conclusion that followed was that the sugar found in deoxyribonucleosides was 2-deoxy-D-ribose.

CHO
OH
OH
CH_2OH

2-deoxy-D-ribose

It was also established that mild acid hydrolysis of DNA in the presence of benzyl mercaptan yielded benzyl mercaptal of 2-deoxy-D-ribose, which can be isolated in crystalline form. As was demonstrated in the fifties by chromatographic analysis, the only sugar in DNAs of plant, animal and bacterial origin is 2-deoxy-D-ribose. In 1954, a full description of 2-deoxy-D-ribose isolated from different types of DNA was provided by comparison with its synthetic analogue (melting point of mixed sample, optical rotation, synthesis of derivatives of known structure).

In the late fifties, an unusual carbohydrate was found in some preparations of RNA extracted from plants and animals, which was identified as 2-*O*-methyl-D-ribose.

CHO
OCH_3
OH
OH
CH_2OH

2-*O*-methyl-D-ribose

As was established later, 2-*O*-methyl-D-ribose is present in very small amounts in certain RNAs, which is why this sugar is referred to as a rare or minor component of RNA.

1.4 Bonding Between Carbohydrate Moiety and Heterocyclic Base

1.4.1 Purine Nucleosides

It is logical to assume that adenines and guanines in nucleosides are linked to the carbohydrate moiety in the same manner. This is why bonding at the oxygen atom is impossible in the case of guanine. The ribosides of adenine and guanine readily lend themselves to deamination in the presence of nitrous acid with the linkage between the base and carbohydrate remaining intact.

$\xrightarrow{HNO_2}$ hypoxantine glycoside

$\xrightarrow{HNO_2}$ xantine glycoside

R-D-ribose residue

Consequently, the possibility of bonding at amino groups is also ruled out. Thus, the sugar moiety may be linked to adenines and guanines at C^8 (C-C bond) or at one of the nitrogens (C-N bond) of the pyrimidine (N^1 or N^3) or imidazole (N^7 or N^9) rings. These two types of bonds must differ markedly in stability under conditions of acid hydrolysis, since it is known that *N*-glycosides undergo hydrolysis rather easily, whereas *C*-glycosides are extremely stable. The ease of hydrolysis of purine nucleosides attests to their being *N*- rather than *C*-glycosides. This is also corroborated by the fact that the adenine and guanine nucleosides can be converted into C^8-substituted derivatives (the substituent at the carbon in the imidazole ring is absent). For example, adenine riboside is readily brominated at the imidazole ring carbon:

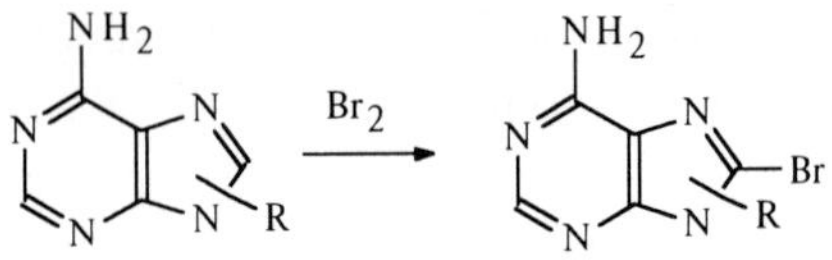

$\xrightarrow{Br_2}$

R-D-ribose residue

Out of the four nitrogens of the purine base those in the pyrimidine ring (N^1 and N^3) are excluded because methylation of the xanthine riboside yielded, as has already been mentioned, by deamination of guanine riboside gives theophylline (1,3-dimethyl xanthine) riboside.

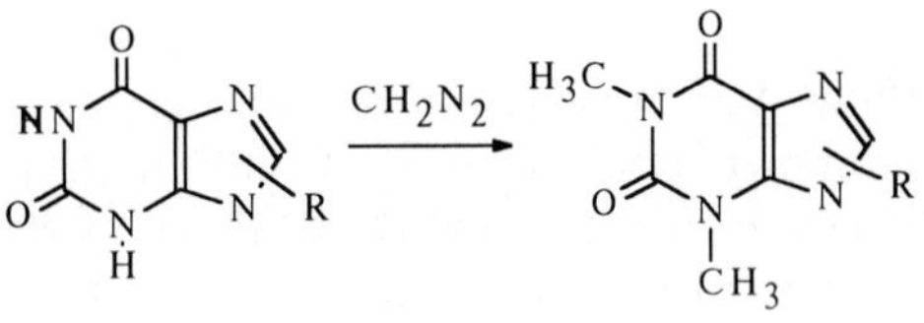

$\xrightarrow{CH_2N_2}$

R-D-ribose residue

The nitrogens at positions 7 and 9 can be regarded as possible binding sites.

The question as regards the site of sugar to base binding was answered once and for all by Gulland in the forties. He was the first to take advantage of the fact that the UV spectra of the purine base containing an alkyl radical and a carbohydrate moiety at the same nitrogen atom are identical and markedly dependent on the position of the substituted nitrogen in the heterocyclic nucleus. Comparison of the UV spectra of adenine riboside with those of 7- and 9-methyladenine showed that the adenine nucleoside has a UV spectrum that is virtually identical with that of 9-methyladenine and bears no resemblance to that of 7-methyladenine.

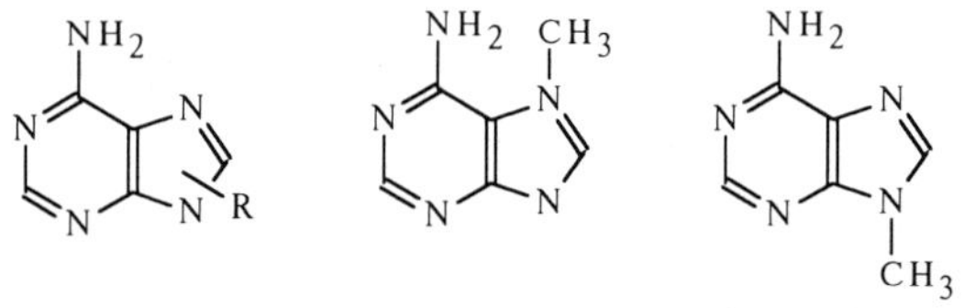

ribosyladenine | 7-methyladenine | 9-methyladenine

Similarly, the UV spectrum of guanine riboside looks like the spectrum of 9-methylguanine and differs from that of 7-methylguanine. These findings clearly indicate that purine nucleosides are 9-ribosylpurines. The structures of adenine and guanine deoxyribosides are very much the same.

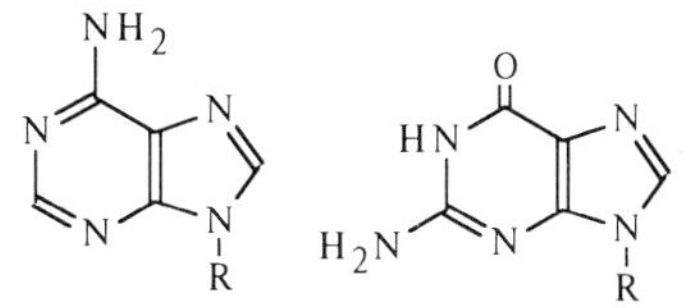

R - D-ribose or 2-deoxy-D-ribose residue

Rigorous proof that the sugar moiety in purine nucleosides is at position 9 was provided by the synthesis of the respective nucleosides, conducted by Todd and coworkers.

1.4.2 Pyrimidine Nucleosides

In the case of pyrimidine nucleosides, the involvement of the substituent at position 4 in the glycosidic bond should be ruled out in view of the fact that deamination of ribosylcytosine may yield ribosyluracil:

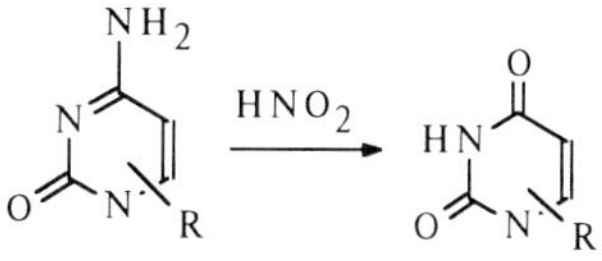

$\xrightarrow{HNO_2}$

R - monosaccharide residue

The greater stability of pyrimidine nucleosides during acid hydrolysis, as compared to *O*-glycosides, suggests that the oxygen at position 2 does not participate in the formation of a bond with the carbohydrate moiety either. The same applies to the carbons at positions 5 and 6, since the substitution products associated with C^5 and C^6 [5-bromo- and 5,6-di(phenylhydrazyl)-nucleosides] have been derived from a natural uracil nucleoside.

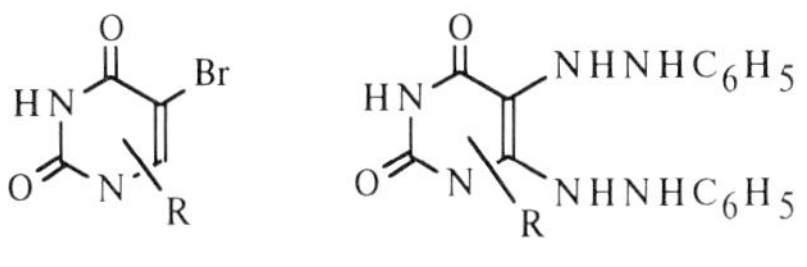

R-D-ribose residue

Consequently, only N^1 and N^3 of the pyrimidine ring can be linked to the sugar. Methylation of ribosyluracil gives mono-*N*-methylribosyluracil (uracil is converted into 1,3-dimethyluracil under the same conditions) which breaks down to 3-methyluracil during acid hydrolysis:

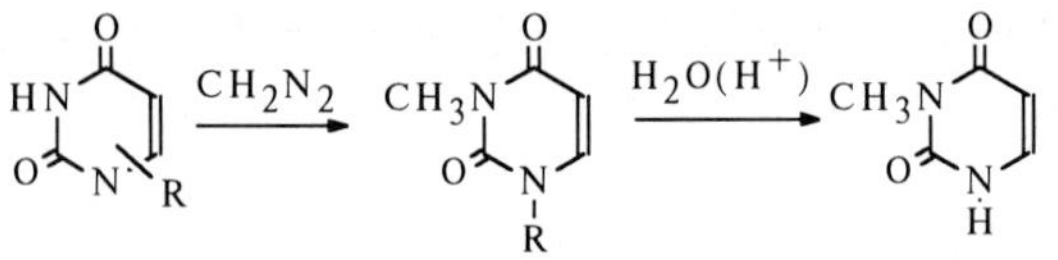

CH_2N_2 $\quad$ $H_2O(H^+)$

R-D-ribose residue

This is indicative of the fact that uracil nucleoside, just like cytosine nucleoside, is 1-riboside.

The similarity between the UV spectra of cytosine riboside and deoxyriboside, as well as the isolation of 3-methylthymine from DNA methylation products, attest to linkage between the sugar of deoxynucleosides and N^1 of the pyrimidine ring.

The conclusions as to the site of ribose binding in nucleosides have been corroborated by X-ray structure analysis.

1.5 Size of the Oxide Ring in the Sugar

The structure of the oxide ring in ribonucleosides has been established by the methylation method. 2,3,5-*O*-Trimethyl-D-ribofuranose has been isolated from an acid hydrolysate of methylated nucleosides, its oxidation yielding at first trimethyl-D-ribonolactone and then *meso*-dimethoxysuccinic acid:

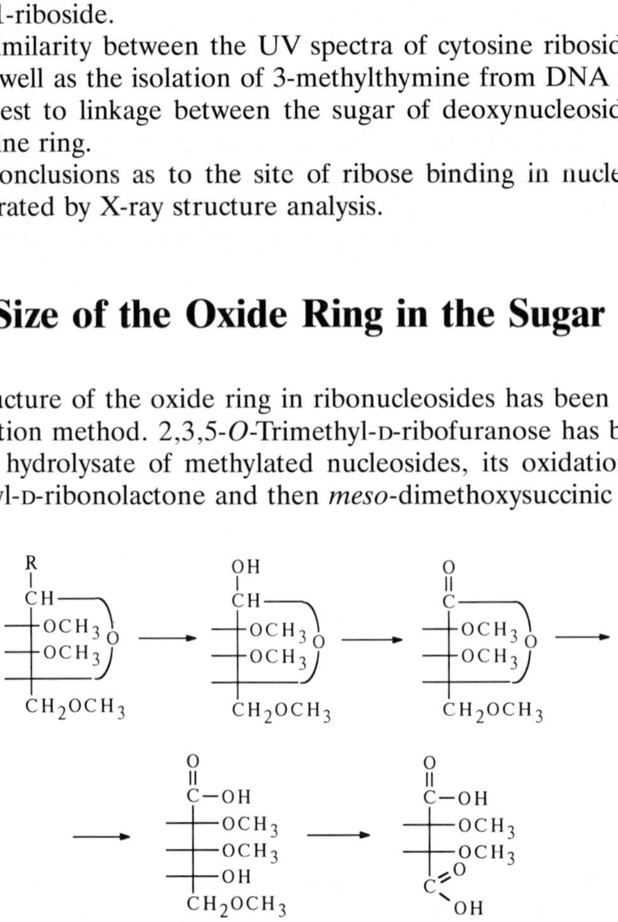

The furanose structure of ribonucleosides is also borne out by their response to triphenylchloromethane which, as is well known, reacts predominantly with the primary hydroxy groups of the carbohydrate, giving triphenylmethyl (trityl) esters. All nucleosides isolated from DNA and RNA form 5′-*O*-trityl derivatives with this reagent.

A convenient and simple way to determine the size of the ring in the sugar is through oxidation with periodic acid. This method which was developed for *O*-glycosides was applied to nucleosides by Todd and coworkers. It is based on the possibility of oxidative cleavage of the C-C bond in 1,2-diols and similar compounds in the presence of periodic acid:

$$-\overset{H}{\underset{OH}{C}}-\overset{H}{\underset{OH}{C}}- \xrightarrow{HIO_4} -C{\overset{O}{\underset{H}{\diagdown}}} + -C{\overset{O}{\underset{H}{\diagdown}}}$$

$$-\overset{H}{\underset{OH}{C}}-\overset{H}{\underset{OH}{C}}-\overset{H}{\underset{OH}{C}}- \xrightarrow{2\ HIO_4} 2-C{\overset{O}{\underset{H}{\diagdown}}} + HCOOH$$

The reaction proceeds quantitatively in aqueous solutions. Measurement of the spent amount of periodic acid by iodometric titration may give an indication whether the glycoside is a furanoside or a pyranoside. Given below are some data concerning the composition of the oxidation products and the oxidizing agent requirements for furanosides and pyranosides of D-ribose and 2-deoxy-D-ribose:

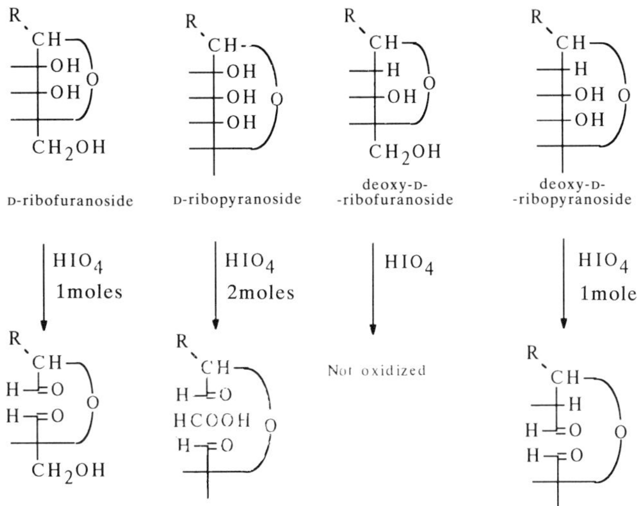

The experimental data on oxidation of ribonucleosides and deoxyribonucleosides indicate that the only structure these compounds can have is that of furanose. The oxidation of nucleosides with periodic acid has provided the basis for a micromethod to determine the size of the carbohydrate ring with chromatographic analysis of the dialdehyde reduction (and hydrolysis) products. The method permits the carbohydrate ring size to be determined at very low concentrations of the substances under investigation.

1.6 Configuration of the Glycoside (Anomeric) Center

After it had been definitively demonstrated (in the early fifties) that nucleosides have the furanose structure, the way their formulas used to be written was revised. Fischer's formulas for carbohydrates are too cumbersome and fail to adequately represent the three-dimensional configuration of their molecules. Instead, use was made of so-called "perspective" formulas proposed by Haworth, in which the convex portion of the carbon chain projects upward from the page, while the oxygen of the furanose ring lies in the background. The molecule is pictured in compliance with the laws of perspective, as is shown below (when the ribofuranose ring is represented in this fashion, it is conventionally assumed to be two-dimensional.

$HOCH_2$ O B

HO OH(H)

It would now be appropriate to consider the interrelation between Fischer's projection formulas and Haworth's perspective ones, which is of paramount importance for writing the formulas of nucleosides in the context of stereochemistry of the glycoside (anomeric) center.

The emergence of a new asymmetric carbon (C^1) when the ribofuranose ring of D-ribose is closed gives rise to two stereoisomeric forms: α-anomer with *cis*-configuration at C^1 and C^2 and β-anomer with *trans*-configuration at the same carbons.

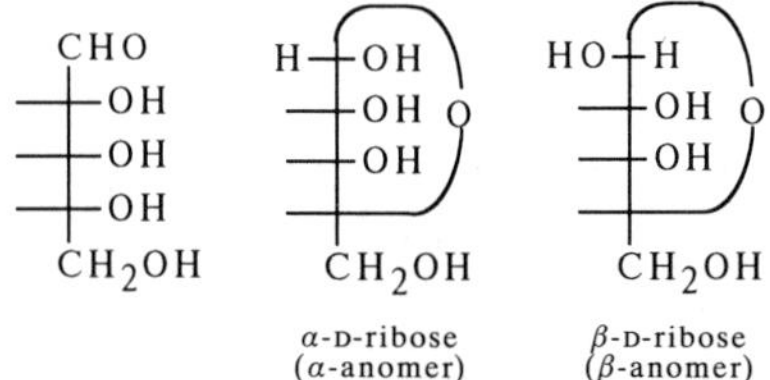

α-D-ribose (α-anomer) β-D-ribose (β-anomer)

Corresponding to these anomers are two series of glycoside derivatives, namely:

R OH O OH CH_2OH R OH O OH CH_2OH

α-D-ribofuranoside β-D-ribofuranoside

R-aglycone residue

To go from Fischer's formulas to perspective ones, in which the oxygen of the ring shares the same plane with the carbon chain C^1-C^4, one must alter the former so as to bring to light the mutual arrangement of the hydroxyls and hydroxymethyl group with respect to the plane of the furanose ring. To this end, one must interchange, in the above formulas of the α- and β-anomers, the substituents at C^4 twice (when pairs of substituents are interchanged an even number of times, the configuration at the asymmetric center remains the same). In the following scheme this operation is illustrated for the α-anomer (at first, the rearrangement involves H and CH_2OH, then H and the oxygen of the ring):

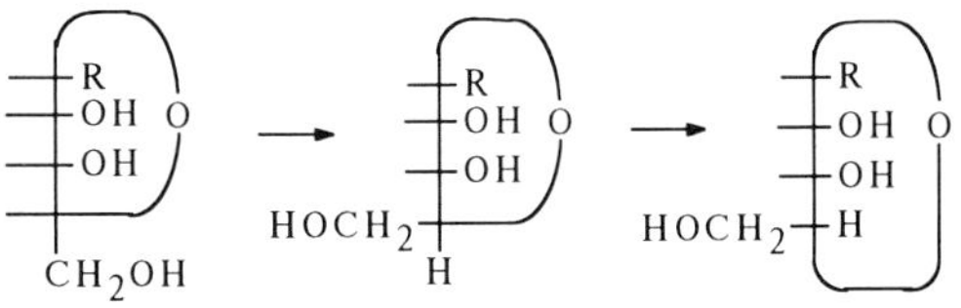

When perspective formulas are now written, the groups to the right of the vertical line should be arranged below the ring plane, whereas those to the left should be arranged above it; the oxygen must be placed at the point farthest away from the page. The C-C bonds projecting from the page are drawn in bold lines to create the effect of perspective. Shown below are the perspective formulas of α- and β-D-ribofuranosides after appropriate alterations (R stands for aglycone):

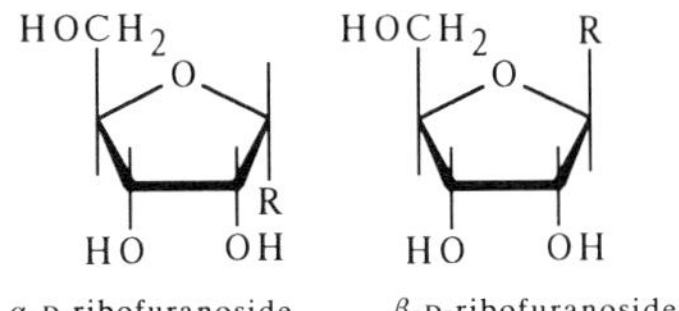

α-D-ribofuranoside β-D-ribofuranoside

Consequently, the aglycone lies below the carbohydrate ring plane in α-glycosides and above the plane in β-glycosides – that is, on the same side with the CH_2OH group.

The configuration of the glycoside center in purine and pyrimidine nucleosides has been established chemically by Todd and coworkers. They have found that 5′-*O*-toluenesulfonyl (tosyl) esters of ribonucleosides rapidly undergo intramolecular alkylation when heated. In the case of 2′,3′-*O*-substituted adenine nucleoside, the *para*-toluene sulfonate of N^3,5′-cycloadenine nucleoside is formed:

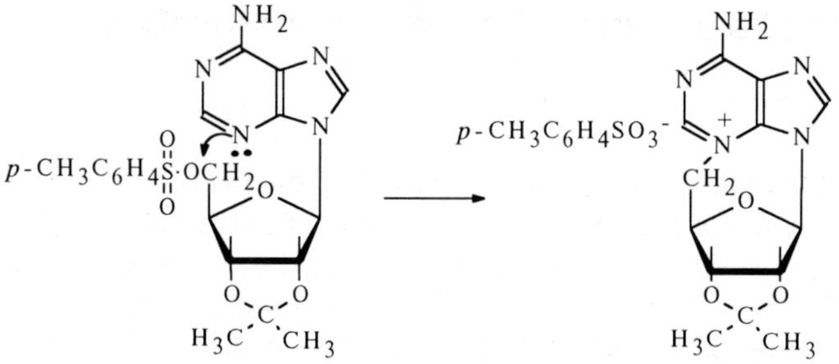

Similarly, O^25'-cyclic nucleoside was derived from cytosine nucleoside:

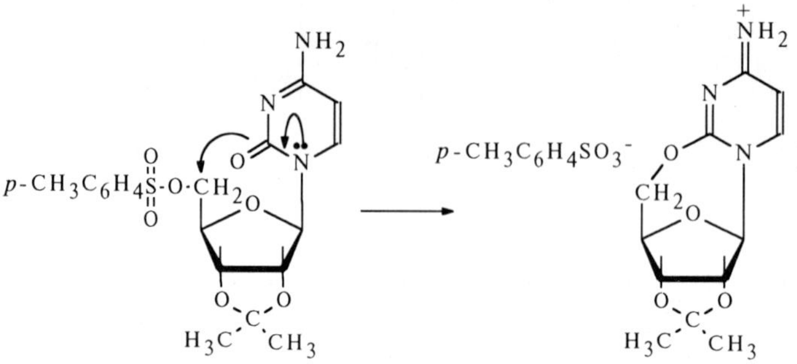

Intramolecular cyclization with elimination of the tosylate ion is possible only provided the heterocyclic base and the CH_2OH group lie on the same side of the carbohydrate ring plane – that is, with the glycoside center in the starting nucleoside having β-configuration. Hence, as was established by Todd and coworkers, naturally occurring adenine and cytosine ribonucleosides are characterized by β-configuration of their glycoside center.

Likewise, cyclization methods have been used to corroborate β-configuration of the glycoside center in guanine and uracil ribonucleosides as well as deoxyribonucleosides. For instance:

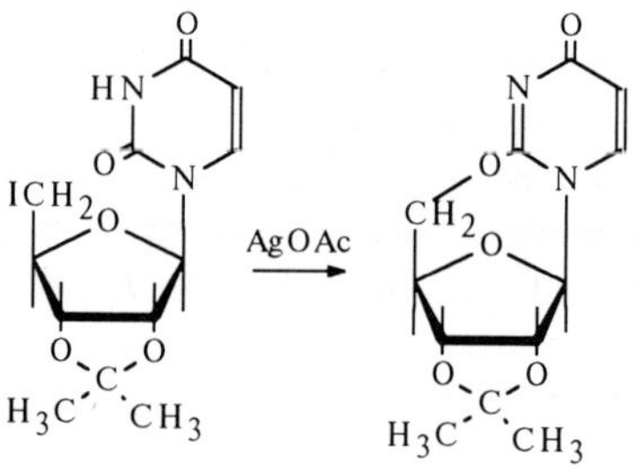

The results of X-ray structural analysis of crystalline cytosine nucleoside have confirmed the β-configuration of its glycoside center.

1.7 Nomenclature and Abridged Formulas of Nucleosides

The available structural data indicate that the purine nucleosides isolated from hydrolysates of nucleic acids are essentially 9-β-D-ribo- or 9-β-2′-deoxy-D-ribofuranosides of adenine and guanine, while the isolated pyrimidine nucleosides are 1-β-D-ribo- or 1-β-2′-deoxy-D-ribofuranosides of cytosine, uracil and thymine. The numbering of atoms in pyrimidine and purine nucleosides is shown in the following formulas:

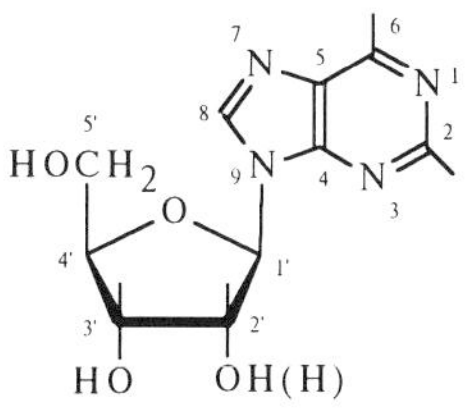

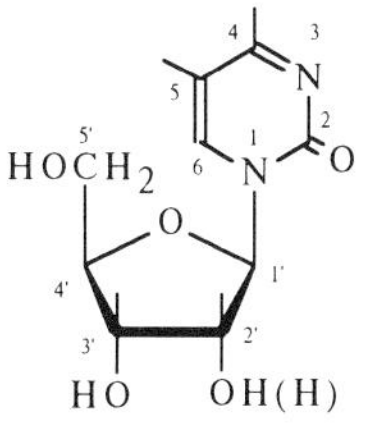

Tables 1-2 and 1-3 list the structural formulas and names of the major nucleosides of RNA and DNA. In view of the fact that difficulties arise when complete structural formulas of nucleosides are used in writing those of nucleic acids and their large fragments, abridged structural formulas are used instead as well as abbreviated (one- or three-letter) symbols which are also listed in Tables 1-2 and 1-3.

If abbreviated letter symbols of nucleosides occur in a text, they normally consist of a single letter, although such symbols coincide with one-letter symbols of the constituent pyrimidines or purines (see Table 1-1). In cases where bases of nucleic acids are mentioned in the same text along with nucleosides without using formulas (complete or abridged), three-letter symbols of nucleosides may be used to avoid confusion.

An abridged structural formula (see Tables 1-2 and 1-3) represent the sugar as a vertical line (a modified borrowing from the chemistry of carbohydrates) with the abbreviated symbol of the base constituting the nucleoside added on top. The hydroxyl groups are represented by horizontal lines with the OH group (the latter is omitted sometimes) – on the right in the case of 2′- and 3′-hydroxyl groups or on the left to denote the 5′-OH group. For the 2′- and 3′-hydroxyl groups to be represented correctly, the vertival line is mentally divided into three parts with a horizontal line associated with the 2′-hydroxyl being connected to the end of the first part and a line associated with the 3′-hydroxyl being connected to the end of the second part.

Table 1-2. Structure, Name and Symbols of Major Pyrimidine and Purine Nucleosides Isolated from RNA.

Structural formula	Name	Abridged structural formula	Symbol
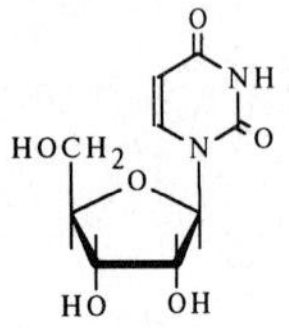 O, NH, O, N, $HOCH_2$, O, HO, OH	Uridine (1-β-D-ribofuranosyluracil)	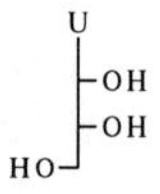 U, –OH, –OH, HO–	U, Urd
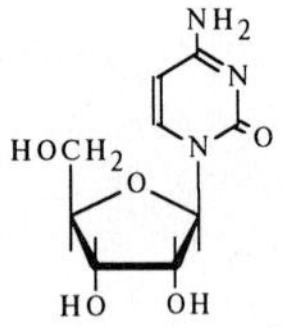 NH_2, N, O, N, $HOCH_2$, O, HO, OH	Cytidine (1-β-D-ribofuranosylcytosine)	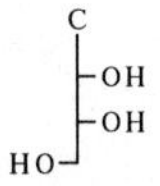 C, –OH, –OH, HO–	C, Cyd
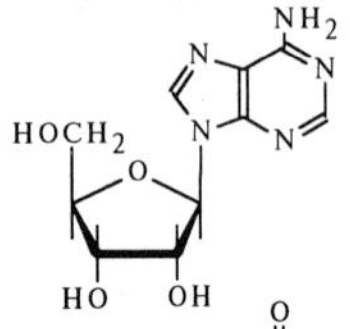 NH_2, N, N, N, N, $HOCH_2$, O, HO, OH	Adenosine (9-β-D-ribofuranosyladenine)	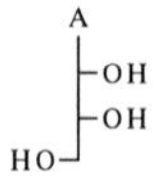 A, –OH, –OH, HO–	A, Ado
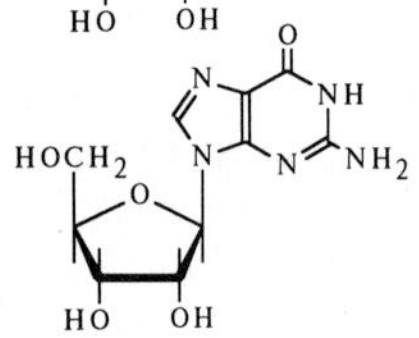 O, N, NH, N, N, NH_2, $HOCH_2$, O, HO, OH	Guanosine (9-β-D-ribofuranosylguanine)	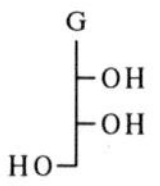 G, –OH, –OH, HO–	G, Guo

The difference between the abridged structural formulas of adenine ribo- and deoxyribonucleosides (adenosine and deoxyadenosine) is in the absence of a horizontal line with an OH group, which indicates that the 2′-hydroxyl group is absent:

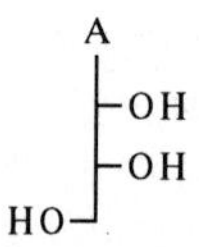

A
–OH
–OH
HO–

adenosine

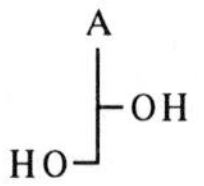

A
–OH
HO–

deoxy-
adenosine

When structural formulas are written in this way, the presence of cyclic oxygen is always implied.

Table 1-4 lists pyrimidine nucleosides of some phage DNAs in which they replace the major nucleosides (see Tables 1-2 and 1-3).

Table 1-3. Structure, Name and Symbols of Major Pyrimidine and Purine Nucleosides Isolated from DNA.

Structural formula	Name	Abridged structural formula	Symbol
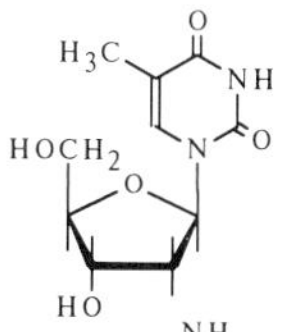	Deoxythymidine (1-β-2′-deoxy-D-ribofuranosylthymine)	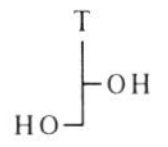	dT, dThd
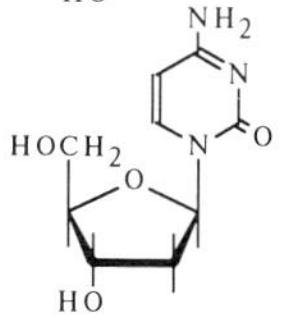	Deoxycytidine (1-β-2′-deoxy-D-ribofuranosylcytosine)	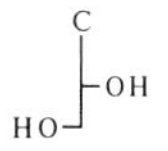	dC, dCyd
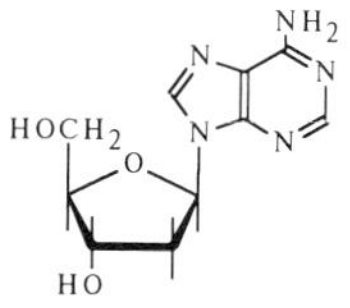	Deoxyadenosine (9-β-2′-deoxy-D-ribofuranosyladenine)	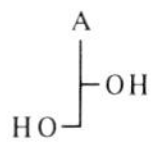	dA, dAdo
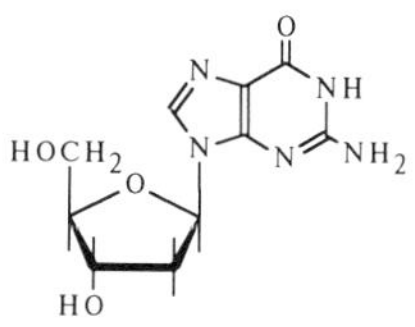	Deoxyguanosine (9-β-2′-deoxy-D-ribofuranosylguanine)	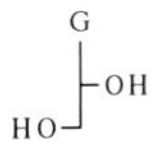	dG, dGuo

Table 1-4. Structure and Name of Pyrimidine Deoxyribonucleosides Constituting DNAs of Some Phages.

Structural formula	Name	Source (substituted nucleoside)
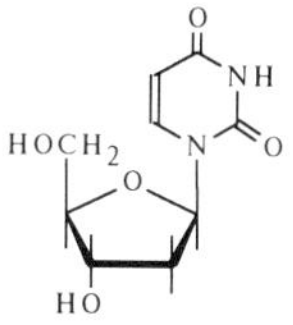	Deoxyuridine (1-β-2′-deoxy-D-ribofuranosyluracil)	Phage PNS 1 (deoxythymidine)
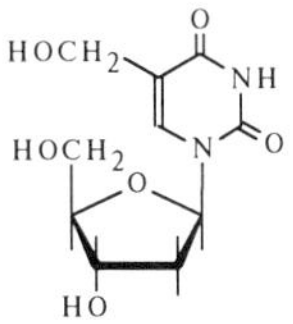	5-Hydroxymethyldeoxyuridine (1-β-2′-deoxy-D-ribofuranosyl--5-hydroxymethyluracil)	Phage SP 8 (deoxythymidine)

Table 1-4. (Continued).

Structural formula	Name	Source (substituted nucleoside)
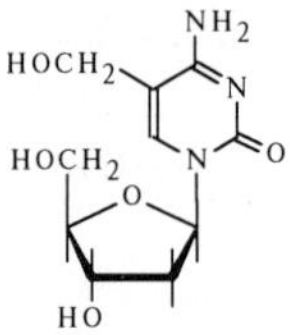 NH_2 $HOCH_2$ N N O $HOCH_2$ O HO	5-Hydroxymethyldeoxycytidine (1-β-2′-deoxy-D-ribofuranosyl- -5-hydroxymethylcytosine)	T-even phages (deoxycytidine)
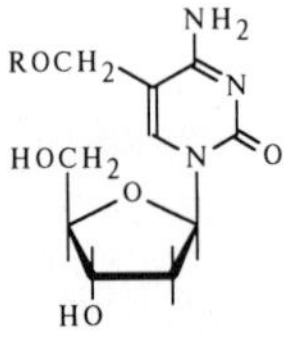 NH_2 $ROCH_2$ N N O $HOCH_2$ O HO	5-Hydroxymethyldeoxcytidine glycosides ($R = \alpha$-D-gluco- pyranosyl-, β-D-glucopyranosyl- or α-gentiobiosyl)	T-even phages (deoxycytidine)

1.8 Minor Nucleosides

In addition to the above-described major nucleosides of wide occurrence, nucleic acids also contain nucleosides that are derivatives of minor pyrimidine or purine bases or 2′-*O*-methylribose. Such nucleosides are referred to as rare or minor by analogy with the constituent bases or sugar. At present, more than 40 such compounds are known.

Minor nucleosides vary widely in chemical composition, and no classification has as yet been developed for them. Whatever classification exists has to do with the type of the pyrimidine or purine base of which the minor nucleoside is a derivative. Accordingly, minor nucleosides differing from major ones by substitution of functional groups in the carbohydrate moiety or base are grouped with minor components related to the major nucleoside. The structure of minor nucleosides was usually identified using the above-described techniques as well as through controlled conversion into nucleosides of a known structure. The structure of minor nucleosides has been corroborated by synthesis. Tables 1-5 and 1-6 give the structural formulas, names and abbreviated symbols of minor nucleosides isolated from RNA and DNA.

Table 1-5. Structure, Name and Symbols of Minor RNA Nucleosides of the General Formula

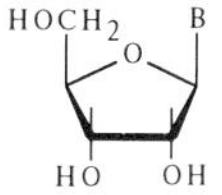

$HOCH_2$ O B HO OH

Structure	Name	Symbol
	Uridine Derivatives	
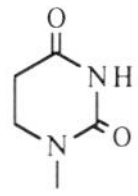 O, NH, N, O	Dihydrouridine (1-β-D-ribofuranosyl--5,6-dihydrouracil)	D or hU
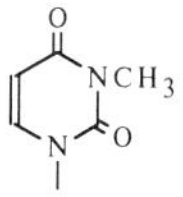 O, NCH_3, N, O	3-Methyluridine (1-β-D-ribofuranosyl--5,6-methyluracil)	m^3 U
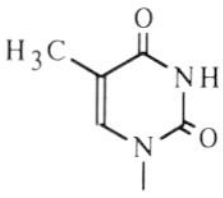 H_3C, O, NH, N, O	Ribothymidine (1-β-D-ribofuranosyl-thymine)	rT or T
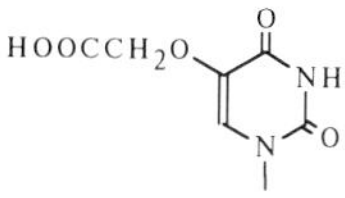 $HOOCCH_2O$, O, NH, N, O	Uridyl-5-hydroxyacetic acid (1-β-D-ribofuranosyl-5-carboxymethylhydroxy-uracil)	V
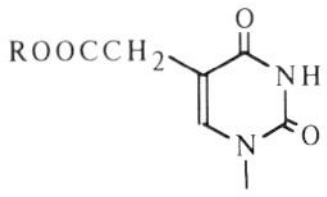 $ROOCCH_2$, O, NH, N, O	(R = H) 5-Carboxymethyluridine (1-β-D-ribofuranosyl-5-carboxymethyluracil) (R = CH_3) 5-Carbmethoxymethyluridine	cm^5 U cmm^5 U
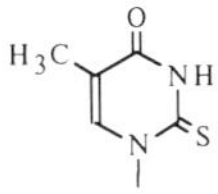 H_3C, O, NH, N, S	2-Thio-5-methyluridine (1-β-D-ribofuranosyl-2-thio-4-hydroxy-5-methylpyrimidine)	s^2m^5 U
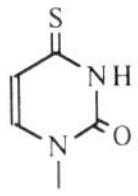 S, NH, N, O	4-Thiouridine (1-β-D-ribofuranosyl--2-hydroxy-4-thiopyrimidine)	s^4 U
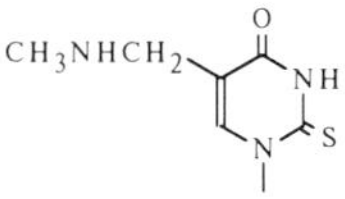 CH_3NHCH_2, O, NH, N, S	5-(Methylaminomethyl)-2-thiouridine	s^2mam^5 U
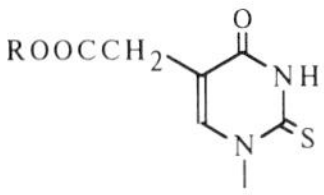 $ROOCCH_2$, O, NH, N, S	(R = H) 2-Thio-5-carboxymethyluridine (R = CH_3) 2-Thiouridyl-5-acetic acid, methyl ester	s^2cm^5 U s^2am^5 U

Table 1-5. (Continued).

Structure	Name	Symbol
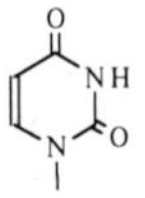 In addition, the proton of hydroxyl at $C^{2'}$ is substituted by CH_3	2′-*O*-Methyluridine (1-β-D-2′-*O*--methylribofuranosyluracil)	Um
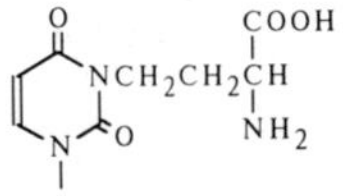	γ-(Uridyl-3)-α-aminobutyric acid	X
	Cytidine Derivatives	
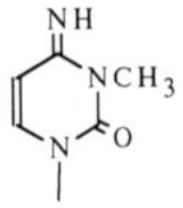	3-Methylcytidine (1-β-D-ribofuranosyl--3-methylcytosine)	m^3 C
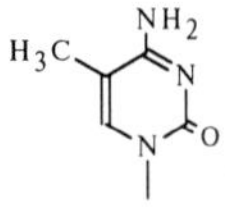	5-Methylcytidine (1-β-D-ribofuranosyl--5-methylcytosine)	m^5 C
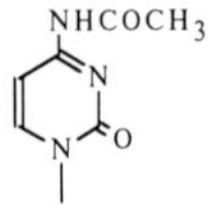	4-N-Acetylcytidine (1-β-D-ribofuranosyl--4-N-acetylcytosine)	ac^4 C
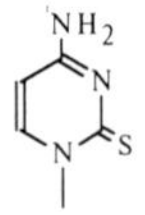	2-Thiocytidine (1-β-D-ribofuranosyl--2-thio-4-aminopyrimidine)	s^2 C
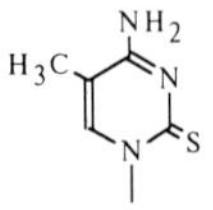	2-Thio-5-methylcytidine (1-β-D-ribofuranosyl--2-thio-4-amino-5-methylpyrimidine)	s^2m^5 C
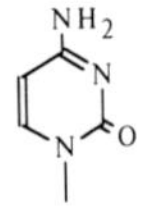 In addition, the proton of hydroxyl at $C^{2'}$ is substituted by CH_3	2′-*O*-Methylcytidine (1-β-D-2′-*O*-methyl ribofuranosylcytosine)	Cm

Table 1-5. (Continued).

Structure	Name	Symbol
	Adenosine Derivatives	
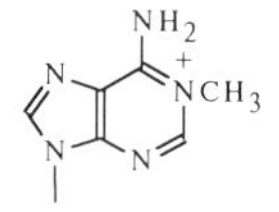 NH_2, $\overset{+}{N}CH_3$	1-Methyladenosine (9-β-D-ribofuranosyl--1-methyladenosine)	m^1 A
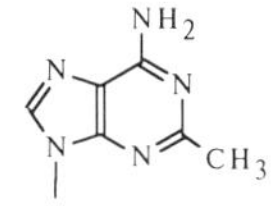 NH_2, CH_3	2-Methyladenosine (9-β-D-ribofuranosyl--2-methyladenosine)	m^2 A
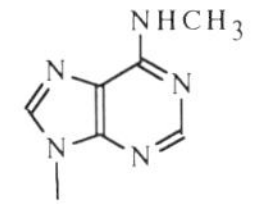 $NHCH_3$	6-*N*-Methyladenosine [9-β-D-ribofuranosyl--6-(methylamino)purine]	m^6 A
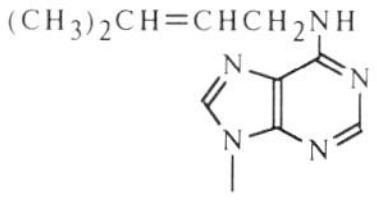 $(CH_3)_2CH{=}CHCH_2NH$	6-*N*-Isopentenyladenosine [9-β-D-ribofuranosyl--6-(γ, γ-dimethylallyl)aminopurine]	i^6 A
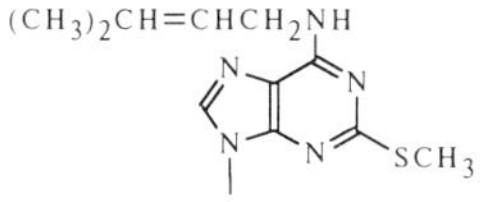 $(CH_3)_2CH{=}CHCH_2NH$, SCH_3	2-Methylthio-6-*N*-(Δ^2-isopentenyl)adenosine [9-β-D-ribofuranosyl-6-(γ, γ-dimethylallyl)-amino-2-methylthiopurine]	ms^2i^6 A
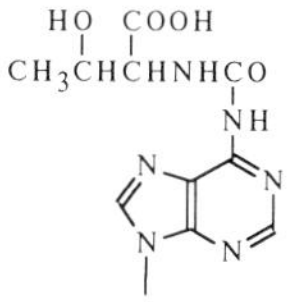 HO COOH, $CH_3CHCHNHCO$, NH	6-*N*-(*N*-Threonylcarbonyl)adenosine	t^6 A
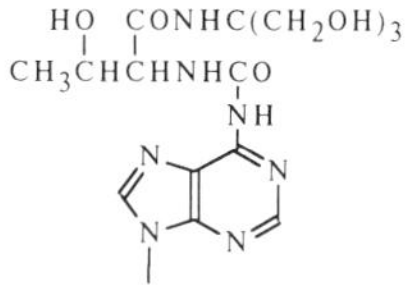 HO $CONHC(CH_2OH)_3$, $CH_3CHCHNHCO$, NH	6-*N*-[*N*-Threonylcarbonyltri(hydroxamethyl)-aminomethyl]adenosine	N^+
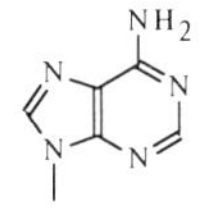 NH_2; In addition, the proton of hydroxyl at $C^{2'}$ is substituted by CH_3	2'-*O*-Methyladenosine (9-β-D-2'-*O*-Methyl ribofuranosyladenine)	Am
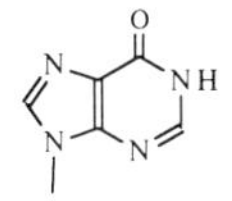 O, NH	Inosine (9-β-D-ribofuranosylhypoxanthine)	I
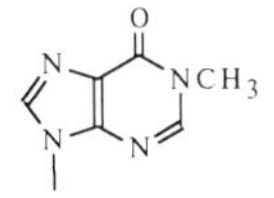 O, NCH_3	1-Methylinosine (9-β-D-ribofuranosyl-1-methylhypoxanthine)	m^1 I

Table 1-5. (Continued).

Structure	Name	Symbol
	Guanosine Derivatives	
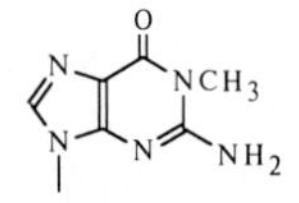	1-Methylguanosine (9-*β*-D-ribofuranosyl-1-methylguanine)	m^1 G
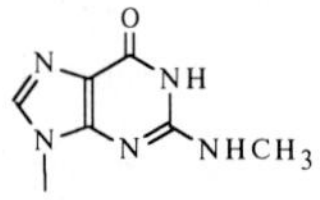	2-*N*-Methylguanosine [9-*β*-D-ribofuranosyl--2-(methylamino-4-hydroxypurine)	m^2 G
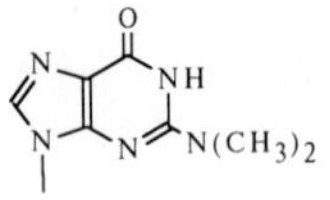	2-*N,N*-Dimethylguanosine (9-*β*-D-ribofuranosyl--2-dimethylamino-4-hydroxypurine)	m^2_2 G
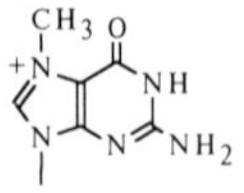	7-Methylguanosine (9-*β*-D-ribofuranosyl-7-methylguanine)	m^7 G
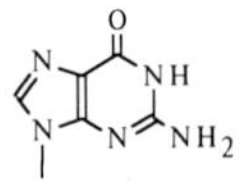 In addition, the proton of hydroxyl at $C^{2'}$ is substituted by CH_3	2′-*O*-Methylguanosine (9-*β*-D-2′-*O*-methyl-ribofuranosylguanine)	Gm
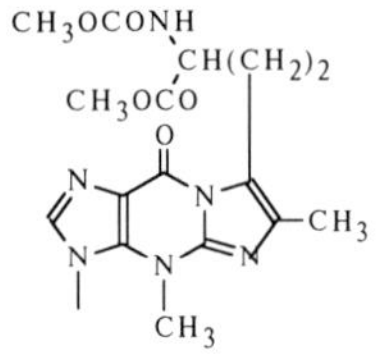	–	Y

Table 1-6. Structure, Name and Symbols of Minor DNA Nucleosides of the General Formula

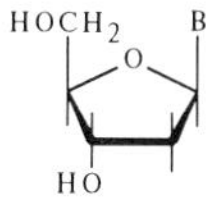

Structure	Name	Symbol
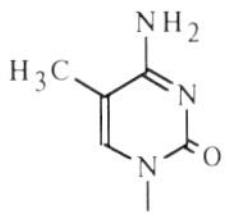	5-Methyldeoxycytidine (1-β-2′-deoxy- -D-ribofuranosyl-5-methylcytosine)	dm^5 C
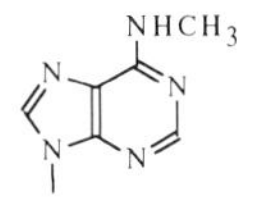	6-*N*-Methyldeoxyadenosine (9-β-2′-deoxy- -D-ribofuranosyl-6-methylaminopurine)	dm^6 A
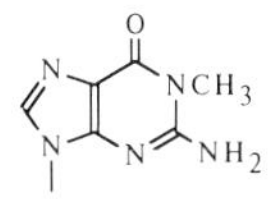	1-Methyldeoxyguanosine (9-β-2′-deoxy- -D-ribofuranosyl-1-methylguanine)	dm^1 G
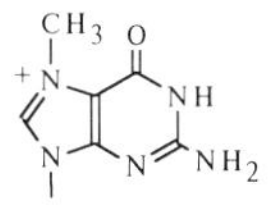	7-Methyldeoxyguanosine (9-β-2′-deoxy- -D-ribofuranosyl-7-methylguanine)	dm^7 G
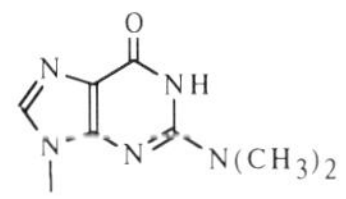	2-*N,N*-Dimethyldeoxyguanosine (9-β-2′- -deoxy-D-ribofuranosyl-2-dimethylamino- -6-hydroxypurine)	dm$_2^2$ G

1.9 Pseudouridine

Apart from major and minor nucleosides with *N*-glycosidic bonds some RNAs also contain sizable amounts of an unusual nucleoside with a different type of glycosidic bond, known as pseudouridine. This unusual, or "fifth", nucleoside was discovered as far back as 1951, but it took ten years to decipher its structure in spite of the fact that the chief methods for establishing the structure of nucleosides had been developed by the early fifties and become available to many laboratories around the world. The conventional methods for determining a structure did not work on pseudouridine. It was soon found out that the unusual nucleoside has the same chemical composition as uridine. However, unlike uridine, it is extremely stable under conditions of acid hydrolysis, which explains why attempts to break it down into components failed for quite some time. The UV absorption spectrum of the compound differed from that of uridine and 1-alkyluracils and was closely similar to that of 5-hydroxymethyluracil. These findings had led to a conclusion that the "fifth nucleoside" is a derivative of uracil, yet, in contrast to the latter, represents a *C*-glycoside with glycosidic bonding at C^5 or C^6 in the pyrimidine ring. This conclusion was supported by the results of UV spectroscopy at different values of pH.

The bathochromic shift of the absorption curve corresponding to the nucleoside of interest in the UV region at $pH > 8.0$, as compared to a more acid medium, suggested that the heterocyclic base of this nucleoside contains two hydroxy groups prone to ionization.

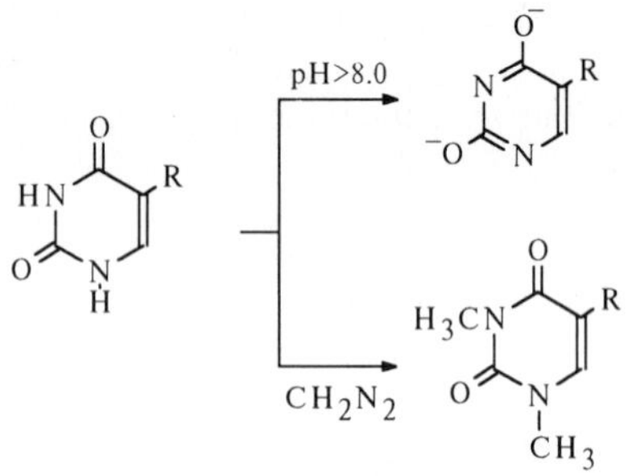

R - ribose residue

Methylation of both nitrogens of the base was significant in demonstrating that they are not substituted in the starting nucleoside. The presence of D-ribose was proved by hydrazine-induced cleavage of the nucleoside. Thus, it was established that the "fifth nucleoside" differs from uridine only by the nature of its glycosidic bond. It is for this reason that the nucleoside was called pseudouridine.

The fact that pseudouridine has the structure of 5-ribofuranosyluracil was established definitively through double oxidation with periodic acid followed by reduction with sodium borohydride – that is, through conversion into 5-hydroxymethyluracil.

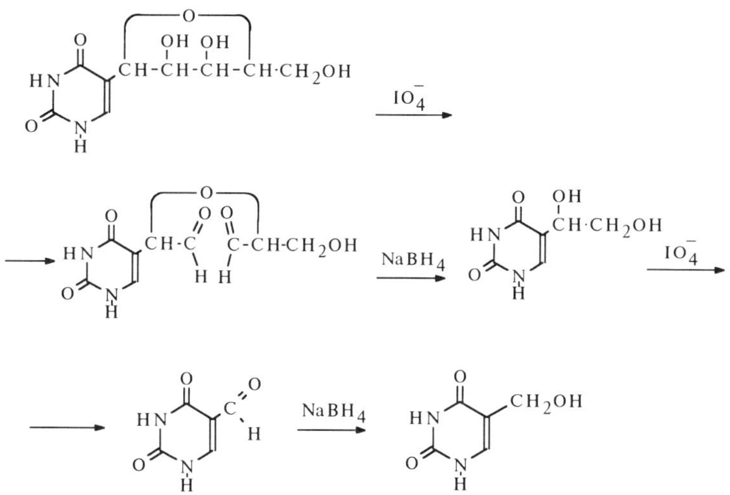

The results of NMR spectroscopy, attesting to presence of a proton at C^6 in pseudouridine and its absence at C^5, corroborated the findings of the preceding studies. The structure of pseudouridine, just as that of 5-D-ribofuranosyluracil, was also confirmed by synthesis. The configuration of the glycoside center was established in the usual manner:

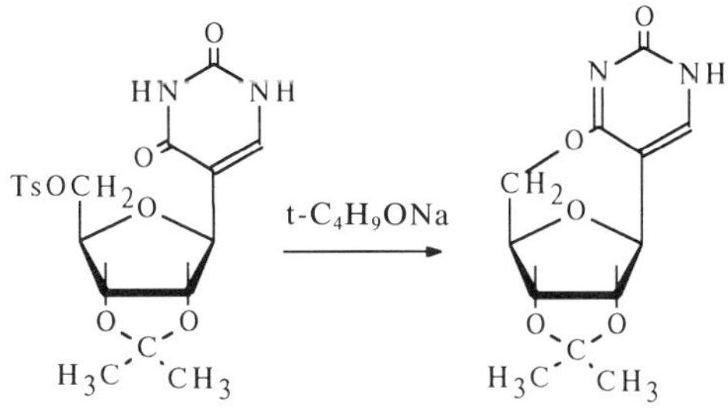

Thus, pseudouridine is a 5-β-D-ribofuranosyluracil. Given below are the commonly recognized structural formula and abbreviated symbols of this nucleoside:

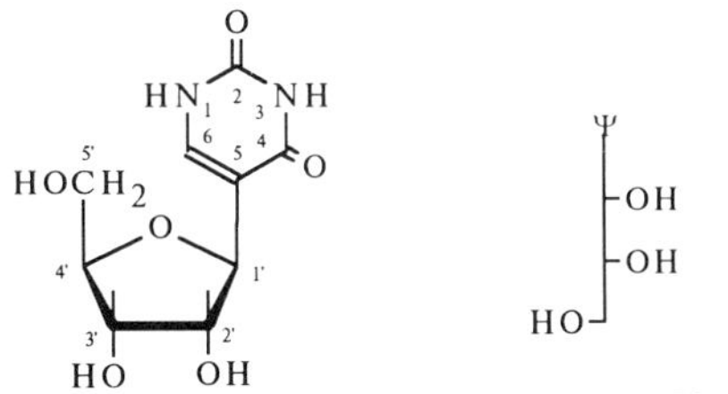

Ψ or ΨU

Some RNAs have also been found to contain 2′-*O*-methylpseudouridine (designated by the abbreviated symbol Ψm).

1.10 Nucleoside Antibiotics

Microorganisms as well as plant and animal tissues have been a source of nucleosides that are not present in nucleic acids but are found in the cell in a free state. Most of these compounds are antibiotics and are widely used in treating malignant tumors and some other diseases. Nucleoside antibiotics also serve as an extremely important tool for studying the mechanisms of biochemical reactions, primarily biosynthesis of nucleic acids and protein.

There are more than 20 nucleoside antibiotics known today, in which the aglycone is a pyrimidine or purine base; they differ from ordinary and minor nucleosides either by the structure of the carbohydrate moiety or that of the aglycone. We shall now consider the structure of the most interesting nucleoside antibiotics.

1.10.1 Purine Nucleosides

Structurally, these nucleosides are similar to adenosine, which is a major component of nucleic acids. This may explain their biological activity.

Puromycin. This antibiotic is obtained from the actinomycete *Streptomyces alboniger* and displays antibacterial activity.

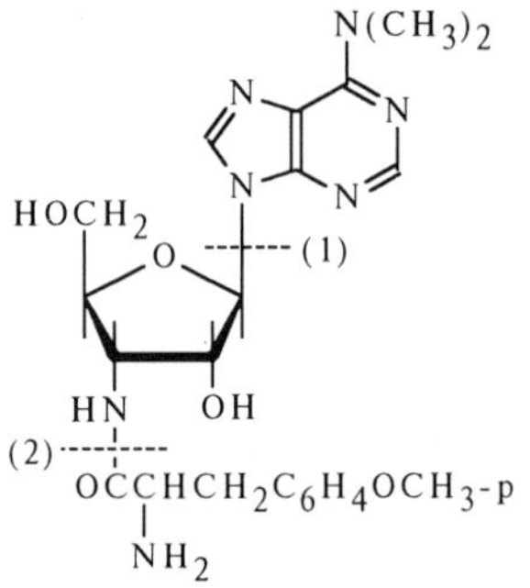

$N(CH_3)_2$
N N N N
$HOCH_2$ O (1)
HN OH
(2)
$OCCHCH_2C_6H_4OCH_3$-p
NH_2

Alcoholysis of puromycin in the presence of hydrogen chloride yields three substances identified as 6-dimethyl-aminopurine [cleavage of glycosidic bond; path (1)], *O*-methyl-L-tyrosine [cleavage of amide bond; path (2)], and 3-amino-3-deoxy-D-ribose.

Puromycin gives a triacetate during acetylation (with two hydroxyls of the carbohydrate moiety being acylated along with the amino group in the amino acid moiety), which is converted into *N*-acetylpuromycin after treatment with ammonia solution in methanol (these conditions are conducive to cleavage of the ester bonds with the amide bond remaining intact). These transformations provide additional proof to the presence of two free hydroxyl groups in puromycin. Treatment of puromycin with phenyl isothiocyanate then with a sodium alcoholate results in an amino acid and a free "amino nucleoside" which is oxidized with periodic acid (1 mole), this being indicative of the furanose structure of the constituent sugar.

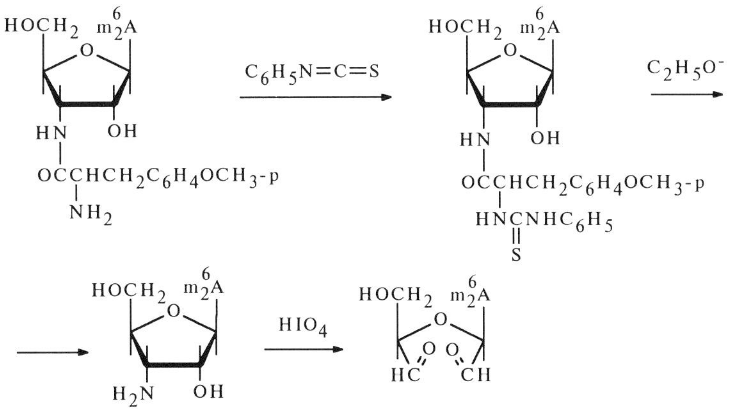

The stability of puromycin itself toward periodic acid also suggests that the amino acid (*O*-methyltyrosine) moiety in its molecule is linked with the amino group of the carbohydrate moiety via an amide bond.

The β-configuration of the glycoside center in puromycin has been established by the usual procedure based on 5′-methylsulfonic acid derivative of the protected "amino nucleoside".

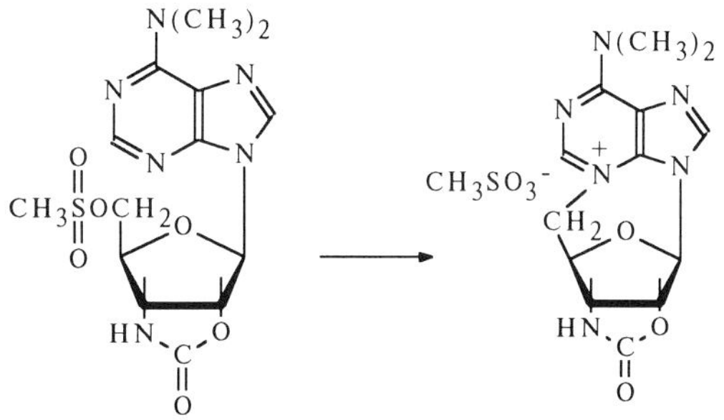

The structure of puromycin was definitively established by synthesis.

Lysylaminoadenosine and homocitrullylaminoadenosine. These two nucleoside antibiotics [(1) and (2), respectively], structurally closely related to puromycin, have been isolated from the mycelium of *Cordyceps militaries.* Their structure was determined as in the case of puromycin.

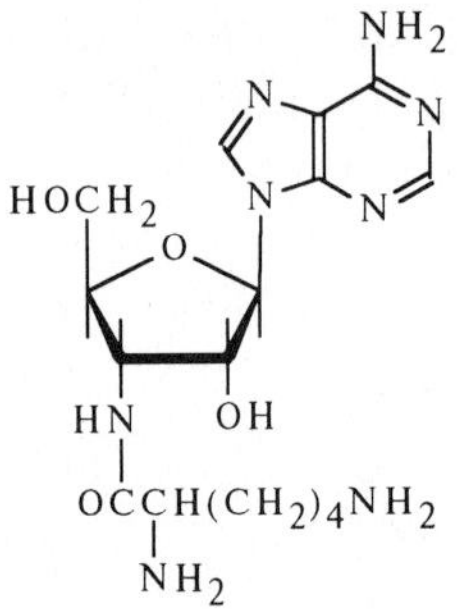

(1) 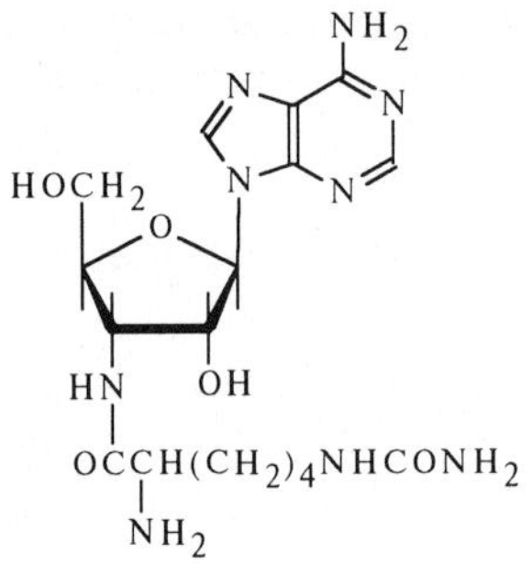 (2)

The presence of an amide bond in lysylaminoadenosine (1) has been demonstrated by isolation of bis(2,4-dinitrophenyl)lysine after treatment with 2,4-dinitrofluorobenzene followed by acid hydrolysis. The amino acids present in both antibiotics have been assumed to be in an L-form.

Cordycepin. This nucleoside has also been obtained from the mycelium of *Cordyceps militaries.* Acid hydrolysis of the antibiotic results in adenine and deoxypentose. According to the spectral characteristics, the glycosyl is at N^9 of adenine.

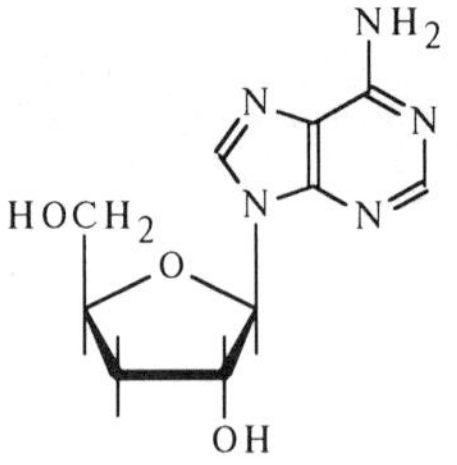

Cordycepin is not oxidized by periodate. Its treatment with *para*-nitrophenylhydrazine gives osazone. The implication is that the sugar has the structure of 3′-deoxyribose.

Nebularine is present in *Agaricus nebularis* and other fungi. When subjected to acid hydrolysis it breaks down into D-ribose and an unsubstituted purine. The results of UV spectroscopy and oxidation with periodic acid indicate that the carbohydrate moiety in the furanose form is associated with N^9 of the purine.

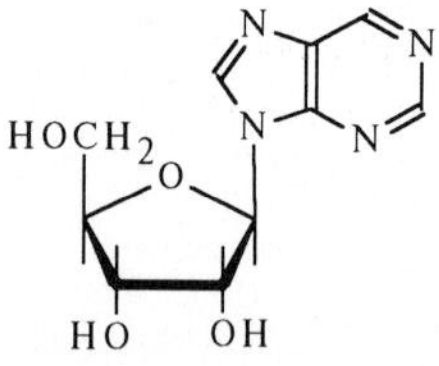

The β-configuration of the glycoside center has been established by its relation to adenosine which, just as nebularine, was derived from synthetic 9-β-D-ribofuranosyl-6-chloropurine:

nebularine ← H_2/Pd — 9-β-D-ribofuranosyl-6-chloropurine — NH_3 → adenosine

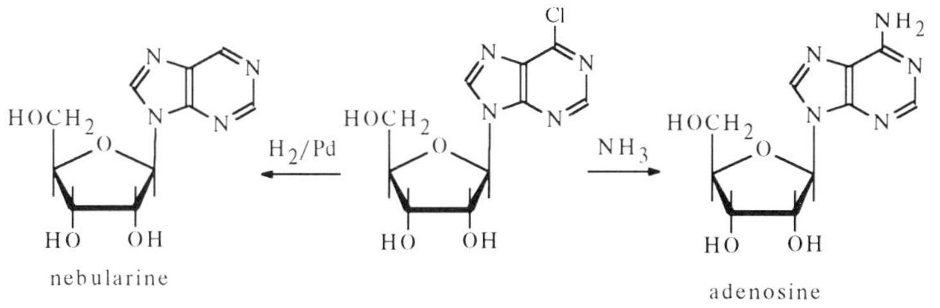

Tubercidin. This nucleoside was isolated from a culture of *Streptomyces tubecidiens.* Its hydrolysis in the presence of the ion–exchange resin Dowex 50 gives 7-deazaadenine (3) and D-ribose.

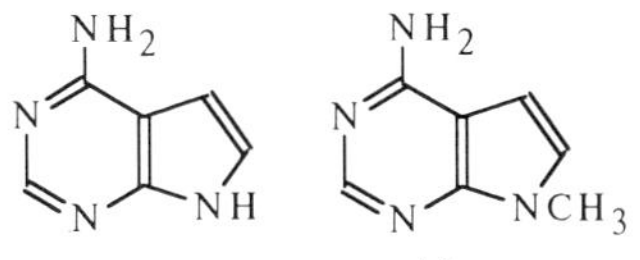

(3) (4)

The absorption spectrum of tubercidin in the UV region coincides with that of synthetic 9-methyl-7-deazaadenine (4), which indicates that the carbohydrate moiety in the tubercidin molecule (5) is linked with the nitrogen in the position 9 (N^9).

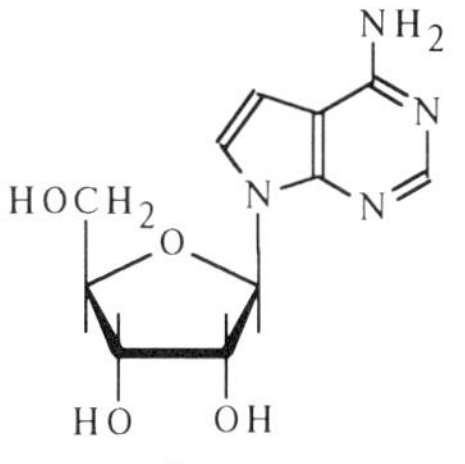

(5)

The ease of conversion of 5′-*O*-tosyl ester of a tubercidin derivative into N^3,5′-cyclic nucleoside (heating in acetone) suggests that the glycoside center in the antibiotic has a β-configuration.

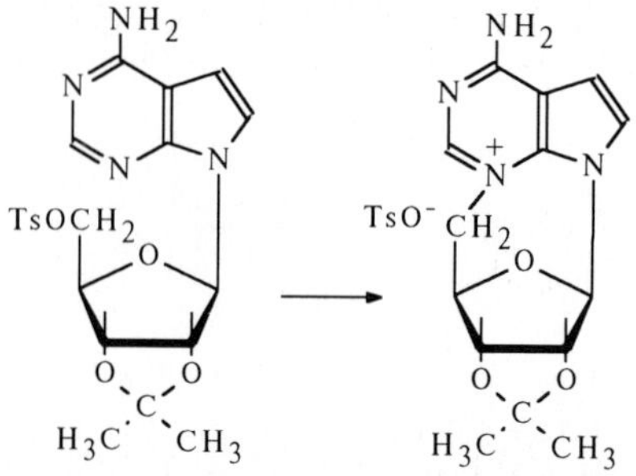

1.10.2 Pyrimidine Nucleosides

All known antibiotics in this category were isolated from *Streptomyces* cultures. They are essentially N^1-glycosides of cytosine, their molecules containing fragments of 4-amino-4-deoxyhexose and an amino acid. Shown below are the formulas of some antibiotics with definitively established structure.

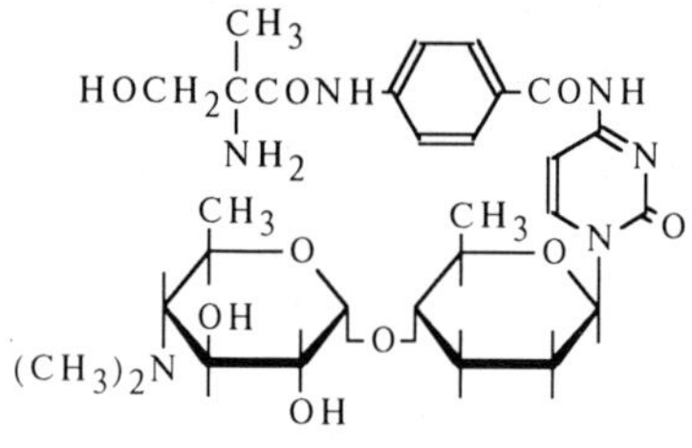

amycetin

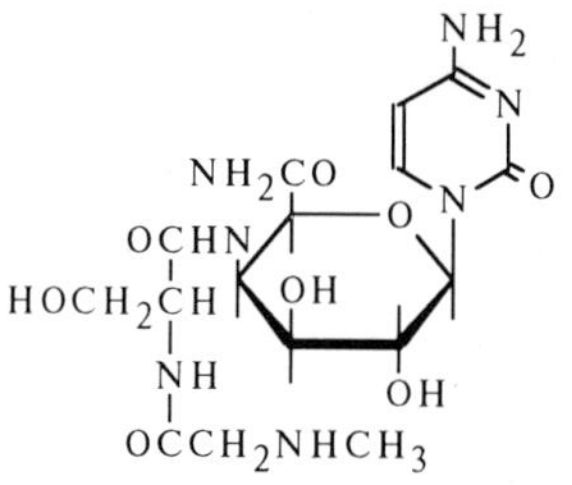

gougerotin

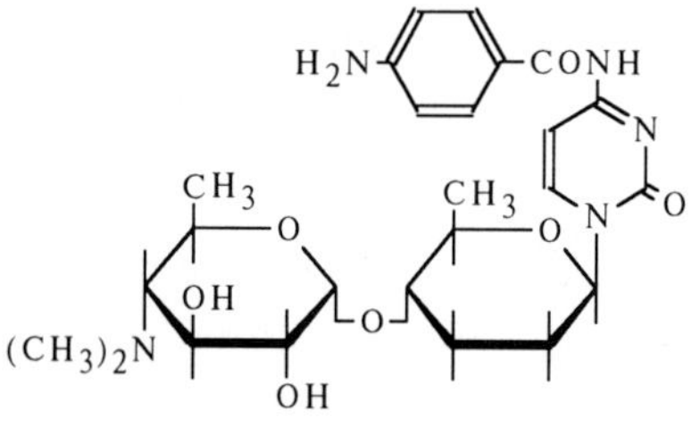

plicacetin

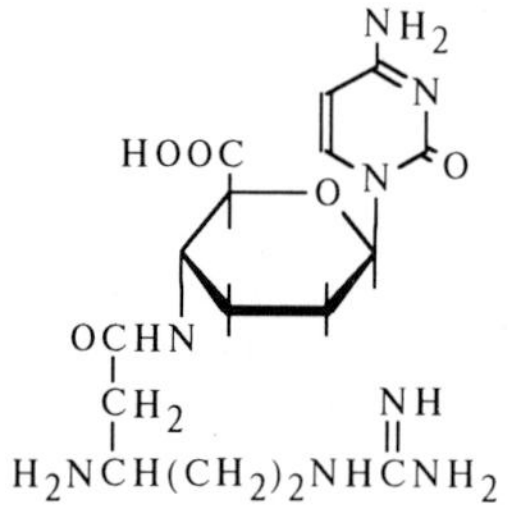

blasticidin

1.11 Other Nucleosides

In addition to the nucleosides produced by microorganisms, a number of free nucleosides have been discovered in plant and animal tissues, whose role in biochemical processes is yet to be elucidated. The structure of some of them is given below:

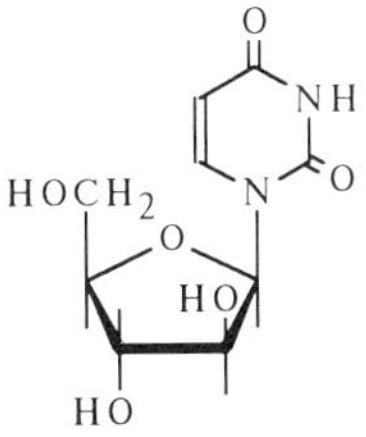

spongouridine

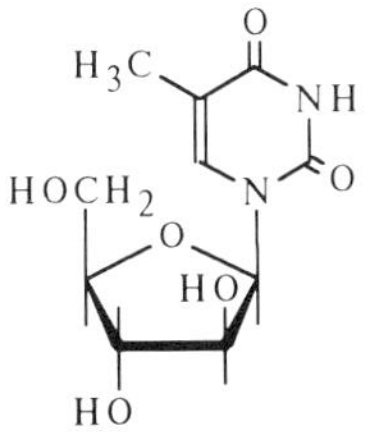

spongothymidine

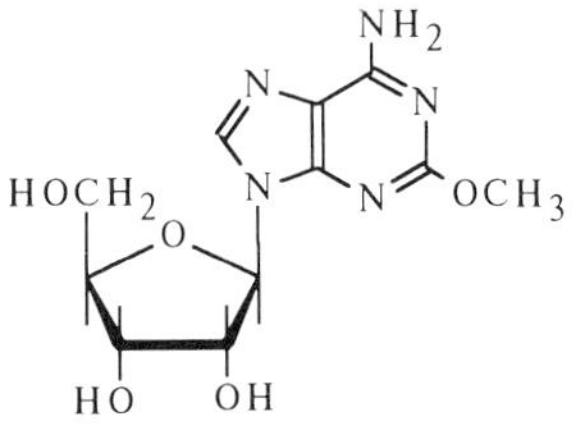

spongosine

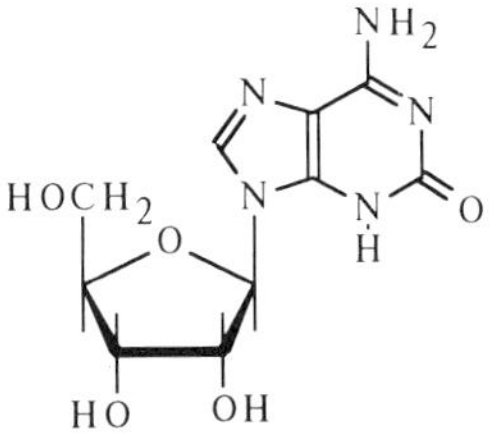

crotonoside

Spongouridine, spongothymidine and spongosine were extracted from the sponges *Cryptotethya crypta* directly with the aid of acetone. The structure of these compounds was established using standard procedures (described above) and corroborated by synthesis. Crotonoside was isolated from beans. When subjected to acid hydrolysis, it yields isoguanine and D-ribose, while its deamination yields xanthosine also obtained when guanosine is exposed to nitrous acid. This is why the structure of 9-β-D-ribofuranosyl-2-hydroxy-6-aminopurine was ascribed to crotonoside and this was subsequently corroborated by synthesis.

References

1. A. M. Michelson, *The Chemistry of Nucleosides and Nucleotides.* Academic Press, N.Y.–London (1963).
2. R. K. Robins, *Purines and Related Cyclic Systems,* in: *Heterocyclic Compounds.* Vol. 8, R. C. Elderfield (Ed.), John Wiley & Sons, Inc., N.Y.–London–Sydney (1967).
3. G. Kenner and A. Todd, *Pyrimidine and Its Derivatives,* in: *Heterocyclic Compounds.* Vol. 6, R. C. Elderfield (Ed.), John Wiley & Sons, Inc., New York (1960).
4. T. L. V. Ulbricht, in: *Comprehensive Biochemistry.* Vol. 8, M. Florkin and E. H. Stotz (Eds.), Elsevier, Amsterdam–London–N.Y., pp. 158–208 (1963).
5. D. O. Jordan, *The Chemistry of Nucleic Acids.* Butterworths, London–Washington, p. 358 (1960).
6. R. W. Chambers, in: *Progress in Nucleic Acid Res. and Mol. Biol.* Vol. 5, J. N. Davidson and W. E. Cohn (Eds.), Academic Press, N.Y.–London, pp. 349–398 (1966).
7. J. J. Fox, K. A. Watanabe, and A. Bloch, *ibid.* pp. 251–313.
8. *Nucleoside Analogues: Chemistry, Biology and Medical Application.* R. T. Walker, E. De Clercq and F. Eckstein (Eds.), Plenum Press, N.Y.–London (1979).
9. R. H. Hall, *Modified Nucleosides in Nucleic Acids.* Columbia University Press, New York, (1971).

2 Properties of Nucleosides

2.1 Heterocyclic Bases

2.1.1 General Concepts

An unsubstituted pyrimidine is an aromatic system in which, just as in pyridine, the carbons are deficient in electrons due to electronegativity of the nitrogens[1]. There are two factors contributing to this, namely, induction and mesomeric effects. The former is especially pronounced at positions next to the nitrogens – that is, 2, 4 and 6. Position 5 is electrondeficient but to a lesser degree.

The mesomeric effect may manifest itself only in one direction, with electrons being shifted toward the nitrogen, otherwise unviable six-membered cyclic structures with a cumulative system of multiple or triple bonding would be formed.

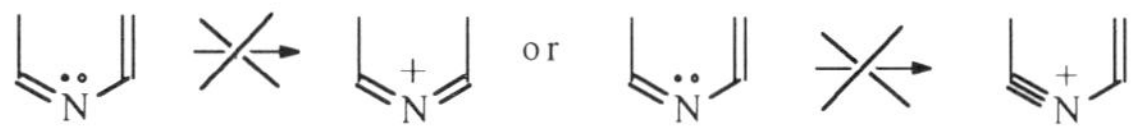

or

The electron-attracting mesomeric effect of the nitrogen atoms N^1 and N^3 may be represented by the following resonance structures[2]:

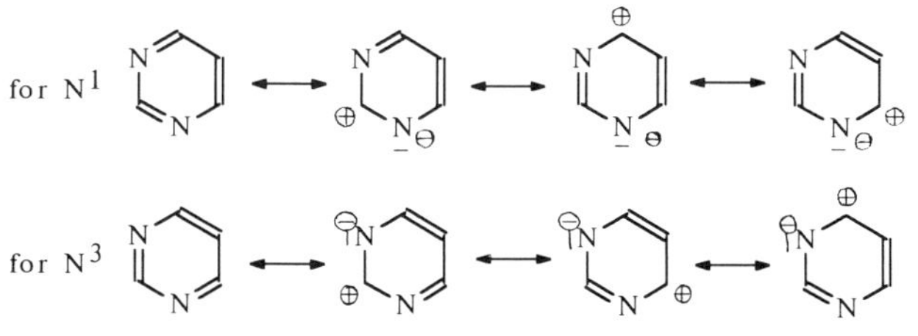

for N^1

for N^3

[1] According to the scale proposed by Pauling, the electronegativity of nitrogen is 3 and that of carbon is 2.4.

[2] It should be borne in mind that these structures do not exist in reality but are merely indicative of the direction taken by electron density shifts.

The above structures show that the mesomeric effect of the nitrogen atoms also leads to electron deficiency in the carbons at positions 2, 4 and 6. Consequently, the aromatic ring in pyrimidine is stable toward electrophilic reagents even to a greater extent than in pyridine. The only site vulnerable to attack by an electrophilic reagent is position 5. Therefore, the latter is often referred to as aromatic. It is also natural that positions 2, 4 and 6 must be reactive in reactions with nucleophilic reagents. These principles also apply to unsubstituted pyrimidine, which is confirmed by experimental evidence, albeit rather scant because this compound is extremely difficult to isolate. Various derivatives of pyrimidine, whose reactions are more complex yet are governed by the same basic principles, have been studied much more comprehensively.

In considering the purine ring (containing an imidazole nucleus in addition to the pyrimidine one) it is appropriate to draw an analogy between the imidazole and pyrrole nuclei.

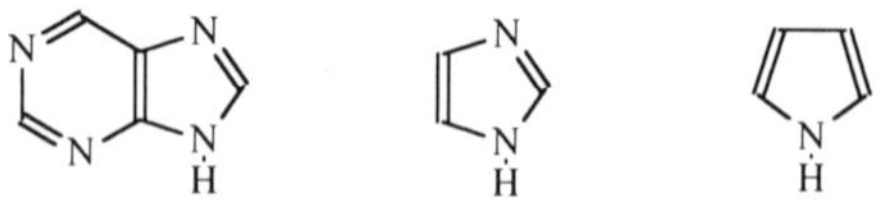

The lone-pair electrons of nitrogen in pyrrole are involved in formation of the aromatic sextet – that is, they become incorporated into the nucleus. The distribution of electron density in the pyrrole nucleus can be represented by the following resonance structures:

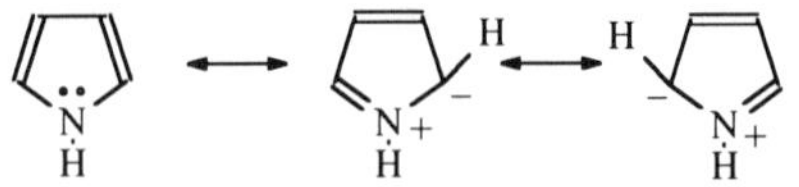

The electronegativity of nitrogen in this case manifests itself only in polarization of the N-H bond, which leads to increased mobility of the proton associated with the nitrogen atom. As a result, electrophilic substitution in the case of pyrrole is facilitated primarily in the α-positions. Thus, the pyrimidine ring in the purine nucleus is marked by surplus of π-electrons. It is expected that positions 2 and 6 in purine are chemically similar to positions 2 and 4 in the pyrimidine ring. Position 8 may be compared to the α-position in pyrrole. It should be taken into account, however, that the π-electron clouds of both monocyclic systems in purines overlap and that the true electron density distribution in each ring may change as a consequence of electron donation by the imidazole ring to the pyrimidine one.

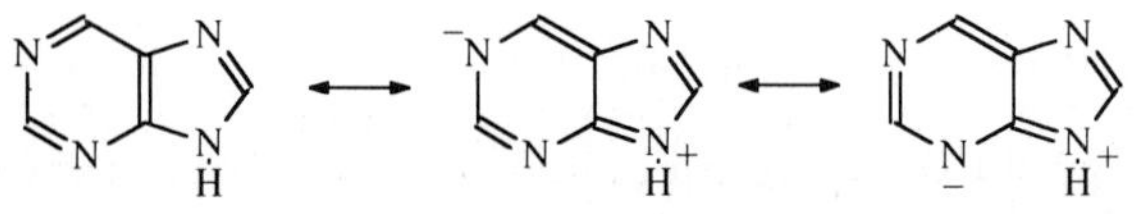

In the presence of electron-donor groups in the pyrimidine moiety, the electron density in the imidazole ring will increase.

The pyrimidine and purine rings, just as those of pyridine and pyrrole, are two-dimensional and all of the hydrogen atoms linked with them lie in the same plane, with the electron sextet in each nucleus, just as in benzene, forming regions of increased electron density above and below the ring plane.

Such are the qualitative concepts[3] usually invoked when discussing the chemical behavior of pyrimidines and purines.

2.1.2 Tautomerism

The ability of heterocyclic bases to undergo tautomeric interconversions markedly influences the three-dimensional structure of nucleic acids as well as the reactivity of the bases themselves, both at the monomer (as part of nucleosides and nucleotides) and polymer levels.

Since nucleic acids contain hydroxy and amino derivatives of purine and pyrimidine, these compounds may display, similarly to hydroxy and amino derivatives of pyridine, two types of tautomerism: lactim-lactam and enamine-ketimine:

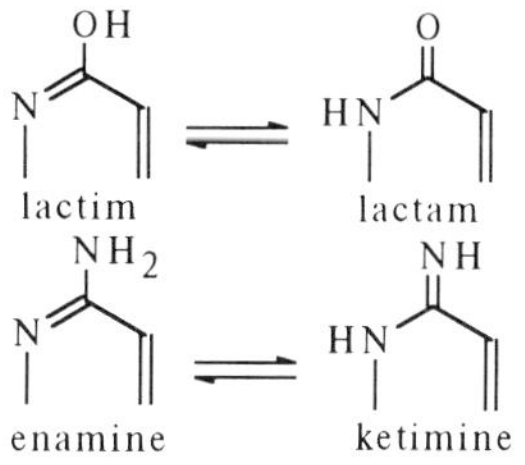

OH, N — lactim ⇌ O, HN — lactam

NH_2, N — enamine ⇌ NH, HN — ketimine

As regards hydroxy derivatives, if they are in the lactam form, the lone-pair electrons of the "amide" nitrogen may lead to formation of a new aromatic system. No rupture of conjugation at the carbonyl carbon takes place because $C{=}C{-}\underset{\|}{C}{-}C{=}C$ systems are characterized by both direct and cross-conjugation. Another argument in favor of the lactam form is the greater affinity to the proton of the nitrogen atom, as compared to the oxygen atom.

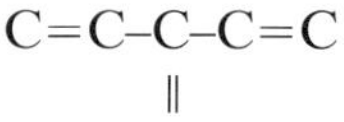

In the case of amino derivatives, it is generally more difficult to chose between the enamine and ketimine forms, although the above assumptions favor the enamine form. Thus, for example, the lactam-enamine form of tautomerism will be more preferable for cytosine and guanine:

[3] Information about the corresponding quantitative calculations can be found in the book "Organic Chemistry of Nucleic Acids" by N. K. Kochetkov and E. I. Budovskii, published in English by Plenum Press (see references at the end of this chapter).

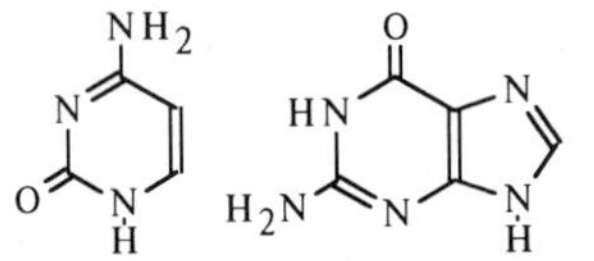

In view of the fact that substituents are present at N^1 of pyrimidine bases and N^9 of purine bases in nucleic acids and their monomer units, the possibilities of tautomeric transitions are limited.

Experimental proof supporting the existence of particular tautomeric forms of pyrimidines and purines in appropriate nucleosides (in aqueous solutions) was provided by UV, IR and NMR spectroscopy. The bases of nucleic acids were compared with substituted pyrimidines and purines in which a particular tautomeric form is possible. In the case of uridine, for instance, N^3-methyluridine and other methylated uracils were used for comparison. At neutral pH values, uracil in uridine may exist in the following three forms:

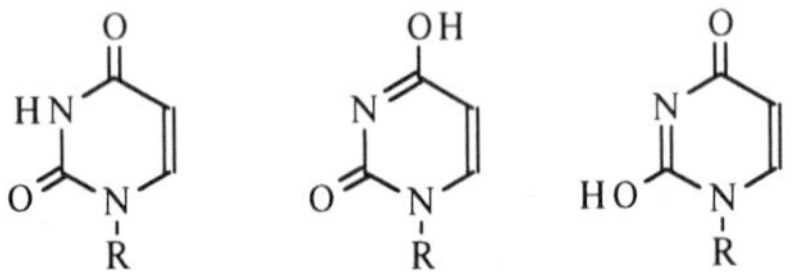

The IR spectra of uridine and N^3-methyluridine incapable of enolyzation turned out to be virtually identical, thereby suggesting that uridine in aqueous solutions exists in a diketo form. This conclusion was also confirmed by the difference between the IR spectra of uridine and its 4-ethoxy derivative which seemed to possess one of the enolic forms. Moreover, it was established that the UV spectrum of uracil resembles more closely that of 1,3-dimethyluracil (1) and differs markedly from that of 1-methyl-4-ethoxydihydropyrimid-2-one (2).

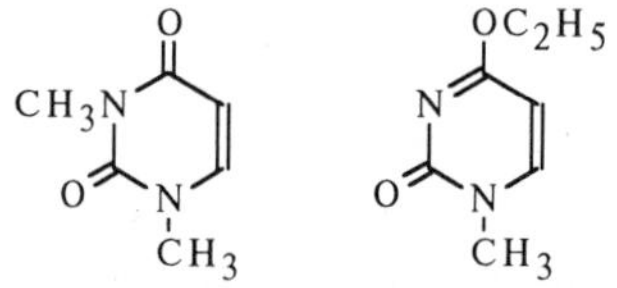

(1) (2)

The NMR spectra of uridine and thymidine in nonaqueous solutions were found to feature a single resonance peak of the proton of the N–H group, which also supports the conclusions drawn from the UV and IR spectra to the effect that uridine and thymidine exist in a diketo form. According to the results of IR spectroscopy and X-ray structural analysis, the diketo form was also ascribed to 5,6-dihydrouridine (a minor nucleoside).

R-ribose residue

The similarity between the UV and IR spectra of *cytidine* (3), *deoxycytidine* (4) or N^1-methylcytosine (5) and their 4-dimethylamino derivatives has led to the conclusion that cytosine nucleosides exist in a keto-amino form:

(3)(R-ribose residue)
(4)(R-deoxyribose residue)
(5)(R=CH_3)

Final proof supporting this conclusion was provided by the finding that the NMR spectrum of cytidine contains a peak typically associated with an aromatic amino group and lacks the peak corresponding to the proton of the N–H group.

Theoretically, *adenosine* may exist both in imino and amino forms. The similarity between the IR spectra of 6-amino and 6-dimethylamino derivatives favors the latter.

R-ribose residue

This conclusion is corroborated by the NMR spectra of adenosine and deoxyadenosine, having a characteristic double-proton peak which corresponds to the amino group, but no peaks corresponding to other protons associated with nitrogen.

Comparison of the IR spectra of *guanosine* (6), 1,9-dimethyl-guanine (7) and 9-β-D-ribofuranosyl-2-amino-6-methoxypurine (8) in D_2O reveals close similarity between the structures of the first two compounds and their difference from the structure of the third compound lacking the carbonyl group (R being ribose residue).

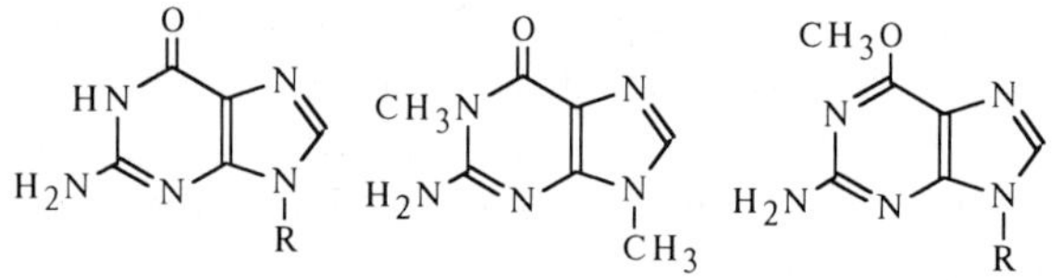

(6) (7) (8)

When the NMR spectra of guanosine and 1,9-dimethylguanine were compared, it was found that the single-proton peak in the spectrum of guanosine corresponds to the proton bound to N^1. The fact that the singlet double-proton preak is not split means that both protons are bound to the same atom which therefore can only be the nitrogen of the 2-amino group. All this is the evidence in support of the keto-amino structure of guanosine.

Comparison of the IR spectra of *inosine* and its derivatives with tautomeric forms (9) and (10) has shown that it has a keto structure (11):

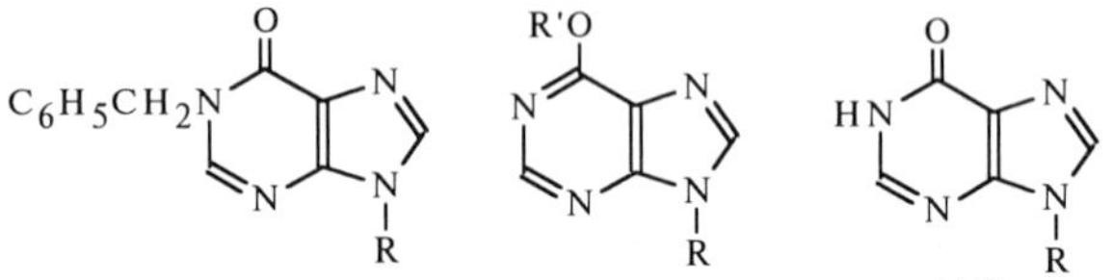

(9) (10) (11)

R-ribose residue; R'=Alk

If one proceeds from the assumption that the electron density in unsubstituted nuclei of pyrimidine and purine is dicontinuous, or split (see above), as well as from the data on the most preferable tautomeric forms of existence of the major pyrimidine and purine bases in nucleosides, one can draw conclusions about the distribution of electron density in the corresponding substituted heterocyclic systems:

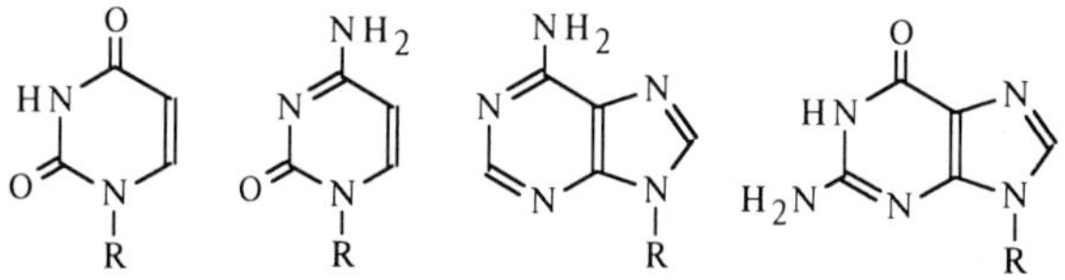

Atoms donating two electrons to the common π-electron system (nitrogen atoms of the pyrrole type, such as N^1 in pyrimidine and N^9 in purines, as well as those of exocyclic amino groups) must exhibit some electron deficiency (δ^+). At the same time, atoms contributing a single electron to the aromatic system and having a pair of free electrons (endocyclic nitrogen atoms of the pyridine type: N^3 of cytosine, N^1 of adenine, and also N^3 and N^7 of adenine and guanine) must exhibit a partial negative charge (δ^-). Accordingly, it is these atoms that must be affected by electrophilic reagents in the first place.

Therefore, all of the carbon atoms, except for C^5 in pyrimidines, must react with nucleophiles. The latter atom is more likely to interact with electrophiles. Such electron density distribution is consistent with calculation results. The reactivity of pyrimidine and purine bases in nucleosides and in nucleic acids also supports the above.

2.1.3 Reactions with Electrophilic Reagents

Substitution Reactions at Carbon Atoms. These reactions include substitution for the protons bound to the carbon atoms at position 5 in pyrimidine bases or position 8 in the purine ring. They lead to modification of the heterocycle in the nucleoside without breaking it. *Halogenation* is the most common of these reactions. Free halogens in a nonaqueous medium trigger direct substitution for the hydrogen (at C^5) in pyrimidine nucleosides:

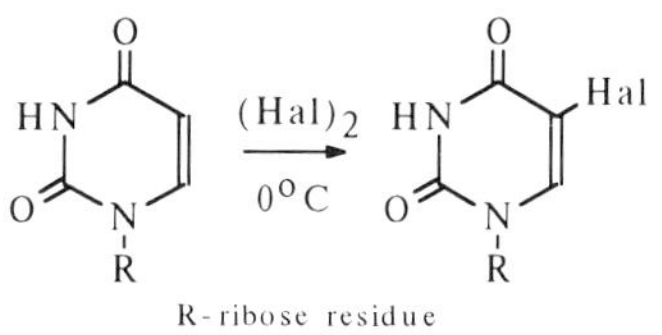

R-ribose residue

The reaction proceeds under mild conditions with a practically quantitative yield. This has been used to obtain 5-chloro and 5-bromo derivatives of uridine and cytidine. Iodination requires much more vigorous conditions – with heating and the presence of nitric acid which seems to be necessary to remove the hydrogen iodide reducing the iodo derivative produced during the reaction. When ICl is used as the iodinating agent, the reaction proceeds under much milder conditions. In the presence of water, addition at the double bond involving C^5 and C^6 takes place in the reaction mixture.

In the case of purine nucleosides, where the competing reaction of addition at the above-mentioned double bond is impossible, halogenation proceeds in both aqueous and nonaqueous solvents:

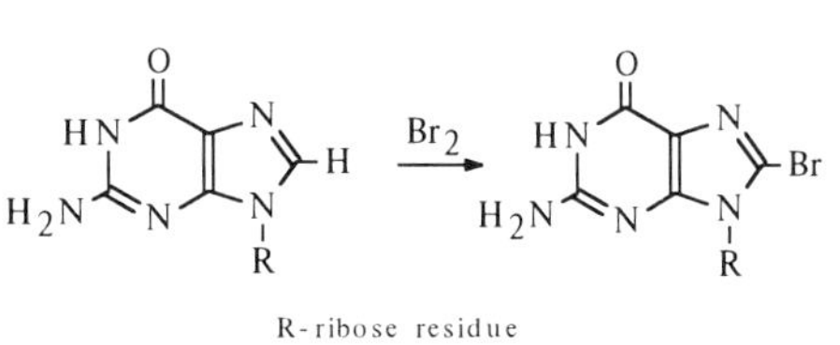

R-ribose residue

Guanosine lends itself more readily to halogenation than adenosine. 8-Halogenpurines are rather unstable compounds and are easily hydrolysed during halogenation with rupture of the imidazole ring.

Nitration of nucleosides requires much more vigorous conditions (higher temperature, nitrating mixture). Therefore, one must protect the hydroxy groups of the carbohydrate moiety in the nucleoside against oxidation.

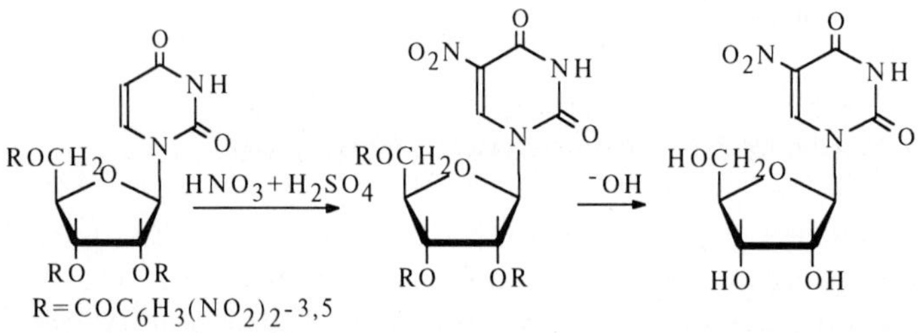
ROCH2
HNO3+H2SO4
RO OR
R=COC6H3(NO2)2-3,5
O2N
ROCH2
RO OR
-OH
HOCH2
HO OH

The corresponding 5-amino nucleosides can be obtained by reduction of the 5-nitro derivatives.

Heating of uridine together with formaldehyde in the presence of hydrochloric acid leads to its 5-hydroxymethylation.

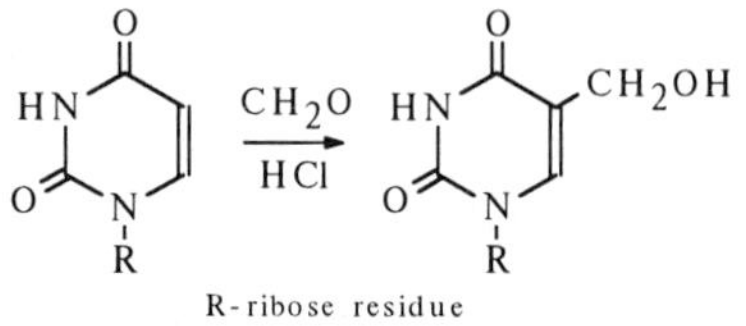
CH2O
HCl
CH2OH
R-ribose residue

Under similar conditions, cytidine does not undergo hydroxymethylation (probably because of protonation of the amino group whose positive charge prevents electrophilic substitution).

5-Hydroxymethyluracil is easily formed when the base is hydroxymethylated in an alkaline medium.

Aminomethylation takes place in the presence of hydrochlorides of secondary amines and formaldehyde.

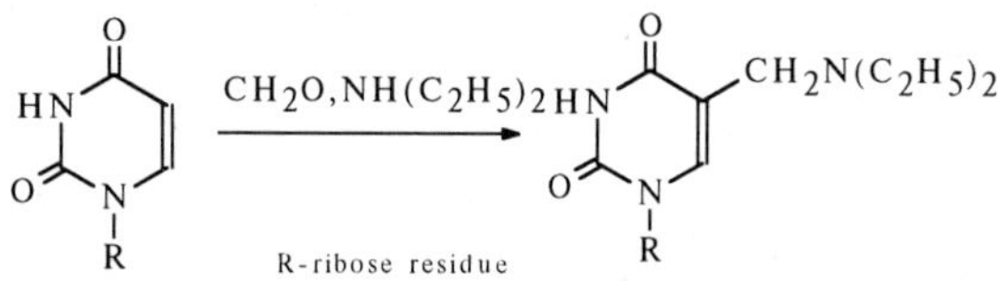
CH2O,NH(C2H5)2
CH2N(C2H5)2
R-ribose residue

5-Hydroxymethyl and 5-aminomethyl derivatives can be converted into the corresponding 5-methylnucleosides through hydrogenation over a platinum or rhodium catalyst.

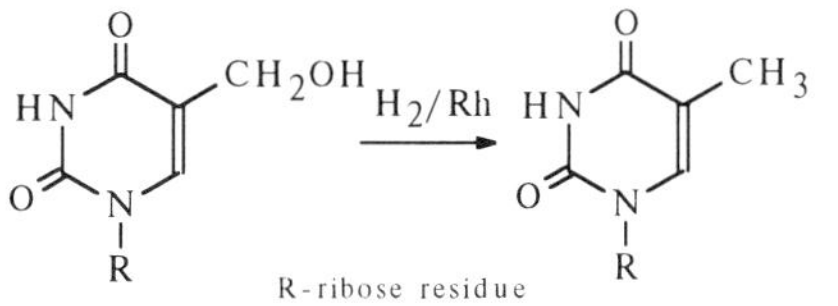

R-ribose residue

These reactions are widely used in nucleoside chemistry for their ability to easily produce 5-methyl and 5-hydroxymethyl derivatives as well as their analogues.

Substitution Reactions at Nitrogen Atoms. The most typical and widely used reactions of this type are those of alkylation with diazomethane, alkyl halides and alkyl sulfates. Other common reactions include those of additon at polarized C=C and C=N bonds and also those yielding N-oxides.

Methylation. The behavior of pyrimidine and purine derivatives during the methylation with diazomethane is largely dependent on the reaction conditions. In the absence of protic solvents the predominant process is substitution of a methyl group for the hydrogen atom exhibiting the most acidic properties – that is, substitution at N^3 in uracil derivatives (this may be equated with the need for preliminary protonation of diazomethane) and at N^1 in guanine derivatives.

The methylation can be represented by the following scheme:

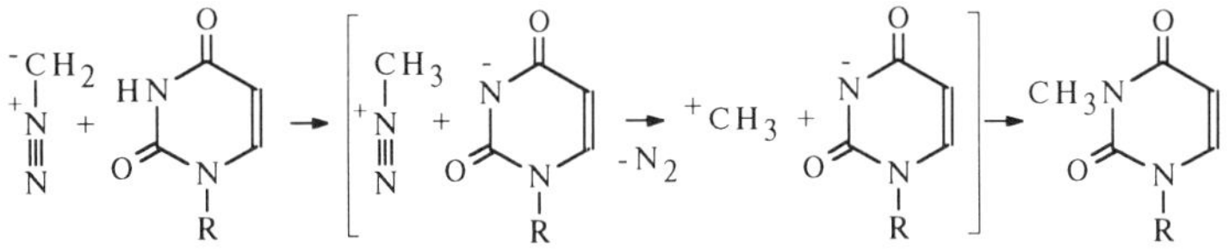

R-ribose residue

If other proton donors are also present in the reaction mixture, the methylation of the pyrimidine or purine ring proceeds at the site with the highest electron density because diazomethane becomes electrophilic in the course of the reaction without participation of the heterocyclic base.

$$^{-}CH_2{-}\overset{+}{N}{\equiv}N + H^+ \longrightarrow \left[CH_3\overset{+}{N}{\equiv}N \longrightarrow {}^{+}CH_3 + N_2\right]$$

For instance, when a guanosine suspension in ether is treated with diazomethane, the reaction yields *N^1*-methylguanosine, whereas the reaction in a water–ether medium gives *N^7*-methylguanosine:

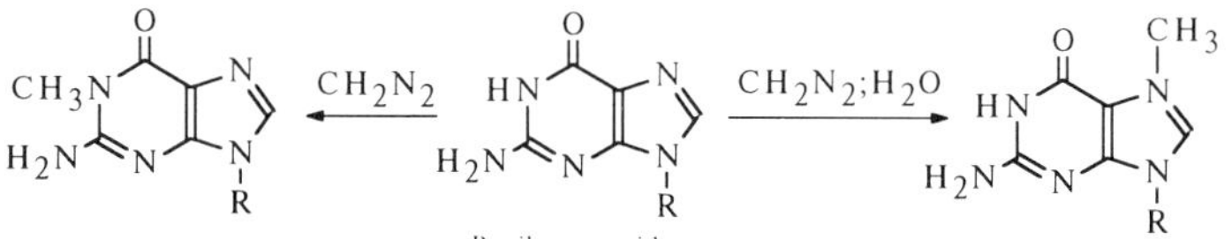

R-ribose residue

The methylation of uracil nucleosides at the heterocycle always involves only N^3; cytosine nucleosides are methylated only in the presence of water, giving N^3-substituents:

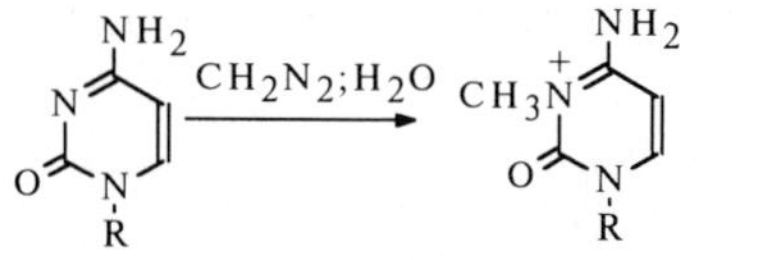

$CH_2N_2; H_2O$

R-ribose and deoxyribose residue

The propensity of nucleosides for interaction with diazomethane in water–ether solutions decreases as follows: guanosine $\approx$ uridine $>$ cytidine $>$ adenosine.

In addition to diazomethane, methyl iodide and dimethyl sulfate are also used for the methylation of bases. The alkyl radical in the molecules of both substances is associated with electronacceptor groups which are substituted, when interacting with sufficiently strong nucleophiles, and act as departing groups. During the reaction, the alkyl (methyl) radical is transferred to the nucleophile and alkylation of the latter occurs. In this particular case, heterocyclic bases act as nucleophiles. When treated with these reagents in a neutral medium, guanosine, adenosine and cytidine undergo methylation, whereas uridine and thymidine do not enter into the reaction.

Guanosine is alkylated giving N^7-methylguanosine; in the presence of potassium carbonate (K_2CO_3), however, the main product of the reaction is N^1-methylguanosine.

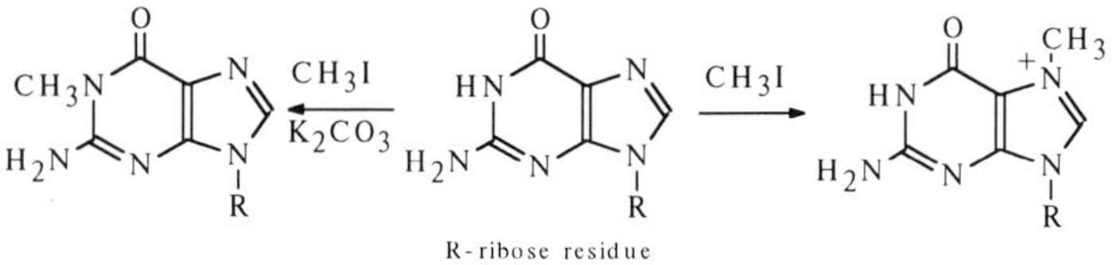

CH_3I, K_2CO_3 ; CH_3I

R-ribose residue

It is most likely that in the presence of K_2CO_3 the –NH–CO group in the heterocyclic ring of guanosine is partially converted into the anionic form –N=C–O$^-$ which (just as most ambidentate ions) reacts with methyl iodide at the most nucleophilic site or, in ohter words, at the nitrogen atom.

Methylation of adenosine proceeds primarily at N^1, and that of cytidine, at N^3.

The ease with which methylation by such agents as methyl iodide and dimethyl sulfate takes place changes in the following decreasing order: guanosine $>$ adenosine $>$ cytidine $\gg$ uridine.

Interaction with Reagents Having C=N and C=C Bonds. The heterocyclic moieties of nucleosides react readily with compounds having C=N and C=C bonds. The mechanisms of such reactions seem to boil down to nucleophilic addition at the polarized double bond.

Examples include reactions with carbodiimides and acrylonitrile. They usually involve heterocycles with –CO–NH– groups, primarily uridine:

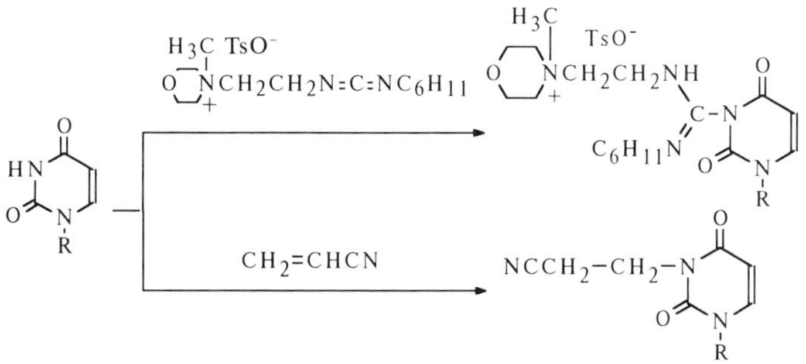

Other substances reacting with carbodiimide include thymidine and guanine nucleosides as well as such minor nucleosides as inosine and pseudouridine.

Oxidation with Peracids. Similarly to many nitrogen-containing heterocycles, heterocyclic bases in nucleosides are oxidized with peracids to yield N-oxides. For example, oxidation of adenosine in the presence of hydrogen peroxide and acetic acid gives N^1-oxide:

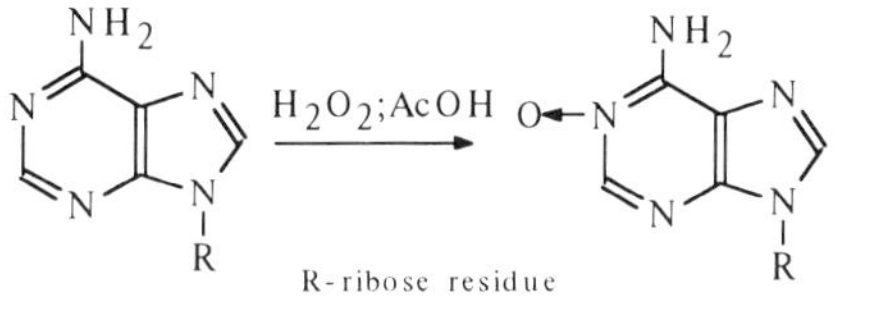

R-ribose residue

Cytidine is oxidized with arylcarboxylic peracids to yield N^3-oxide:

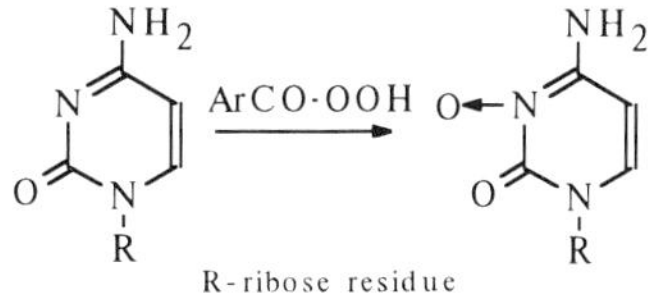

R-ribose residue

Isotopic Exchange of Hydrogen Atoms. The substitution of deuterium and tritium for hydrogen atoms in the purine and pyrimidine rings may proceed, depending on the reaction conditions, as electrophilic or protophilic exchange. The hydrogen atoms linked to nitrogen are exchanged most rapidly, while in the case of those bound to carbons the exchange is much slower. The exchange of the isotope back for hydrogen is also rapid.

Of particular interest for nucleoside studies are samples containing a more or less strongly bound hydrogen isotope and, in this connection, proton exchange reactions at the carbons of the heterocyclic base. If we take protons of low mobility, those bound to the carbon of the imidazole ring (C^8) in purine bases are exchanged more easily than others, and the exchange rate increases with increasing pH of the medium and with temperature.

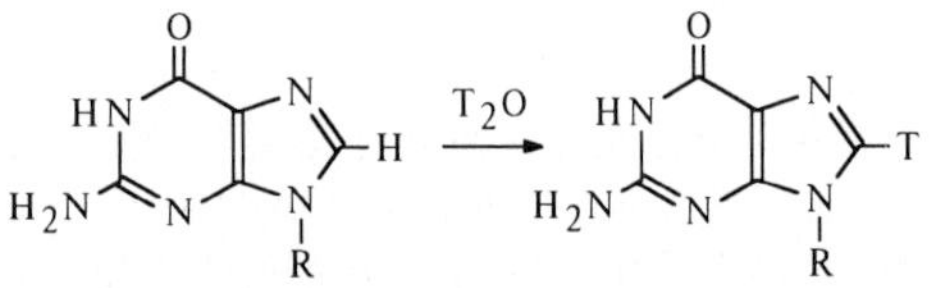

R-ribose and deoxyribose residue

Nucleosides labeled in this fashion are employed in biological experiments that are usually conducted at neutral pH values and temperatures not exceeding 40 °C – that is, conditions ruling out the loss of the label. The exchange in pyrimidine derivatives proceeds under more vigorous conditions, the hydrogen at C^5 being exchanged more easily than at C^6.

2.1.4 Reactions with Nucleophilic Reagents

Reactions of heterocyclic bases with nucleophilic reagents bring about substitution of functional groups (amino groups as a rule) in the pyrimidine or purine nucleus. This is why they arouse a great deal of interest for modification of nucleic acids. The reactions with hydroxylamine, *O*-alkylhydroxylamines as well as hydrazine and its derivatives are the most common.

Depending on pH, hydroxylamine may react selectively with different bases of nucleic acids. At acid and neutral pH values, it reacts with cytosine:

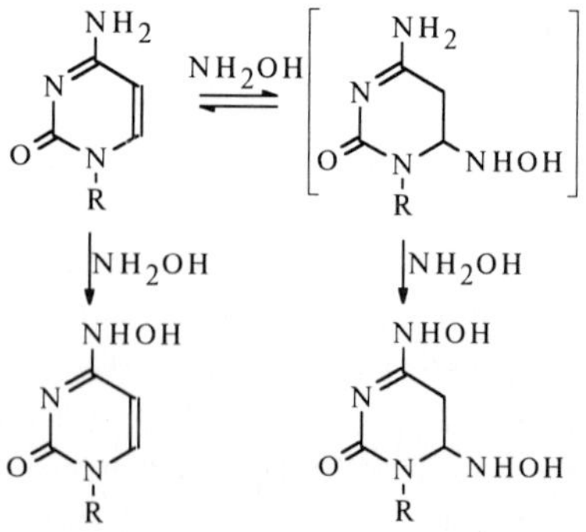

R-ribose residue

The fastest step of the reaction is the nucleophilic addition of hydroxylamine at the double bond involving C^5 and C^6. The subsequent nucleophilic substitution at C^4 gives both di- and monohydroxyamino derivatives of cytosine. *O*-Alkylhydroxylamines react in a similar manner. The reaction rate reaches its maximum at pH 5–6, which indicates that cytidine enters into the reaction in a protonated form, while hydroxylamine (pK_a 6.5) does so as a free base.

The reaction of hydroxylamine and *O*-methylhydroxylamine with adenosine is markedly slower (optimal pH value is 4 or 5). It yields exclusively products of substitution of a hydroxylamino or *O*-methylhydroxyamino group for the amino one.

The amino group in guanosine does not seem to be modified by hydroxylamine and its *O*-alkyl derivative. Only hydroxylamine reacts with uridine and only in an alkaline medium. This reaction leads to the rupture of the heterocycle. This is why it will not be considered here, although the reaction is extensively used for specific modification of nucleic acids at uracil units.

Another group of nucleophilic reagents is represented by hydrazine and its derivatives. The reactions involving them are conducted in an aqueous medium, their mechanism being heavily dependent on the pH of the latter. Nucleophilic substitution occurs only at pH 7 with only cytosine nucleoside entering into the reaction; the reaction products are mono- and disubstituted hydrazines:

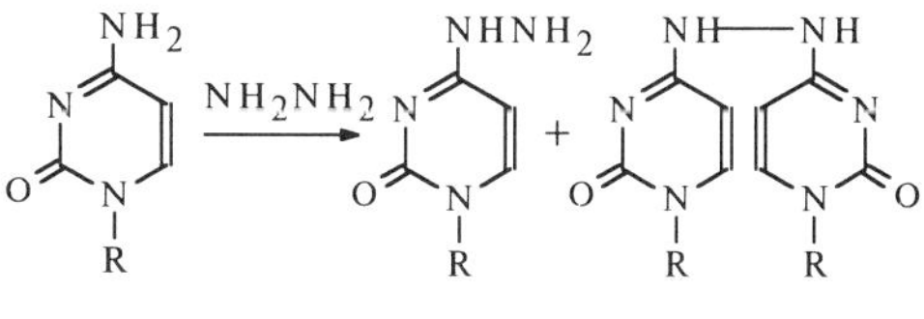

R-ribose and deoxyribose residue

The reaction with monomethylhydrazine proceeds under similar conditions and yields a single product:

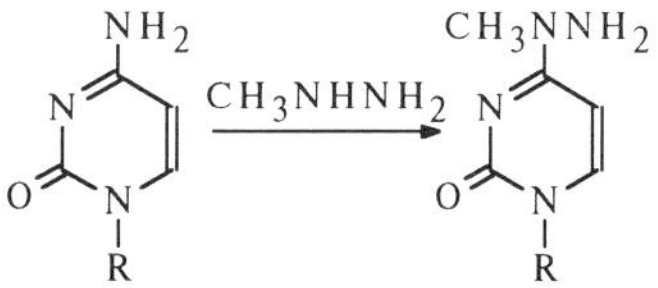

R-ribose residue

Purine nucleosides as well as uridine and thymidine do not react with hydrazine and its derivatives under the above conditions.

In an alkaline medium or nonaqueous solvents hydrazine breaks down uracil and cytosine derivatives. In contrast, it has virtually no effect on purine nucleosides at all.

Apart from the reactions discussed above, nucleophilic substitution for amino groups in heterocyclic bases proceeds in the presence of an alkali and amines. Cytosine nucleosides are the most reactive in this case.

It is konwn that C^5, C^6-dihydrocytidine undergoes deamination most easily in the presence of an alkali; this reaction seems to proceed as follows:

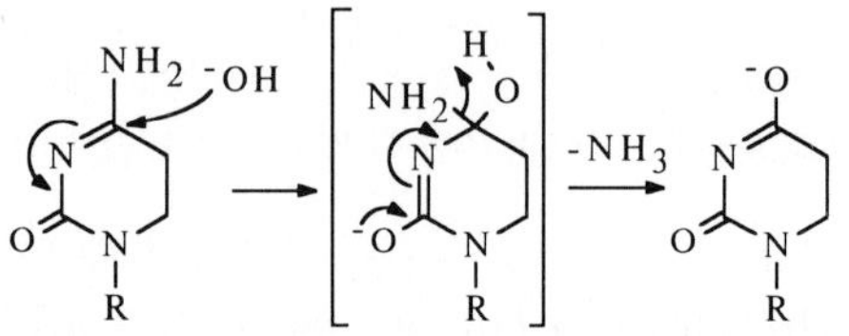

R-ribose residue

In the case of cytidine, the deamination seems to be preceded by hydration giving the corresponding dihydro derivative which is first deaminated as shown above, then undergoes dehydration yielding uridine.

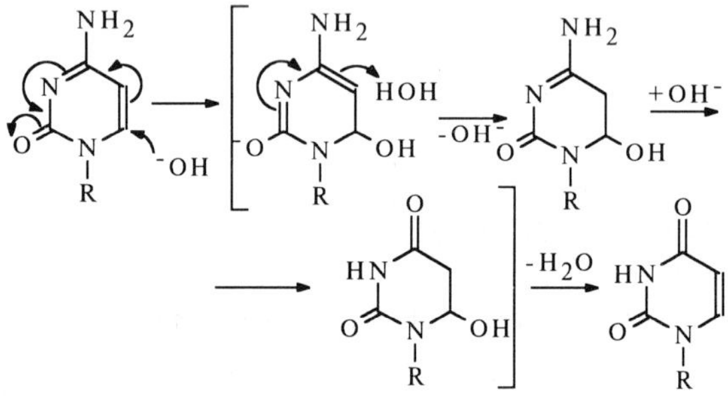

R-ribose residue

Under similar conditions, 5-methyldeoxycytidine reacts at a slower rate, as compared to deoxycytidine, which is quite consistent with the reaction mechanism. Attempts to make adenosine enter into this reaction are successful only under vigorous conditions.

Cytidine and deoxycytidine undergo reamination in aqueous solutions:

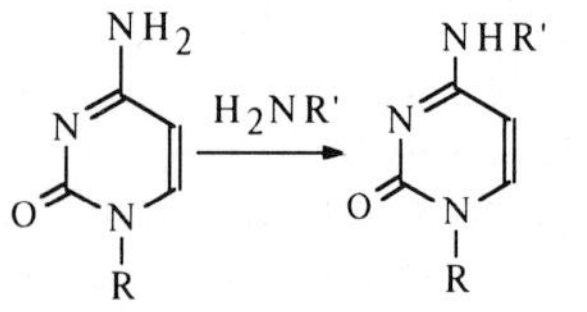

R-ribose and deoxyribose residue; R'=Ar

The reactions of amino group substitution in nucleosides, especially in the case of cytosine derivatives, are widely used in nucleic acid chemistry as a means for controlled modification of bases. Nucleophilic substitution reactions also find use in synthetic chemistry of nucleosides. The substitution of sulfur for the carbonyl oxygen (e. g., in uracil derivatives) also belongs to nucleophilic substitution reactions. For example:

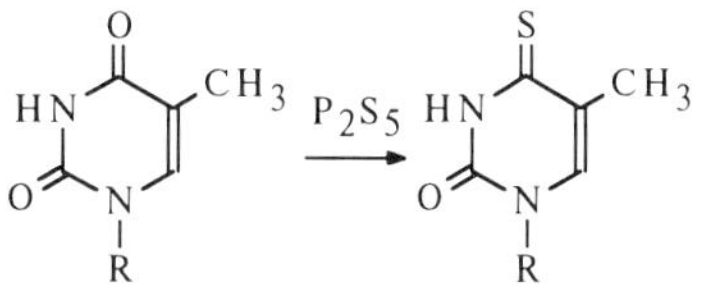

P_2S_5

R-ribose residue

As a result of incorporation of thio groups into ribo- and deoxyribopyrimidine nucleosides, reactive intermediates are formed, from which both major natural nucleosides and their various analogues can be obtained. For instance:

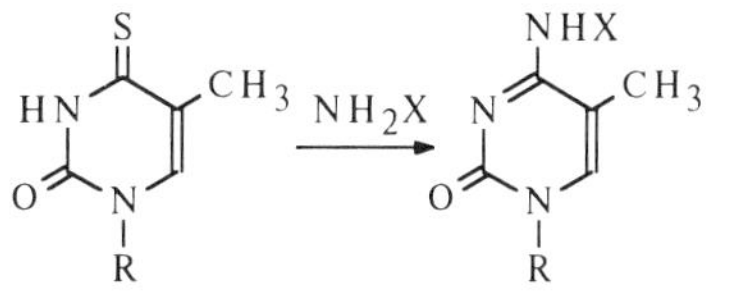

NH_2X

R-ribose and deoxyribose residue; X=H,OH,NH_2

These and similar reactions of nucleophilic substitution have become quite popular in the modification of pyrimidine and purine bases which are subsequently subjected to ribosylation as part of synthesis of natural nucleosides and their analogues with 2-, 4- and 6-substituents.

2.1.5 Addition Reactions

These reactions are typical of pyrimidine derivatives with a reactive C^5-C^6 double bond. They have become instrumental as a means for imparting greater lability to the glycosidic bond in pyrimidine nucleosides when analyzing nucleoside structure as well as during the synthesis of various analogues of nucleosides as part of studies into the mechanisms of their action.

Halogenation of uracil and cytosine nucleosides in an aqueous medium has long been the most fully studied addition reaction. It proceeds in several steps and results, under certain conditions, in addition of two bromine atoms and a hydroxyl (R is ribose or deoxyribose residue).

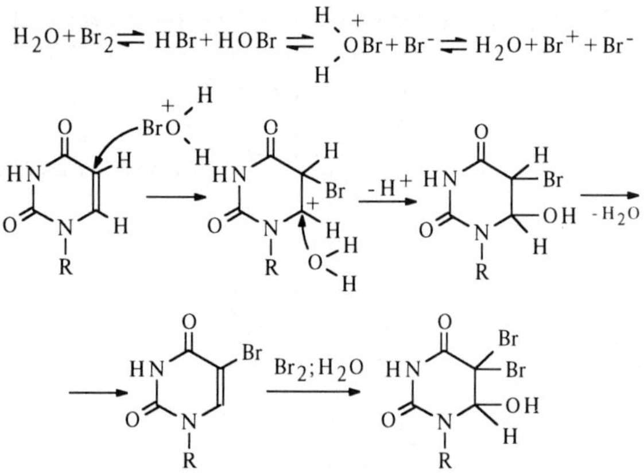

$$H_2O + Br_2 \rightleftharpoons HBr + HOBr \rightleftharpoons H_2O^{+}Br + Br^- \rightleftharpoons H_2O + Br^+ + Br^-$$

The reaction rates differ from one step to another, and the 5-bromo-6-hydroxy-6,6-dihydro derivative can also be obtained with a good yield.

Similarly, a bisulfite ion is added at the C^5-C^6 double bond. This reaction has been studied especially well on nucleotides.

The addition of osmium tetraoxide at the C^5-C^6 double bond causes hydroxylation of the pyrimidine ring.

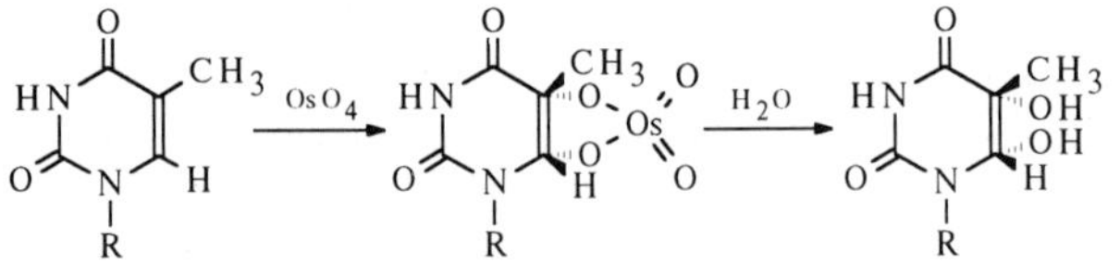

R-deoxyribose residue

The oxidation of thymidine is much easier than that of uracil and cytosine ribo- and deoxyribonucleosides.

The family of furocumarin derivatives known as psoralens has been actively investigated with regard to the capacity of these compounds to act both as dermal photosensitising agents and as probes

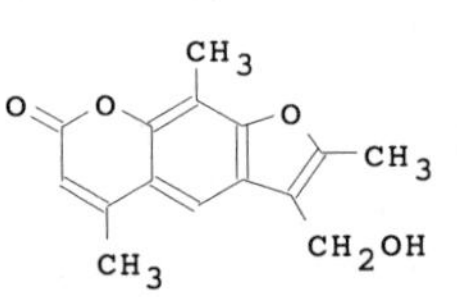

of nucleic acids structure and function. The biological activity of psoralens is primarily the result of covalent binding which involves three discrete steps: (1) nonbonding, intercalating bonding to the DNA; (2) upon irradiation at 365 nm, formation of a monoaddition product between the psoralen and a DNA base, probably a pyrimidine; (3) absorption of a second photon by some monoadducts, which results in interstrand cross-linking. The native

DNA were examined. Five nucleoside-HMT monoaddition products were isolated and characterised; three were deoxythymidine adducts and two were deoxyuridine adducts from deoxycytidine-HMT adducts. Two of these adducts are given below:

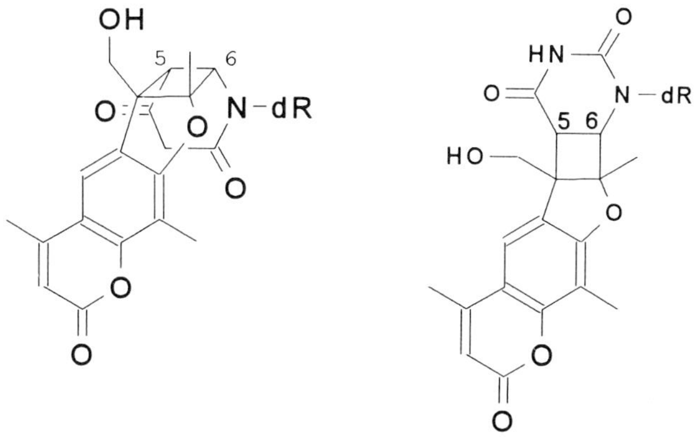

The results of their study indicated that (1) a limited number of nucleoside-psoralen adducts are formed with native dsDNA, and (2) the stereochemistry of the adducts is apparently determined by the geometry of the noncovalent intercalated complex formed by HMT and DNA prior to irradiation.

The double bond in pyrimidine nucleosides lends itself readily to hydrogenation. 5,6-Dihydro derivatives are obtained with a good yield in catalytic hydrogenation in the presence of rhodium on aluminum oxide and also when platinum or palladium catalysts are employed.

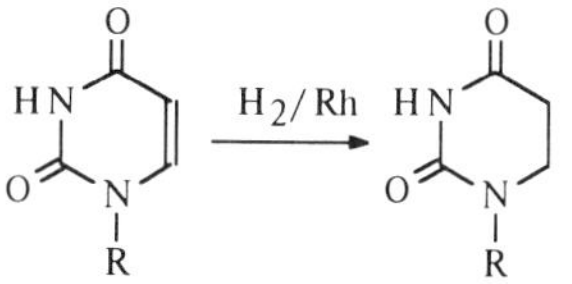

R-ribose residue

In the case of cytidine, the reaction does not stop at the step of addition of one mole of hydrogen, which is followed by a subsequent reduction step. Therefore, when 5,6-dihydrocytidine is the desired product, the process must be carefully controlled. Reduction of pyrimidine nucleosides is also possible using sodium in liquid ammonia or a sodium amalgam in water, however, complex mixtures of reduction products are formed in such cases.

Purine nucleosides are immune to the action of the above-mentioned reducing agents.

2.1.6 Reactions Involving Exocyclic Amino Groups

From the standpoint of amino group properties, cytosine, adenine and guanine closely resemble aromatic amines containing electron-acceptor substituents (e. g., *para*-nitroaniline) in their rings. The pK_a values of the amino groups of the above heterocycles range from 2.5 to 4. None the less, reactions with electrophilic reagents proceed readily at the amino groups of the bases. They are universally used in the synthesis of various derivatives of nucleosides and for their modification at the monomer and polymer levels. The most typical reaction in this category is amino group acylation (the acylation also involving the hydroxy groups of the carbohydrate moiety). Of the three major nucleosides containing amino groups in the heterocyclic nucleus, cytidine enters into acylation reactions most easily. For selective *N*-acetylation the reaction is carried out in an alcohol solution with boiling in the presence of an equimolar amount of acetic anhydride.

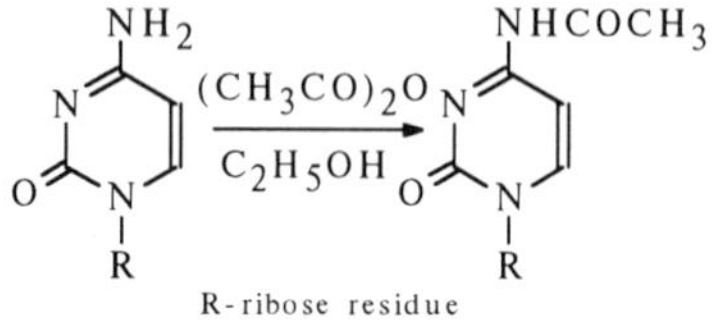

R-ribose residue

The structure of the acetylation products has been established by UV, IR and NMR spectroscopy. When the reaction is conducted under vigorous conditions (surplus of the acylating agent, high temperature), the acylation involves not only the amino group of the heterocycle but also the heterocyclic nitrogen (to say nothing of the hydroxy groups of the carbohydrate moiety), for example:

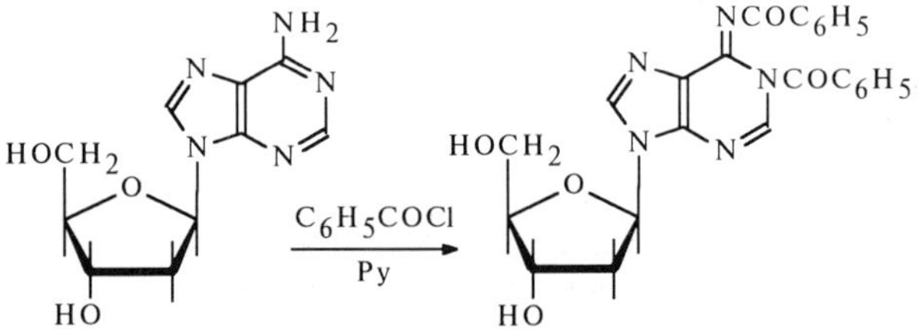

Benzoylation of deoxyguanosine (and guanosine) under the same conditions gives a similar derivative, for example:

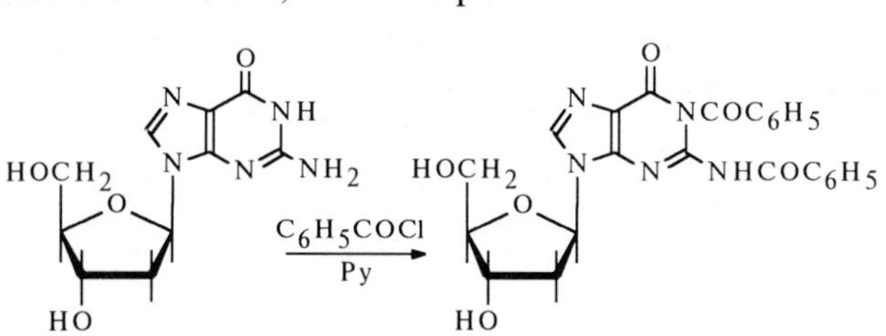

An important reaction also involving the amino group is that between respective nucleosides and aldehydes. A reaction with formaldehyde in aqueous solutions is most likely to yield *N*-methylol derivatives of $RNHCH_2OH$. Only guanosine and its derivatives can enter into a reaction with glyoxal:

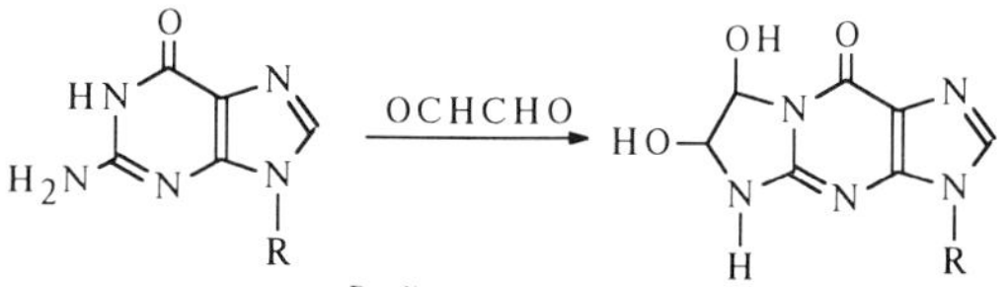

R-ribose residue

The selectivity of this reaction indicates how important the $-NH-C(NH_2)=N-$ group is for its course. 3-Ethoxybutan-2-on-1-al reacts just as guanine nucleosides do:

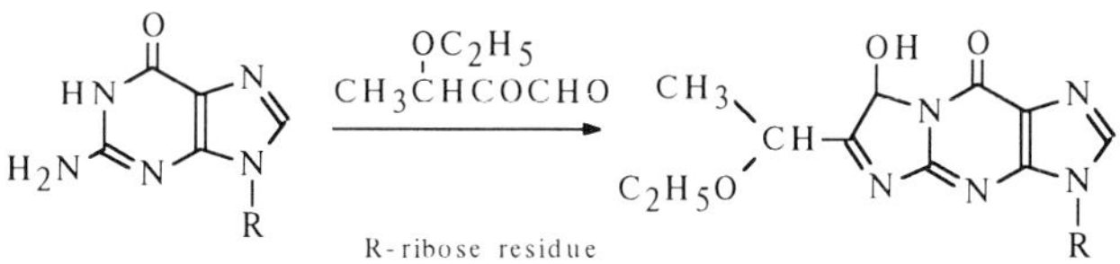

R-ribose residue

Nucleosides containing amino groups in the heterocycle typically enter into a reaction with nitrous acid. A case in point is adenosine which is converted into inosine in the presence of nitrous acid:

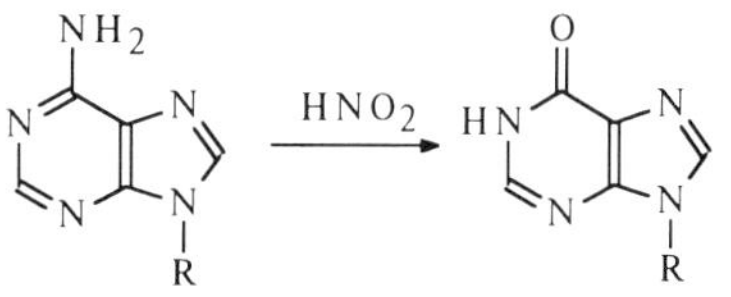

R-ribose residue

Deamination of guanosine and cytidine yields, respectively, xanthosine and uridine. The reaction seems to result in the corresponding diazonium salt in which no delocalization of the positive charge takes place and which readily dissociates, just as diazoalkyls, in an aqueous medium giving a hydroxy derivative. For instance:

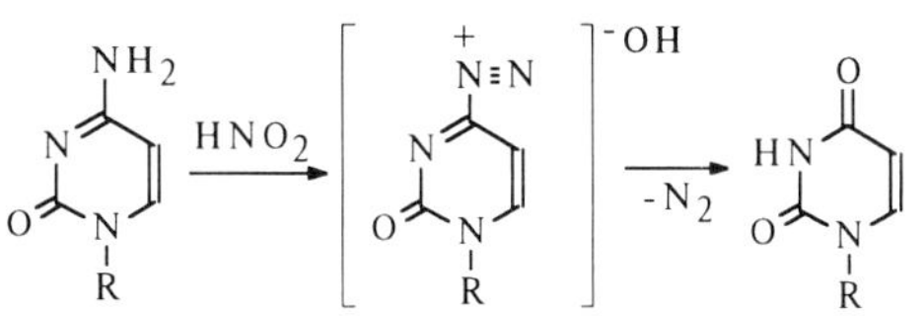

R-ribose residue

2.2 Reactions at the Carbohydrate Moiety

The pentose moiety of nucleosides is typically involved in three types of reactions: substitution for hydrogen atoms in hydroxyl groups, oxidation, and substitution for hydroxyl groups.

2.2.1 Substitution for Hydrogen Atoms in Hydroxyl Groups

This category includes such reactions as acylation and alkylation of sugars. They are used in the synthesis of nucleotides and their derivatives to block the hydroxyl groups of the carbohydrate moiety, therefore, this section deals primarily with the conditions under which the blocking groups are removed.

Acylation. Nucleosides react with anhydrides and acid chlorides in anhydrous pyridine at room temperature, giving nucleosides completely acylated in the carbohydrate moiety; for instance, acetylation may proceed as follows:

In the case of cytidine, the amino group of the base also undergoes acylation under the same conditions. The primary hydroxyl in deoxynucleosides lends itself more readily to acylation, which makes it possible to obtain 5′-monoacetyl derivatives under mild conditions:

Acetylation of 5′-substituents of adenosine and uridine yields a mixture of 2′,5′- and 3′,5′-diacetyl derivatives. Heating of the latter with anhydrous pyridine or water brings about 2′ → 3′ isomerization as a result of which 3′,5′-di-*O*-acetylnucleoside is obtained with an almost quantitative yield:

Alkaline hydrolysis of *O*-acylnucleosides permits acyl groups to be removed under mild conditions.

The ester bond is also easily cleaved when *O*-acylnucleosides are treated with an alkaline solution of hydroxylamine:

AcOCH$_2$ B, AcO OAc $\xrightarrow{NH_2OH;\, pH>7}$ HOCH$_2$ B, HO OH + 3AcNHOH

Aminoacylation of ribonucleosides has been studied most thoroughly. The synthesis on which all methods are based resides in a reaction of 5′-protected nucleosides with amino acids protected (or deprotected) at the amino group and activated at the carboxyl:

YOCH$_2$ B, HO OH $\xrightarrow{\text{Z-NHCH(R)CO-X}}$ YOCH$_2$ B, Z-NHCH(R)C(O)O OH + 2'-O-Isomer $\longrightarrow$ HOCH$_2$ B, NH$_2$CH(R)C(O)O OH + 2'-O-Isomer

The methods for activating amino acids (the nature of X) and their protective groups (Z) have been borrowed from peptide synthesis; Y is usually a trityl group or its analogue.

An investigation of this reaction has led to two conclusions as regards the reactivity of nucleosides. Firstly, the activation of amino acids for the reaction with ribose hydroxyls must be sufficiently effective (only anhydrides and imidazolides of amino acids are capable of reacting). Secondly, in absolute solvents only the NH_2 group of cytosine, if the hydroxyls of ribose are excepted, reacts with anhydrides, imidazolides and active esters of *N*-protected amino acids.

Similarly to the synthesis of 3′(2′)-*O*-aminoacylnucleosides, the method based on the use of dicyclohexylcarbodiimide (the synthesis seems to involve formation of amino acid anhydrides as intermediates) makes it possible to introduce *N*-protected peptides into nucleosides. *O*-Aminoacyl-nucleosides can be isolated from these peptides by eliminating the protective groups. The hydroxyl groups of the carbohydrate moiety in nucleosides may also be acylated with derivatives of inorganic acids. Phosphorylation of nucleosides has become routine. Among the most widely used reactions of this type are also those with aryl(alkyl)sulfonyl chlorides, yielding esters of sulfonic acids.

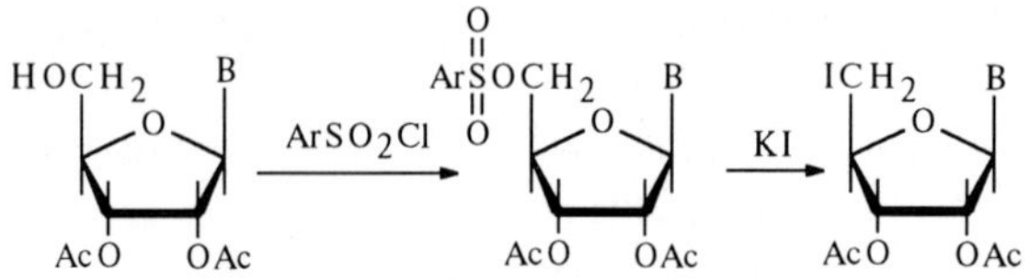

The arylsulfo group in such derivatives may be replaced by iodine. If a deprotected nucleoside is introduced into the reaction with aryl(alkyl)sulfonyl chloride, a nucleoside with a fully substituted ribose moiety can be obtained, for example:

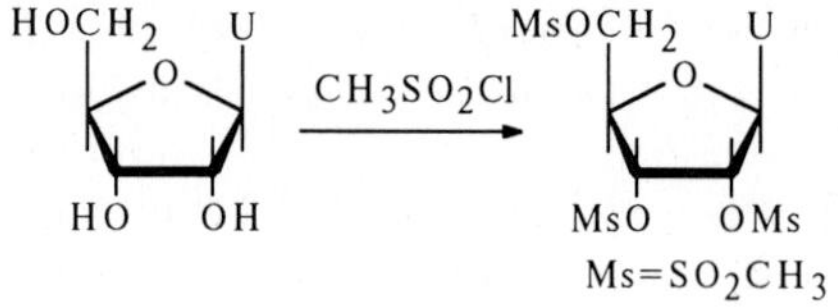

$Ms = SO_2CH_3$

Alkylation. As has already been mentioned, in addition to *N*-alkyl derivatives a reaction between alkylating agents (diazomethane, methyl iodide) and nucleosides also yields *N,O*-dialkylnucleosides, the 2′-hydroxyl group undergoing alkylation with the greatest ease.

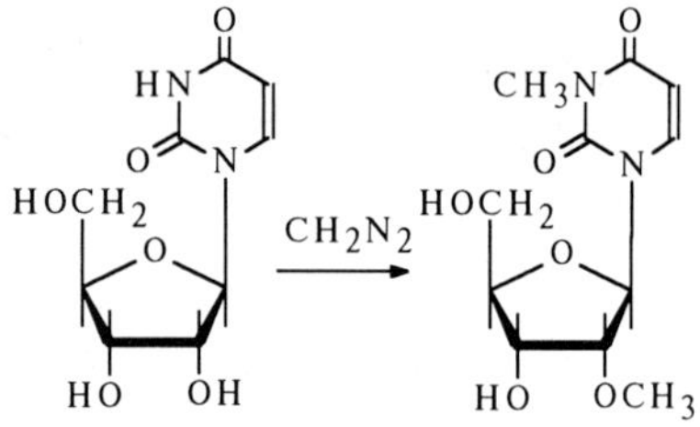

Selective 2′-*O*-methylation can be conducted successfully when adenosine or cytidine is heated together with diazomethane in an aqueous solution of 1,2-dimethoxyethane. For example:

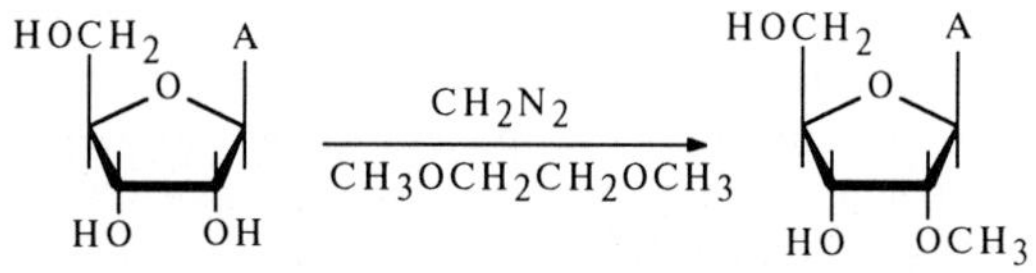

Alkylation of nucleosides with triphenylchloromethane proceeds in a selective manner, at the primary hydroxyl group.

HOCH2 B / HO OH(H) —TrCl→ TrOCH2 B / HO OH(H)

$Tr=(C_6H_5)_3C$

It is only in the case of uridine that 2′,5,-*O*-ditrityluridine is obtained along with the 5′-*O*-monotrityl [5′-*O*-mono(triphenylmethyl)] derivative:

HOCH2 U / HO OH —TrCl→ TrOCH2 U / HO OH + TrOCH2 U / HO OTr

Triphenylmethyl esters are hydrolysed (elimination of the blocking trityl group) by the action of dilute hydrochloric or acetic acid. Incorporation of a methoxy group into the phenyl ring of trityl ester enhances the lability of the ether bond. The mono(methoxy)triphenylmethyl (MeOTr) group, for example, is completely removed in the presence of 80% acetic acid at 20 °C within two hours, whereas removal of the trityl group under the same conditions requires several days.

$p\text{-}CH_3OC_6H_4(C_6H_5)_2C\text{-}OCH_2$ B / HO OH(H) —80% AcOH, 20°C, 2 hrs→ HOCH2 B / HO OH(H)

The reaction with *para*-mono(methoxy)triphenylchloromethane, MeOTrCl, is widely used in the synthesis of nucleoside derivates (especially in the synthesis of polynucleotides) by virtue of its high selectivity (only the primary hydroxyl group is involved) and the ease of removal of the protective group. Another rather popular reaction is acetal-yielding alkylation involving vinyl ethers.

In the first description of this reaction, uridine-3′5′-cyclic phosphate reacted with dihydropyrane:

OCH2 U / O=P / HO O OH —(dihydropyran), H+→ OCH2 U / O=P / HO O O-(tetrahydropyranyl)

The reaction is conducted in organic solvents in the presence of dry hydrogen chloride. The tetrahydropyranyl groups is removed by mild acid hydrolysis.

Similar derivatives are yielded by reactions between nucleosides and 2-methoxypropene:

AcOCH2 B O AcO OH → ($CH_3C(OCH_3)=CH_2$) AcOCH2 B O AcO $OC(CH_3)_2$ OCH_3

Reactions with Carbonyl Compounds. Ribonucleosides rather easily react with aldehydes and ketones in the presence of acid catalysts to yield cyclic acetals or ketals. The reaction with acetone is by far the most common:

HOCH2 B O HO OH → ($(CH_3)_2CO$) HOCH2 B O O O C H_3C CH_3

In the case of benzaldehyde, a similar reaction gives rise to a new asymmetric center with the result that two diastereomers are formed:

HOCH2 B O HO OH → (C_6H_5CHO) HOCH2 B O O O C* H C_6H_5

Ketals and acetals are easily hydrolysed in a weakly acid medium. The specificity of the interaction of aldehydes and ketones with the *cis*-glycol group and the ease of hydrolysis of the forming compounds are reasons why this reaction is extensively used for the selective protection of 2′- and 3′-hydroxyl groups in ribonucleosides.

2.2.2 Oxidation

It has already been mentioned above that nucleosides possessing a 2′,3′-*cis*-glycol group (ribofuranosides) readily undergo oxidation with periodic acid (X is phosphoric acid, its ester or another group):

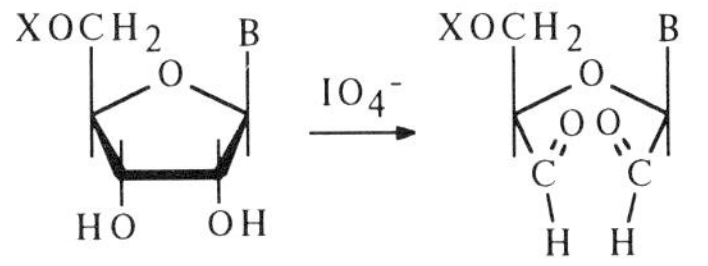

The reaction is rather specific and proceeds in an aqueous medium under extremely mild conditions. The oxidation products – dialdehydes – exist in aqueous solutions in the form of hydrates. Dialdehydes interact most readily with the usual reagents sensitive to aldehyde groups. For instance, their treatment with phenylhydrazine gives bis(phenylhydrazones):

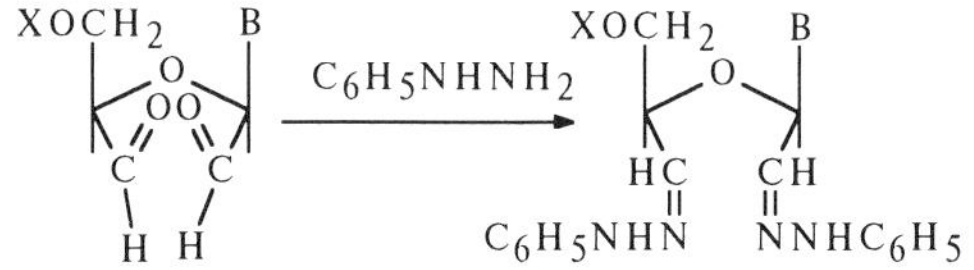

Dialdehydes and their derivatives are very easily oxidized in an alkaline medium with simultaneous cleavage of the glycosidic bond and detachment of XO^- (β-elimination) (here, X is phosphoric acid or its ester):

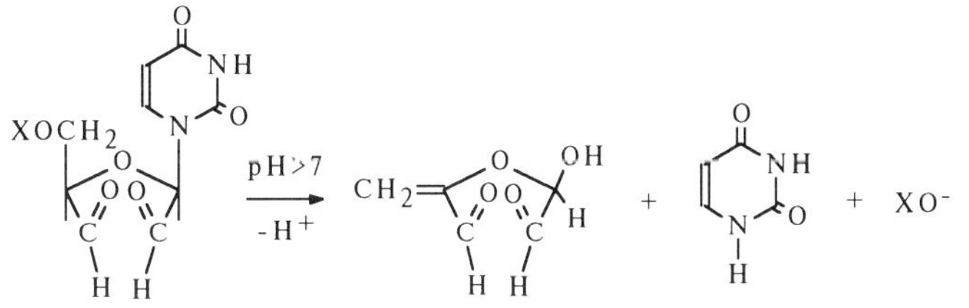

The products of periodate oxidation of ribonucleosides are readily reduced to the corresponding triols in the presence of sodium borohydride which is relatively stable in aqueous media:

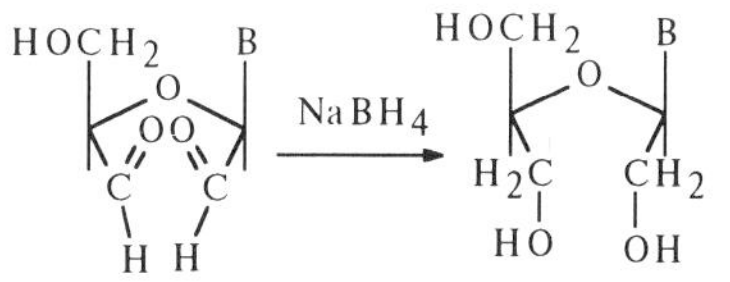

By using sodium borotritide, $NaBT_4$, in this reaction one can easily label the oxidized nucleoside molecule.

The above-described oxidation reaction and its combination with reduction are widely used in various structural studies in the chemistry of nucleosides and nucleic acids.

Treatment of ribonucleosides with oxygen over a platinum catalyst leads to selective oxidation of the primary hydroxyl group with formation of uronic acid derivatives:

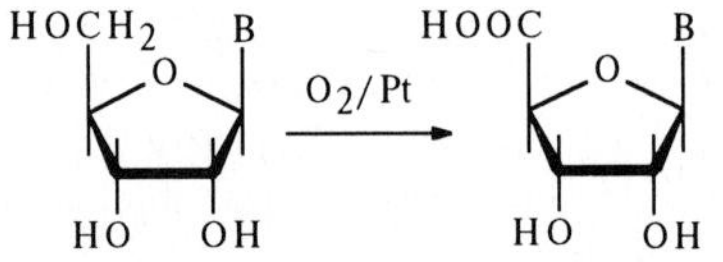

2.3 Reactions Involving Heterocyclic Bases and the Carbohydrate Moiety

A typical reaction in this category is the above-mentioned intramolecular alkylation of nucleosides, giving rise to a new ring, for example:

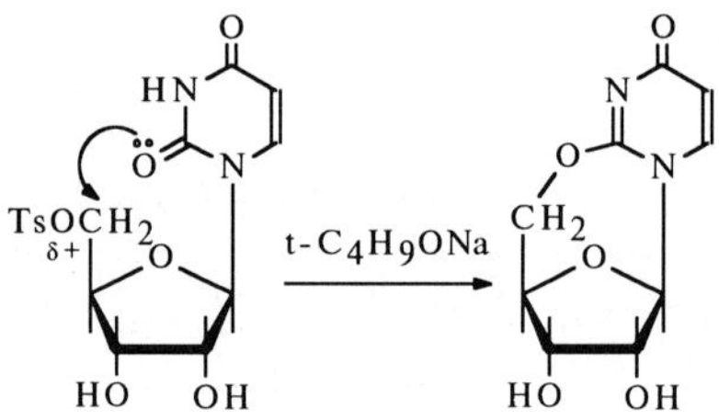

The starting compounds for obtaining cyclonucleosides may be aryl- or alkylsulfonates of nucleosides, containing both primary and secondary hydroxyl groups:

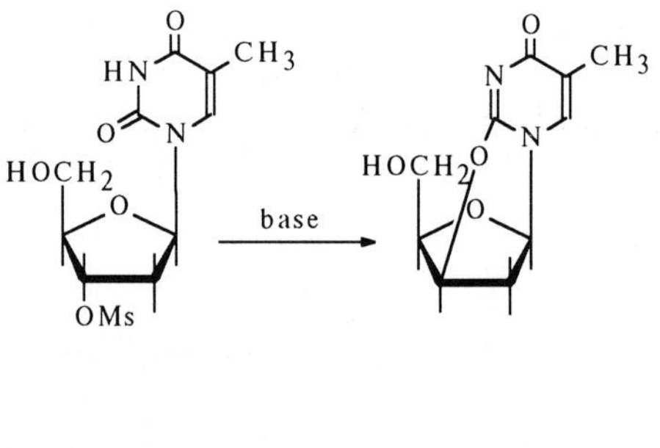

Similar transformations are also undergone by the corresponding 2′-derivatives of nucleosides:

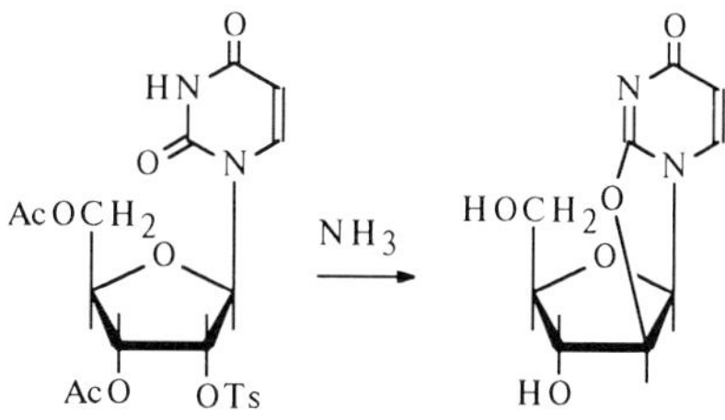

Uracil and thymine cyclonucleosides can also be derived from appropriate iododeoxynucleosides in the presence of silver salts:

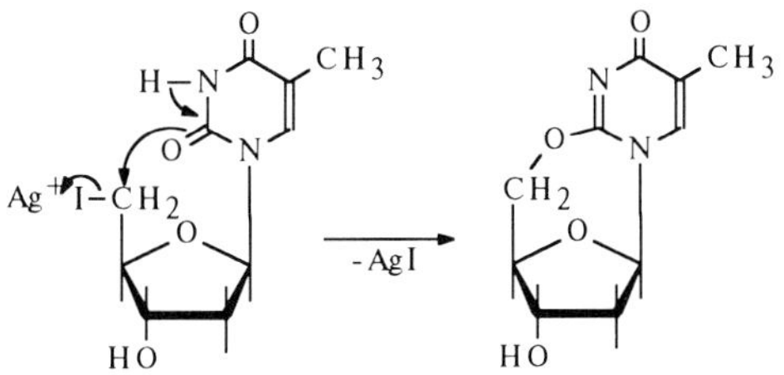

Depending on the carbon atom (5′, 3′, or 2′) of the carbohydrate moiety involved in the formation of the oxygen-containing bridge with the uracil (thymine) nucleus, distinction is made between O^2,5′-, O^2,3′- and O^2,2′-cyclic nucleosides. If we take a look at the ease of their formation, cyclic nucleosides can be arranged in the following order: O^2,2′-cyclic nucleoside > O^2,3′-cyclic nucleoside > O^2,5′-cyclic nucleoside.

Various cyclic nucleosides obtained from thymidine, uridine and 5-substituents of the latter are widely used in the synthesis of nucleoside analogues with a modified carbohydrate moiety and pyrimidine.

In addition to the intramolecular alkylation giving rise to new cyclic structures, other reactions have already been described, including those of alkylation and acylation, which involve both heterocyclic bases and the carbohydrate moiety.

2.4 Stability of *N*-Glycosidic Bonds

The *N*-glycosidic bonds in major nucleosides isolated from RNA and DNA are usually highly stable in neutral and alkaline media and prone to hydrolysis in the presence of mineral and organic (HCOOH, CH_3COOH, CCl_3COOH) acids. The hydrolysis rate depends on the hydrogen ion concentration; the reaction proceeds according to the general scheme:

$HOCH_2$ N / O / HO OH(H) $\xrightarrow{H_2O, H^+}$ $HOCH_2$ O ∿OH / HO OH(H) + N–H

The ease of hydrolysis is largely dependent on the nature of the heterocyclic base and sugar moiety.

2.4.1 Effect of the Heterocyclic Base Species

Acid hydrolysis of purine derivatives is much faster than of pyrimidine ones, which becomes evident from the following kinetic data on nucleoside hydrolysis (pH 1.0, 37 °C):

Compound	k_1, s^{-1}
Adenosine	3.6 10^{-7}
Guanosine	9.36 10^{-7}
Cytidine	1 10^{-9}
Uridine	1 10^{-9}
Deoxyadenosine	4.3 10^{-4}
Deoxyguanosine	8.3 10^{-4}
Deoxycytidine	1.1 10^{-7}
Deoxyuridine	1 10^{-7}

For hydrolysis of purine ribonucleosides, heating with 0.1 N hydrochloric acid (100 °C, 1 hour) is quite sufficient, in contrast, pyrimidine nucleosides requires boiling with 3 N HCl (125 °, 4 hours).

The hydrolysis rate is also markedly affected by substituents in the heterocyclic ring.

Glycosidic bonds in guanine derivatives are more sensitive to acids, as compared to those in adenine derivatives. Substitution of hydroxy groups for the amino ones in these nucleosides leads to a perceptible labilization of the glycosidic bonds. For instance, derivatives of such a hydroxy analogue of guanine as xanthine are hydrolysed more easily than those of hypoxanthine which is a hydroxy analogue of adenine.

Introduction of alkyl groups at positions 3 and 7 of the purine ring (alkylation of nitrogens N^3 and N^7) also causes a tangible decrease in the glycosidic bond strength. A case in point is 7-methyldeoxyguanosine

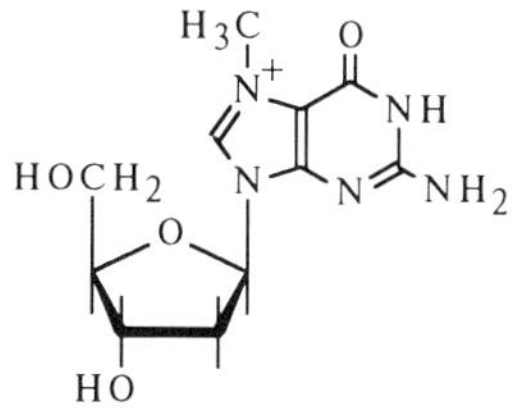

whose hydrolysis under mild conditions proceeds at a rate ten thousand times faster than in the case of deoxyguanosine. N^3- and N^7-methyl derivatives of deoxyadenosine are much less stable than the latter.

In the case of pyrimidine nucleosides, as opposed to purine ones, substitution of a hydroxy group for the amino group at position 4 renders the glycosidic bond much more stable. As can be inferred from the date that follow, under conditions where deoxyuridine remains virtually unchanged most of deoxycytidine is hydrolysed (hydrolysis with 5% trichloroacetic acid, 100 °C, 30 min):

Compound	Degree of cleavage, %
Deoxycytidine	76
Deoxyuridine	3
Deoxythymidine	4.3
5-Bromodeoxyuridine	16

Introduction of a methyl group at position 5 of the pyrimidine ring does not produce any material effect on the hydrolysis rate (deoxythymidine is practically as stable as deoxyuridine). However, an electron-acceptor group at the same position makes cleavage of the glycosidic bond much easier: the 5-bromo derivative is hydrolysed more easily than the starting deoxyuridine (see above). The stability of the glycosidic bond in cytosine nucleosides is markedly affected by introduction of an acyl into the amino group of the heterocycle. For instance, 4-*N*-acyl derivatives of cytidine and deoxycytidine are hydrolysed in an acid medium much more readily than the starting nucleosides.

The glycosidic bond becomes even less stable in C^5,C^6-dihydropyrimidine nucleosides. The stability of such compounds in an acid medium is comparable to that of ordinary glycoside amines.

The lower stability of the glycosidic bond as a result of upset aromaticity of the pyrimidine ring is widely used while determining the structure of nucleosides. For example, to isolate the sugar or base from a pyrimidine nucleoside, it is first subjected to hydrogenation, then the 5,6-dihydronucleoside formed is hydrolysed under very mild conditions (0.1 N HCl, 100 °C, 15 min).

2.4.2 Effect of Substituents in the Carbohydrate Moiety

The stability of the glycosidic bond is strongly dependent on the nature of the substituent at positions 2′ and 3′ in the carbohydrate moiety of the nucleoside. Ribonucleosides are much more stable (100–1000 times) toward hydrolysis than the corresponding deoxynucleosides (see above data on kinetics of hydrolysis).

Substitution of an electronegative hydroxy group for hydrogen in the furanose cycle leads to a marked increase in stability of the glycosidic bond, as can be inferred from the following kinetic data (hydrolysis with 1 N HCl, 100 °C):

Compound	Formula	Half-life, min
2′,3′-Dideoxyuridine	(12)	8.2
2′-Deoxyuridine	(13)	104
Uridine	(14)	not hydrolysed

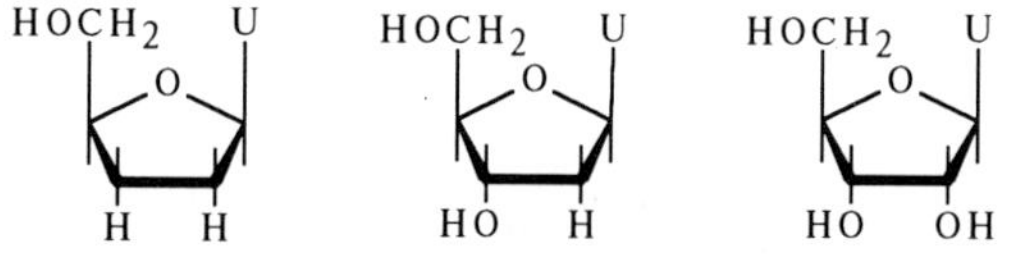

HOCH2 U O H H (12) — HOCH2 U O HO H (13) — HOCH2 U O HO OH (14)

The glycosidic bond becomes even more stable when iodine is substituted for the hydroxy group or when such electron-acceptor groups as toluenesulfo- or 2,4-dinitrobenzoyls are inserted into the ribose.

2.4.3 Mechanism of Hydrolysis of *N*-Glycosidic Bonds

The linear relationship between the rate of acid hydrolysis of the *N*-glycosidic bond in nucleosides and the pH value of the reaction mixture suggests that the cleavage of this bond is preceded by protonation of the heterocyclic base, which, consequently, must be the step determining the rate of the reaction.

It is currently believed that hydrolysis of the *N*-glycosidic bond is not accompanied by opening of the furanose ring, as was previously assumed.

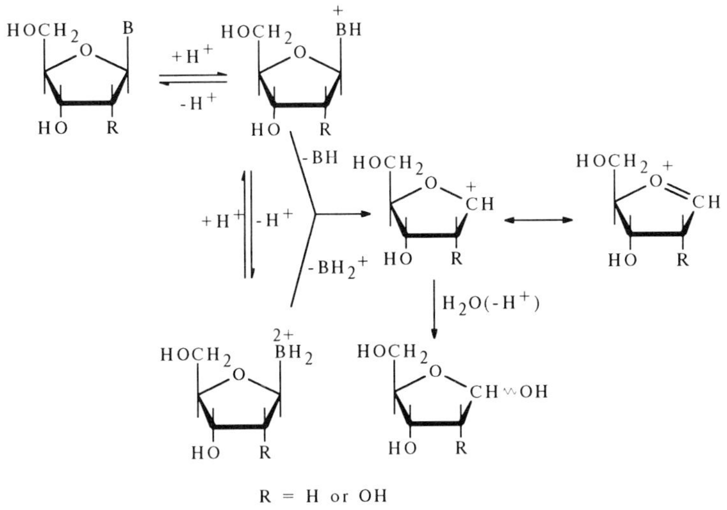

R = H or OH

The above scheme shows clearly that in the course of the reaction, which seems to be based on a unimolecular mechanism, the base is detached from the carbohydrate moiety in a mono- or diprotonated state. For example, hydrolysis of the glycosidic bond in deoxycytidine may be written as follows:

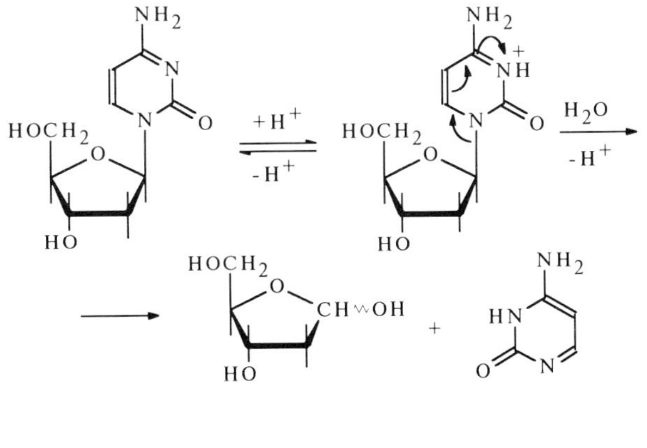

The assumption is that during hydrolysis of deoxythymidine the base is protonated at the oxygen atom. For instance:

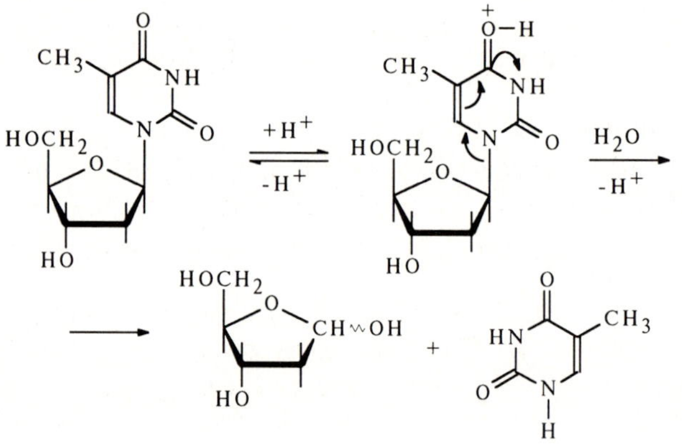

In the case of purine nucleosides, the protonation seems to proceed at the nitrogen of the imidazole ring, N^7, which (at least in guanine derivatives) is the site of maximum electron density.

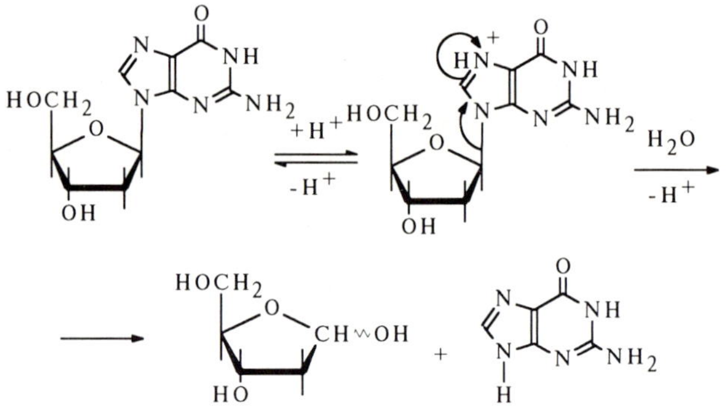

Such a mechanism for the cleavage of the *N*-glycosidic bonds in purines is further supported by the extremely high lability of nucleosides methylated at position 7 or 3. In this case, the positive charge at the respective nitrogen atom is fixed strongly enough. Therefore, such nucleosides are much less stable toward acid hydrolysis than the corresponding non-methylated analogues; moreover, they can also be hydrolysed in an alkaline medium. For example, the acid hydrolysis rate for 7-methyldeoxyguanosine is several orders of magnitudes above that for deoxyguanosine and is independent of the pH value of the reaction mixture. In addition this nucleoside is easily hydrolysed in an alkaline medium.

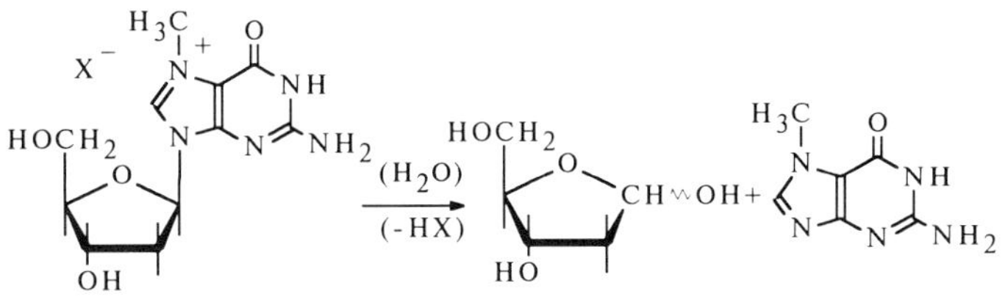

3-Methyldeoxyguanosine is even more labile in a weakly acid medium.

Thus, cleavage of the *N*-glycosidic bond in purine nucleosides usually involves conversion of at least one of the nitrogen atoms in the corresponding base into the onium form.

In conclusion, it should be pointed out that nucleosides are not always stable in neutral and alkaline media. There are some exceptions when the *N*-glycosidic bond is hydrolysed with relative ease in such media. For example, deoxythymidine, which is more or less stable in acid media, is hydrolysed rather rapidly at alkaline pH values close to neutral.

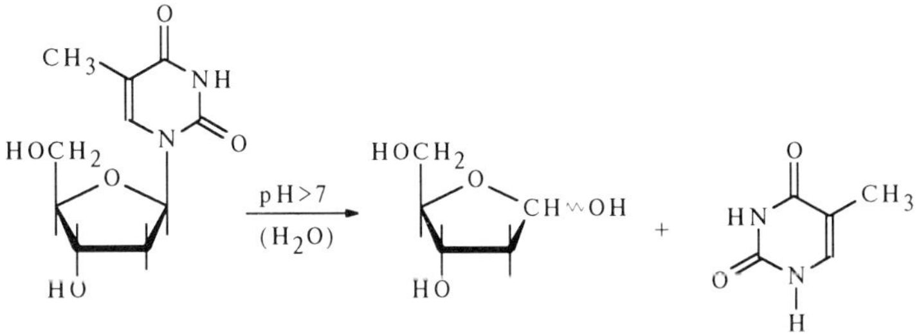

Adenosine and deoxyadenosine are hydrolysed, when heated with 1 N NaOH (100 °C, 1 hour), by about 25 per cent to yield adenine, whereas guanine nucleosides are stable under the same conditions.

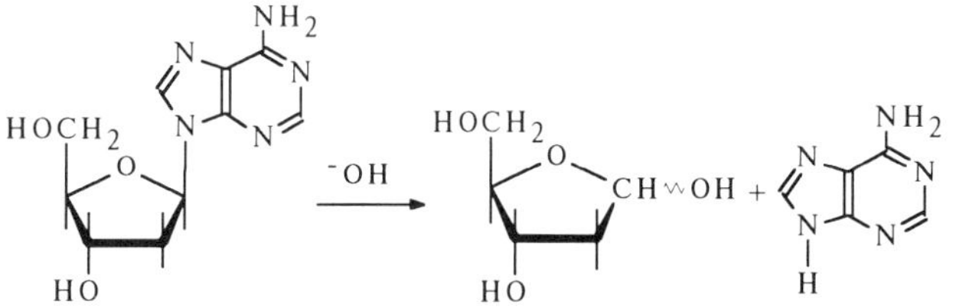

Notably, the pattern established for acid hydrolysis (in an acid medium guanosine nucleosides are hydrolysed more easily than adenosine ones) is not observed in this case. Unfortunately, the available experimental data are too scant to throw more light on the reasons for such difference.

2.5 Properties of Pseudouridine

The glycosidic C–C bond linking the ribose and uracil in pseudouridine exhibits properties differing widely from those of the ordinary *N*-glycosidic bond. It is highly stable in an acid medium, which is why, as has already been mentioned, serious difficulties were encountered in establishing the structure of pseudouridine by usual methods: it took a long time before the structure of the carbohydrate and pyrimidine moieties were elucidated, along with the linkage between the two. Treatment with an acid under vigorous conditions leads to isomerization rarely observed in *N*-glycosides. The unusual properties of the glycosidic bond stem from the allyl nature of the glycosidic carbon, which facilitates heterolytic rupture of the oxygen–glycosidic carbon bond.

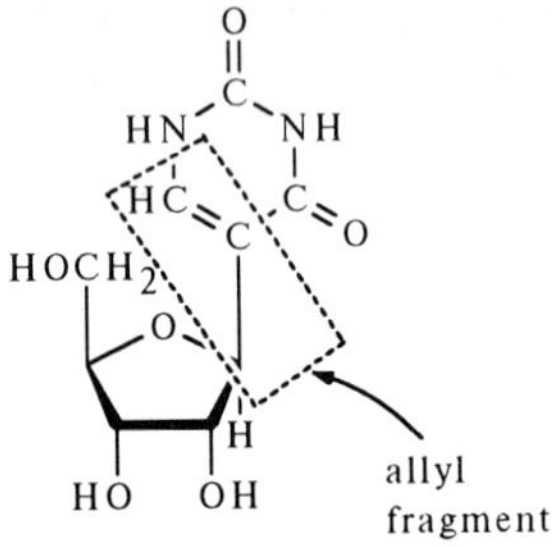

Among the factors contributing to the delocalization of the positive charge in ion (15) is also the N^1 atom, which may be represented using the following structures:

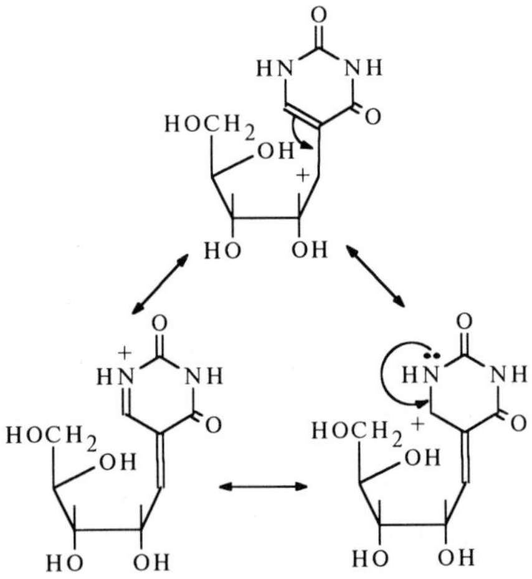

It is the relative ease with which the glycosidic carbon–oxygen bond in pseudouridine undergoes a heterolytic rupture with subsequent transformations of the emerging ion (15) that accounts for the tendency for isomerization, displayed by this nucleoside, and the structure of the resulting compounds.

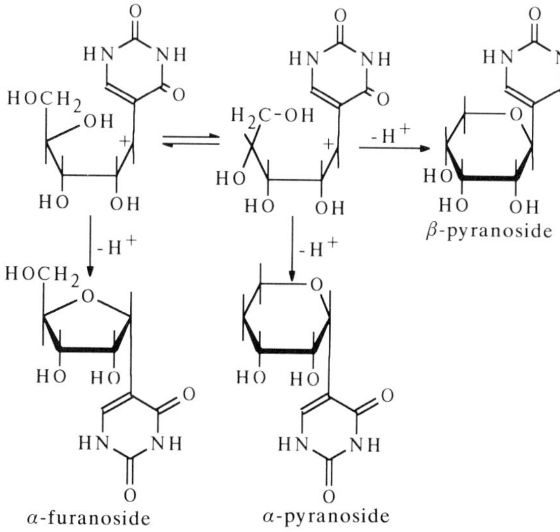

β-pyranoside

α-furanoside

α-pyranoside

The formation of α-furanoside is understandable if it is taken into account that the carbonium center (at the glycosidic carbon) in ion (15) is planar and may be subjected to nucleophilic attack by the oxygen of the secondary hydroxyl group on either side of the plane. The result is, in one case, the starting pseudouridine having a β-configuration and, in another, its anomer.

Similar factors are also involved in the formation of anomeric pyranosides, except that in this instance it is the oxygen of the primary hydroxyl group that acts as the nucleophilic agent.

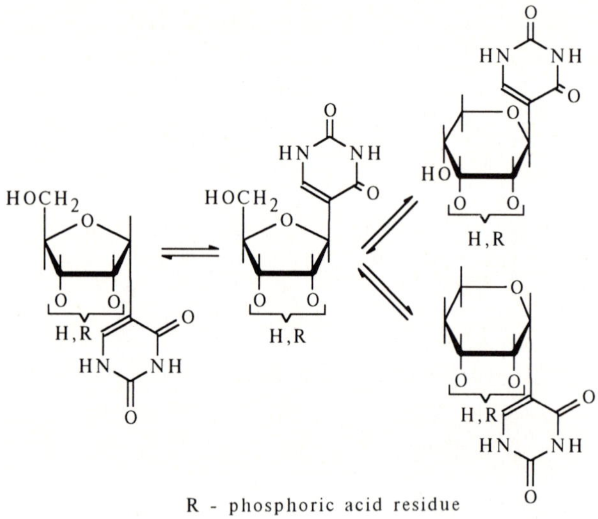

R - phosphoric acid residue

The above isomerization mechanism is borne out by the fact that a similar process is observed in cases where the 2′- or 3′-hydroxyl groups are protected.

At the same time, pseudouridine derivatives at the 5′-hydroxyl group are isomerized in an acid medium to yield only α-furanosides (without formation of α- and β-pyranosides):

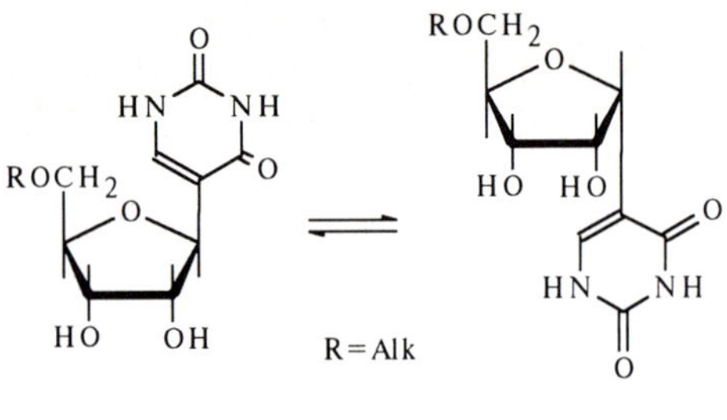

R=Alk

Pseudouridine also undergoes isomerization in alkaline media with formation of all possible isomers. The mechanism of this process can be written as follows:

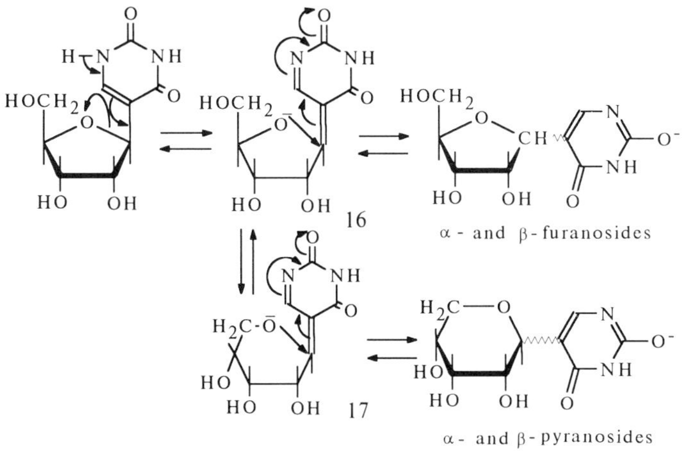

α- and β-furanosides

α- and β-pyranosides

The reaction proceeds as addition of alcohols to α,β-unsaturated carbonyl compounds in the presence of bases, for example:

$$CH_2{=}CHCHO \xrightarrow[\text{base}]{C_2H_5OH} C_2H_5OCH_2CH_2CHO$$

Addition at the multiple C–C bond may involve both primary and secondary alcohol groups [in forms (16) and (17), respectively]. Since the glycoside center in forms (16) and (17) has a two-dimensional configuration (being of ethylene type), addition of the primary hydroxyl gives α-and β-pyranosides, whereas that of the secondary hydroxyl results in α- and β-furanosides.

Just as pyrimidine nucleosides, pseudouridine is hydrogenated over a rhodium catalyst. However, in contrast to N-glycosides typically subject to addition of hydrogen at the double C^5–C^6 bond, it is the ether bond that undergoes hydrogenolysis in the case of pseudouridine, the double C^5–C^6 bond not being affected at all:

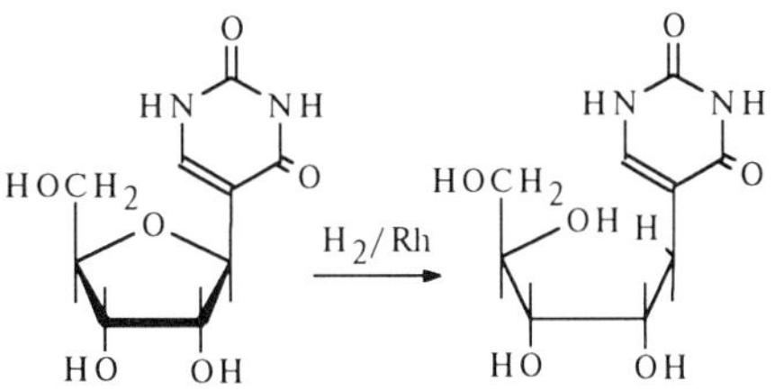

$$\xrightarrow{H_2/Rh}$$

This process is similar to hydrogenolysis of allyl and benzyl ethers:

$$CH_2{=}CHCH_2OR \xrightarrow{H_2/Pd} CH_2{=}CHCH_3 + ROH$$

$$C_6H_5{-}CH_2OR \xrightarrow{H_2/Pd} C_6H_5{-}CH_3 + ROH$$

Consequently, even in this case the behavior of pseudouridine closely resembles that of compounds of the allyl type.

References

1. N. K. Kochetkov and E. I. Budovskii, *Organic Chemistry of Nucleic Adids.* Plenum Press, London–N.Y. (1972).
2. A. M. Michelson, *The Chemistry of Nucleosides and Nucleotides.* Academic Press, N.Y.–London (1963).
3. T. L. V. Ulbricht, in: *Comprehensive Biochemistry.* Vo. 8, M. Florkin and E. H. Stotz (Eds.), Elsevier, Amsterdam–London–N.Y., pp. 158–208 (1963).
4. D. O. Jordan, *The Chemistry of Nucleic Acids.* Butterworths, London-Washington, p. 358 (1960).
5. R. W. Chambers, in: *Progress in Nucleic Acid Res. and Mol. Biol.* Vol. 5, J. N. Davidson and W. E. Cohn (Eds.), Academic Press, N.Y.–London, pp. 349–398 (1966).
6. R. Shapiro, in: *Progress in Nucleic Acid Res. and Mol. Biol.* Vol. 8, J. N. Davidson and W. E. Cohn (Eds.), Academic Press, London, pp. 73–112 (1968).

3 Structure of Nucleotides

3.1 Introduction

As can be inferred from hydrolysis of nucleic acids, a monomer unit of these biopolymers includes a phosphoric acid residue in addition to that of a nucleoside. Heterocyclic base, pentose and phosphoric acid residues are present in any nucleic acid in strictly equimolar amounts; hence, in a monomer, too, there is a mole of phosphoric acid for every mole of the nucleoside. Since acid hydrolysis of monomers under certain conditions gives pentose phosphates and a heterocyclic base, the linkage between the phosphate group and nucleoside is evidently via pentose. The ester type of bonding between the phosphate group and pentose seems to be the most probable of all possible ones, with the phosphate group being linked to one of the hydroxyls, such as 5′-hydroxyl.

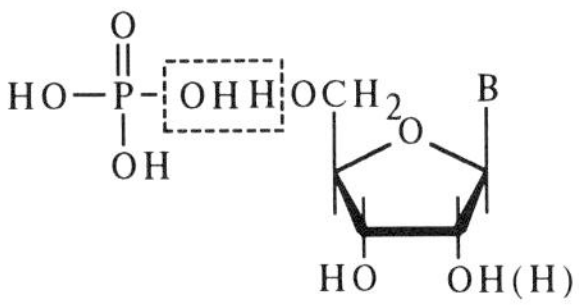

O
HO—P-OH H OCH2 B
OH O
HO OH(H)

The phosphomonoester type of bonding in nucleic acid monomeres is corroborated both by their chemical properties and by their behavior in the presence of enzymes specifically cleaving phosphomonoester bonds. Enzymes are known, namely, phosphomonoesterases (PME), that break down phosphonucleosides to phosphoric acid and nucleosides. It has already been mentioned that acid hydrolysis of a phosphonucleoside gives pentose phosphate, but if hydrolysis is conducted under mild conditions at pH 4 (phosphomonoester bonds at this pH value are most labile as a rule), dissociation to a nucleoside and phosphoric acid takes place. Shown below are the corresponding transformations involving adenosine phosphate.

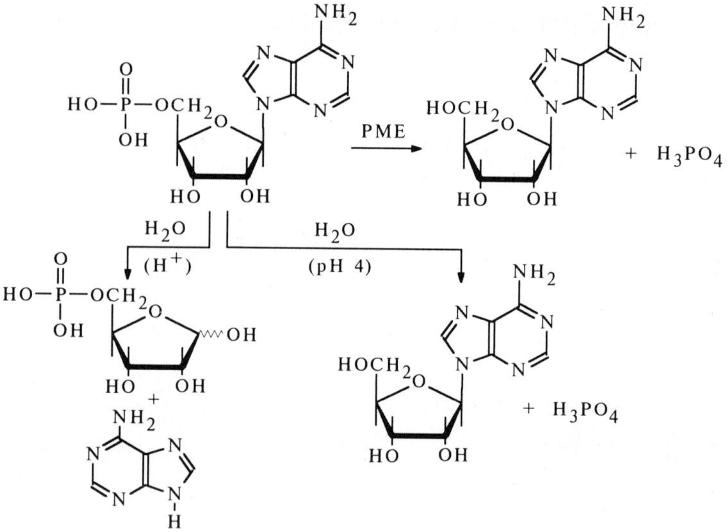

Phosphates of nucleosides are referred to as *nucleotides.* Depending on the structure of the pentose involved, distinction is made between *ribonucleotides* (monomer units of RNA) and *deoxyribonucleotides* (monomer units of DNA). According to the structure of the heterocyclic base constituting the nucleotide, there are purine and pyrimidine nucleotides.

Nucleotides, just as nucleosides, are usually isolated from nucleic acids after chemical or enzymatic hydrolysis. Ribonucleic acids break down to a mixture of ribonucleoside 3′- and 2′-phosphates after the treatment with a 0.3 N potassium hydroxide solution (37 °C, 18 hours). Boiling with 0.1 N HCl gives pyrimidine nucleoside 3′(2′)-phosphates (hydrolysis of purine nucleotides under these conditions proceeds at the glycosidic bond). Ribonucleoside 5′-phosphates result from treatment of RNA with the enzyme phosphodiesterase (PDE) obtained from snake venom, yeasts, and other sources. Deoxyribonucleotides are obtained only through enzymatic hydrolysis (digestion) of DNA, because chemical hydrolysis proceeds with complications; usually, PDE from snake venom is needed for breaking down DNA to deoxynucleoside 5′-phosphates. Under the action of PDE isolated from spleen, DNA as well as RNA yields the corresponding nucleoside 3′-phosphates. Separation of nucleotides is possible by ion-exchange and thin-layer chromatography, paper chromatography, and electrophoresis. Nucleotides are identified with the aid of UV spectra and also qualitative reactions with the phosphate group.

3.2 Nomenclature and Isomerism

Depending on the positions occupied by the phosphate group in ribo- or deoxyribonucleoside, distinction is made between three types of (isomeric) nucleotides:

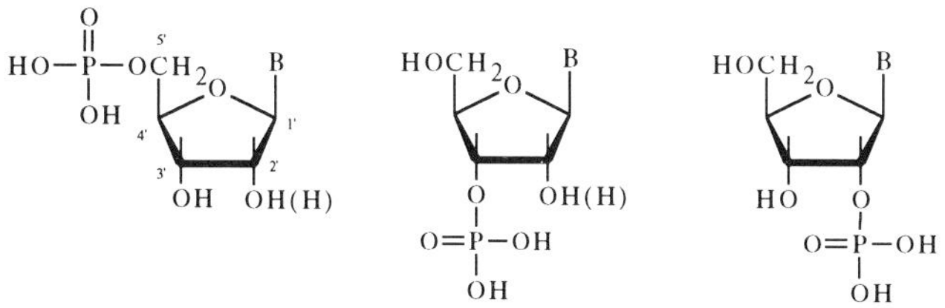

nucleoside-5'-phosphates nucleoside-3'-phosphates nucleoside-2'-phosphates

A nucleotide's name is usually based on that of the associated nucleoside. In the most common version, it consists of the nucleoside's name and position of the phosphate. For example, uridine 5′-phosphate:

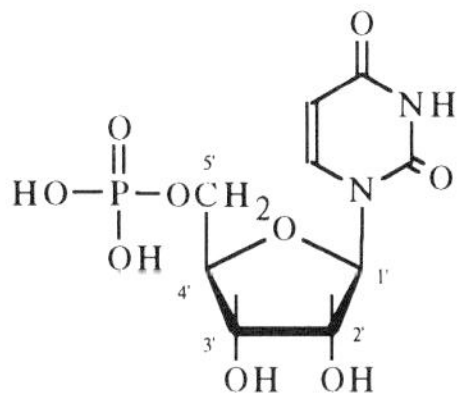

An alternative name of this nucleotide is 5-uridylic acid. In the case of pyrimidine nucleotides, the nomenclature consists of the name of the corresponding nucleoside (uridyl, cytidyl, etc.) plus the ending "ic" and the word "acid". For example: uridyl + ic acid = uridylic acid. Purine nucleotides have similar nomenclature with the difference that the basic component is the name of the corresponding base (guanyl, adenyl, etc.). For example: guanyl + ic acid = guanylic acid. The composite name is always preceded by the position of the phosphate in the nucleotide[1)]. Adenosine 3′-phosphate, for instance, may also be called 3′-adenylic acid.

Table 3-1 lists the names of the major nucleotides isolated from natural nucleic acids (DNA and RNA). Also listed are abridged formulas and abbreviated symbols, which, in turn, are based on abridged formulas and abbreviated symbols for the corresponding nucleosides plus an abbreviated symbol of the phosphate group. According to the rules of IUPAC and IUB, the phos-

1) Different notation is sometimes also used, for example: uridylic-3′ acid.

phate is designated as "p". The abridged formulas for the above nucleotides are as follows:

U / OH / OH / p— or pU

uridine 5'-phosphate
(5'-uridylic acid)

A / OH / p / HO— or Ap

adenosine 3'-phosphate
(3'-adenylic acid)

It can be seen that even more abbreviated symbols consisting of only two letters are used for nucleotides, the capital letter corresponding to the name of the nucleoside (see Table 3-1) and the small letter (p) on the left indicating that there is a phosphate group at position 5′. If the phosphate group is linked to the carbohydrate moiety at position 3′, the letter "p" is placed to the right of the capital letter (symbolyzing the nucleoside). The presence of a phosphate group at position 2′, in guanosine 2′-phosphate, for example, is indicated as follows:

G / p / OH / HO— G2'p

In the case of deoxynucleotides, the letter "d" is added before the symbol.

The names and abbreviated symbols of nucleotides that are derivatives of minor nucleosides (see Tables 1-5 and 1-6) are based on the same principle. For instance, pseudouridine 5′-phosphate (5′-pseudouridylic acid) has the abbreviated symbol pΨ, while deoxyuridine 3′-phosphate (3′-deoxyuridylic acid) has the symbol dUp.

Their abridged formulas are as follows:

Ψ / OH / OH / p—

U / p / HO—

Table 3-1. Structure, Names and Symbols of the Major Nucleotides Isolated from RNA and DNA.

Nucleoside in the nucleotide*	Nucleotide name	Abridged formula	Abbreviated symbol
Uridine	Uridine 5′-phosphate (5′-uridylic acid)	U / —OH / —OH / p—	pU
	Uridine 3′-phosphate (3′-uridylic acid)	U / —OH / —p / HO—	Up
	Uridine 2′-phosphate (2′-uridylic acid)	U / —p / —OH / HO—	U2′p
Cytidine	Cytidine 5′-phosphate (5′-cytidylic acid)	C / —OH / —OH / p—	pC
	Cytidine 3′-phosphate (3′-cytidylic acid)	C / —OH / —p / HO—	Cp
	Cytidine 2′-phosphate (2′-cytidylic acid)	C / —p / —OH / HO—	C2′p
Adenosine	Adenosine 5′-phosphate (5′-adenylic acid)	A / —OH / —OH / p—	pA
	Adenosine 3′-phosphate (3′-adenylic acid)	A / —OH / —p / HO—	Ap
	Adenosine 2′-phosphate (2′-adenylic acid)	A / —p / —OH / HO—	A2′p

Table 3-1. (Continued).

Nucleoside in the nucleotide*	Nucleotide name	Abridged formula	Abbreviated symbol
Guanosine	Guanosine 5′-phosphate (5′-guanylic acid)	G; –OH; –OH; p–	pG
	Guanosine 3′-phosphate (3′-guanylic acid)	G; –OH; –p; HO–	Gp
	Guanosine 2′-phosphate (2′-guanylic acid)	G; –p; –OH; HO–	G2′p
Deoxythymidine	Deoxythymidine 5′-phosphate (5′-deoxythymidylic acid)	T; –OH; p–	pdT
	Deoxythymidine 3′-phosphate (3′-deoxythymidylic acid)	T; –p; HO–	dTp
Deoxycytidine	Deoxycytidine 5′-phosphate (5′-deoxycytidylic acid)	C; –OH; p–	pdC
	Deoxycytidine 3′-phosphate (3′-deoxycytidylic acid)	C; –p; HO–	dCp
Deoxyadenosine	Deoxyadenosine 5′-phosphate (5′-deoxyadenylic acid)	A; –OH; p–	pdA
	Deoxyadenosine 3′-phosphate (3′-deoxyadenylic acid)	A; –p; HO–	dAp
Deoxyguanosine	Deoxyguanosine 5′-phosphate (5′-deoxyguanylic acid)	G; –OH; p–	pdG
	Deoxyguanosine 3′-phosphate (3′-deoxyguanylic acid)	G; –p; HO–	dGp

* For complete structural formulas of nucleosides as well as their abridged formulas and abbreviated symbols, see Tables 1-2 and 1-3.

3.3 Structure of Nucleotides

It has already been pointed out that nucleotides, which are essentially monomer units of nucleic acids, represent nucleosides monophosphorylated at the sugar moiety. Since the structure of nuleosides was discussed in the preceding chapter, here we shall deal only with those aspects of nucleotide structure which have to do with the position of the phosphate group in their molecule.

3.3.1 Nucleoside 5′-Phosphates

Nucleoside 5′-phosphates result from hydrolysis of RNA and DNA in the presence of the enzyme phosphodiesterase (PDE) isolated from snake venom as well as from actinomyces and yeasts.

$$\text{RNA} \xrightarrow{\text{PDE}} \text{pA} + \text{pG} + \text{pU} + \text{pC}$$

$$\text{DNA} \xrightarrow{\text{PDE}} \text{pdA} + \text{pdG} + \text{pdT} + \text{pdC}$$

Nucleotides of this type were isolated for the first time from muscle extracts rather than nucleic acid. As far back as 1847, Liebig isolated a nucleotide which was named muscle inosinic acid. This nucleotide seems to result from enzymatic deamination of 5′-adenylic acid (adenosine 5′-phosphate). The latter is known to be present in muscles in a free state as an intermediate product of biosynthesis.

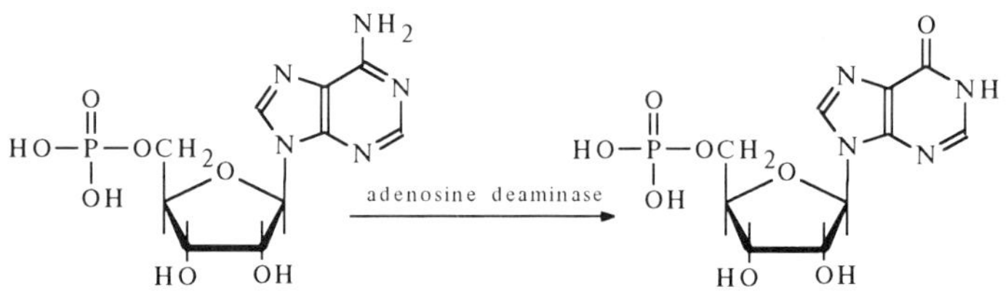

adenosine deaminase

As has been established in recent years, inosinic acid in the form of phosphate of a minor nucleoside, inosine to be exact (see Table 1-5), is present in some RNAs and plays an important role in their functioning.

In spite of the fact that inosinic acid was isolated long before the discovery of nucleic acids, its structure was established only in 1909 when it was demonstrated that this nucleotide is hydrolysed in a weakly acid medium to inosine and phosphoric acid, whereas its acid hydrolysis yields ribose 5-phosphate and hypoxanthine:

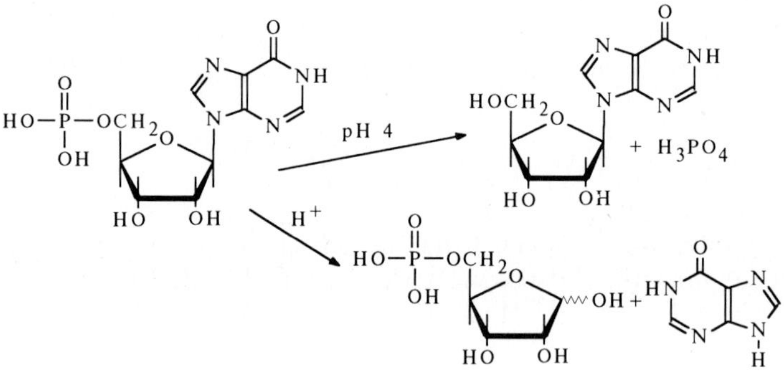

pH 4

$+ H_3PO_4$

H^+

The structure of ribose 5-phosphate was established by converting it, during oxidation with dilute nitric acid, into 5-phospho-D-ribonic acid (1):

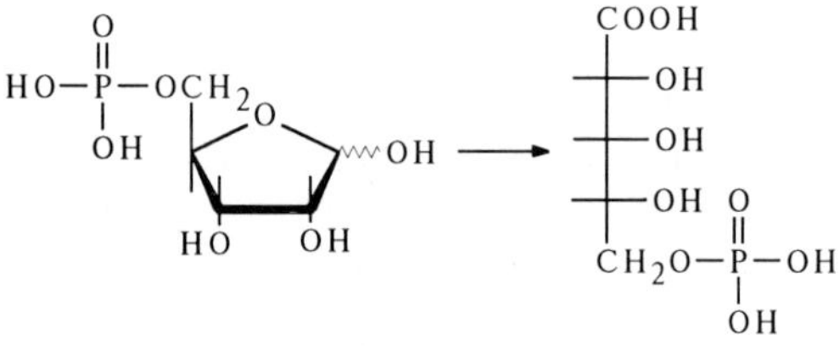

(1)

If the phosphate group in the ribose phosphate under investigation had been linked to the 2- or 3-hydroxyl group of ribose, the oxidation would have resulted in phosphotrihydroxyglutaric acid (2):

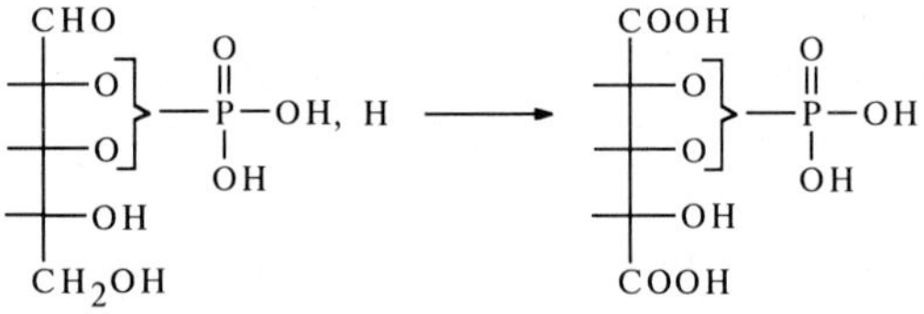

(2)

However, this acid has never been found among the oxidation products of the corresponding ribose phosphastes.

Ribose 5-phosphate was then converted into a methyl glycoside whose reduction produced an optically active ribite 5-phosphate:

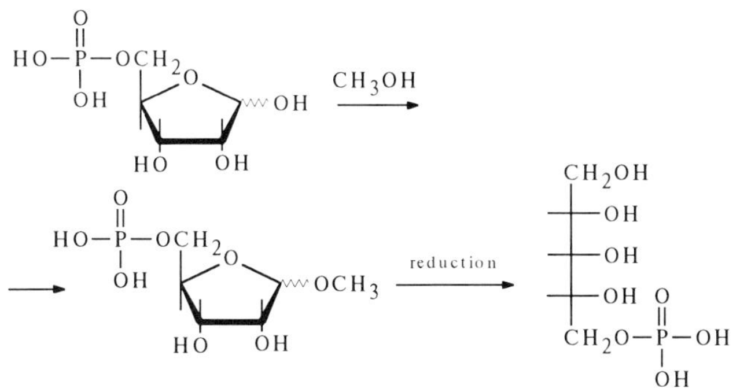

Final evidence to support the structure of ribose 5-phosphate was provided by its synthesis from 2,3-isopropylidene-1-*O*-methylribofuranoside according to the following scheme:

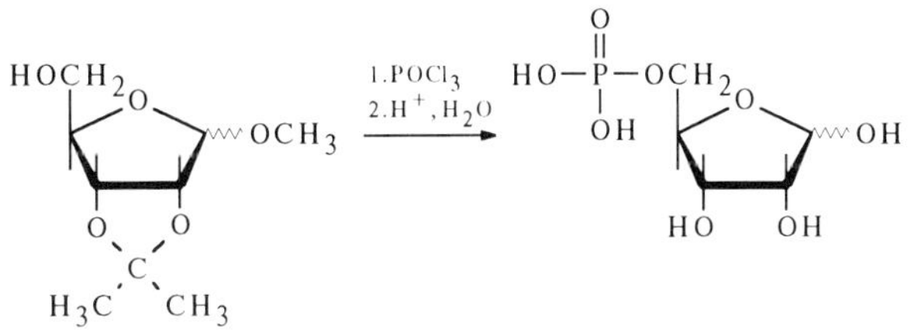

The ribosse phosphate isolated after hydrolysis of inosinic acid and the synthesized ribose 5-phosphate turned out to be identical. This finding has suggested that the phosphate group in inosinic acid and, consequently, in muscle adenylic acid, occupies position 5′ in the corresponding nucleoside. The adenosine phosphate isolated from RNA after its hydrolysis with snake venom PDE was found to be identical with muscle adenylic acid or, in other words, adenosine 5′-phosphate. Its structure was also corroborated by synthesis from 2′, 3′-*O*-isopropylideneadenosine:

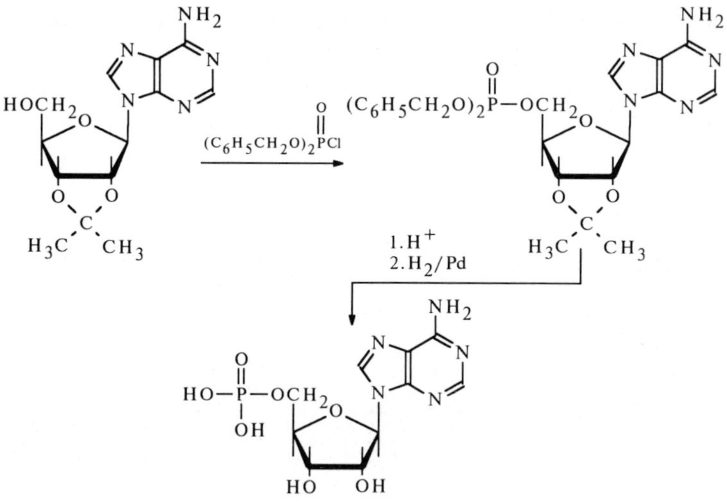

The structure of other ribonucleoside 5′-phosphates (pG, pU and pC), formed during PDE hydrolysis of RNA has been established in a similar manner and supported by syntheses from guanosine, cytidine and uridine derivatives, conducted using the above procedure. It should be noted that the phosphorylating agent in the synthesis of guanosine 5′-phosphate was tetra-*para*-nitrophenyl pyrophosphate:

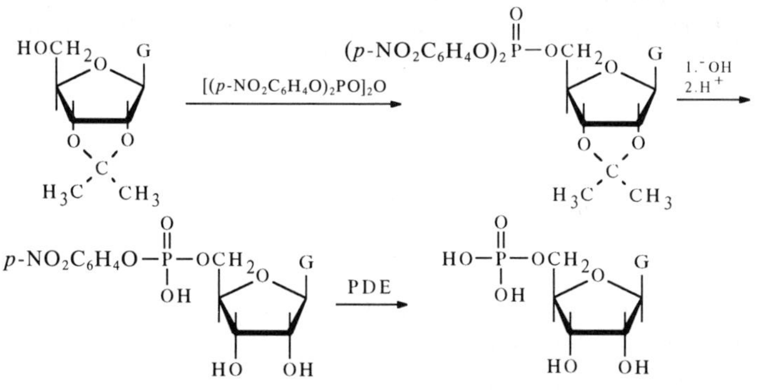

During mild alkaline hydrolysis of a completely protected nucleoside, one of the *para*-nitrophenyl groups was removed; the isopropylidene group was removed as usual by acid treatment. The other *para*-nitrophenyl group was removed by PDE hydrolysis. Comparison of each of the synthesized nucleoside 5′-phosphates (pA, pG, pC and pU) with the nucleotides isolated from RNA has shown them to be identical with each other.

Ribonucleoside 5′-phosphates with a free *cis*-glycol group (as opposed to nucleoside 2′- and 3′-phosphates) are easily oxidized with periodic acid:

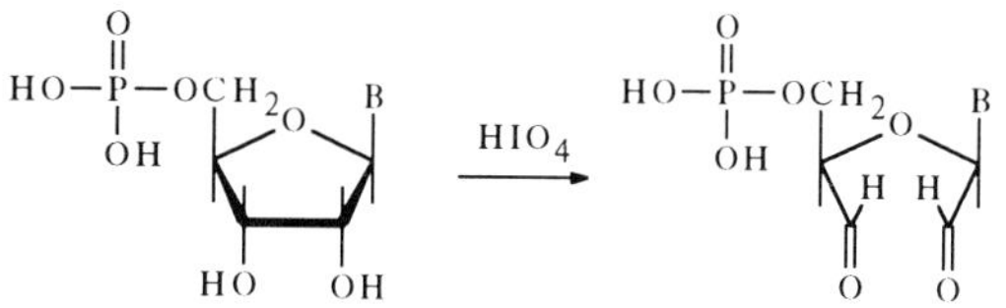

The oxidation requires 1 mole of periodic acid, without formation of formic acid in the process.

The results of oxidation with periodic acid have confirmed the structure of the ribonucleoside 5′-phosphates isolated from RNA. Natural ribonucleoside 3′(2′)-phosphates are not oxidized under these conditions. It is for this reason that the periodate oxidation method is widely used to determine the position of the phosphate group in ribonucleoside phosphates.

Synthetic and naturally occurring ribonucleoside 5′-phosphates are easily dephosphorylated in the presence of snake venom PME also known as 5′-nucleotidase:

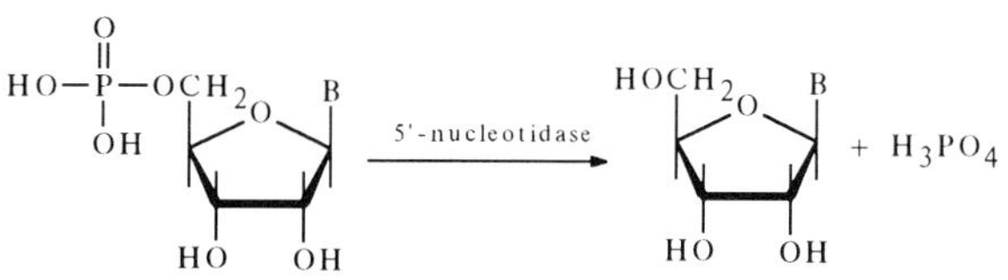

At present, in order to determine the position of the phosphate group in nucleotides of unknown structure extensive use is made of this enzyme because it does not break phosphomonoester bonds in nucleoside 3′- and 2′-phosphates. The ability of a nucleotide to break down to a nucleoside and phosphoric acid under the action of 5′-nucleotidase is considered to be solid proof that its phosphate group is at position 5′. The crystalline deoxyribonucleotides isolated in 1935 during enzymatic[2] hydrolysis of DNA also turned out to be 5′-phosphates. The determination of the phosphate group position in these compounds was based on the fact that their breakdown under the effect of 5′-nucleotidase resulted in equimolar quantities of the corresponding deoxynucleosides and phosphate:

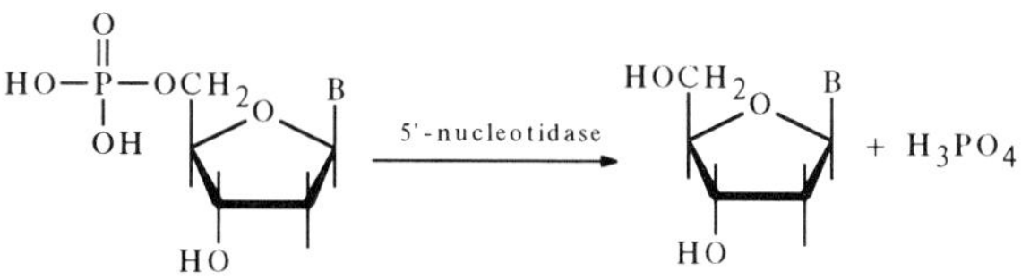

[2] Intestinal PDE was used for hydrolysis. At present, snake venom PDE is employed for the same purpose.

The structure of the deoxynucleoside 5′-phosphates derived from DNA was corroborated by synthesis. Given below by way of example is the synthesis of deoxythymidine 5′-phosphate from 3′-*O*-acetyldeoxythymidine obtained, in turn, from 5′-*O*-trityldeoxythymidine.

$TrOCH_2$ ... T, HO $\xrightarrow{Ac_2O}$ $TrOCH_2$ ... T, AcO $\xrightarrow{H^+}$

$\rightarrow$ $HOCH_2$ ... T, AcO $\xrightarrow{(C_6H_5CH_2O)_2POCl}$ $(C_6H_5CH_2O)_2P(O)OCH_2$ ... T, AcO $\xrightarrow{^-OH}$

$\rightarrow$ $C_6H_5CH_2O-P(O)(OH)-OCH_2$ ... T, HO $\xrightarrow{H_2/Pd}$ $HO-P(O)(OH)-OCH_2$ ... T, HO

pdT

The protective groups in the fully protected nucleotide were eliminated by alkaline hydrolysis during which the acetyl and one of the benzyl groups were removed. The second benzyl group was removed through hydrogenolysis over a palladium catalyst. Similarly, pdG, pdA and pdC were synthesized from the corresponding 3′-acetyldeoxynucleosides.

In view of the high reactivity of the amino group in cytosine nucleosides, use was made, as the starting compound in the synthesis, of a nucleoside also protected at the amino group. It was prepared by exhaustive acetylation of 5′-trityldeoxycytidine with subsequent removal of the trityl group through acid hydrolysis. The phosphorylation of the resulting diacetyl derivative and isolation of free pdC were conducted in exactly the same way as during the synthesis of pdT (see above).

NHCOCH$_3$ TrOCH$_2$ HO AcO NH$_2$ N O Ac_2O H^+ HOCH$_2$ 1.$(BzlO)_2POCl$ 2.^-OH 3.H_2/Pd HO—P—OCH$_2$ OH pdC

The deoxynucleoside 5′-phosphates synthesized by the methods described above have turned out to be identical with those isolated from DNA after PDE hydrolysis.

3.3.2 Nucleoside 3′- and 2′-Phosphates

Alkaline hydrolysis of RNA yields a mixture of ribonucleoside 3′- and 2′-phosphates of the four major nucleosides constituting RNA.

Evidence of their belonging to monophosphates has been supplied by hydrolysis with aqueous ammonia, which resulted in equimolar amounts of the corresponding nucleosides and phosphate.

Given below are data to support the structure of adenosine 3′- and 2′-phosphates isolated individually by ion-exchange chromatography. These compounds are not oxidized with periodic acid, which is indicative of substitution of a phosphate group for the hydrogen of one of the hydroxyls in the *cis*-1,2-diol group of adenosine. When kept in an acid medium, each of the above compounds became transformed to an equilibrium mixture of adenosine 2′- and 3′-phosphates. In contrast, adenosine 5′-phosphate did not undergo any transformation under the same conditions. Each of the isomers (2′- and 3′-) was transformed, when treated with trifluoroacetic anhydride, to adenoside 2′,3′- cyclic phosphate. The scheme that follows illustrates the transformations involving only the 2′- and 3′-hydroxyl groups:

$$\text{HOCH}_2\text{-adenosine 3'-phosphate} \xrightarrow{(CF_3CO)_2O} \left[\text{3'-}O\text{-P(O)(OH)OCOCF}_3\right] \xrightarrow{-CF_3COOH} \text{adenosine 2',3'-cyclic phosphate}$$

A similar explanation can be provided for the formation of a cyclic phosphate from the 2′-isomer as well.

These transformations indicated that the phosphoric acid residue in isomeric adenosine phosphates is at position 2′ or 3′. It was an extremely difficult task to make sure that each adenosine phosphate isolated from the alkaline hydrolysate of RNA belongs to 2′- or 3′-monophosphates. To this end, each isomer was subjected to acid hydrolysis under very mild conditions[3] to cleave the glycosidic bond. The resulting ribose phosphate was reduced to ribitol phosphate, and its optical activity was determined. The following scheme shows the conversion of one of the isomers of optically active ribitol 2-phosphate, the results of which attest to the fact that the starting nucleoside has the structure of adenosine 2′-phosphate.

$$\text{adenosine 2'-phosphate} \xrightarrow{H^+} \text{ribose 2-phosphate} \xrightarrow{[H]} \text{ribitol 2-phosphate } (CH_2OH\text{–}CH(OPO(OH)_2)\text{–}CHOH\text{–}CHOH\text{–}CH_2OH)$$

At the same time ribitol 3-phosphate

$$CH_2OH\text{–}CHOH\text{–}CHOPO(OH)_2\text{–}CHOH\text{–}CH_2OH$$

cannot be optically active because of the symmetry of its structure (*meso*-form). Indeed, when the other isomer was treated in the same manner, an optically inactive monophosphate was formed, which gave every reason to ascribe the adenosine 3′-phosphate structure to the nucleotide.

[3] The hydrolysis time was kept short to minimize the migration of the phosphate group.

Adenosine 2′- and 3′-phosphates display different mobility during ion-exchange chromatography, paper chromatography, and electrophoresis under certain conditions, which makes their identification possible.

A similar approach was used to establish the structure of isomeric phosphates of guanidine, cytidine and uridine. However, in the case of pyrimidine, nucleosides whose glycosidic bond is known to be hydrolysed under relatively vigorous conditions, the double C^5–C^6 bond was subjected to hydrogenation at first. The reduced nucleotides were successfully hydrolysed to ribose phosphates whose structure was established by the methods described above. Here is how such conversion occurs in the case of uridine 3′-phosphate:

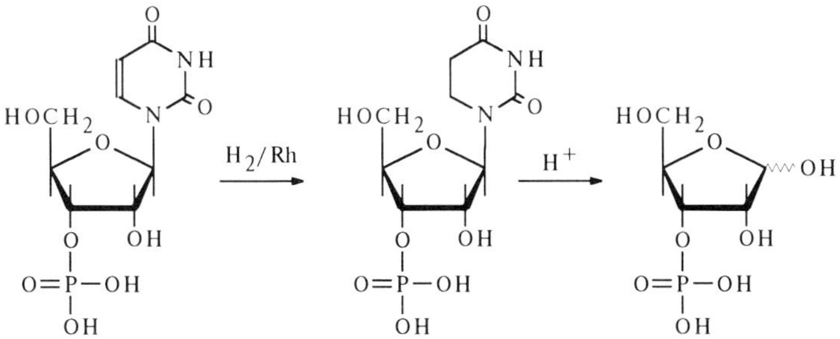

The structure of uridine 3′-phosphate was confirmed by X-ray structural analysis.

The structure of deoxyribonucleoside 3′-phosphates resulting from hydrolysis of DNA in the presence of spleen PDE was corroborated by comparison with isomers of a known structure, namely, the corresponding deoxynucleoside 5′-phosphates.

In contrast to deoxyribonucleoside 5′-phosphates, the 3′-isomers are not cleaved by 5′-nucleotidase. This fact is put to practical use when it is necessary to determine the position of the phosphate group in deoxyribonucleosides. The structure of all deoxynucleoside 3′-phosphates was definitively established by syntheses. The following example represents synthesis of deoxythymidine 3′-phosphate from 5′-*O*-trityldeoxythymidine:

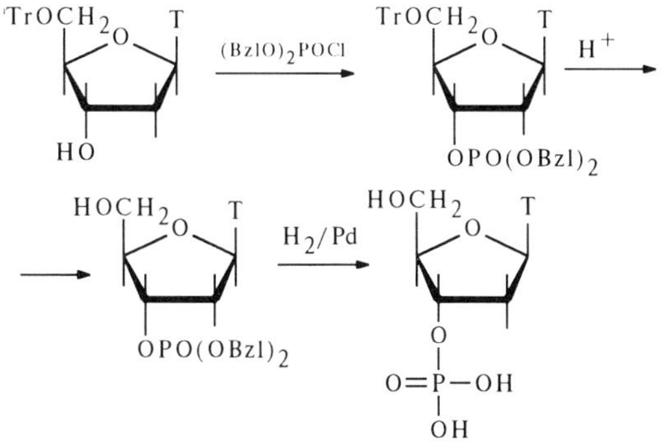

In addition to phosphates of the major nucleosides, hydrolysates of RNA and DNA have also yielded those of minor nucleosides (see Tables 1-5 and 1-6); their structure was estasblished by the methods, described above.

3.3.3 Nucleoside Cyclic Phosphates

When ribonucleic acids are treated with pyrimidyl ribonuclease (its abbreviated name is pyrimidyl RNase), pyrimidine nucleotides are formed as intermediates, their phosphate being linked both to the 2′- and 3′-hydroxyl groups; they are known as cyclic phosphates.

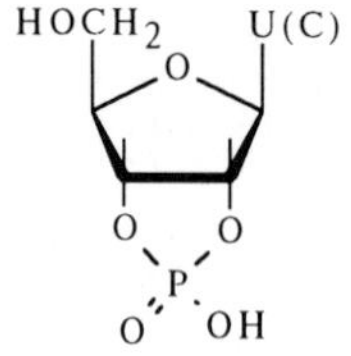

HOCH2 U(C) O O P O OH

pyrimidine nucleoside 2′,3′-cyclic phosphate

These phosphates are further hydrolysed in the presence of pyrimidyl RNase to the corresponding nucleoside 3′-phosphates (without formation of the 2′-isomer):

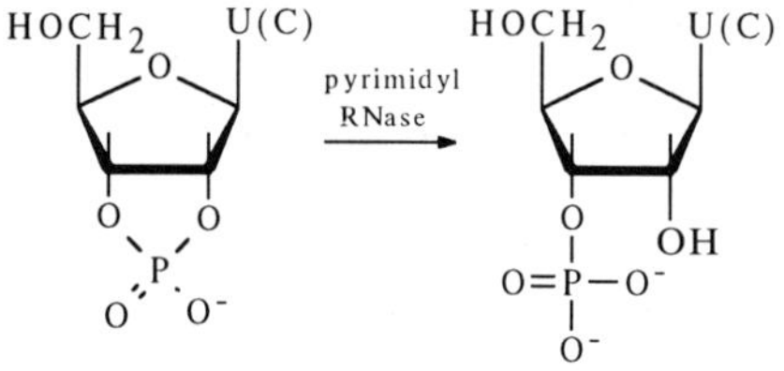

HOCH2 U(C) O O P O O- —pyrimidyl RNase→ HOCH2 U(C) O OH O=P—O- O-

Pyrimidine and purine nucleoside 2′,3′-cyclic phosphates are also formed when a solution of RNA in formamide with ammonia is heated or during hydrolysis of RNA in the presence of barium carbonate. The structure of these compounds has been determined from titration data (with only one hydroxyl group of the phosphoric acid residue being titrated), their behavior during electrophoresis, as well as acid or alkaline hydrolysis yielding a mixture of 3′- and 2′-phosphates.

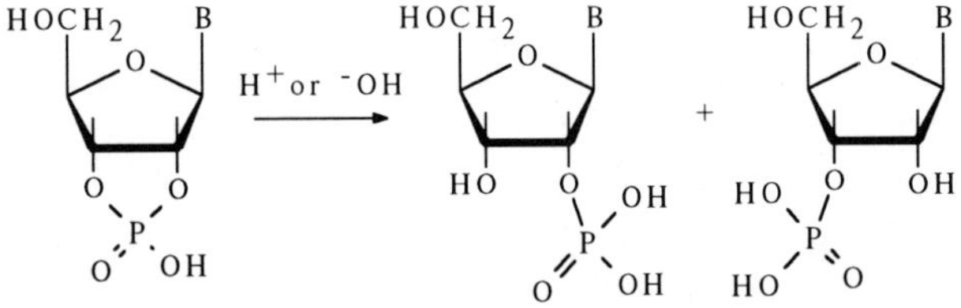

HOCH2 B O O P O OH —H+ or -OH→ HOCH2 B HO O P OH O OH + HOCH2 B HO O P OH HO O

The following abridged formulas and abbreviated symbols have been adopted for nucleoside 2′,3′-cyclic phosphates

N>p

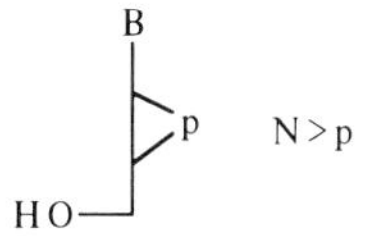

(it should be remembered that N stands for nucleoside). For example, uridine 2′,3′-cyclic phosphate has the following abbreviated symbol:

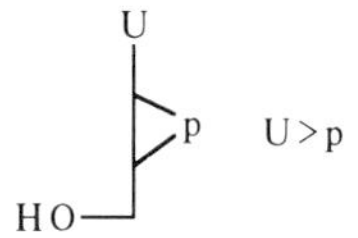

U>p

Natural subtrates have also provided a source of adenosine 3′,5′-cyclic phosphate (3) which is produced in the cell from adenosine 5′-triphosphate.

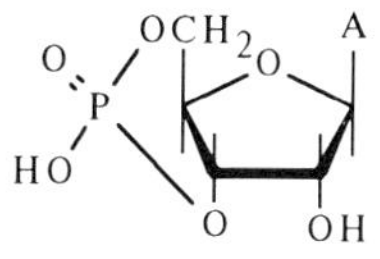

(3)

Just as in the case of nucleoside 2′,3′-cyclic phosphates, only one hydroxyl of the phosphate group is titrated in compound (3). The phosphodiester nature of the nucleotide is also confirmed by electrophoretic mobility. Adenosine 3′,5′-cyclic phosphate is not oxidized with periodic acid, which is indicative of a substituent at one of the hydroxyls of the *cis*-diol group. Under the action of diesterases (enzymes hydrolysing diphosphates), adenosine 3′,5′-cyclic phosphate breaks down to a mixture of adenosine 3′- and 5′-phosphates (adenosine 2′-phosphate has not been found in the hydrolysates):

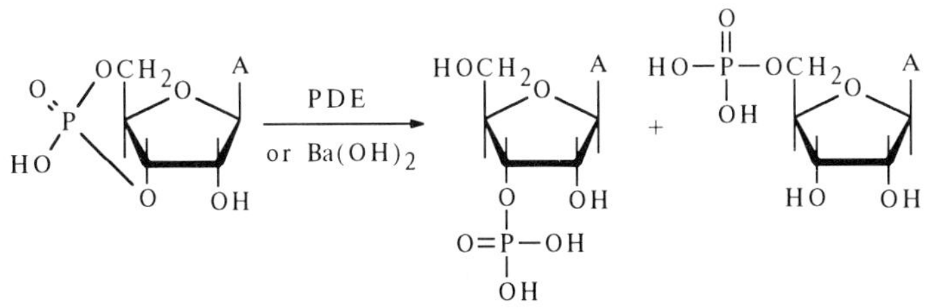

PDE or $Ba(OH)_2$

Hydrolysis of adenosine 3′,5′-cyclic phosphate with a dilute $Ba(OH)_2$ solution also yields a mixture of adenosine 3′- and 5′-phosphates (with a 5:1 ratio); no 2′-phosphate is formed in this case as well.

Adenosine 3′,5′-cyclic phosphate plays an extremely important role as a regulator of biosynthetic processes occurring in the cell.

3.3.4 Nucleoside 3′(2′),5′-Diphosphates

Nucleosides containing two phosphate groups are also called nucleotides. 3′,5′-Diphosphates of pyrimidine deoxynucleosides, which are produced in significant amounts during acid hydrolysis of DNA (in purine nucleotides the glycosidic bond is hydrolysed under such conditions) have been studied more thoroughly. The structure of one of them namely, thymidine 3′,5′-diphosphate, was established by acid hydrolysis to such known compounds as thymine and 2-deoxy-D-ribose diphosphate. First, the double C^5–C^6 bond in the nucleotide was reduced through hydrogenation over a rhodium catalyst with the result that cleavage of the glycosidic bond became possible under mild conditions:

$\xrightarrow{H_2/Rh}$ $\xrightarrow{H^+}$

3′,5′-diphosphoribose

A mixture of isomeric ribonucleoside 3′,5′- and 2′,5′-diphosphates is formed during alkaline hydrolysis of RNA.

Diphosphates of ribonucleosides are formed during enzymatic or chemical hydrolysis of RNA only from its terminal groups, therefore, their quantity is very small compared to other hydrolysis products.

The usual approaches are employed in assigning abbreviated symbols to nucleoside diphosphates. For example, thymidine 3′,5′-diphosphate and adenosine 2′,5′-diphosphate are written as follows:

or pdTp

or pA2′p

3.4 General Comments Regarding the Structure of Monomer Units in Nucleic Acids

Structurally, all constituent nucleotides in nucleic acids share a number of common features.

(1) The phosphoric acid residue both in the ribonucleotide (monomer unit of RNA) and in deoxyribonucleotide (monomer unit of DNA) is always linked to pentose:

Phosphoric acid residue	—	D-ribose (RNA) or 2-deoxy-D-ribose (DNA)	—	Pyrimidine or purine base

Thus, pentose phosphate is the common constituent of monomer units in both RNA and DNA. It does not seem to impart any unique properties to the monomer unit of the corresponding nucleic acid; quite on the contrary, its role boils down to giving the unit a universal trait inherent in every monomer.

(2) The only component of ribo- and deoxyribonucleotides that may vary is the heterocyclic base. A characteristic structural feature of this component stems from the fact that it belongs to two similar yet markedly distinct classes of heterocyclic compounds – pyrimidine and purine bases. Both belong to aromatic heterocycles containing several nitrogen atoms in the ring, as well as exocyclic oxo and/or amino groups. These heterocyclic bases, however, differ in size, electron density distribution in the molecule, and also in the nature of the nitrogen involved in the linkage with the universal component of the monomer. As can be seen from Figure 3-1, heterocyclic bases and, consequently, monomer units in RNA and DNA form a rather limited set. In both cases, it consists of two pyrimidine and two purine bases. The difference between the constituent bases in both nucleic acids is minor and resides only in that RNA contains uracil instead of 5-methyluracil (thymine) which is present in DNA.

Hence, the distinctive features of the monomer unit in a particular nucleic acid must be determined by the heterocyclic base, and the role (function) to be performed by the monomer unit in question must be a manifestation of the special qualities inherent in the base – that is, its structure and properties.

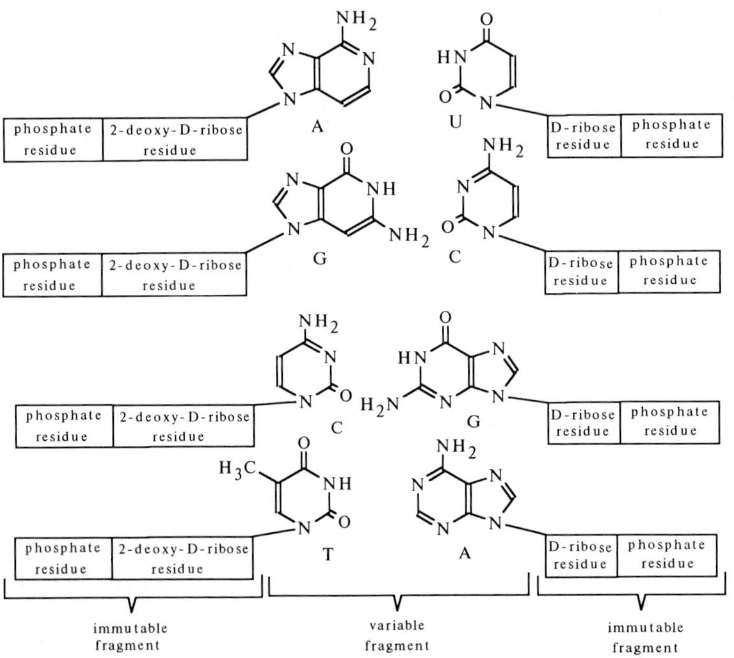

Fig. 3-1. Structural units of DNA and RNA.

(3) All monomer units of nucleic acids have some common structural features: their permanent (carbohydrate phosphate) and variable (heterocyclic base) components are linked via an *N*-β-glycosidic bond, whereas pentose (ribose or deoxyribopentose) always belongs to the D-series and have a β-configuration.

It is interesting to compare the typical common features of these monomer units and those of proteins which are, just as nucleic acids, highly important biopolymers.

The monomer units of a protein are reprensented by organic compounds of a single type, namely, amino acids. Their structural component that varies is the side chain radical R which determines whether the amino acid is hydrophobic or hydrophilic.

immutable fragment

···—NHCHCO—NHCHCO—NHCHCO—NHCHCO— ···

variable fragment → R'   R''   R'''   R''''

The monomer unit of nucleic acids is made up of three components with widely differing structure and properties. Two of them are organic compounds belonging to entirely diverse categories (carbohydrates and herterocyclic bases), while the third one is essentially a residue of phosphoric acid exhibiting, unlike many inorganic acids, a number of unique properties (it is tribasic, virtually lacks oxidative and reductive capacity, and its esters display, as will be shown in what follows, dual reactivity). Whereas the invariable and variable components of the protein's monomer unit, amino acids to be precise, form a framework of considerable strength, the nucleic acid's monomer unit, or nucleotide, is held together by weak linkages, namely, glycosidic and phosphoester bonds, which may be selectively ruptured (in enzymatic or chemical hydrolysis, etc.). The monomer unit of protein lacks such "hot spots".

The most striking difference is in the number of monomer units: a protein typically contains at least twenty different units (20 different "ordinary" amino acids may be encountered), while each nucleic acid (DNA or RNA) has only four[4].

---

[4] We are speaking only about major, and not minor (rare), pyrimidine and purine nucleotides.

References

1. A.M. Michelson, *The Chemistry of Nucleosides and Nucleotides.* Academic Press, N.Y.–London (1963).
2. D.M. Brown, in: *Comprehensive Biochemistry.* Vol. 8, M. Florkin and E.H. Stotz (Eds.), Elsevier, Amsterdam–London–N.Y., pp. 209–269 (1963).
3. D.O. Jordan, *The Chemistry of Nucleic Acids.* Butterworths, London–Washington, p. 358 (1960).

4 Properties of Nucleotides

The bases and carbohydrate moiety in nucleotides retain the properties they normally exhibit in nucleosides. The chemical behavior of nucleotides is largely determined by the presence of a phosphate group. For instance, they are strong acids and readily soluble in water; the phosphate group in nucleotides is almost as reactive as monoalkyl phosphates, although in certain ways it is distinct from the latter.

Unfortunately, the properties of nucleotides, determined by the presence of the phosphate group, have not been studied completely enough, and specialists sometimes have to resort to analogies with simpler compounds.

4.1 Acid-Base Behavior

Each nucleotide can give away or accept a proton or, in other words, is capable of ionization. Table 4-1 lists experimentally established ionization constants for heterocyclic bases and pentose, both in nucleosides and in nucleotides (data for free bases are also included for comparison) as well as for phosphate groups.

4.1.1 Ionization of Bases

The capacity of the heterocyclic bases in nucleosides and nucleotides for ionization – that is, to accept (basic properties) or give up (acid properties) a proton – is dependent on their structure.

Basic properties are displayed by pyrimidines and purines containing an amino group at position 4 (6), such as cytosine and adenine. However, according to the UV and IR spectra, it is not the amino group that serves as the site of proton attachment, as was long believed, but rather the neighboring cyclic nitrogen (N^3 in cytosine and N^1 in adenine), which agrees well with

Table 4-1. pK_a Values for Nucleotides and Their Components.

Compound	pK_a values for			
	base	pentose	phosphate group (pK_1)	phosphate group (pK_2)
Adenine	4.25			
Adenosine	3.63	12.35		
Deoxyadenosine	3.8			
Adenosine 5′-phosphate	3.74	13.06	0.9	6.05
Adenosine 2′- and 3′-phosphates (mixture)	3.7		0.9	6.1
Guanine*	3.0; 9.32			
Guanosine	2.1; 9.33	12.3		
Deoxyguanosine	2.4; 9.33			
Guanosine 5′-phosphate	2.9; 9.6		0.7	6.3
Guanosine 2′- and 3′-phosphates (mixture)	2.4; 9.8		0.7	6.0
Uracil	7.48; 11.48			
Uridine	9.25	12.59		
Deoxyuridine	9.3			
Uridine 5′-phosphate	9.5		1.0	6.4
Uridine 2′- and 5′-phosphates (mixture)	9.96		1.0	5.9
Cytosine**	4.6; 12.2			
Cytidine	4.1	12.24		
Deoxycytidine	4.25			
Cytidine 5′-phosphate	4.5		0.8	6.3
Cytidine 2′-phosphate	4.30		0.8	6.19
Cytidine 3′-phosphate	4.16		0.8	6.04
Thymine	9.94			
Deoxythymidine	9.8	12.85		
Deoxythymidine 5′-phosphate	10.0		1.6	6.5

* In the case of guanine derivatives, two pK_a values are given – for the amino and hydroxyl groups, respectively.

** The pK_a values are given, respectively, for the amino and hydroxyl groups of cytosine; in cytidine and its phosphates, the base is in the tautomeric oxo form, and pK_a in this case is for its amino group only.

the results of electron density calculations and can be explained by mesomerism. The protonation of the cyclic nitrogen atoms of cytosine and adenine is also facilitated by possible delocalization of the positive charge:

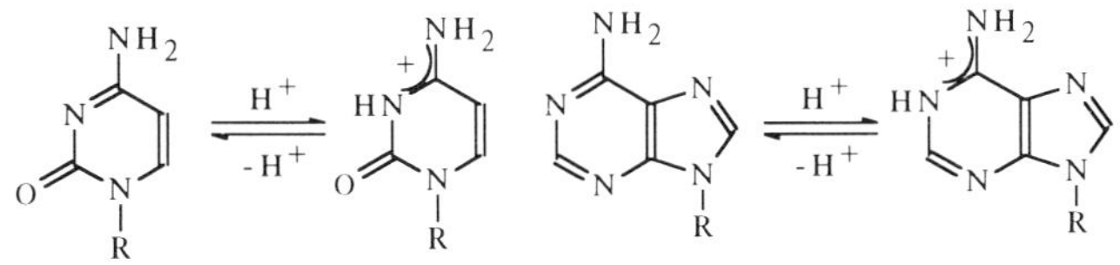

R - hydrogen or pentose phosphate residue

It is evident from Table 4-1 that adenine and cytosine as nucleoside and nucleotide constituents are weaker bases than aniline (pK_a 4.6).

Bases which can exist in nucleosides or nucleotides in a tautomeric hydroxy form and lack amino groups in the heterocyclic nucleus (uracil, thymine) easily give away their proton in an alkaline medium; that is, they exhibit acid properties:

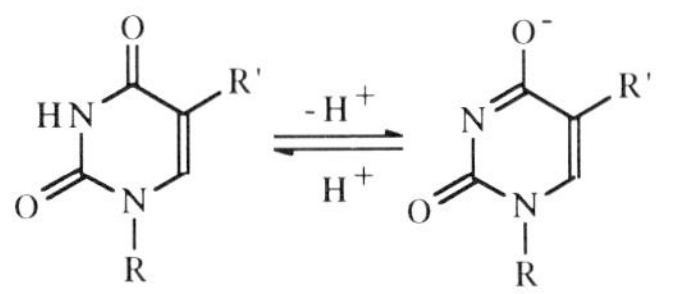

R - pentose or pentose phosphate residue;
R' = H or CH_3

As can be seen from Table 4-1, uracil and thymine are close to phenol in terms of protonation capacity (pK_a 9.99).

Guanine can display both basic and acid properties: it accepts a proton in an acid medium and loses it in an alkaline one. Just as in the above cases, the site of proton attachment is the nitrogen of the heterocyclic nucleus, rather than the amino group. Calculations have shown that the electron density is maximum at N^7. Protonation of the latter is confirmed by spectroscopic data and also by increased lability of the glycosidic bond in the cation of guanosine, caused by delocalization of the positive charge at both nitrogens in the imidazole ring:

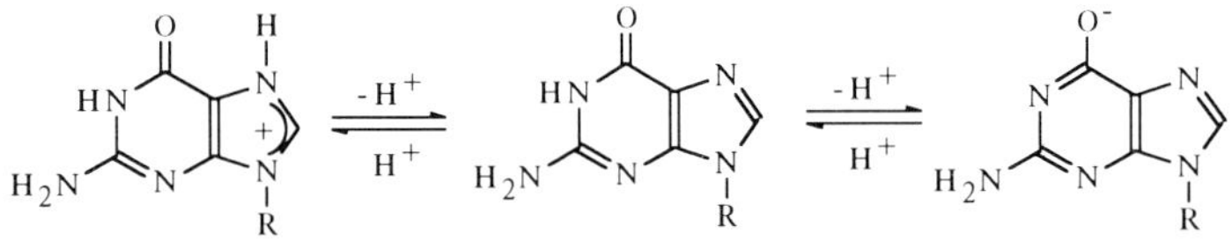

R - pentose or pentose phosphate residue

The pK_a values for guanine (see Table 4-1) indicate that it is a weaker base, as compared to cytosine and adenine, and closer to uracil and thymine in acid behavior (all bases are nucleoside or nucleotide constituents). The nature of the pentose moiety exerts only a minor influence on the ionization constant. The observed decrease in pK_a from the base to deoxyribonucleoside and further to ribonucleoside can be explained by the negative inductive effect of pentose, which is weaker in the case of deoxyribose, as opposed to ribose.

The pK_a value for the base is strongly affected by the presence of phosphate groups in nucleotides, albeit in a complicated manner. Since it might appear that the inductive effect of the phosphate group would lead to a further decrease in the pK_a value for the base, whereas in fact the opposite is true (it is greater in nucleotides than in nucleosides), one may assume that the effect of the phosphate group manifests itself in steric interactions. Proceeding from the structural formulas of nucleosides, it would be logical to assume that such interactions between the base and phosphate group must be especially pronounced in the case of nucleoside 5′-phosphates whose phosphate groups may be sterically close to the base. The 2′(3′)-phosphate group which is in *trans*-position with respect to the heterocycle must exert a weaker influence on the properties of the base.

4.1.2 Ionization of Hydroxyl Groups in Pentose

The dissociation observed in all nucleosides at a pH value close to 12 (see Table 4-1) is ascribed to ionization of the hydroxyl groups in the sugar.

4.1.3 Ionization of the Phosphate Group

The phosphoric acid residue in a nucleotide exhibits the properties of dibasic acids and, accordingly, has two dissociation constants.

$$RO-\overset{\overset{\displaystyle O}{\|}}{\underset{\underset{\displaystyle OH}{|}}{P}}-OH \xrightleftharpoons{k_1} RO-\overset{\overset{\displaystyle O}{\|}}{\underset{\underset{\displaystyle OH}{|}}{P}}-O^- \xrightleftharpoons{k_2} RO-\overset{\overset{\displaystyle O}{\|}}{\underset{\underset{\displaystyle O^-}{|}}{P}}-O^-$$

The pK_a values corresponding to the first and second steps of ionization of the phosphate groups in nucleotides are given in Table 4-1. As can be inferred from the tabulated data, the pK_a value corresponding to the first step of ionization ranges from 0.7 to 0.9 and that corresponding to the second step ranges from 6.0 to 6.4.

Nucleoside cyclic phosphates are essentially monobasic acids with a pK_a ≈ 0.7.

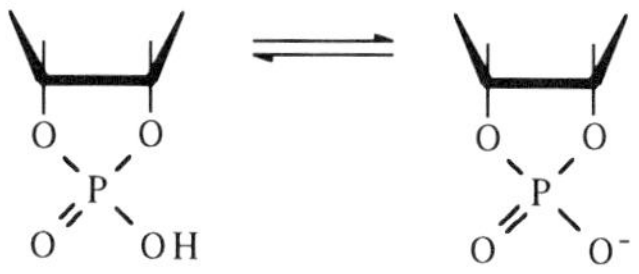

The pK_a decreases in the order uridine 5′-phosphate > adenosine 5′-phosphate > cytidine 5′-phosphate > guanosine 5′-phosphate which is indicative of the influence exerted by the cation emerging during protonation of the base on the first step of ionization of the phosphate group. Among the possible reasons for this phenomenon is intramolecular hydrogen bonding, for example:

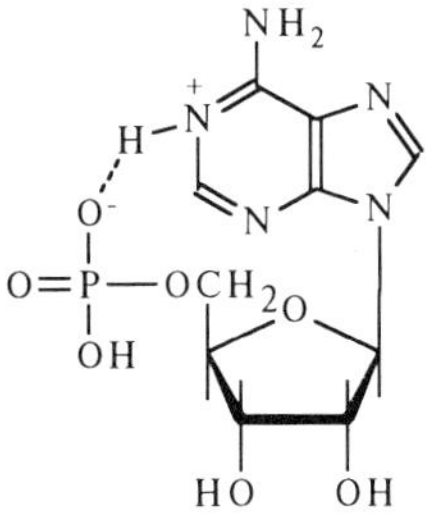

Comparison of the pK_a values for the phosphate group of nucleotides (0.7–1.0) with that for methyl phosphate (1.5) shows that in nucleotides the first step of dissociation of the phosphate group is strongly affected by the nucleoside.

Comparative pK_a values for the bases and phosphate groups in four major nucleotides are represented in Figure 4-1 as dissociation curves at pH values ranging from 0 to 8.

Proceeding from the differences in the pK_a values for the amino group, which, as can be seen from Figure 4-1, become manifest in the pH interval of 2 to 5, one can segregate a mixture of the four major nucleotides[1] into indi-

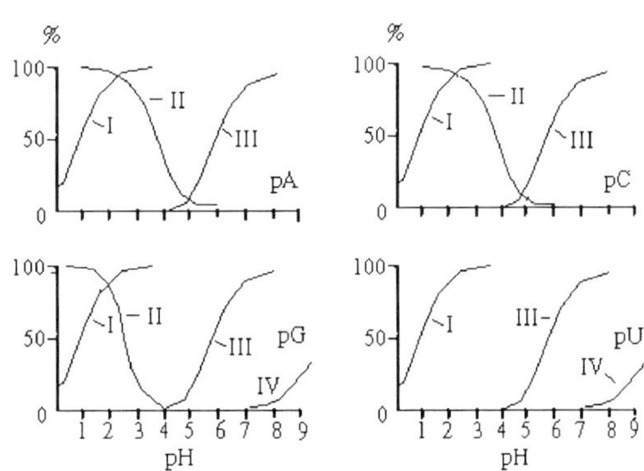

Fig. 4-1. Dissociation curves for ionized groups of nucleoside 5′-phosphates I and III – first and second steps of dissociation of the phosphate group, respectively; II – dissociation associated with the presence of an amino group; IV – dissociation associated with the presence of a hydroxy group in the carbohydrate moiety.

[1] Such mixtures are obtained, for example, after hydrolysis of nucleic acids.

vidual compounds. In the above pH interval, the phosphate group is completely dissociated at the first hydroxyl group, whereas at the second one it is still deprived of any charge whatsoever (noticeable ionization begins after pH 5). The widest differences in the net charges of nucleotides are observed at pH 3.5, when the positive charge at the bases (see curves in Fig. 4-1) equals 0.54 for adenosine 5′-phosphate, 0.05 for guanosine 5′-phosphate, 0.84 for cytidine 5′-phosphate, and 0.0 for uridine 5′-phosphate. Since each nucleotide has a negative charge equal to unity at pH 3.5, the free negative charges (net charges) are, respectively, 0.46 for adenosine 5′-phosphate, 0.95 for guanosine 5′-phosphate, 0.16 for cytidine 5′-phosphate, and 1.0 for uridine 5′-phosphate. The molecules of these four nucleotides are almost of the same size and, consequently, this factor does not produce any decisive effect on their mobility. Thus, the above values of net charges indicate that the electrophoretic mobility of the molecules is relatively high at pH 3.5. The dissociation curves can be used to calculate the relative mobility of nucleotides for any other pH value, which is a widely used technique in the separation of various nucleotide mixtures both by electrophoresis and ion-exchange chromatography.

Thus, the nucleotide molecule as a whole is capable of displaying both base and acid behavior at virtually any pH value. The implication is that, depending on local proton concentrations, a nucleotide can act as their donor and also as their acceptor.

4.2 Formation of Derivatives with Metals

Nucleotides form salts with metal ions, including salts of the chelate type. The possible interactions between various metals and nucleotides are summarized below (with respect to the site of binding with the nucleotide).

Binding site	Metal ions
Phosphate group	Li^+, Na^+, K^+, Rb^+, Cs^+, Mg^{2+}, Ca^{2+}, Sr^{2+}, Ba^{2+}
Phosphate group and ribose	B^{3+}, UO_2^{2+}
Phosphate group and heterocyclic base	Co^{2+}, Ni^{2+}, Mn^{2+}, Zn^{2+}, Cd^{2+}, Pb^{2+}, Cu^{2+}
Ribose and heterocyclic base	Co^{3+}
Heterocyclic base	Ag^+, Hg^{2+}

Depending on the metal species, it may become associated with the nucleotide both through proton substitution in groups capable of ionization and through linkage with the heterocyclic nitrogen atoms marked by high electron density (donor–acceptor bond). As regards metals incapable of complexing, such as alkali and alkaline-earths, nucleotides combine with them to form salts at the phosphate group. Nucleotides are usually isolated from nucleic acid hydrolysates and stored in the form of such salts. If nucleotides or nucleosides have a free *cis*-diol group, they can form complexes with boric acid. Depending on the diol concentration (in nucleotide or nucleoside), one molecule of boric acid may be coordinated with one or two diol molecules.

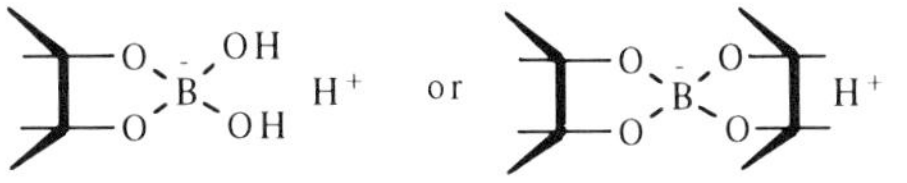

The formation of such complexes can be ascertained electrophoretically, this approach being often used in structural studies for determining 2′-3′-*cis*-diol groups.

Nucleotides also form chelate salts with ions of bivalent copper. In this case, the bonding with the metal involves the phosphate group and the nitrogen of the heterocyclic base.

In the absence of a phosphate group – that is, in the case of nucleosides – interaction with metal ions capable of complexing may involve the primary hydroxyl group of ribose and the amino group of the base.

The complex-producing interaction between adenosine and bivalent mercury salts involves only the nitrogens of the base, without pentose. It has been speculated that the metal is linked to the base at the nitrogen atoms at positions 1 and 7. Significantly, unsubstituted pyrimidine and purine bases interacting with ions of silver and bivalent mercury yield stable salts that have long been used for nucleoside synthesis.

The site of metal binding to nucleosides and nucleotides (see above) was determined using various physicochemical methods, such as UV, IR and NMR spectroscopy, as well as thermodynamic characteristics. It should be pointed out in conclusion that complexes of metals with nucleotides play a critical role in the latter's functioning at the monomer (in biosynthetic processes) and polymer levels.

4.3 Reactions at Heterocyclic Bases and Pentose

Nucleotides as well as nucleosides are known to enter into reactions involving pyrimidine and purine bases and also carbohydrate moieties. Reactions of the first type include electrophilic substitution of protons in the heterocyclic nucleus. A case in point is hydroxymethylation of cytidine 5′-phosphate:

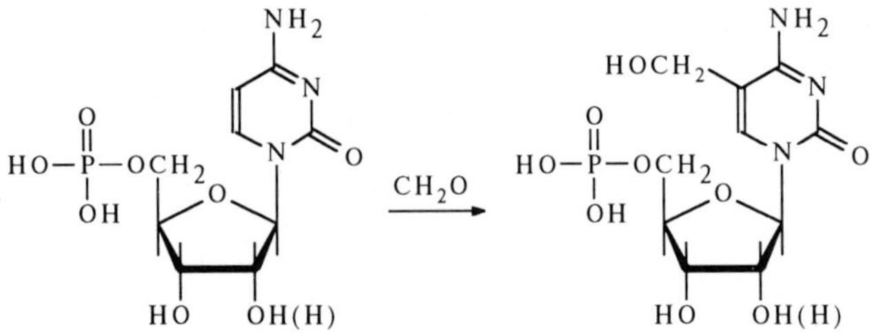

Hydroxymethylation of the uracil and cytosine nuclei in their ribo- and deoxyribonucleotides yields the corresponding 5-hydroxymethyl derivatives that are otherwise difficult to obtain.

A typical reaction for pyrimidine nucleotides is addition at the double bond C^5–C^6. Such addition at the C^5–C^6 bond of pyrimidine derivatives of the bisulfite ion is widely used to modify bases in nucleotides and nucleic acids[2]. The reaction is conducted in neutral aqueous solutions – that is, similarly to modification in the presence of hydroxylamine and halogens. The reaction with bisulfite proceeds at a much faster rate than with other nucleophilic agents and results, in the case of uracil derivatives, in saturation of the double bond C^5–C^6.

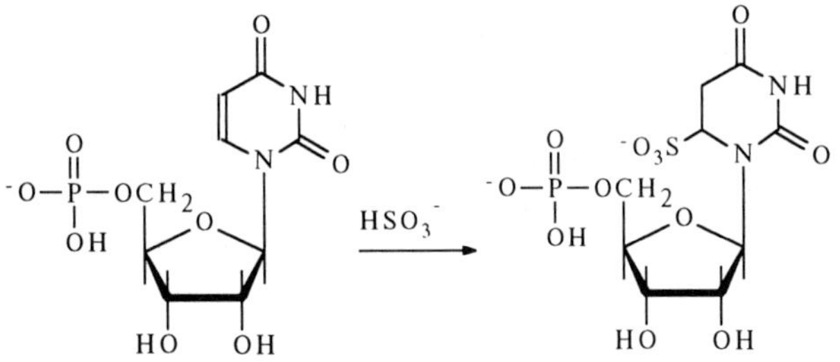

The reaction with cytosine derivatives in the absence of nucleophilic agents proceeds in a similar manner, but if the reaction mixture contains a nucleophile capable of attacking C^4, substitution for the amino group occurs at the same time. In the presence of methylhydroxylamine, for example, the following stable derivative is formed:

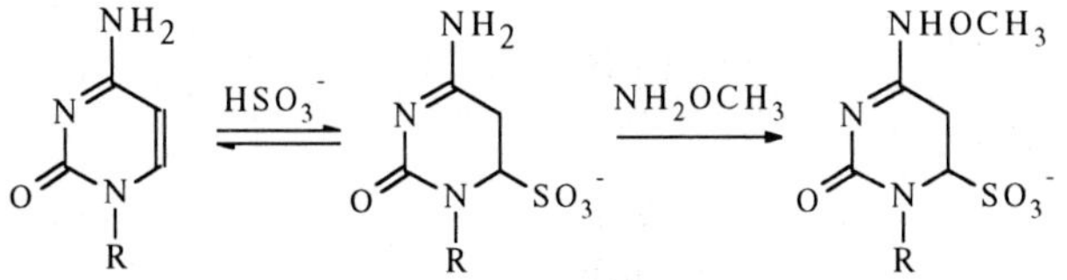

R - pentose phosphate residue

Practical uses of this and similar reactions are described elsewhere.

[2] This reaction has not been studied as thoroughly at the nucleoside level.

When hydrogenated over a rhodium catalyst, uridine 5′-phosphate is easily converted into the corresponding 5,6-dihydrouracil derivative:

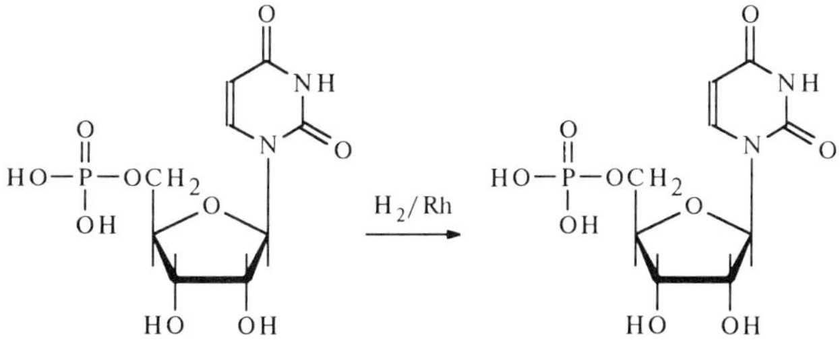

$$\xrightarrow{H_2/Rh}$$

Methylation reactions at the nucleotide level proceed both at the heterocycle and at the phosphate group. For example, methylation of uridine 5′-phosphate with diazomethane yields a mixture of N^3-methyluridine 5′-phosphate and uridine 5′-methyl phosphate with a 3:1 ratio.

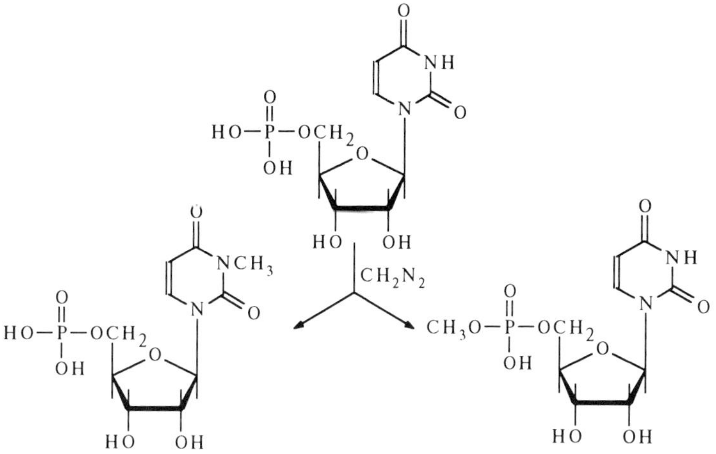

$$CH_2N_2$$

A similar picture is observed in the case of guanosine 5′-phosphate.

Alkylation of nucleotides with alkyl halides or dimethyl sulfate at pH ~ 4, when nucleotides exist in the form of a monoanion less nucleophilic than the dianion (which exists at pH 7.0), involves the heterocycle primarily. For instance, adenosine 5′-phosphate is smoothly converted into N^1-methyladenosine 5′-phosphate under such conditions:

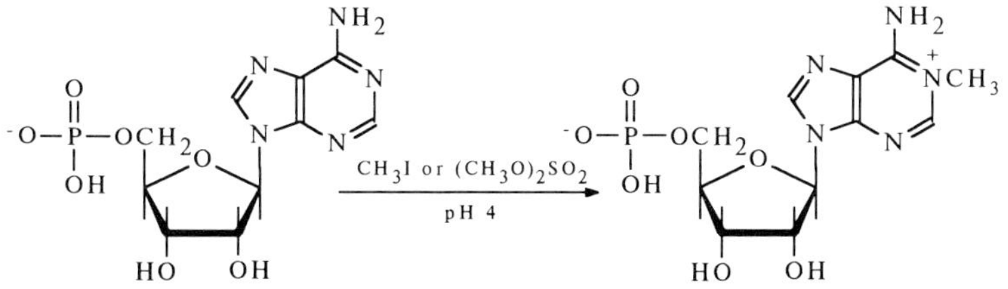

$$\xrightarrow[\text{pH 4}]{CH_3I \text{ or } (CH_3O)_2SO_2}$$

Other nucleotides are alkylated in very much the same fashion, the alkyl group occupying the same positions in the heterocycle as in the case of nucleosides.

Treatment of nucleotides with anhydrides or chlorides of organic acids results in substitution of acyl groups for all mobile hydrogen atoms in the pentose and acylation of the amino groups of the heterocyclic bases. If the latter lack amino groups, only derivatives at the carbohydrate moiety can be obtained. In this instance, a mixed anhydride with a phosphate group seems to emerge as an intermediate, which, however, is easily hydrolysed when the reaction mixture is treated with water. In the case of 5′-phosphates, this reaction proceeds without any complications; for example:

HO—P(=O)(OH)—OCH$_2$ (T) HO $\xrightarrow[Py]{Ac_2O}$ [AcO—P(=O)(OH)—OCH$_2$ (T) AcO] $\xrightarrow{H_2O}$ HO—P(=O)(O$^-$)—OCH$_2$ (T) AcO

The acylation of 3′-phosphates is less smooth. For instance, the main acylation product of uridine 3′-phosphate is 5′-acetyluridine 2′,3′-cyclic phosphate, while the yield of the 2′,5′-diacetyl derivative is very low (this side reaction naturally does not take place at all in the case of deoxynucleotides):

HOCH$_2$ (U) O—P(=O)(OH)O$^-$, OH $\xrightarrow[Py]{Ac_2O}$ AcOCH$_2$ (U) O, O—P(O)O$^-$

Such conversion is again caused by formation of a mixed anhydride as an intermediate, whose phosphorus atom carries an excess positive charge and, therefore, is prone to nucleophilic attack by the adjacent hydroxyl group:

O—P(=O)(O$^-$)—O—C(=O)—CH$_3$, :OH $\longrightarrow$ O, O—P(O)O$^-$ + CH_3CO^- (C=O) + H^+

Introduction of a nucleophilic agent more active than the hydroxyl group into the reaction mixture suppresses the side reaction and no cyclic phosphate is formed. For example, acetylation of uridine 3′-phosphate with acetic anhydride in the presence of tetraethylammonium acetate gives 2′,5′-*O*-diacetyluridine 3′-phosphate with a good yield:

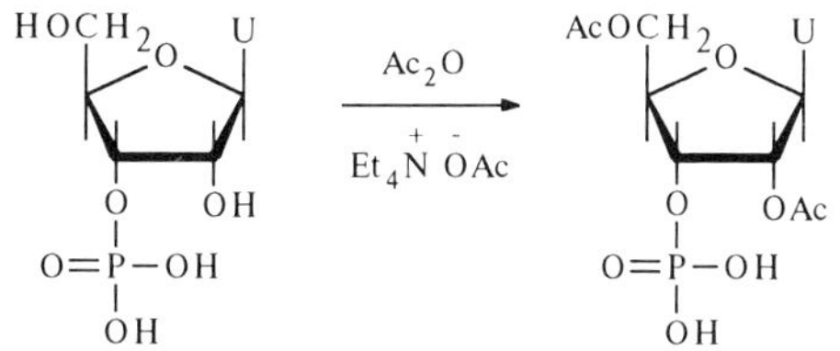

Acetylation of nucleotides containing amino groups in the nucleus of the heterocyclic base also yields the corresponding amides. Deoxycytidine 5′-phosphate, for example, is converted into 3′,N^4-dianisoyldeoxycytidine 5′-phosphate, when treated with *para*-anysoyl chloride:

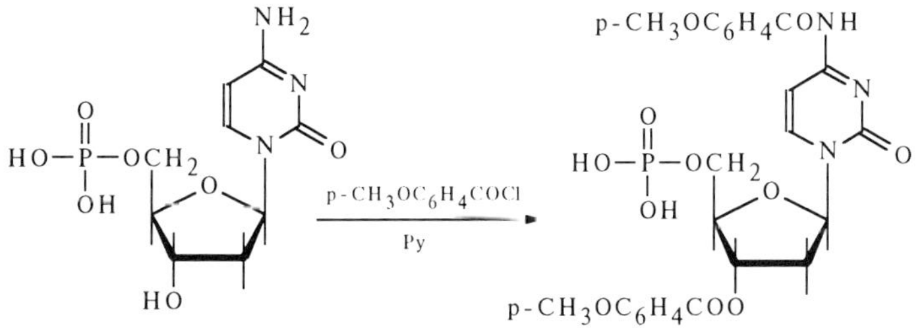

Acylation of adenine and guanine nucleotides also proceeds at the amino group and pentose. The acyl derivatives thus obtained are widely used in the synthetic chemistry of nucleotides (e. g., in polynucleotide synthesis). Studies into the properties of these compounds have shown that their amide bond is easily ruptured in the presence of ammonia but remains stable in alkaline media (pH > 12). The ester bonds in the carbohydrate moiety are easily broken both by alkalis and ammonia. Such difference seems to stem from the fact that the amide nitrogen in acyl aminopyrimidine (or -purine) behaves as phthalimide nitrogen or, in other words, exhibits acid properties:

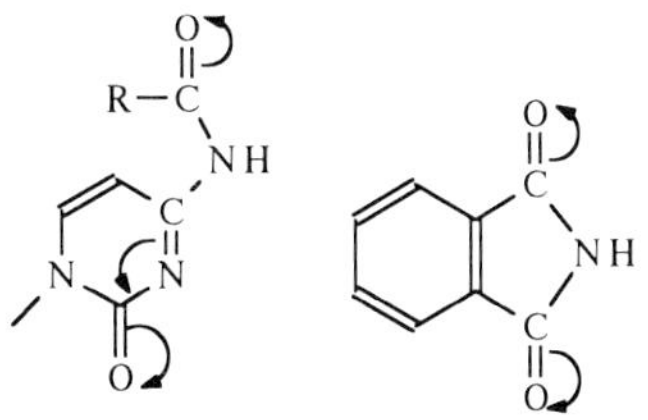

The detachment of the proton from the imide nitrogen in a highly alkaline medium results in its becoming negatively charged. Naturally, the resulting charge is delocalized, which can be expressed by the following boundary structures:

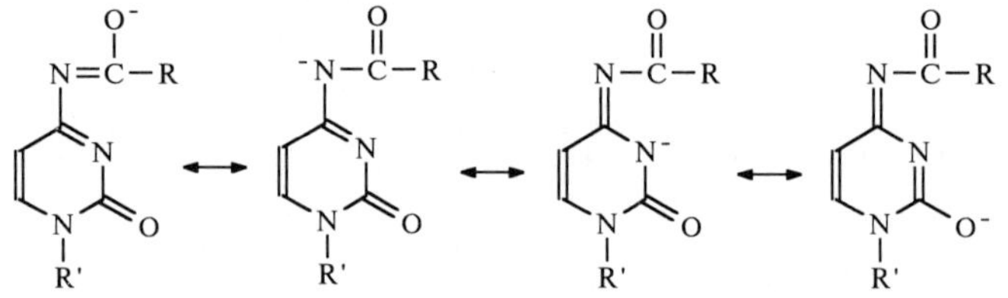

R'-pentose residue

Alkaline hydrolysis of such an amide bond (nucleophilic attack of the carbonyl carbon by the hydroxyl anion) must proceed at a very slow rate if the concentration of the deprotonated form is sufficiently high.

Treatment of such amides with concentrated aqueous ammonia solutions gives rise to ammonolysis. Ammonia attack may involve both the carbonyl carbon and that of the heterocycle (the uncharged NH_3 group must attack both carbon atoms with greater ease than the hydroxyl ion, provided the amide nitrogen is deprotonated under these conditions as well). As a consequence, the acyl group is eliminated in both cases.

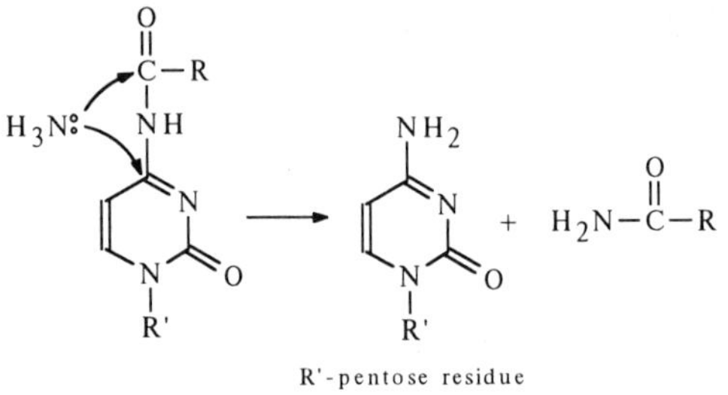

R'-pentose residue

Alkaline hydrolysis of the ester bond in acylated nucleotides follows the usual pattern, its rate being proportional to the hydroxyl ion concentration.

It appears that the ester bond undergoes both hydrolysis and ammonolysis when nucleotides acylated at the carbohydrate moiety are treated with aqueous ammonia solutions.

The different properties of the amide and ester bonds are put to practical use in selective removal of the acyl group linked to pentose, for example, to obtain deoxynucleotides in which only the amino group of the heterocyclic base is blocked. To this end, the deoxynucleotide acylated at pentose as well as the heterocycle is treated with an alkali (usually 1 N NaOH), and an *N*-acylated nucleotide is produced with a good yield. For instance, N^2-isobutyryldeoxyguanosine 5′-phosphate has been obtained from N^2,3′-diisobutyryldeoxyguanosine 5′-phosphate:

$^{-}O-P(O)(O^{-})-OCH_2$ … $NHCOCH(CH_3)_2$, $(CH_3)_2CHCOO$ $\xrightarrow{NaOH}$ $^{-}O-P(O)(O^{-})-OCH_2$ … $NHCOCH(CH_3)_2$, HO

The stability of amide bonds in *N*-acylnucleotides under the conditions described above depends on the structure of the heterocyclic base and the nature of the acyl. In the case of acetic acid derivatives, the amide bond stability diminishes in the series: guanosine > adenosine > cytidine. In N^4-acylated cytidine 5′-phosphates it decreases with varying acyl group in the following sequence: anisoyl- > benzoyl- > acetyl-. With this in view, different acyl groups are introduced into nucleotides (and nucleosides) for selective blocking. The less stable the *N*-acyl derivative forming a given heterocycle, the more stabilizing the acyl group is to be introduced into it. For example, anisoylation is used to block the amino group in cytosine nucleotides; in the case of guanosine, acylation with chlorides or anhydrides of fatty acids is used for the same purpose. The best way to selectively block the amino group of the heterocycle is through its benzoylation. Using excessively labile derivatives (such as *N*-acetylcytidines) or excessively stable ones (such as *N*-anisoylguanosines) makes no sense because, in the former case, the protective groups can be easily and uncontrollably eliminated in the presence of many nucleophilic agents and, in the latter, difficulties arise when attempts are made to selectively remove protective groups at the final stages of chemical transformations. The reactions involving all nucleotides of nucleic acids are discussed elsewhere.

4.4 Some Properties of the Phosphate Group (General Concepts)

For a better understanding of those properties of nucleotides which stem from the presence of a phosphate group in their molecules, we must first of all review the currently adopted concepts regarding the structure of the phosphate group and also discuss, using such simple analogues of nucleotides as alkyl phosphates and similar compounds as examples, the mechanism of nucleophilic substitution at the phosphorus atom as a reaction most typical of phosphates. This will not only throw more light on certain regularities in the properties of nucleotides, but also make it possible to know more about the special properties imparted to the phosphate group by the nucleoside. In this section, the structure and properties of the phosphate group are, wherever necessary, compared with those of its formal analogue – the carboxyl group, which allows us to emphasize the peculiar behavior of organic phosphoric acid derivatives.

Moreover, alkyl phosphates are used as examples to illustrate the transformations of crucial importance in the chemistry of nucleotides and nucleic acids, namely, hydrolysis and reactions involving the neighboring groups.

4.4.1 Structure of the Phosphate Group and Mechanism of Nucleophilic Substitution at the Phosphorus Atom

As is known, the electron configuration of phosphorus is $1s^2 2s^2 2p^6 3s^2 3p^3$. Its valence is determined by the structure of the outermost shell which, characteristically, includes not only $3s$ and $3p$ but also $3d$ orbitals (Fig. 4-2).

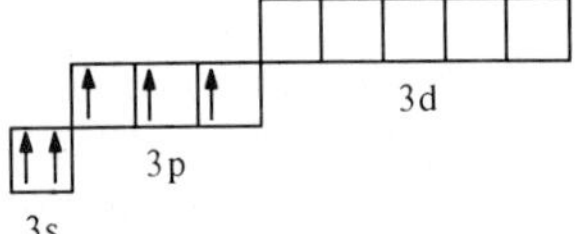

Fig. 4-2. Structure of the outermost electron shell of the phosphorus atom.

Phosphoric acid and its esters belong to compounds with a four-coordinated phosphorus atom. Such molecules are tetrahedral. The four σ-bonds with sp^3 hybridization of the electron orbitals are directed toward the apices of the tetrahedron:

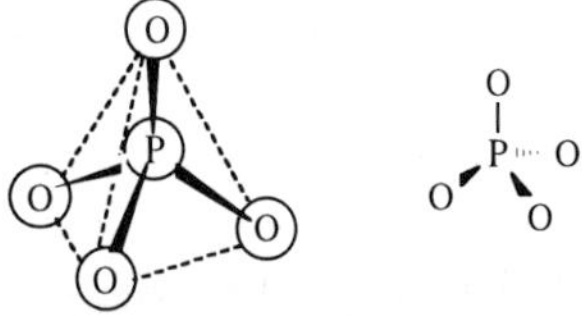

Thus, the phosphate group structure differs widely from that of the carboxyl group in carboxylic acids. It is well known that the latter group has a two-dimensional structure: the formation of the three σ-bonds in it involves the sp^2-hybrid orbitals of the carbon atom (they lie in the same plane and form 120° angles), while the fourth, π-bond results from overlapping of the p orbitals of the carbon and oxygen atoms at a right angle to this plane. The π-bond is polarized because of the pronounced electronegativity of the oxygen atom, as compared to the carbon, the degree of polarization being dependent on the inductive effect and mesomerism of the substituents at the carbonyl carbon.

At present, there is no consensus regarding the nature of bonding in compounds with a tetrahedral phosphorus atom. None the less, all assumptions to this effect are based on the general concepts concerning the structure and properties of *p* and *d* orbitals. It has been speculated that the π-bonding in the phosphoryl (P=O) group may involve vacant 3*d* orbitals[3]. Since there are five such orbitals (see Fig. 4-2) and each of them can, in principle, receive the fifth electron, the phosphorus atom can interact with that of oxygen in several ways to form a π-bond in the phosphoryl group. As a result of partial overlapping of the 3*d* and *p* orbitals of all four oxygen atoms in the phosphate group, double bonding is not localized in the phosphoryl group but is distributed, albeit unevenly, among all oxygen atoms of the phosphate group. For example, knowledge of the lengths of the P–O bonds (X-ray structural data) in dibenzyl phosphate, $(C_6H_5CH_2O)_2P(O)OH$, has led to a conclusion that each of the single P–O bonds accounts for 10% of double bonding and the P=O bond in the phosphoryl group accounts for 70%. Besides, the difference between double bonds in the C=O and P=O groups lies in the fact that the latter bond is polarized to a much greater extent (the phosphorus atom readily loses electrons from the *d* orbitals and probably for this reason is less electronegative than carbon). Such structural features of the phosphate group manifest themselves in the greater nucleophilicity of the phosphoryl oxygen and electrophilicity of the phosphorus atom. The transfer of electronic effects of the substituents, especially the conjugation effect with participation of the phosphoryl group, in phosphate esters and their derivatives is much less pronounced than in carboxylic acid derivatives because the substituents do not lie in the same plane.

At the same time, the phosphate and carboxyl groups share a common feature; namely, their central atom carries a partial positive charge and is therefore capable of being attacked by nucleophilic agents. When the phosphorus atom undergoes such an attack, nucleophilic substitution of one of the ligands (as the substituents at the phosphorus atom are usually called) can take place in phosphoric acid derivatives.

This process may be based on one of the two mechanisms: associative (A) and dissociative (B). In the former case, the substrate is coordinated with the nucleophile at the step determining the reaction rate. The resulting association is rather stable and may be regarded as an intermediate compound. This process is usually represented by the symbol S_N2. The dissociative mechanism is characterized by the fact that the substitution rate determining step involves

[3] Detailed information about the 3*d* orbitals of the phosphorus atom (angle functions, configuration and notation) and their involvement in π-bonding in the ground state of the molecule with a tetrahedral phosphorus atom as well as σ- and π-bonding in the transient state can be found in R. Hudson's and A. Kirby and S. Warren's books (see references at the end of this chapter).

departure of one of the substituents and formation of an active phosphorus compound with a smaller number of substituents at the central atom. This type of transformation is represented by the symbol S_N1. In some cases, however, the substitution event is a synchronous biomolecular process and no intermediate compound is formed, in contrast to S_N2 substitution at the carbon atom. The mechanism of such reactions is referred to as intermediate and denoted $I_a(S_Ni)$ ("a" indicates that the substitution process is still associative in essence). The first two mechanisms are discussed in what follows here in this section; mechanism I_a is treated in the section dealing with the properties of phosphoric acid amides whose transformations led to its discovery in the first place.

Mechanism S_N2 of nucleophilic substitution at the tetrahedral phosphorus atom has been corroborated by numerous kinetic experiments with hydrolysis and alcoholysis of halogen phosphates. It has been speculated that, formally, the reaction proceeds in very much the same manner as nucleophilic substitution at the saturated carbon: the nucleophile Nu attacks the reaction site from "behind", with respect to the leaving group X.

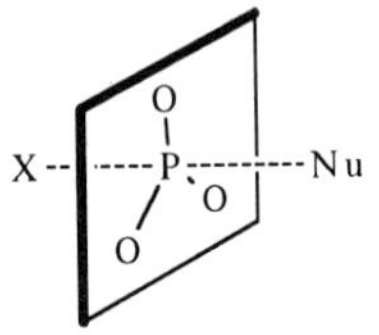

After departure of the substituent X, the configuration is turned inside out. A Walden inversion during nucleophilic substitution at the tetrahedral phosphorus atom has been demonstrated in many experiments aimed at elucidating the reactivity of optically active phosphorus-containing compounds.

It is believed that the most likely structure of the intermediate compound is a trigonal bipyramid in which the entering and leaving (substituted) groups (Nu and X, respectively, in Fig. 4-3A) occupy axial positions[4].

Geometrically, this structure is reminiscent of the transition state arising during an S_N2 reaction at the saturated carbon atom (Fig. 4-3B). In the latter case, however, the five bonds differ markedly in strength: the three bonds at the base of the bipyramid are quite strong, while the two others (C–X and C–Nu) are close to ionic bonds. At the same time, by virtue of the ability of the phosphorus atom to accept electrons at the vacant 3*d* orbitals (see Fig. 4-2), it can form compounds with five covalent bonds of comparable, if not identical, strength. In intermediate compounds of type A, for example (see Fig. 4-3), the three bonds at the bipyramid base (equatorial bonds), formed with the participation of sp^2-hybrid orbitals, are stronger. The two other (axial) bonds involving *pd*-hybrid orbitals are weaker.

[4] These positions are sometimes referred to as apical.

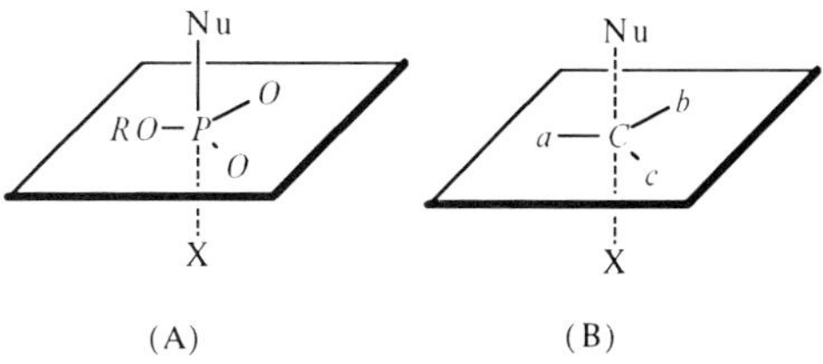

Fig. 4-3. Structures of the intermediate compound arising during nucleophilic substitution at the phosphorus atom in the phosphate group (A) and the transient state during nucleophilic bimolecular substitution at the saturated carbon (B)
Nu and X – entering and leaving groups (atoms) in an axial position; a, b, c – different residues.

Consequently, we are dealing here with an important unique property of the phosphorus atom, namely, its ability to form additional bonds involving its vacant 3*d* orbitals. In this case, the octet rule is usually not observed, and the phosphorus atom in similar structures can be surrounded by a shell of ten electrons.

The mechanism of nucleophilic substitution at the phosphorus atom, described in the context of the phosphate group, differs substantially from that of nucleophilic substitution at the carbonyl carbon in carboxylic acid derivatives, where formation of the intermediate structure is ensured by rupture of the double C=O bond.

The nucleophilic attack of the carbon in the carbonyl group proceeds at a right angle to the plane accommodating the RCOX molecule. The structure emerging after the breaking of the C=O bond is characterized by sp^3 hybridization of the electron orbitals in the carbon atom (with the four σ-bonds being directed toward the apices of the tetrahedron).

The rate of nucleophilic substitution at the tetrahedral phosphorus atom is determined by the following three major factors: (1) substituent species, (2) species of the leaving group, and (3) species of the nucleophilic agent.

Substituents. The ease of nucleophilic substitution at the tetrahedral phosphorus atom may be dependent both on the donor–acceptor properties of the substituents and on their three-dimensional structure.

One should begin discussion of the effect exerted by alkoxy groups in phosphates by comparing them with analogous derivatives of carboxylic acids. It is a well known fact that alkoxy groups may produce negative inductive and positive mesomeric effects on the associated atoms or groups of atoms. In cases where conjugation is possible, the positive mesomeric effect is much more pronounced than the negative inductive one.

It has already been mentioned above that carbonyl-containing compounds have two-dimensional structure. This is why the lone-pair electrons of some atoms adjacent to the carbonyl group interact with it rather strongly (are in conjugation):

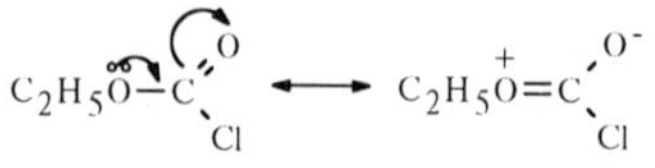

$$C_2H_5\ddot{O}-C(=O)Cl \longleftrightarrow C_2H_5\overset{+}{O}=C(O^-)Cl$$

A similar conjugation in phosphates of tetrahedral configuration is not as manifest (for structural features of the phosphoryl group, see above). Nevertheless, they may also display conjugation of the oxygen and phosphorus atoms due to the vacant 3*d* orbitals of the latter (so-called) $p_\pi - d_\pi$ conjugation). The lone-pair electrons of the oxygen in the alkoxy group may be transferred to the 3*d* orbitals of the phosphorus atom, which, in turn, must give rise to further hindrances for interaction between the phosphorus and the nucleophile. The presence and degree of such conjugation are confirmed by the results of hydrolysis of carboxylic and phosphoric chlorides:

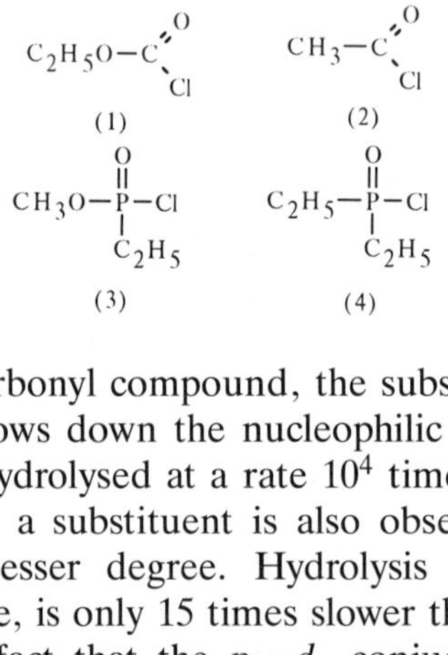

$$C_2H_5O-C(=O)Cl \quad (1) \qquad CH_3-C(=O)Cl \quad (2)$$

$$CH_3O-P(=O)(C_2H_5)-Cl \quad (3) \qquad C_2H_5-P(=O)(C_2H_5)-Cl \quad (4)$$

In a carbonyl compound, the substituent exhibiting *p*-donor properties drastically slows down the nucleophilic attack. For example, ethyl chlorocarbonate (1) is hydrolysed at a rate 10^4 times slower than acetylchloride (2). The effect of such a substituent is also observed in a phosphoryl compound, but to a much lesser degree. Hydrolysis of methoxyphosphoryl derivative (3), for example, is only 15 times slower than that of its alkyl analogue (4).

The fact that the $p_\pi - d_\pi$ conjugation of substituents with the phosphorus atom affects the activity of the latter in nucleophilic substitution reactions is confirmed by well known examples of decreasing reactivity as alkyl radicals in compounds of a similar structure are substituted by alkoxy or amide groups. It has been found that reactivity during alkaline hydrolysis decreases perceptibly in the following order: $(C_2H_5)_2POCl > CH_3O(C_2H_5)POCl > (CH_3O)_2POCl$. The steric hindrances arising during hydrolysis of these compounds are more or less the same. Just as in the previous case, the changes in reactivity are ascribed to the fact that the lone-pair electrons of the oxygen in the alkoxy group are shifted to the 3*d* orbitals of the phosphorus atom. Since the 3*d* orbitals of the phosphorus atom are rather fully occupied as a result of such conjugation which also leads to a decrease in its surplus positive charge, incorporation of the lone-pair electrons of the nucleophilic agent into the 3*d* orbitals of the phosphorus atom is hindered significantly in the transition state with an associated drop in the rate of nucleophilic substitution.

An even more pronounced inhibiting effect on nucleophilic substitution at the phosphorus atom is exerted by an alkylamide (or dialkylamide) group. For instance, alkaline hydrolysis of the diamide $[(CH_3)_2N]_2P(O)F$ proceeds at a rate slower by a factor of tens of thousands than in the case of the methylmethoxy derivative $CH_3O(CH_3)P(O)F$. This influence of the amino group in phosphamides is attributed to the fact that by virtue of its more pronounced positive mesomeric effect and less pronounced negative inductive one, as compared to the alkoxy group, it weakens the electrophilic properties of the associated phosphorus atom to a much greater degree (due to the $p_\pi - d_\pi$ conjugation).

In halogen phosphates $(RO)_2P(O)X$, the halogens (X = F and Cl) may also be in $p_\pi - d_\pi$ conjugation with the phosphorus atom. Experimental evidence attests to a lesser degree of involvement of the chlorine atom in such conjugation, in comparison with fluorine, the greater electronegativity of the latter notwithstanding (acid chlorides are much more reactive than acid fluorides). Thus, the degree of conjugation seems to be dependent not only on electronegativity, but also on the structure of the outermost shell of the atom associated with phosphorus.

A similar relationship has also been observed in the case of trisubstituted phosphates and thiophosphates, $(RO)_2P(O)OR'$ and $(RO)_2P(O)SR'$, differing in the structure of one substituent. Dialkyl thiophosphates are tens of thousands of times more reactive than trialkyl phosphates.

The exact course taken by nucleophilic substitution at the tetrahedral phosphorus atom is strongly affected not only by the capacity of substituents for conjugation, but also by their size. In this case, steric hindrances are more influential than in the series of carboxylic compounds with planar structure at the carboxyl group. In esters of the $(CH_3O)_2P(O)X$ type, replacement of the methyl group by a *tert*-butyl one leads to a rather sharp drop in the rate of alkaline hydrolysis.

Leaving Group. This factor is as important in nucleophilic substitution reactions as the electrophilicity of the phosphorus atom. The effect of the leaving group is rather simple. The greater the ease with which its linkage to the reaction site is broken, the more reactive the substance. Since the leaving group departs taking its electron pair along, all factors conducive to its greater electronegativity usually facilitate the course of the reaction. In compounds of the $(RO)_2P(O)X$ type, X is always the leaving group if it is associated with the phosphorus atom via a "weaker" bond than that linking the alkoxy group with the latter. We have already seen examples in which the X group was represented by a chlorine atom, RS and other groups. If the function of the leaving group is performed by a phosphoric acid anion, when nucleophilic substitution takes place in pyrophosphates, for example, the attack ends in liberation of a more stable anion – that is, anion of a stronger acid. For instance, asymmetrical dibenzyldiphenyl pyrophosphate reacts with alcohols to yield exclusively dibenzyl phosphates (diphenylphosphoric acid is stronger than the dibenzylphosphoric one).

$$(C_6H_5O)_2\overset{O}{\overset{\|}{P}}-O-\overset{O}{\overset{\|}{P}}(OCH_2C_6H_5)_2 \xrightarrow{\;HOR\;} (C_6H_5O)_2\overset{O}{\overset{\|}{P}}-O^- H^+ + RO-\overset{O}{\overset{\|}{P}}(OCH_2C_6H_5)_2$$

Nucleophilic Agent. In the course of nucleophilic substitution, the nucleophile displaces the group being substituted from the molecule of the substrate (in our case, phosphate). As has already been pointed out, the ease with which the nucleophilic substance is attached to the phosphorus atom acting as the reaction site depends on a number of factors. One of them is *nucleophilicity* or, in other words, capacity of a reagent to give up its electron pair for combining with the reaction site. Nucleophilicity is determined by the mobility of electrons in the nucleophilic particle, its polarizability, and steric factors. The nucleophilic agents reacting readily with the tetrahedral phosphorus atom can be arranged as follows [with respect to chlorophosphinates of the $R_2P(O)Cl$ type]: $HO^- > C_6H_5O^- > C_2H_5OH > C_2H_5S^-$, CH_3COO^-. Much more pronounced nucleophilicity is displayed by hydroxylamine and hydroxamic acids. The increased activity of these compounds stems from the fact that the atom involved in the nucleophilic attack is found next to another having lone-pair electrons ("α-effect"). This seems to ensure greater polarizability of the nucleophilic agent. Naturally, depending on how fully the 3*d* orbitals of the phosphorus atom are occupied as a result of $p_\pi - d_\pi$ conjugation with the substituents, the basicity and polarizability of nucleophiles may influence the reaction rate differently. It has been demonstrated that the more pronounced the conjugation of substituents, for example, alkoxy and dialkylamide groups with 3*d* orbitals, the less significant the effect of basicity of the nucleophile on the reaction rate and the greater the importance of polarizability.

It has already been mentioned that during substitution based on the S_N2 mechanism in the transition state, the bonds of the saturated carbon atom (see Fig. 4-3B) differ widely in strength, and the transition state is not likely to exist for a more or less long time. At the same time, the corresponding state for the phosphorus atom in phosphates (see Fig. 4-3A) is characterized by a much smaller difference in bond strength and is comparatively more stable. It is these specific features that make restructuring of the transition state (type A in Fig. 4-3) possible, which may ultimately affect the end result of the substitution. Such restructuring is not accompanied by cleavage of the phosphorus–substituent bonds and boils down to changes in bond angles and lengths in a sufficiently persistent transient state, as represented by the trigonal bipyramid. This kind of restructuring is schematically shown below.

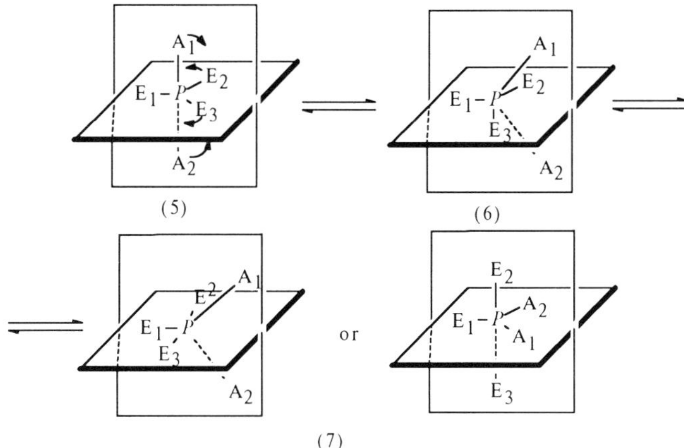

(5) (6) (7)

Let substituents E_1, E_2 and E_3 in the starting trigonal bipyramid occupy equatorial positions, with substituents A_1 and A_2 being in axial ones.

At a certain point in time, the substituents A_1 and A_2 start moving toward each other in the plane of the page. Simultaneously, the substituents E_2 and E_3 move in a plane perpendicular to that of the page so that E_3 moves upward from the page, while E_2 moves in the opposite direction. As a result, a state (6) arises in which the substituents A_1, A_2 and E_3 form the base of a tetragonal bipyramid with apex E_1. The subsequent changes in angles, proceeding in the same directions, lead to a state (7) where the substituents E_3 and E_2 lie on the same straight line, whereas E_1, A_1 and A_2 are arranged in the plane of the page at an angle of 120°. In the resulting trigonal bipyramid, the axial positions are now occupied by the substituents E_3 and E_2 and the equatorial positions are occupied by E_1, A_1 and A_2. Such restructuring of the transient state, (5) → (7), which, of course, is reversible, is called pseudorotation. The line connecting the central atom with the substituent E_1 is known as axis of rotation. The pseudorotation phenomenon is observed in such compounds where the energies of the phosphorus–substituent bonds are reasonably close. In the state (6), the bond lengths are practically equalized (rehybridization of the three sp^2 equatorial and two *pd* axial orbitals, leading to five sp^3d hybrid orbitals), and the function of the axis of rotation can be performed by virtually any phosphorus–substituent bond, provided it does not differ markedly from others. Some examples of pseudorotation are considered below in section 4.4.4 dealing with cyclic phosphates.

Mechanism S_N1. Some reactions of nucleophilic substitution at a tetrahedral phosphorus atom are believed to involve a mechanism different from S_N2 described above. The reaction apparently proceeds in two steps, in this case: first, one of the substituents (X) in the phosphate group is eliminated with intermediate formation of the corresponding metaphosphate, then, the nucleophilic agent (Y^-) is attached to it:

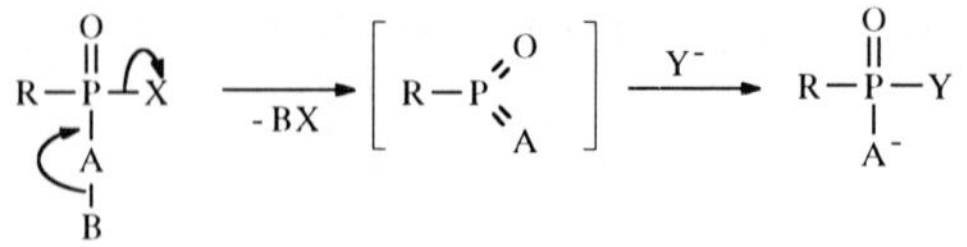

This is what is known as mechanism S_N1 of phosphorus substitution. However, the existing proof of the latter cannot yet be considered convincing enough. Now follow examples of reactions believed to be based on this mechanism.

Salicyl phosphate (8) is hydrolysed much more rapidly than unsubstituted phenyl phosphate or *para*-carboxyphenyl phosphate (the hydrolysis rate is maximum at pH 5). It was assumed earlier that nucleophilic catalysis takes place and salicyloyl phosphate (9) undergoes hydrolysis. As was established later, the reaction takes a different course: salicyl phosphate (8) is hydrolysed more rapidly than salicyloyl phosphate (9) and is essentially an intermediate compound forming during hydrolysis of the latter:

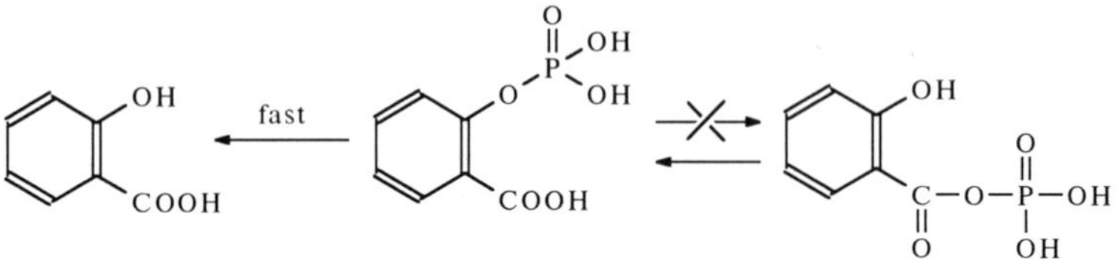

(8) (9)

The reason for a faster reaction rate in going from (9) to (8) seems to be intramolecular acid catalysis leading to formation of metaphosphate $[PO_3^-]$:

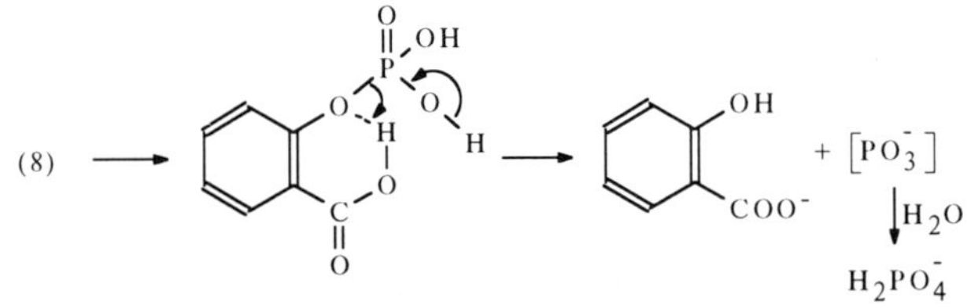

1,8-Dihydroxynaphthalene monophosphate is hydrolysed in a similar manner:

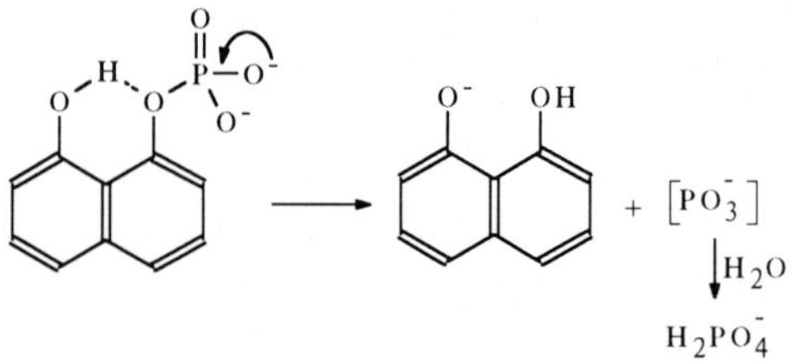

The hydrolysis of acyl phosphates and pyrophosphates also seems to involve the S_N1 mechamism:

$$R{-}C(=O){-}O{-}PO_3H^- \xrightarrow{\text{slowly}} R{-}C(=O)OH + [PO_3^-] \xrightarrow{H_2O} H_2PO_4^-$$

$$HO{-}P(=O)(O^-){-}O{-}P(=O)(OH){-}O^- \longrightarrow H_2PO_4^- + [PO_3^-] \xrightarrow{H_2O} H_2PO_4^-$$

More substantial evidence in favor of the S_N1 mechanism of substitution at the phosphorus atom was provided in studies of tetraethyl pyrophosphate hydrolysis. The reaction proceeds at a fast rate at pH values exceeding 9 and is catalyzed by phosphate ions. If the catalyst is a phosphate ion containing the oxygen isotope ^{18}O, the label is detected in the molecule of the emerging diethyl phosphate. This is indicative of intermediate formation of a metaphosphate anion in this reaction.

This conclusion is borne out by formation of methyl phosphate (11) when the reaction is conducted in methanol and also trimetaphosphate (10) in a concentrated aqueous solution.

$$HO{-}P(={}^{18}O)({}^{18}O^-)_2 + (C_2H_5O)_2P(=O){-}O{-}P(=O)(OC_2H_5)_2 \longrightarrow$$

$$\longrightarrow (C_2H_5O)_2PO_2^- + (C_2H_5O)_2P(=O){-}O{-}P(={}^{18}O)({}^{18}O^-)_2 \longrightarrow$$

$$\longrightarrow (C_2H_5O)_2P(=O){-}{}^{18}O^- + [{}^{18}O{=}P({=}{}^{18}O){-}{}^{18}O^-] \xrightarrow{H_2O} HO{-}P(={}^{18}O)({}^{18}O^-)({}^{18}OH)$$

$$[{}^{18}O{=}P({=}{}^{18}O){-}{}^{18}O^-] \longrightarrow \text{(10)}; \quad [{}^{18}O{=}P({=}{}^{18}O){-}{}^{18}O^-] \xrightarrow{CH_3OH} CH_3O{-}P(={}^{18}O)({}^{18}O^-)({}^{18}OH) \text{ (11)}$$

(10) (11)

4.4.2 Catalysis of Nucleophilic Substitution at the Phosphorus Atom

Nucleophilic Catalysis (increasing the reactivity of the phosphorus atom). Many reactions with the participation of phosphate derivatives have been found to involve nucleophilic catalysis by tertiary amines. Alcoholysis of tetrabenzyl pyrophosphate, for example, is accelerated by imidazole:

$$(BzlO)_2P(=O)-O-P(=O)(OBzl)_2 \rightleftharpoons {}^{-}O-P(=O)(OBzl)_2 + HN\overset{+}{\frown}N-P(=O)(OBzl)_2 \xrightarrow[\text{(fast)}]{C_3H_7OH} C_3H_7O-P(=O)(OBzl)_2$$

A similar effect is produced by pyridine, however, sterically hindered amines, such as 2,6-lutidine, do not catalyze such reactions.

Catalysis with a nucleophilic agent resides in that substitution of the leaving group with formation of an intermediate compound and subsequent replacement of the catalyst by the nucleophile whose entry is facilitated by it proceed at much faster rates than in a non-catalyzed reaction. This has the following explanation: tertiary amine is a stronger nucleophile than alcohol, and the forming intermediate with an ammonium group at the phosphorus atom is more reactive than pyrophosphate because the p_π–d_π conjugation between the phosphorus and substituent containing the ammonium nitrogen is completely upset and the latter exerts a strong negative induction effect. An example of unstable compounds belonging to this category is provided by pyridinium derivatives of alkyl phosphates:

$$C_5H_5\overset{+}{N}-P(=O)(OR)_2$$

General Basic Catalysis. What sets this type of catalysis apart from the nucleophilic one is the deuterium isotope effect. It is known that reactions whose rate is determined by cleavage of the bonds formed by deuterium atoms proceed five to eight times more slowly than the corresponding reactions involving atoms of protium (instead of deuterium). Alcoholysis of pyrophosphates in the presence of bases is slowed down from ROH to ROD. This indicates that during the step preceding nucleophilic attack at the phosphorus atom the alcohol is activated by the base, the activation boiling down to "removal" of the proton:

$$RO-P(=O)(OR)-O-P(=O)(OR)-OR \longrightarrow RO-P(=O)(OR')-OR + {}^{-}O-P(=O)(OR)_2$$

(D)H—Ö—R'

B (2,6-lutidine) $\longrightarrow BH^+$

Since in this case the base serving as the catalyst does not participate in the nucleophilic attack at the phosphorus atom, its steric features do not affect the reaction rate in any significant manner. For instance, phosphorylation of alcohols with pyrophosphates (see above) is catalyzed by tertiary amines, including such sterically hindered ones as, say, 2,6-lutidine.

General Acid Catalysis. A good illustration of acid catalysis of nucleophilic substitution at the phosphorus atom is provided by reactions between sarin (isopropyl methylphosphonofluoridate) with phenols. For example, the pyrocatechol monoanion reacts with sarin at a much faster rate than the phenolate ion.

iso-C_3H_7—O, O---HO; H_3C, F, $^-$O

or

iso-C_3H_7—O, O, $^-$O; H_3C, F----HO

$\xrightarrow{(-F^-)}$ iso-C_3H_7—O, O, P, HO, H_3C, O

A possible reason is hydrogen bonding between the nondissociated hydroxyl group of pyrocatechol and the phosphoryl oxygen or sarin fluorine atoms, disturbing the oxygen–phosphorus conjugation with the result that the phosphorus atom is more readily attacked by the nucleophile, or phenoxy ion to be precise. Catalysis along similar lines seems to be likely for reactions in which the function of the catalytically active group is performed by the hydrogen-containing ammonium group.

CH_2, $\overset{+}{N}$, CH_3, CH_3, H, O^-, O, H_3C—P, F, iso-C_3H_7O $\xrightarrow{(-F^-)}$ CH_2, $\overset{+}{N}$, CH_3, CH_3, H, O, H_3C—P=O, OC_3H_7-iso

The exceptionally high rate of acid hydrolysis of a sarin analogue containing a trialkyl amino group can be ascribed to intramolecular catalysis.

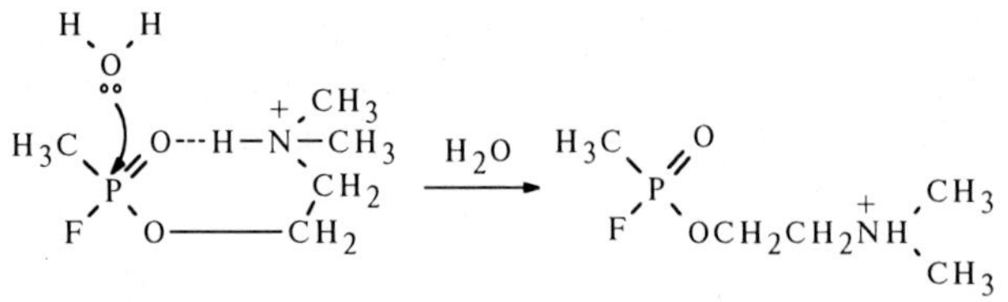

$$H_3C(F)P(=O)OCH_2CH_2\overset{+}{N}H(CH_3)_2 \cdots H_2O \xrightarrow{H_2O} H_3C(F)P(=O)OCH_2CH_2\overset{+}{N}H(CH_3)_2$$

The general acid catalysis may involve nondissociated carboxyl groups. Diethyl phenylphosphonate which contains a carboxyl group in the *ortho* position is easily hydrolysed, when heated with water, to a free phosphate:

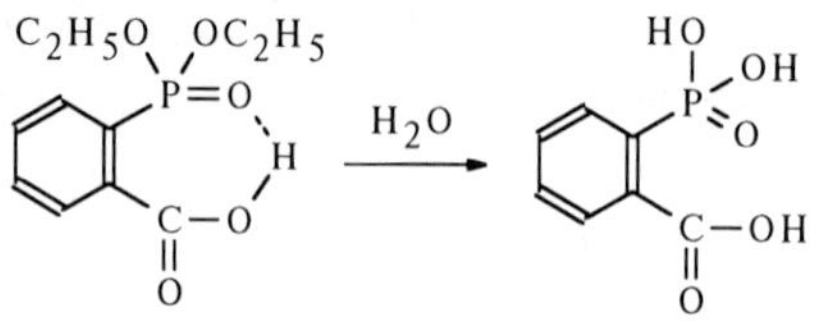

$$o\text{-}C_6H_4[P(=O)(OC_2H_5)_2](COOH) \xrightarrow{H_2O} o\text{-}C_6H_4[P(=O)(OH)_2](COOH)$$

The stability toward hydrolysis under the same conditions of the sodium salt (12) and alkyl esters (13) of the same acid as well as the corresponding *para*-isomer (14) provides additional proof to support intramolecular acid catalysis with the participation of an *ortho*-carboxyl group.

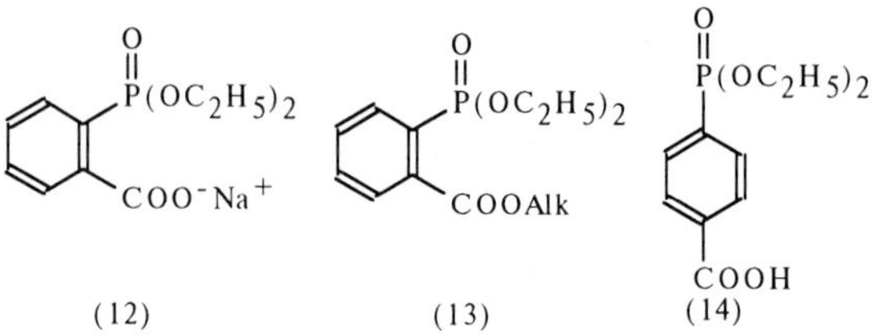

$o\text{-}C_6H_4[P(O)(OC_2H_5)_2](COO^-Na^+)$ (12)

$o\text{-}C_6H_4[P(O)(OC_2H_5)_2](COOAlk)$ (13)

$p\text{-}C_6H_4[P(O)(OC_2H_5)_2](COOH)$ (14)

Electrophilic Catalysis. Many examples may be used to illustrate catalysis of phosphate hydrolysis by metal ions. A case in point is alcoholysis of tetrabenzyl pyrophosphate with a propyl alcohol and tertiary amine mixture, the rate of which substantially increases in the presence of lithium or calcium ions.

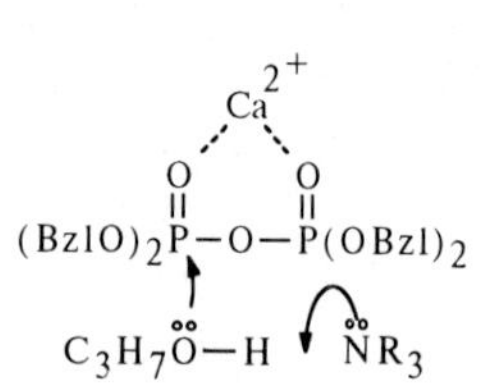

$$Ca^{2+} \cdots (BzlO)_2P(=O)-O-P(=O)(OBzl)_2 \leftarrow C_3H_7\ddot{O}-H \;\; \ddot{N}R_3$$

This effect also seems to be a consequence of enhanced electrophilic properties of the phosphorus atom after drawing electrons from the phosphoryl oxygen. Hence, the mechanism of catalysis in this case is close to the acid one, with the difference that the role of the proton is played by the metal ion. Similarly, diisopropyl fluorophosphate and related compounds are rapidly hydrolysed in the presence of complexes formed by bivalent copper and α,α'-dipyridyl:

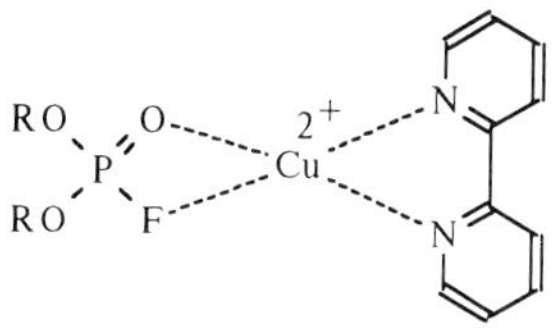

4.4.3 Hydrolysis of Alkyl Phosphates

Since nucleotides and polynucleotides remained for a long time relatively little known compounds, most conclusions made as regards the mechanisms of their transformations associated with the phosphate group more often than not were based on analogies with the corresponding reactions involving alkyl phosphates. In the early fifties, for example, evidence was provided, by the mechanisms of hydrolysis of mono- and dialkyl phosphates as well as phosphates of simple polyols, to explain the difference in hydrolytic stability between RNA and DNA. The same findings made it possible to predict and explain many properties of nucleotide derivatives from the structure of the associated nucleoside and the phosphate group position in the latter. The facts accumulated so far are not yet sufficient for detailed analysis of the mechanisms of many reactions involving nucleotides and their derivatives, especially enzymatic reactions with participation of the phosphate group. There is no doubt, however, that a major contribution to investigation of these processes will be made by the thoroughly studied reactions involving such analogues of nucleotides as alkyl phosphates and phosphates of polyols. Accordingly, this section deals with reactions involving mono-, di- and trialkyl phosphates and also those of polyols. The reactions discussed here are of great importance in the chemistry of nucleotides and their derivatives. Particular attention is given to processes with participation of the neighboring groups – that is, accompanied by intramolecular catalysis. Most of the reactions considered in this section are based on the mechanism of nucleophilic substitution at the tetrahedral phosphorus atom. However, depending on the exact form in which the phosphate group takes part in the reaction (anion, neutral molecule), the latter may take a different course as well. The general concepts as regards such reactions are also discussed.

Hydrolysis of Monoalkyl Phosphates. The rate of hydrolysis of most simple monoalkyl phosphates passes through a maximum lying in the pH range of 3 to 5. A typical curve showing the alkyl phosphate hydrolysis rate versus pH of the medium is given in Figure 4-4.

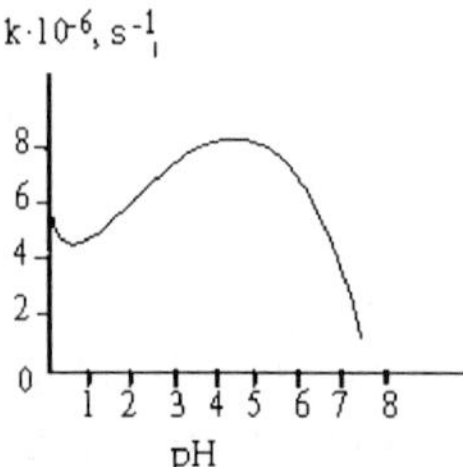

Fig. 4-4. Monomethyl phosphate hydrolysis rate versus pH of the medium at 100 °C.

At alkaline pH values, the hydrolysis rate drops sharply. At acid pH values, a minimum is observed (at pH 1), then the hydrolysis rate increases again with acidity of the medium. Since the pK_a values for the first and second steps of dissociation of methyl phosphate are 1.54 and 6.31, respectively, it can be assumed that the hydrolysis rate maximum at pH 4 corresponds to the highest monoanion concentration in the solution. The decreasing hydrolysis rate in an alkaline medium suggests that the dianion is not active (apparently because the electrostatic repulsion hinders the attack of the dianion by the hydroxyl ion). A nondissociated acid is less active than its monoanion. At pH < 1, one can observe acid-catalyzed hydrolysis with formation of a conjugate acid:

$$H^+ + ROPO_3H_2 \rightleftharpoons RO\overset{+}{P}O_3H_3$$

The mechanism of hydrolysis of the methyl phosphate monoanion has been at the center of protracted debates which ended in rejection of the bimolecular substitution hypothesis (S_N2). The methyl phosphate monoanion hydrolysis rate is much higher than that for the dimethyl phosphate monoanion (see below). This has led to speculations that for a molecule to display some kind of "special" reactivity, in a substituted phosphate a nondissociated hydroxyl must coexist with a negatively charged oxygen atom. Two very similar S_N1 mechanisms of general intramolecular acid catalysis with participation of the nondissociated hydroxyl group in the monoanion have been discussed. It has been assumed that the proton can be transferred to the leaving group either with involvement of water, through emergence of a six-membered cyclic transition state, or without such involvement, through intramolecular migration:

$$R{-}O{-}PO_3H^- \cdot H_2O \longrightarrow R{-}OH + H_2O + [PO_3^-] \xrightarrow{H_2O} H_2PO_4^-$$

$$R{-}O{-}P(O)(O^-){-}OH \rightleftharpoons R{-}O^+(H){-}P(O)(O^-)_2 \longrightarrow R{-}OH + [PO_3^-] \xrightarrow{H_2O} H_2PO_4^-$$

Both mechanisms include formation of a hypothetical metaphosphate. There is no direct proof of metaphosphate formation during hydrolysis of monoalkyl phosphates at pH $\approx$ 4, yet the hypothesis is supported by some indirect evidence. The stability of monophosphate dianions, if their hydrolysis is assumed to be based on a similar S_N1 mechanism, seems to stem from the fact that many alkoxy groups are eliminated but with difficulty and can "depart" only if they are protonated in advance, as is the case with monoanions. Dissociation in an alkaline medium with formation of an RO^- anion occurs only in the case of esters containing strong electron-acceptor groups for example:

$$(^-O)_2P(O)OC_6H_3(NO_2)_2\text{-}2,4 \longrightarrow [PO_3^-] + {}^-OC_6H_3(NO_2)_2\text{-}2,4$$

In spite of the many difficulties arising in the investigation of hydrolysis of the neutral form of monoalkyl phosphates, $ROPO_3H_2$, it was found to be accompanied by rupture of the oxygen–carbon bond in the R radical. Thus, in this case the monophosphate acts as an alkylating agent, just as alkyl sulfates. The substitution proceeds as an S_N2-type reaction at the saturated carbon atom:

$$H_2O + RCH_2O{-}P(O)(OH)_2 \longrightarrow H_2O\cdots C(H)(H)(R){-}O{-}P(O)(OH)_2 \longrightarrow RCH_2OH + {}^-O{-}P(O)(OH)_2$$

As regards *tert*-butyl phosphate and α-D-glucose 1-phosphate, whose neutral forms are hydrolysed at a rate 10^5 times faster[5] than methyl phosphate but also with cleavage of the C–O bond, their hydrolysis is believed to be based on the S_N1 mechanism.

The mechanism of hydrolysis of such esters seems to be determined by the ease of formation of the intermediate carbocation.

[5] The hydrolysis rate versus pH curve for these compounds has no minimum at pH 1.

$$\text{glucopyranosyl}-O-P(=O)(OH)_2 \longrightarrow \left[\text{glucopyranosyl cation}\ (C^{+}-H)\right] + {}^{-}O-P(=O)(OH)_2$$

$$\downarrow H_2O$$

$$\text{glucopyranose}\ (\sim OH)$$

The hydrolysis of monophosphates of primary alcohols at pH 1 is based on a bimolecular mechanism. Experiments with $H_2{}^{18}O$ have shown that two parallel processes occur. The first process involves cleavage of the C–O bond (73 %), while in the second process it is the P–O bond that is broken (27 %):

$$1)\ H_2^{18}O + CH_3-\overset{+}{O}(H)-P(=O)(OH)_2 \longrightarrow H_2^{18}O^{\delta+}\cdots C(H)(H)\cdots O(H)-P(=O)(OH)_2 \xrightarrow{-H^+} CH_3^{18}OH + HO-P(=O)(OH)_2$$

$$2)\ H_2^{18}O + (HO)(OH)P(=O)-\overset{+}{O}(H)-CH_3 \longrightarrow \left[H_2^{18}O\cdots P(=O)(HO)(OH)\cdots O(H)-CH_3\right]^{+} \xrightarrow{-H^+} H^{18}O-P(=O)(OH)-OH + CH_3OH$$

Monophosphates formed by secondary and tertiary alcohols are hydrolysed in an acid medium by a monomolecular mechanism with cleavage of the C–O bond only, for example:

$$(CH_3)_3C-\overset{+}{O}(H)-P(=O)(OH)_2 \longrightarrow (CH_3)_3C^{+} + O{=}P(OH)_3$$

$$(CH_3)_3C^{+} \xrightarrow{H_2O,\ -H^+} (CH_3)_3COH$$

Table 4-2 is a brief summary of the effect of molecular structure of monoalkyl phosphates on their behavior during hydrolysis.

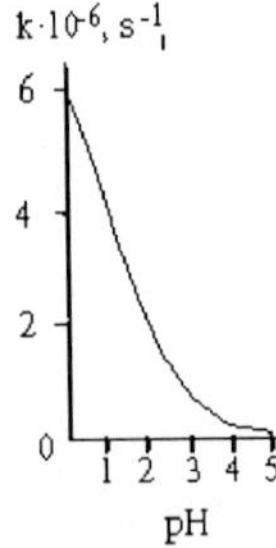

Fig. 4-5. Dimethyl phosphate hydrolysis rate versus pH of the medium at 100 °C.

Table 4-2. Effect of Molecular Structure of Monoalkyl Phosphates on Their Behavior During Hydrolysis.

Monophosphate structure	Radical	Molecularity of the reaction	k_1 (at 100 °C, $\mu = 0$)	Bond broken
Dianion	Any alkyl	Bimolecular	Very slow	P–O
	Phenyl	Monomolecular	$1.8 \cdot 10^{-2}$	P–O
Monoanion	Any alkyl	Monomolecular	$8.23 \cdot 10^{-6}$	P–O
Neutral	*n*-Alkyl	Bimolecular	$5 \cdot 10^{-8}$	C–O
	tert-Alykl	Monomolecular	$4.8 \cdot 10^{-3}$	C–O
	Glycoside	Monomolecular	$3 \cdot 10^{-2}$	C–O
Conjugate acid	*n*-Alkyl	Bimolecular	$2 \cdot 10^{-6}$	C–O
$ROPO_3HA^+$	*sec*-Alkyl	Monomolecular	$1.08 \cdot 10^{-6}$	P–O
	tert-Alykl	Monomolecular	$1.05 \cdot 10^{-8}$	C–O

Hydrolysis of Dialkyl Phosphates. Dialkyl phosphates are among the least reactive esters[6]. The curve representing their hydrolysis rate constant as a function of pH has no maximum (Fig. 4-5), which is indicative of the extremely slow hydrolysis of the dialkyl phosphate monoanion.

Hydrolysis of the diester in its neutral form proceeds predominantly (70–80 %) with cleavage of the C–O bond [equation (1)], while the P–O is broken only in 20 to 30 per cent of the cases [equation (2)].

$$H_2\ddot{O} + CH_3O{-}P(=O)(OH){-}OCH_3 \longrightarrow H_2O{\cdots}CH(H)(H){\cdots}O{-}P(=O)(OH){-}OCH_3 \longrightarrow$$

$$\longrightarrow H_2\overset{+}{O}{-}CH_3 + {}^{-}O{-}P(=O)(OH){-}OCH_3 \longrightarrow CH_3OH + CH_3O{-}P(=O)(OH){-}OH \quad (1)$$

$$H_2\ddot{O} + (HO)P(=O)(OCH_3)(OCH_3) \longrightarrow H_2O{\cdots}P(O)(HO)(OCH_3){\cdots}OCH_3 \longrightarrow HO{-}P(=O)(OH){-}OCH_3 + CH_3OH \quad (2)$$

[6] Exceptions include acid phosphates of 1,2-diols, containing a phosphate group at the adjacent carbon atoms, as well as cyclic diphosphates – so-called cyclic phosphates (see below).

As the medium becomes more acidic, the dialkyl phosphate hydrolysis rate increases with acid concentration. The protonated form of the diester is hydrolysed, primarily, at the C–O bond (up to 90%), just as the neutral form, the P–O bond being affected insignificantly (10%).

$$H_2\ddot{O} + CH_3\overset{+}{O}(H)-P(=O)(OH)-OCH_3 \longrightarrow \left[H_2O\cdots C(H)(H)(H)\cdots O(H)-P(=O)(OH)-OCH_3\right]^+ \longrightarrow$$

$$\longrightarrow H_2\overset{+}{O}-CH_3 + HO-P(=O)(OH)-OCH_3$$

Thus, the widest difference in the hydrolysis rate of di- and monophosphates is observed in the case of the monoanionic form. The differences in the rates of hydrolysis of other ionic forms are not as pronounced.

Hydrolysis of Trialkyl Phosphates. Triphosphates undergo hydrolysis in both acid and alkaline media. At alkaline pH values, the mechanism of the reaction is bimolecular and includes a nucleophilic attack of the tetrahedral phosphorus atom by a hydroxyl ion (experiments with $H_2{}^{18}O$ have shown that the P–O bond is usually broken in this case):

$$\underset{(\text{from } H_2{}^{18}O)}{H^{18}O^-} + (RO)_2P(=O)-OR \longrightarrow \left[H^{18}O\cdots P(=O)(OR)(OR)\cdots OR\right]^- \longrightarrow {}^{18}O^--P(=O)(OR)_2 + ROH$$

The geometric structure of the transition state is a trigonal bipyramid in which the incoming and leaving groups occupy axial positions. This is corroborated by the fact that alkaline hydrolysis of optically active triphosphates is usually accompanied by inversion. Formation of the transition state involves polarization of bonds – the nucleophile (anion) donates electrons to the phosphorus atom. When the substituent departs in anionic form, the phosphorus loses these electrons and passes into the initial state. Due to the high sensitivity of the conjugation toward charge reversal, it is more pronounced in the transition, as opposed to the ground, state. In the presence of an RO group with a negative inductive effect in the triester molecule, two effects opposite in sign are observed, namely: inductive effect, leading to displacement of σ electrons from the phosphorus atom along the P–O bond (in the P–O–R group), and an increase in the p_π–d_π conjugation, with displacement of the lone-pair electrons of the oxygen in the RO group to the 3*d* orbitals of the phosphorus atom. As the electron-donor property of the radicals (R) in the triphosphate becomes more pronounced, the positive mesomerism of the RO groups increases, which leads to shorter P–O bonds in the P–O–R groups and a much slower rate of alkaline hydrolysis.

In neutral and acid solutions, the hydrolysis rate for triphosphates is weakly dependent on the proton concentration, which suggests that the unprotonated form of the phosphate is hydrolysed at a faster rate. Experiments with $H_2{}^{18}O$ have shown that it is the C–O bond that is broken predominantly in an acid medium:

$$H_2{}^{18}O + CH_3{-}O{-}P(=O)(OCH_3){-}OCH_3 \longrightarrow H_2{}^{18}O\cdots CH_3\cdots O{-}P(=O)(OCH_3){-}OCH_3 \longrightarrow CH_3{}^{18}OH + HO{-}P(=O)(OCH_3){-}OCH_3$$

4.4.4 Cyclic Phosphates

Hydrolysis of Cyclic Phosphates. Unlike simple dialkyl phosphates, five-membered cyclic diphosphates are extremely reactive. Ethylene phosphate, for example, is hydrolysed with extreme ease in both acid and alkaline media. Under such conditions, the hydrolysis rate for this compound is about 10^7 or 10^8 times faster than that for dimethyl phosphate (see above), and only the P–O bond is broken.

The high reactivity of cyclic phosphates, as compared to non-cyclic ones, can be explained by the ring being strained. The straining seems to occur only in five-membered cyclic phosphates, in view of the fact that the reactivity of compounds with a larger (six- and seven-membered) ring equals that of dimethyl phosphate or is even lower. The exact nature of the conjugation in five-membered cyclic phosphates is yet to be elucidated, however, the greater ease of their hydrolysis, in comparison with acyclic derivatives, suggests that formation of the transition state is accompanied by elimination of the strained state[7]. This hypothesis is supported by the fact that treatment of ethylene phosphate with ^{18}O-labelled water in an acid medium along with incorporation of the isotope into the phosphate group as a result of hydrolysis also leads to oxygen exchange without breaking of the ring, the latter reaction proceeding at the same rate as hydrolysis ($k_{hydr}/k_{exch} = 5$).

$$(CH_2O)_2P(={}^{18}O){}^{18}OH \xleftarrow[k_{exc}]{H_2{}^{18}O} (CH_2O)_2P(=O)OH \xrightarrow[k_{hydr}]{H_2{}^{18}O} HOCH_2CH_2O{-}P(=O)(OH){-}{}^{18}OH$$

Consequently, emergence of a transition state has been postulated for both reactions, having the structure of a trigonal bipyramid (15) (see below) in

[7] There is an alternative hypothesis accounting for the greater ease of hydrolysis of five-membered cyclic phosphates. It has been proposed that the strained state of the ring in these compounds leads to a situation where the cyclic oxygen atoms are removed from positions conductive to p_π–d_π conjugation with the phosphorus atom. As a result, the latter becomes more electrophilic and prone to attack by water (hydrolysis).

which the P–O bonds of the five-membered ring occupy an axial position and an equatorial one. In this nonplanar ring, the O–P–O angle is 90°, as opposed to 99° in the starting cyclic phosphate.

Calculations of the angular straining have shown that the gain in energy, resulting from elimination of the state of strain during emergence of the transition state (15), ranges from 12 to 25 kJ/mole (3–6 kcal/mole). In accordance with the rules of nucleophilic substitution at the tetrahedral phosphorus atom, the incoming and leaving groups occupy axial positions. Consequently, hydrolysis of ethylene cyclic phosphate must be accompanied by cleavage of the P–O bond in the axial position as well as by formation of hydroxyethylene phosphate containing an oxygen isotope in the phosphate group:

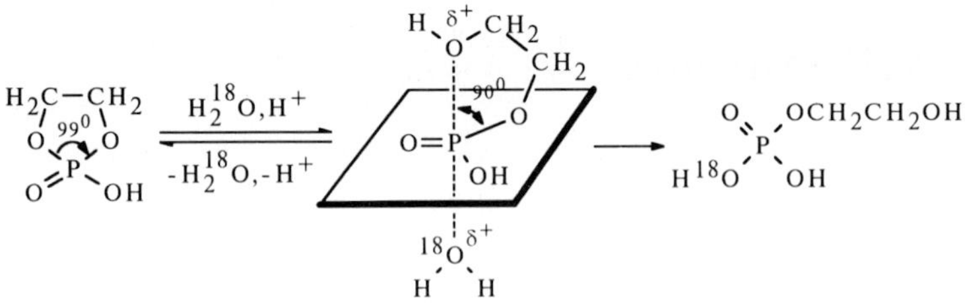

$H_2^{18}O, H^+$ ⇌ $-H_2^{18}O, -H^+$

(15)

On the other hand, oxygen exchange in ethylene phosphate with the five-membered ring remaining intact must also occur in keeping with the general rule – that is, departure of the group being substituted (in this case, hydroxyl) from the axial position. Hence, in the transition state of type (15), resulting from nucleophilic substitution in the ethylene phosphate molecule, the axial positions may be occupied not only by $-O-CH_2-$ groups but also hydroxyl ones with labelled and unlabelled oxygen atoms. This may also occur as a result of restructuring of the transition state, known as pseudorotation. In the case under consideration, the corresponding transformation can be written as follows:

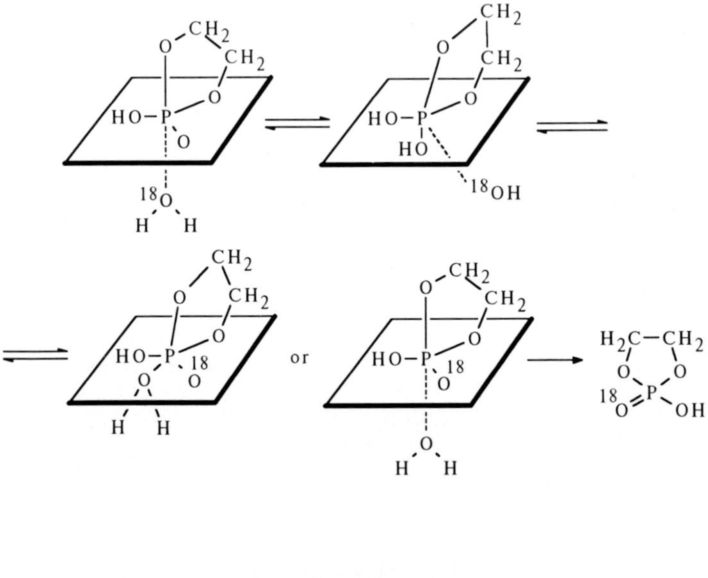

or

It can be seen that the oxygen exchange may occur without opening of the ring. The probability of another possible restructuring of the starting trigonal bipyramid, during which the O–P–O group finds itself in the equatorial position, is low because in such an event the O–P–O angle would be equal to 120° and that would render the system much more strained.

A similar mechanism also seems to be involved in hydrolysis of methylethylene cyclic phosphate which has been found to undergo not only rupture of the cycle and oxygen exchange but also rapid hydrolysis of the exocyclic group with the ring remaining intact.

pseudorotation (rotation axis P-OCH_3)

pseudorotation (rotation axis P=O)

oxygen exchange

hydrolysis with ring opening

hydrolysis of exocyclic phosphodiester bond

Reactions Accompanied by Formation of Cyclic Structures Including a Phosphate Group. It is known that in an acid medium the phosphate group in *N*-phosphorylated hydroxyamino acids undergoes N → O migration. For example:

$$CH_3OCO\text{–}CH(CH_2OH)\text{–}HN\text{–}P(O)(OC_3H_7\text{-iso})_2 \xrightarrow{H^+} CH_3OCO\text{–}CH(\overset{+}{N}H_3)\text{–}CH_2\text{–}O\text{–}P(O)(OH)_2$$

In this case, one might expect emergence of a transition state incorporating a five-membered ring:

X=NH; Y=OH

However, the three constituent atoms in the resulting five-membered ring must lie on the same straight line (the leaving group X and incoming group Y must be in an axial position or, in other words, the X–P–Y angle must be 180°), which is impossible. Since such migration causes detachment of both isopropyl groups, it is more likely to expect a sequence of steps including hydrolysis and cyclization with departure of the isopropoxy groups and opening of the ring:

$$CH_3O\text{–}C(O)\text{–}CH(CH_2OH)\text{–}HN\text{–}P(O)(OC_3H_7)_2 \xrightarrow[H^+]{H_2O} CH_3O\text{–}C(O)\text{–}CH(CH_2OH)\text{–}HN\text{–}P(O)(OH)(OC_3H_7) \longrightarrow$$

$$\longrightarrow CH_3O\text{–}C(O)\text{–}\underset{\text{(cyclic: HC–CH}_2\text{–O–P(O)(OH)–NH)}}{HC} \xrightarrow[H^+]{H_2O} CH_3O\text{–}C(O)\text{–}CH(NH_2)\text{–}CH_2\text{–}O\text{–}P(O)(OH)\text{–}OH$$

A similar migration of the phosphate group was observed during alkaline treatment of glycero-1-methyl phosphate (16):

The formation of equal amounts of glycerol 1- and 2-phosphates is due to emergence of a five-membered cyclic phosphate as a result of the nucleophilic attack on the phosphorus atom by the adjacent hydroxyl. As has already been mentioned, such compounds are highly unstable and undergo hydrolysis at virtually the same rate along both possible paths (a) and (b).

Methyl(β-hydroxyethyl) phosphate is fully converted into hydroxyethyl phosphate in a similar manner:

Under the same conditions, methyl(β-methoxyethyl) phosphate, which cannot be transformed to a cyclic structure, is not hydrolysed. The above findings suggest that hydrolysis of diphosphates rather stable in an alkaline medium (see above) is substantially facilitated in cases where cyclic phosphates can form as intermediate products of hydrolysis.

The fact that alkaline hydrolysis of glycero-1-methyl phosphate fails to produce glycero-2-methyl phosphate, the main process being elimination of the methoxy group to yield the corresponding cyclic phosphate, is indicative of a possible mechanism of this transformation. The transition state seems to emerge as a result of the phosphorus atom being attacked by the 2-hydroxyl group of glycerol. In addition to the incoming group, the methoxyl group also finds itself in the axial position. The reaction ends in departure of the latter group.

$$H_2C{-}CH{-}CH_2OH \text{ (cyclic O-P(O)(O^-)-OCH}_3\text{)} \longrightarrow CH_3O\text{-----}P\text{-----}OH \xrightarrow{-CH_3OH} \text{cyclic glycerol phosphate } H_2C{-}CH{-}CH_2OH, \; O{-}P(=O)(O^-){-}O$$

If the intermediate structure (17) had undergone pseudorotation [formation of structure (18)], the reaction products would have included glycero-2-methyl phosphate; the latter's absence indicates that no pseudorotation takes place in this case (this phenomenon will be discussed at greater length in what follows).

$$\text{(17)} \not\longrightarrow \text{(18)} \longrightarrow HOCH_2{-}CH(O{-}P(=O)(O^-){-}OCH_3){-}CH_2OH$$

Migration of the phosphate group can occur in the case of non-alkylated glycol monophosphates:

$$CH_2O{-}P(=O)(OH)_2,\ CHOH,\ CH_2OH \xrightarrow{H^+} [\text{cyclic } CH_2{-}O{-}P(=O)(OH){-}O{-}CH,\ CH_2OH] \longrightarrow CH_2O{-}P(=O)(OH)_2,\ CHOH,\ CH_2OH + CH_2OH,\ CH{-}O{-}P(=O)(OH)_2,\ CH_2OH$$

This process is possible only in an acid medium, which agrees well with the above-described mechanism of nucleophilic substitution at the tetrahedral phosphorus atom (it is the water molecule that is the leaving group in an acid medium).

Glycero-1 and glycero-2 phosphates are not capable of interconversion in alkaline media. Consequently, no cyclic phosphate is formed in this case. The reason is the lack of a suitable leaving group. It is most likely that although a cyclic structure similar to the one emerging in the transition state is formed, it is not capable of undergoing pseudorotations, which also rules out the possibility of isomerization.

$$HOCH_2-CH(OH)-CH_2O-PO_3^{2-} \rightleftharpoons \text{cyclic intermediate} \nrightarrow \text{cyclic intermediate} \rightleftharpoons HOCH_2-CH(OPO_3^{2-})-CH_2OH$$

4.4.5 β-Elimination Reactions

Alkyl phosphates containing such electronegative substituents as C≡N, COOH, COOR and others at the β-position in the alkyl group readily undergo β-elimination of the phosphate group in an alkaline medium. If ethyl phosphate practically does not break down during mild alkaline hydrolysis, β-cyanethyl phosphate and its esters do so easily under the same conditions.

$$HO^- + H-CH(C{\equiv}N)-CH_2-O-P(O)(O^-)-OR \xrightarrow{-H_2O} CH_2{=}CH-C{\equiv}N + {}^-O-P(O)(O^-)-OR$$

$R = H$ or Alk

β-Elimination of the phosphate group is always accompanied by cleavage of the C–O bond and formation of the corresponding unsaturated compound. The latter undergoes subsequent transformations in some instances. For example, during β-elimination of the phosphate group, serine phosphate is converted into pyruvic acid:

$$H_2N-C(COO^-)(H)-CH_2-O-P(O)(O^-)-O^- \xrightarrow[HO^-]{-H_2O} CH_2{=}C(COO^-)(NH_2) + {}^-O-P(O)(O^-)-O^-$$

$$CH_2{=}C(COO^-)(NH_2) \longrightarrow CH_3-C(COO^-){=}NH \xrightarrow[-NH_3]{H_2O} CH_3-C(O)-COO^-$$

4.5 Reactions of Nucleotides Involving the Phosphate Group

4.5.1 Chemical and Enzymatic Dephosphorylation

Cleavage of the phosphomonoester bond in nucleotides gives rise to nucleosides:

$$RO-\overset{\overset{\displaystyle O}{\|}}{\underset{\underset{\displaystyle OH}{|}}{P}}-OH \xrightarrow{H_2O} ROH + H_3PO_4$$

R-nucleoside residue

This reaction, initiated and sustained both by chemical reagents and by enzymes cleaving phosphomonoester bonds of a particular structure, is widely used in the synthesis of nucleosides and also in structural studies.

Chemical methods for hydrolysing nucleotides to nucleosides have not been studied completely enough in spite of the fact that the course of hydrolysis of the phosphomonoester bond in simple monophosphates is known in every detail. The general method of chemical hydrolysis of nucleotides to nucleosides is based on heating at 100 °C in a buffer (usually ammonium formiate solution) at pH 4. As was demonstrated in experiments with simple alkyl phosphates, such conditions are optimal for cleavage of phosphomonoester bonds. The hydrolysis of nucleotides at the P–O bond seems to proceed just as in the case of simple monophosphates.

The rate of detachment of phosphate groups depends but insignificantly on their position in a nucleoside, a more important factor being the structure of the heterocyclic base. Purine nucleotides are hydrolysed faster than pyrimidine ones. For example, guanosine 5′-phosphate is hydrolysed to guanosine (pH 4, 100 °C) at a rate twice that of hydrolysis of uridine 5′-phosphate to uridine.

Heating nucleotides in aqueous solutions at pH < 1 also leads to cleavage of the phosphomonoester bonds. Under such conditions, however, it is only pyrimidine nucleotides that are hydrolysed to nucleosides (cytidine being markedly deaminated in this case). When the same conditions are maintained for purine nucleotides, hydrolysis of *N*-glycosidic bonds proceeds at a perceptibly higher rate. Ribonucleoside 3′(2′)-phosphates are readily split to nucleosides in a neutral medium in the presence of salts and hydroxides of thorium, zirconium, lead, lanthanum and some other metals. For instance, uridine 3′(2′)-phosphate is almost quantitatively converted into a nucleoside when boiled for 20 min with lanthanum hydroxide:

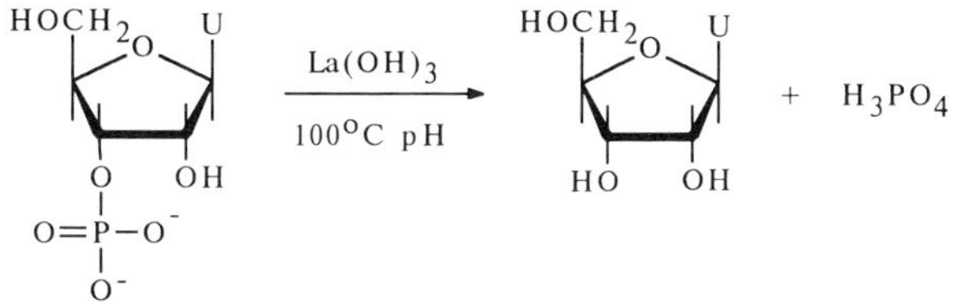

It takes two to four hours at pH 6 (100 °C) for ribonucleoside 3′(2′)-phosphates to be dephosphorylated in the presence of bivalent lead ions. In the presence of thorium salts at 37 °C, nucleoside 3′(2′)-phosphates lose inorganic phosphate both in acid (pH 3.6) and alkaline (pH 8.3) media.

Deoxyribonucleotides are not dephosphorylated under such conditions. Apparently, hydrolysis requires a hydroxyl group next to the phosphate group. The role of this group is yet to be elucidated. A possible reason for the catalysis of hydrolysis of the monoester bond by heavy metal compounds is their ability to combine with phosphate groups, which renders the phosphorus atom more electrophilic.

Enzymatic dephosphorylation of nucleotides may be mediated both by nonspecific phosphomonoesterases (phosphatases) and by phosphomonoesterases acting only on nucleotides. The nonspecific phosphomonoesterases (PME) include the alkaline phosphatase from *E. coli,* acid phosphatase from the prostate gland, alkaline phosphatase from wheat, and some others. The rate at which nucleotides are hydrolysed by these enzymes (to nucleosides) is usually the same as that in the case of simple monophosphates. The above phosphatases hydrolyse all mononucleotides irrespective of the phosphate group position.

Specific PME selectively hydrolyse nucleotides of a particular structure. Among the best known and widely used enzymes is that hydrolysing nucleoside 5′-phosphates. By virtue of its specific action, this enzyme has been named 5′-nucleotidase:

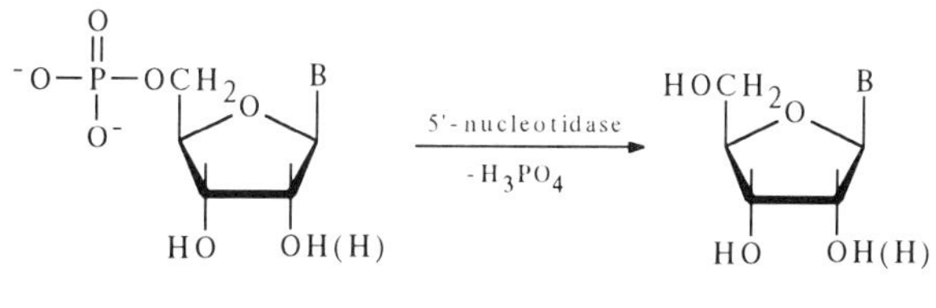

PME from snake venom is generally employed for hydrolysis of nucleoside 5′-phosphates in determining the structure of nucleotides or synthesis of nucleosides.

The phosphatases hydrolysing only nucleoside 3′-phosphates are known as 3′-nucleotidases. Such enzymes are isolated from various sources of plant origin (wheat, soybean, takadiastase, etc.). Nucleoside 2′- and 5′-phosphates are not hydrolysed by these enzymes.

4.5.2 Migration of the Phosphate Group

In the case of ribonucleoside 3′(2′)-phosphates, which contain a hydroxyl next to the phosphate group, migration of the phosphate group is observed along with hydrolysis of the P–O bond in an acid medium. This process passes, just as with 1- and 2-phosphates of glycerol, through an intermediate five-membered cyclic phosphate resulting from nucleophilic substitution at the tetrahedral phosphorus atom with the participation of the neighboring hydroxyl group. Nucleoside 2′,3′-cyclic phosphate is not stable in acid media and readily undergoes hydrolysis yielding a mixture of nucleoside 3′-phosphate and nucleoside 2′-phosphate:

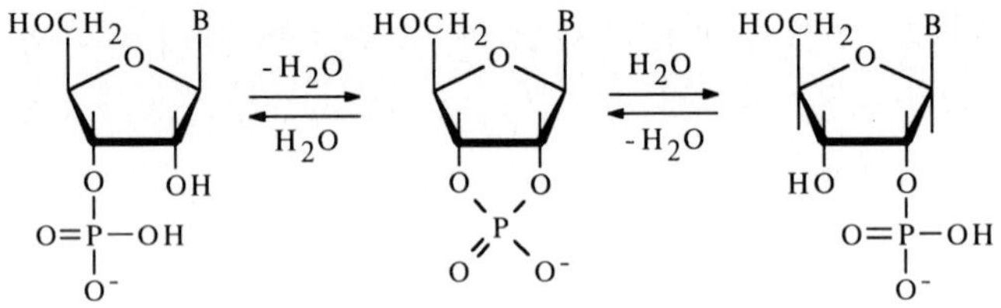

HOCH2 B O; O OH; O=P—OH; O⁻ — $-H_2O$ / H_2O ⇌ — HOCH2 B O; O O; P; O O⁻ — H_2O / $-H_2O$ ⇌ — HOCH2 B O; HO O; O=P—OH; O⁻

No migration takes place in alkaline media, just as in the case of glycerophosphates. To preserve pure isomers (2′- or 3′-phosphates), the ribonucleotide solution is usually alkalized with ammonia. When a mixture of isomers is separated by ion-exchange chromatography, care should be taken that the pH of the solution is above 7.

4.5.3 Alkylation of the Phosphate Group

Alkylation of nucleoside 5′-phosphates in a neutral aqueous solution with diazomethane (at 20 °C) yields the corresponding methyl esters of nucleotides, for example (for adenine and cytosine derivatives):

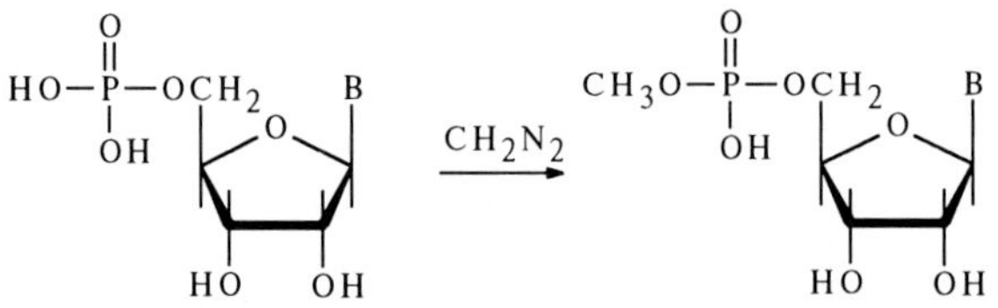

O; HO—P—OCH2 B; OH O; HO OH — CH_2N_2 → — O; CH_3O—P—OCH2 B; OH O; HO OH

Under similar conditions, uridine and guanosine 5′-phosphates also undergo methylation at the base with N^3- and N^7-methyl derivatives forming along with methyl esters, for example:

$$\text{Uridine 5'-phosphate } (HO-P(O)(OH)-OCH_2-) \xrightarrow{CH_2N_2} CH_3O-P(O)(OH)-OCH_2- \text{ (N}^3\text{-methyl uridine derivative, } NCH_3)$$

For selective alkylation of the phosphate group in nucleotides, use is made of active nucleotide derivatives.

4.5.4 Activation of the Phosphate Group in Nucleotides. Synthesis of Some Derivatives with Respect to the Phosphate Group

In reactions involving nucleotides, the heterocyclic base, as well as hydroxyl groups of pentose and the phosphate group may be affected. For reactions to proceed selectively with respect to the phosphate group, the latter must be activated. To this end, use is made of carbodiimides, aryl sulfochlorides, and diphenylchlorophosphate. The corresponding "active" derivatives can be obtained in the case of any nucleoside 5′-phosphates and deoxyribonucleoside 3′-phosphates. Ribonucleoside 3′(2′)-phosphates are converted, during such activation, into the corresponding ribonucleoside 2′,3′-cyclic phosphates.

The resulting active nucleotide derivatives cannot be isolated because of their instability. Once formed, they are put to immediate use in the synthesis of esters, amides, anhydrides, and so forth.

Activation by Carbodiimides. Formally, carbodiimides are dehydrating agents.

$$R'O-\overset{O}{\overset{\|}{\underset{O^-}{\underset{|}{P}}}}-OH + R''XH \xrightarrow{R-N=C=N-R} R'O-\overset{O}{\overset{\|}{\underset{O^-}{\underset{|}{P}}}}-XR'' + (RNH)_2C=O$$

R - Alk or Ar; R' - nucleoside residue;
R''X - alkohol or amine residue

The mechanism of such reactions is yet to be completely understood, but some experimental evidence gives reason to believe that they yield a carbodiimide and nucleotide adduct (19).

$$R'O-P(=O)(OH)(O^-) + R-N{=}C{=}N-R \longrightarrow R-NH-C(=N-R)-O-P(=O)(OH)(OR') \;(19) \xrightarrow{H^+}$$

$$\longrightarrow R-NH-\overset{+}{C}{=}NHR\,[-O-P(=O)(O^-)(OR')] \;\; R''-X-H \;(20) \longrightarrow R'O-P(=O)(OH)-XR'' + (RNH)_2C{=}O$$

The adduct (19) subsequently receives a proton and turns into an active derivative (20) in which the phosphorus atom becomes more electrophilic. The next step is nucleophilic substitution at the phosphorus atom with formation of the corresponding derivatives. NMR spectroscopic studies into the transformations undergone by nucleotides in the presence of dicyclohexylcarbodiimide (DCC), carried out by D. G. Knorre and coworkers, have shown that the reaction intitially yields pyrophosphates then tri- and tetraphosphates[8]. Nucleophilic substitution in the active derivative (20) seems to take place, the function of the nucleophile R″XH being performed by the nucleotide itself.

$$(20) + R'O-P(=O)(O^-)-O^- \xrightarrow[-(RNH)_2CO]{} R'O-P(=O)(O^-)-O-P(=O)(O^-)-OR'$$

Then, the resulting pyrophosphate is activated by DCC to form a compound similar to (20), which can be converted into tri- as well as tetraphosphates.

$$R-NH-\overset{+}{C}{=}NHR\,[-O-P(=O)(OR')-O-P(=O)(O^-)-OR']$$

$$+\; R'O-P(=O)(O^-)-O-P(=O)(O^-)-OR' \longrightarrow {}^-O-P(=O)(OR')-O-P(=O)(OR')-O-P(=O)(OR')-O-P(=O)(OR')-O^-$$

$$+\; R'O-P(=O)(O^-)-O^- \longrightarrow {}^-O-P(=O)(OR')-O-P(=O)(OR')-O-P(=O)(OR')-O^-$$

---

[8] Thus, the earlier speculations that the reaction involves intermediate metaphosphate formation have not been borne out.

When other nucleophiles are present in the reaction mixture, the above tri- and tetraphosphates act as phosphorylating agents. The proposed mechanism is supported by additional experimental evidence. For instance, addition to the reaction mixture (absolute pyridine) of trialkylamines, which are strong bases and, therefore, prevent formation of protonated adducts of the (20) type, usually slows down and sometimes even inhibits the reaction. Enhancement of the nucleophilic properties of reagent R″XH speeds up the reaction perceptibly; monoalkyl phosphates (in the form of dianions), for example, enter into the reaction under consideration much more readily than dialkyl phosphates (monoanions).

Dicyclohexyl carbodiimide is often used for activation of the phosphate group. The nucleotide activated in this manner reacts smoothly with primary and secondary amines as well as various acids, giving the corresponding derivatives with respect to the phosphate group. The following scheme illustrates some examples of synthesis of nucleoside 5′-phosphate derivatives in the presence of DCC (similar transformations may also involve deoxyribonucleoside 3′-phosphates).

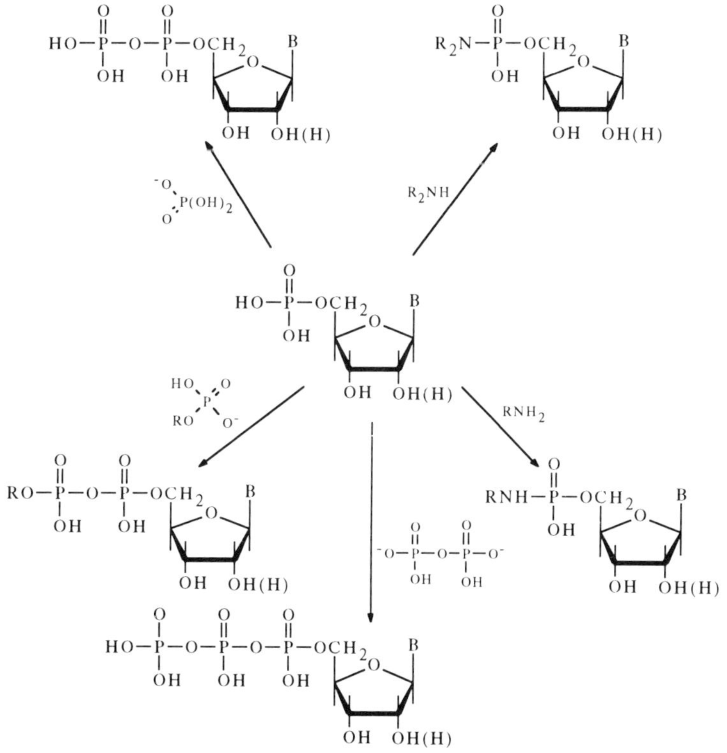

Nucleotides activated by carbodiimide can react with alcohols. In this case, in order to avoid side self-condensation reactions, the hydroxy groups of pentose and the amino groups of the heterocyclic base must be protected in advance; after condensation, the protective groups are removed with the aid of ammonia.

NHBz, HO–P(=O)(OH)–OCH$_2$, AcO; 1. DCC, 2. ROH, 3. NH_3 → RO–P(=O)(OH)–OCH$_2$, HO, NH_2

Activation by Aryl Sulfochlorides. Another method for activating nucleotides resides in their treatment with chlorides of arylsulfonic acids in absolute pyridine where nucleotides (in the form of salts with trialkylamines) are readily soluble. The first step of the process seems to be formation of a mixed anhydride:

$$RO{-}P(O)(O^-){-}OH + Cl{-}SO_2{-}Ar \xrightarrow[-HCl]{Py} RO{-}P(O)(O^-){-}O{-}SO_2{-}Ar$$

R-nucleoside residue; Ar=$C_6H_4CH_3$-p

The formation of such anhydrides, however, defies detection. Knorre and coworkers have been successful in demonstrating, with the aid of pulsed NMR spectroscopy on ^{31}P nuclei, that the early stage of the process is marked by accumulation of pyrophosphate in the reaction mixture, along with linear trimetaphosphate:

$$[RO{-}P(O)(O^-){-}O{-}SO_2{-}Ar \xrightarrow{-ArSO_3^-} RO{-}PO_2] \xrightarrow{RO{-}P(O)(OH)O^-} {}^-O{-}P(O)(OR){-}O{-}P(O)(OR){-}O^- \xrightarrow{[RO{-}PO_2]} {}^-O{-}P(O)(OR){-}O{-}P(O)(OR){-}O{-}P(O)(OR){-}O^-$$

R-nucleoside residue

During the subsequent interaction with aryl sulfochloride, the pyro- and polyphosphate become converted into the corresponding monomeric nucleoside metaphosphates stabilized with pyridine (their NMR-^{31}P spectra feature a singlet with δ near 5 ppm), for example:

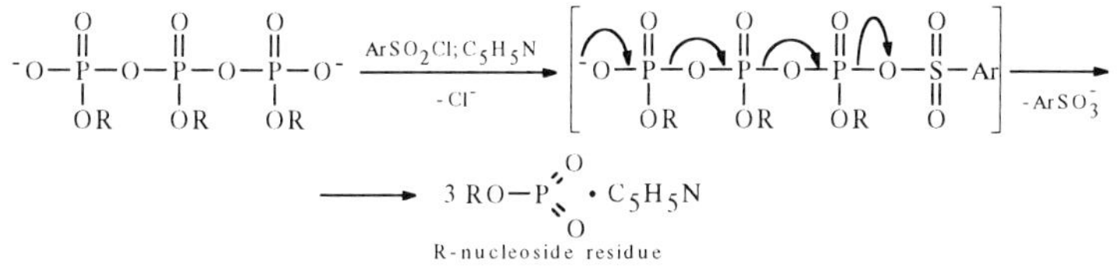

$$3\,RO{-}PO_2 \cdot C_5H_5N$$

R-nucleoside residue

The use of an aryl sulfochloride fixed on a polymer substrate has made it possible to obtain solutions containing only a monomeric metaphosphate (the polymeric sulfonate is easily removed by simple filtration).

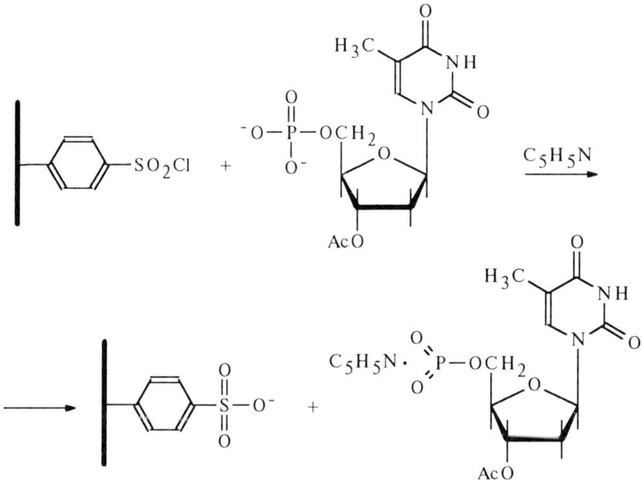

All of the reactions mentioned above are conducted under conditions of complete absence of moisture whose contact with the monomeric nucleoside metaphosphate would immediately transform it into a nucleotide:

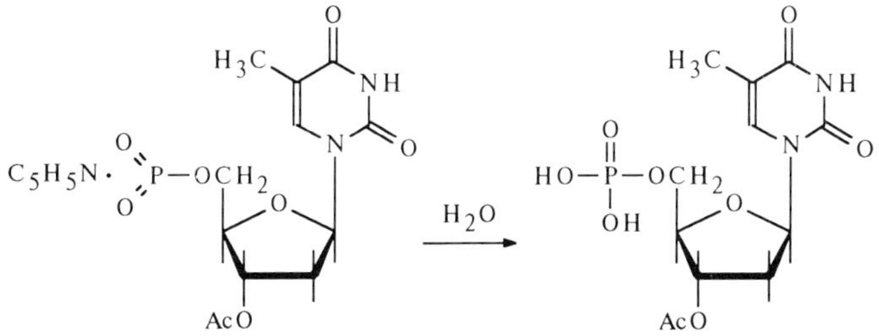

The nucleoside metaphosphates derived in this fashion can be used to obtain nucleotide derivatives with respect to the phosphate group. The following scheme represents some reactions involving 3′-acetyldeoxythymidine 5′-metaphosphate.

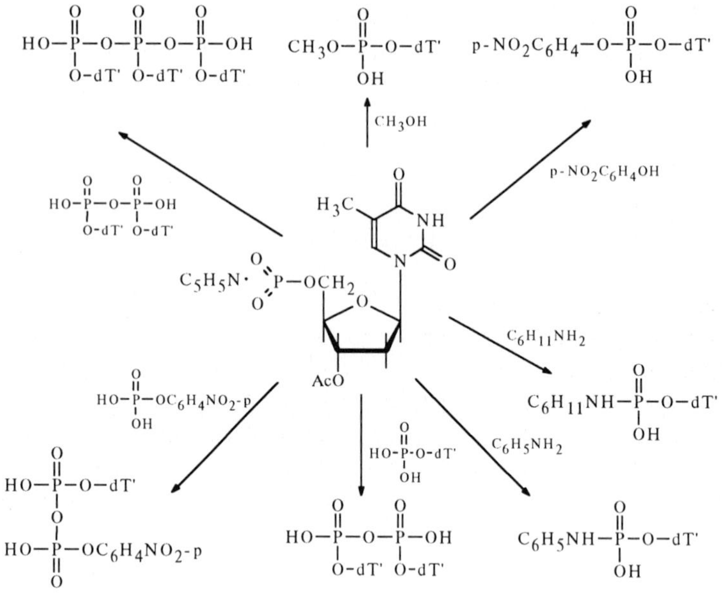

dT' - 3'-O-acetyldeoxythymidine residue

The interaction between metaphosphates and nucleosides containing a deprotected hydroxyl group in the carbohydrate moiety results in internucleotide linkage. This reaction is widely used for synthesizing oligonucleotides.

To activate nucleotides use is made of both simple aryl sulfochlorides, such as *para*-toluene sulfochloride, and sterically hindered ones, such as mesitylene sulfochloride (21) and 2,4,6-triisopropylphenyl sulfochloride (22):

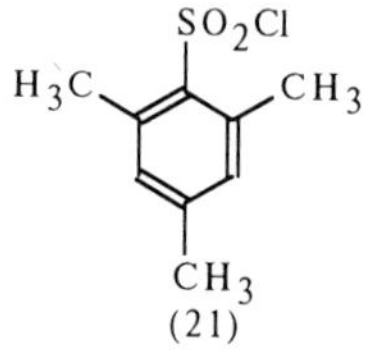

(21) 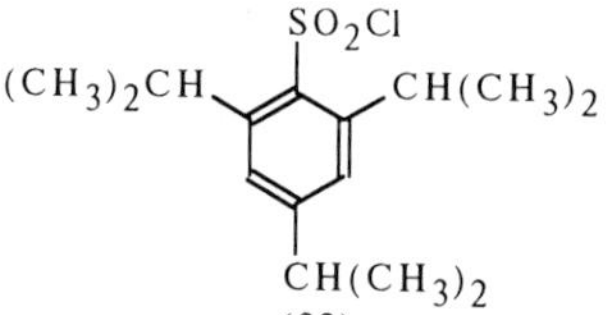 (22)

Sulfochloride (22) gives particularly good results. The reason seems to be that the leaving group forming during nucleotide activation is essentially an anion of an arylsulfonic acid which is stronger than, say, *para*-toluenesulfonic acid (the delocalization of the positive charge at the sulfur atom occurs virtually without participation of the aromatic ring so that the sulfo group is displaced from the plane of conjugation with the latter).

Activation by Diphenyl Chlorophosphate. The phosphate group in nucleotides is also activated with the aid of diphenyl chlorophosphate. The reaction is conducted in absolute dioxane where the forming pyrophosphate is rather stable.

$$RO{-}P(=O)(OH){-}O^- + (C_6H_5O)_2P(=O)Cl \longrightarrow RO{-}P(=O)(O^-){-}O{-}P(=O)(OC_6H_5)_2$$

R - nucleoside residue

Since diphenylphosphoric acid is stronger than the monophosphate (nucleotide) participating in the formation of an anhydride by the second hydroxyl of the phosphate group ($pK_2 \sim 6.0$), in reactions with nucleophilic agents such anhydrides serve as a donor of the nucleotide (the leaving group usually is the diphenylphosphoric acid anion). Reactions of this kind proceed with particular ease in pyridine. Given below are some examples illustrating the use of a respective mixed anhydride of 5′-adenylic acid for synthesis of its esters with phenols, amides, mixed anhydrides with carboxylic acids, as well as derivatives with phosphoric acid, phosphates and similar compounds.

$$(C_6H_5O)_2P(=O){-}O{-}P(=O)(OH){-}OCH_2\text{-adenosine}$$

$$\xrightarrow{H_3PO_4} HO{-}P(=O)(OH){-}O{-}P(=O)(OH){-}O{-}A$$

$$\xrightarrow{H_2SO_4} HO{-}S(=O)_2{-}O{-}P(=O)(OH){-}O{-}A$$

$$\xrightarrow{\text{morpholine (O NH)}} \text{O N}{-}P(=O)(OH){-}O{-}A$$

$$\xrightarrow{O_2N{-}C_6H_3(NO_2){-}OH} O_2N{-}C_6H_3(NO_2){-}O{-}P(=O)(OH){-}O{-}A$$

$$\xrightarrow{RO{-}P(=O)(OH){-}OH} RO{-}P(=O)(OH){-}O{-}P(=O)(OH){-}O{-}A$$

$$\xrightarrow{RCOOH} RC(=O){-}O{-}P(=O)(OH){-}O{-}A$$

A - 5′-adenosine residue; R = Alk

When such mixed anhydrides are employed, the hydroxyl groups of pentose are not protected because the side process of their phosphorylation both by the nucleotide and diphenyl phosphate is practically at a standstill.

Synthesis of Nucleoside Cyclic Phosphates. The above procedures for activating the phosphate group are extensively used also for intramolecular phosphorylation, for instance, synthesis of ribonucleoside 2′,3′-cyclic phosphates.

When DCC is added to a solution of ribonucleoside 3′(2′)-phosphate in pyridine in the presence of trialkylamine, ribonucleoside 2′,3′-cyclic phosphate is produced practically with a quantitative yield:

$HOCH_2$ B ... $O=P-O^-$ ... $C_6H_{11}NH-C{=}NHC_6H_{11}$ → $HOCH_2$ B ... $+ (C_6H_{11}NH)_2CO$

The preference of intramolecular phosphorylation leading to formation of a cyclic phosphate has been proved experimentally. It was found, for example, that when the above reaction is conducted with a large surplus of methyl alcohol (such conditions are created to synthesize methyl esters of nucleotides), no methyl ester of nucleoside 3′(2′)-phosphate is formed and the only product is nucleoside 2′,3′-cyclic phosphate.

Methyl esters of nucleotides, just as all diphosphates, are stable under the conditions described above and do not react even with carbodiimide.

The phosphate group can also be activated, during cyclic phosphate synthesis, by the already mentioned diphenyl chlorophosphate as well as anhydrides and chlorides of carboxylic acids, such as trifluoroacetic anhydride, chlorocarbonate, and so on. For example, addition of chlorocarbonate to an aqueous solution of uridine 2′(3′)-phosphate gives uridine 2′,3′-cyclic phosphate with a quantitative yield:

$HOCH_2$ U ... OH ... $O=P-O^-$ OH $\xrightarrow{C_2H_5O-C(=O)Cl}$ $HOCH_2$ U ... $O=P-O^-$ ... $C_2H_5O-C=O$ →

→ $HOCH_2$ B ... $+ C_2H_5OH + CO_2$

The ease of conversion of ribonucleoside 2′(3′)-phosphate into a five-membered cyclic phosphate is due to the fact that the hydroxyl group in its molecule is positioned conveniently for attacking the activated phosphorus atom. During nucleophilic substitution, the hydroxyl and electron-acceptor groups can find themselves in an axial position, which will lead to formation of a new P–O bond with simultaneous departure of the electron-acceptor group:

X - leaving group

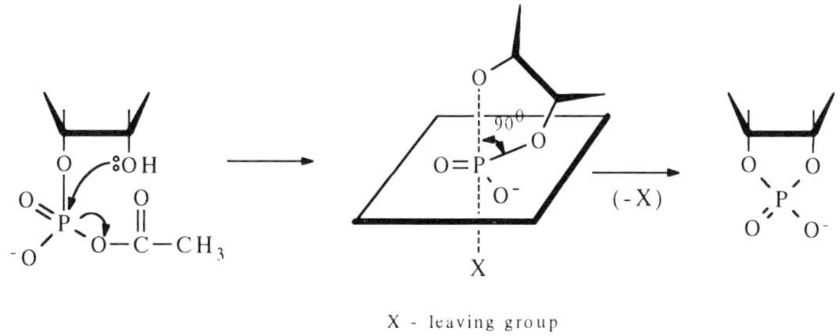

When the position of the hydroxyl group is not favorable for the intramolecular reaction, for instance, in the case of activation of nucleoside 5′-phosphates, the main process is formation of a symmetrical pyrophosphate [path (*a*)]:

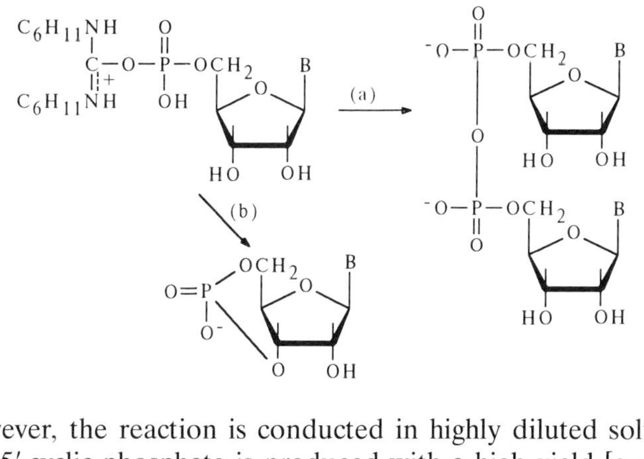

If, however, the reaction is conducted in highly diluted solutions, ribonucleoside 3′,5′-cyclic phosphate is produced with a high yield [e. g., path (*b*)]. Thus, even a poorly positioned hydroxyl group (in *trans*-position) can attack the phosphorus atom to form a six-membered ring.

4.5.5 Acylation of the Phosphate Group

Mixed anhydrides of nucleoside 5′-phosphates with carboxylic and amino acids play a major role in many biosynthetic processes. Accordingly, methods for their synthesis have been developed. As can be seen from the scheme in section 4.5.4, by using anhydrides of nucleoside 5′-phosphates and diphenylphosphoric acid one can obtain mixed anhydrides of the corresponding nucleotides and carboxylic acids, the latter being used directly as the starting compounds. A simple and effective way to synthesize mixed anhydrides of nucleoside 5′-phosphates and carboxylic acids is based on a reaction between a nucleotide and carboxylic anhydride or chloride (for behavior of ribonucleoside 3′-phosphates in this reaction, see 4.5.4), for example:

$$X{-}\overset{O}{\overset{\|}{C}}R \;+\; {}^{-}O{-}\overset{O}{\overset{\|}{\underset{OH}{\underset{|}{P}}}}{-}OCH_2(\text{ribose-}A,\ 2'\text{-}OH,\ 3'\text{-}OH) \xrightarrow{(-X)} RCO{-}O{-}\overset{O}{\overset{\|}{\underset{OH}{\underset{|}{P}}}}{-}OCH_2(\text{ribose-}A,\ 2'\text{-}OH,\ 3'\text{-}OH)$$

$R = CH_3, C_2H_5, C_6H_5, C_6H_2(CH_3)_3$-1,3,5; $X = RCOO$ or Cl

If more than twice the excess amount of adenosine 5′-phosphate is made to react with acetic anhydride, the reaction does not terminate at the step of mixed anhydride, or acetyl adenylate, formation but yields diadenosine 5′-pyrophosphate. The latter results from nucleophilic substitution of the acetic acid anion ($pK_a \approx 4.8$) in the intermediate acetyl adenylate by that of a weaker acid, namely, adenosine 5′-phosphate ($pK_a \approx 6$).

$$RO{-}\overset{O}{\overset{\|}{\underset{O^-}{\underset{|}{P}}}}{-}O^- + (CH_3CO)_2O \xrightarrow[-CH_3COO^-]{} \left[RO{-}\overset{O}{\overset{\|}{\underset{O^-}{\underset{|}{P}}}}{-}O{-}\overset{O}{\overset{\|}{C}}{-}CH_3\right] \xrightarrow{RO-PO_3^{2-}}$$

$$\longrightarrow RO{-}\overset{O}{\overset{\|}{\underset{O^-}{\underset{|}{P}}}}{-}O{-}\overset{O}{\overset{\|}{\underset{O^-}{\underset{|}{P}}}}{-}OR + CH_3COO^-$$

R - 5′-adenosine residue

Two methods are used primarily for synthesis of mixed anhydrides of nucleotides and amino acids. The first method, similar to the one described above, resides in making a nucleotide to react with a mixed anhydride of a carbobenzoxy derivative of (Cbz)-amino acid and ethyl carbonate:

$$\mathrm{R{-}CH(NH{\cdot}Cbz){-}COOH \xrightarrow[\textit{tert}\text{-amine}]{C_2H_5OCOCl} R{-}CH(NH{\cdot}Cbz){-}\overset{O}{\overset{\|}{C}}{-}O{-}\overset{O}{\overset{\|}{C}}{-}OC_2H_5 + {}^{-}O{-}\overset{O}{\overset{\|}{P}}(O^-){-}OR' \xrightarrow{-CO_2;\,-C_2H_5OH}}$$

$$\mathrm{\longrightarrow R{-}CH(NH{\cdot}Cbz){-}\overset{O}{\overset{\|}{C}}{-}O{-}\overset{O}{\overset{\|}{P}}(O^-){-}OR' \xrightarrow[-CO_2;\,-C_6H_5CH_3]{H_2/Pd} R{-}CH(NH_3^+){-}\overset{O}{\overset{\|}{C}}{-}O{-}\overset{O}{\overset{\|}{P}}(O^-){-}OR'}$$

$R = CH_3, C_2H_5, C_6H_5$ etc; R′ = 5′-nucleoside residue; $Cbz = C_6H_5CH_2O{-}\overset{O}{\overset{\|}{C}}{-}$

After the carbobenzoxy group is removed by hydrogenation over a palladium catalyst, the corresponding aminoacyl nucleotide is formed.

To produce mixed anhydrides of amino acids and adenosine 5′-phosphate, an alternative procedure may be used – the carboxyl group is activated in carbobenzoxy-amino acid with the aid of DCC. The resulting adduct is attacked by a nucleotide anion to give the desired mixed anhydride:

$$\mathrm{R{-}CH(NH{\cdot}Cbz){-}COOH \xrightarrow{DCC,\,H^+} R{-}CH(NH{\cdot}Cbz){-}\overset{O}{\overset{\|}{C}}{-}O{-}\overset{+}{C}(NHC_6H_{11}){=}NHC_6H_{11} \xrightarrow{R'O{-}P(=O)(O^-)OH}}$$

$$\mathrm{\longrightarrow R'O{-}\overset{O}{\overset{\|}{P}}(O^-){-}O{-}\overset{O}{\overset{\|}{C}}{-}CH(NH{\cdot}Cbz){-}R \xrightarrow[-CO_2;\,C_6H_5CH_3]{H_2/Pd} R'O{-}\overset{O}{\overset{\|}{P}}(O^-){-}O{-}\overset{O}{\overset{\|}{C}}{-}CH(NH_3){-}R}$$

R - amino acid radical; R′ - 5′-adenosine residue

This is how a number of peptide derivatives of nucleoside 5′-phosphates were also synthesized. Moreover, DCC was also instrumental in producing anhydrides of fatty acids and adenosine 5′-phosphate.

4.6 Properties of Nucleotide Derivatives with Substituents in the Phosphate Group

Advances in methods for analyzing and synthesizing nucleic acids as well as studies into the mechanism of action of nucleotide metabolism enzymes have in recent years brought the properties of the phosphate group in nucleotides substituted at the phosphorus atom to the fore. This is why we have decided to devote this section to the properties of some compounds belonging to this category.

4.6.1 Nucleoside Cyclic Phosphates

Nucleoside cyclic phosphates comprise an important class of nucleotide derivatives with respect to the phosphate group. They can be regarded as internal diesters of nucleosides. As has already been pointed out, five-membered ribonucleoside 2′,3′-cyclic phosphates (23) are formed when RNA is dissociated under conditions of chemical or enzymatic hydrolysis and also as intermediate compounds during isomerization of ribonucleoside 2′- and 3′-phosphates in an acid medium. The methods for synthesizing these compounds were described in section 4.5.4.

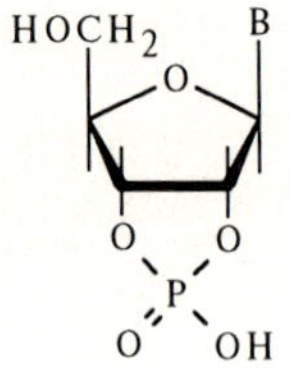

(23) 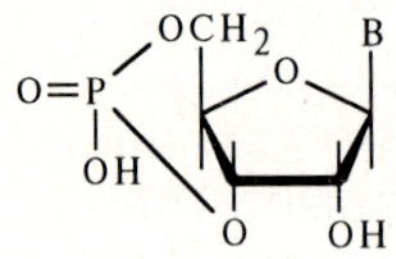 (24)

The cell has also been found to contain a six-membered cyclic phosphate, namely, adenosine 3′,5′-cyclic phosphate [belonging to compounds of type (24)]; it exists in a free state and plays a key role in the regulation of many biochemical processes.

Hydrolysis. Similarly to simple five-membered cyclic phosphates, ribonucleoside 2′,3′-cyclic phosphates are highly labile compounds and can exist only at neutral pH values and low temperatures. They readily undergo hydrolysis in alkaline and acid media to yield nucleoside 2′- and 3′-phosphates in practically equal amounts:

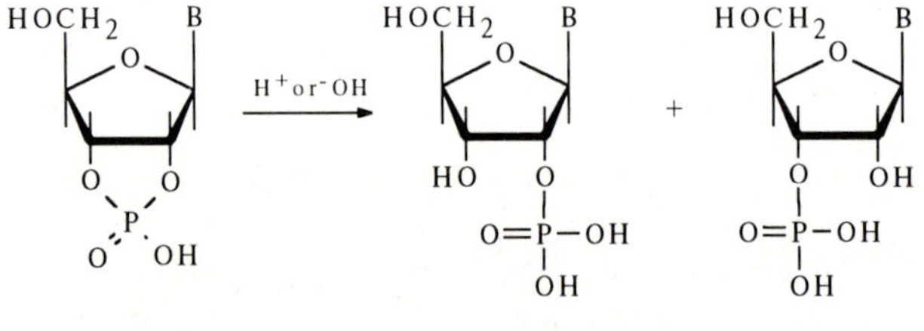

In an acid medium, hydrolysis of the P–O bond in nucleoside 2′,3′-cyclic phosphates proceeds at a rate 10^6 to 10^7 times faster than in the case of simple dialkyl phosphates.

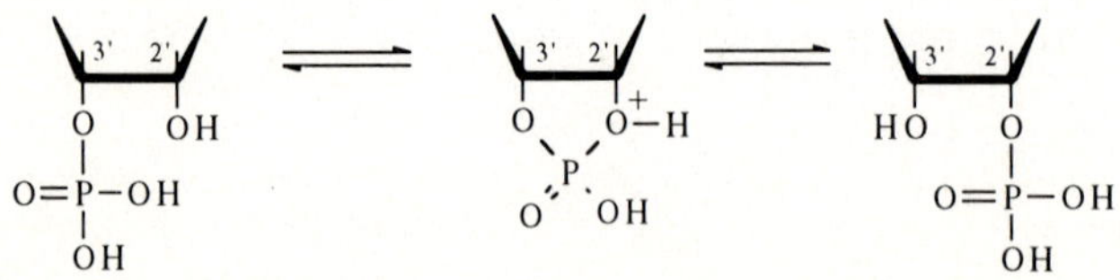

The reaction is reversible, although the rate of the backward process is much slower. A certain influence on the hydrolysis rate is exerted by the nature of the nucleotide's constituent base. Pyrimidine nucleoside 2′,3′-cyclic phosphates are hydrolysed faster than the purine ones; the most labile compound is uridine 2′,3′-cyclic phosphate. Alkaline hydrolysis of ribonucleoside 2′,3′-cyclic phosphates is slower than acid hydrolysis, the reaction in this case being irreversible:

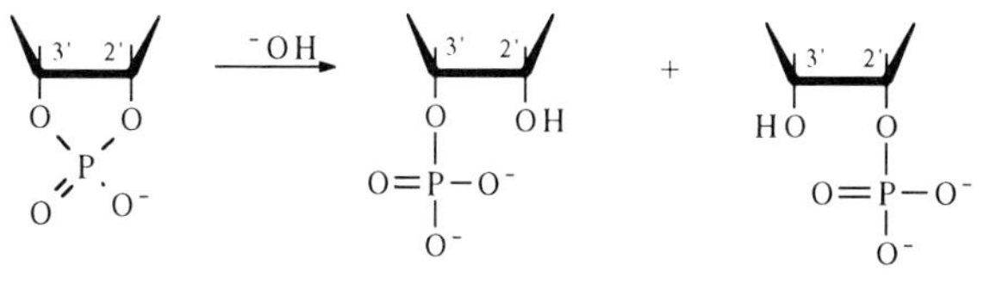

The greatest lability in an alkaline medium is displayed by uridine 2′,3′-cyclic phosphate, and the greatest stability, by adenosine 2′,3′-cyclic phosphate.

In weakly acidic and neutral media at 100 °C, nucleoside 2′,3′-cyclic phosphates are converted into the corresponding nucleosides. The reaction rate-determining step is hydrolysis of the phosphodiester bond, followed by rapid hydrolysis of the phosphomonoester one.

The ease of cleavage of the five-membered cycle in ribonucleoside 2′,3′-cyclic phosphates seems to be due (by analogy with simpler five-membered cyclic phosphates) to the ring being less strained in the transition state.

The formation of equal quantities of nucleoside 2′- and 3′-phosphates during acid as well as alkaline hydrolysis can be explained by the reaction passing with the same probability through two transition states differing in the nature of the hydroxyl which occupies an axial position (2′ or 3′); in acid hydrolysis, any of the transition states readily undergoes pseudorotation.

Just as would be expected from comparison of the stabilities of five- and six-membered cyclic phosphates or simple glycols, ribonucleoside 3′,5′- and deoxyribonucleoside 3′,5′-cyclic phosphates are much less labile than the 2′,3′-isomers. As a rule, cleavage of the P–O bond during acid hydrolysis [path (*a*)] is accompanied by that of the glycosidic bond [path (*b*)]:

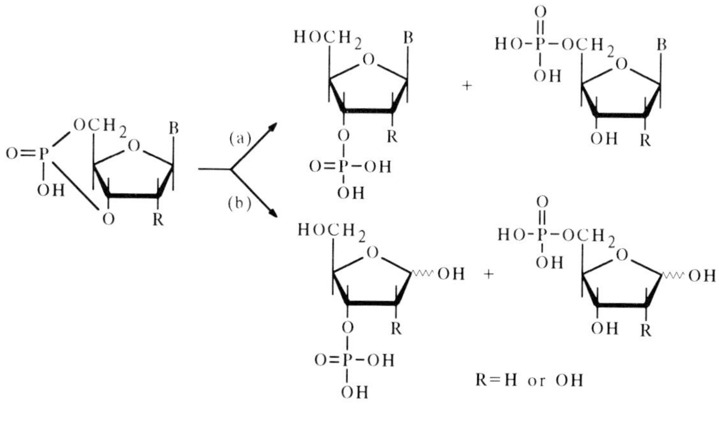

R = H or OH

Hydrolysis along path (*a*) yields a mixture of nucleoside 3′- and 5′-phosphates (with the 3′-isomer being predominant), while hydrolysis along path (*b*) gives a mixture of the corresponding ribose or deoxyribose phosphates.

The main products of alkaline hydrolysis of nucleoside 3′,5′-cyclic phosphates are nucleoside 3′- and 5′-phosphates forming at a 5:1 ratio [path (*a*)]. This process is particularly smooth in the presence of barium hydroxide, which seems to stem from the catalytic effect of barium ions capable of coordinating with the phosphate group.

Significantly, the rate of hydrolysis of the phosphodiester bonds in the six-membered ring of nucleoside 3′,5′-cyclic phosphate is still higher than in the case of ordinary six-membered cyclic phosphates, such as propanediol 1,3-cyclic phosphate. This has to do with the peculiarities of the conformation of nucleoside 3′,5′-cyclic phosphates in which the six-membered ring is *trans*-fused with the five-membered ribo- or deoxyribofuranose ring. X-ray structural analysis has shown that adenosine 3′,5′-cyclic phosphate (25) has a structure with an unusual conformation of ribose in which C^4 does not lie in the ring plane.

In the case of uridine 3′,5′-cyclic phosphate (26), a conformation of an envelope inwardly bent at $C^{3'}$ of ribose is assumed.

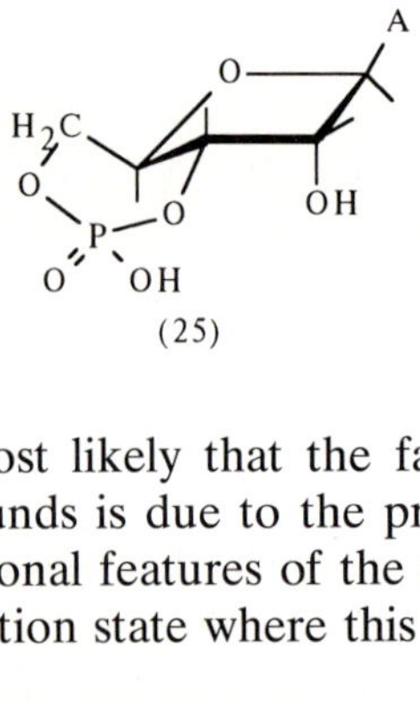

(25) 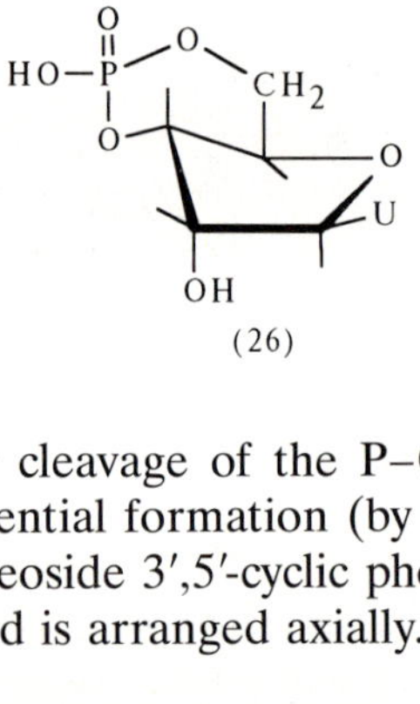 (26)

It is most likely that the faster cleavage of the P–O bond (at $C^{5'}$) in these compounds is due to the preferential formation (by virtue of the special conformational features of the nucleoside 3′,5′-cyclic phosphate structure) of such a transition state where this bond is arranged axially.

Enzymatic Cleavage. A large number of enzymes are known which cleave phosphoester bonds in nucleoside cyclic phosphates. The best known of them is pyrimidyl RNase, the specificity of its action residing in that it hydrolyses the P–O bond formed by the 2′-hydroxyl in pyrimidine nucleoside 2′,3′-phosphates:

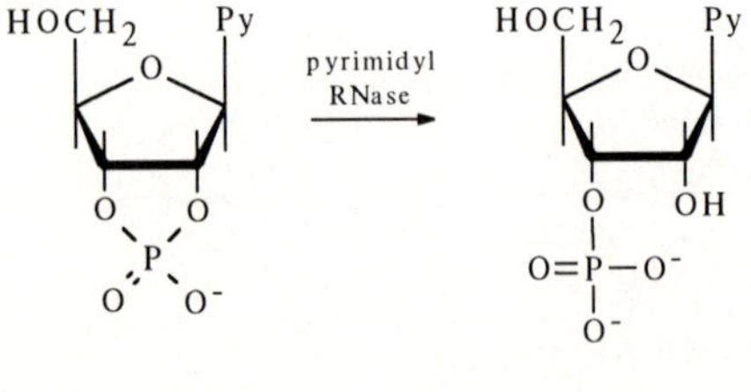

pyrimidyl RNase

No nucleoside 2′-phosphates are formed in this case. This means that the P–O bond at $C^{3'}$ in the five-membered ring of pyrimidyl RNase is not affected. Purine nucleoside 2′,3′-cyclic phosphates are hydrolysed by other enzymes. For instance, guanyl RNase isolated from takadiastase hydrolyses guanosine 2′,3′-cyclic phosphates to guanosine 3′-phosphates, with the P–O bond at $C^{3'}$ remaining intact:

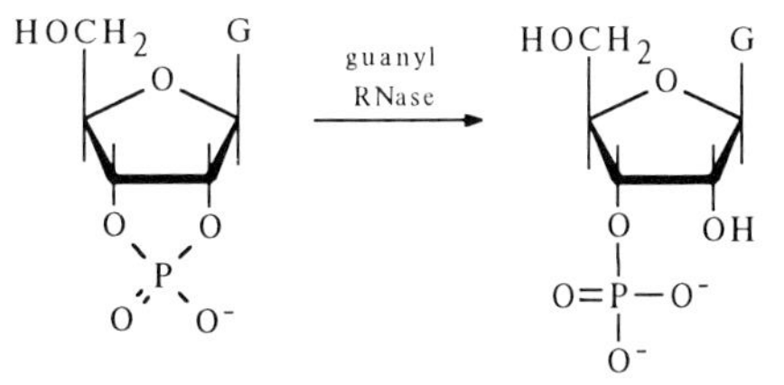

guanyl RNase

Similar specificity of action is observed in exonuclease A_5 isolated from actinomyces.

Phosphodiesterase (PDE) from *E. coli* hydrolyses nucleoside 2′,3′-cyclic phosphates to nucleoside 3′-phosphates irrespective of the nature of the constituent base:

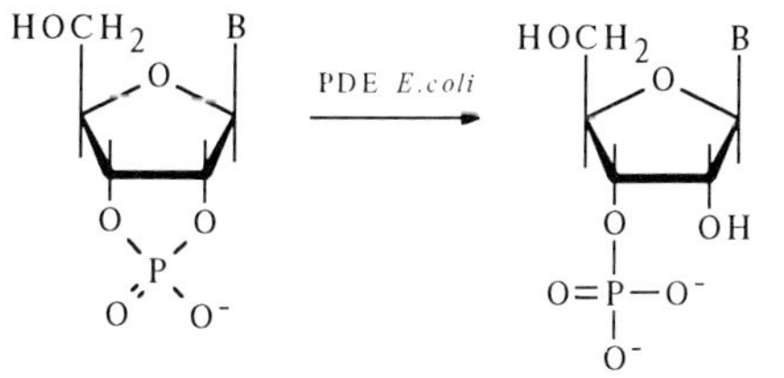

PDE *E.coli*

Mammalian tissues have been found to contain PDE hydrolysing nucleoside 3′,5′-cyclic phosphates. For example, PDEs isolated from the bovine heart hydrolyse adenosine 3′,5′- and uridine 3′,5′-cyclic phosphate to the corresponding 5′-phosphates with particular ease.

Other Reactions. In anhydrous alcohols, ribonucleoside 2′,3′-cyclic phosphates are rapidly converted, in the presence of hydrogen ions, into a mixture of ribonucleoside 2′- and 3′-alkyl phosphates:

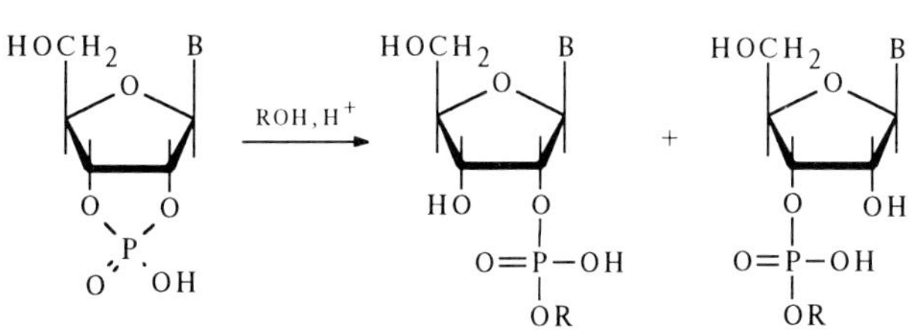

ROH, H^+

R= CH_3, C_2H_5, C_3H_7, *tert*-C_4H_9 etc.

Alcoholates of alkali metals in alcoholic media also hydrolyse ribonucleoside 2′,3′-cyclic phosphates to ribonucleoside 2′(3′)-alkyl phosphates. This reaction, however, proceeds at a much slower rate, as compared to a similar acid-catalyzed reaction with alcohols, and is accompanied by side processes. For example, treatment of adenosine 2′,3′-cyclic phosphate with sodium benzylate in benzyl alcohol results in adenosine 3′- and 2′-benzyl phosphates with a sizable admixture of adenosine 3′- and 2′-phosphates.

The mechanism of these reactions seems to be similar to that proposed for hydrolysis of nucleoside cyclic phosphates, catalyzed by acids and bases, with the difference that the nucleophile attacking the tetrahedral phosphorus atom in this case is an alcohol or alkoxy-ion molecule.

Ribonucleoside 2′- and 3′-alkyl phosphates are easily formed from the corresponding nucleoside 2′,3′-cyclic phosphates if a mixed anhydride is first obtained with diphenylphosphoric acid, for example, then treated with excess alcohol. The following conversion mechanism has been proposed:

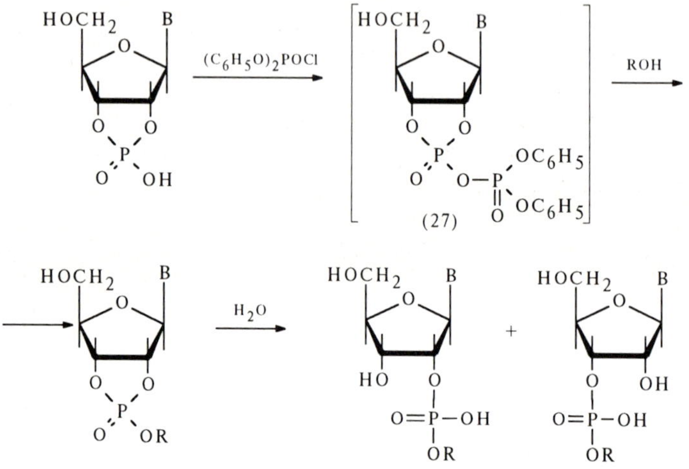

The extremely unstable compound (28) (the lability of its phosphotriester bonds is enhanced by the cyclic nature of the compound) is readily hydrolysed to form nucleoside 2′- and 3-alkyl phosphates. Thus, it is assumed that in the nucleophilic substitution at the phosphorus atom forming part of the five-membered ring of anhydride (27) the leaving group is the diphenylphosphoric acid anion [path (*a*)]. However, this mechanism lacks corroborating evidence.

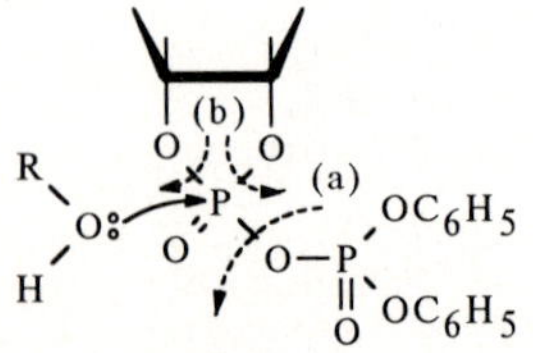

If one proceeds from the general concepts concerning nucleophilic substitution at the tetrahedral phosphorus atom in cyclic phosphates, this mechanism seems unlikely because in this case the O–P–O group of the five-membered ring must occupy an equatorial position in the transition state (axial positions being occupied by the incoming alcohol molecule and leaving diphenyl phosphate group), which will lead to a drastic increase in strain and almost never materializes. Apparently, it is the 2′- or 3′-hydroxyl group of ribose that occupies an axial position and becomes the leaving entity. Thus, when anhydride (27) is attacked by an alcohol (see scheme), two isomeric mixed anhydrides must form:

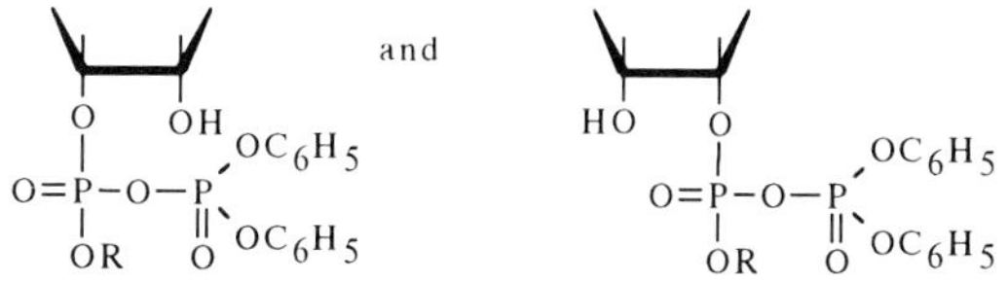

$$\text{O}=\text{P}(\text{OR})(\text{O-sugar, 2'-OH})-\text{O}-\text{P}(=\text{O})(\text{OC}_6\text{H}_5)_2 \quad \text{and} \quad \text{O}=\text{P}(\text{OR})(\text{O-sugar, 3'-HO})-\text{O}-\text{P}(=\text{O})(\text{OC}_6\text{H}_5)_2$$

Subsequently, these anhydrides must be easily converted into a cyclic triester (28) as a result of intramolecular nucleophilic substitution at the phosphorus atom. Hence, the conversion of anhydride (27) into cyclic triester (28) seems to involve intermediate anhydrides (29); during this step [(27) → (28)] the role of the diphenyl phosphate group boils down to enhancing the electrophilicity of the nucleotide phosphorus atom.

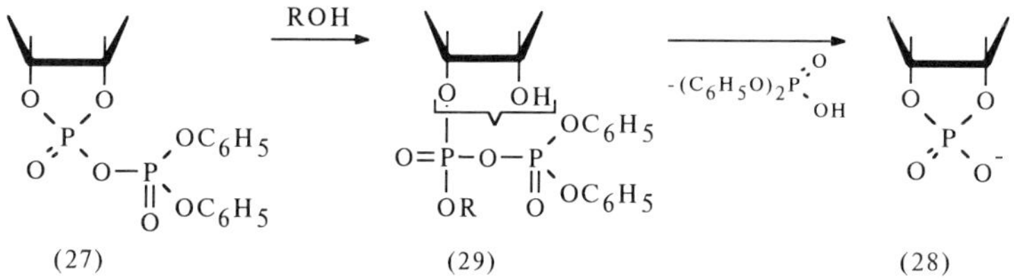

$$\underset{(27)}{\text{cyclic }\text{O}_2\text{P}(=\text{O})-\text{O}-\text{P}(=\text{O})(\text{OC}_6\text{H}_5)_2} \xrightarrow{\text{ROH}} \underset{(29)}{\text{O}=\text{P}(\text{OR})(\text{O})-\text{O}-\text{P}(=\text{O})(\text{OC}_6\text{H}_5)_2} \xrightarrow{-(\text{C}_6\text{H}_5\text{O})_2\text{P(O)OH}} \underset{(28)}{\text{cyclic }\text{O}_2\text{P}(=\text{O})\text{O}^-}$$

More stable substituted cyclic phosphates are produced from nucleoside 3′,5′-cyclic phosphates. For example, treatment of uridine 3′,5′-cyclic phosphate with diphenyl phosphochloride followed by addition of an amine results in an amide of uridine 3′,5′-cyclic phosphate, which has been isolated and characterized:

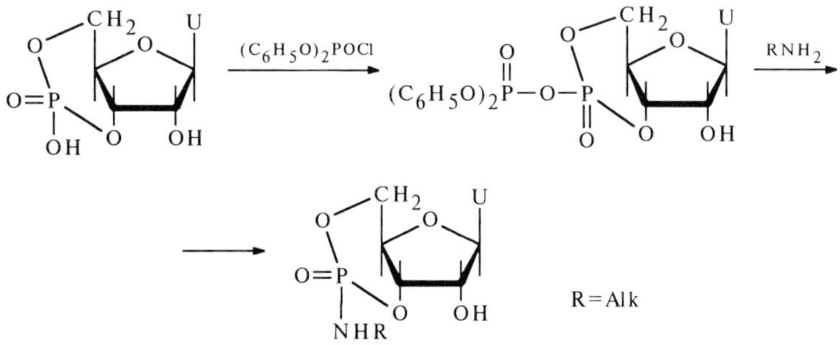

$$\text{Urd-3',5'-O}_2\text{P(O)OH} \xrightarrow{(\text{C}_6\text{H}_5\text{O})_2\text{POCl}} (\text{C}_6\text{H}_5\text{O})_2\text{P(O)}-\text{O}-\text{P(O)-3',5'-Urd} \xrightarrow{\text{RNH}_2} \text{Urd-3',5'-O}_2\text{P(O)NHR}$$

R = Alk

The greater stability of such a trisubstituted cyclic phosphate, as compared to the above-described nucleoside 2′,3′-alkyl cyclic phosphates, stems from both the greater stability of the six-membered ring and the presence of an amide group at the phosphorus atom.

4.6.2 Alkyl Esters of Nucleotides

The reactions leading to cleavage of phosphoester bonds occupy a special place among the chemical transformations involving alkyl esters of nucleotides. These reactions are widely used in the chemistry of nucleic acids, where they form the basis of the analytical procedures employed to determine composition and structure.

Two principal types of reactions leading to cleavage of the phosphodiester bonds in alkyl esters of nucleotides are known. Those of the first type include hydrolysis catalyzed by acids, alkalis and bases and proceeding as nucleophilic substitution at the phosphorus atom with cleavage of the P–O bond. Alkyl esters of nucleotides, with a carbonyl group at the β-position of the carbohydrate moiety with respect to the alkyl phosphate, are characterized by β-elimination of the latter, accompanied by cleavage the C–O bond. The reactions of both types will be discussed at greater length in what follows.

Hydrolysis of Phosphodiester Bonds. Alkyl esters of nucleotides form two distinct groups in terms of phosphodiester bond stability. The first group includes compounds in which the alkyl phosphate group is linked with one of the hydroxyls of the *cis*-glycol group; that is, ribonucleoside 3′- or 2′-alkyl phosphates (30):

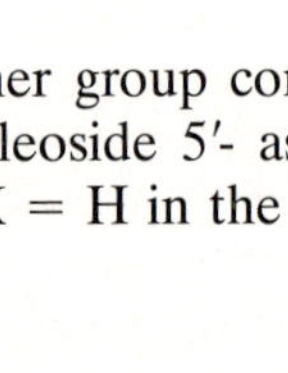

(30) R=Alk

The other group comprises the rest of the alkyl esters of nucleotides; that is ribonucleoside 5′- as well as deoxyribonucleoside 3′- and 5′-alkyl phosphates [(31); X = H in the second compound]:

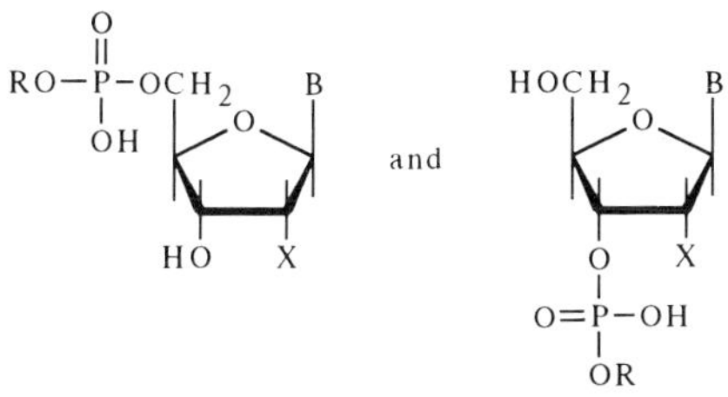

(31) R=Alk; X=H or OH

The sharp difference in stability of the phosphodiester bonds in compounds belonging to both groups stems from the presence or absence of a hydroxy group at the α-position with respect to the phosphate group.

Ribonucleoside 3′(2′)-alkyl phosphates (30) have a strong resemblance, in terms of their properties, to alkyl phosphoglycols – that is, they are stable only at neutral pH values. Under conditions of alkaline or acid catalysis, these compounds readily dissociate to free nucleotides. Therewith, the cleavage of phosphodiester bonds proceeds, just as in the case of alkyl phosphoglycols, as transphosphorylation, which is to say that the neighboring hydroxyl group of ribose is involved in this instance. The initially formed ribonucleoside 2′,3′-cyclic phosphate yields a mixture of nucleoside 3′- and 2′-phosphates during hydrolysis.

The catalytic function of bases B resides in that by forming a hydrogen bond with the hydroxyl group adjacent to the phosphodiester one (with 2′-OH in the scheme below) they render this hydroxyl group more nucleophilic. The result is promotion of the intramolecular nucleophilic attack at the phosphorus atom.

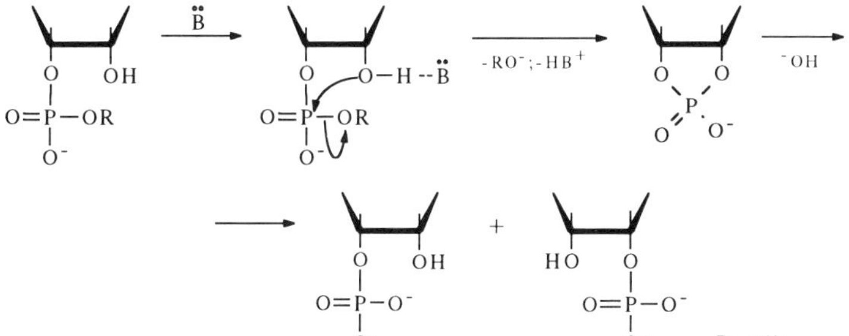

R=Alk

The fact that no cytidine 2′-benzyl phosphate has been found among the products of incomplete alkaline hydrolysis of cytidine 3′-benzyl phosphate suggests that the transition state associated with the nucleophilic substitution at the phosphorus atom, involving the 2′-hydroxyl group, has the structure of a trigonal bipyramid in which axial positions are occupied by the attacking 2′-hydroxyl and leaving benzyl groups. The nucleophilic attack ends in formation of nucleoside 2′,3′-cyclic phosphate. In this case, there is no pseudorotation which may result in axial arrangement of the P–O bond at $C^{3'}$ with subsequent departure of the 3′-hydroxyl group and formation of nucleoside 2′-benzyl phosphate.

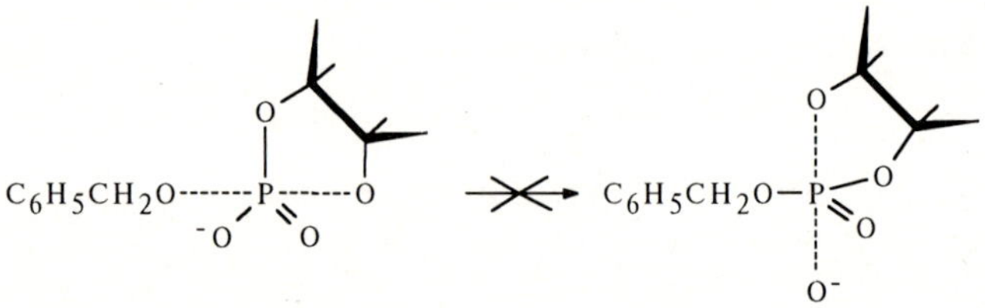

After the cyclic phosphate has formed, it undergoes hydrolysis with extreme ease, giving a mixture of nucleoside 2′- and 3′-phosphates without migration of the phosphate group.

During acid hydrolysis of nucleoside 3′(2′)-alkyl phosphates (30), migration of the alkyl phosphate group is observed along with cleavage of the phosphodiester bond. For example, during incomplete (by 50%) acid hydrolysis of cytidine 3′-benzyl phosphate, the reaction mixture was found to contain cytidine 2′-benzyl phosphate (25%) in addition to cytidine 2′- and 3′-phosphates as well as the unreacted ester.

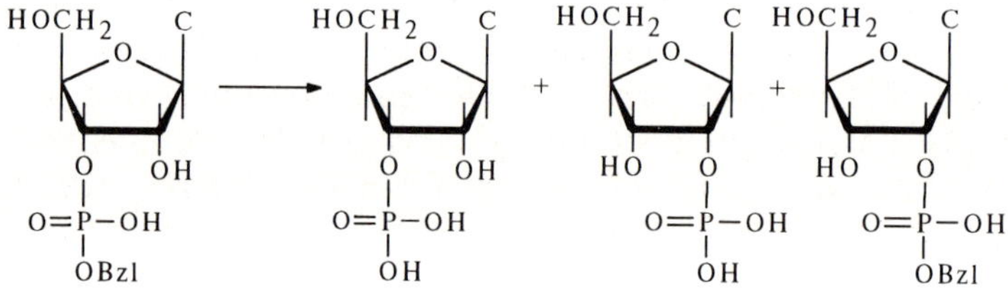

The formation of cytidine 2′-benzyl phosphate may be due to the pseudorotation in the transition state arising in the course of the substitution.

The compounds of structure (31), in which the nucleophilic properties of the neighboring hydroxyl group cannot influence the course of the reaction, behave like dialkyl phosphates or, in other words, are stable toward acid and alkaline hydrolysis. The most stable are such derivatives of adenosine as adenosine 5′- as well as deoxyadenosine 5′- and 3′-benzyl phosphates.

During acid hydrolysis of esters of type (31) under vigorous conditions, the cleavage of phosphodiester bonds is complicated by that of the glycosidic bond (especially in the case of purine nucleotides). The cleavage of phosphodiester bonds, if it is made possible, occurs not only between the oxygen and phosphorus atoms, as was the case with phosphodiesters of structure (30), but also primarily between the oxygen and carbon.

Alkaline hydrolysis of nucleoside phosphate esters of type (31) is facilitated by incorporation of electron-acceptor substituents into the residue R. Interestingly, nucleoside 3′,5′-cyclic phosphates may form at the same time, although the *trans*-position of the 3′-hydroxyl and CH_2OH groups hampers this reaction. For instance, deoxythymidine 3′-(*para*-nitrophenyl phosphate) is converted into deoxythymidine 3′,5′-cyclic phosphate in an alkaline medium.

$HOCH_2$ T O O OH $O=P-O^-$ $OC_6H_4NO_2$-p

$\xrightarrow[-\,p\text{-}O_2NC_6H_4]{^-OH}$

CH_2 T O O $O=P$ O^- O OH

The cyclization (nucleophilic substitution at the phosphorus atom with the participation of a sterically hindered 5′-hydroxyl group) is in this case facilitated by the ease of departure of the stable *para*-nitrophenolate anion.

Two conclusions should be drawn from the finding that ribonucleoside 3′(2′)-alkyl phosphates arc highly labile. Firstly, the intramolecular nucleophilic attack of the phosphorus atom in nucleoside 3′(2′)-phosphates is facilitated by the 2′(3′)-hydroxyl group being sterically close to the phosphate group. In the presence of such a hydroxyl group in a nucleotide molecule, the nucleophilic attack of the phosphorus atom by water or a hydroxyl anion proceeds at a slower rate. The intramolecular nucleophilic attack results in an intermediate state with a five-coordinated phosphorus atom, in which one of the *cis*-hydroxyl groups of ribose (the attacking one) is arranged axially and the other (linked to the phosphate group), equatorially. Another axially arranged substituent is always an alkoxy group. The departure of this group (the main direction of nucleophilic substitution at the phosphorus atom) is followed by formation of a cyclic diphosphate which is easily hydrolysed, unlike dialkyl phosphates, to a mixture of 2′- and 3′-phosphonucleosides. During this step, the attacking group is a water molecule. Thus, if a nucleoside 3′(2′)-phosphate contains a 2′(3′)-hydroxyl group sterically close to the phosphate group, both emergence of the transition state and the subsequent hydrolysis of the intermediately forming cyclic phosphate are accelerated. Consequently, the neighboring hydroxyl group catalyzes the hydrolysis of nucleoside 3′(2′)-alkyl phosphates (a similar pattern is observed in the case of polyol phosphates as well).

The other important aspect of hydrolysis of ribonucleoside 3′(2′)-alkyl phosphates is the fact that it involves different mechanisms depending on whether the medium is acid and alkaline.

If the process is catalyzed by an acid, the intermediate state (32) arising from ribonucleoside 3′-alkyl phosphate undergoes two transformations: conversion into cyclic phosphate (33) and pseudorotation as a result of which the nondissociated hydroxyl of the phosphate group and that of ribose, associated with the alkyl phosphate proup prior to the reaction [structure (36)] occupy axial positions. If the hydroxyl group of ribose "leaves" such a state, migration of the alkyl phosphate group takes place and the emerging isomer (34) again undergoes the same transformations involving the hydroxyl group adjacent to the phosphodiester bond and the end products are the nucleoside 2′- and 3′-phosphates forming in equal amounts.

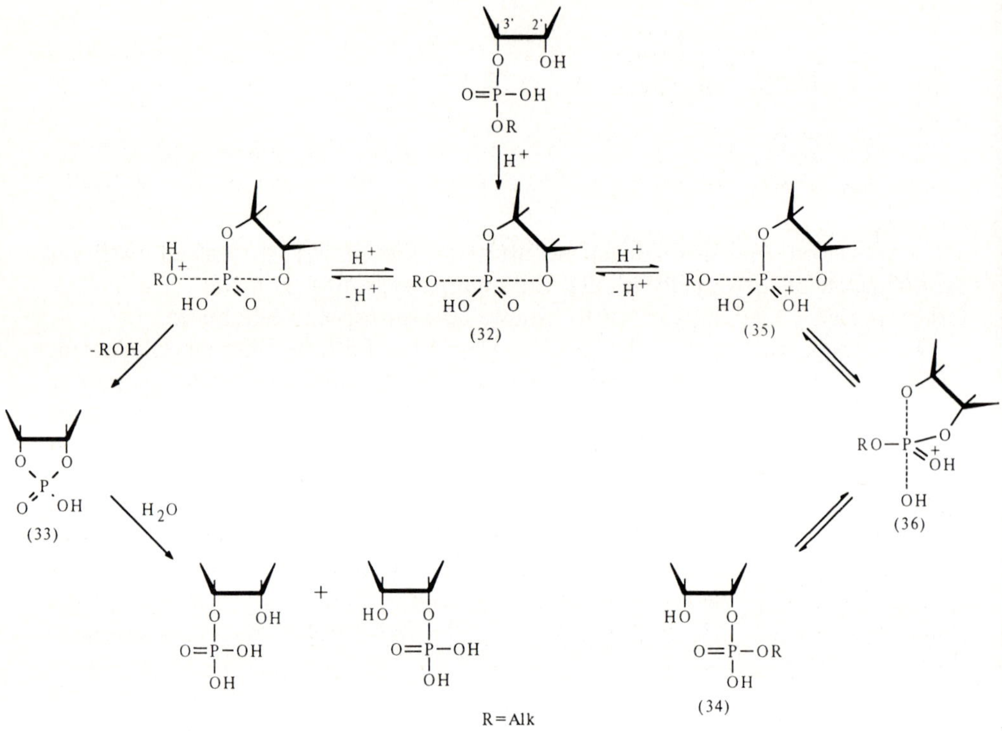

Pseudorotation in intermediate state (32) is impossible because of the absence of two sufficiently electronegative groups in the equatorial position, which could "stretch" the equatorial P–O bonds to the length of axial ones. After protonation, structure (32) may undergo conversion into structure (35) where such groups are already present (hydroxyl and 3′-O-CH< group). Therefore, pseudorotation and subsequent transition toward compound (34) become possible.

During alkaline hydrolysis, the P–O bonds cannot be equalized in length with subsequent pseudorotation in transition state (32) because the bonds linking the phosphoryl oxygen atom and that of the oxygen resulting from the dissociation ($P{-}OH \rightarrow P{-}O^-$) to the central phosphorus atom remain much stronger than the rest of the P–O bonds.

This is why the transformations catalyzed by bases are not accompanied by migration of the alkyl phosphate group. The products of alkaline hydrolysis of nucleoside 3′(2′)-alkyl phosphates at any step of the process are nucleoside 3′- and 2′-phosphates.

Hydrolysis of Phosphotriester Bonds. Triesters, which are analogues of diesters of types (30) and (31), also differ in stability of the phosphotriester bonds.

Triesters in which the nucleophilic behavior of the neighboring hydroxyl cannot affect the course of the reaction are in many ways similar to trialkyl phosphates.

As regards stability of phosphotriester bonds, compounds containing a hydroxyl next to the phosphotriester group stand distinctly apart:

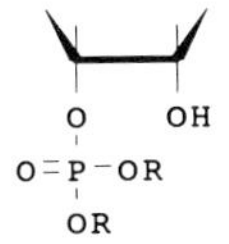

O OH
O=P−OR
OR

Such triesters should behave like protonated diesters. They correspond to an activated form (30, H = R) that initiates diester cleavage and are relatively unstable. Assuming an S_N2 mechanism, one can expect formation of a cyclic triester as an intermediate product after the elimination of an OR anion. This intermediate will be hydrolysed by a water molecule to a mixture of 2′- and 3′-phosphodiesters:

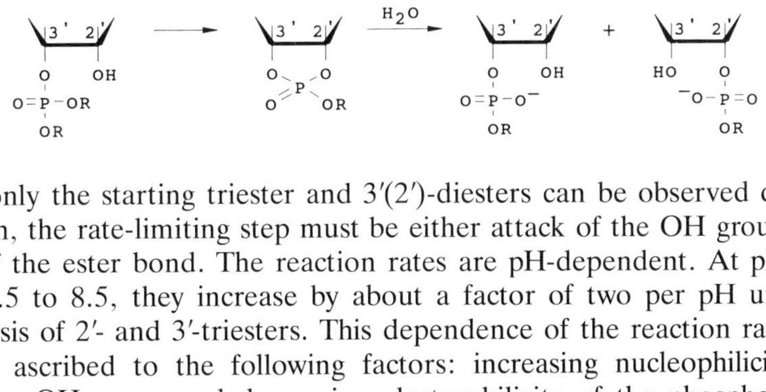

3′ 2′ → 3′ 2′ —H_2O→ 3′ 2′ + 3′ 2′

Since only the starting triester and 3′(2′)-diesters can be observed during the reaction, the rate-limiting step must be either attack of the OH group or cleavage of the ester bond. The reaction rates are pH-dependent. At pH ranging from 4.5 to 8.5, they increase by about a factor of two per pH unit during hydrolysis of 2′- and 3′-triesters. This dependence of the reaction rates on pH can be ascribed to the following factors: increasing nucleophilicity of the attacking OH group and decreasing electrophilicity of the phosphorus (both factors influencing the rate of the attack).

Enzymatic Cleavage. Some of the enzymes hydrolysing phosphodiester bonds in nucleoside cyclic phosphates also break down noncyclic esters of nucleotides. A case in point is pyrimidyl RNase which catalyzes the hydrolysis of pyrimidine nucleoside 3′-alkyl phosphates, 2-ketopyrimidine derivatives without any substituent at N^3. The result is the corresponding nucleoside 3′-phosphates or their derivatives. Nucleoside 2′,3′-cyclic phosphates emerge as intermediates. For example:

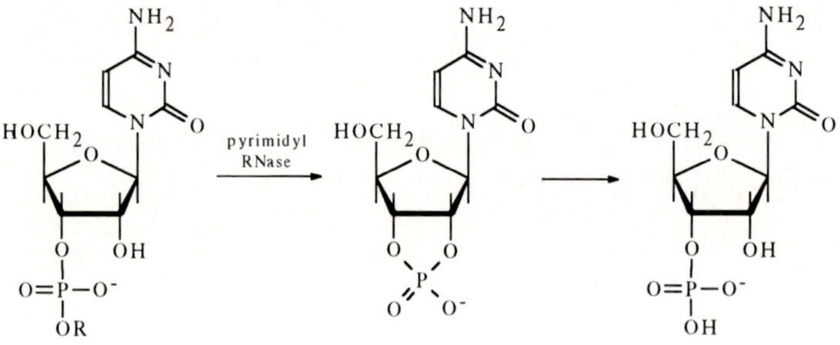

pyrimidyl RNase

R = Alk

Pyrimidyl RNase cleaves the coresponding phosphodiester bonds formed by uridine and cytidine derivatives, as well as some minor components of RNA, such as pseudouridine, 5,6-dihydrouridine, and ribothymidine.

The enzyme does not hydrolyse pyrimidine ribonucleoside 2′- and 5′-alkyl phosphates, nor does it break down any purine ribonucleoside alkyl phosphates. Pyrimidine and purine deoxyribonucleoside 3′(5′)-alkyl phosphates are not hydrolysed by pyrimidyl RNase either.

This enzyme with its capacity for breaking phosphodiester bonds in pyrimidine ribonucleotides only with a particular position of the alkyl phosphate group is widely used in structural studies because it permits one to determine the structure of nucleoside alkyl phosphates as well as enables controlled hydrolysis of ribonucleic acids.

Guanyl RNase (derived from takadiastase or actinomyces[9]) catalyzes the hydrolysis of ribonucleoside 3′-phosphates, derived from 6-ketopurines unsubstituted at N^1 and N^7. They include guanosine among the major RNA components and inosine, N^2-methyl guanosine and N^2-dimethyl guanosine among the minor ones (see Table 1-5). Just as in the above-described cases, this reaction involves a cyclic phosphate formation step, for example.

[9] These two preparations of guanyl RNAse differ in that the activity of one of them (isolated from actinomyces) is less dependent on the presence of substituent at N^1 and N^7 in the guanine nucleus of the subastrate.

guanyl RNase

R = Alk

In some instances, the second step of the reaction, namely, hydrolysis of nucleoside 2′,3′-cyclic phosphates, proceeds at a very slow rate so that the latter accumulate in sizable amounts. Guanyl RNase does not catalyze the hydrolysis of alkyl phosphates of other ribo- and deoxyribonucleosides.

By virtue of its highly specific action, guanyl RNase is the major enzyme used in determining the RNA structure.

In addition to the nucleases considered above, ther is a number of enzymes that are less specific. Worthy of mention among the enzymes most frequently used for breaking phosphodiester bonds are phosphodiesterases (PDEs) isolated from the spleen and micrococci, which catalyze the breaking of virtually any nucleoside 3′-alkyl phosphates down to the corresponding nucleoside 3′-phosphates. Another commonly used enzyme is PDE from snake venom, which, in contrast to the PDEs mentioned above, catalyzes the hydrolysis of nucleoside 5′-alkyl phosphates to the corresponding nucleoside 5′-phosphates.

β-Elimination reactions. The cleavage of phosphodiester bonds in alkyl esters of nucleotides may be accompanied by breaking not only of P–O but also C–O bonds. The latter must, in this case, be activated by the β-carbonyl group in pentose. The incorporation of such a carbonyl group may be achieved either through oxidation of the hydroxyl groups in the carbohydrate moiety, which leads to activation of the 5′- and 3′-phosphoester bonds, or through elimination of the heterocyclic base, which leads to activation of the 3′-phosphoester bond.

Notably, the products of periodate oxidation of nucleoside 5′-alkyl phosphates lend themselves readily to hydrolysis in an alkaline medium. This entails cleavage of the C–O bond with detachment of the phosphomonoester:

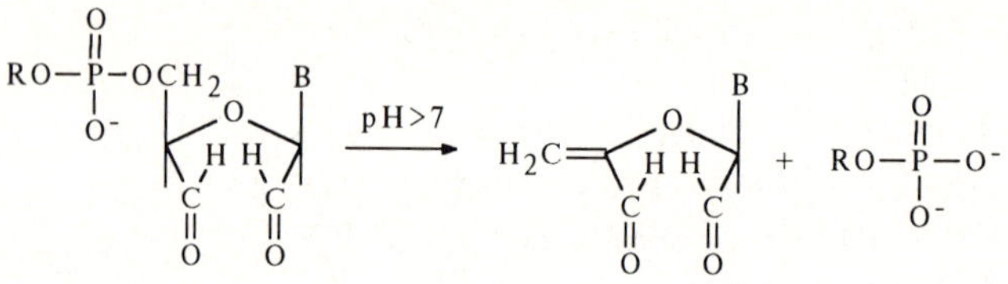

Along with β-elimination of the alkyl phosphate group, the oxidized nucleoside in an alkaline medium often undergoes cleavage of the *N*-glycosidic bond with separation of the heterocyclic base. Primary amines serve as effective catalysts of such cleavage.

The 5′-alkyl phosphate group may also be eliminated in the corresponding deoxynucleotide derivatives after oxidation of their secondary hydroxyl group. For instance, when solutions of deoxynucleoside 5′-alkyl phosphates are heated in a mixture of acetic anhydride and pyridine with dimethyl sulfoxide, deoxyribose is oxidized with subsequent β-elimination of the alkyl phosphate group and simultaneous cleavage of the *N*-glycosidic bond:

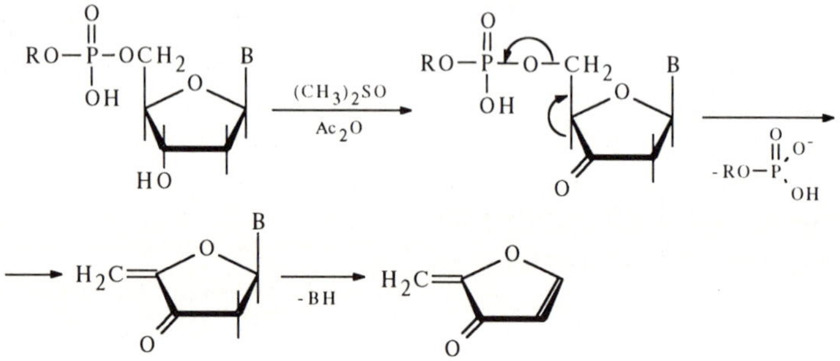

In the case of deoxynucleoside 3′-alkyl phosphates, a similar cleavage of the phosphodiester bond can be easily achieved in an alkaline medium after oxidation of the primary hydroxyl group with atmospheric oxygen over a platinum catalyst but with preliminary conversion of the carboxyl into an amide group:

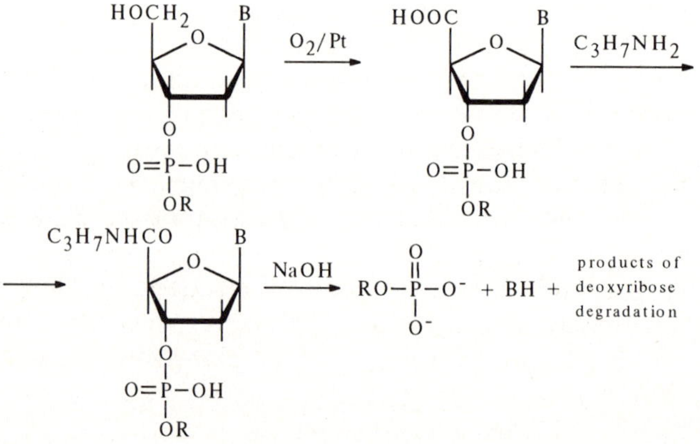

4.6.3 Mixed Anhydrides of Nucleotides

As has already been mentioned, mixed anhydrides of nucleotides occur widely in nature and play a major role in many biosynthetic processes. The most important of them are phosphoric acid derivatives known as nucleotide coenzymes or nucleotide anhydrides[10)].

Anhydrides with Phosphoric Acids. It has already been mentioned that anhydrides of nucleotides and diphenylphosphoric acid are highly reactive, their instability making it impossible to isolate these compounds, but they can be made to react with various nucleophilic agents in an absolute solvent. In such cases, the nucleophile replaces the diphenylphosphate group to yield the corresponding nucleotide derivatives. Anhydrides of nucleotides with monoesters of phosphoric or unsubstituted phosphoric and pyrophosphoric acids are less reactive. Such anhydrides are stable in neutral media but are hydrolysed in acid ones. At alkaline pH values, they exist in the form of di- or trianions, which is why it is difficult for the hydroxyl anion to approach them because of electric repulsion. As a result, they are relatively stable toward alkaline hydrolysis. Cleavage of the pyrophosphate bond is easy in an alkaline medium as well if the pyrophosphate group is next to a hydroxyl one; in this case, the reaction is accompanied by formation of a cyclic phosphate. For instance, treatment of cytidine 5′-diphosphoglycerol with an aqueous solution of ammonia gives cytidine 5′-phosphate and glycero-1,2-cyclic phosphate:

$$CH_2(OH)-CH(OH)-CH_2O-P(=O)(O^-)-O-P(=O)(O^-)-OCH_2\text{-}[\text{ribose}(HO,\,OH)]\text{-}C \xrightarrow{NH_4OH}$$

$$\longrightarrow CH_2(OH)-CH(O)-CH_2(O)\,[P(=O)O^-] \;+\; {}^-O-P(=O)(O^-)-OCH_2\text{-}[\text{ribose}(HO,\,OH)]\text{-}C$$

Anhydride of cytidine 5′-phosphate with phosphoric or pyrophosphoric acid is stable under the same conditions. The role of the neighboring hydroxyl in such anhydrides is reminiscent of the involvement of 2′(3′)-hydroxyl of ribose in alkaline hydrolysis of ribonucleoside 3′(2′)-alkyl phosphates.

The effect of the hydroxyl group that may lead to formation of a six-membered cyclic phosphate is less pronounced. For example, coenzyme A (37) partially dissociates to a cyclic phosphate when treated with an alkali.

---

10) The chemistry of these compounds is treated in the well known monograph by A. M. Michelson (see references at the end of this chapter).

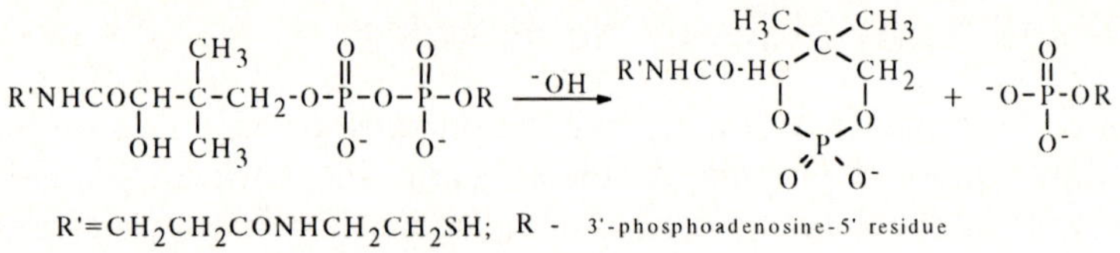

R'=$CH_2CH_2CONHCH_2CH_2SH$; R - 3'-phosphoadenosine-5' residue

When anhydride of deoxythymidine 5′-phosphate is heated in pyridine, the pyrophosphate undergoes partial conversion into deoxythymidine 3′,5′-cyclic phosphate:

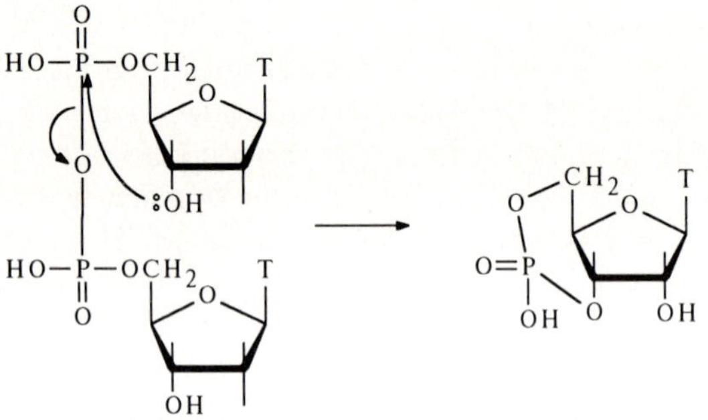

Descriptions of intermolecular reactions between such anhydrides and nucleophilic agents are scant. Only a pyrophosphorolysis reaction involving nucleotide anhydrides is more or less well known. For example, anhydride of adenosine 5′-phosphate reacts with pyrophosphoric acid in pyridine to yield adenosine 5′-triphosphate:

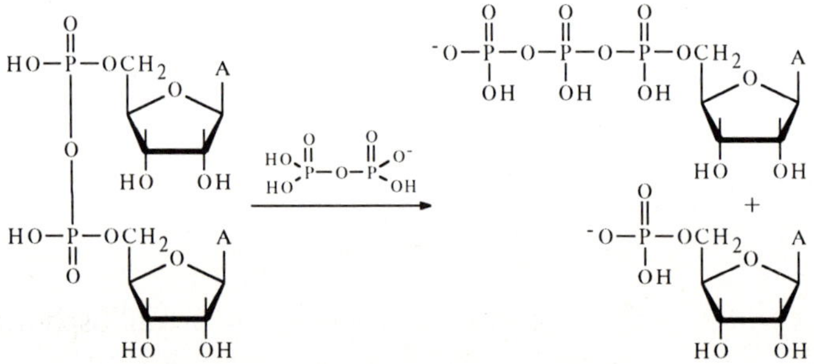

Moreover, the reaction mixture was found to contain adenosine 5′-diphosphate which seems to result from nucleophilic substitution for the terminal phosphate group in the triphosphate:

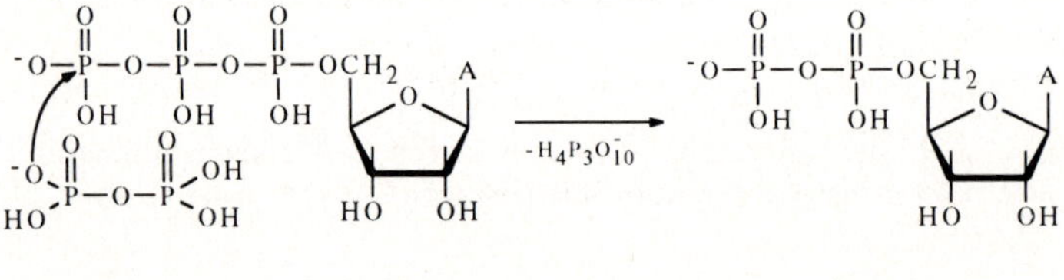

Anhydrides with Carboxylic Acids. Mixed anhydrides of nucleotides and carboxylic acids belong to the most important biologically active nucleotide derivatives. Formation of a mixed anhydride of adenosine 5′-phosphate and an amino acid marks the beginning of activation of the latter, necessary for its incorporation into the peptide chain during protein biosynthesis in ribosomes. This reaction, proceeding in the presence of the enzyme aminoacyl-tRNA ligase specific for each amino acid, resides in nucleophilic attack of the α-phosphate link in adenosine 5′-triphosphate by a free amino acid.

A similar activation mechanism has been established for acetic, butyric and other carboxylic acids participating in the biosynthesis of steroids and other biologically important compounds.

$$^{-}O-P(=O)(O^-)-O-P(=O)(O^-)-O-P(=O)(O^-)-OCH_2\text{-[ribose(HO, OH)]-A} + {}^{+}NH_3CH(R)C(=O)O^- \xrightarrow{\text{enzyme}}$$

$$\longrightarrow {}^{+}NH_3CH(R)C(=O)-O-P(=O)(O^-)-OCH_2\text{-[ribose(HO, OH)]-A} + {}^{-}O-P(=O)(O^-)-O-P(=O)(O^-)-O^-$$

Before discussing the properties of mixed anhydrides of nucleotides with carboxylic acids, we must take a brief look at those of their simpler analogues – acyl phosphates.

General Concepts About Acyl Phosphates. Acyl phosphates feature two reactive sites that can be attacked by nucleophiles – phosphorus in the phosphate group and carbon in the carbonyl.

$$R-C(=O)-O-P(=O)(O^-)-OR'$$

However, due to differences in the structure of carboxyl and phosphate groups, the machanisms of nucleophilic substitution at the above-mentioned atoms are widely dissimilar. The effectiveness of the attack at the carbonyl carbon is determined by the ease of linkage with the nucleophile, whereas the decisive factor for the nucleophilic attack at the phosphate phosphorus is the nature of bonding with the group being substituted (the course of both reactions is also materially affected by the nature of the substituents, nucleophilic agent and solvent). In order to demonstrate the peculiar behavior of acyl phosphates in nucleophilic substitution reactions, it would be appropriate at this juncture to compare the behavior of asymmetric mixed anhydrides of two carboxylic acids and asymmetric pyrophosphates.

$$R-\underset{\underset{O}{\parallel}}{C}-O-\underset{\underset{O}{\parallel}}{C}-R' \qquad (R''O)_2\overset{\overset{O}{\parallel}}{P}-O-\overset{\overset{O}{\parallel}}{P}(OR''')_2$$

The nucleophilic agent Nu attacks the mixed anhydrides of carboxylic acids at the most electrophilic site – that is, at the carbonyl carbon of a stronger acid (RCOOH). The result is the corresponding derivative of the latter, and the function of the leaving group is performed by the anion of the weaker acid ($R'COO^-$).

$$R'C(=O)-O-C(=O)R + {}^-Nu \longrightarrow R'C(=O)Nu + RC(=O)O^-$$

The mixed anhydride of two phosphoric acids, or asymmetric pyrophosphate, is attacked by the nucleophile at both phosphorus atoms; the effectiveness of the attack in this case, however, is determined not by the magnitude of the partially positive charge at these atoms but by the nature of the leaving group: the more stable the departing anion (i. e., the stronger the acid corresponding to the anion), the easier the nucleophilic substitution. Thus, if the first of the two phosphoric acids $(R''O)_2P(O)OH$ and $(R'''O)_2P(O)OH$ corresponding to the asymmetric pyrophosphate shown above is stronger, the attack at the phosphorus of the second acid will be more effective:

$$(R''O)_2\underset{\underset{O}{\parallel}}{P}-O-\underset{\underset{O}{\parallel}}{P}(OR''')_2 + \bar{N}u \longrightarrow (R''O)_2\underset{\underset{O}{\parallel}}{P}-O^- + Nu-\underset{\underset{O}{\parallel}}{P}(OR''')_2$$

Consequently, the nucleophile receives the residue of the weaker phosphoric acid.

Acyl phosphates are most likely to exhibit properties of both carboxylic anhydrides and pyrophosphates. Hence, it becomes clear that they can serve as both acylating and phosphorylating agents. The path of the reaction is determined by many factors, including the nature of the nucleophilic agent, the degree of screening of the reactive sites (carbonyl carbon and phosphorus of the phosphate group), the type of solvent, the thermodynamic stability of the reaction products, and some others. More often than not, reactions between acyl phosphates and nucleophilic agents in the absence of enzymes result in transfer of the acyl group to the nucleophile (yielding the corresponding carboxylic acid derivative).

Acyl phosphates of the general formula

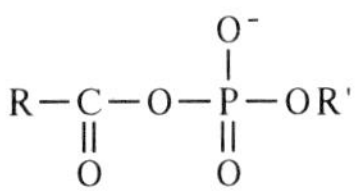

$$\mathrm{R{-}\underset{\underset{O}{\|}}{C}{-}O{-}\overset{\overset{O^-}{|}}{\underset{\underset{O}{\|}}{P}}{-}OR'}$$

act as acylating agents, provided the carboxylic acid is stronger than the correspoding phosphoric one (pK of the secondary hydroxyl of the phosphate group, involved in anhydride bonding, is in the neighborhood of 6.0, whereas pK of the carboxyl group in carboxylic acids usually ranges from 2 to 4). The path of the reaction seems to be determined by the greater electrophilicity of the carboxyl carbon and its accessibility as a result of two-dimensional structure of the carboxyl group.

The interaction of acyl phosphates with amines usually leads to acylation of the latter. The following reaction between acyl phosphates and hydroxylamine is especially well known:

$$\mathrm{R{-}\underset{\underset{O}{\|}}{C}{-}O{-}\overset{\overset{O^-}{|}}{\underset{\underset{O}{\|}}{P}}{-}OR' + NH_2OH \longrightarrow R{-}\underset{\underset{O}{\|}}{C}{-}NHOH + HO{-}\overset{\overset{O^-}{|}}{\underset{\underset{O}{\|}}{P}}{-}OR'}$$

The reaction proceeds with extreme ease. At pH 5–7, for example, the quantitative formation of hydroxamic acid takes ten minutes; the latter gives colored complexes with iron salts, which are useful in quantitative estimation of the acyl phosphate content in the reaction mixture. The hydroxamic reaction is known to be the standard of reactivity of carboxylic acid derivatives.

The interaction of acetyl phosphate with tertiary amines (as opposed to primary and secondary ones) can lead not only to acetylation but also phosphorylation of the amine:

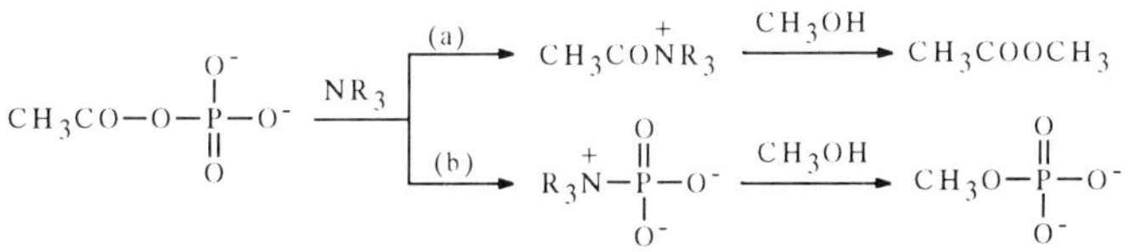

$$\mathrm{CH_3CO{-}O{-}\overset{\overset{O^-}{|}}{\underset{\underset{O}{\|}}{P}}{-}O^-} \xrightarrow{NR_3} \begin{cases} \xrightarrow{(a)} \mathrm{CH_3CO\overset{+}{N}R_3} \xrightarrow{CH_3OH} \mathrm{CH_3COOCH_3} \\ \xrightarrow{(b)} \mathrm{R_3\overset{+}{N}{-}\overset{\overset{O}{\|}}{\underset{\underset{O^-}{|}}{P}}{-}O^-} \xrightarrow{CH_3OH} \mathrm{CH_3O{-}\overset{\overset{O}{\|}}{\underset{\underset{O^-}{|}}{P}}{-}O^-} \end{cases}$$

Depending on the structure of the tertiary amine, the reaction may proceed predominantly along each of the indicated paths (*a*) and (*b*). For instance, the acylation of imidazole and trimethylamine proceeds along either path with equal probability, while in the case of pyridine phosphorylation is the predominant path [path (*b*)].

In contrast to acetyl phosphate, the reactions of acyl phosphates with primary, secondary and tertiary amines yield acylated amines:

$$\mathrm{R{-}\underset{\underset{O}{\|}}{C}{-}O{-}\overset{\overset{O^-}{|}}{\underset{\underset{O}{\|}}{P}}{-}OR'} \xrightarrow{\;>N-\;} \mathrm{R{-}\underset{\underset{O}{\|}}{C}{-}N<}$$

The rates of such reactions are usually an order of magnitude higher than in the case of simple acetyl phosphate.

The mild conditions of such reactions and the high yields of acylated amines form the basis of one of the peptide synthesis methods:

$$\mathrm{Cbz\text{-}NH\underset{\underset{R}{|}}{C}H\underset{\underset{O}{\|}}{C}{-}O{-}\underset{\underset{O}{\|}}{P}(OR')_2} + \mathrm{NH_2\underset{\underset{R''}{|}}{C}HCOOH} \longrightarrow \mathrm{Cbz\text{-}NH\underset{\underset{R}{|}}{C}HCO{-}NH\underset{\underset{R''}{|}}{C}HCOOH}$$

When the nucleophilic attack of the carbonyl carbon in an acyl phosphate is hindered, phosphorylation becomes the predominant reaction path, for example:

$$\mathrm{C_6Cl_5{-}\underset{\underset{O}{\|}}{C}{-}O{-}\underset{\underset{O}{\|}}{P}(OC_2H_5)_2} \xrightarrow[\mathrm{-C_6Cl_5COO^-}]{\mathrm{NH_2C_6H_5}} \mathrm{C_6H_5NH{-}\underset{\underset{O}{\|}}{P}(OC_2H_5)_2}$$

Let us now consider acyl phosphates of more complex structure, in which the function of the phosphate moiety is performed by nucleotides.

Anhydrides of Nucleotides and Carboxylic Acids. As has already been mentioned, such anhydrides can be synthesized and isolated individually. Some nucleotide anhydrides belonging to the class of acyl phosphates may be stored for long periods of time. Among the most labile nucleotide derivatives of this type are anhydrides with acetic and other fatty acids as well as with amino acids.

$$\mathrm{R\underset{\underset{O}{\|}}{C}{-}O{-}\overset{\overset{O^-}{|}}{\underset{\underset{O}{\|}}{P}}{-}OCH_2\text{-}(ribose/deoxyribose, B; HO, OH(H))}$$

R - alkyl or alkylamino group

These compounds are easily hydrolysed as a rule and react with hydroxylamine and amines. All these reactions lead to transfer of the acyl group to the nucleophile. The reaction with hydroxylamine (hydroxamic reaction) is used, just as in the case of simple acyl phosphates (see above), for quantitative determination of such anhydrides. For instance:

$$RC(O)\text{—O—}P(O)(O^-)\text{—OCH}_2\text{-nucleoside(B)} \xrightarrow{NH_2OH} RC(O)NHOH + {}^-O\text{—}P(O)(O^-)\text{—OCH}_2\text{-nucleoside(B)}$$

R - alkyl or alkylamino group

The reaction with phosphoric acid in pyridine proceeds in a similar manner. For example:

$$CH_3C(O)\text{—O—}P(O)(O^-)\text{—OCH}_2\text{-nucleoside(T)} \xrightarrow{HPO_4^{2-}} CH_3C(O)\text{—O—}P(O)(O^-)\text{—OH} + {}^-O\text{—}P(O)(O^-)\text{—OCH}_2\text{-nucleoside(T)}$$

The resulting acetyl phosphate reacts further with the next phosphoric acid molecule to yield an inorganic pyrophosphate:

$$CH_3CO\text{—O—}P(O)(O^-)\text{—O}^- + {}^-O\text{—}P(O)(O^-)\text{—OH} \longrightarrow {}^-O\text{—}P(O)(OH)\text{—O—}P(O)(OH)\text{—O}^- + CH_3COO^-$$

The reverse reaction in pyridine is virtually nil, and the equilibrium shifts toward formation of pyro- and polyphosphates. The partial conversion of nucleotides into symmetric pyrophosphates is usually observed when they are exposed to carboxylic anhydrides in order to incorporate protective acyl groups (at the hydroxyl groups of pentose or into the nucleus of the heterocycle).

Special properties are displayed by anhydrides of ribonucleoside 3′(2′)-phosphates and carboxylic acids, which are easily converted, due to proximity of the 2′(3′)-hydroxyl group of ribose, into the corresponding nucleoside 2′,3′-cyclic phosphates. Thus, unlike anhydrides of ribonucleoside 5′-phosphates with carboxylic acids, their 3′-isomers are essentially phosphorylating rather than acylating agents. A similar change in properties from 5′- to 3′-derivatives seems to be caused by that the phosphate group in the latter is sterically close to the hydroxyl one, which results in some kind of intramolecular catalysis. Consequently, it may be assumed that anhydrides of nucleoside 5′-phosphates with carboxylic acids can act as phosphorylating agents if they are formed at intermediate steps of biosynthetic processes or if an enzyme ensures approach of the nucleophilic agent to their phosphate, and not carboxyl, group.

There is no reason why the possibility of such reactions should be ruled out, although they are yet to be observed.

The most stable of the known anhydrides of nucleotides and carboxylic acids are mesitylenecarboxylic acid derivatives. They can be isolated individually and stored for a long period ot time. Studies into the reactions of these anhydrides with nucleophilic agents have shown that their acylating function is inhibited. In reactions with nucleophilic agents, anhydrides of nucleotides and mesitylenecarboxylic acid behave like phosphorylating agents.

If, however, the reaction with amines proceeds rather easily in aqueous solutions, in the case of weaker nucleophiles, such as mercaptans and alcohols, better results are obtained in organic solvents.

Thus, mixed anhydrides of nucleotides with mesitylenecarboxylic acid were the first representatives of "active" nucleotide derivatives that could be isolated individually and used in further reactions. What makes this class of acyl nucleotides promising is that nucleotide or polynucleotides can be linked covalently with their aid to the nucleophilic sites of various substances, the appropriate reaction being conducted in aqueous media (anhydrides of nucleotides with mesitylenecarboxylic acid are not hydrolysed at neutral pH values and at room temperature). It may be assumed, in particular, that these anhydrides can be used as inhibitors of nucleotide metabolism enzymes at least when the active site of the enzyme is occupied by a sufficiently active nucleophilic group (imidazoles, ε-amino group of lysine, SH group).

NH2—P(O)(O-)—OCH2 B HO OH(H)
RNH—P(O)(O-)—OCH2 B HO OH(H)
NH3
RNH2
CH3 CH3 CH3 C(O)—O—P(O)(O-)—OCH2 B HO OH(H)
RSH
ROH
N NH
N N—P(O)(O-)—OCH2 B HO OH(H)
RS—P(O)(O-)—OCH2 B HO OH(H)
RO—P(O)(O-)—OCH2 B HO OH(H)

4.6.4 Amides of Nucleotides

Amides of nucleotides – derivatives of aliphatic and aromatic amines – usually are rather stable. The presence of an amide moiety in the nucleotide molecule naturally gives rise to some specific properties standing distinctly apart from those of the corresponding alkyl esters. In view of the important role of amides of nucleotides in synthesis, along with relatively scant information about them, discussed in what follows are simple amidophosphates whose properties are in many ways similar to those of nucleotide amides.

General Concepts About Amidophosphates. The reactivity of amidophosphates is strongly dependent on the form in which they react: that of a conjugated acid (38), neutral molecule (39), or anion (40):

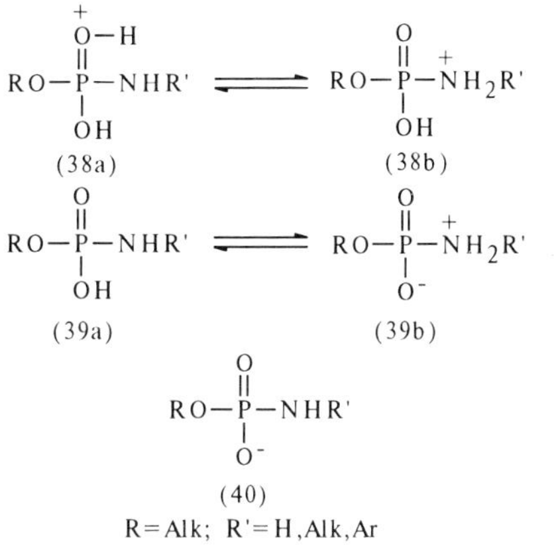

$$\underset{(38a)}{RO-\overset{\overset{+}{O}-H}{\overset{\|}{P}}(OH)-NHR'} \rightleftharpoons \underset{(38b)}{RO-\overset{O}{\overset{\|}{P}}(OH)-\overset{+}{N}H_2R'}$$

$$\underset{(39a)}{RO-\overset{O}{\overset{\|}{P}}(OH)-NHR'} \rightleftharpoons \underset{(39b)}{RO-\overset{O}{\overset{\|}{P}}(O^-)-\overset{+}{N}H_2R'}$$

$$\underset{(40)}{RO-\overset{O}{\overset{\|}{P}}(O^-)-NHR'}$$

$$R = Alk;\ R' = H, Alk, Ar$$

The conjugated acid (38) seems to represent a mixture of two forms differing in the site of proton attachment: form (38a) is marked by protonation of the phosphoryl oxygen, while form (38b) is marked by that of the amide nitrogen. The proton in the neutral form (39) may also occur either at the oxygen (39a) or at the nitrogen (39b). The anionic form of the amidomonophosphate carries a negative charge concentrated primarily at the oxygen atom.

The hydrolysis rate of amidophosphates is to a great extent dependent on the pH of the medium and structure of the amide moiety. Aliphatic amine derivatives are highly unstable in acid media (the hydrolysis rate in this case varying with the acid concentration), hydrolysed slowly at pH 4–6 and virtually stable at alkaline pH values. Hence, the conjugated acid (38b) is the most reactive form. It is quite possible that hydrolysis of the neutral form is determined by the presence of molecules protonated at the nitrogen (39b). Amidophosphates that are derivatives of aromatic amines are hydrolysed in acid media at a slower rate than aliphatic amine derivatives, which seems to stem from the difficulties in protonation of the amide nitrogen whose lone-pair electrons are involved in conjugation with the aromatic nucleus. The relationship between the hydrolysis rate and capacity of the amide nitrogen for protonation is supported by the results of hydrolysis of *N*-acyl amides.

Such compounds are hydrolysed more easily at pH 4, whereas at lower and higher pH values the hydrolysis rate drops sharply, which is strongly reminiscent of the behavior of monophosphates. In this connection, the following hydrolysis mechanism (through an intermediate hydration step) has been proposed for them:

$$\text{(cyclic intermediate: } O_2P(O^-)\text{–NHAc, } H_2O\text{)} \xrightarrow{\text{fast}} \left[O_2\overset{-}{P}{=}O \right] + H_2O + NH_2Ac; \quad \left[PO_3^- \right] \xrightarrow{H_2O} H_2PO_4^-$$

Proceeding from the above data and similar evidence concerning the behavior of unsubstituted amidophosphate, the following mechanisms of hydrolysis and solvolysis of amidophosphates have been proposed.

The *bimolecular (associative) mechanism* is similar to the S_N2 mechanism of nucleophilic substitution for other phosphoric acid derivatives containing a suitable leaving group. In accordance with this mechanism, the stability of the monoanionic form (40) toward nucleophilic substitution is due to the strong $p\pi$–$d\pi$ conjugation of the phosphorus and nitrogen atoms, which leads to occupation of the 3*d* orbitals of the phosphorus and hinders the formation of an intermediate compound of the trigonal bipyramid type during the nucleophilic attack. Moreover, departure of the amide group in such a form is disadvantageous. In the case of the protonated form (39b), the situation is altogether different because the 3*d* orbitals become vacant as a result of the upset $p\pi$–$d\pi$ conjugation and the protonated amide group is apparently ready to "depart" in the form of amine. Still more reactive for the same reason is the form (38b) of conjugated acid, as the conjugation of the phosphorus and oxygen atoms is less pronounced because of no negative charge being present at the latter.

As regards hydrolysis and solvolysis of the neutral form of amidomonophosphates, the bimolecular mechanism is most likely. For example, methylphosphate cyclohexylamide is converted into methylethyl phosphate by less than ten per cent during solvolysis in 50% ethanol.

$$CH_3O{-}P(O)(O^-){-}\overset{+}{N}H_2C_6H_{11} \xrightarrow[-C_6H_{11}NH_2]{50\%\ C_2H_5OH} CH_3O{-}P(O)(OH){-}OC_2H_5 + CH_3O{-}P(O)(OH){-}OH$$

If the nucleophilic substitution for this amide had been based on the S_N1 type mechanism (if only the substrate molecule were involved in the rate-determining step), no selectivity with respect to one of the nucleophilic agents would have been observed because the rates of the reaction with a molecule of water and alcohol would have been the same.

The *monomolecular (dissociative) mechanism* is analogous to the S_N1 mechanism of nucleophilic substitution at the phosphorus atom, which boils down to elimination of the labile metaphosphate, promoted, just as in the case of the S_N2 mechanism, by protonation of the amide nitrogen:

$$RO-P(=O)(O^-)-\overset{+}{N}H_2R' \longrightarrow [RO-PO_2] + R'NH_2$$
$$[RO-PO_2] \xrightarrow{X} RO-P(=O)(O^-)-X$$

The distinguishing feature of this mechanism, as opposed to the bimolecular one, is lack of selectivity toward the nucleophile.

Such a mechanism has been proposed for the above-described reaction of solvolysis of methylphosphate cyclohexylamide in 50% ethanol, but in the presence of pyridine, since the yield of methylethyl phosphate in this case practically corresponds to the ethanol content in the reaction mixture.

$$CH_3O-P(=O)(O^-)-\overset{+}{N}H_2C_6H_{11} \xrightarrow[-C_6H_{11}NH_2]{C_5H_5N} [CH_3O-P(=O)(O^-)-\overset{+}{N}C_5H_5] \xrightarrow[-C_5H_5N]{}$$
$$\longrightarrow [CH_3O-PO_2] \xrightarrow{C_2H_5OH+H_2O} CH_3O-P(=O)(OH)-OC_2H_5 + CH_3O-P(=O)(OH)-OH$$

The intermediate compound in this case seems to be methylphosphorylpyridinium which is then converted into methylmetaphosphate. A monomolecular mechanism has also been proposed for solvolysis of the monoanionic form of *N*-arylamidophosphates.

Intermediate Mechanism (Ia). Some reactions are believed to involve a transition state intermediate between trigonal bipyramid and planar metaphosphate. Such a transition state is characterized by a lengthening of the P–N bond and a shortening of the bond which links the phosphorus atom with the incoming nucleophile. Transformations of this kind possess features of both mono- and bimolecular reactions (as regards the effect produced by the nucleophile). A corresponding transition state is believed to be involved in solvolysis of the monoanionic form of the unsubstituted amidophosphate (40).

This hypothesis is based on the fact that hydrolysis of the amidophosphate is 10^3 times faster, as compared to that of *para*-nitrophenyl phosphate, although if the reaction with water were based on an S_N2 type mechanism, the unsubstituted amidophosphate would be expected to react at a slower rate than the aromatic ester. On the other hand, if the reaction of solvolysis in aqueous alcohol proceeded through formation of a metaphosphate, then alkyl phosphate and phosphoric acid would be yielded in the same ratios as during solvolysis of *para*-nitrophenyl phosphate, whose monomolecular mechanism is universally accepted. However, the amidophosphate yields twice as much alkyl phosphate as *para*-nitrophenyl phosphate. These findings suggest both mono- and bimolecular mechanisms of solvolysis of the unsubstituted amidophosphate monoanion.

In addition to hydrolysis and alcoholysis, amidophosphates also enter into other reactions of nucleophilic substitution for the amino group. Most typical in this respect are reactions with such nucleophiles as phosphoric acid and its esters, carboxylic acids, and amines. During hydrolysis of benzyl amidophosphate, the anion of benzylphosphoric acid acts as an amino group substituting nucleophile. In the presence of two moles of water, this reaction yields an ammonium salt of symmetrical dibenzyl pyrophosphate:

$$H_2O + C_6H_5CH_2O-P(=O)(O^-)-\overset{+}{N}H_3 \longrightarrow C_6H_5CH_2O-P(=O)(HO)(O^- \overset{+}{N}H_4)$$

$$C_6H_5CH_2O-P(=O)(HO)(O^- \overset{+}{N}H_4) + C_6H_5CH_2O-P(=O)(^-O)(\overset{+}{N}H_3) \xrightarrow{H_2O} C_6H_5CH_2O-P(=O)(H_4\overset{+}{N}\,\overset{-}{O})-O-P(=O)(OCH_2C_6H_5)(O^- \overset{+}{N}H_4)$$

In the absence of water, monoamide of dibenzyl pyrophosphate is formed in inert solvents. In this case, an amidophosphate molecule protonates the amino group of the other molecule to give a nucleophile (amidophosphate anion) and a substrate containing a sufficiently electrophilic phosphorus atom and an appropriate leaving group (amido-phosphate protonated at the nitrogen atom):

$$2\ C_6H_5CH_2O-P(=O)(H_2N)(OH) \longrightarrow C_6H_5CH_2O-P(=O)(H_2N)(O^-) + C_6H_5CH_2O-P(=O)(HO)(\overset{+}{N}H_3) \longrightarrow$$

$$\longrightarrow C_6H_5CH_2O-P(=O)(H_2N)-O-P(=O)(OCH_2C_6H_5)(O^{-}\,{}^{+}NH_4)$$

This reaction was studied on many substances and now underlies the preparative method of synthesis of nucleoside 5′-pyrophosphates and related compounds.

Amidomonophosphates also react with carboxylic acids to yield mixed anhydrides. For instance, benzyl amidophosphate dissolved in glacial acetic acid gives acetylbenzyl phosphate. Apparently, in this case, too, protonation of benzylamidophosphate takes place during the first step (see equation above). The protonated form of amidophosphate (at the nitrogen atom) is then nucleophilically attacked not by the like anion, as in the above-described case, but by excess acetic acid.

$$(C_6H_5CH_2O)(HO)P(=O)\overset{+}{N}H_3 + CH_3COOH \longrightarrow (C_6H_5CH_2O)(H_4N^{+}\,{}^{-}O)P(=O)OCOCH_3$$

Similarly, the function of nucleophiles can be performed by amines. In this case, one amino group at the phosphorus atom is replaced by another (such reactions are referred to as transamination). Just as the above-considered reactions of substitution of the amino group in alkylamidomonophosphates, transamination seems to involve the protonated form of amidophosphates. A case in point is reactions involving methylphosphate cyclohexylamide.

$$CH_3O-\overset{O}{\overset{\|}{\underset{OH}{\underset{|}{P}}}}-NHC_6H_{11} \longrightarrow CH_3O-\overset{O}{\overset{\|}{\underset{O^-}{\underset{|}{P}}}}-\overset{+}{N}H_2C_6H_{11} \xrightarrow[-NH_2C_6H_{11}]{NH_2R} CH_3O-\overset{O}{\overset{\|}{\underset{OH}{\underset{|}{P}}}}-NHR$$

$$R = \alpha\text{-}C_{10}H_7,\ OCH_3,\ CH_2CF_3$$

Notably, used as nucleophiles in this case are sufficiently weak amines which do not produce stable ammonium forms in the course of the reaction and, therefore, retain their nucleophilic properties.

The most reactive amides of phosphoric acid and its esters are imidazolides. Imidazolides of mono- and dialkyl phosphates react with phosphoric acid and its esters to give pyrophosphates with a quantitative yield. They also react readily with carboxylic acids, amines and alcohols:

$$RO-P(O)(OH)-NHR' \xleftarrow{R'NH_2} RO-P(O)(OH)-N(\text{imidazolyl}) \xrightarrow{R'OH} RO-P(O)(OH)-OR'$$

$$RO-P(O)(OH)-O-C(O)-R' \xleftarrow{R'COOH} RO-P(O)(OH)-N(\text{imidazolyl}) \xrightarrow{R'OP(O)(OH)_2} RO-P(O)(OH)-O-P(O)(OH)-OR'$$

R and R' = Alk

The high reactivity of phosphoimidazoles with respect to nucleophiles is most likely to stem from the ease of intramolecular protonation of the second nitrogen in the imidazole, resulting in an active form in which the phosphorus is linked with the positively charged nitrogen:

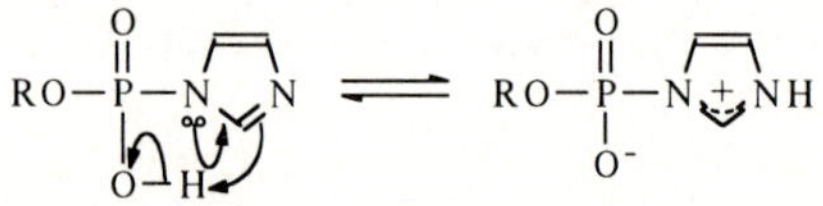

$$RO-\overset{O}{\overset{\|}{\underset{O^-\,H}{\underset{|}{P}}}}-N\langle\text{imidazole}\rangle N \rightleftharpoons RO-\overset{O}{\overset{\|}{\underset{O^-}{\underset{|}{P}}}}-N^{+}\langle\text{imidazole}\rangle NH$$

The ease of formation of an active ammonium form in the case of imidazolides seems to be due to the protonation of the nitrogen not linked with the phosphorus and having its basic properties unaffected by the proximity of the phosphate group.

The evidence discussed in this section suggests that amidomonophosphates are stable and rather reactive compounds.

Amides of Nucleotides. Amides of nucleotides are essentially amidomonophosphates as well. It would be natural to assume that many properties of these compounds are more or less very similar. As will be shown here, this is true in some respects, and the many regularities established for amidophosphates also apply to the corresponding nucleotide derivatives. It is for this reason that comparison of the reactivities of these two classes of compounds makes it possible to put knowledge about the properties of the most thoroughly studied simple derivatives to practical use in the chemistry of nucleotide amides.

Insofar as their stability toward hydrolysis is concerned, nucleotide amides resemble closely simple amides of alkyl phosphates: they are stable in aqueous solutions at neutral and alkaline pH values and lend themselves readily to hydrolysis in acid media. As can be inferred from comparative data on the hydrolysis rate of amide derivatives of uridine 5′-phosphate, the rate of the reaction (0.05 N HCl, 37°C)

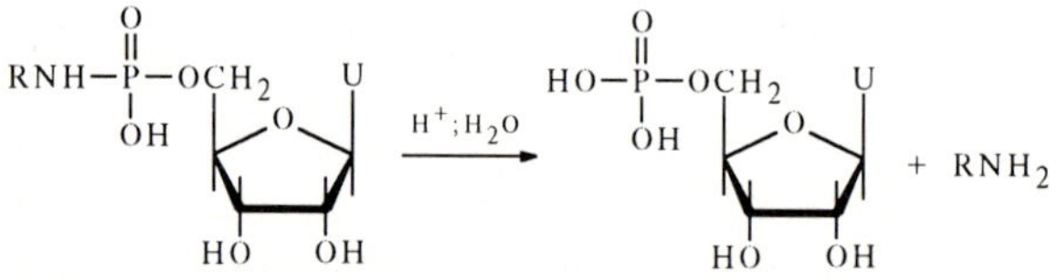

$$RNH-\overset{O}{\overset{\|}{\underset{OH}{\underset{|}{P}}}}-OCH_2\text{(uridine)} \xrightarrow{H^+;\,H_2O} HO-\overset{O}{\overset{\|}{\underset{OH}{\underset{|}{P}}}}-OCH_2\text{(uridine)} + RNH_2$$

depends on the corresponding amine species, the relationship between the two being rather complex:

Initial amide for amidonucleotide	pK_a of amine	Half-life, min	$k \cdot 10^2$ min^{-1}
para-Nitroaniline	1.00	160	0.43
meta-Chloroaniline	3.52	118	0.59
Aniline	4.57	70	0.99
para-Anisidine	5.29	45	1.53
2,2,2-Trifluoroethylamine	5.70	<5	–
Morpholine	8.36	15	4.78
β-Phenylisopropylamine	10.00	46	1.50
n-Butylamine	10.43	84	0.83
Cyclohexylamine	10.64	144	0.48

The highest stability toward hydrolysis is displayed by amides – amine derivatives exhibiting weak as well as strong basic properties. Since, as has already been mentioned, hydrolysis of amidophosphates, similarly to reactions of nucleophilic amino group (or substituted amino group) substitution, begins with protonation of the amide nitrogen, it is natural to assume that the rate of the process will depend precisely on this reaction step. This assumption may explain the above regularity in view of the fact that in both extreme cases (when the corresponding amine is either a very weak or a very strong base) the amide nitrogen becomes electron-deficient due to the electron-acceptor influence of the radical at the nitrogen, in the case of amides derived from amines with extremely weak basic properties, or due to the substantial increase in the p_π–d_π conjugation between the nitrogen and phosphorus atoms, in the case of amides containing strong electron-donor substituents at the same position (which are, consequently, derivatives of a strong base).

Amidonucleotides, just as simple amidophosphates, react rather easily with phosphate anions. Uridine 5′-amidophosphate, for example, yields the corresponding pyrophosphate while reacting with phosphoric acid and its esters in pyridine.

The reactions apparently involve the same mechanism as in the case of alkyl amidophosphates (see above).

Various amidonucleotides differing in basicity of the starting amine have been used in reactions with phosphate anions: anisidides, morpholides, piperidides, and cyclohexylamides. Nucleotide morpholides are most commonly used in the synthesis of nucleoside 5′-pyrophosphates and analogous compounds. One of the reasons for the reactivity of morpholides, just as in the case of acid hydrolysis, may be the ease of emergence of the activated protonated form. Reactions of morpholides of uridine 5′-, guanosine 5′-, adenosine 5′- and cytidine 5′-monophosphates with phosphoric and pyrophosphoric acids as well as phosphates have given various pyro- and triphosphates belonging to the class of nucleotide coenzymes with yields ranging from 50 to 70 per cent. Represented below by way of example is synthesis of coenzyme A from the morpholide of 2′,3′-cyclophosphoadenosine 5′-phosphate and pantothenoyl phosphate.

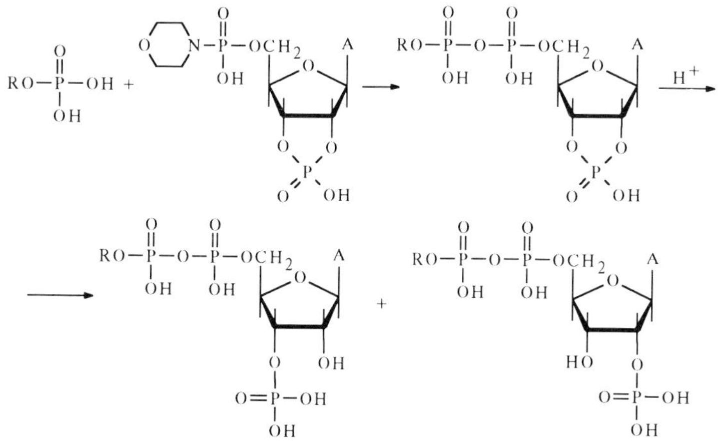

(41) coenzyme A

(42)

R = $CH_2C(CH_3)_2CH(OH)CONHCH_2CH_2CONHCH_2CH_2SH$

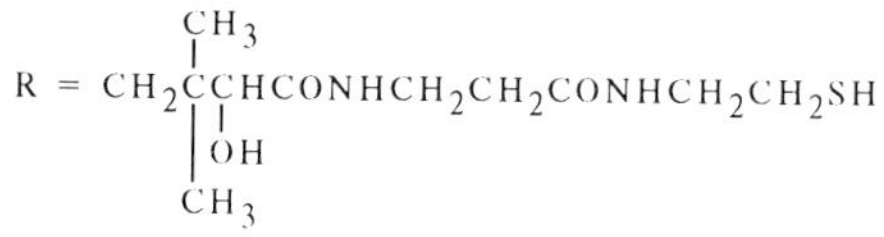

The overall yield of coenzyme (41) and isocoenzyme A (42), which are formed in equal amounts, is 65 per cent in terms of the morpholide entered into the reaction. The individually isolated coenzyme A has turned out to be identical with its natural counterpart.

In spite of the spectacular advances in employing amidonucleotides in the synthesis of nucleotide 5′-pyrophosphates and similar compounds, attempts to use them in reactions with other nucleophilic agents have not been successful. The same is true as regards making alcohols to react with amidonucleotides to yield nucleotide esters. Only when the reaction between the morpholide of adenosine 5′-phosphate with methanol taken in a markedly (50-fold) surplus amount is conducted in pyridine in the presence of an equimolar quantity of dry HCl can a low yield of adenosine 5′-methylphosphate be achieved:

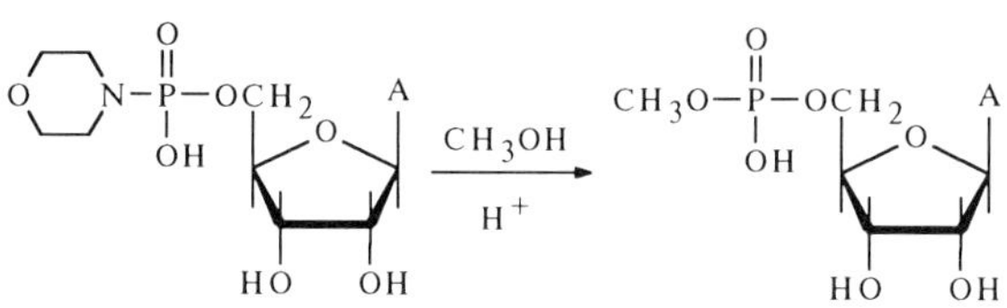

$\xrightarrow[H^+]{CH_3OH}$

Attempts to use other alcohols in this reaction have failed. The most promising starting compounds for the synthesis of nucleotide esters are imidazolides of the latter. They have been derived from nucleotides and *N,N′*-carbonyl diimidazole:

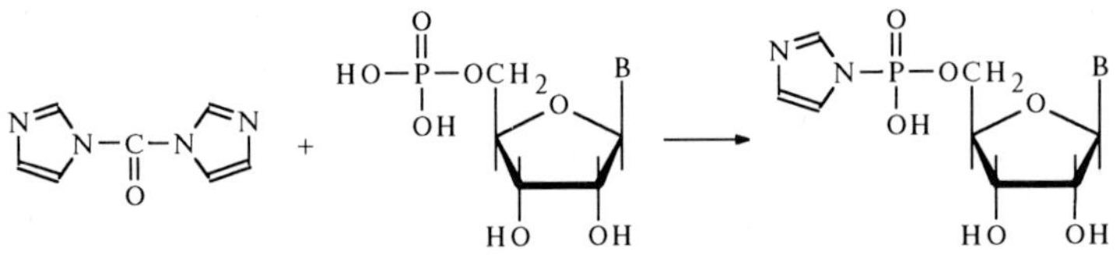

As in the case of simple phosphoimidazoles (see above), imidazolide of adenosine 5′-phosphate reacts rather easily with primary alcohols and *para*-nitrophenol. The reaction with tertiary alcohols proceeds with greater difficulty.

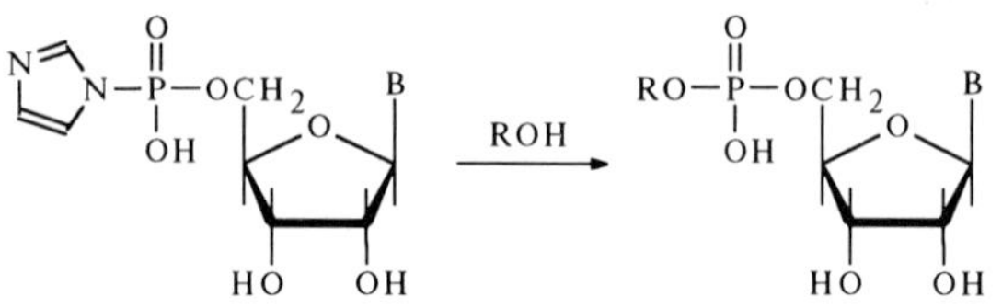

4.7 Hydrolysis of *N*-Glycosidic Bonds

The *N*-glycosidic bonds in nucleotides, as well as nucleosides, are rather stable at neutral and alkaline pH values, but prone to acid hydrolysis. The *N*-glycosidic bonds in purine nucleotides are hydrolysed with much greater ease than in pyrimidine ones.

The mechanism of acid hydrolysis of *N*-glycosidic bonds and the influence of various factors on their stability were discussed at length in Chapter 2. The described regularities apply to nucleotides as well. This section deals only with the effect of the phosphate group on the stability of *N*-glycosidic bonds. The rates of acid hydrolysis of *N*-glycosidic bonds in purine nucleotides and the corresponding nucleosides differ but insignificantly, especially in the series of adenine derivatives (pH 1.0, 37 °C):

Compound	k_1, s^{-1}
Adenosine 5′-phosphate	$3.8 \cdot 10^{-7}$
Adenosine 3′(2′)-phosphate	$3.3 \cdot 10^{-7}$
Guanosine 3′(2′)-phosphate	$6.6 \cdot 10^{-7}$
Deoxyadenosine 5′-phosphate	$3.1 \cdot 10^{-4}$

Compound (Continued)	k_1, s^{-1}
Deoxyguanosine 5′-phosphate	$1.8 \cdot 10^{-4}$
Deoxycytidine 5′-phosphate	$2.0 \cdot 10^{-8}$
Deoxythymidine 5′-phosphate	$2.0 \cdot 10^{-8}$
Adenosine	$3.6 \cdot 10^{-7}$
Guanosine	$9.36 \cdot 10^{-7}$
Cytidine	$1.0 \cdot 10^{-9}$
Uridine	$1.0 \cdot 10^{-9}$
Deoxyadenosine	$4.3 \cdot 10^{-4}$
Deoxyguanosine	$8.3 \cdot 10^{-4}$
Deoxycytidine	$1.1 \cdot 10^{-7}$

Similarly, minor differences in hydrolysis rate are observed in the case of adenosine 3′(2′)-phosphate and guanosine 3′(2′)-phosphate as well as the corresponding nucleosides in 6 N hydrochloric acid at 100 °C.

As regards pyrimidine deoxyribo derivatives, incorporation of the phosphate group into a nucleoside renders the glycosidic bond more stable. This stability increases from mono- to diphosphates (0.4 N H_2SO_4, 100 °C):

Compound	k_1, s^{-1}
Deoxycytidine	3.7
Deoxycytidine 5′-phosphate	0.6
Deoxycytidine 3′,5′-diphosphate	0.28
Deoxythymidine	0.34
Deoxythymidine 5′-phosphate	0.23
Deoxythymidine 3′,5′-diphosphate	0,12

This trend seems to stem from the fact that undissociated (under conditions of acid hydrolysis) phosphate groups are strong electron-acceptor substituents because their phosphorus atom carries a partial positive charge. The effect of the phosphate group on the stability of the glycosidic bond is comparable with that of the toluenesulfo or 2,4-dinitrobenzoyl group.

An opposite effect on the glycosidic bond stability in uridine derivatives is exerted by the 3′,5′-cyclic phosphate group. Uridine 3′,5′-cyclic phosphate loses its base very rapidly when treated with 1 N hydrochloric acid (the half-life at 100 °C being only 8 min). Uridine 5′- and 3′-phosphates are rather stable under such conditions. Similar lability is displayed by other pyrimidine nucleoside 3′,5′-cyclic phosphates as well.

In contrast, the *N*-glycosidic bonds in purine ribonucleoside 3′,5′-cyclic phosphates are hydrolysed at a much slower rate than in the corresponding monophosphates.

The marked labilizytion of the *N*-glycosidic bonds in pyrimidine nucleoside 3′,5′-cyclic phosphates and, on the other hand, their increasing stability in analogous purine derivatives (as compared to ribonucleoside 5′- and 3′-phosphates) are most likely due to the special conformation of pentose in their molecules.

References

1. N. K. Kochetkov and E. I. Budovskii, *Organic Chemistry of Nucleic Acids.* Plenum Press, London–N.Y. (1972).
2. R. F. Hudson, *Structure and Mechanism in Organo-Phosphorus Chemistry.* Academic Press, London–N.Y. (1965).
3. A. Kirby and S. Warren, *Organic Chemistry of Phosphorus* Elsevier, Amsterdam (1967).
4. H. G. Khorona, *Some Recent Developments in the Chemistry of Phosphate Esters of Biological Interest.* John Wiley & Sons, Inc., N.Y.–London (1961).
5. A. M. Michelson, *The Chemistry of Nucleosides and Nucleotides.* Academic Press, N.Y.–London (1963).
6. G. R. Cox and O. B. Ramsey, *Mechanism of Nucleophilic Substitution in Phosphate Esters.* Chem. Revs., pp, 317–352 (1964).
7. H. Goldwhite, *Introduction to Phosphorus Chemistry.* Cambridge University Press, Cambridge–London (1981).
8. G. Lowe, *Chiral [^{16}O, ^{17}O, ^{18}O] Phosphate Esters.* Acc. Chem. Res., Vol. 16, pp. 244–251 (1983).
9. T. Ueda and J. J. Fox, In: *Advances in Carbohydrate Chemistry.* Vol. 22, M. L. Wolfrom and R. S. Tipson (Eds.), Academic Press, N.Y.–London, pp. 307–419 (1967).
10. R. Shapiro, in: *Progress in Nucleic Acid Res. and Mol. Biol.* Vol. 8, J. N. Davidson and W. E. Cohn (Eds.), Academic Press, London, pp. 73–112 (1968).
11. *Nucleic Acids in Chemistry and Biology.* G. M. Blackburn and M. J. Gait (Eds.), IRL Press, Oxford–N.Y.–Tokyo (1990).
12. S. H. Westheimer, *Pseudo-rotation in the Hydrolysis of Phosphate Esters. Acc. Chem. Res.,* Vol. 1, pp. 70–78 (1968).

5 Primary Structure of Nucleic Acids

5.1 Introduction

Nucleic acids are high-molecular weight compounds – polycondensation products of nucleotides. The monomer units of a nucleic acid molecule usually are either only ribonucleotides or only deoxyribonucleotides. Depending on the type of the constituent monomer units, distinction is made between ribonucleic (RNA with ribonucleotides as the monomer units) and deoxyribonucleic (DNA with deoxyribonucleotides as the monomer units) acids.

Both polynucleotide types, DNA and RNA, are found in all living cellular organisms, except for viruses which contain polymers of one type only. The structural differences between the monomer units in polynucleotides of the two types account for their different three-dimensional structure (or macrostructure). As a consequence, the chemical properties of both polymers differ markedly.

DNA and RNA are also dissimilar biologically. They perform different functions and, consequently, occupy different positions in the cell. It is extremely important to know the structure of nucleic acids if the mechanism of their functioning is to be understood. A distriction is made between primary structure and macromolecular (dimensional) structure. By primary structure is meant the sequence of monomer units linked by covalent bonds into a continuous polynucleotide chain. In determining the primary structure of nucleic acids one is actually trying to establish the type of the intermonomer linkage and the sequence of monomer units in the polymer chain. The macromolecular structure of a nucleic acid is defined as the spatial arrangement of polynucleotide chains in the molecule.

5.2 Major Types of Nucleic Acids, Their Isolation and Characteristics

5.2.1 DNA

DNA is an essential component of every living cell as well as some viruses and bacteriophages. Distinction is made between cellular and viral (phage) DNAs.

Cellular DNAs. The DNAs of prokaryotic and eukaryotic cells differ in both molecular weight and location in the cell. In simply structured prokaryotic cells with a single membrane (bacteria[1]), cyanobacteria and some other organisms) and a single chromosome, virtually all of DNA is present in the form of just one macromolecule with a molecular weight exceeding $2 \cdot 10^9$. In bacteria, the DNA molecule (accounting for about 1 wt % of the cell) occupies the nuclear zone and does not seem to be bound to proteins. The cytoplasm of bacteria has been found to contain extrachromosomal DNA molecules. Such molecules are known as *plasmids* (episomes). The molecular weight of plasmid DNA ranges from $1 \cdot 10^6$ to $75 \cdot 10^6$. Eukaryotic cells (cells of higher organisms, fungi, protozoa, and most algae) contain a nucleus surrounded by the membrane and have a few or many chromosomes. In such cells, the DNA is almost entirely concentrated in the nucleus and distributed among the chromosomes in which it is bound to proteins (histones and other nuclear proteins). The molecular weight of such DNAs is twice that of the corresponding prokaryotic DNAs. In addition to nuclear DNA, eukaryotic cells contain a small amount of cytoplasmic DNA. This category includes *satellite* DNA as well as mitochondrial and chloroplast DNA's. Their molecular weights are much lower, as compared to chromosomal DNAs (ranging from one to tens of millions). Analogues of plasmid DNAs in bacterial cells have been found among cytoplasmic DNAs in eukaryotes.

Viral (Phage) DNAs. Many viruses and phages contain DNAs differing in molecular weight and in three-dimensional structure. Single-stranded DNA of bacteriophage ψ X 174 is a polynucleotide chain with a molecular weight of $1.7 \cdot 10^6$. The polyoma virus contains a double-stranded DNA having a molecular weight of $3 \cdot 10^6$. A similar molecular weight is observed in virus SV 40 DNA, while the molecular weight of the phage λ DNA is greater by one order of magnitude.

[1] Colon bacillus (*Escherichia coli*) is a well known representative of prokaryotic cells.

5.2.2 RNA

Just as DNA, RNA is an essential component of all living cells and also many viruses and bacteriophages. A distinction is made between cellular and viral (phage) RNAs respectively.

Cellular RNAs. Three main types of cellular RNAs are known: messenger (mRNA), ribosomal (rRNA) and transfer (tRNA). Table 5-1 lists some characteristics of each type of RNA as well as their percentage of the total cellular RNA in *E. coli.*

The RNA content of most cells exceeds that of DNA many times. In bacterial cells, almost all of RNA is found in the cytoplasm. More than 10 per cent of all RNAs in animal cells (e.g., liver) is found in the nucleus (mRNA), about 15 per cent are in mitochondria (rRNA and tRNA), and more than 50 per cent are in ribosomes (rRNA). Each ribosome is made up of three rRNA molecules with sedimentation constants of 23S, 16S and 5S for *E. coli.* Transfer RNAs differ but insignificantly in molecular weight. The greatest molecular weight differences are observed in mRNAs, which has to do with their function as carriers of genetic information.

Viral RNAs. RNA is a constituent of most plant viruses. Among the most thoroughly studied is the tobacco mosaic virus which contains RNA having a molecular weight of $2 \cdot 10^6$. Many bacteriophages, such as MS2 and R17, also contain RNA (with a molecular weight in the neighborhood of a million). Also known are RNA-containing animal viruses, such a s polio virus. All of the above viruses contain a single-stranded RNA molecule per viral particle.

Table 5-1. Some Characteristics of Different Types of RNA *(E. Coli).*

RNA type	Sedimentation constant*	Molecular weight	Number of nucleotides	Percentage of total cellular RNA
mRNA	6S–25S	25,000–1,500,000	75–3000	2
rRNA	5S	35,000	100	82
	16S	550,000	1500	
	23S	1,000,000	3100	
tRNA	4S	23,000–30,000	75–90	16

* The distribution is given according to the sedimentation coefficient of the nucleic acid sample in a solution of moderate ionic strength (0.1) expressed in Svedberg units (S).

5.3 Nature of Internucleotide Linkages

Studies into internucleotide linkages in polymeric nucleic acid molecules were completed successfully in the early fifties, immediately after the structure of nucleotides had been established and some properties of their derivatives (primarily esters) had been elucidated. By the same time, methods for isolating and purifying DNA and RNA had been developed so that the nature of intermonomer linkages was studied on pure, albeit markedly degraded, nucleic acid preparations.

The earliest information about the type of intermonomer or, as has become more common to say, internucleotide linkage came from potentiometric titration experiments. Their results indicated that RNA as well as DNA contains a single hydroxyl at each phosphate group ($pK_a \sim 1$). The conclusion that followed was that each nucleic acid contains a disubstituted phosphate group as a structural unit.

It was natural to assume that phosphate groups join nucleosides by crosslinks involving two of their hydroxyls, whereas one hydroxyl remains free. It only remained to find out which parts of nucleoside moieties participate in the linkage with phosphate groups.

Since nucleic acids can be deaminated in the presence of nitrous acid, it is obvious that the amino groups of pyrimidine and purine bases are not involved in internucleotide linkage. Moreover, potentiometric titration has shown that the oxo(hydroxy) groups of guanines and uracils in nucleic acids are free as well. This finding led to the conclusion that internucleotide linkages are formed by the phosphate group and the hydroxyls of carbohydrate moieties (i.e., they are essentially phosphodiester bonds), which are consequently responsible for formation of the polymer chain (nucleic acid). Thus, what is generally referred to as internucleotide linkage is, in fact, a system of bonds:

$$\mathrm{C{-}O{-}\overset{\overset{\displaystyle O}{\|}}{\underset{\underset{\displaystyle OH}{|}}{P}}{-}O{-}C}$$

(where C is the primary or secondary carbon of the sugar). As has already been mentioned, depending on the reaction conditions, hydrolysis of DNA and RNA yields nucleotides with different positions of the phosphate group:

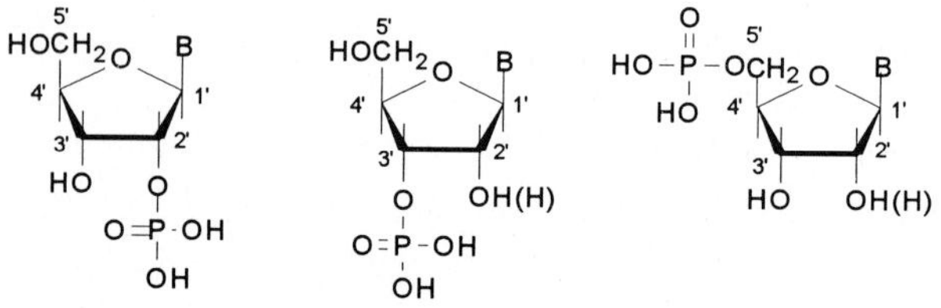

If it is assumed that all internucleotide linkages in nucleic acids are identical, then, in addition to the phosphate group, they may contain only the 3′-hydroxyl of one nucleoside moiety and the 5′-hydroxyl of the other (3′-5′-bond). And if they are different, the polymer chain of DNA can accommodate three types of bonds at a time: 3′-5′, 3′-3′ and 5′-5′. In the case of RNA, the number of bond types must be greater because of involvement of the 2′-hydroxyl as well.

The true nature of internucleotide linkages in native DNA and RNA was established as a consequence of directed chemical and enzymatic hydrolysis of the biopolymers with subsequent isolation and identification of the resulting fragments.

5.3.1 Internucleotide Linkage in DNA

Chemical hydrolysis of DNA as a method for degrading the polymer to establish the nature of internucleotide linkage has turned out to be useless for all practical purposes. DNA is not hydrolysed at alkaline pH values, which is quite consistent with the assumption that the internucleotide linkage is phosphodiester in nature (the stability of dialkylphosphates in alkaline media has already been discussed). When treated with acid even under mild conditions, DNA is hydrolysed both at phosphodiester bonds and at *N*-glycosidic bonds formed by purine bases. Consequently, hydrolysis of the polymer does not yield consistent results, yet it has been possible to isolate, among products of acid DNA hydrolysis, diphosphates of pyrimidine deoxynucleosides, which are identical with synthetic deoxycytidine and deoxythymidine 3′,5′-diphosphates:

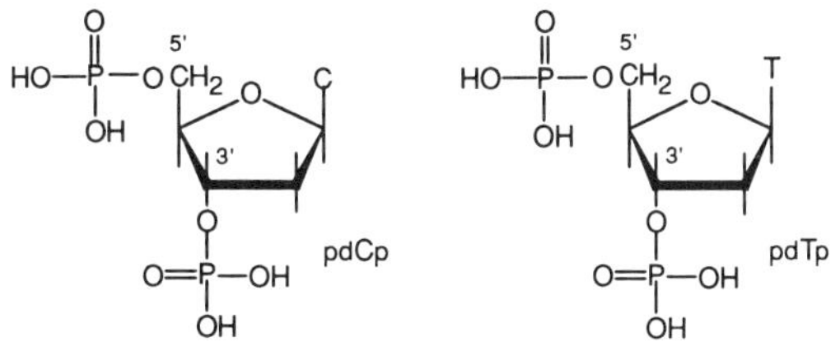

pdCp pdTp

The mechanism of pdCp and pdTp formation will be discussed later; at this juncture, it is important to point out that the presence of these compounds among DNA degradation products is indicative of involvement of both hydroxyl groups, at least those of pyrimidine monomer units, in formation of the internucleotide linkage.

Enzymatic hydrolysis of DNA has turned out to be more specific. When DNA preparations were treated with snake venom phosphodiesterase (PDE), the polymer was almost completely hydrolysed to deoxynucleoside 5′-phosphates whose structure was established through comparison with the corresponding synthetic nucleotides.

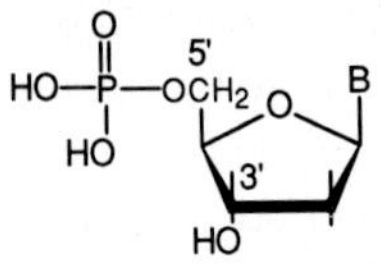

B- A,G,T and C residues

These findings suggest involvement of the 5′-hydroxyls of all four deoxynucleosides constituting DNA in internucleotide linkage. DNA in the presence of micrococcal or spleen PDE is hydrolysed in a similar manner but to 3′-phosphates.

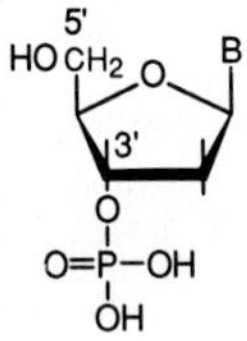

B- A,G,T and C residues

The results of DNA hydrolysis in the presence of phosphodiesterases of different specificity clearly indicate that nucleosides in DNA are linked through the medium of the phosphate group which also esterifies the hydroxyl at the secondary carbon (position 3′) of one nucleoside moiety and that at the primary carbon (position 5′) of another.

Thus, it has been conclusively proved that internucleotide linkage in DNA involves the phosphates group as well as 3′- and 5′-hydroxyls of the nucleosides [(a) and (b) show the directions of hydrolysis of the DNA polynucleotide chain in the presence of snake venom and spleen or micrococcal PDE, respectively]:

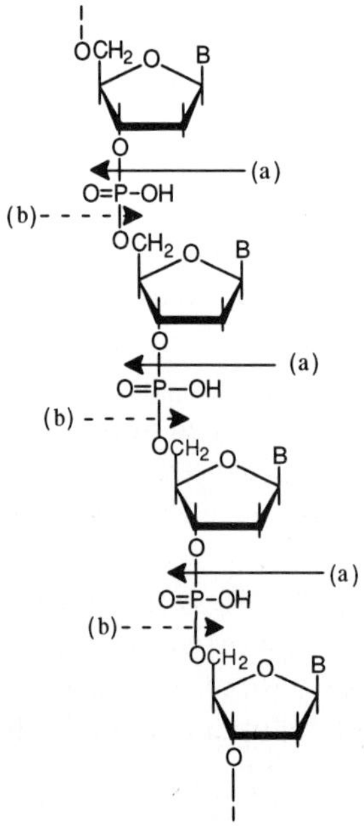

The possibility of a different polymer structure with alternating 3′-3′ and 5′-5′ bonds between nucleosides was ruled out because such an assumption was at variance with experimental data. For instance, such a polymer should not have been hydrolysed completely (to monomers) in the presence of snake venom PDE which selectively breaks down only nucleoside 5′-alkylphosphates. The same applies to spleen PDE which selectively hydrolyses nucleoside 3′-alkylphosphates.

5.3.2 Internucleotide Linkage in RNA

The nature of internucleotide linkage in RNA has turned out to be more challenging. Already in the earliest studies into the structure of RNA it was found to be extremely unstable during alkaline hydrolysis. The main products of alkaline hydrolysis of RNA are ribonucleoside 2′- and 3′-phosphates yielded in practically equal amounts:

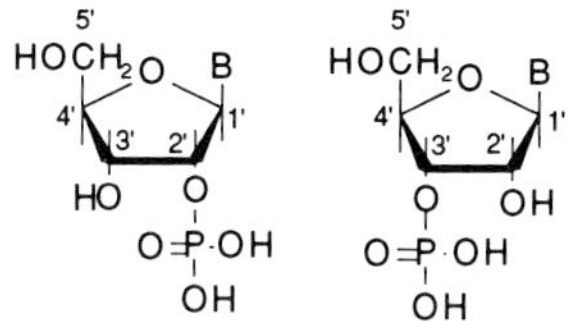

B-residues A,G,T and C

No ribonucleoside 5′-phosphates are formed in the process. These findings were hard to reconcile with the assumed phosphodiester nature of internucleotide linkage in RNA (as has already been mentioned, ordinary dialkylphosphates are relatively stable in alkaline media) and called for more comprehensive studies. An extremely important role in such studies, carried out in the early fifties by Todd and coworkers, was played by synthetic ribonucleotide alkylphosphates, which were obtained to simulate a particular type of phosphodiester bonds. The stability of such model compounds under conditions of alkaline hydrolysis has already been treated at length.

The results obtained by Todd and collaborators to elucidate the mechanisms of the transformation undergone by ribonucleotide alkylphosphates suggest that the internucleotide linkage in RNA as well as DNA involves the phosphate group and 3′- and 5′-hydroxyls of the carbohydrate moieties. A similar bond in RNA must be readily hydrolysed in an alkaline medium because the neighboring 2′-hydroxyl must catalyze this process at pH > 10, which marks the onset of ionization of the ribose hydroxyls. It should be strongly emphasized that the intermediate compounds forming during alkaline hydrolysis must be all of the four ribonucleoside 2′,3′-cyclic phosphates, while the end products are the ribonucleoside 3′- and 2′-phosphates (four pairs of isomers) resulting from their hydrolysis. The following scheme illustrates alkaline hydrolysis of an RNA fragment with a random nucleotide sequence in an alkaline medium:

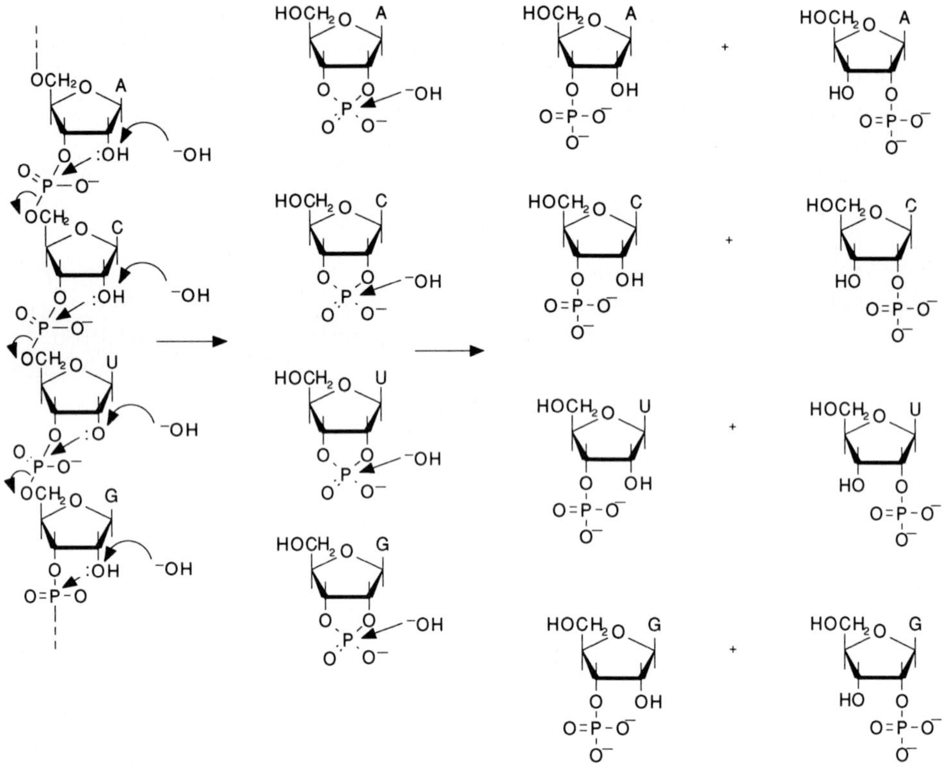

Of course, other types of internucleotide linkage could not be excluded at that stage of investigations, namely, homogeneous 2′–5′ and mixed 3′–5′ and 2′–3′ bonds which must break in an alkaline medium to yield the same intermediates and end products. The hypothesis stating that the 3′–5′ (2′–5′) type of internucleotide linkage must exist in RNA was supported by two findings. When treated with an alkali, RNA is immediately hydrolysed to low-molecular fragments, which is indicative of virtually simultaneous cleavage of internucleotide linkages over the entire length of the polymer chain, rather than step by step, proceeding from the end of the chain on as α–hydroxyls become free (such a cleavage mechanism might be expected if the monomers were linked by a 2′–3′ or similar bond). Besides, mild alkaline treatment of RNA was found to inhibit the cleavage at the stage of formation of the nucleoside 2′,3′-cyclic phosphates which had been isolated and identified.

The absence of dimers with 5′–5′ bonding among the RNA hydrolysis products provided additional evidence in favor of the 3′–5′ or 2′–5′ nature of internucleotide linkage in RNA.

Thus, the results of alkaline hydrolysis showed that the number of possible internucleotide linkage types in RNA is limited but did not elucidated the structure of this polymer.

More conclusive evidence as regards the type of internucleotide linkade in RNA was provided, just as in the case of DNA, by enzymatic hydrolysis, or digestion.

RNA hydrolysis in the presence of snake venom PDE to ribonucleoside 5′-phosphates yielded a direct proof that 5′-hydroxyls participate in the phosphodiester bonding between monomer units. Later this was definitively established as a result of discovery of RNA phosphorolysis in the presence of the enzyme polynucleotide phosphorylase (PNPase), yielding ribonucleoside 5′-pyrophosphates:

$$\text{RNA} + H_3PO_4 \xrightleftharpoons{\text{PNPase}} \text{nucleoside 5}'\text{-pyrophosphates}$$

Thus, it only remained to elucidate the nature of the second hydroxyl involved in the formation of internucleotide linkage. This problem was partially solved with the aid of yet another enzyme used for controlled RNA digestion, namely, pyrimidyl ribonuclease (RNase).

As was shown earlier, this enzyme breaks down only pyrimidine ribonucleoside 3′-alkylphosphates to ribonucleoside 3′-phosphates (via the intermediate ribonucleoside 2′,3′-cyclic phosphate). It was found that the enzyme acts on RNA in a similar fashion. The following example with a random nucleotide sequence illustrates how this reaction in combination with other already known ones can be used in structural studies. Experiments with any purified RNA samples have shown that the quantity of the phosphoric acid formed when the polymer is treated with pyrimidyl RNase and then with phosphomonoesterase (PME), as well as the amount of the periodic acid spent in the subsequent oxidation, is equivalent to that of the pyrimidines in the RNA sample. This suggested that at least pyrimidine nucleotides in RNA are linked with the neighboring ones only by a 3′–5′ bond. This conclusion is borne out by the results of alkaline treatment of enzymatic hydrolysates of RNA after it has been exposed to RNase: in an alkaline medium, migration of the phosphate group in ribonucleoside 3′- and 2′-phosphates is impossible, and the presence of only pyrimidine ribonucleoside 3′-phosphates in the corresponding hydrolysates is a clear indication of the 3′–5′ type of the internucleotide linkage in the case of pyrimidine nucleotides.

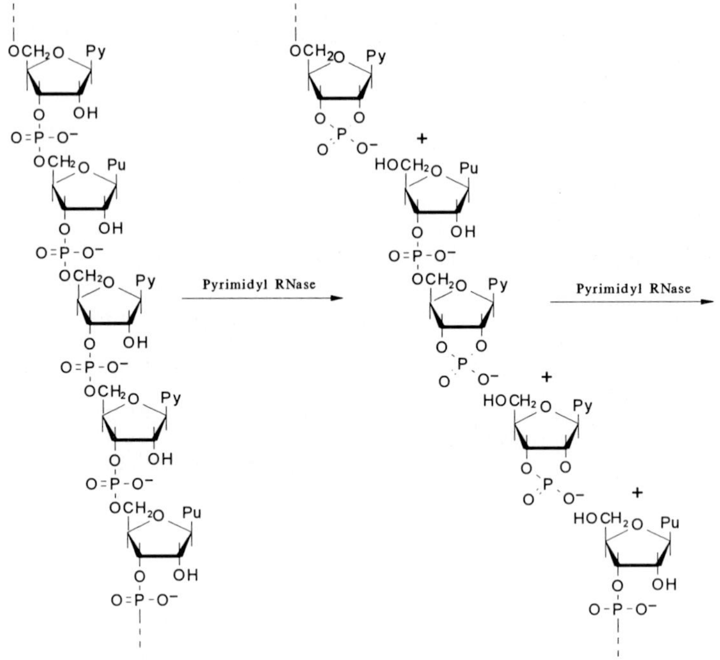

The logical assumption that the 3′–5′ bond must be the only intermonomer linkage in RNA was soon corroborated in experiments where spleen PDE was used to digest RNA. Just as in the case of DNA, this enzyme catalyzed hydrolysis of the polynucleotide chain to nucleoside 3′-phosphates, the entire RNA being broken down to nucleotides.

It should be pointed out that guanyl (T_1) RNase and also other enzymes are currently used for directed cleavage of 3′–5′ bonds in RNA.

In short, it has been found that the internucleotide linkage in RNA as well as DNA is the same over the entire linear polymer chain and involves the phosphate group and 3′- and 5′-hydroxyls of the pentose. Consequently, the molecular backbone of both nucleic acids represents a carbohydrate-phosphate chain with D-ribose (RNA) or 2′-deoxy-D-ribose (DNA) and phosphate groups alternating at regular intervals. In both cases, the polymer chain contains a regularly recurring group of atoms consisting of three carbons, two oxygens and a phosphorus

$$HO[-C^{5'}-C^{4'}-C^{3'}-O-P-O]_n-C^{5'}-C^{4'}-C^{3'}-OH$$

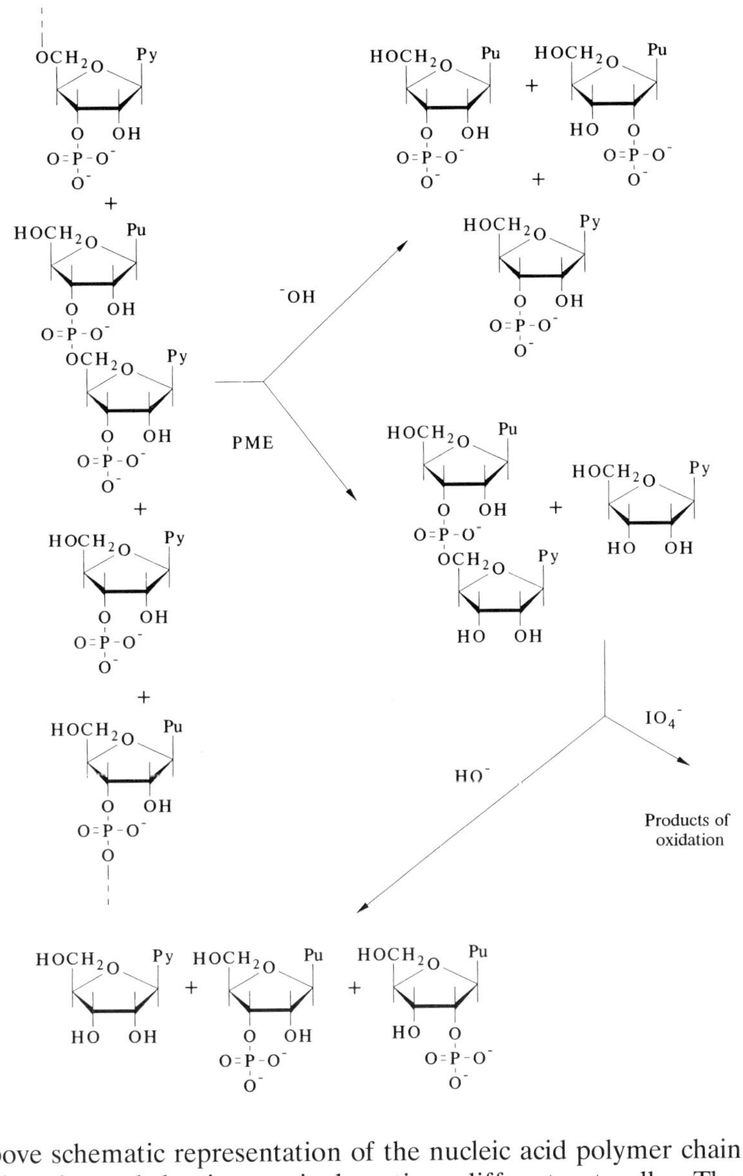

The above schematic representation of the nucleic acid polymer chain suggests that it is polar and that its terminal portions differ structurally. The terminal fragment with the free 5′-hydroxyl (or 5′-phosphate) not involved in the internucleotide linkage is known as the 5′ end (or 5′ terminus) of nucleic acids, while the terminal fragment with the free 3′-hydroxyl or 3′-phosphate is known as the 3′ end (or 3′ terminus):

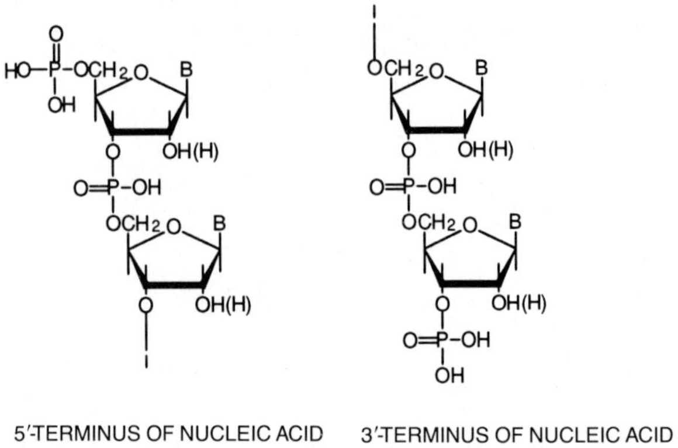

5′-TERMINUS OF NUCLEIC ACID 3′-TERMINUS OF NUCLEIC ACID

The structural formulas of DNA and RNA can be illustrated by the following scheme showing fragments of polymer chains with a random sequence of monomer units:

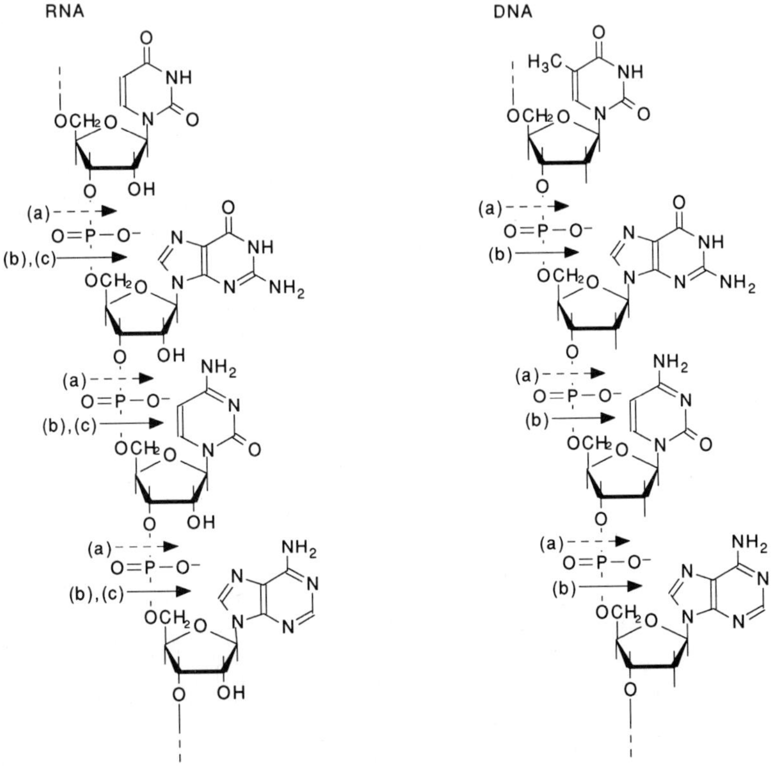

RNA DNA

The scheme shows the sites where internucleotide linkages are cleaved in the presence of snake venom PDE [direction (a)], spleen PDE [direction (b)] and alkali [direction (c)].

5.4 Nomenclature, Abridged Formulas and Abbreviations

The fragments resulting from enzymatic or chemical hydrolysis of nucleic acids and their synthetic analogues are usually termed oligonucleotides if the molecule contains a small number of monomers. The prefix "oligo" comes from carbohydrate chemistry where oligomers (from Greek "oligos" – few + "meros" – part) are compounds occupying an intermediate position, in terms of size, between monomers and polymers. The natural and synthetic nucleotide polymers comprising more than ten monomer units are usually called polynucleotides. Synthetic poly(oligo)nucleotides may have internucleotide linkages other than those of the 3′–5′ type (such as 5′–5′ or 2′–5′). According to the exact number of monomer units in the chain, oligonucleotides can be referred to as dinucleotides, trinucleotides, and so on. Similar terms are used to denote polynucleotides with a particular chain length. The names defining the number of monomer units in the chain have Greek or Latin roots. For example, a 15-membered nucleotide is known as pentadecanucleotide and a 20-membered one as icosanucleotide.

In accordance with the commonly accepted nomenclature of polypeptides, a polynucleotide is regarded as a chain in which every nucleotide esterifies the hydroxyl of the next nucleotide, rather than a set of nucleosides linked by phosphodiester groups. Therefore, the complete name of an oligo(poly)nucleotide is also composed as that of a polypeptide; that is, it includes the names of the nucleotide residue, separated, in parentheses, by the type of the internucleotide linkage between the corresponding monomer units.

The name of a polynucleotide usually begins with the 5′ end (the residue of the 5′-terminal nucleotide being named first). In such cases, the internucleotide linkage is denoted as (3′→5′), the arrow indicating the direction in which the name is composed (from the 5′ toward the 3′ end of the chain).

The name of a polynucleotide chain may also be composed in reverse order (from the 3′ to the 5′ end). Then, the internucleotide linkage should be denoted as (5′→3′). Given below by way of example are the names of a tetradeoxyribonucleotide and a triribonucleotide.

Tetradeoxyribonucleotide

Structural formula and full names | Abridged formulas and abbreviations

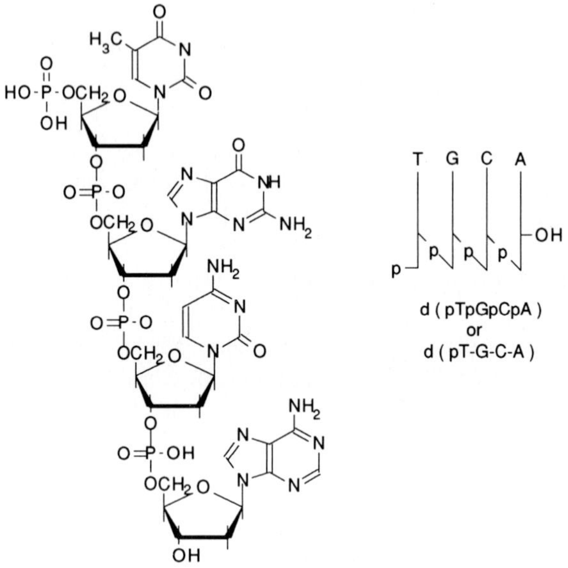

d (pTpGpCpA)
or
d (pT-G-C-A)

5′-phosphothymidylyl (3′ → 5′) deoxyguanylyl (5′ → 3′) deoxyguanilil (3′ → 5′) deoxyadenozine

deoxyadenylyl (5′ → 3′) deoxyctidylic – (5′ → 3′) deoxyguandylic (5′ – 3′) deoxythymidylic – 5′-acid (or-thymidine-5′-phosphate)

The choice of the direction in which the name is composed is usually determined by the nature of the terminal groups of the chain. Poly(oligo)nucleotides are written in a direction from the 5′ to the 3′ end, as a rule. The abbreviations in most cases replacing the cumbersome complete structural formulas are formed following the same rules that govern the writing of abridged formulas of monomer components. The only addition to these rules is the abbreviation of the internucleotide linkage (internucleotide phosphodiester group). Such linkage is symbolized by a curved or straight sloping (diagonal) line interrupted by lowercase "p". Thus, the internucleotide phosphorus atom is denoted in exactly the same way as the phosphate group in a monomer component or that at the end of the polynucleotide chain, while its phosphodiester nature is expressed as bonds with the 3′-hydroxyl (left diagonal line) and 5′-hydroxyl (right diagonal line).

In spite of the fact that abridged formulas greatly facilitate the writing of polynucleotides, they become useless when the structural formulas of nucleic acids are to be written (e.g., tRNAs consisting of 75–85 monomer units). In such instances, use is made of abbreviated letter symbols according to the rules that apply to nucleotides. In this case, too, the internucleotide phospho-

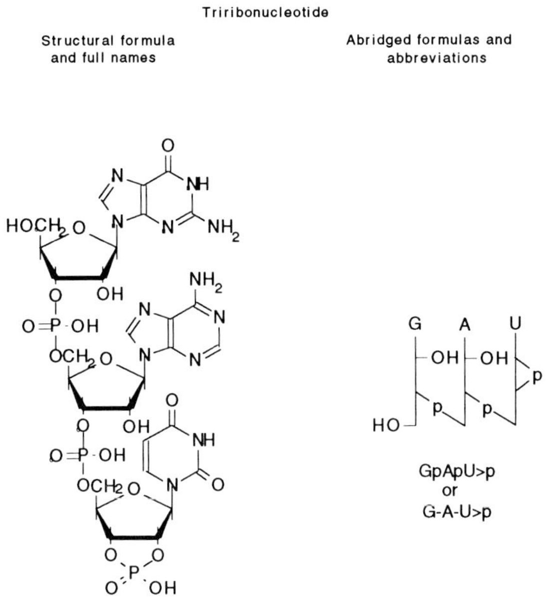

Triribonucleotide

Structural formula and full names	Abridged formulas and abbreviations
guanylyl (3′ → 5′) adenylyl (3′ → 5′) – uridine-2′,3′-cyclophosphate or 2′,3′-cyclophosphouridyl – (5′ → 3′) adenylyl (5′ → 3′) guanosine	GpApU>p or G-A-U>p

rus is denoted by lowercase "p", just as in the above scheme, inserted between the uppercase letters standing for the corresponding nucleotides. The letter "p" to the left of the uppercase letter denotes the 5′-phosphomonoester group, while the same letter to the right stands for the 3′-phosphomonoester group. The symbol of the internucleotide phosphorus ("p") can be replaced by a hyphen, which further simplifies the structural formulas of nucleic acids, or even dropped altogether. In this case, only the symbols of the terminal phosphate groups ("p") are left. In such abridged notation, deoxyribopolynucleotides differ from ribopolynucleotides by the letter "d" added before the abbreviation in parentheses. For example, decaribonucleotide and decadeoxyribonucleotide can be written as follows:

Decaribonucleotide	Decadeoxyribonucleotide
pUpGpCpCpApUpGpApGpU	d(pTpGpCpCpApTpGpApGpT)
or	or
pU-G-C-C-A-U-G-A-G-U	d(pT-G-C-C-A-T-G-A-G-T)
or	or
pUGCCAUGAGU	d(pTGCCATGAGT)

In the case of homopolynucleotides – that is, polynucleotides consisting only of one monomer, the formulas become even simpler. The formula of oligothymidylic acid, for example, can be written as follows:

$$d(pT)_n, \text{ where } n \leq 10$$

As can be seen from this formula, oligothymidylic acid contains ten or fewer nucleotides, has a 5′-terminal phosphate group and a free 3′-hydroxyl in pentose.

Sometimes, yet simpler formulas are used. For linstance, the above oligomer can also be written as $d(T) \leq 10$, or $d(T)_n$, or oligo(dT).

High-molecular weight homopolymers are written using a single letter (nucleoside symbol) following the prefix "poly". For example, the polymer of deoxyriboadenylic acid is written as poly(dA) and that of riboadenylic acid is written as poly(A).

5.5 Nucleotide Composition

An important characteristic of nucleic acids is their nucleotide composition or, in other words, composition and ratio of the consituent monomer units. In the late forties and early fifties, when such research tools as paper chromatography and UV spectroscopy came into being, many analyses of the composition of nucleic acids were carried out (Chargaff, Belozersky). Their results provided a decisive argument for rejecting the older notions of nucleic acids as polymers containing recurring tetranucleotide sequences (so-called tetranucleotide theory of nucleic acid structure reigned supreme in the thirties and forties) and paved the way toward modern concepts not only of the primary structure of DNA and RNA but also their macromolecular structure and functions.

The method for determining the composition of nucleic acids is based on analysis of the products of their enzymatic or chemical degradation. Three chemical methods are usually employed. Acid hydrolysis under vigorous conditions (70% perchloric acid, 100°C, one hour or 100% formic acid, 175°C, two hours), conducted for analysis of both DNA and RNA, leads to cleavage of all *N*-glycosidic bonds and formation of a mixture of purine and pyrimidine bases. In the case of RNA, use can be made of mild acid hydrolysis (1 N hydrochloric acid, 100°C, one hour), resulting in purine bases and pyrimidine nucleoside 2′(3′)-phosphates, as well as alkaline hydrolysis (0.3 N KOH, 37°C, 20 hours), yielding a mixture of nucleoside 2′(3′)-phosphates.

Since the number of nucleotides of each type in nucleic acids is equal to that of the corresponding bases, to establish the nucleotide composition of a given nucleic acid one can simply determine the quantitative ratio of the bases. To this end, individual compounds are separated from the hydrolysates by paper chromatography or electrophoresis (when the hydrolysis yields nucleotides). Irrespective of whether it is associated with the carbohydrate moiety or not, each base displays a characteristic absorption maximum in the UV spectrum, whose intensity depends on concentration. Therefore, the UV spectra of the separated compounds can be used to find the quantitative ratio of the bases and, consequently, determine the nucleotide composition of the starting nucleic acid.

Quantitative analysis of minor nucleotides, especially such unstable ones as dihydrouridylic acid, is based on enzymatic hydrolysis (snake venom and spleen PDE).

Experience with the above analytic techniques has shown that nucleic acids of different origin consist, with rate exceptions, of four major nucleotides, whereas minor nucleotides may vary widely.

As will be shown later, analyses of the nucleotide composition of DNA have made it possible to establish its three-dimensional structure.

5.5.1 Composition of DNA

While studying the nucleotide composition of native DNAs differing in origin, Chargaff noticed the following regularities.

1. All DNAs, regardless of their origin, contain equal numbers of purine and pyrimidine bases. Hence, in any DNA there is a pyrimidine nucleotide for each purine one.
2. Any DNA always contains equal amounts of adenine and thymine, guanine and cytosine, in pairs usually denoted A=T and G=C. These two regularities give rise to a third one.
3. The number of bases containing amino groups in positions 4 of the pyrimidine ring and 6 of the purine ring (cytosine and adenine) equals that of bases containing an oxo group in each of the same positions (guanine and thymine); that is, $A + C = G + T$. These regularities have become known as Chargaff's rules. It was also established that for each type of DNA the sum of guanine and cytosine is not equal to the sum of adenine and thymine, which is to say that the ratio $(G + C)/(A + T)$ usually differs from unity (it may be greater or less than unity). This serves as an indicator to distinguish between the two main types of DNA: the $A \cdot T$ type with adenine and thymine being predominant and the $G \cdot C$ type with prevalence of guanine and cytosine.

The ratio of the sum of guanine and cytosine to that of adenine and thymine, characterizing the nucleotide composition of a given DNA, is known as the coefficient of specificity. Each DNA has a characteristic coefficient of specificity, which may vary from 0.3 to 2.8. In calculating this coefficient, one must take into account the content of minor bases and replacement of the major bases by their derivatives. For example, when the coefficient of specificity is calculated for DNA from wheat embryos, containing six per cent of 5-methylcytosine, the latter forms part of the sum of guanine (22.7%) and cytosine (16.8). The meaning of Chargaff's rules for DNA became understood after its three-dimensional structure had been established.

5.5.2 Composition of RNA

The first indications of the nucleotide composition of RNA came from the analysis of RNA preparations representing mixtures of cellular RNAs (ribosomal, messenger and transfer RNA, known as total RNA fraction). Chargaff's rules are not obeyed in this case, although there is a certain relationship between the content of guanine and cytosine on the one hand and that of adenine and uracil on the other.

The results of RNA analyses conducted in recent years indicate that individual RNAs do not obey Chargaff's rules either. However, the differences in the contents of each pair of bases are insignificant for most RNAs so that we can say that in general Chargaff's rules hold for RNA as well. This is due to the macrostructural features of RNA.

As has already been mentioned, minor bases are characteristic structural features of some RNAs. The corresponding nucleotides are usually present in transfer and some other RNAs in very small amounts, which is why determination of the nucleotide composition of such RNAs in its entirety is sometimes rather difficult.

In conclusion of this section it should be emphasized once again that knowledge of the nucleotide composition of DNA and RNA has been fundamental in elucidating the primary structure of these biopolymers. What is more, it has prepared the groundwork for concepts of the macromolecular structure of DNA and RNA as well as their functions. Yet the nucleotide composition ist not fully representative even of the primary biopolymer structure. To establish the primary structure of nucleic acids one must know the sequence in which nucleotides occur in individual nucleic acid molecules.

5.6 Sequence of Nucleotide Units

The order in which nucleotide units are arranged in a nucleic acid polymer chain is commonly known as nucleotide sequence. Determination of the nucleotide sequence in nucleic acids or their fragments is the primary task of nucleic acid chemistry and molecular biology. This is due to the fact that the sequence of monomer units in nucleic acids determines the content of their inherent genetic information, the ability to carry it, as well as the three-dimensional structure and chemical properties of the polymer. Without knowing the primery structure of nucleic acids one cannot study the fine mechanisms underlying the functions of DNA and RNA. Knowledge of the primary structure of individual portions of nucleic acids makes it possible to construct synthetic models which serve as rather useful research tools in molecular biology, molecular genetics, enzymology, and many other sciences concerned with the functions of living organisms.

Systematic studies aimed at establishing the primary structure of nucleic acids began after methods for isolating some of them on an individual basis had been developed.

The first objects of such studies were transfer RNAs for which reliable isolation and purification techniques were elaborated in the early sixties. Since the molecular weight of tRNAs is relatively low, they provided a good basis for developing standard analytical procedures. At present, the primary structure of hundreds of individual tRNAs and some high-molecular weight RNAs is a matter of common knowledge.

The general approach currently used to determine the nucleotide sequence is basically the same as that proposed by Sanger to determine the amino acid sequence in proteins and known as the unit method. It consists of controlled fragmentation of biopolymer chains, separation of the resulting fragments (oligomers), and determination of the sequence of monomer units in these fragments. To reconstruct the original structure of the polymer, the analysis based on the above-described scheme should be repeated at least once, the second fragmentation involving other parts of the chain so that two sets of fragments have partially identical monomer unit sequences. Such an approach to reconstruction (establishment of the primary structure) can be schematically represented as follows:

Sites of chain cleavage during the 1st analysis

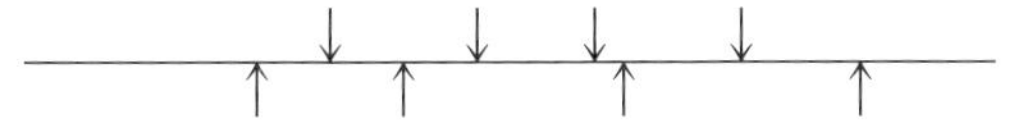

Sites of chain cleavage during the 2nd analysis

Comparison of the two sets of oligomers reveals overlapping sequences, which allows the fragments to be properly arranged one with respect to another. This is what is known as the overlapping unit method.

In the case of DNAs whose molecular weight is usually very high, the first analysis also involves the overlapping unit method but each unit is characterized by its molecular weight. It is this characteristic that permits the mutual arrangement of the units in the molecule to be determined using the above scheme (so-called mapping).

Until recently, the most commonly used technique for controlled cleavage of internucleotide bonds, as part of the unit method of analyzing the primary nucleic acid structure, has been enzymatic hydrolysis or digestion.

In the mid seventies, radically new approaches were developed for determining the nucleotide sequence in nucleic acids, different from the unit method and allowing the rapid and reliable sequencing of high-molecular weight compounds. These methods, currently applicable to DNA and RNA and widely used for nucleic acid sequencing, will be discussed at greater length in Chapter 6.

The present chapter will deal with sequencing only in the context of oligonucleotides which are still valuable as effective means for determining the structure of synthetic oligonucleotides.

In what follows, we shall first of all discuss the general aspects of determining the primary structure of oligonucleotides, including the specific action of nucleases – enzymes breaking the internucleotide bonds in RNA and DNA.

5.6.1 Internucleotide Bond-Breaking Enzymes (Nucleases)

To determine the nucleotide sequence in oligo(poly)nucleotides and nucleic acids use is made of various enzymes breaking certain intermonomer bonds.

We shall first discuss enzymes of nucleic acid metabolism, which catalyze hydrolysis of polynucleotides and nucleis acids. Such enzymes are known as nucleases and classed with phosphodiesterases because they break internucleotide phosphodiester bonds. Used for structural studies in addition to phosphodiesterases, are also phosphomonoesterases which catalyze cleavage of the phosphomonoester bond with release of the phosphate from oligo(poly)-nucleotides having a terminal phosphate group. The substrates of phosphomonoesterases are, as a rule, products of nucleic acid hydrolysis by nucleases.

The classification of nucleases is usually based on three characteristics. The first, fundamental characteristic is the substrate specificity of the enzyme – that is, the ability of nuclease to hydrolyse RNA, DNA, or both. Hence such names as ribonuclease (abbreviated as RNase), deoxyribonuclease (DNase) and nonspecific nucleases whose abridged form, in the case of phosphodiesterases, is PDE plus the source from which it is isolated. The second characteristic is the way in which the polymer is digested by the enzyme or, in other words, its ability to catalyze hydrolysis of the polymer chain endolytically – that is, at a particular site inside the chain, or exolytically – that is, beginning from the end of the chain and breaking step by step the terminal internucleotide bonds. This is how the terms endonucleases and exonucleases came into being. Finally, the third characteristic is the type of cleavage of a particular internucleotide linkage which may be hydrolysed in two ways depending on its structure. For instance, cleavage of the internucleotide linkage in the shown fragments is possible both between the 3′-hydroxyl and phosphorus atom, giving a 5′-phosphate group and freeing the 3′-hydroxyl, and between the phosphorus and 5′-hydroxyl, giving terminal 3′-phosphate and 5′-hydroxyl groups:

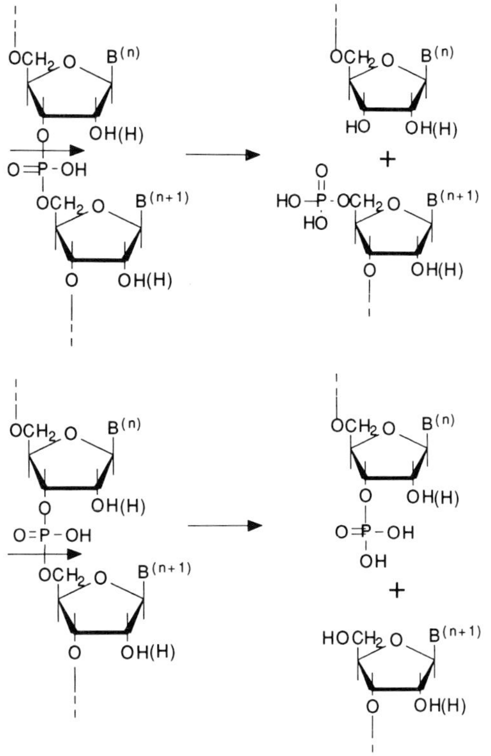

This characteristic is usually ignored in the name of the enzyme, the assumption being that the mechanism of phosphodiester bond cleavage by the enzyme, whose name takes into account the first and/or second characteristics, is known. We shall now consider the mechanism of action of the nucleases most widely employed in studying the structure of oligo(poly)nucleotides and nucleic acids.

5.6.1.1 Ribonucleases

Pyrimidyl Ribonuclease. The most fully studied enzyme in this category is pyrimidyl nuclease (abbreviated as RNase A). It was discovered in 1920 in the pancreas and was crystallized by Kunitz in 1940 (*J. Gen. Physiol.* **24**, 15). The molecular weight of the enzyme is 13,700; it is stable in a broad pH range and very stable when heated in weakly acid solutions. The activity is maximum at pH 7.7 (65 °C). The enzyme has no effect on DNA. Pyrimidyl RNase exhibits high specificity of endolytic action: it catalyzes hydrolysis of any internucleotide 3′–5′ bonds formed by the 3′-hydroxyl of the pyrimidine nucleotide. The nucleotide acting as donor of the 5′-hydroxyl may be either a pyrimidine or purine one. What makes the action of this enzyme special is that it splits RNA only over single-stranded portions without affecting the double-stranded ones.

The cleavage of internucleotide linkages proceeds in two steps. The first step is transphosphorylation (intramolecular nucleophilic substitution involving the 2′-hydroxyl) to yield an oligonucleotide terminating in a 2′,3′-cyclic phosphate group of the pyrimidine nucleotide[2)]. The second step is hydrolysis of the cyclic phosphate group (intermolecular nucleophilic substitution involving a water molecule) to yield an oligonucleotide with the terminal 3′-phosphate group of the pyrimidine nucleotide. The general scheme of RNA cleavage by pyrimidyl ribonuclease is given below:

The products of short-term treatment of RNA with pyrimidyl ribonuclease are invariably pyrimidine ribonucleoside 2′,3′-cyclic phosphates and purine oligoribonucleotides with pyrimidine nucleoside 2′,3′-cyclic phosphate at the 3′ end of the oligonucleotide chain.

2) The first step is reversible. The reverse reaction is used in the synthesis of oligonucleotides with pyrimidine ribonucleoside 2′,3′-cyclic phosphates serving as the substrates.

Thus, pyrimidyl ribonuclease is essentially a highly specific phosphodiesterase catalyzing the degradation of diphosphates in which one alcohol group must be represented by the 3′-hydroxyl of pyrimidine ribonucleoside. The other alcohol group does not have to be a nucleoside. As demonstrated in experiments on model substrates, pyrimidyl RNase also hydrolyses pyrimidine ribonucleoside 3′-alkylphosphates.

Synthetic pyrimidine ribonucleoside 2′,3′-cyclic phosphates are also easily hydrolysed in the presence of pyrimidyl RNase.

To illustrate the action of the enzyme on polyribonucleotides, here is a scheme showing the cleavage of an 18-membered fragment forming part of a tRNA:

↓ ↓ ↓ ↓ ↓ ↓ ↓ ↓ ↓

A–Ψ–C–U–G–C–Ψ–U–I–A–C–A–C–G–C–A–G–A

↓ pyrimidyl RNase

A–Ψp + G–Cp + I–A–Cp + A–Cp + G–Cp + A–G–A + 2UP + ΨP + Cp

As a result of such digestion, four dinucleotides are formed along with two trinucleotides and three pyrimidine nucleotides. A distinctive structural feature of the hydrolysates is the presence of pyrimidine nucleotides at their 3′ end. Adenosine turned out to be 3′-terminal only in one of the trinucleotides, which immediately indicates that this oligonucleotide occupies the 3′ end of the polynucleotide chain. As can be inferred from the above scheme, the internucleotide linkage involving the pseudouridine 3′-phosphate group is broken just as similar linkages formed by uridine and cytidine 3′-phosphates. It should also be noted that the specificity of pyrimidyl RNase with respect to pyrimidine nucleotides is not absolute. The enzyme hydrolyses the internucleotide bonds formed by the adenosine 3′-phosphate group, albeit at a much slower rate (by about two orders of magnitude).

Guanyl Ribonuclease. The second enzyme of this group, isolated from so-called takadiastase which is an extract from the mold Aspergillus oryzae is known as ribonuclease T_1 (RNase T_1). Similar enzymes have been isolated from actinomyces and other sources. The specificity of ribonuclease T_1 is narrower than that of pyrimidyl RNase, because it breaks only one type of internucleotide linkage, namely, bonds formed by the 3′-phosphate group of guanosine (and its derivatives) and the 5′-hydroxyl of any neighboring nucleotide. By virtue of the specificity of its action, RNase T_1 is often referred to as guanyl ribonuclease. The enzyme cleaves only single-stranded portions of RNA.

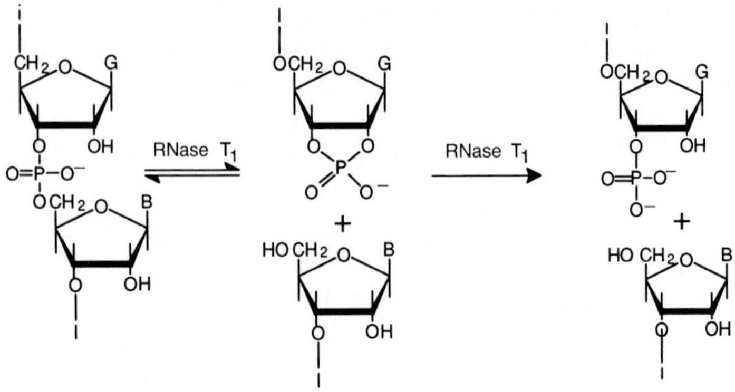

Just as in the case of pyrimidyl RNase, hydrolysis proceeds in two steps with intermediate formation of guanosine 2′,3′-cyclic phosphate or an oligonucleotide with a 3′-terminal guanosine 2′,3′-cyclic phosphate. The first step is reversible; the cyclic phosphate hydrolysis rate is much slower than that of transphosphorylation. And as with pyrimidine RNase, the specificity of ribonuclease T_1 is not absolute. The enzyme hydrolyses the internucleotide bonds formed by the inosine 3′-phosphates methylated at the pyrimidine ring, guanosine 3′-phosphates, and also xanthosine 3′-phosphates (in the latter case, hydrolysis is much slower). Now follows a scheme showing how ribonuclease T_1 splits the same 18-membered polynucleotide which was used to illustrate the action of pyrimidyl RNase:

$$\text{A–Ψ–C–U–}\overset{\downarrow}{\text{G}}\text{–C–Ψ–U–}\overset{\downarrow}{\text{I}}\text{–A–C–A–C–}\overset{\downarrow}{\text{G}}\text{–C–A–}\overset{\downarrow}{\text{G}}\text{–A}$$

$$\downarrow \text{RNase } T_1$$

$$\text{A–Ψ–C–U–Gp + C–Ψ–U–Ip + A–C–A–C–Gp + C–A–Gp + A}$$

Comparison of the mechanisms of action of pyrimidyl RNase and RNase T_1 shows that the oligonucleotide units resulting from treatment with the second enzyme are much larger. A characteristic feature of RNase T_1 hydrolysates is the presence of guanosine 3′- or minor inosine 3′-phosphates at the 3′ end of the oligonucleotides. Studies into hydrolysis of model compounds with RNase T_1 have shown that this enzyme has many potential uses in nucleotide chemistry. Guanyl RNase is one of the enzymes that have played a decisive role in structural investigations of RNA.

5.6.1.2 Deoxyribonucleases

The first enzymes of this type were isolated from the pancreas (pancreatic DNase or DNase I) and spleen (DNase II). Both have turned out to be endonucleases.

Pancreatic Deoxyribonuclease (DNase I). This enzyme has been found to lack pronounced specificity toward bases as it breaks some internal linkages. At cleavage sites, the phosphate group remains linked to the 5′-hydroxyl:

DNase I

PDE of serpent venom

The resulting oligonucleotides are good substrates for snake venom PDE (see below), which becomes useful in preparative hydrolysis of DNA to deoxyribonucleoside 5′-phosphates.

Spleen deoxyribonuclease (DNase II). This enzyme, just as DNase I, is not very specific with respect to bases and cleaves internucleotide bonds in DNA in such a manner that the phosphate group remains at the 3′-hydroxyl:

DNase II

PDE of spleen

The products of DNA hydrolysis in the presence of DNase II serve as substrates for spleen PDE (see below). Such DNA hydrolysis to nucleotides is used in structural studies but seldom applied to preparative isolation of the corresponding deoxyribonucleotides.

In addition to the DNases just described, there are many other deoxyribonucleases (both endo- and exonucleases). These enzymes have yet to be completely studied and sometimes are not readily available. Restriction enzymes and endonuclease IV should be mentioned among the DNases already employed in structural studies.

Endonuclease IV. This enzyme was isolated for the first time in 1969 from *E. coli* infected with phage T4. The enzyme most effectively catalyzes the cleavage of the internucleotide linkage formed by the 3′-hydroxyl of deoxythymidine and 5′-phosphate group of the deoxycytidylic acid and acts only on single-stranded DNA. The result is oligonucleotides one of which contains the 3′-terminal deoxythymidine and the other has a deoxycytidine 5′-phosphate at the 5′ end:

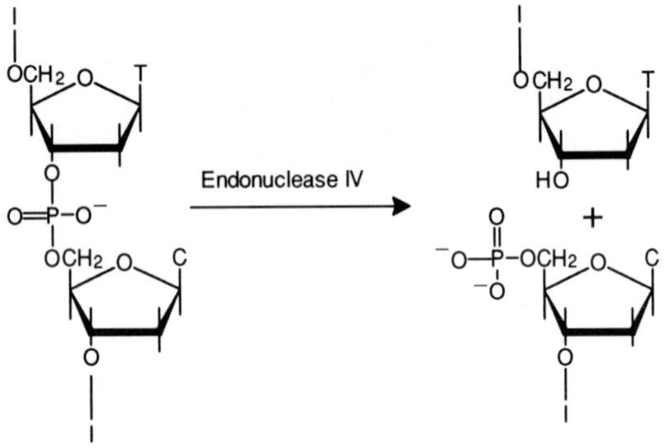

As regards other internucleotide linkages, endonuclease IV cleaves also the dGpdC bond, but at a much slower rate.

5.6.1.3 Nonspecific Nucleases

Phosphomonoesterases (PME). The enzymes belonging to this group are sometimes referred to as phosphatases. They cleave phosphomonoester bonds in nucleotides or in ribo- and deoxyribooligo(poly)nucleotides with a 5′- or 3′-terminal phosphate group.

Because they break any phosphomonoester bonds, phosphomonoesterases are known as nonspecific. This category of PME includes alkaline phosphatase from *E. coli*, acid phosphatase from the prostate, and alkaline phosphatase from wheat. In their presence, nucleotides are degraded to nucleosides, while oligonucleotides with terminal phosphate groups are converted into those with free 3′- and 5′-hydroxyls; a molecule of inorganic phosphate is removed in the course of the reaction:

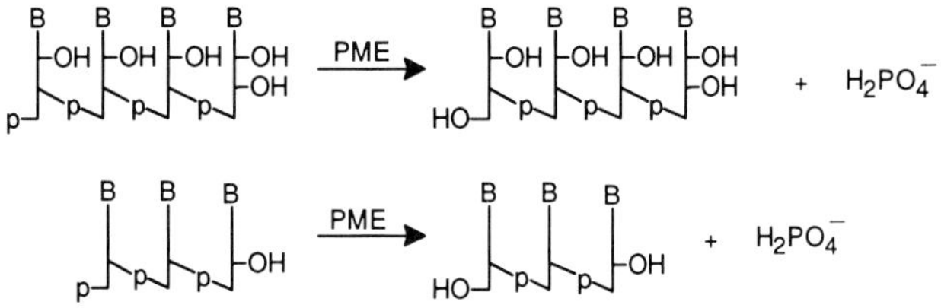

Specific phosphomonoesterases are treated elsewhere. In addition to these PMEs hydrolysing only nucleotides mention should be made of 3′-nucleotidase found in *E. coli*. This enzyme removes 3′-terminal phosphate groups in RNA and DNA.

Snake Venom Phosphodiesterase. The enzyme is isolated from the venom of cobras, vipers, and so on. It cleaves RNA and DNA step by step beginning from the 3′ end with removal of nucleoside 5′-phosphates.

When hydrolysis is conducted under conditions when not all internucleotide linkages have time to be cleaved (limited amount of the enzyme, low temperature), a statistical set of molecules is formed including both starting molecules and those shortened from the 3′ end by one, two, three and more nucleotide units.

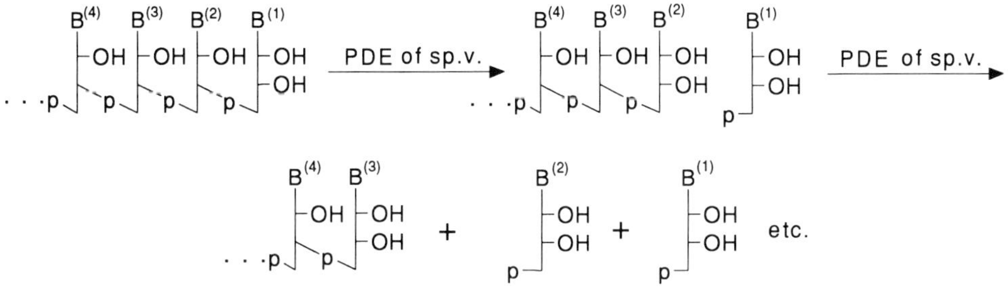

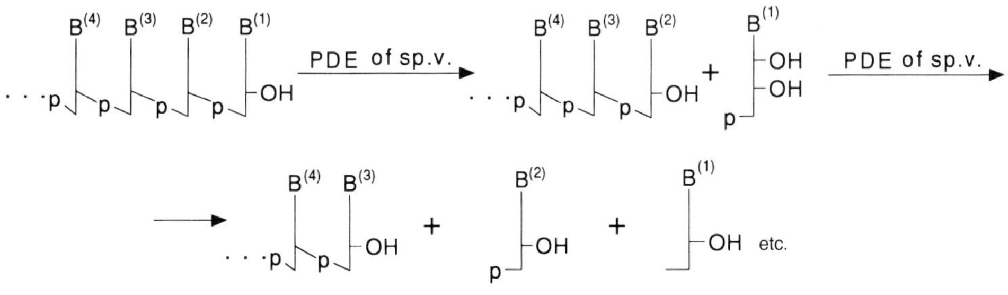

The presence of a phosphate group at the 3′ end of an oligo(poly)nucleotide in the substrate inhibits the enzyme. In such cases, snake venom PDE is used after treating the oligo(poly)nucleotide with phosphomonoesterase. This enzyme is employed for preparative separation of deoxyribonucleoside and ribonucleoside 5′-phosphates, respectively, from DNA and RNA.

Spleen Phosphodiesterase. This enzyme catalyzes cleavage of oligo(poly)nucleotides from 5′ end of the polymer chain. Each catalytic event breaks only the terminal internucleotide linkage with removal of nucleoside 3′-phosphate. When hydrolysis is not complete, a statistical set of molecules shortened from the 5′ end by one, two, three and more nucleotide units is formed, just as in the case of snake venom PDE.

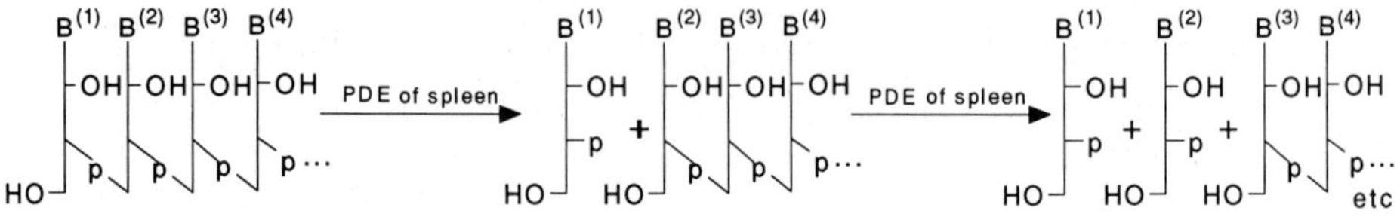

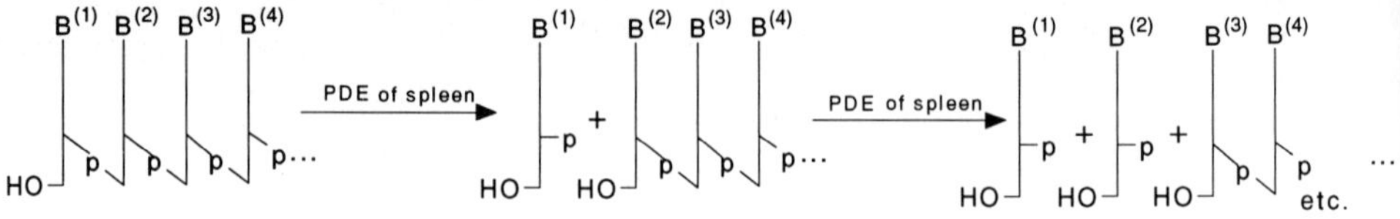

The enzyme does not cleave polynucleotides containing a phosphate group at the 5′ end.

5.6.2 Methods for Determining the Nucleotide Sequence in Oligonucleotides

As has already been mentioned, determination of the nucleotide sequence in nucleic acids is a rather difficult task and must be accomplished in stages: at first, the polynucleotide chain must be split into units or, to be more precise, oligonucleotides, and then one must determine the nucleotide sequence in each oligonucleotide chain. Analysis of chains consisting of three or four monomer units is not difficult, but as the chain becomes longer, the difficulties increase progressively. Should this be the case, a lot depends on the nature of individual nucleotide units and their mutual arrangement in the oligomer chain.

Oligoribonucleotides. The procedure of analyzing the structure of oligoribonucleotides comprises several steps, the nucleotide composition being determined first, then the terminal fragments and, finally, the sequence of monomer units.

The easiest way to determine the nucleotide composition of oligoribonucleotides is through alkaline hydrolysis with subsequent identification of the corresponding ribonucleoside 3′(2′)-phosphates forming as a result; for example:

A U U C G
HO–ApUpUpCpGp $\xrightarrow{^{-}OH}$ Ap + 2Up + Cp + Gp + Corresponding 2′ - isomers

Exonucleases, such as snake venom PDE, can be used for the same purpose:

U G G A C
pUpGpGpApC $\xrightarrow{\text{PDE of sp. venom}}$ pU + 2pG + pA + pC

The determination of the terminal fragments of the molecules (so-called terminal analysis) may also involve chemical and enzymatic hydrolysis. Alkaline hydrolysis makes it possible to determine the nature of the terminal fragments within a single run, provided the terminal phosphate group is linked to the 5′-hydroxyl (the scheme does not show the corresponding 2′-isomeric nucleotide products):

A U G U
pApUpGpU $\xrightarrow{^{-}OH}$ pAp + HO–Up + HO–Gp + HO–U

In addition to the usual products of alkaline hydrolysis, ribonucleoside 3′(2′)-phosphates, the hydrolysate in this case also contains 3′,5′-diphosphonucleoside (5′-terminal fragment) and a free nucleoside (3′-terminal fragment), which can be separated and identified by paper chromatography and electrophoresis.

The nature of the terminal nucleotides may also be determined by way of complete hydrolysis in the presence of snake venom and spleen PDE, if the terminal phosphate groups are removed in advance. For instance:

A C C G
pApCpCpGp $\xrightarrow{\text{PME}}$ HO–ApCpCpG

$\xrightarrow{\text{PDE of sp.vn.}}$ HO–A + 2 pC + pG

$\xrightarrow{\text{PDE of spleen}}$ HO–Ap + 2 HO–Cp + HO–G

It can be seen that as a result of treatment with snake venom PDE the nucleoside is formed from the 5′-terminal fragment of the molecule, and if spleen PDE is used, the source of the nucleoside is the 3′-terminal fragment. Thus, by using spleen PDE one can get the same information as after alkaline treatment, which is why alkaline hydrolysis is preferred for determining the 3′-terminal groups in oligoribonucleotides – alone or in combination with enzymatic hydrolysis.

Terminal analysis of polynucleotides may be used to determine the length of the polymer chain (it is the numerical ratio between terminal fragments and the rest of nucleotides that is determined).

If it is necessary to find out whether the polymer has phosphate groups at its ends, it is treated with PME. Appearance of a phosphate in the hydrolysate indicates that terminal phosphate groups are present.

In determining the nucleotide sequence in short oligonucleotides it is sometimes quite sufficient to resort to alkaline hydrolysis and/or digestion with snake venom PDE. This is illustrated by the above example and the examples that follow (also formed are the corresponding 2′-isomeric nucleotides which are not shown):

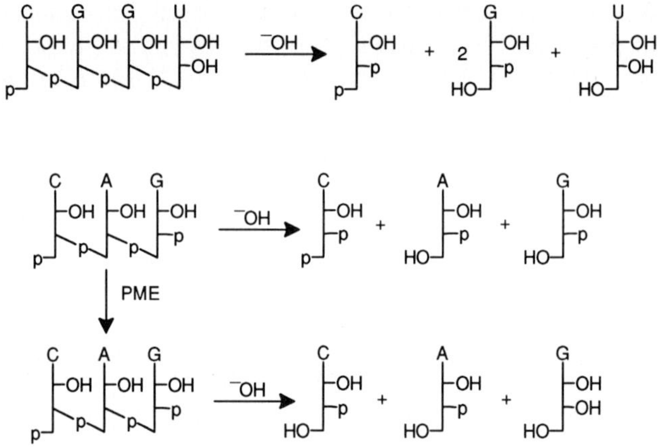

Analysis of longer oligoribonucleotides can be based on two approaches: consecutive removal of nucleotides from one of the ends of the oligonucleotide or its cleavage at the internucleotide linkages using enzymes of different specificity with subsequent establishment of the order in which monomers alternate in the units and reconstruction of the starting oligonucleotide (method of overlapping units).

The consecutive removal of monomer units by a chemical method was for the first time performed in Todd's laboratory. The procedure begins from the 3′ end after periodate oxidation of the 2′,3′-*cis*-diol group in the oligoribonucleotide. The resulting dialdehyde is selectively cleaved under mild conditions in a β-elimination process (the emerging carbonyl group facilitates detachment of the proton at $C^{4'}$) to give an oligonucleotide shortened by one monomer, a dialdehyde fragment, and a base.

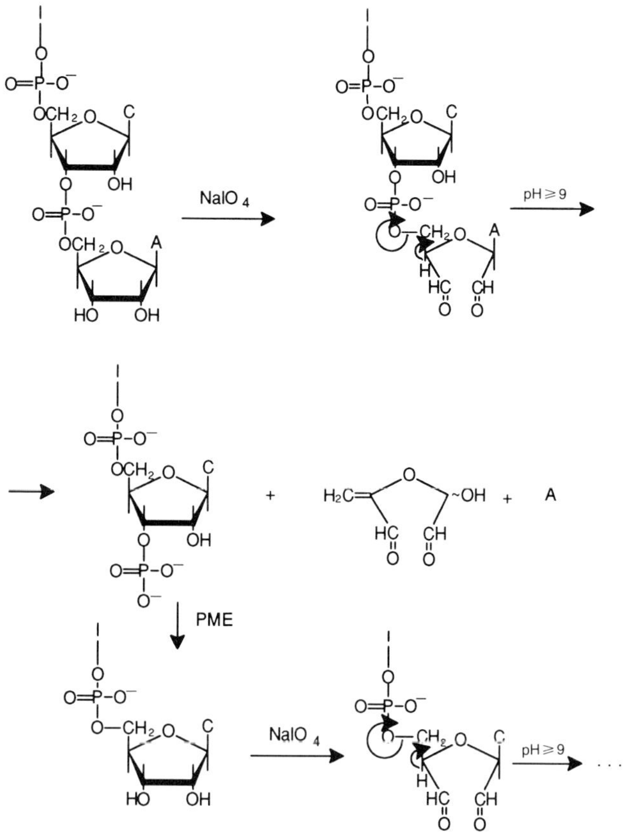

Since reliable methods for analyzing bases are available, one can easily determine the nucleotide with the 3′-terminal group which was removed.

Enzymatic dephosphorylation in the presence of phosphomonoesterase yields an oligonucleotide shortened by one monomer, again containing a 2′,3′-diol group prone to oxidation, which can undergo the same sequence of transformations. A prerequisite for successful application of this method is completeness of each of the three consecutive reactions (oxidation, β-elimination, and dephosphorylation) as well as absence of nuclease traces in the phosphomonoesterase preparation used. If these conditions are not met, by-products rapidly accumulate in the reaction mixture and analysis becomes complicated.

The above procedure makes it possible to determine the nucleotide sequence in oligoribonucleotides rather reliably. It has also been used to determine the 3′-terminal sequence in tRNA. A case in point is establishment of the 3′-terminal sequence in phenylalanine tRNA from *E. coli* having a length of 19 nucleotides.

The consecutive removal of nucleotides from the ends of an oligonucleotide can also be achieved in the presence of exonucleases. Short-term treatment of an oligoribonucleotide with snake venom PDE gives all possible oligonucleotides – a statistical set (as opposed to the above-mentioned step-by-step shortening of the chain) of molecules undergoing no changes and shortened by one, two, three and more monomer units. For example:

pApUpUpCpCpG
pApUpUpCpC
pApUpUpC
pApUpU
pApU

Nucleoside 5′-phosphate

The resulting oligonucleotides are separated by ion-exchange chromatography and 3′-terminal nucleotides are determined in each of them. It can be easily seen that the enumeration of 3′-terminal nucleotides in the unchanged oligonucleotide and then in that shortened by one monomer unit and so on all the way to the dinucleotide (the two nucleotides are enumerated in the latter) will represent the primary structure of the oligomer under analysis, beginning from the 3′ end. This approach is widely used at present to determine the nucleotide sequence in oligodeoxyribonucleotides.

The enzymatic method of determining the nucleotide sequence, involving removal of the terminal nucleotides with the aid of PDE, has its limitations similarly to the chemical method.

For instance, when long oligonucleotides are treated with an enzyme, individual cleavage of internucleotide linkages are observed in the middle of the chain because of the presence of trace amounts of other enzymes. As a result, attempts to determine sequences of eight to ten monomer units in this fashion have seldom been successful. Another approach most widely used to determine the nucleotide sequence in oligoribonucleotides involves cleavage at internucleotide linkages with subsequent separation of short oligonucleotides, establishment of their structure, and reconstruction of the starting molecule. Here are some simple examples illustrating the method under discussion (in both cases, the 3′-terminal nucleotide is known[3]).

[3] Both oligonucleotides are products of polynucleotide hydrolysis with pyrimidyl and guanyl RNases, respectively.

$$\text{ApGpApApCp} \xrightarrow{\text{RNase T}_1} \text{ApGp} + \text{ApApCp}$$
$$\text{ApGp} \xrightarrow{\text{1. PME; 2. }^{-}\text{OH}} \text{Ap} + \text{G} \qquad \text{ApApCp} \xrightarrow{\text{1. PME; 2. }^{-}\text{OH}} 2\text{Ap} + \text{C}$$

$$\text{ApCpApCpGp} \xrightarrow{\text{pyrimidyl RNase}} 2\text{ApCp} + \text{Gp}$$
$$\text{ApCp} \xrightarrow{\text{1. PME; 2. }^{-}\text{OH}} \text{Ap} + \text{C}$$

A partial hydrolysis method has been developed for analysis of oligonucleotides containing polypyrimidine sequences whose structure cannot be determined by complete hydrolysis with pyrimidyl RNase. During such hydrolysis under mild conditions, some internucleotide linkages normally cleaved during complete hydrolysis may remain intact, which makes it possible to isolate oligopyrimidine units. Subsequent analysis of partial degradation products allows one to easily determine their structure and then the structure of the starting oligonucleotide. The following example may illustrate such an approach. The following abridged formula has been written for the heptanucleotide forming a part of valine tRNA (A. A. Baev, T. V. Venkstern *et al.*) after determination of the general nucleotide composition and terminal nucleotides (remember that nucleoside symbols separated by commas in parentheses stand for an unknown sequence):

$$\text{A (A,U,}\Psi\text{,C,C)Gp}$$

After partial hydrolysis of this heptanucleotide in the presence of pyrimidyl RNase two trinucleotides were separated, their structure being determined rather easily:

$$\text{A(A,U,}\Psi\text{,C,C)Gp} \xrightarrow{\text{pyrimidyl RNase}} \text{ApApCp} + (\text{U,}\Psi\text{,C})\text{p} + \text{Gp}$$
$$(\text{U,}\Psi\text{,C})\text{p} \xrightarrow{\text{PME}} (\text{U,}\Psi\text{,C})$$
$$\Psi + \text{pU} + \text{pC} \xleftarrow{\text{snake venom PDE}} (\text{U,}\Psi\text{,C}) \xrightarrow{\text{pyrimidyl RNase}} \Psi\text{p} + \text{Up} + \text{C}$$

These results give unambiguously the following nucleotide sequence in the starting heptanucleotide:

$$\text{ApApCp}\Psi\text{pUpCpGp}$$

Another example (also from the same source) shows how one can gradually approach the true structure in the case of the hexanucleotide (A,U,Ψ,C,C)Gp for which complete enzymatic hydrolysis does not provide an unambiguous answer. The data necessary for determining the structure of this hexanucleotide are listed in Table 5-2.

The above examples show that determination of the oligoribonucleotide structure is a more or less difficult problem which must be solved in a particular way for each individual case.

Oligodeoxyribonucleotides. When the nucleotide sequence is determined in oligodeoxyribonucleotides, the composition and terminal fragments are analyzed first, just as in the case of oligoribonucleotides, and only then the entire sequence.

Table 5-2. Determination of the Structure of Hexanucleotide CpApΨpCpUpGp.

Hydrolysis conditions	Hydrolysis products	Possible structures	Number of possible structures
1. PME 2. –OH	Ap + Up + Ψp + 2Cp + G	(A,U,Ψ,C,C)Gp	60
1. PME 2. Snake venom PDE	C + pA + pU + pΨ + pC + pG	C(A,U,Ψ,C)Gp	24
Pyrimidyl RNase (complete hydrolysis)	2 Cp + ApΨp + Up + Gp	C[U,C,(ApΨ)]Gp	6
Pyrimidyl RNase (partial hydrolysis	→ Pentanucleotide C(U,ΨA,C) ↓ 1. PME 2. Pyrimidyl RNase 2Cp + ApΨp + U i. e. C[(ApΨ),C]Up	CpAp pCpUpGp or CpCpApΨpUpGp	2
	→ Dinucleotide (C,U) ↓ 1. PME 2. –OH Cp + U i. e. CpUp →	CpApΨpCpUpGp	1

The nucleotide sequence may be determined by resorting to acid hydrolysis under vigorous conditions (72 % perchloric acid, 100 °C, or 85 % formic acid, 175 °C), during which a mixture of pyrimidine and purines bases is formed as a result of cleavage of *N*-glycosidic bonds. In general, however, the nucleotide composition is analyzed together with the terminal fragments by way of complete enzymatic hydrolysis with exonucleases – snake venom and spleen PDE. Again, as in the case of oligoribonucleotides, the terminal phosphate is removed with the aid of PME, then treated with PDE under conditions when all internucleotide linkages are broken, for instance:

T G C A
p–Tp–Gp–Cp–A–OH → (PME) → HO–Tp–Gp–Cp–A–OH

PDE of sp. venom: HO–T–OH + pG–OH + pC–OH + pA–OH

PDE of spleen: HO–Tp + HO–Gp + HO–Cp + HO–A–OH

The composition of the nucleosides found in both hydrolysates reveals the nature of the 5′- and 3′-terminal units of the oligodeoxyribonucleotide, while the ratio between the amounts of the emerging nucleoside and nucleotides[4] gives an indication of the chain length and composition of the starting oligodeoxyribonucleotide.

The main difficulty in determining the primary structure of oligodeoxyribonucleotides stems from the fact that until recently no enzymes have been found to be as specific as pyrimidyl and guanyl RNases with respect to oligo(poly)nucleotides[5].

Only one approach has been elaborated for determining the primary structure of oligodeoxyribonucleotides, namely, partial hydrolysis with phosphodiesterases. Just as with oligoribonucleotides, the problem boils down to separation of the oligomer mixture resulting from digestion with the number of nucleotide units gradually going down to two. The analysis of terminal nucleotides in each emerging oligomer makes it possible to determine the order in which nucleotides alternate in the starting molecule.

4) This ratio is usually determined after chromatographic separation of the nucleoside by comparing the absorption intensity in the UV spectrum of the nucleoside and nucleotide fractions.

5) Endonuclease IV is an enzyme that is too specific and scarce.

It should be pointed out that the most difficult task is to find the conditions when the hydrolysate would contain the entire set of nucleotides from the unreacted starting one to the dinucleotide. The oligodeoxyribonucleotide under analysis is first treated with PME, then either with snake venom or spleen PDE. In the former case, the set of oligodeoxyribonucleotides is formed as a consequence of removal of the monomeric 5′-phosphates from the 3′ end of the oligomer chain, and in the latter, as a result of removal of the monomeric 3′-phosphates from the 5′ end. Given below are the sets of oligonucleotides resulting from incomplete degradation of a dodecadeoxyribonucleotide, aided with snake venom and spleen PDE:

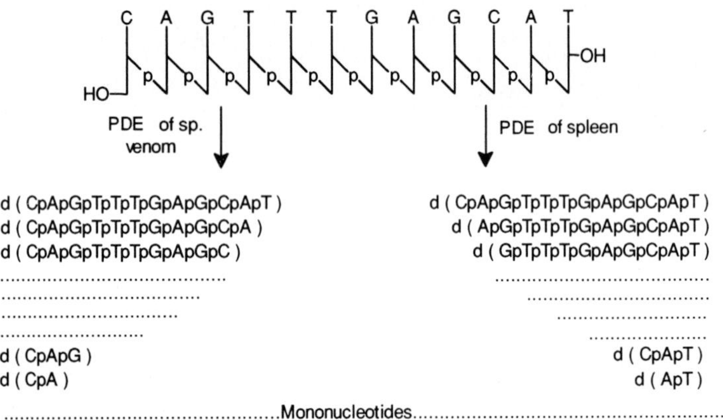

In the beginning of analysis one may obtain any one of the above sets from the starting oligonucleotide. Then, each individual component (except for the mononucleotides) is isolated to determine which nucleotide is terminal. If a set of oligonucleotides is obtained using snake venom PDE, the 3′-terminal fragments are determined in each by separating them in the form of nucleosides under the action of spleen PDE. For example:

C A G T T T G A G C A T
HO– –OH
PDE of spleen

2 C + 3 A + 3 G + 3 T + T (–OH)

3′ - terminal fragment

C A G
HO– –OH
PDE of spleen

C + A + G (–OH)

3′ - terminal fragment

And if a set of oligonucleotides is derived through hydrolysis of the starting oligomer in the presence of spleen PDE, then the 5′-terminal fragments are determined (in the form of nucleosides) in each isolated oligonucleotide using snake venom PDE. For example:

C A G T T T G A G C A T
HO–…p…–OH —PDE of sp.venom→

→ C (HO–, –OH; 5' - terminal fragment) +3 A (p–, –OH) +3 G (p–, –OH) +4 T (p–, –OH) + C (p–, –OH)

C A G
HO–…p…–OH —PDE of sp.venom→ C (HO–, –OH; 5' - terminal fragment) + A (p–, –OH) + G (p–, –OH)

The coincidence of the results obtained in the two ways described above is indicative of their reliability.

However, the need to isolate each oligonucleotide resulting from hydrolysis of the sample under investigation and subject each of them to terminal analysis renders the method too cumbersome, especially when we are dealing with longer oligomers.

With this in view, Sanger recently developed a method, sometimes called the “wandering-spot” method, which allows the nucleotide sequence in oligodeoxyribonucleotides to be determined without isolation of each individual oligonucleotide resulting from partial digestion of the sample under analysis. It includes preparation of so-called nucleotide maps (see below), which are essentially the results of two-dimensional separation of oligonucleotides obtained through enzymatic hydrolysis (first, in the presence of PME for removing the terminal phosphate groups, then, as has already been described, snake venom or spleen PDE). A ^{32}P-labeled terminal phosphate group is introduced into the oligonucleotides under analysis so that one could work with minimal amounts of the substances. The 5′-terminal hydroxyls undergo phosphorylation. The reaction is conducted in the presence of the enzyme polynucleotide kinase and γ-^{32}P-ATP:

$$\mathrm{d(NpN'p\ldots)} \xrightarrow[\text{5'-polynucleotide kinase}]{\overset{*}{\mathrm{p}}\mathrm{ppA}} \mathrm{d(\overset{*}{p}NpN'p\ldots)}$$

Nucleotides containing labeled phosphate (indicated $\overset{*}{\mathrm{p}}$ in the above scheme) are easily detectable when sensitive X-ray films are used (autoradiography).

Two-dimensional separation of a mixture of 5′-phosphorylated oligodeoxyribonucleotides obtained in the above-described fashion proceeds in two steps: first, the mixture is subjected to electrophoresis on a narrow strip of cellulose acetate at pH 3.5 in one direction; then, the result is placed on a DEAE-cellulose plate so that it serves as the starting line in thin-layer chromatography with an aqueous solution of the oligoribonucleotide mixture produced during alkaline hydrolysis of RNA being used as the eluent (this process is known as "homochromatography"; oligonucleotides without labeled phosphorus do not impede the determination). The method offers a clear separation of labeled oligonucleotides differing in length just by a single unit.

In order to understand the results of two-dimensional separation or, in other words, be able to "read the nucleotide map" one must know the extent to which the mobility of the oligonucleotide changes in the course of electrophoresis and chromatography as a result of removal of a mononucleotide unit. The mobility of oligonucleotides during electrophoresis depends both on the net negative charge of the molecule and on its size: the greater the charge, the greater the mobility, the latter dropping as the molecular weight increases. Therefore, a substantial decrease in charge even with simultaneously diminishing molecular weight will result in a particle with reduced electrophoretic mobility, whereas a small decrease in charge with a simultaneous decrease in molecular weight will give a particle with increased mobility. The difference in total charges of the nucleotides stems from the fact that cytosine, adenine, and guanine exhibit different basicity – that is, protonation capacity and, consequently, ability to carry a positive charge. At pH 3.5, the phosphate groups in nucleotides carry only a negative charge, and the bases differ in protonation capacity. This is why electrophoresis should be conducted at pH 3.5. In this case, the net charges of deoxycytidine, deoxyadenosine, deoxyguanosine and deoxythymidine phosphates approach –0.15, –0.45, –0.9 and –1, respectively.

Consider two extreme cases. If thymidine phosphate is removed from an oligodeoxyribonucleotide or, in other words, the molecule loses a negative charge, and in spite of the marked decrease in molecular weight, whatever is left of the oligonucleotide becomes electrophoretically less mobile; and if cytidine phosphate with a net charge of only –0.15 is removed and the molecular weight is still considerable, the electrophoretic mobility of the oligonucleotide with reduced molecular weight and volume, but not charge, turns out to be greater, as compared to the starting oligodeoxyribonucleotide. The mobility of oligonucleotides during chromatography under preset conditions always increases inversely with the length of the oligonucleotide chain. Thus, the oligonucleotide becomes more mobile as it loses mononucleotide units.

Let us now see how nucleotide maps will be affected by the above regularities. The removal of deoxycytidine and deoxythymidine phosphates will change the nucleotide map as shown below on the top.

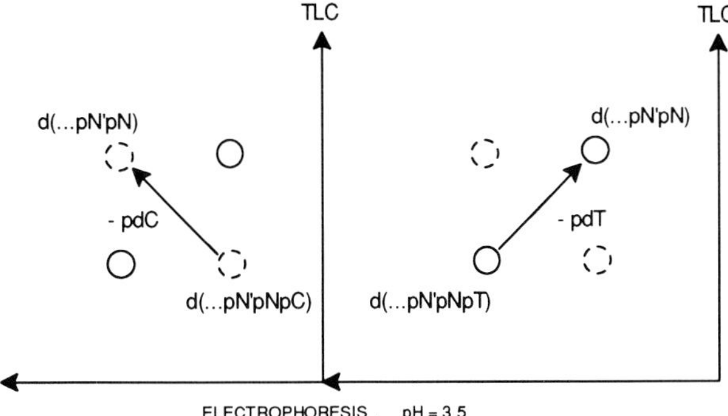

It can be seen that the two-dimensional movement of the oligonucleotide shortened by one unit is given, depending on the nature of the removed unit, by the angle formed with the perpendicular to the starting line of the unshortened nucleotide.

- pdC - pdT

The more chromatographically mobile of every two adjacent spots on the two-dimensional map is a product of removal of the terminal unit from the less mobile precursor. If it is also more mobile electrophoretically, the removed nucleotide is deoxycytidylic acid, and if it is less mobile, then we are dealing with deoxythymidylic acid. Given below by way of example is a two-dimensional map produced during analysis of the undecadeoxyribonucleotide $d(C_5T_6)$ sequence after incomplete digestion with snake venom PDE:

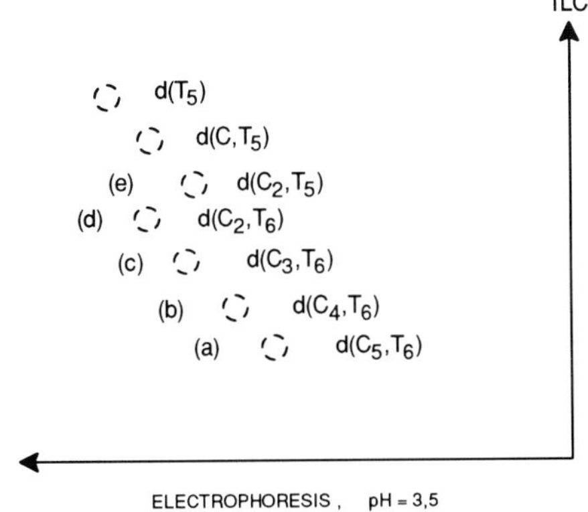

Spot (a) corresponds to the starting oligodeoxyribonucleotide d(C_5T_6). The next spot (b), which is more mobile chromatographically, is a product of removal of a nucleotide from the 3′ end of the starting oligonucleotide. Since compound (b) is also more mobile electrophoretically, pdC is the detached unit. Thus, it may be assumed that the 3′-terminal unit in d(C_5T_6) is pdC. The two following spots (c) and (d) seem to be products of sequential removal of two more deoxycytidine 5′-phosphate groups; that is, the formula of the starting oligonucleotide can be written as follows:

d[(2C,6T)pCpCpC]

Compound (e) is electrophoretically less mobile than (d), yet, as can be inferred from chromatographic mobility, it is formed from compound (d) as a result of removal of a nucleotide. The implication is that compound (e) is formed from (d) as a consequence of removal of deoxythymidylic acid; that is, the formula of the oligonucleotide can be written as

d[(2C,5T)pTpCpCpC]

Complete analysis of the map gives the 3′-terminal sequence of the oligonucleotide as

d(. . . pCpCpTpCpCpC);

that is, the structure of the starting compound is

d(pTpTpTpTdpTpCpCpTpCpCpC).

This is how the nucleotide sequence in many oligodeoxyribonucleotides containing pyrimidine nucleotides has been determined.

Two-dimensional (chromatography-electrophoresis) maps of oligonucleotides also containing purine deoxyribonucleotides reveal the same pattern: removal of all nucleotides (both pyrimidine and purine ones) leads to higher chromatographic mobility of the fragment, yet during electrophoresis its mobility remains virtually the same after removal of deoxyadenylic acid whose loss brings about a decrease in the net charge (by –0.5) along with a tangible molecular weight reduction. Removal of pdG results in lower electrophoretic mobility of the fragment (the net charge of the molecule is reduced by –0.9) but to a lesser extent than in the case of pdT removal.

The change in mobility of an oligodeoxyribonucleotide during two-dimensional chromatography-electrophoresis as a result of removal of pyrimidine and purine nucleotides can be shown schematically as follows:

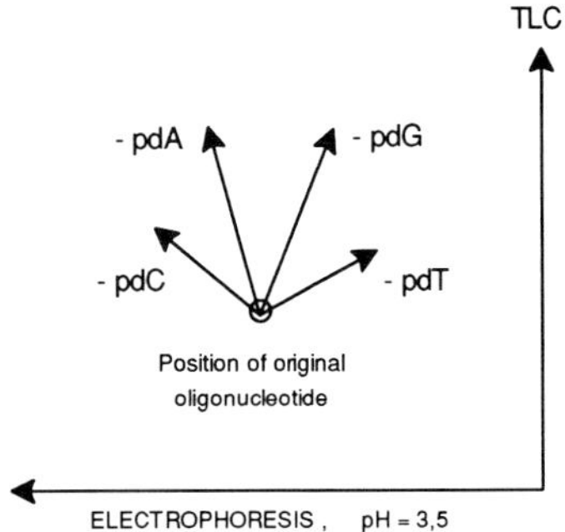

Note that removal of purine nucleotides increases the chromatographic mobility of the remaining oligonucleotide to a much greater extent than that of pyrimidine ones.

Evidently, this phenomenon cannot be explained by the fact that purine nucleotides have a greater molecular weight because the difference in the latter is relatively small (pdC 227, pdT 242, pdA 251, pdG 267); a more likely reason is the pronounced adsorption capacity of purine bases (on DEAE-cellulose in the case under consideration).

Shown below is a nucleotide map of d(C_2,T_4,A_2,G_4) – a dodecanucleotide containing all four deoxyribonucleotides. After partial digestion with spleen PDE, all eleven compounds became so arranged that it was possible to determine the initial nucleotide sequence:

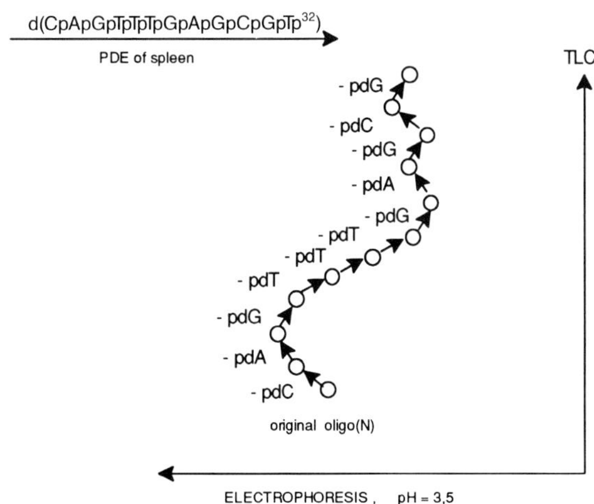

This approach was used to determine the nucleotide sequence of many oligodeoxyribonucleotides isolated from hydrolysates of phage and other DNAs.

Among the drawbacks of the method is its inability to obviate the difficulties that arise while selecting the right conditions for partial digestion of the oligonucleotide under analysis with phosphodiesterases, which would make it possible to obtain a complete set of oligonucleotides – from the intact starting one to that consisting of two units. Moreover, the method may involve errors stemming from small changes in mobility during analysis of rather long oligonucleotides. The best results were obtained while studying oligonucleotides with no more than ten monomer units.

At present, the Maxam–Gilbert method is also used to determine the primary structure of oligonucleotides.

References

1. S. Mandeles, *Nucleic Acid Sequence Analysis.* Columbia University Press, N.Y.–London (1972).
2. *Oligonucleotide Synthesis: A Practical Approach.* M. J. Gait (Ed.), IRL Press, Oxford–Washington, D. C., pp. 135-151 (1984).
3. Y. Mizuno, *The Organic Chemistry of Nucleic Acids.* Elsevier, Amsterdam–Oxford–N.Y.–Tokyo, pp. 255–295 (1986).
4. *Synthesis and Application of DNA and RNA.* S. A. Narang (Ed.), Academic Press, Orlando–San Diego–N.Y., etc. pp. 137–180 (1987).

6 Determination of the Primary Structure of Nucleic Acids

6.1 Introduction

One of the most spectacular achievements of molecular biology in the past decade was the development of very rapid methods for determining the primary structure of DNA and, later, RNA, which are known as sequencing – that is, determining the sequence of nucleotides in these biopolymers. Although the currently used sequencing techniques appeared as recently as 1977, their origin is to be found in the various branches of nucleic acid chemistry and enzymology, which were already well developed in the sixties.

In the early seventies, all those involved in sequencing could not help being fascinated by what became known as the block method of RNA sequencing. Developed in the mid sixties to determine the nucleotide sequence in tRNA, this method had made it possible to establish the structure of hundreds of tRNAs, many 5S RNAs, and even some high-molecular weight RNAs within mere five to eight years.

Naturally, the first DNA sequencing methods (early seventies) were closely similar to the block method. Radical changes had taken place by the mid seventies, when it became clear that the block method of RNA sequencing could not be applied to DNA. The decisive factor was the absence of enzymes hydrolysing DNA at one or only two of the four bases (analogs of guanyl and pyrimidyl RNases). The cul-de-sac into which scientists had worked themselves forced them to think about drastically new approaches. The problem had boiled down to developing a new concept of sequencing. Soviet scientists were involved in its elaboration at the very outset (see, e. g., the paper by Ye. D. Sverdlov and coworkers, published in FEBS Lett. in 1973). The new principle of determining the primary structure of DNA may be formulated as follows: label the ends, cleave the long polynucleotide statistically at one of the four nucleotides, fractionate the fragments, and measure the length of each labeled fragment. However, before this concept was embodied in a working method for sequencing high-molecular weigth DNAs then RNAs, two methodological difficulties had to be obviated. First of all, it was necessary to develop a technique of fractionation along the length of oligo- and polynucleotides differing by a single nucleotide link (or a single negative charge). Sets of such fragments are formed when the new sequencing concept is put to practical use. Although homochromatography was used as recently as the late sixties to separate small oligonucleotides, the problem was solved only with the advent of such a powerful technique as polyacrylamide gel electrophoresis (PAGE).

The other problem was finding a way to isolate individual DNA fragments (individual high-molecular weight RNAs in the case of RNA) necessary for analysis. Used initially for the purpose were bacterial plasmids, bacteriophages, or animal viruses (e. g., SV40) containing homogeneous DNAs, which made it possible to easily produce several hundreds of micrograms or 100 picamoles of DNA for analysis. However, it was only advances in genetic engineering, especially in molecular cloning, that permitted experimenters to isolate in any amounts and easily identify any fragments of genomic DNAs under investigation. State-of-the-art methods allow one to sequence nucleic acid fragments consisting of no more than 500 nucleotides within a single experiment. Such a limitation is imposed by the current possibilities of gel electrophoresis. This is why molecules containing a thousand and more pairs of nucleotides are usually split into so-called restriction fragments with the aid of restriction enzymes.

The methodological difficulties were obviated along with elaboration of the new sequencing concept. The first seminal work outlining the polymerase copying method was published by F. Sanger in 1975 under the title "A Rapid Method for Determining Sequences in DNA by Primed Synthesis with DNA Polymerase" (the updated sequencing technique was described in 1977). It was followed by another work of equal importance, based on chemical destruction of DNA, which appeared in 1977: "A New Method for Sequencing DNA" by A. Maxam and W. Gilbert.

Thus, the DNA sequencing methods belong essentially to two types of techniques differing in the exact way to obtain statistical sets of labeled oligo- and polynucleotides, which allow the nucleotide sequence in the DNA under investigation to be "read" immediately after fractionation. One of the techniques (Sanger's method) resides in enzymatic synthesis of DNA copies (cDNAs) terminating at one of the four nucleotides; the other technique is based on chemical cleavage of DNA at one or only two of the four nucleotides (Maxam–Gilbert method). In spite of the Sanger's method appearing before that developed by Maxam and Gilbert, the chemical method will be considered first in keeping with the general purpose of this textbook to cover as fully as possible recent developments in the chemistry of nucleic acids.

The revolutionary impact of the new methods on DNA sequencing was felt immediately by those involved in the sequencing of high-molecular weight RNAs.

The traditional block method for determining the primary structure of RNA had reached its limit by the late seventies during sequencing of single-stranded RNA of phage MS-2, containing 3500 nucleotides. It became evident at that time that rapid sequencing methods must be applied to RNA as well. This book also covers the rapid RNA sequencing methods. The solid-phase technique for sequencing DNA and RNA will be discussed in what follows along with the use of computers in this area.

6.2 DNA Mapping

A prerequisite condition for sequencing high-molecular weight DNAs is a preparation of a restriction map – that is, representing the mutual arrangement of sites of cleavage by particular restriction endonucleases over the DNA fragment of interest and approximate distances between them.

The mutual arrangement of the large blocks resulting from the cleavage of the DNA chain by restriction enzymes can be determined without any prior information about the entire sequence of nucleotides in these blocks. This is possible owing to the specific action of restriction enzymes. One can resort for this purpose both to incomplete cleavage by the same restriction enzyme and to another restriction enzyme. The method is based on the sharp difference between the molecular weights of the fragments resulting from complete and incomplete cleavage, as well as the possibility of distinct separation of the cleavage products by means of electrophoresis. The latter is performed in a polyacrylamide or agarose gel. What happens in the process is separation of DNA fragments by molecular weight. Having plotted mobility versus molecular weight (for which purpose low-molecular weight DNAs of some phages with a known molecular weight are employed), one can find the molecular weight of the DNA under analysis from its electrophoretic mobility.

An example of such an approach is provided by the following scheme of analysis of SV40 virus DNA with a molecular weight of $3 \cdot 10^6$.

Complete cleavage of this DNA, which is essentially a cyclic double-stranded polydeoxyribonucleotide, by a mixture of restriction enzymes Hind isolated from *Haemophilus influenzae* R_d gives 11 fragments equal in length to 400 to 2500 nucleotides; they are shown on the diagram as fragments A through K. They are separated electrophoretically in polyacrylamide gel. At the same time, the initial DNA is treated with the same enzyme under mild conditions to obtain, then to separate by the same method, polynucleotides with a molecular weight higher than in the first case. After complete cleavage of these blocks the investigator determines which of the fragments A through K they contain and in what sequence (see Fig. 6-1).

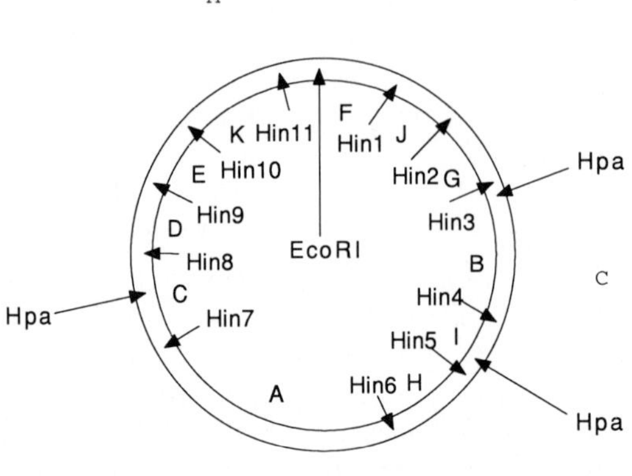

Fig. 6-1. DNA mapping.

By using the overlapping block method one can determine the structure of the DNA of interest. The data for the SV40 virus DNA are shown in Table 6-1.

Table 6-1. Results of Cleaving SV40 Virus DNA by a Mixture of Restriction Endonucleases Hind.

Reaction No.*	Composition of blocks resulting from incomplete cleavage, according to complete hydrolysis data**	Overlapping of blocks, according to electrophoresis data
1	G, J	G J
2	H, I	(H, I)
3	F, K	F K
4	E, K	K E
5	J, F	J F
6	F, G, J	G J F
7	C, D	(C, D)
8	B, G	B G
9	B, G, J	B G J
10	B, F, G, J	B G J F
11	B, F, G, J, K	B G J F K
12	A, C, D	(C, D) A
13	B, F, G, H, I, J	(H, I) B G J F
14	B, F, G, H, I, J, K	(H, I) B G J F K
15	A, C, D, E	E (C, D) A
16	A, B, G, H, I, J, K	A (H, I) B G J

* The fractions are arranged in decreasing order of their mobility during PAGE.
**The products of complete hydrolysis are denoted by capital letters in decreasing order of their size, A being the largest.

It can be seen that the fragments occur in the following sequence:

E(C,D)A(H,I)BGJFK

The position of fragments C and D as well as H and I cannot be determined from the above data. The location of all fragments was determined by using a mixture of other restriction enzymes Hpa (see above), which cleave SV40 DNA into three fragments: Hpa A, Hpa B and Hpa C with chains 4200, 3400 and 2000 nucleotides long, respectively, each fragment being subsequently treated by a mixture of enzymes Hind. Under the action of the latter, Hpa A was cleaved into fragments D, E, F, G, J, K and a protion of C; Hpa B was cleaved into fragments A, H and a portion of C; and Hpa C was cleaved into fragments B and I. By comparing these results one can distribute all eleven fragments A through K as shown in the scheme. The results also make it possible to define the mutual arrangement of the three fragments produced by the Hpa-mediated cleavage.

The next section covers some restriction enzymes widely used in genetic engineering, including DNA mapping and fragmentation.

6.3 Restriction Endonucleases

Restriction endonucleases cleave at specific sequences of double-straded DNAs. These enzymes were discovered in various bacteria in the early seventies. At present, more than 800 restriction endonucleases have been isolated from different (hundreds of) bacterial strains. All restriction endonucleases belong to three major classes (I, II and III). Of particular interest to gene engineers are restriction endonucleases of type II, recognizing 4- to 8-nucleotide sequences. In such segments, they cleave the double-stranded DNA chain at linkages located either in the middle or at the edges of the recognition site. Usually, but not always, the sequences recognized by restriction endonucleases are symmetrical (have a second-order axis of symmetry). The cleavage of each chain proceeds at identical internucleotide linkages with respect to the axis of symmetry. Shown schematically below are nucleotide sequences (commonly called sites) recognized by some restriction endonucleases. The arrows indicate the internucleotide linkages that undergo cleavage.

It can be seen that certain enzymes cleave DNA in such a way that the duplex portions at the cleavage sites have protruding 5′ (*Eco*RI) or 3′ (*Hha*I, *Pst*I) ends. In other instances (*Hae*III), products with blunt ends are formed. The recognized tetranucleotide sequences occur at intervals equal to $4^4 = 256$ nucleotides, on the average, while the hexanucleotide ones occur at intervals equal to $4^6 = 4096$ nucleotides. Consequently, the cleavage of DNA by a res-

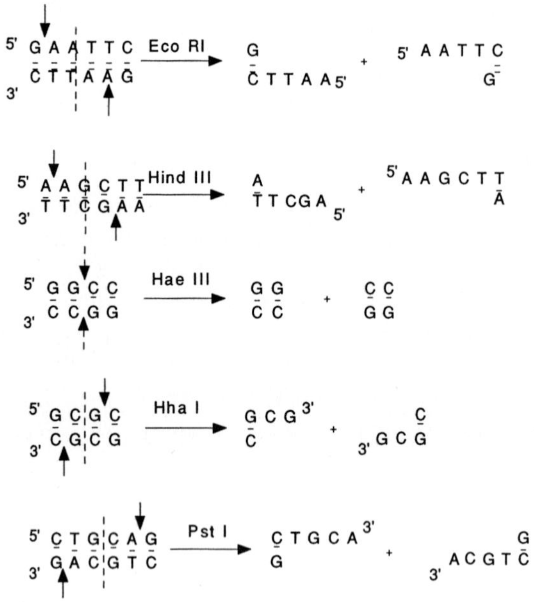

5' G A A T T C — Eco RI → G + 5' A A T T C
3' C T T A A G — C T T A A 5' + G

5' A A G C T T — Hind III → A + 5' A A G C T T
3' T T C G A A — T T C G A 5' + A

5' G G C C — Hae III → G G + C C
3' C C G G — C C + G G

5' G C G C — Hha I → G C G 3' + C
3' C G C G — C + 3' G C G

5' C T G C A G — Pst I → C T G C A 3' + G
3' G A C G T C — G + 3' A C G T C

triction endonuclease recognizing hexanucleotide sites produces longer fragments. Significantly, since restriction endonucleases recognize strictly defined portions of DNA, they thus provide a highly precise tool for controlled cleavage of DNA into fragments of predetermined lengths having, in turn, strictly defined termini. Summarized in Table 6-2 are some restriction endonucleases used in structural studies.

Table 6-2. Some Restriction Enzymes Used for Analyzing the Primary DNA Structure. The enzymes recognizing the same nucleotide sequences are called isoschizomers. The arrows indicate the internucleotide linkages that undergo cleavage. In the case of double-stranded oligodeoxyribonucleotide sequences, the polarity of the ends is always 5′–GGCC–3′

. . . .

3′–CCGG–5′

The same applies to other sequences as well.

Abridged enzyme name	Recognized nucleotide sequence in DNA	Number of cleavage *sites in DNA of*		Source strain of the enzyme
		phage λ	SV40 virus	
	↓			
*Hae*III	GG CC	>50	18	*Haemophilus aegyptus*
	· · · ·			
*Blu*II	CC GG	>50	18	*Brevibacterium luteum*
	↑			

Table 6-2. (Continued).

Abridged enzyme name	Recognized nucleotide sequence in DNA	Number of cleavage *sites in DNA of* phage λ	SV40 virus	Source strain of the enzyme
*Hap*II	↓ C CG G · ·· ·	>50	1	*Haemophilus aphrophilus*
*Hpa*II	G GC C ↑	>50	1	*Haemophilus parainfluenzae*
*Alu*I	↓ AG CT ·· ·· TC GA ↑	>50	32	*Arthrobacter luteus*
*Eco*RII	↓ CCTGG · · · · · GGACC ↑	>35	16	*Escherichia coli*
*Eco*RI	↓ G AATT C · · · · · · C TTAA G ↑	5	1	*Escherichia coli*
*Hind*III	↓ A AGCT T · · · · · · T TCGA A ↑	6	6	*Haemophilus influenzae R_d*
*Hinc*II	↓ GTPy PuAC · · · · · ·	34	7	*Haemophilus influenzae R_c*
*Hind*II	CAPu PyTG ↑	34	7	*Haemophilus influenzae R_d*
*Hpa*I	↓ GTt AAC · · · · · ·	11	5	*Haemophilus parainfluenzae*
*Apo*I	CAA TTG ↑	11	5	*Arthrobacter polychromogenes*
*Bam*HI	↓ G GATC C · · · · · · C CTAG G ↑	5	1	*Bacillus amyloliquefaciens H*

Table 6-2. (Continued).

Abridged enzyme name	Recognized nucleotide sequence in DNA	Number of cleavage *sites in DNA of* phage λ	SV40 virus	Source strain of the enzyme
*Bal*I	↓ TGG CCA · · · · · · ACC GGT ↑	15	0	*Brevibacterium albidum*
*Xma*I	↓ C CCGG G · · · · · · G GGCC C ↑	3	0	*Xanthomonas malvacearum*
*Sma*I	↓ CCC GGG · · · · · · GGG CCC ↑	3	0	*Serratia maccescens* S_b
*Pst*I *Xma*II	↓ C TGCA G · · · · · · G ACGT C ↑	18	2	*Providencia stuartii* *Xanthomonas malvaceaurm*

6.4 Controlled Chemical DNA Cleavage Method (Chemical Sequencing)

6.4.1 Basic Principle of the Method

The method was proposed in 1977 by the American scientists A. Maxam and W. Gilbert. It was immediately recognized as very simple, fast and reliable and is now most extensively used worldwide. Its simplicity is determined not only by the sequencing procedure itself, but also by the preparation of the starting individual DNA sample. All is needed, as has already been mentioned, are restriction endonucleases and enzymes enabling incorporation of label into the 5′ or 3′ end of DNA.

The basic principle of the method is location of each of the four bases along the polynucleotide chain relative to one (5′- or 3′-) terminal nucleotide which thus becomes a reference point. To this end, one resorts to statistical chemical modification in four different reactions involving one or two of the four DNA bases to permit subsequent quantitative cleavage of the sugarphosphate backbone at the modification sites. The ideal result is obtained when the base in question in each 200- to 300-nucleotide DNA fragment under investigation is modified only in one particular position. Cleavage of the polymer gives rise to a plurality of molecules differing in length according to the position of the base with respect to the reference point. If, for example, the reference point is a ^{32}P-labeled 5′-terminal nucleotide, then only molecules containing the 5′-terminal sequence can be "seen". To determine the length of the resulting fragments use is made of polyacrylamide gel electrophoresis (PAGE) under denaturating conditions (in the presence of urea), when the products of partial DNA cleavage are sorted by size. When electrophoresis is over, its result is fixed by placing the gel on an X-ray film. After exposure and development, the film (autoradiograph) shows dark bands corresponding to the position of ^{32}P-labeled oligonucleotides in the gel. Naturally, unlabeled oligonucleotides, also formed during chemical degradation, do not show on the autoradiograph. While reading the sequence out from the latter, the experimenter simply notes down the base-specific reagent that has cleaved the chain at the next nucleotide.

Shown schematically below is a sequence of nucleotides in a 20-unit polydeoxyribonucleotide and the process of degradation of the polynucleotide chain at the cytosine (C) sites, as well as the resulting labeled and unlabeled oligonucleotides.

```
20-unit DNA
fragment ³²P-             5         10        15        20
labeled at      *pG-A-C-T-A-C-C-G-T-A-C-C-T-A-G-C-G-T-G-G
the 5' end

Degradation products:

labeled         *pG-Ap
                *pG-A-C-T-Ap
                *pG-A-C-T-A-Cp
                *pG-A-C-T-A-C-C-G-T-Ap
                *pG-A-C-T-A-C-C-G-T-A-Cp
                *pG-A-C-T-A-C-C-G-T-A-C-C-T-A-Gp

unlabeled               pT-Ap
                        pT-A-Cp
                        pT-A-C-C-G-T-Ap
                        pT-A-C-C-G-T-A-Cp
                        pT-A-C-C-G-T-A-C-C-T-A-Gp
                        pT-A-C-C-G-T-A-C-C-T-A-G-C-G-T-G-Gp
                              pC-G-T-Ap
                              pC-G-T-A-Cp
                              pC-G-T-A-C-C-T-A-G-Gp
                              pC-G-T-A-C-C-T-A-G-C-G-T-G-Gp
                                pG-T-Ap
                                pG-T-A-Cp
                                pG-T-A-C-C-T-A-Gp
                                pG-T-A-C-C-T-A-G-C-G-T-G-Gp
                                        pC-T-A-Gp
                                        pC-T-A-G-C-G-T-G-Gp
                                          pT-A-Gp
                                          pT-A-G-C-G-T-G-Gp
```

*pCpApGpTpGpCpApGpTpGpTpTp
*pCpApGpTpGpCpApGpTpGpTpTpGpAp
*pCpApGpTpGpCpApGpTpGpTpTpGpApGpCpC

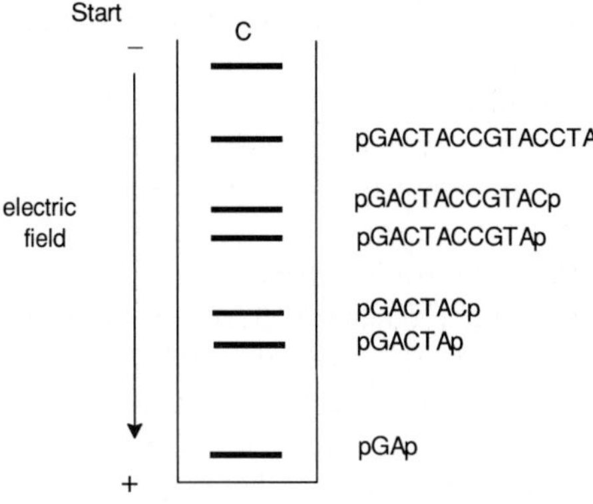

Fig. 6-2. Autoradiogram of a mixture of oligonucleotides after statistical degradation of the 20-unit DNA fragment at C (see text) and separation by PAGE. Shown on the right is the primary structure of the corresponding labeled oligonucleotides.

Figure 6-2 is an autoradiogram of this oligonucleotide mixture after PAGE, used to "read" the sequence of the cytosine nucleotides along the icosanucleotide chain from the 5′ end.

After such an experiment, the formula of the icosanucleotide under investigation can be written as follows:

$$\begin{array}{cccccccccccccccccccc} &&&&5&&&&&10&&&&&15&&&&&20\\ N&N&C&N&N&C&C&N&N&N&C&C&N&N&N&C&N&N&N&N \end{array}$$

N–N–C–N–N–C–C–N–N–N–C–C–N–N–N–C–N–N–N–N

where N is the unknown nucleoside.

The other letter symbols – A, G and T standing for the corresponding nucleotides – are arranged in the same fashion along the chain of the DNA fragment of interest.

Thus, by combining the PAGE technique with determination of radioactive phosphate A. Maxam and W. Gilbert successfully developed a rapid method for determining the primary structure of DNA through its chemical degradation. It is now hard to imagine that as recently as the mid seventies it took years to determine the primary structure of a 30- to 40-nucleotide DNA fragment, whereas already by the early eighties analysis of structures containing a thousand and more nucleotides had become a matter of just a few days work.

6.4.2 Chemical Methods for Specific Cleavage of Polydeoxyribonucleotide Chain

Statistical (partial) specific chemical cleavage of internucleotide linkages in DNA at each of the four nucleotides in four different reactions conducted simultaneously forms the basis of sequencing.

Practically, every chemical cleavage of a particular internucleotide linkage comprises three consecutive steps: modification of the heterocyclic base, detachment of the modified base from the sugar, and β-elimination of the internucleotide phosphate groups from the 3′ then 5′ positions of deoxyribose of the polynucleotide which "lost" the base during the second step. Figure 6-3 illustrates these steps in the context of modification of guanines by their methylation with dimethyl sulfate, followed by cleavage of the *N*-glycosidic bond an β-elimination.

Emphasis should be placed on the intimate realtionship between the reaction at each step of the process and the preceding ones. The cleavage of internucleotide linkages in DNA occurs only at the sugar where the base has been detached. The specificity of such degradation of the polynucleotide chain is ensured at the first step, when the base is modified, and the reaction is conducted under conditions "mild" enough to keep the degradation within the desired limits. The subsequent reactions, namely, removal of the base and β-elimination, proceed in a quantitative manner. Hence, the crucial factor is the selection of the modifying reagents ensuring specificity of the modification (involvement of only one base in the reaction) and, consequently, the site at which the polynucleotide chain is to undergo cleavage. The main modifying agents are dimethyl sulfate and hydrazine. Dimethyl sulfate easily methylates the nitrogens of the purine bases in DNA. Methylation in double-stranded DNA proceeds perceptibly only in two positions – at N^7 of guanine and N^3 of adenine. It should be pointed out that the methylation of guanine in DNA proceeds at a rate almost an order of magnitude greater, as compared to adenine ones. This is precisely what makes cleavage possible during sequencing. In both cases, the methylation of purines leads to the destabilization of the *N*-glycosidic bond linking the base and the sugar.

Given below is the scheme of guanine base methylation in an oligonucleotide chain. The positive charge emerging during methylation of N^7 is delocalized in the imidazole ring among the atoms N^7, C^8 and N^9. This weakens the *N*-glycosidic bond, and the latter is broken in a broad pH range at a rate several orders of magnitude greater than in the case of unsubstituted deoxyguanosine.

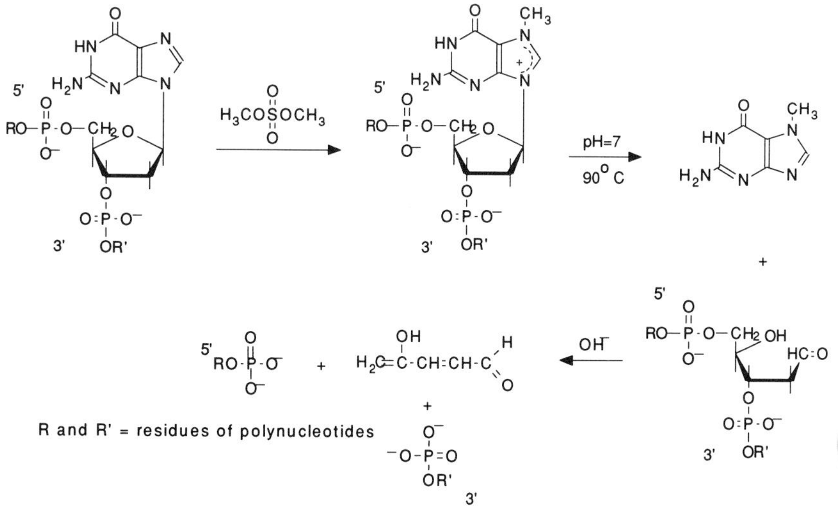

R and R' = residues of polynucleotides

The resulting "base-free" deoxyribose readily passes from the cyclic form into an open one characterized by reactions of β-elimination of the 3′-phosphate group. Such reactions are usually catalyzed by alkalis or organic bases. The unsaturated sugar emerging after "departure" of the phosphate dianion associated with the 3′-terminal polynucleotide sequence being handled again undergoes β-elimination, but this time the 5′-phosphate group is involved. A consequence of such a double β-elimination is the degradation of the polynucleotide chain at the modified deoxyguanosine link.

If such a transformation in different molecules of a polynucleotide being sequenced affects only one of the many deoxyguanosine links (the ideal case of limited modification), the reaction mixture accumulates all possible oligonucleotides – products of polymer chain cleavage at sites previously occupied by deoxyguanosine.

As has already been mentioned, the methylation of N^3 in adenines also takes place under the same conditions, albeit at a rate one order of magnitude slower, with the result that the degradation of DNA occurs at sites previously occupied by adenine nucleosides as well.

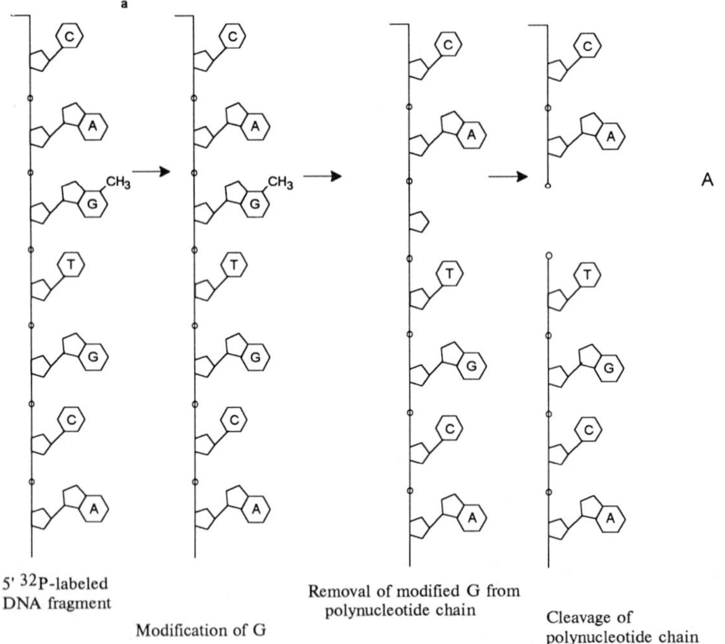

Fig. 6-3. Sequencing of 5′-labeled DNA fragment by limited (statistical) methylation with dimethyl sulfate, followed by removal of N^7-methylguanine and β-elimination of the phosphate groups linked with this nucleoside in the polynucleotide chain: (a) chemical reactions at the modification site; (b) structure of the fragments resulting from limited modification of guanines in the 18-unit oligodeoxyribonucleotide, having a ^{32}P-labeled 5′-terminal phosphate.

The positive charge emerging at N^3 during the methylation of the pyrimidine fragment of the adenine ring doubles the lability of the glycosidic bond, as compared to N^7-methyl deoxyguanosine. The subsequent cleavage of the polynucleotide chain at the sites of N^3-methyl deoxyadenosines proceeds at neutral pH values (ca. 7), its mechanism being similar to that described above for deoxyguanosine links.

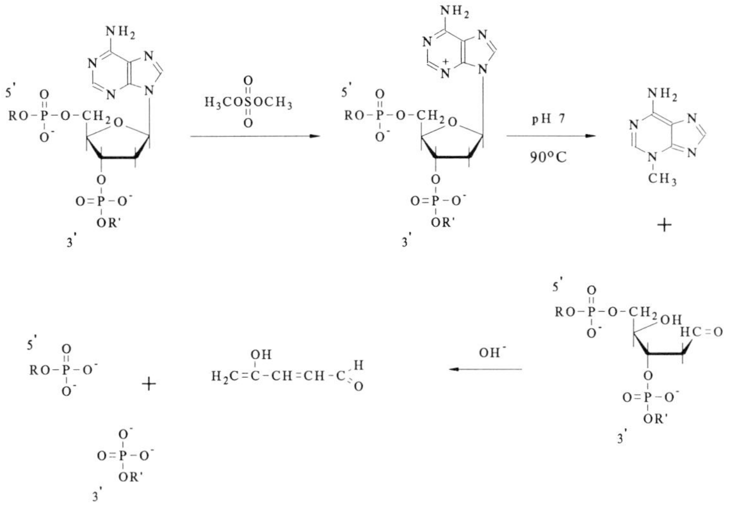

When this process is conducted on end-labeled DNA with subsequent electrophoretic separation of the degradation products and autoradiography, the positions of guanine and adenine nucleotides in the polynucleotide can be identified. However, since the methylation of N^7-guanine in this reaction proceeds at a much faster rate, most DNA cleavages will occur at guanines. The autoradiograph will show that the bands corresponding to cleavage at G are always more intense than those corresponding to cleavage at A (G > A). And if the reaction of base detachment is conducted at pH 2 under conditions of much milder hydrolysis of *N*-glycosidic bonds in 3-methyladenine links, followed by the usual alkaline treatment, the A bands on the autoradiograph will be more intense than the G bands (A > G) (see Table 6-3). These differences in the rates of methylation and detachment of methylated purines were used to distinguish between the constituent guanines and adenines of DNA. In practical experiments, these differences did not always manifest themselves to the same degree, which is why the main efforts were aimed at providing the right conditions for locating adenines in reactions not involving guanines. One of such reactions turned out to be degradation of DNA with a 1 N alkali at 90 °C, opening the adenine ring and, after subsequent treatment with piperidine, breaking the polynucleotide chain at modified A sites. These conditions are conducive to partial degradation at C sites as well (A > C). Provision for three different sets of conditions for DNA modification (G > A, A > G, A > C) has made it possible to reliably determine the positions of G and A in the polynucleotides under investigation.

Later, the right conditions for cleavage only at guanine links were found.

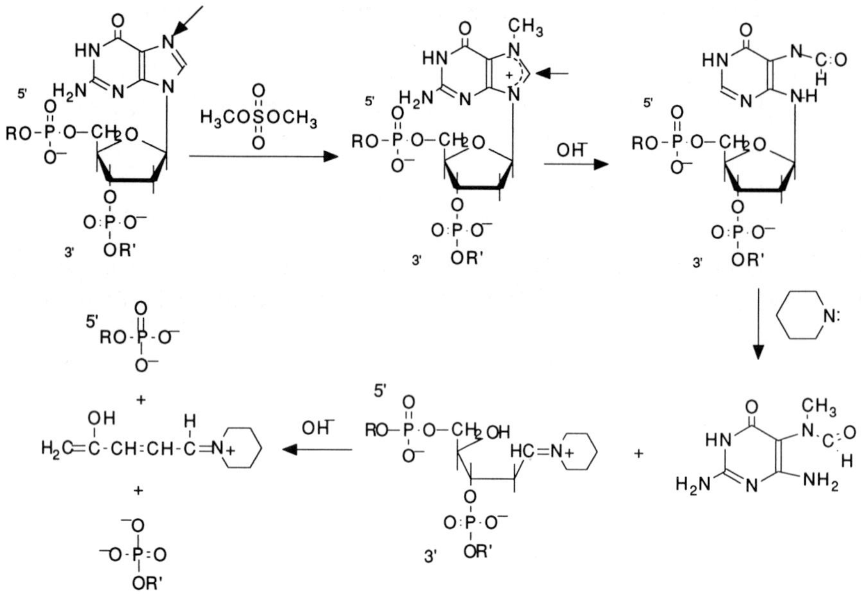

G-specific cleavage is conducted as follows: after methylation with dimethyl sulfate, the reaction mixture is treated with piperidine in free-base form, whose presence allows only guanines to be removed from the polynucleotide chain. It is well known that the imidazole ring in N^7-methylguanine nucleoside is opened by mild alkaline treatment (pH $>$ 10). The resulting *N*-glycoside formed with the participation of the primary amine is converted into azomethine which is typically involved in β-elimination reactions, just as the aldehyde form of the depurinated deoxyribose.

The G-specific reaction has made it possible to simplify the procedure of locating purine nucleotides. The G+A cleavage started being conducted in an acid medium.

Acid depurination also leads to DNA cleavage at guanine and adenine, but no distinction between the two is possible. The mechanism of acid hydrolysis of *N*-glycosidic bonds in purine nucleosides was treated at length elsewhere. As is currently believed, only protonation of N^7 in guanine and N^3 in adeninen renders the *N*-glycosidic bond labile. The corresponding bonds in the pyrimidine links of DNA at low pH values ($<$5) are much more stable. In the latest version of sequencing (A+G modification), limited hydrolysis with formic acid is used.

To determine the position of pyrimidine bases in the polynucleotide chain, they are modified with hydrazine. The most reactive fragment in pyrimidines of nucleic acids is the double bond $C^5–C^6$. In a reaction with hydrazine, one of its molecules is added at the above-mentioned double bond with the result that the strong nucleophilic group of hydrazine becomes linked to the most electrophilic carbon C^6.

NH_2NH_2

R=H or CH_3

The pyrimidine ring is no longer planar and loses its aromatic properties. Notably, the presence of the methyl group at position 5 (thymine) enhances the stability of this pyrimidine base toward nucleophiles. This is why thymines enter into such reactions with greater difficulty as a rule. Naturally, such transformations are unusual for purine bases because the double C–C bond in the purine nucleus, corresponding to $C^5–C^6$ in pyrimidines, is also included in the aromatic system of imidazole.

After addition of the hydrazine molecule at the double bond $C^5–C^6$, the heterocyclic ring of pyrimidines is opened as a result of intramolecular nucleophilic substitution involving the added hydrazine, for example:

NH_2NH_2

R=chain of polynucleotides

The new five-membered ring now includes the hydrazine nitrogens, whereas those of cytosine are beyond the ring. These and subsequent transformations of one of the cytosines in the polynucleotide chain are shown below.

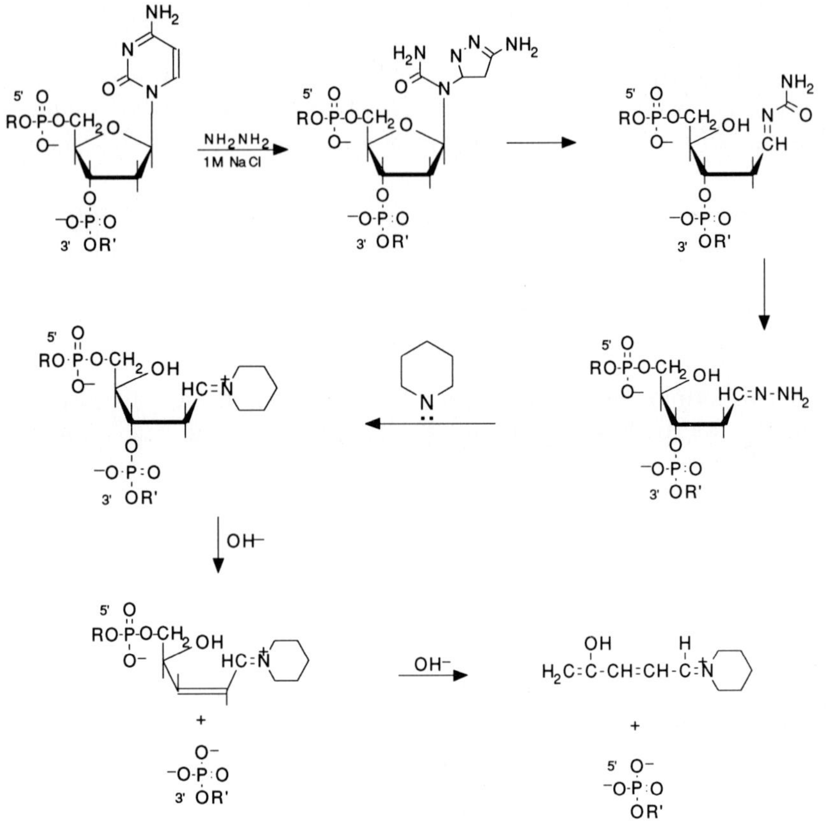

The available data on the mechanism of the reactions between pyrimidine bases and hydrazine in nucleic acids suggest that one of the following groups may remain linked to the sugar at the modification site: the glycosidic nitrogen of urea in the form of a secondary amine; the glycosidic nitrogen in the form of tertiary amine; or hydrazine itself. In all structures of this type the bond linking the sugar to the nitrogen is sufficiently reactive. Treatment with piperidine easily gives the corresponding azomethine with a fixed positive charge, characterized by β-elimination of the 3′- then 5′-phosphate, described above in the context of structures with sugar in an open (aldehyde) form, resulting from depurination. The above-described mechanism of DNA cleavage at cytosines is still hypothetical and needs additional studies. However, the experimental evidence available so far (hydrazinolysis with 17–18 M hydrazine followed by treatment with 1 M piperidine, 90 °C, 30 min.) indicates that the right conditions for cleavage of the internucleotide linkages in the cytidines of the polynucleotide chain have been chosen.

The modification of thymines in DNA with hydrazine proceeds in a similar fashion:

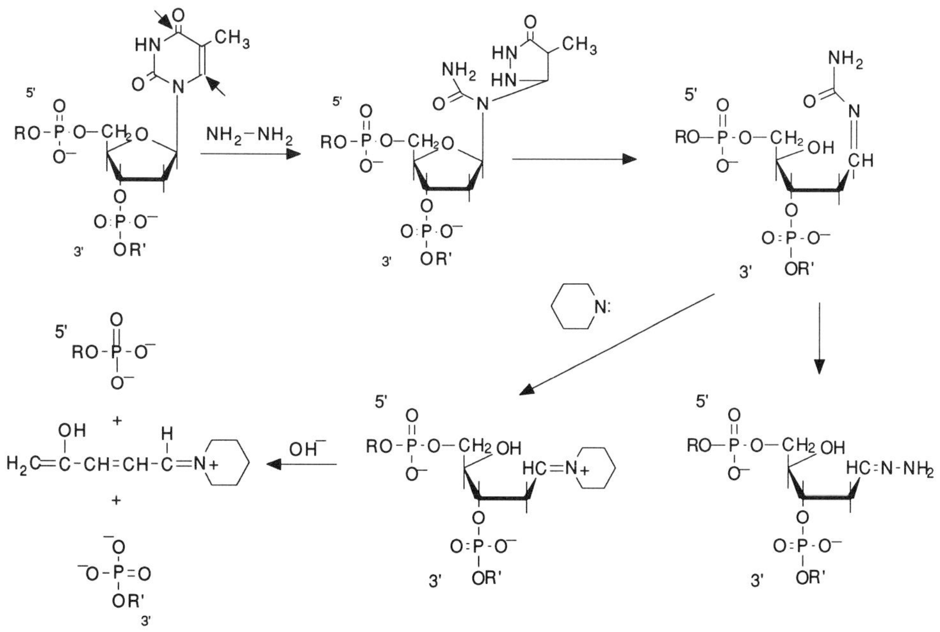

In this case, too, the treatment of the DNA modified with piperidine at T is followed by its degradation at the same units (the same β-elimination mechanism is involved).

Thus, the above conditions are conducive to modification and cleavage of the polynucleotide chain at both pyrimidines (modification at C+T). By accident, while refining the method, A. Maxam noticed that addition of a salt (1 M NaCl) to the experimental solution drastically slows down the reactions at thymines. This finding became useful in elaborating a method for specific DNA cleavage only at cytidines (modification at C).

All these studies into statistical chemical modification and degradation of DNA have resulted in an optimal procedure for specific cleavage of the polynucleotide chain during sequencing, namely: (G); (G+A); (C+T); (C). The advantages of this procedure are as follows.

Firstly, all of the four reactions involve piperidine (for breaking the sugar-phosphate backbone), which is removed by evaporation, while the salt-free cleaved DNA can be very easily dissolved in a few microliters of formamide for application on a thin (0.3 mm) slab gel for sequencing.

Secondly, the side processes normally occurring during the degradation have been minimized in the four reactions, which ensures good separation during PAGE and permits hundreds of nucleotides to be sequenced from the labeled end of DNA by this method.

Thirdly, the sequencing scheme (G); (G+A); (C+T); (C) makes the autoradiograph rather easy to "read" by virtue of its inherent 1:2:2:1 symmetry (see Fig. 6–7). The central columns being flanked on both sides by the G+A

Table 6-3.

NN	Cleavage specificity	Modifying agent	Modified base displacement conditions	Cleavage conditions
1	G	Dimethyl sulfate	Piperidine	Piperidine
2	G+A	Acid	Acid	Piperidine
3	C+T	Hydrazine	Piperidine	Piperidine
4	C	Hydrazine + NaCl	Piperidine	Piperidine
5	G>A	Dimethyl sulfate	Heating at pH 7	NaOH or piperidine
6	A>G	Dimethyl sulfate	Acid	NaOH
7	A>C	NaOH	Piperidine	Piperidine
8	G	Methylene blue	Piperidine	Piperidine
9	T	OsO_4 or $KMnO_4$	Piperidine	Piperidine
10	C	$NH_2OH \cdot HCl$ pH 6	Piperidine	Piperidine
11	A	Diethyl pyrocarbonate	Piperidine	Piperidine

and C+T ones each containing a base from the corresponding adjacent pair makes the sequencing picture most vivid.

None the less, Table 6–3 below lists the main base-specific DNA cleavage reactions used in sequencing.

Reactions 5 through 11 are conducted only when it becomes necessary to verify the results obtained the usual way.

6.4.3 Obtaining Individual DNA Fragments and ^{32}P-Labeling

The starting material for sequencing usually includes certain duplex DNA fragments isolated after treatment of linear DNAs with restriction endonucleases or similar restriction of a cyclic DNA (from plasmids, phages, viruses) at a single site. What is more, if the DNA duplex is long enough and contains, say, more than a thousand nucleotide pairs, it must be broken into shorter fragments because in a single sequencing experiment only 200 to 300 nucleotides can be handled. To this end, the restriction sites are mapped and DNA is cleaved at these sites. Such procedures result in double-stranded DNAs with unique terminal sequences (see, e.g., Tables 6-1 and 6-2) which can be labeled at the 5′ or 3′ end, depending on which end of the strand is important for sequencing. In each case, the end-labeled sample represents a duplex DNA. Since only one labeled strand is necessary for analysis, the next step is either separation of the strands (Fig. 6-4a) or treatment with a restriction nuclease with a specificity different from the initial one (Fig. 6-4b).

The state-of-the-art cloning techniques using plasmids and phages as vectors make it possible to obtain any DNA fragments in amounts of hundreds of picomoles, which is quite sufficient for sequencing of a chain up to five thousand base pairs long.

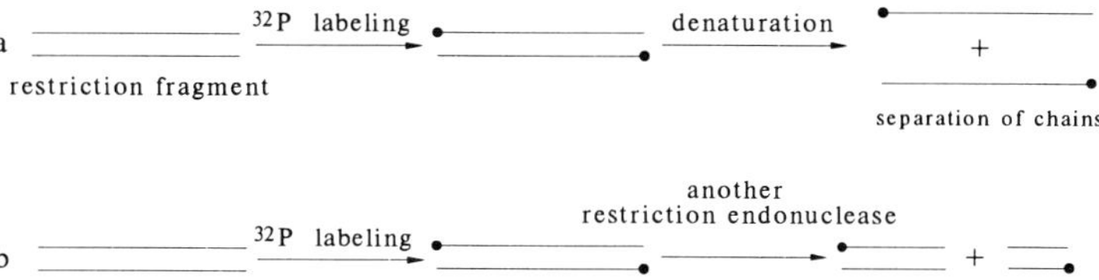

Fig. 6-4. Two ways of preparing DNA fragments with a single label: (a) DNA labeled at two strands is subjected, after denaturation, to PAGE for separation of the strands; (b) DNA labeled at two strands is cleaved by a second restrictase (E) to give duplexes labeled at one strand only.

As has already been mentioned, one of the basic principles of the chemical sequencing method is the presence, at the 5′ or 3′ end of the DNA under analysis, of an atom or a group of atoms that would enable detection of the chemical degradation products after the separation. More often than not, a ^{32}P-phospate group or ^{32}P-labeled nucleotide incorporated into the DNA with the aid of appropriate enzymes is used for this purpose. The labeled terminal nucleotide must be the same in all molecules of the DNA under investigation. Since fragments obtained by controlled cleavage of a native DNA using restriction endonucleases are most commonly involved in the sequencing of naturally occurring DNAs, the terminal nucleotides are identified by the specificity of the corresponding enzyme. To incorporate the label use is made of three enzymes and three different ^{32}P-precursors:

- polynucleotide kinase and [γ-^{32}P] ATP – for labeling the 5′ ends;
- terminal transferase and [α-^{32}P] ribonucleoside 5′-triphosphate – for labeling the 3′ ends;
- DNA polymerase and [α-^{32}P] deoxyribonucleoside 5′-triphosphate – also for labeling the 3′ ends complementary to the protruding 5′ ends of the duplex DNA. The use of nucleoside 5′-triphosphate preparations with a high specific activity (300–6000 Ci/mole) facilitates sequencing DNA fragments amounting to one or two picomoles.

Labeling the 5′ Ends. For direct phosphorylation of the ends of restriction fragments to take place, their 5′-phosphate must be removed in advance with the aid of alkaline phosphatase. Double-stranded DNAs with protruding 5′ ends (see Table 6-2) are more readily phosphorylated by polynucleotide kinase, as compared to DNAs with blunt ends or a protruding 3′ end. Table 6-4 lists some of the restriction endonucleases which leave protruding 5′ ends after DNA cleavage.

Figure 6-5 shows schematically the procedure of labeling double-stranded fragments at the 5′ ends. After dephosphorylation, alkaline phosphatase is removed by way of phenolic extraction and the precipitated DNA is treated with a polynucleotide kinase and [γ-^{32}P] ATP with a specific activity of 1000 Ci/mole.

Table 6-4. Restriction Endonucleases Cleaving DNA into Fragments with a Protruding 5′ End.

Restriction endonuclease	Recognized nucleotide sequence	Fragment end structure after restriction
*Eco*RI	↓ . . . NGAATTCN . . . . . . NCTTAAGN . . . ↑	pAATTCN . . . GN . . .
*Hind*III	↓ . . . NAAGCTTN . . . . . . NTTCGAAN . . . ↑	pAGCTTN . . . AN . . .
*Sal*I	↓ . . . NGTCGACN . . . . . . NCAGCTGN . . . ↑	pTCGACN . . . GN . . .
*Xma*I	↓ . . . NCCCGGGN . . . . . . NGGGCCCN . . . ↑	pCCGGGN . . . CN . . .
*Eco*RII	↓ . . . NCCAGGN . . . . . . NGGTCCN . . . ↑	pCCAGGN . . . N . . .

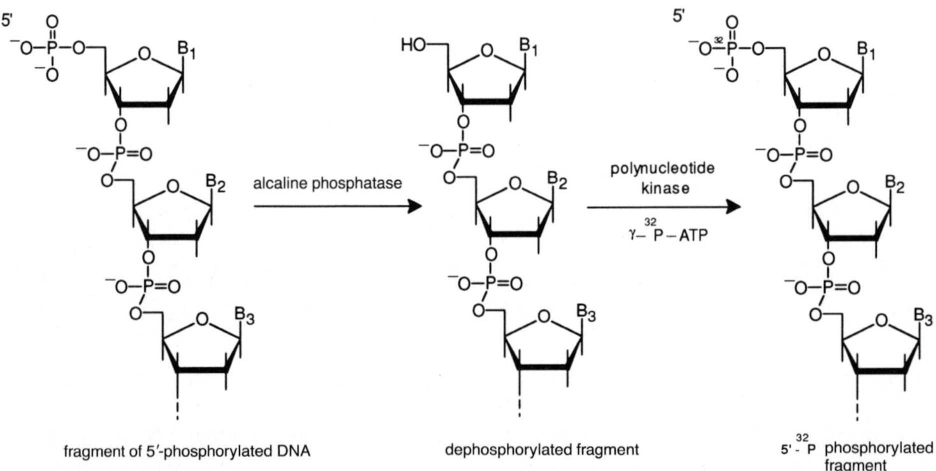

Fig. 6-5. ^{32}P-labeling of the protruding 5′ end of the restriction endonuclease-treated DNA fragment.

An alternative method is known whereby the 5′-terminal phosphate of DNA is exchanged directly for a labeled phosphate with the aid of polynucleotide kinases and [γ-^{32}P] ATP.

Shown below (Fig. 6-6) is the standard procedure for preparing a set of duplex DNA fragments with one of the strands being labeled at the 5′ end. The starting DNA, which may be a large restriction fragment or else a plasmid, phage or viral DNA, is exposed to a restriction endonuclease (E_1). As a result of such restriction with subsequent phosphorylation short DNA duplexes labeled at both 5′ ends are formed with the aid of a polynucleotide kinase, from which a labeled single-stranded DNA can be isolated after denaturation followed by gel electrophoresis. The mixture of labeled duplexes can also be treated with a second restriction endonuclease (E_2). The result of this treatment is subjected to electrophoresis for isolating products with only one labeled strand.

The procedure is then repeated using the same restriction endonucleases (E_1 and E_2) but in the reverse order (Fig. 6-6, top). The illustrated hypothetical DNA fragment has three sites recognized by restriction endonuclease E_1 and three sites recognized by E_2. The restriction by means of a mixture of E_1 and E_2 gives seven duplexes (1 through 7 in Fig. 6-6). It can be seen that in this case only fragments 1 through 4, 6 and 7 are labeled at one strand. As regards fragment 5, it is either labeled at both strands or is not labeled at all (after treatment with E_1 and then E_2). Fragments 1 and 7 have only one strand labeled if they were treated with E_2 at first and then with E_1 (Fig. 6-6, top). Fragments 2, 3, 4 and 6 are labeled at the opposite strands, depending on the sequence of treatment with E_1 and E_2.

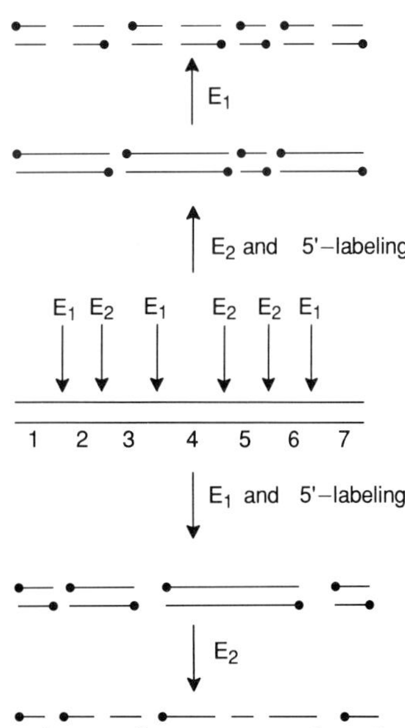

Fig. 6-6. Preparation of duplex DNAs 5′-end labeled at one of the strands only.

Labeling the 3′ Ends with the Aid of Terminal Transferase. Terminal transferase extracted from calf thymus polymerizes ribonucleotides in a template-independent reaction at the 3′ ends of DNA strands. The function of substrates is performed by ribonucleoside 5′-triphosphates and, if the latter contain an α-labeled phosphate, such as [α-^{32}P] ATP, the 3′ ends of DNA receive a polyribonucleotide labeled at the internucleotide phosphate.

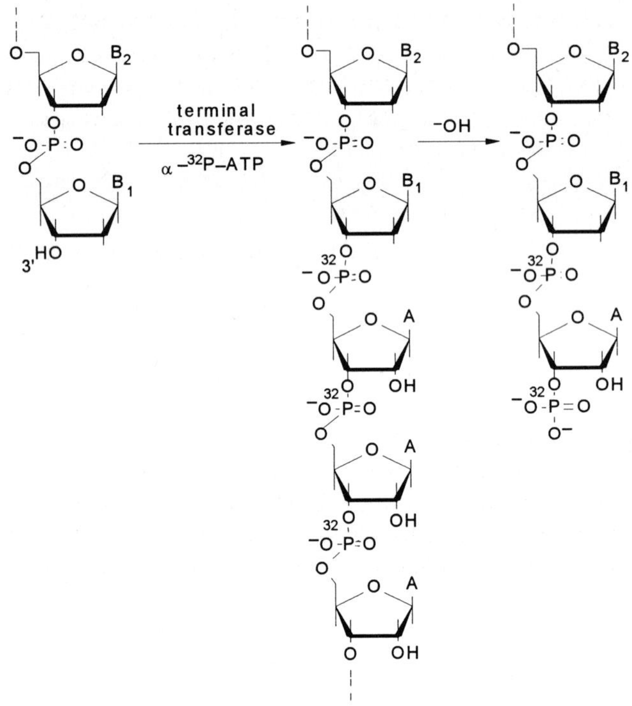

Alkaline treatment of such a mixed DNA-oligo-A-polymer gives a DNA with a "double" label at the 3′ end: the 3′-terminal phosphate and the internucleotide one next to it are labeled. The mechanism of alkaline hydrolysis (the hydrolysis involves all internucleotide phosphate groups adjacent to the 2′-hydroxyl groups; i.e., only in the riboadenyl links) is represented schematically below:

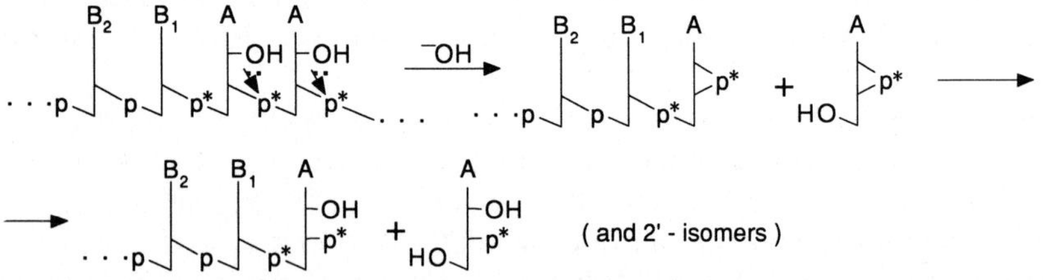

The first step of the alkaline treatment includes nucleophilic substitution at the internucleotide phosphorus with the participation of a 2′-hydroxyl. The resulting cyclic phosphates are then easily hydrolysed to a mixture of 2′- and 3′-phosphates. This process results in incorporation of two ^{32}P-labeled phosphate groups into the 3′ end of the DNA. 3′ end labeling may be used in both single- and double-stranded DNAs. If pyrimidine nucleoside 5′-triphosphates are used as [α-^{32}P]-labeled nucleotides, alkaline treatment may be replaced by pyrimidyl RNase (the hydrolysis mechanism remaining the same).

An alternative way to label the 3′ end is extending the latter into duplex DNA with a protruding 5′ end with the aid of an *E. coli* or phage T_4 DNA polymerase. Here is a schematic representation of such a reaction in which the restriction fragment *Hin*dIII is extended:

$$\begin{array}{l} \text{5}'\text{ pAGCTTNN} \text{———} \\ \qquad\quad\;\; \text{ANN} \text{———} \end{array} \xrightarrow[\text{DNA polymerase}]{\text{dATP, dGTP} \atop \text{dCTp, }[\alpha\text{-}^{32}\text{P}]\text{dTTP}} \begin{array}{l} \text{pA–GCTTNN} \text{———} \\ \text{Tp*CGAANN} \text{———} \end{array}$$

By using only dTTP containing an α-labeled phosphate, one can incorporate the label into the 3′-terminal nucleotide of DNA.

An important step in extracting DNA from gel is its elution ensuring a high yield. Moreover, it is important to avoid degradation of DNA in the process. The extraction method must ensure that DNA is highly concentrated and free of the electrophoretic buffer, enzyme inhibitors and low-molecular weight polyacrylamide fragments. The best way is diffusion of DNA from the degraded gel in a salt sulution with subsequent filtration and ethanol-assisted precipitation of the DNA from the filtrate. This is how DNA fragments varying in length from tens to thousands of nucleotides are extracted from polyacrylamide gel.

6.4.4 Polyacrylamide Gel Electrophoresis: High-Resolution System for Separating Oligo(poly)nucleotides According to Chain Length

The third basic principle of the sequencing method is separation of oligo- and polynucleotide chains according to size by way of gel electrophoresis. All labeled fragments resulting from the chemical degradation of DNA have one end in common and another end varying in length. Each fragment in the array contains all the nucleotides present in the preceding smaller fragment plus one more at the variable end. At pH > 7, two adjacent fragments differ by a small increment of charge and mass, one of the four mononucleotides. During PAGE, the larger fragment moves more slowly than the smaller one because it enters into greater noncovalent interaction with the gel matrix.

Currently used techniques permit separation of such DNA strands differing by a single nucleotide at lengths ranging from several to hundreds of nucleotides. Since these oligo(poly)nucleotides differing in length by a mere nucleotide appear on the X-ray film as bands, resolution is determined by two parameters: thickness of the bands and the distance between their centers. One must make sure that the concentration of the sample applied onto the gel is sufficiently high for the gel to be uniform, for diffusion to be weak, and for scattering of the radioactive emission to which the film is exposed to be modest. Then the bands on the film will be thin. The center-to-center distance between adjacent bands is determined by the "retarding" action of the gel matrix. Use is commonly made of thin 8 % polyacrylamide gels containing 50 % urea at pH 8.3 (8.3 M). Thin sequencing gels have important advantages: electrophoresis on such gels can be rather fast (at high voltage), and because scattering during autoradiography is reduced, the bands come out sharper on the film. Usually, within a single run of four standard reaction mixtures (G, G+A, C+T, C) anywhere between 100 to 150 nucleotides can be "read" in 8 % polyacrylamide gel. 20 % polyacrylamide gel is more suitable for "reading" the first 30 nucleotides from the labeled end.

For the purposes of autoradiography, the gel on the top of a glass plate is covered with thin plastic wrap and the plate is placed on X-ray film inside a light-tight holder. To slow down the "smearing out" of the bands during exposure, such holders are stacked in a freezer.

6.4.5 Direct Reading of Nucleotide Sequence from Autoradiogram

The great success of the chemical and Sanger's enzymatic sequencing techniques stems from the fact that the nucleotide sequence determined within a single analysis (an average of 300 nucleotides, the exact amount depending on the resolution of the gel electrophoresis) can be read directly from the autoradiogram. It would not be an exaggeration to say that the contribution of these techniques to molecular biology is more important than that of any other method or theory proposed since the dicovery of the secondary DNA structure by Watson and Crick. Let us now see how a sequence of nucleotides is read. A typical autoradiogram of a sequencing gel is represented in Figure 6-7.

The autoradiogram shows vertical ladders of horizontal bands. Band at the bottom correspond to cleavage products near the labeled end or, in other words, short oligonucleotides. The upper bands correspond to cleavages progressively more distant from the labeled end. The spacing between the bands decreases with increasing length of the oligo(poly)nucleotides. The dark band interrupting the vertical row corresponds to the full-length DNA fragment not cleaved in all four reaction mixtures.

Fig. 6-7. Autoradiogram of a sequencing gel used in analysis of DNA ^{32}P-labeled at the 5′ end and its interpretation. The four vertical ladders correspond to reaction mixtures separated in 8 % polyacrylamide gel after limited cleavage at guanines (G), guanines and adenines (G+A), cytosines and thymines (C+T) and cytosines (C), using reactions 1-4 (Table 6-3).

In order to read the sequence of nucleotides (bases in the sugar-phosphate backbone of the polynucleotide) from these bands, let us look first at the bottom portion of the autoradiogram. The bands in the very bottom correspond to the short labeled DNA fragments. So we must move now from the bottom-most band up, along the G+A and C+T ladders and interpret each band. Note that these two ladders together contain all bands arising from partial cleavage of end-labeled DNA. If a band occurs on the G+A ladder and falls under G further to the left, it is a result of cleavage at guanine. If the G ladder is empty, then the band under G+A must result from cleavage at adenine. Similarly, a band on the C+T ladder to the right of the center line corresponds to cleavage at a pyrimidine base, whereas the presence or absence of a band under C suggests that the cleavage has taken place at cytosine or thymine, respectively. Thus, going from one band to another from the bottom upward allows one to immediately write down the sequence of nucleotides from the 5′ end (which is labeled) of the DNA. The pattern on the film must be continuous, and the bands on the ladders must be sharply resolved and consistent with one DNA sequence. However, some bands tend to be doubled up, blurred, too light, or even missing. As a result, reading the DNA sequence becomes impossible. Such aberrations normally stem from improperly conducted chemical (insufficient specificity of modification reactions or incomplete β-elimination) and enzymatic reactions described earlier. Other reasons may be the presence of spurious fragments contaminating the main DNA fragment, DNA nicks, and microheterogeneous labeled DNA ends. A detailed description of the causes leading to all possible aberrations and ways to eliminate them can be found listed in a review by A. Maxam and W. Gilbert (see references).

6.4.6 Double-Stranded DNA Sequencing Strategy

There are two main approaches to DNA sequencing. The first approach is used when it becomes necessary to determine the primary structure of a particular functionally significant DNA segment (a promoter, replication origin, structural gene, etc). It boils down to separation of the segment in question, after treatment of the sample, by means of a restriction enzyme and preparative PAGE. In this way, hundreds of picomoles of the DNA of interest can be obtained, which corresponds to about 1 mg DNA containing about 5000 pairs of bases. After electrophoresis, the gel is exposed to UV light against a fluorescent background. When this is done, the DNA bands appear dark. Once the desired band has been identified, the DNA fragment is extracted and precipitated with the aid of alcohol. The strands of such a fragment are most commonly separated by way of denaturation at 90 °C with subsequent electrophoreses and sequencing after ^{32}P-labeling of the ends. It is still not clear what makes the complementary strands go apart during electrophoresis. In all likelihood, an important role is played by the configuration of each strand, which is dependent on the primary structure. The possibility to separate the strands permits one to clearify the primary DNA structure at the site of secondary restriction and to confirm the sequence when the second DNA strand is decoded. To separate strands up to 500 (and sometimes even more) nucleotides in length, use is made of 5 % PAGE. This technique is employed routinely when the DNA portion to be decoded is small and precisely mapped, as well as for separating cloned fragments of eukaryotic chromosomes from their bacterial vectors.

The second approach is used when it is necessary to sequence plasmid, phage or viral DNA. In this case, one can do without a detailed restriction map and a stockpile of isolated DNA fragments. At first, DNA is cleaved into fragments that are easy to label at the 5′ or 3′ ends (this purpose will best be served by restriction endonucleases which leave 5′ ends extended). Then, the 5′ or 3′ ends of all these fragments are labeled at once. The subsequent analysis may proceed along two paths: (1) denaturation for strand separation and (2) digestion with a second restriction enzyme (see Figs. 6-4 and 6-6), followed by the separation of fragments labeled only at a single strand. If the scheme of Figure 6-6 makes use of the same two restrictases in reverse order, the result is the same set of fragments labeled at their opposite strand. Figure 6-8 illustrates the procedure of preparing fragments for sequencing of a cloned gene. It shows separation of fragments from a recombinant DNA (RV), based on the second path.

All fragments in the right ladder (RV) are labeled at both strands (because no treatment with a second enzyme was used), which is why they cannot be sequenced. All fragments in the left ladder ($V_{E1,\ E2}$) lack inserts of the cloned gene under investigation. Only the bands (4, 5 and 6) in the central ladder ($RV_{E1,\ E2}$) are fragments with a single label, include gene inserts and can be sequenced. Complete sequencing of both strands of a particular restriction

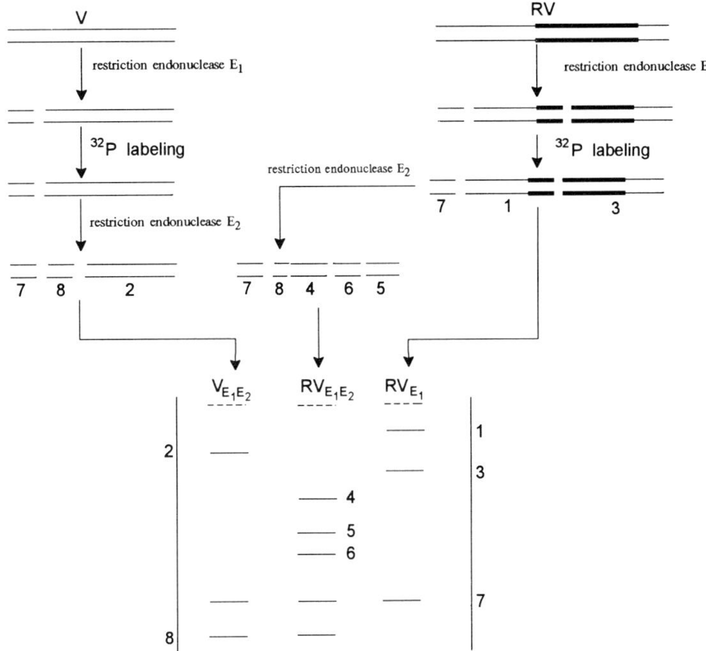

Fig. 6-8. Separation of fragments of a cloned gene after treatment with two restriction endonucleases E_1 and E_2. The same treatment was undergone by the vector (V) used to obtain the recombinant DNA (RV).

fragment is sometimes difficult by the above method. There are several reasons for this: the length of the strands may exceed the resolution of the sequencing genes; the separation of the strands may not be complete; difficulties may arise with determination of the terminal nucleotides. In such cases, the complementary strand is sequenced from the labeled 3′ end, rather than the 5′ end, or vice versa. One can also obtain fragments through digestion with another restriction enzyme and use newly emerging inner ends for sequencing. Such a method of sequencing over the entire length and along both strands is illustrated below.

The top portion of the scheme shows a DNA sequence containing sites of recognition by restriction endonucleases *Alu*I, *Hae*III, and a third enzyme. The sequence of this double-stranded DNA can be decoded as follows: treatment with restriction endonuclease *Alu*I gives rise to a fragment that can be ^{32}P-labeled at both 5′ ends with the aid of polynucleotide kinase or at both 3′ ends with the aid of terminal transferase. Then, restriction endonuclease *Hae*III is used for asymmetric cleavage of the *Alu*I-fragment, followed by sequencing of the resulting fragments from the labeled 5′ (1 and 2) or 3′ ends (3 and 4). It is also possible to cleave the strands of the *Alu*I fragment and sequence the intact ones beginning from the labeled 5′ (5 and 6) or 3′ ends

```
Enzyme III  AluI             HaeIII      AluI            Enzyme III
    ↓        ↓                  ↓           ↓                  ↓
  5'...GTCAG CTCGCCCTTATGG CCACACTAG CTAATGGGCATGACT..
     ..CAGTC GAGCGGGAATACC GGTGTGATC GATTACCCGTACTGA...5'
         (1)CTCGCGCTTATGG
                                         AluI/kinase/HaeIII/gel
                          GGTGTGATC(2)
                          CCACACTAG(3)
                                         AluI/polymerase/HaeIII/gel
         (4)GAGCGGGAATACC
         (5)CTCGCCCTTATGGCCACACTAG
                                         AluI/kinase/separation of strands
            GAGCGGGAATACCGGTGTGATC(6)
            CTCGCCCTTATGGCCACACTAG(7)
                                         AluI/polymerase/separation of strands
         (8)GAGCGGGAATACCGGTGTGATC
                       (9)CCACACTAGCTA..
                                           HaeI/kinase/enzyme III/gel
     ..AGTCGAGCGGGAATACC(10)
     ..TCAGCTCGCCCTTATGG(11)
                                         HaeI/polymerase/enzyme III/gel
                      (12)GGTGTGATCGAT
```

(7 and 8). Moreover, the starting DNA may first be treated with restrictase *Hae*III, labelled at the 5′ or 3′ ends, then treated with a third restriction enzyme cleaving the strand from both sides, and sequence in opposite directions from the *Hae*III site, bypassing the *Alu*I sites (9-12). All odd-numbered sequences validate completely or partially the even-numbered complementary sequence and vice versa. The results of all independent sequencing procedures are then represented by arrows pointing to the left and to the right, each arrow indicating which strand has been sequenced and which restriction site has been used for the purpose.

In conclusion, here is an example showing the results of sequencing of plasmid pBR322 (Fig. 6-9).

The transverse bands on top correspond to the sites of DNA pBR322 cleavage with restriction endonucleases *Hinf*I and *Ava*II, and the restriction maps

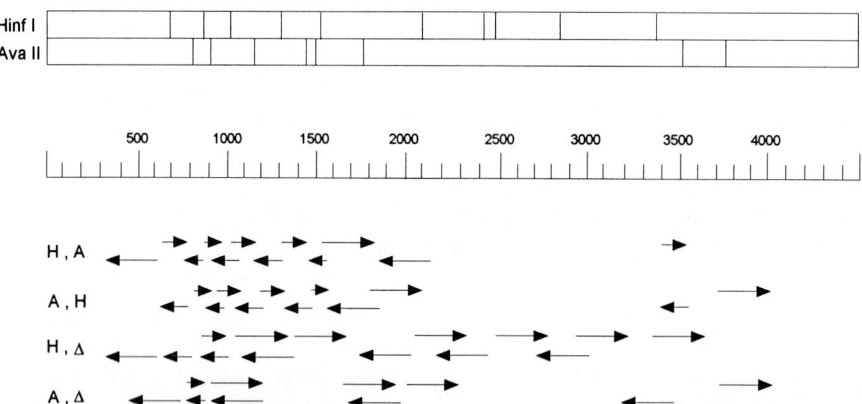

Fig. 6-9. Position of sequenced fragments of plasmid pBR322 labeled at one end (using two restriction endonucleases and polynucleotide kinase).

are arranged one above the other. The scale underneath represents the distances from the unique site of pBR322 cleavage with restriction endonuclease *Eco*RI in nucleotide pairs. The plasmid is 4362 base pairs long. Shown below the scale are rows of arrows. Each arrow begins from the ^{32}P-labeled 5′ end resulting from the cleavage with restriction endonucleases *Hinf*I or *Ava*II, and its length equals that of the corresponding fragment, or 250 base pairs (resolution of gel electrophoresis) if the fragment is longer. The arrows mark the DNA sites next to the labeled ends, which can be decoded without any particular difficulty. The arrows in the row designated H,A correspond to the fragments labeled at one end and resulting from the following sequence of reactions: *Hinf*I → kinase → *Ava*II. The arrows in row A,H correspond to the fragments resulting from the reactions in reverse sequence: *Ava*II → kinase → *Hinf*I. Both experiments have made it possible to sequence a total of 1700 base pairs, with both strands being read so as to verify and more unambiguously determine the sequence of each strand. The two bottom rows of arrows are designated H, Δ and A, Δ. They represent, respectively, the DNA fragments resulting from the following treatments: *Hinf*I → kinase → melting and separation of strands (H,Δ) and *Ava*II → kinase → melting and separation of strands (A,Δ). Both experiments, carried out together with the previous two, have made it possible to determine the structure of the plasmid over 70 per cent (3000 b.p.) of its length.

The vector most commonly used for cloning DNA fragments is plasmid pBR322, which contains six unique sites recognized by restriction endonucleases *Pst*I, *Sal*I, *Ava*I, *Hind*III, *Bam*HI and *Eco*RI, occupying a small region of the *lac*Z gene. The DNAs inserted at these sites of plasmid pBR322 and then cloned can be sequenced by the Maxam–Gilbert method without complicated fragment treatment and separation procedures.

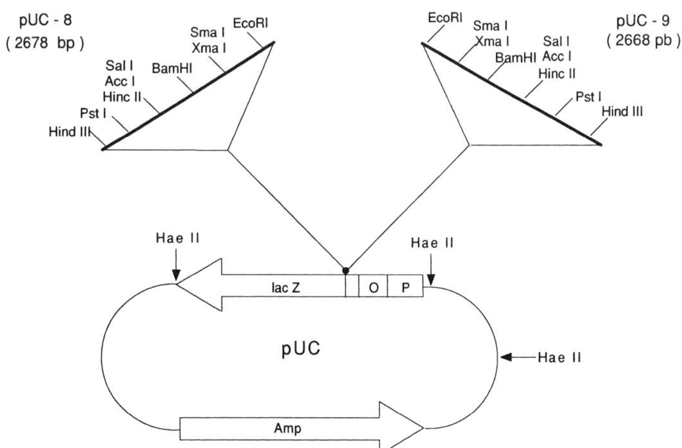

Fig. 6-10. Commercially available vectors pUC-8 and pUC-9.

The company "Pharmacia" has been proposing vectors pUC for this purpose. They are essentially plasmids constructed on the basis of plasmid pBR322 and contain sites, over a portion of the *lac*Z gene, recognized by unique restriction endonucleases. Two such commercially available vectors (pUC-8 and pUC-9) are represented in Fig. 6-10.

6.5 DNA Sequencing by Polymerase Copying Method (Enzymatic Sequencing Method)

The polymerase method (often called enzymatic sequencing method, as opposed to chemical) was virtually the first tool proposed for sequencing large DNAs. It is based on the work done by F. Sanger, who spent more than ten years after the determination of the primary structure of insulin (1963) elaborating nucleic acid sequencing techniques. Having developed successful sequencing procedures for small oligonucleotides, Sanger's laboratory came up with what became known as the "plus and minus method" (1975). It took a mere two years for Sanger to published, along with the presentation of the chemical method (1977), a description of his modified method (DNA sequencing using chain terminating inhibitors), which became known as the "dideoxy method". This sequencing technique, based on template-directed and primed synthesis of single-stranded DNA with the aid of DNA polymerase, provides a reliable and simple way to sequence DNA, along with the above-described Maxam–Gilbert method. The potential of the method was obvious already in 1976, when Sanger and coworkers determined the primary structure of the genome of phage ΦX174. Although the genome contained "only" 5375 nucleotides, the possibility to decode it completely and thus gain access to all the genetic information contained in the genome marked a breakthrough in this field. It is precisely this phage that demonstrated the possibility of encoding a gene within a gene.

6.5.1 Basic Principle of the Method: Priming and Termination of Enzymatic Synthesis of DNA Copies

Conceptually, both sequencing methods – chemical and enzymatic – are identical. However, if the Maxam–Gilbert method is based on specific DNA cleavage determined by the nature of the constituent bases, the underlying principle of Sanger's method is statistical DNA synthesis terminating at one of the four nucleotides. Thus, both methods are aimed at obtaining a complete (statistical) set of DNA fragments terminating at each of the four nucleotides. It was found that synthesis with subsequent PAGE and autoradiography provides an effective means for producing such sets, as opposed to degradation.

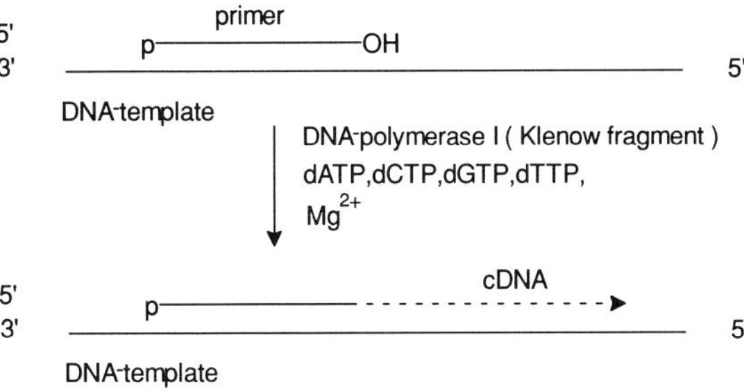

Fig. 6-11. Synthesis of a DNA (cDNA) copy using a single-stranded template DNA with a primer, which is an oligonucleotide complementary to a particular site, in the presence of all four deoxynucleoside 5′-triphosphates. The synthesis is catalyzed by DNA polymerase I, or Klenow fragment, lacking 5′→3′ exonuclease activity.

The substrate in the polymerase copying method is a single-stranded fragment, or a DNA template (which is essentially the molecule of interest), annealed to a short complementary fragment, or primer. Figure 6-11 shows schematically the standard procedure for copy DNA (cDNA) synthesis in the presence of Klenow fragment (DNA polymerase I), illustrating the enzymatic process underlying the method. The chemical reaction of enzymatic synthesis of internucleotide linkages, as part of cDNA synthesis, boils down to phosphorylation of the primer's 3′-hydroxyl in the template-primer complex with the next complementary deoxynucleoside 5′-triphosphate. The α-phosphate group of the latter in such an enzyme-substrate complex is attacked by the 3′-hydroxyl of the primer after departure of the pyrophosphate group (see Fig. 6-12).

The motive force of the copying process is formation of a productive enzyme-substrate complex: template-primer-complementary deoxynucleoside 5′-triphosphate-DNA polymerase. Naturally, with the lack of structural consis-

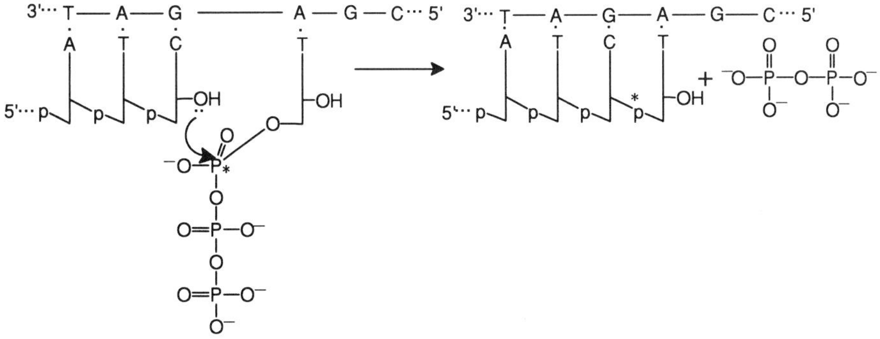

Fig. 6-12. Formation of an internucleotide linkage during attachment of a complementary nucleotide (dpT) in polymerase copying. If nucleoside 5′-triphosphate ^{32}P-labeled at the α-phosphate group is used in the reaction, the cDNA receives a label instrumental in identifying the synthesized polynucleotide by autoradiography. dNTP ^{35}S-labeled at the α-phosphate can also be used. In this case, bands on the autoradiogram are more distinct (because of the β-particle emission energy). However, the specific activity is usually lower and the gel has to be exposed over a longer period of time.

tency in such an enzyme-substrate complex, no internucleotide linkage is formed; consequently, the polymerization (cDNA formation) is inhibited. It is precisely this factor that is used in sequencing by the polymerase copying method, when employed as a polymerization inhibitor is one of the four nucleoside 5′-triphosphates – the one at which copying must be partially (statistically) stopped. For this purpose analogues of ordinary dNTP substrates without a hydroxyl group at the 3′ position are used i.e., 2′,3′-dideoxyribonucleoside 5′-triphosphates, abridged as ddNTP):

ddNTP
B=A,G,C and T

The primer is extended in the presence of all the four usual deoxyribonucleoside 5′-triphosphates and the 2′,3′-dideoxy analogue (ddNTP) of one of them.

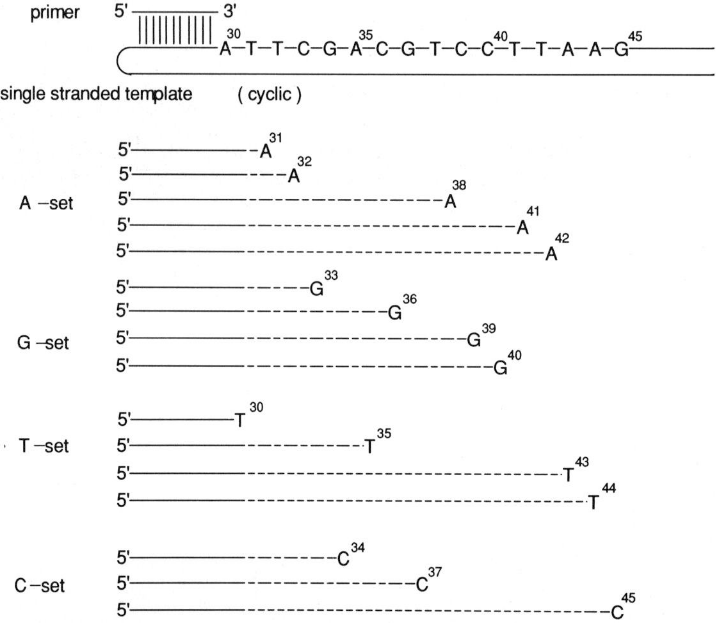

Fig. 6-13. The basic principle of chain termination during DNA sequencing by the polymerase copying method. When such a substrate as, for example, ddATP is introduced into the reaction mixture along with the usual dNTP ones, the synthesis of some chains discontinues at positions T in the template. This gives an A set consisting of all possible oligonucleotides terminating in A (positions 31, 32 and 38 corresponding to T in the template). When ddGTP, ddTTP and ddCTP substrates are used, G, T and C sets are obtained, respectively, the numerals indicating the corresponding chainterminating nucleotides.

The analogue is inserted into the growing strand, the enzyme recognizes it, and another internucleotide linkage is formed with its participation. However, further extension of the complementary strand complex in this position is impossible (see Fig. 6-12) because the inserted analogue lacks a 3′-hydroxyl group (polymerase substrate). One of the four analogues is used to obtain a corresponding set of template copying products having a common origin, which is the primer, and terminating at a nucleotide of appropriate type.

Figure 6-13 is a schematic demonstration of this sequencing procedure, from which it becomes evident that the polymerization in four simultaneous reactions, each involving statistical insertion of the corresponding dideoxynucleoside 5′-phosphate, yields four sets of oligonucleotides (A, G, C an T sets). Each such set comprises oligonucleotides terminating at the 3′ end with a nucleotide (A, G, T or C) which is complementary to its counterpart in the template. If a single dNTP with a ^{32}P-labeled α-phosphate group (usually α-^{32}P-ATP) is introduced into a reaction mixture, all oligonucleotides in the A, G, T and C sets become labeled. The separation of these sets in PAGE by length makes it possible, just as in chemical sequencing, to read the initial sequence from the band pattern on the autoradiogram. Figure 6-14 is an autoradiogram of the reaction mixtures (A, G, T and C sets). In order to obtain a good "statistical" set of cDNA, one must vary the ratio between the terminating and the corresponding ordinary nucleoside 5-triphosphates. At a dNTP/ddNTP ratio equal to 1:100, it is possible to obtain a well resolved set of bands and read as many as 200 and more nucleotides. A lower ddNTP concentration gives risc to larger copying products, which, in combination with a lower concentration of acrylamide, increases the number of nucleotides that can be read off the same gel slab to 300.

In addition to 2′,3′-dideoxynucleoside 5′-triphosphates, the corresponding arabinonucleoside 5′-triphosphates can be used as terminators of enzymatic synthesis of cDNA, such as

$$^{-}O-\overset{O}{\overset{\|}{P}}(O^{-})-O-\overset{O}{\overset{\|}{P}}(O^{-})-O-\overset{O}{\overset{\|}{P}}(O^{-})-OCH_2 \; (\text{arabinofuranose ring: HO, HO}) \; B$$

B=A,G,C and T

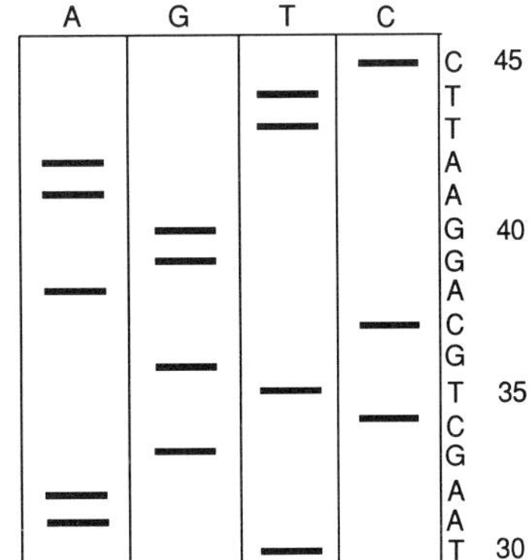

Fig. 6-14. Autoradiogram showing the A, G, T and C sets of Fig. 6-13. The cDNA sequence read off on the right begins after the primer or, in other words, at position 30 and ends at position 45 from the 5′ end. This allows one to write the complementary sequence or the primary structure of the template between nucleotides 30 and 45 from the 3′ end:

30 35 40 45
3′. . . ATTCGACGTCCTTAAG . . . 5′

which are inserted into the growing cDNA strand but inhibit subsequent formation of internucleotide linkage. The inability of "arabinose" fragments to participate in the latter seems to stem from the unfavorable conformation of the 3′-hydroxyl group. Perhaps, this also has to do with the configuration of the 2′-hydroxyl. An alternative type of terminators includes 3′-azido derivatives of 2′-deoxyribonucleoside 5′-triphosphates, introduced into sequencing by A. A. Kraevsky and coworkers:

$$^{-}O-\overset{O}{\overset{\|}{\underset{O^-}{\underset{|}{P}}}}-O-\overset{O}{\overset{\|}{\underset{O^-}{\underset{|}{P}}}}-O-\overset{O}{\overset{\|}{\underset{O^-}{\underset{|}{P}}}}-OCH_2 \;(\text{furanose ring, } 1'\text{-}B,\ 3'\text{-}N_3)$$

B=A,G,C and T

Naturally, termination in this case involves an azido group, instead of the 3′-hydroxyl, in the substrate of DNA polymerase I. Interestingly, insertion of such modified nucleotides into the strand takes place none the less. Important sites of recognition by this enzyme of its substrates include the 5′-triphosphate group and the heterocyclic base. The primers in the polymerase copying method are either single-stranded synthetic oligonucleotides or the restriction fragments of the DNA under analysis (i.e., fragments resulting from hydrolysis of the DNA under analysis with a specific restriction endonuclease).

The polymerase copying method is commonly used to determine reliably anywhere from 15 to 200 (and sometimes more) nucleotides beginning from the primer. Figure 6-15 is an autoradiogram of a typical sequencing gel, produced by the enzymatic method. The primary DNA structure is represented by the shown sequence. For good separation of fragments ranging in size

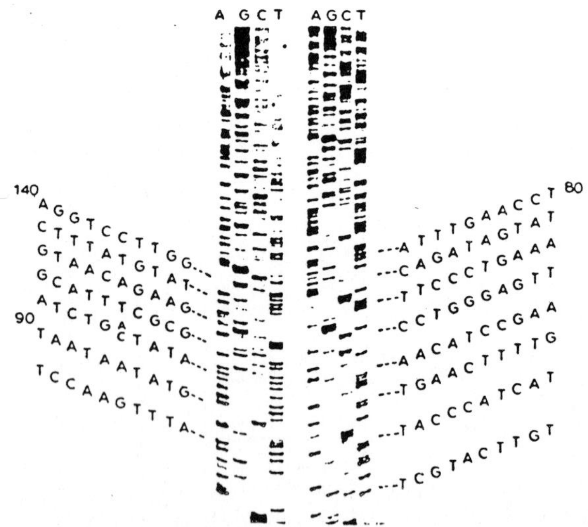

Fig. 6-15. Autoradiogram of a sequencing gel produced by the enzymatic method. Written on either side is the sequence of nucleotides read off from this autoradiogram.

from 50 to 200 nucleotides a 8% polyacrylamide gel is used. In the case of smaller fragments with less than 100 nucleotides, 12% polyacrylamide gel is the right choice. On the other hand, reliable separation of larger fragments (with more than 250 nucleotides) is achieved with the aid of a 6% gel.

6.5.2 Strategy for Sequencing Large Single-Stranded DNAs by the Enzymatic Method

The enzymatic dideoxy sequencing method makes it possible to determine the primary structure of any DNA that can be obtained in a single-stranded form. There are several procedures for preparing single-stranded DNAs, cloning in phage M13 being the most widely used.

Any double-stranded DNA fragment, irrespective of its primary structure, can be inserted into a vector molecule to obtain a recombinant DNA. Separation of DNA fragment mixtures is performed using the cloning technique. A mixture of fragments of any degree of complexity can be separated by plating bacteria transformed by a mixture of recombinant DNAs. In order to find all fragments of the starting mixture, every clone must be used.

To simplify the clone screening procedure, vectors have been developed with a special structure permitting one to distinguish between bacterial colonies carrying a vector with an insert and an "empty" vector by the color of the colonies in plates with an indicator medium. The first such vectors were developed for producing single-stranded DNAs or, in other words, templates for the enzymatic sequencing method. Figure 6-16 illustrates the basic principle of using vector M13 (replicative form of phage M13 DNA) to separate the strands of any DNA fragment.

The starting fragment is incorporated into the replicative double-stranded form of M13 DNA at the sticky ends of the *Eco*RI site occupying a functionally insignificant portion of the phage. Two orientations are possible for inserting a fragment. The prepared mixture of molecules is used to transform a bacterial cell and isolate a single-stranded form from the mature phage DNA, containing either strand A or strand B of the starting DNA fragment. Used for the sequencing of such DNA fragments (A or B) is the same short synthesized or isolated (with the aid of restriction endonucleases) primer forming a complex with the vector DNA near the site of insertion of the starting DNA (see Fig. 6-17).

Here is a brief description of the vector – phage M13 – used for obtaining, in a simple procedure, single-stranded DNAs ready for sequencing. M13 is a filamentous coliphage having a single-stranded genome with a relative molecular weight of $2 \cdot 10^6$. After inoculation with *E. coli* the single-stranded DNA becomes double-stranded, or converted into a replicative form, and during amplification produces about 300 copies per cell. The inoculated cells are not lysed but continue to grow slowly, discharging the phage into the medium.

The filamentous phage M13, just as some other analogous phages (f1, fd), are characterized by three factors extremely important for use in DNA cloning, namely:

(a) the phage DNA can be isolated as a double-stranded molecule;
(b) the DNA "harbored" by the phage may be longer than the phage DNA;
(c) the phage does not lyse the host cell (as opposed, for example, to the lambda phage).

The M13 genome has a small region (about 100 b.p.) which, as has already been mentioned, is functionally insignificant for development of the phage; it is precisely into this region that the foreign DNA can be inserted. Phage M13mp2, for example, has the following structure making it a suitable vector. The above region receives a *lac* promoter and a fragment of the *lacZ* gene encoding the first 146 amino acids of the enzyme β-galactosidase, so that the *N*-terminal fragment of this protein is synthesized in the *E. coli* cells inoculated with phage M13mp2. If use is made in this case of the *E. coli* strain *gal*$^-$ having mutations at the *lac*Z gene segment in question and, therefore, syn-

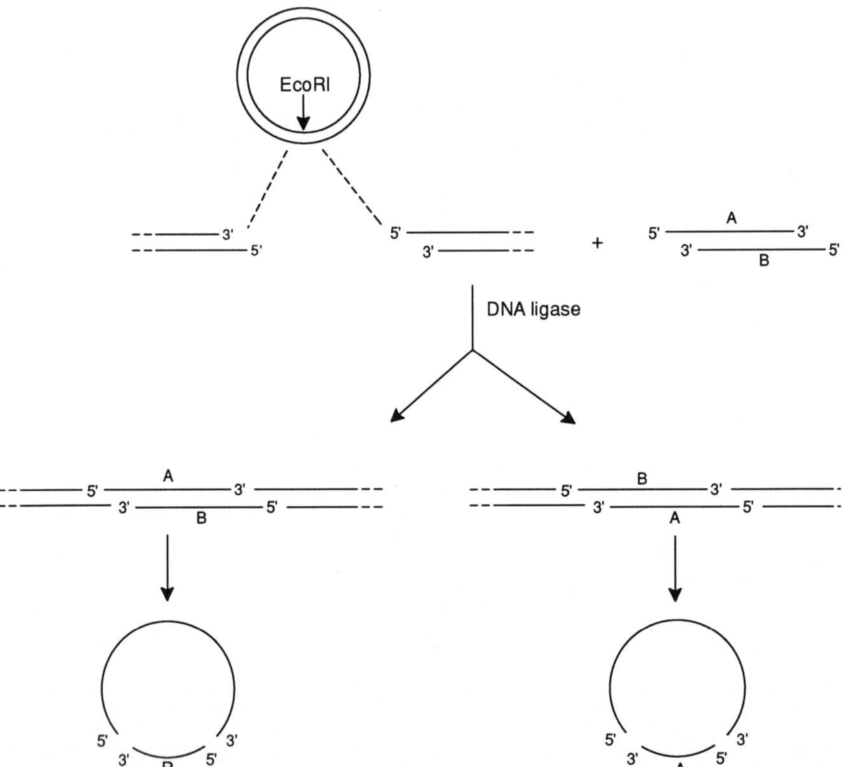

Fig. 6-16. Separation of DNA strands (A and B) by cloning in a vector based on the filamentous bacteriophage M13.

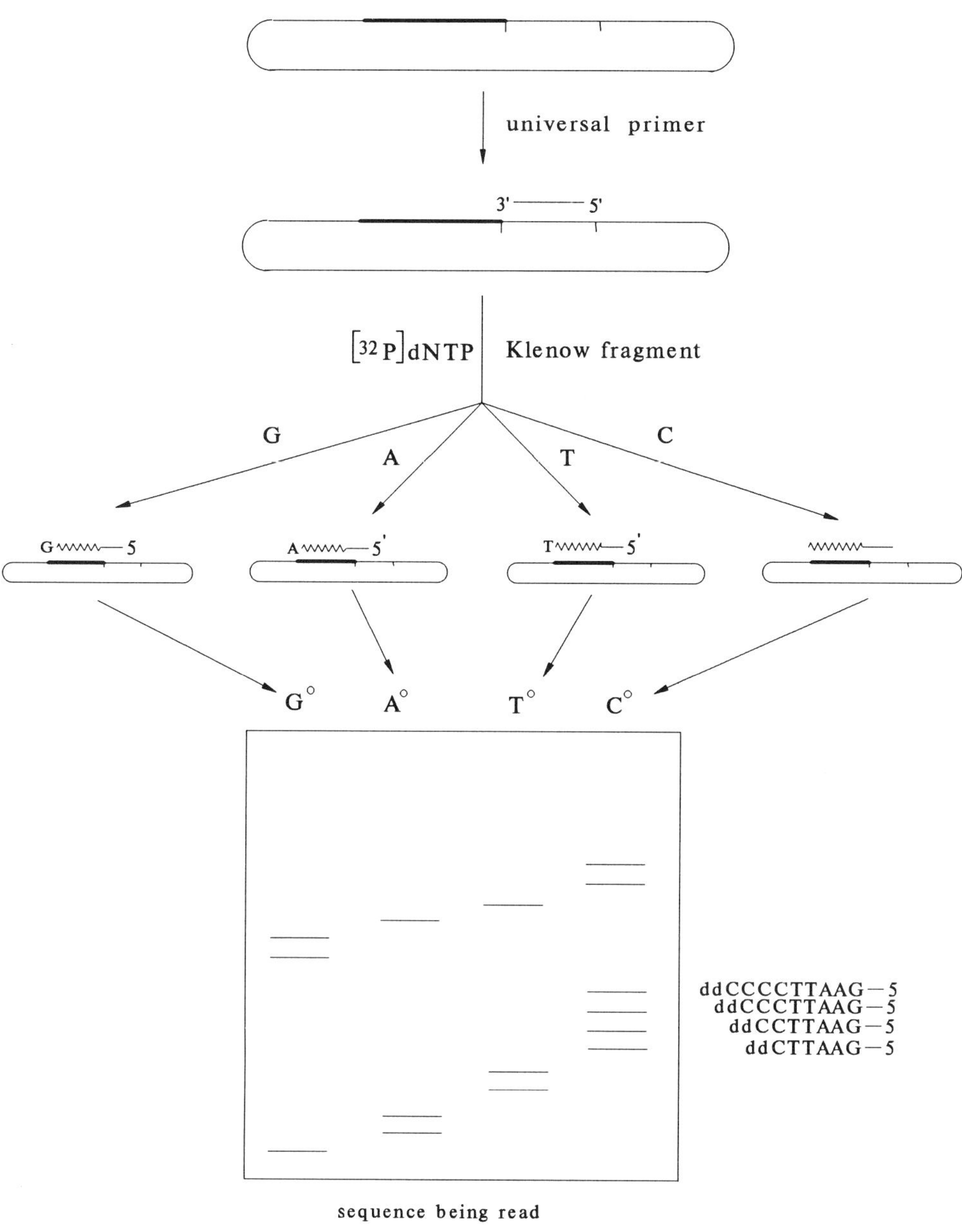

Fig. 6-17. Sequencing of a single-stranded DNA fragment in phage M13 DNA with the aid of a universal primer.

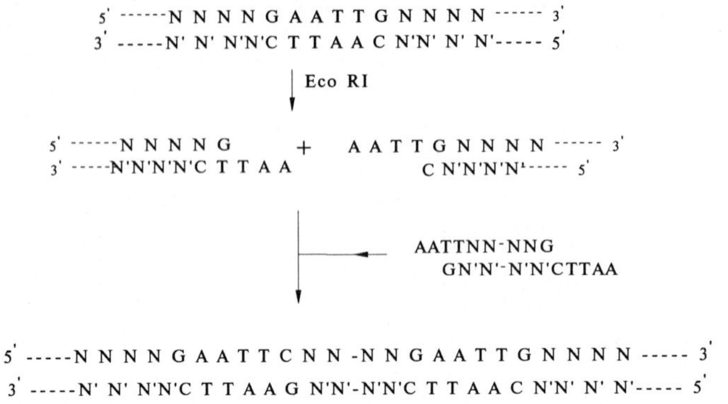

5' ------N N N N G A A T T G N N N N ------ 3'
3' -----N' N' N'N'C T T A A C N'N' N' N'----- 5'
Eco RI
5' ------N N N N G
3' -----N'N'N'N'C T T A A
+
A A T T G N N N N ------ 3'
C N'N'N'N'----- 5'
AATTNN-NNG
GN'N'-N'N'CTTAA
5' -----N N N N G A A T T C N N -N N G A A T T G N N N N ----- 3'
3' -----N' N' N'N'C T T A A G N'N'-N'N'C T T A A C N'N' N' N'----- 5'

Fig. 6-18. Insertion of a foreign DNA at the *Eco*RI site of M13.

thesizing inactive β-galactosidase, its inoculation with the phage produces cells of the *gal*$^+$ phenotype, capable of hydrolysing β-galactosides. This is a result of association of the mutant enzyme encoded by the bacterial DNA with its complete *N*-terminal fragment encoded by the phage DNA, giving rise to a complex exhibiting β-galactosidase activity. The *gal*$^+$ cells can be easily identified because their presence leads to hydrolysis of the chromogenic β-galactosides added to the medium so that the latter turns dark blue. In M13mp2, the fifth and sixth amino acid codons of the β-galactosidase gene correspond to the *Eco*RI restriction endonuclease site (just as in the subsequent generations of this vector; see, e.g., M13mp7 and other vectors in Figs. 6-18 and 6-20). Insertion of a foreign DNA (whose primary structure is to be determined) at this site of M13mp2 usually precludes the above-discussed transformation of *gal*$^-$ cells into *gal*$^+$, and the colonies remain colorless. This serves as an indicator for selecting clones containing the phage with recombinant DNA.

—ACGAATTCA—
M13mp2
—ACGAATTCCCCGGATCCGGGGAATTCA
—M13mp2/Bam HI
—ACGAATTCCCCGGATCCGTCGACCTGCAGGTCGACGGATCCGGGGAATTCA
—M13mp7
EcoRI
BamHI
Sal I
Acc I
Pst I
Sal I
BamHI
EcoRI

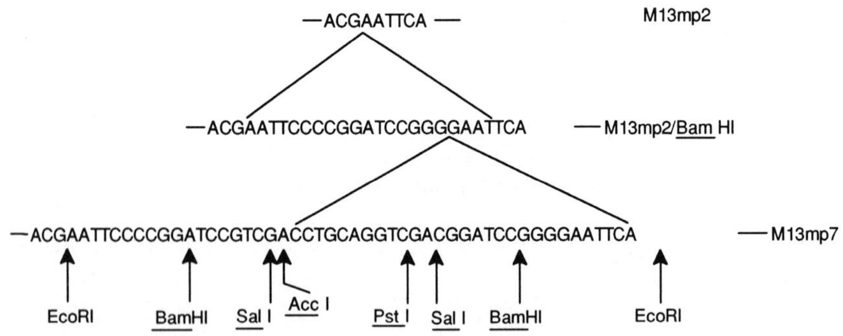

Fig. 6-19. Primary structure of a segment of vector M13mp2 near the *Eco*RI site and the corresponding region in vectors M13mp2/*Bam*HI and M13mp7 constructed on its basis.

As has already been mentioned, a universal cloning vector, M13mp7, was constructed at a later date (see Fig. 6-19). To this end, inserted into the M13mp2 segment was a synthetic DNA containing unique sites of some restriction endonucleases (see Table 6-5). Listed on the left are sequences recognized by restriction endonucleases whose sites are located in vector M13mp7. The right column shows a set of enzymes capable of forming, during DNA hydrolysis, sticky ends corresponding to the sites in the left column.

Shown schematically below is a complete procedure for obtaining a single-stranded DNA suitable for enzymatic sequencing. Figure 6-20 represents the structure of the last-generation vector phage M13mp18 with a decoded polylinker at which a foreign DNA is inserted.

Table 6-5.

Restrictases used for cleavage of replicative form of phage M13mp7	Cleaved sequence	Primary structure of ends of potentially cloned DNA	
*Bam*HI	G↓GATCC	*Bam*HI	G↓GATCC
		*Bal*II	A↓GATCT
		*Bcl*I	T↓GATCA
		*Sau*3A *MbO*I	N↓GATCN
*Eco*RI	G↓AATTC	*Eco*RI	G↓AATTC
*Pst*I	CTGCA↓G	*Pst*I	CTGCA↓G
*Sal*I	G↓TCGAC	*Sal*I	G↓TCGAC
		*Xho*I	C↓TCGAG
*Acc*I	GT↓(CT/AG)AC	*Acc*I	GT↓CGAC
		*Cla*I	AT↓CGAT
		*Hpa*II	NC↓CGGN
		*Taq*I	NT↓CGAN
*Hinc*II *Hind*II	GTPy↓PuAC		

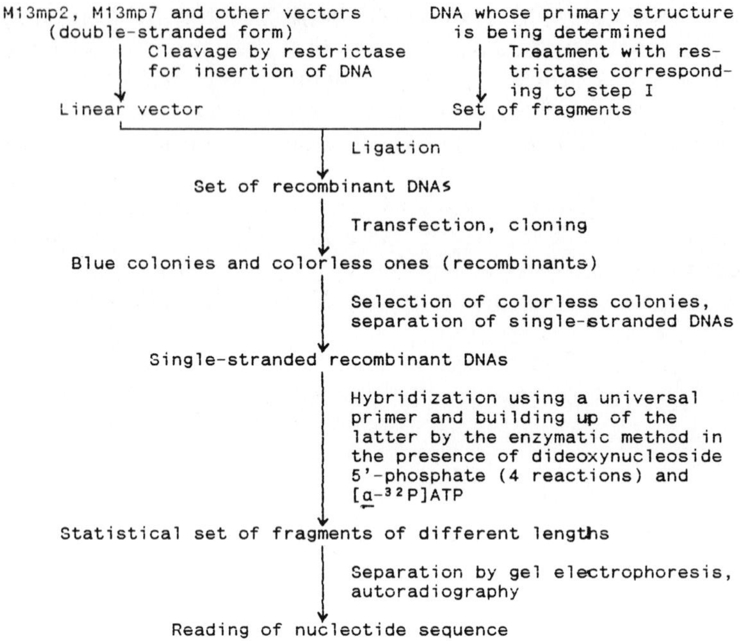

Figure 6-21 illustrates the primary structure of polylinker regions inserted into new generations of vectors based on phage M13. As a consequence of engineering of universal vectors and improvement of methods for cloning and isolation of single-stranded DNAs suitable for sequencing, determination of the primary structure of high-molecular weight DNAs is now possible without restriction endonucleases and their fragmentation can be done using such nonspecific agents as DNase I or by sonication. An advantage offered by such an approach is that a restriction map of the DNA removal under analysis is no longer necessary, which simplifies the experimental procedure significantly. The sequencing involves a fraction containing fragments of limited length (up to 1000 b.p.), their mixture being inserted into the vector by way of blunt- or sticky-end ligation (in the latter case, linkers must be attached to the fragment ends). The result is a mixture of different recombinant DNAs structurally corresponding to the replicative form of phage M13 DNA, each containing a particular fragment of the starting DNA. Then, these DNAs are cloned into *E. coli,* the single-stranded phage DNAs are isolated from individual clones, and the inserts they contain are sequenced by the above-described method using a universal primer (Fig. 6-22). The results are run through a computer to reconstruct the primary structure of the DNA being investigated as a whole. Such a technique (so-called blind sequencing) is by far the fastest, ensuring a sequencing rate of 5000 nucleotides per day on the average. Table 6-6 summarizes the algorithm of this sequencing approach.

A prerequisite for such a sequencing procedure is availability of the appropriate software packages and a nucleotide sequence data base used at the data processing stage.

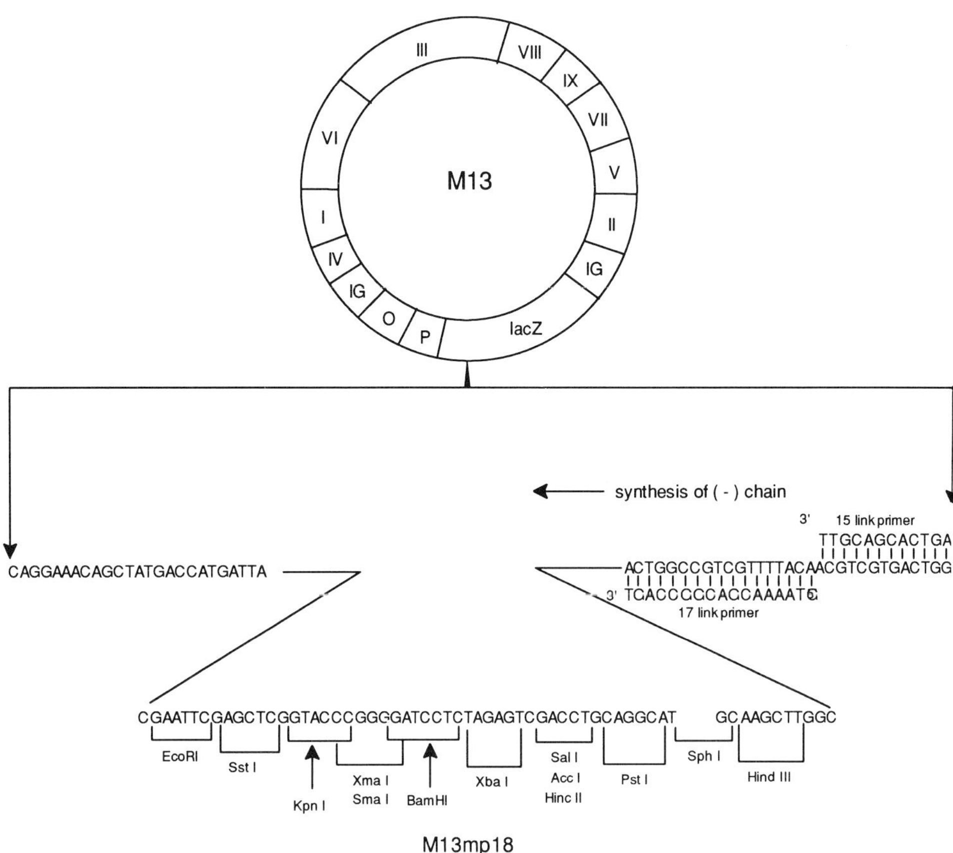

Fig. 6-20. Structure of vector M13mp18. Shown below is the primary structure of the polylinker and the adjacent segments as well as the 15- and 17-unit primers initiating the synthesis of cDNA during sequencing by Sanger's method.

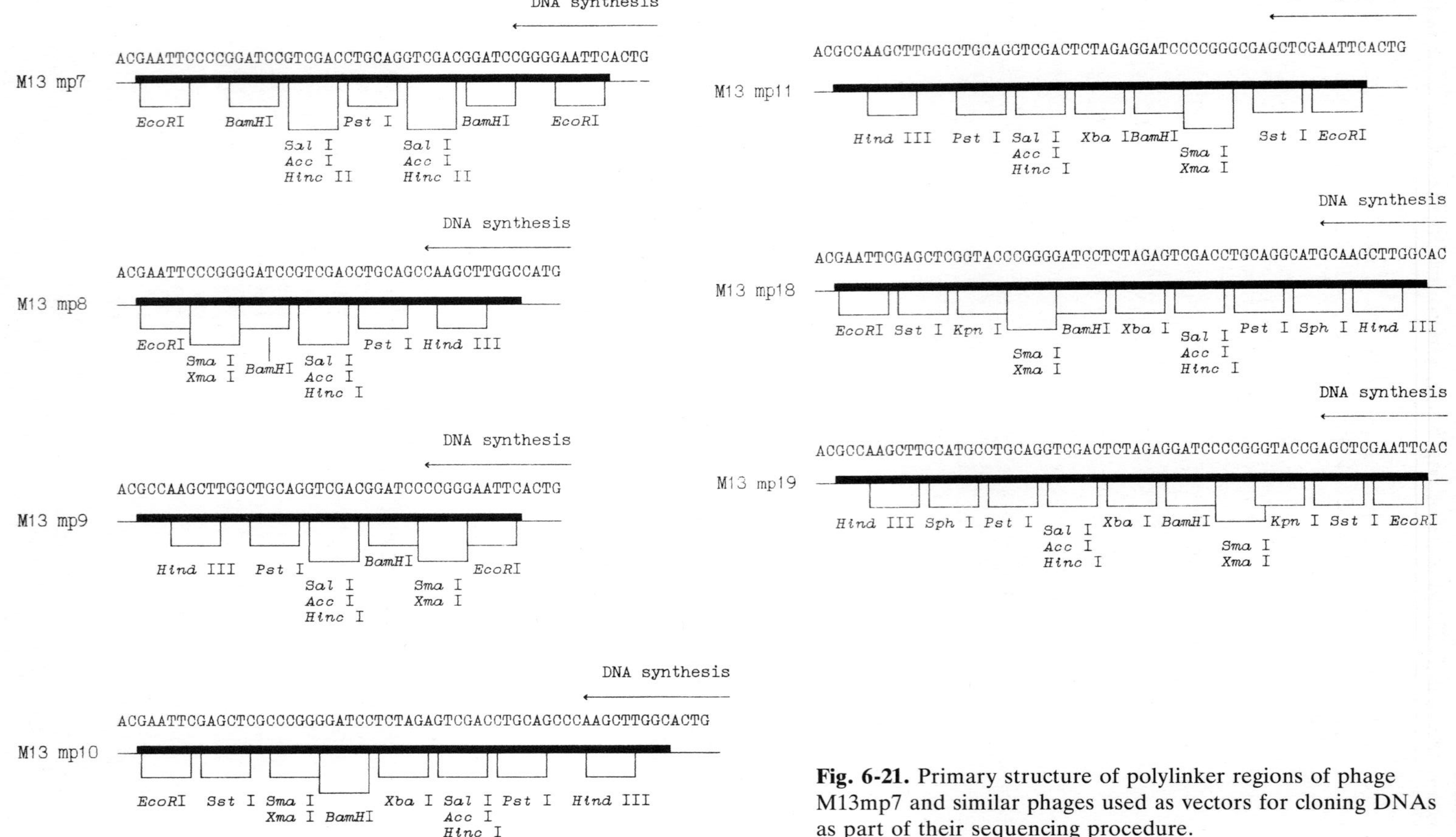

Fig. 6-21. Primary structure of polylinker regions of phage M13mp7 and similar phages used as vectors for cloning DNAs as part of their sequencing procedure.

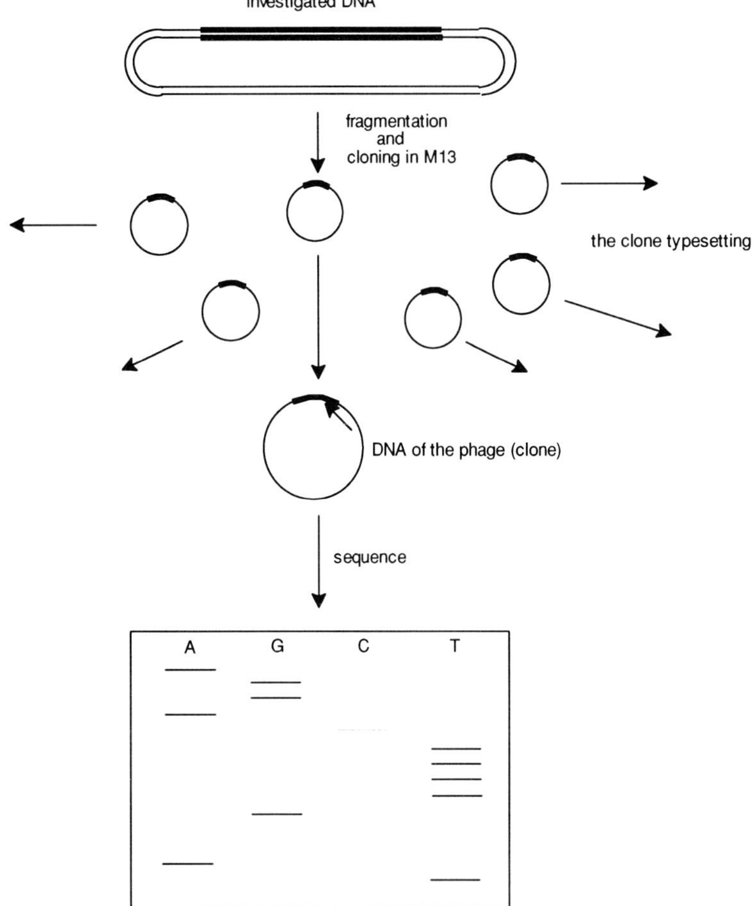

Fig. 6-22. Sequencing of high-molecular weight DNAs by the enzymatic method. Cloning in M13 of its fragments (resulting from nonspecific cleavage), separation of clones, and subsequent use of a universal primer for each clone (the primer is shown as the arrow indicating the direction in which the template is copied).

Table 6-6.

Operation	Mode of its execution
Fragmentation and subcloning	Manual, using standard method
Sequencing of clones	Manual, using standard method and reagents
Data input	Semi-automatic
Integration of partial sequences	Automatic
Search for errors and missing regions	Interactive
Additional sequencing	Manual, using standard method and based on computer data
Additional data input	Semi-automatic
Analysis of resulting structure	Using standard software packages

6.6 Comparison of Chemical and Enzymatic DNA Sequencing Methods

In comparing and evaluating the chemical and enzymatic DNA sequencing methods, one must find answers to the following questions:
(1) how easy is it ot obtain the DNA in the form suitable for sequencing?
(2) what are the number of DNAs necessary for sequencing and the specific activity of the labeled sample?
(3) how time-consuming is the method?
(4) how accurate is the method?

The chemical method is simpler when the DNA under analysis is not too large (200–500 b.p.). When it comes to sequencing a high-molecular weight DNA, preference should be given to the polymerase copying method in order to avoid the restriction procedure yielding individual fragments. In enzymatic sequencing of large single-stranded DNAs (e.g., bacteriophages), one can use a set of priming oligonucleotides whose synthesis today does not call for any significant time and labor expenditures. The ideal method for double-stranded high-molecular weight DNAs is blind enzymatic sequencing using a universal primer (currently offered by many suppliers) and a computer for data processing. The chemical method can also be used, but then one must cut out the DNA fragments of interest from the vector, which makes the entire procedure too complicated. The next criterion is the amount and specific activity of the DNA to be analyzed. One of the reasons for the great popularity of the chemical degradation method is that it allows DNA to be sequenced from any end after the incorporation of a labeled (^{32}P) terminal phosphate. The terminator is kept standard by using restriction endonucleases while preparing the sample for analysis. A drawback of this approach is the impossibility of introducing a highly active label (only one molecule of labeled phosphate per DNA molecule can be inserted). One of the attractions of the enzymatic method is the insertion of a multiple label into the strands being synthesized, which means that a much smaller amount of DNA is necessary for analysis, as compared to chemical sequencing. Simple calculations have shown that for sequencing a 200-membered polynucleotide and obtaining autoradiogram of the same clarity at the same exposure time, 2 picomoles of DNA are needed if the sequencing is done chemically and only 0.1 picomole in the enzymatic case. The chemical method also calls for gamma-labeled ATP with an activity of 3000 Ci/mole, whereas in the enzymatic method the precursor of the label may have a specific activity of 300 Ci/mole, which is more economical.

As regards the labor requirements, both methods seem to be comparable. From the purely experimental standpoint, chemical sequencing is more time-consuming (because of the repetivitve alcohol precipitation, lyophilization and other procedures). In the case of sequencing by the polymerase copying method, on the other hand, the most time-consuming operations are gene engineering of the sample and preparation of the primer.

In terms of accuracy and effectiveness, the close similarity of both methods has been proved by numerous experiments. In both cases, the two characteristics are largely dependent on the structure of extended single-stranded DNA fragments that may form hairpins and have varying mobility during PAGE. Short fragments tend to "run away" from the sequencing gel, which is why one must always have the possibility to check the results of separation in polyacrylamide gel by a cross experiment – that is, sequencing the same portions in other fragments of the starting DNA or sequencing the complementary strand. The latter is usually a must in both methods. Good results in gel electrophoresis of high-molecular weight fragments are obtained at elevated temperatures because they render formation of double-stranded structures difficult. Significantly, good possibilities are being opened up at present for increasing the sequencing rate by either method. This is borne out by new developments in solid-phase chemical sequencing and advent of automatic sequencers based on the enzymatic method. Moreover, the role of computers in processing experimental data is steadily increasing. Today it is hard to make any predictions as regards sequencing rates. The mere fact that work aimed at complete sequencing of the human genome which contains nearly three billion nucleotides is already under way testifies to the spectacular strides being made in this area.

6.7 Determination of Nucleotide Sequences in RNA

The determination of the primary structure of RNA is the first step in elucidating its functional capabilities. The earliest approach used to sequence rather large polynucleotides was the so-called block method developed in the mid sixties for transfer RNAs. The method is based on a combination of complete and partial hydrolyses of tRNAs with pyrimidyl and guanyl RNases followed by isolation of oligonucleotides and their sequencing. By obtaining overlapping blocks one can reconstruct the initial molecular structure. This approach has been treated at length elsewhere. Since this is all history now, we shall not dwell on this method at all but discuss the fast RNA sequencing techniques extensively used at present. It would be natural to apply the chemical degradation and polymerase copying procedures to RNAs. As a matter of fact, both methods started being developed at the same time (1977–80) and, in spite of the difficulties stemming from the different structures of DNA and RNA, RNA sequencing techniques similar to those for DNA have been elaborated.

6.7.1 Direct Chemical Sequencing of RNAs Labeled at the 3′ End

RNA sequencing based on statistical chemical degradation of the polyribonucleotide chain is identical with the method described above for DNA. True, attempts to create adequate conditions for β-elimination stumbled against a serious difficulty arising from the presence of a 2′-hydroxyl group in RNA monomers, or ribonucleotides, and the associated high lability of the internucleotide linkages in the presence of bases.

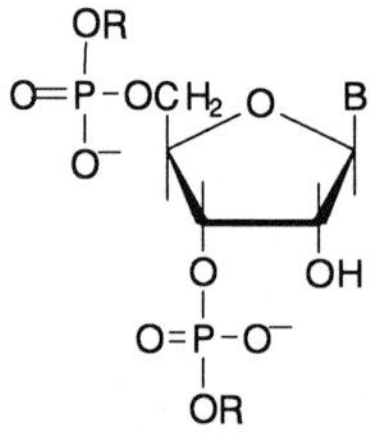

B=A,G,C or U
R= polyribonucleotide chain

This is why it was impossible to use piperidine (pK 11.2) which would lead to nonspecific cleavage of any internucleotide linkage.

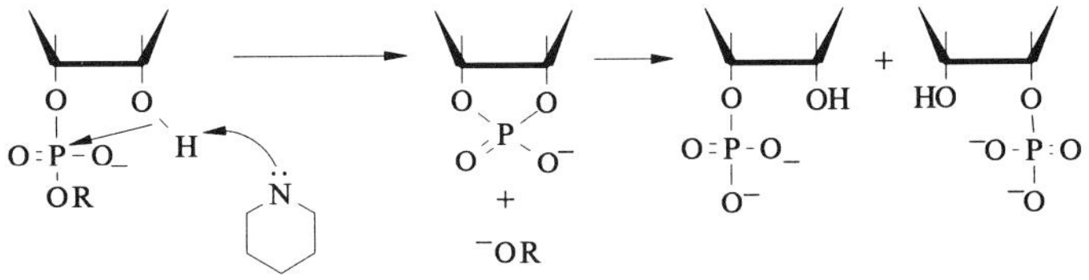

Consequently, different conditions for β-elimination had to be provided. To this end, piperidine was replaced by a weaker base, namely, aniline (pK 4.5). Besides, the presence of a 2′-hydroxyl affects the β-elimination process itself. The phosphate group is easily eliminated from the 3′ position of the sugar moiety, deprived of the heterocyclic base, which is most probably due to the sufficient mobility of the α-proton. As a consequence, the polyribonucleotide chain is cleaved at the site of the removed base. However, the modified sugar moiety in the emerging polyribonucleotide fragment contains a methylene group (as a result of enol-to-ketone conversion) at the former 3′ position of ribose. In such a compound (as well as after its conversion into a Schiff's base), the proton can be removed only from the α-position or, in other words, from the metyhlene group, but not from the β-position (former 4′ position of ribose). Consequently, the second phosphate group is not removed (from the former 5′ position of the sugar), and at their 3′ end the 5′ fragments of the polyribonucleotide contain, after such transformations, an esterified phosphate rather than a free phosphate group (as in the case of DNA). Therewith, the former ribose, which is essentially an α-diketone derivative, may undergo different transformations. Naturally, this may affect the band pattern after PAGE if 5′-labeled RNA is used for analysis. This is precisely why RNA labeling prior to sequencing at the 3′ end (see below) has become extremely popular. In this case, RNA fragments have identical monophosphate groups at the 5′ end.

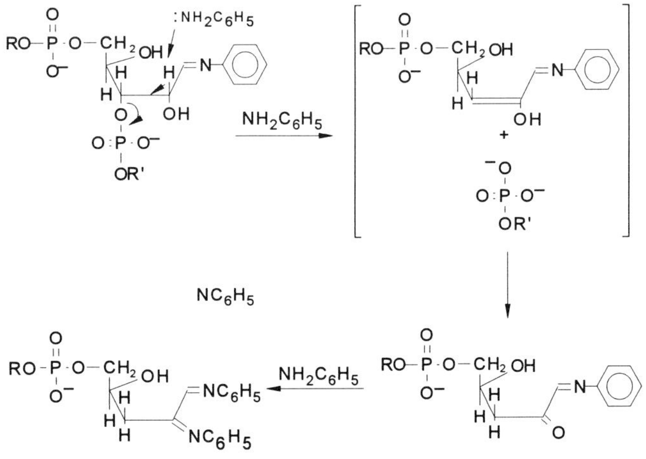

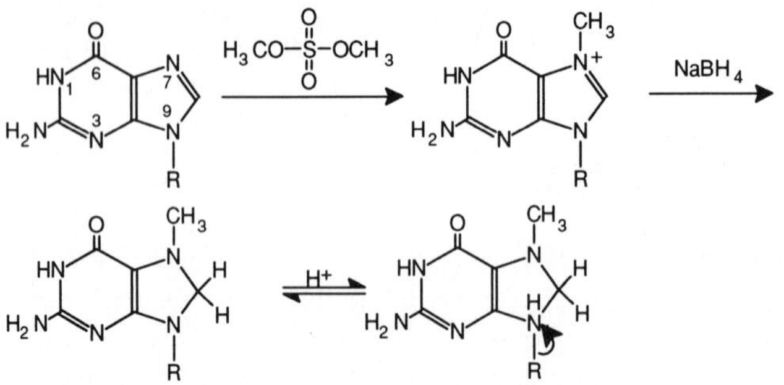

Fig. 6-23. Methylation of the guanine at N^7 with subsequent reduction with sodium borohydride.

The methods for modifying heterocyclic bases, used in the context of DNA, have undergone some changes, when applied to RNA, due to the pronounced lability of the internucleotide linkages.

Guanine Reaction. The modification of guanines in RNA, as well as in DNA, is performed with the aid of dimethyl sulfate (see Fig. 6-23). In this case, the methylation involves predominantly the nitrogen at position 7. However, such an N^7-methylated guanosine link in RNA is stable at neutral pH values. Therefore, for guanine to be removed, the double bond in the imidazole ring is reduced with sodium borohydride (the reduction proceeds easily under rather mild conditions), and the glycosidic bond in the resulting dihydro derivative of 7-methylguanosine becomes highly labile and easily hydrolysed when treated with aniline. The limited methylation of RNA with dimethyl sulfate followed by reduction with $NaBH_4$ and aniline hydrolysis yields polyribonucleotide fragments specifically cleaved at G.

Adenine Reaction (A > G). When diethyl pyrocarbonate reacts with nucleic acids in aqueous solutions, the purine base nitrogens undergo carboethoxylation. It is primarily the nitrogens at position 7 of adenine and guanine that are involved in this case, and the imidazole ring is opened. As a result, the

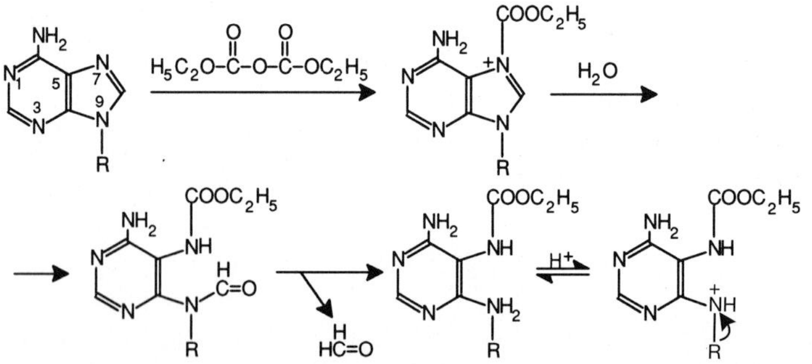

Fig. 6-24. Reaction of diethyl pyrocarbonate with adenine in RNA, which leads to opening of the imidazole ring.

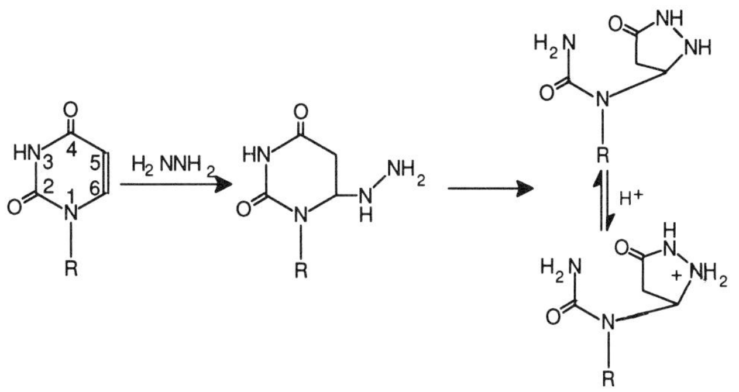

Fig. 6-25. Hydrazinolysis of uracils in RNA, resulting in a labile *N*-glycosidic bond broken after treatment with aniline.

chain can be cleaved with aniline at that position. Since the reaction at adenines proceeds at a rate seven times faster than in the case of guanines, it is used to determine the A positions in polyribonucleotides (see Fig. 6-24).

Uracil Reaction. The mechanism by which unprotonated molecules of hydrazine react with uridine has already been described in the context of cytidine and thymidine. The rate of this reaction, when uracil is involved, is much faster than in the case of C and T ($U \gg C > T$). Once the pyrazolone ring is formed (see Fig. 6-25), it is readily detached from the glycosidic carbon, and an attack with aniline yields a Schiff's base, which leads to β-elimination of the 3′-phosphate group and, consequently, cleavage of the RNA chain at that position.

Cytosine Reaction (C > U). Anhydrous hydrazine reacts primarily at cytosines in the presence of 3 M NaCl (just as in the case of DNA, the reaction is predominantly at C in the presence of 2 M NaCl). The mechanism of the reaction at cytosines has already been described.

Figure 6-26 shows schematically the chemical degradation of a polyribonucleotide at the position where the above-described modification of a particular base has taken place in the presence of aniline. The glycosidic hydroxyl resulting from hydrolysis of the *N*-glycosidic bond (at the modified base site) gives rise to a tautomeric aldehyde form of the sugar moiety. This form reacts with aniline to yield an aldimine (Schiff's base), and, as has already been mentioned, β-elimination and, consequently, RNA cleavage takes place. Aniline titrated with acetic acid to pH 4.6 ensures an optimal pH value for β-elimination and sufficiently mild conditions to avoid nonspecific cleavage of internucleotide linkages in RNA.

As has already been pointed out, for the sequencing results to be unambiguous, the label is usually incorporated into the 3′ end of RNA. This is achieved through a reaction between RNA and [5′-^{32}P]pCp in the presence of RNA ligase:

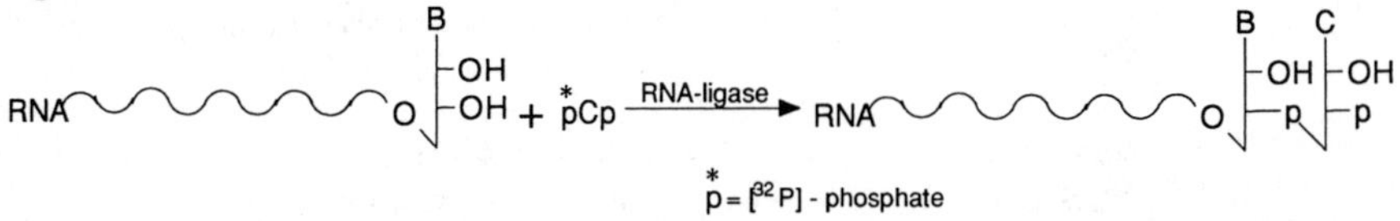

$$RNA\text{–}B(OH)(OH) + \overset{*}{p}Cp \xrightarrow{\text{RNA-ligase}} RNA\text{–}B\text{–}\overset{*}{p}\text{–}C\text{–}p$$

$\overset{*}{p}$ = [^{32}P] - phosphate

Such an RNA labeled at the 3′ end is separated by PAGE then sequenced.

However, the 3′ end may also be labeled after the chemical modifications. This saves a lot of time when several RNAs have to be sequenced in the same experiment. For example, as can be seen from Figure 6-27, four chemical reactions may be conducted with a mixture of several labeled RNAs (RNAs may also be labeled after the chemical modification), followed by fractionation in polyacrylamide gel, elution, and pelleting of an individual RNA.

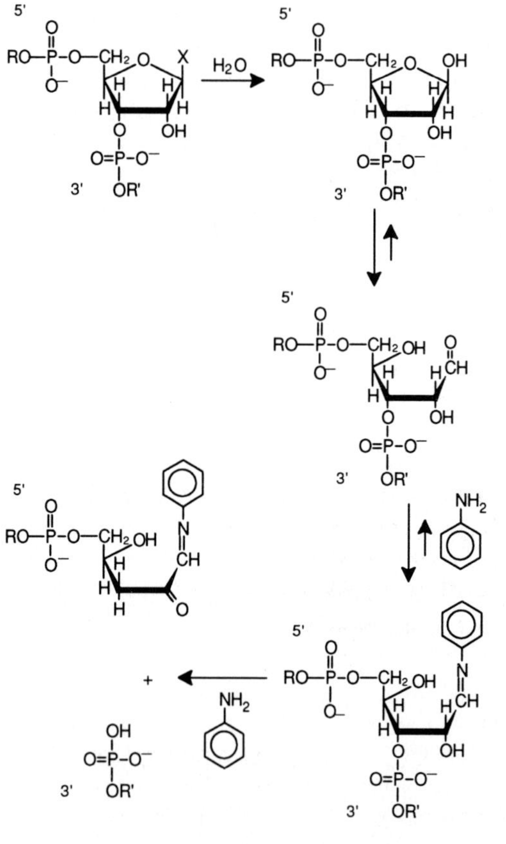

Fig. 6-26. Degradation of RNA at modified bases, yielding a set of 3′-terminal fragments with 5′-phosphate.

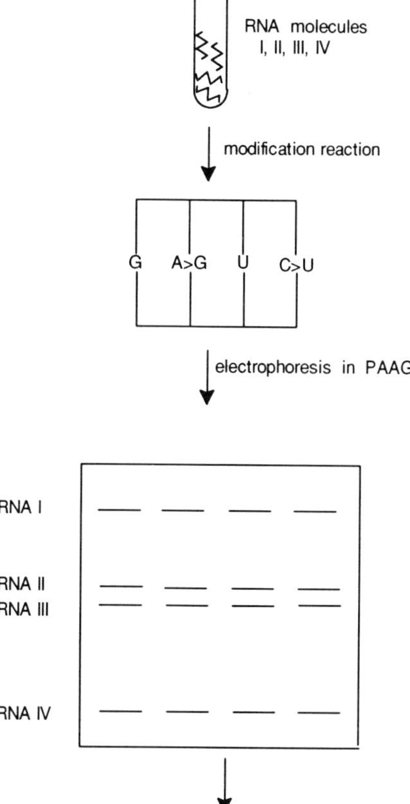

Fig. 6-27. Simultaneous sequencing of several RNAs.

Each modified RNA separated in this fashion is treated with aniline (with cleavage of the polynucleotide chain), then subjected to PAGE. Figure 6-27 shows that RNA molecules I, II, III and IV are separated by PAGE after restricted chemical modification (at G, A > G, C > U and U). As a result, only four chemical reactions are conducted instead of sixteen. The aniline treatment is carried out immediately after the elution from polyacrylamide gel and pelleting. The pellets are then applied onto a standard gel to be subjected to electrophoresis and autoradiographed.

Figure 6-28 is an autoradiogram of the sequencing gel illustrating the primary structure of 5S RNA (the structure is read from the 121st nucleotide; i.e., from the 3′ end). The gel electrophoresis was performed under standard conditions in the presence of 7 M urea.

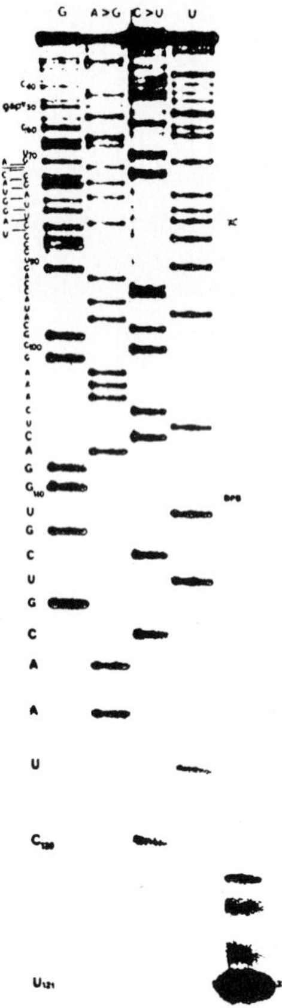

Fig. 6-28. Autoradiogram of the sequencing gel of 5S RNA labeled at the 3′ end with the aid of [5′-^{32}P]pCp and RNA ligase.

6.7.2 Direct Enzymatic Sequencing of RNA

A modification of direct RNA sequencing is the enzymatic method of statistical cleavage of internucleotide linkages with the aid of RNases selectively breaking the phosphodiester bonds formed by pyrimidine and purine nucleoside 3′-phosphates.

As was pointed out in the introduction, the presence of RNases A (pyrimidyl RNase) and T_1 (guanyl RNase) whose substrate specificity has been carefully elucidated has made it possible to develop the block method of RNA sequencing, used already in the mid sixties for determining the sequence of rather large molecules of individual RNAs. These RNases, plus some new ones whose substrate specificity will be described in the following, were used for statistical cleavage of polyribonucleotides instead of their chemical degradation. Four enzymes are usually employed for the purpose. The following schemes illustrate the cleavage of internucleotide linkages by each of them.

RNase T_1 cleaves internucleotide linkages formed by guanosine 3′-phosphates in the polyribonucleotide:

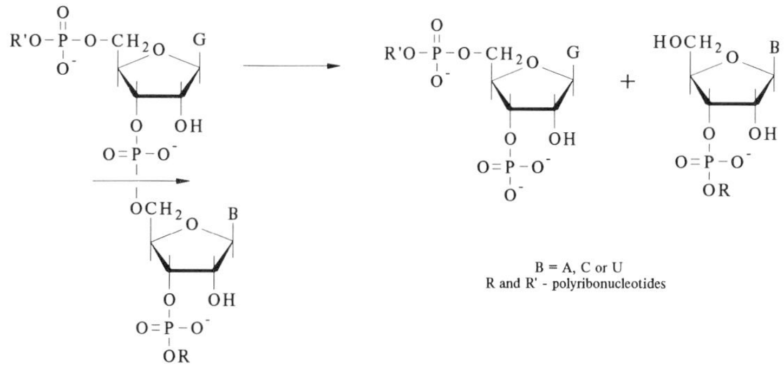

B = A, C or U
R and R' - polyribonucleotides

Hence, RNase T_1 serves as a means for determining the position of G during RNA sequencing.

RNase U_2: The specificity of this enzyme is similar to that of the previous one with the difference that this RNase cleaves internucleotide linkages formed by adenosine 3′-phosphate and is used to determine the position of A in the RNA:

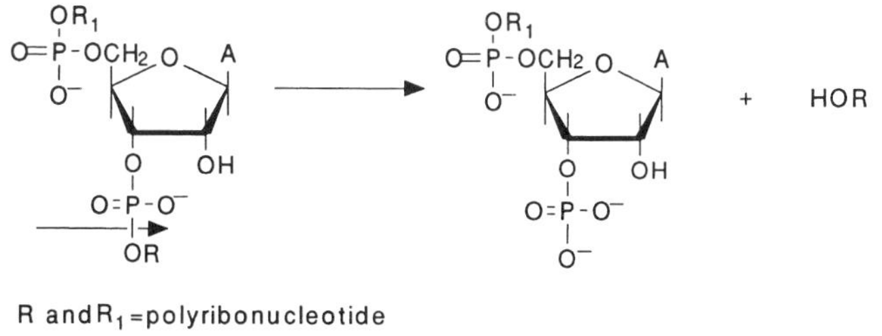

R and R_1 = polyribonucleotide

RNase Phv M (from *Phyzarium polycephalum*) cleaves internucleotide linkages by the same mechanism[1] but involves bonds of two types: thoses formed by adenosine 3′- and uridine 3′-phosphates:

RNase BC (from *Bacillus cereus*) hydrolyses internucleotide linkages formed by pyrimidine nucleoside 3′-phosphates similarly to RNase A (pyrimidyl RNase):

Like in the previous cases, the emerging 3′-terminal RNA fragments lack phosphate groups at the 5′ end (in contrast to the 3′-terminal fragments resulting from chemical degradation of RNA). In addition to the main ribonucleases use is also made of RNase CL3 isolated from chicken liver. This enzyme is highly specific, digesting primarily internucleotide linkages formed by cytidine 3′-phosphates: Cp ↓ N. There is yet another, less specific ribonuclease used for RNA sequencing, namely, nuclease M_1 hydrolysing linkages N ↓ pA, N ↓ pU and N ↓ pG. In this case, the 3′-terminal RNA fragments already have phosphate groups at the 5′ end.

All reactions in which RNAs are digested by these enzymes usually take 12 minutes at 55 °C with subsequent application onto polyacrylamide gel and electrophoresis. Here, just as in direct chemical degradation, 3′-labeled RNA is used. However, during enzymatic digestion the 5′ end of RNA may also be labeled with the aid of polynucleotide kinase and [γ-^{32}P] ATP, because all 5′-terminal fragments resulting from statistical enzymatic hydrolysis with RNases T_1, U_2, Phv M, BC and CL3 have a 3′-terminal 2′,3′-cyclic phosphate group. Statistical hydrolysis of 3 mg RNA under standard conditions requires a single activity unit from each enzyme. Along with enzymatic digestion, which makes it possible to identify the positions of the four nucleotides along the RNA chain from the 5′ or 3′ end, it has also become common practice to resort to incomplete alkaline hydrolysis of the RNA under analysis. Application of alkaline hydrolysis (the autoradiogram usually shows all RNA fragments differing by a single nucleotide) simplifies the reading of the gel. Figures 6-29 and 6-30 are autoradiograms of sequencing gels after sequencing of *E. coli* 5S

[1] The mechanism of internucleotide linkage hydrolysis is dual.

RNA with the aid of enzymes produced by "Pharmacia". The only difference in the sequencing procedures was use of 5′-labeled RNA (Fig. 6-29) and 3′-labeled RNA (Fig. 6-30).

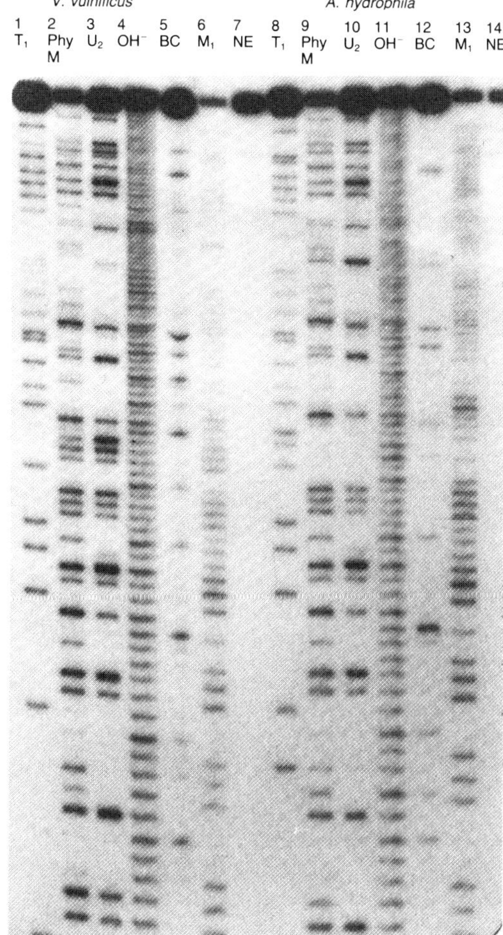

Fig. 6-29. Autoradiogram of the sequencing gel of *E. coli* 5S RNA labeled at the 5′ end. 20% polyacrylamide gel containing 7M urea is used. Channel 1 NE is control (without enzyme). Channel OH corresponds to alkaline hydrolysate of *E. coli* 5S RNA. The other channels correspond to hydrolysates involving the above-described enzymes. Channels 10–12 correspond to hydrolysates involving RNase V_1 isolated from cobra venom and cleaving RNA only over double-stranded portions.

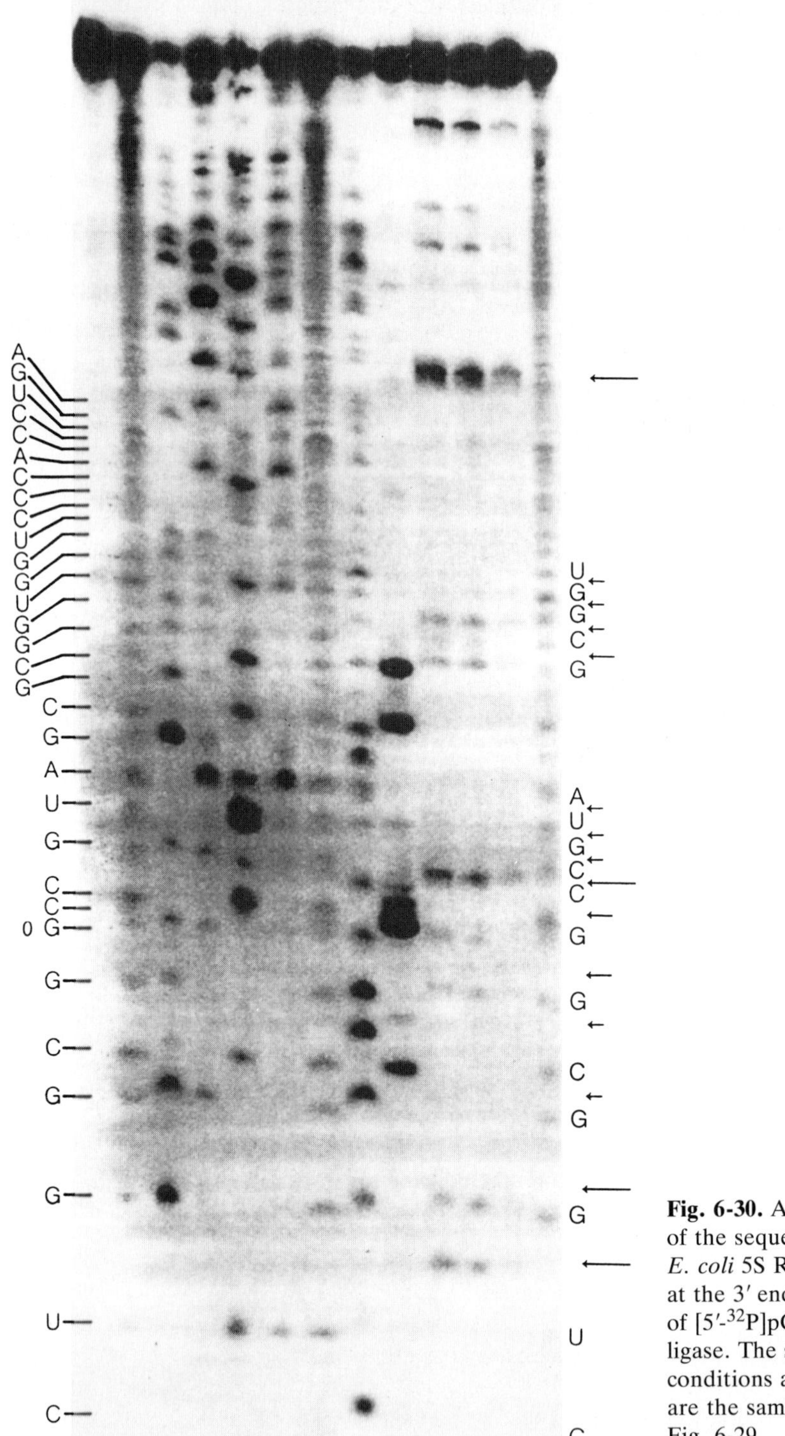

Fig. 6-30. Autoradiogram of the sequencing gel of *E. coli* 5S RNA labeled at the 3′ end with the aid of [5′-^{32}P]pCp and RNA ligase. The sequencing conditions and channels are the same as in Fig. 6-29.

6.7.3 Sequencing of High-Molecular Weight RNAs by Polymerase Copying with the Aid of Reverse Transcriptase

The possibility of DNA copying by way of reverse transcription is widely used for sequencing high-molecular weight RNAs. Two main procedures of RNA sequencing in this manner exist. The first procedure is based on obtaining a single-stranded cDNA with separation in polyacrylamide gel before its direct sequencing. The other calls for obtaining a double-stranded cDNA through reverse transcription and cloning in a suitable vector with subsequent sequencing by any method.

Since the RNA sequencing method described here ultimately boils down to DNA sequencing, we will consider a few examples illustrating only procedures for obtaining individual cDNAs.

The first procedure was used to sequence interferon mRNA. In order to obtain a transcript of interferon mRNA, the latter was not separated from a mixture of polyadenylated mRNAs isolated from human fibroblasts. Instead a primer was added to the complex mixture of mRNA, namely synthetic 15-membered oligodeoxyribonucleotide complementary with respect to the 3′-terminal portion of the interferon mRNA. The nucleotide sequence of the primer, which was expected to produce a complementary complex only with the interferon mRNA, was determined from the known initial structure of the C-terminal region of the protein – fibroblast interferon to be precise. In this case, account is taken of the codons from human genes used predominantly for polycodon amino acids. The primer engineering procedure is shown in below.

C-terminal amino acid sequence of human fibroblast interferon	–MET–SER–TYR–ASN–LEU
Codons of the corresponding amino acids. Most frequently used codons of the corresponding amino acids in human genes	AUG AG$^{U}_{C}$UA$^{A}_{C}$AA$^{U}_{C}$CUN AUG AGC UAC AAC CUG
Structure of synthesized primer	3′-TAC TCG ATG TTG GAC-5′
Formation of complementary complex: 5′(^{32}P)-primer-mRNA of human fibroblast interferon. Reverse transcription. Separation of cDNA and sequencing by the Maxam-Gilbert method	

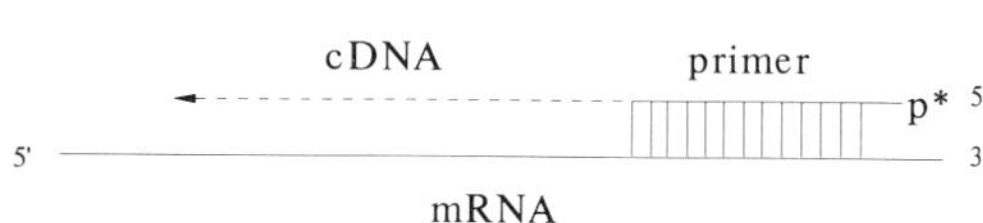

After synthesis, 5′-terminal phosphate is inserted into the 15-linked oligonucleotide with the aid of [γ-^{32}P] ATP and polynucleotide kinase. The labeled primer forms a complementary complex only with interferon mRNA. Following incubation with reverse transcriptase, cDNA is separated by PAGE and sequenced by the Maxam-Gilbert method.

The use of synthetic primers opens up wide possibilities for direct sequencing of rare mRNAs without their separation from complex mRNA mixtures. It also helps to eliminate the highly laborious procedure of mRNA isolation and purification as well as cloning. Among the disadvantages of the method is prefact that only a sequence of 250 to 300 nucleotides can be determined and the fact that structure can be determined only from a single DNA strand. As a matter of fact, the first difficulty is easily circumvented at present by creating new primers in the course of sequencing (it takes a maximum of one day to synthesize and purify a 15-unit oligonucleotide), which are complementary to the RNA portions whose primary structure is already established.

The other approach to RNA sequencing is based on the synthesis of a double-stranded DNA copy. This is how the primary structure of the influenza virus RNA was established. Figure 6-31 shows schematically the transformation of the influenza virus mRNA to a double-stranded transcript, aided by a synthetic primer. After the double-stranded DNA copy had been treated with nuclease S_1, which digests single-stranded portions of DNA, and also with restriction endonucleases, fragments of the RNA transcript were cloned and

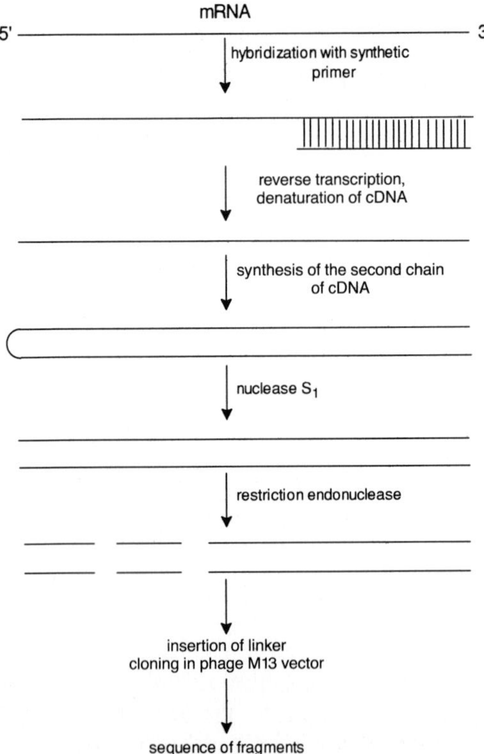

Fig. 6-31. Cloning of fragments of a double-stranded DNA produced by reverse transcription of mRNA and synthetic priming, suitable for sequencing.

sequenced by the Sanger method. The successful implementation of this RNA sequencing approach is associated primarily with automation of oligodeoxynucleotide synthesis. Synthetic oligonucleotides are employed not only as primers during reverse transcription but also as probes in identification of RNA and DNA.

6.8 Solid-Phase Sequencing of DNA and RNA

Oligo- and polynucleotides can be sequenced by the Maxam–Gilbert method with a much lower expenditure of time and reagents if the DNAs and RNAs of interest are first immobilized on modified filter paper.

The solid-phase sequencing principle can best be illustrated by determination of the primary structure of DNA fragments. Used for immobilization is Whatman 540 paper activated with cyanuric chloride and then converted with 2-bromoethylamine in the presence of triethylamine.

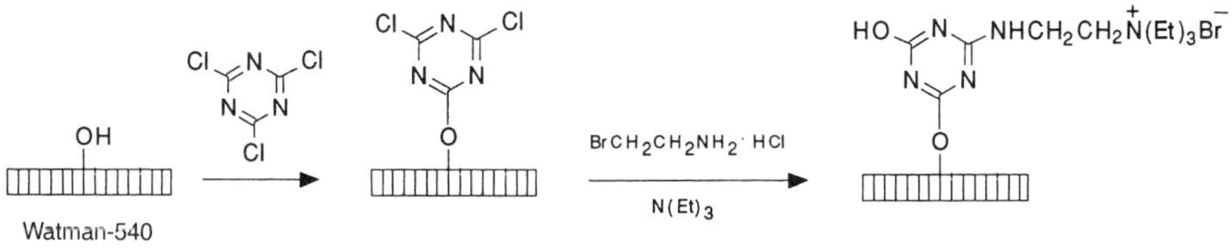

The immobilization of DNA on paper prepared in the above manner is achieved through ion exchange. 120 mg of DNA can be immobilized per square centimeter of paper. Depending on the number of DNA samples to be sequenced, 4n pieces of paper (2–4 mm^2, see Fig. 6-32, left) are cut and marked with a pencil. Paper strips can also be used (Fig. 6-32, right). The DNA to be analyzed (0.5–1.0 μl of sample labeled at the 5′ or 3′ end) is applied onto the prepared paper. When paper strips are used, the DNA application points must be spaced 2 to 5 mm apart. The paper (all pieces or strips) is placed in four reaction vessels to conduct the reactions (G, G+A, T, C).

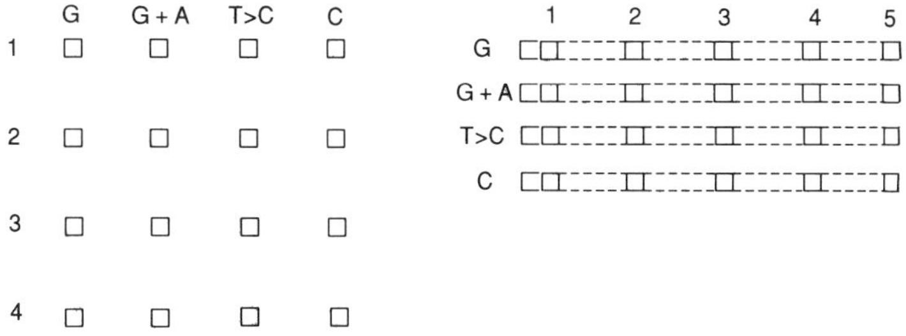

Fig. 6-32. Preparation of ion-exchange paper for immobilization of 5 DNA samples.

The base modification conditions have been changed slightly in comparison with the standard reactions normally used for sequencing in solution. Direct use of the latter for solid-phase sequencing has turned out to be impossible because of the sizable DNA losses during the modifications (hydrazinolysis, alkaline treatment). Dimethyl sulfate is used to modify DNA at the guanines; the position of adenine and guanines (A+G) is determined by treating the samples with 66–88 % formic acid or diethyl pyrocarbonate at 90 °C (A>G). The following scheme illustrates the transformations undergone by adenine in the presence of diethyl pyrocarbonate with subsequent opening of the imidazole ring. The first step is acylation of the nitrogen (N^7) in imidazole.

$$\text{Adenine (R)} \xrightarrow{(C_2H_5OC(O))_2O} N^7\text{-}COOC_2H_5\ \text{adenine}^{+} + \left[{}^{-}O\text{-}C(O)\text{-}OC_2H_5\right] \xrightarrow[-CO_2]{} {}^{-}OC_2H_5 \xrightarrow[H_2O]{} {}^{-}OH + C_2H_5OH$$

Subsequently, the hydroxyl anion (strong nucleophile) immediately attacks the carbon at position 8 of the imidazole ring, which leads to cleavage of the C_8–C_9 bond:

$$\text{NH}_2,\ \text{COOC}_2\text{H}_5,\ {}^{+}\text{N},\ {}^{-}\text{OH} \longrightarrow \text{CH-O-H} \longrightarrow \text{NH-COOC}_2\text{H}_5,\ \text{NCHO-R} \xrightarrow{\text{water, NH}_3} \text{NH-COOC}_2\text{H}_5,\ \text{NH-R}$$

To modify cytosines in DNA hydroxylamine is used (pH 6) instead of hydrazine. It is believed that hydroxylamine is added at the double C_5–C_6 bond and the amine group is substituted. The base modified in this manner is detached in the presence of piperidine from the polynucleotide chain which is then cleaved as usual.

To determine the position of thymines (T) during solid-phase sequencing use is made of oxidation with potassium permanganate in aqueous solutions. The intermediate products of this reaction are 5,6-dihydroxy-5,6-dihydro derivatives.

(Structure: 5,6-dihydroxy-5,6-dihydrothymine — O, CH_3, HN, OH, OH, O, N, H)

These compounds undergo fast opening of the pyrimidine ring at the C_5–C_6 bond, and the sequence of transformations results in formation of *N*-substituted ureas NH_2–CO–NH–R (where R is a polynucleotide). Similar transformations are undergone by cytosines. The oxidation rate of thymines is much higher than that of cytosines. Therefore, the reaction with permanganate makes it possible to determine the position of the former in the first place.

Treatment with piperidine (conditions under which β-elimination takes place) is performed separately for each DNA sample (if paper strips are used, they must be cut). The paper is treated with 10 % aqueous piperidine (30 min, 90 °C), which leads to elution of the DNA. The subsequent treatment, PAGE and autoradiography are carried out under standard conditions.

It should be noted that the solid-phase method allows one to analyze a large number of DNA samples at a time, which cuts down the sequencing time considerably.

For solid-phase RNA sequencing ion-exchange Whatman DE 81 paper is used. 3′-Labeled RNA is immobilized on 4n 2–4 mm^2 pieces of paper, depending on the number of samples, which are then marked with a pencil and placed in four vessels to perform the corresponding modifications. The following conditions are maintained to determine the position of bases in the RNA chain:

G: dimethyl sulfate (pH 5.5, 90 °, 1 min) with subsequent $NaBH_4$ treatment (pH 7, 0 °, 30 min)
A>G: diethyl pyrocarbonate (pH 4.5, 90 °, 10 min)
U+C: 50 % aqueous hydrazine (0 °, 20 min)
C: 3 M NaCl in anhydrous hydrazine (0 °, 10 min).

After modification all paper segments are treated together with 1 M aniline–acetate buffer (pH 4.5, 60 °C, 20 min). For desorption of RNA each paper segment is treated with 2 M $NaClO_4$. RNA is precipitated with an alcohol and subjected to PAGE. The entire procedure takes two hours – that is, the analysis in this case is much quicker than RNA sequencing in solution. Time saving is the greatest when a large number of RNA samples are analyzed, and solid-phase sequencing allows many RNA fragments to be processed simultaneously.

6.9 Polymerase Chain Reactions in Analysis of the Primary Structure of Nucleic Acids

The enzymatic method of DNA sequencing has found broad application primarily because it can be combined with isolation and multiplication of fragments of the DNA under analysis by way of molecular cloning. It was noticed that these fragments are accidentally entrained by a vector from a complex mixture obtained when the polymeric DNA is cleaved in one or another way. In recent years, however, an effective procedure has been developed to produce strictly defined fragments of virtually any DNA, including the total human genome DNA. It is based on DNA-polymerase chain reactions (PCR) which immediately found extensive applications in different fields of molecular biology and gene engineering, including analysis of the primary structure of nucleic acids.

A PCR is based on the DNA polymerase-mediated buildup of the primer oligonucleotide on a single-stranded DNA template (Fig. 6-11). The potential of such a reaction has been drastically expanded by using in it Taq DNA polymerase isolated from the thermophilic bacteria *Thermus aquaticus.* The activity of this enzyme (recently, other enzymes of a similar type have been found as well) is at its peak at 70 °C and it can withstand heating to 95 °C for a short period of time. By virtue of such properties it has become possible to copy and multiple any selected DNA fragment by means of Taq polymerase and two primers oriented toward each other, as is shown in Figure 6-33. The process boils down to running repeated cycles, including denaturation of the DNA at 90–94 °C, hybridization with primers at 50–60 °C, and polymerase-mediated buildup at 70 °C. This gives rise to a chain reaction in which the newly synthesized DNA serves as a template for synthesis of yet another one during the next cycle. Thus, exponential multiplication of a DNA fragment takes place in the course of the PCR. The fragment can be multiplied by a factor of tens of millions over 25 to 30 cycles and produced in microgram quantities even if only a few copies were present in the starting DNA (which may be both double- and single-stranded). The length of the fragment may range anywhere from a hundred to several thousands of base pairs; the termini are determined by the nucleotide sequence of the primers or, in other words, can be defined in advance. The fact that primers of 20 to 25 units long (their synthesis has already become routine) can be used and the PCR can be conducted at an elevated temperature means that the reaction is highly specific and may practically yield a single product. The development of thermocyclers capable of changing the temperature of a large number of samples using a preset program has made it possible to fully automate such reactions.

It should be pointed out that the potential of PCRs is not limited to mere copying of selected DNA or RNA fragments (the latter is possible if a mixture of an RNA and two primers is first treated with reverse transcriptase, then a PCR is conducted to multiply the fragment of the resulting cDNA). PCR modifications have already been elaborated, whereby predetermined changes can be introduced into DNA and selected fragments of different DNAs can be interlinked, thus opening up greater possibilities for gene engineering.

The double-stranded DNA fragments yielded by the PCR can be easily sequenced by the enzymatic method already described. The primer oligonucleotide used in this case (it can, for example, be one of the PCR primers) can be readily hybridized with the corresponding strand of a duplex DNA. To this end, the DNA and primer mixture is heated to 90 °C then cooled rapidly, after which nucleoside triphosphates and DNA polymerase are immediately added. Sometimes, a single-stranded DNA is produced specifically for subsequent sequencing (as well as other purposes) in what has become known as asymmetrical PCR. In this case, two primers are introduced into the PCR in different quantities to give a mixture of a double-stranded DNA and a single-stranded product of primer build-up, taken in an excess amount. It should also be noted that the thermophilic DNA Taq polymerase can be used directly for sequencing as well. Then, the polymerase copying with termination is conducted at a higher temperature, which makes it possible to obviate the complications associated with the tendency of certain nucleotide sequences toward intramolecular complementary interactions.

By resorting to a PCR during sequencing of the genome DNA one can replace the time-consuming molecular cloning by a simple automated process. A particularly attractive feature of this reaction is its applicability to analysis of variable and mutant segments of the genome in comparative and evolutionary studies. By synthesizing primer pairs complementary with respect to the conservative regions confining such segments, one can analyze, within a single experiment, DNA preparations for a great number of individuals at a time. Moreover, analysis with a PCR provides access to DNAs in any form and in smallest amounts. PCR copying is also applicable to DNA in cells and tissues (e.g., hairs, preparations and smears from collections and herbaria, ancient and mummified samples), which opens up a brand new field of genetic research.

As can be seen from the general scheme of PCR (Fig. 6-33), in order to copy an unknown DNA fragment the latter must be preceded and followed by known sequences with which the primers are hybridized. In practice, however, it often becomes necessary to copy and sequence DNA fragments flanking a small known segment on either side. This can be done by means of an inverse PCR. In this case, the DNA under analysis is first cleaved with the aid of a restriction endonuclease, then the resulting fragments are cyclized using DNA ligase. Subsequently, two primers complementary to the known segment and oriented in two opposite directions toward unknown sequences are used for PCR copying of the resulting mixture. As a consequence, only the cyclized fragment containing the known sequence is linearized and multiplied. This sequence is arranged as two segments flanking the obtained product and can serve as a template for sequencing primer oligonucleotides as described above. Other modifications of the PCR have also been developed and are being used.

In conclusion, it should be pointed out that the PCR is ideally suited to automation of the sequencing procedure. By its nature this reaction already provides automation of most of the sample preparation process. As regards the automatic sequencers described below, they can handle PCR products just as successfully as those of molecular cloning.

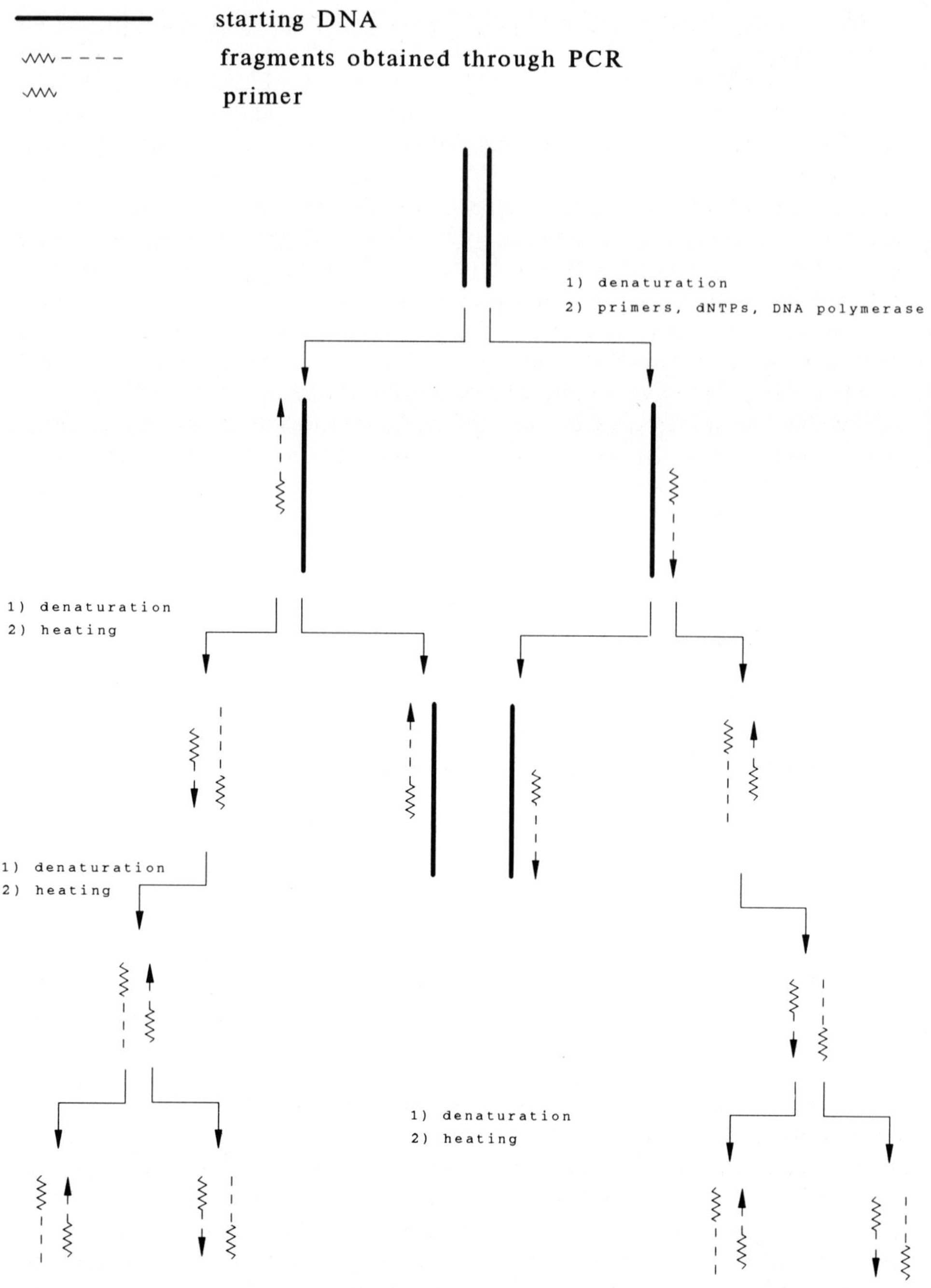

Fig. 6-33. Polymerase chain reaction (first cycles).

6.10 Automation of the Nucleic Acid Sequencing Process

In 1986–87, the first papers were published describing successful attempts to automate DNA sequencing. The major factor of success was the automation of the gel autoradiogram reading procedure. In this case, fluorescent labels are used instead of radioactive ones, which drastically cuts down the time necessary to detect the oligo(poly)nucleotides separated in the sequencing gel. Moreover, macromolecules are detected in the gel during electrophoresis, and all the acquired data are immediately sent to the computer for storage. In other words, all manual operations to dry the gel, place the X-ray film, expose it and analyze the autoradiograms are eliminated.

The chemistry underlying the method under consideration will be first illustrated using a very simple technique recommended for enzymatic sequencing as an example. The primer in the M13 phage system is universal, with an amino-fluorescein-containing fragment being incorporated into the terminal 5′-phosphate in several steps[2].

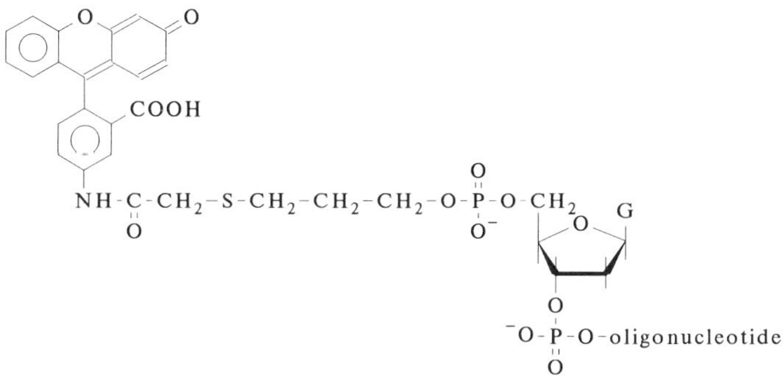

COOH

$$NH{-}\overset{\displaystyle O}{\overset{\|}{C}}{-}CH_2{-}S{-}CH_2{-}CH_2{-}CH_2{-}O{-}\overset{\displaystyle O}{\overset{\|}{P}}(O^-){-}O{-}CH_2 \quad G$$

$$^-O{-}\overset{\displaystyle O}{\overset{|}{P}}(=O){-}O{-}\text{oligonucleotide}$$

The resulting derivative of the primer oligonucleotide, containing the fluorescent label, is purified by means of ion-exchange high performance liquid chromatography (HPLC), desalted by dialysis and stored in the dark at –20 °C.

The sequencing is performed as usual but with a slightly altered dNTP:ddNTP ratio. Table 6-7 lists the corresponding quantities (in μl) of the dNTP and ddNTP solutions introduced into the reaction in the case of preparation of terminating mixtures (T′, C′, G′ and A′) for determining the positions of T, C, G and A, respectively, in the DNA under analysis.

The presence of a rather sizable fluorescent group in the primer does not affect the process of hybridization with the single-stranded M13 phage DNA.

---

[2] The corresponding transformation includes incorporation of mercaptopropanol into the oligonucleotide and a reaction with aminofluorescein iodoacetamide.

Table 6-7.

Molarity of the solution	Nucleoside 5'--triphosphate	T'	C'	G'	A'
0.5 M	dTTP	25	500	500	500
0.5 M	dCTP	500	25	500	500
0.5 M	dGTP	500	500	25	500
0.5 M	dATP	500	500	500	25
10.0 mM	ddTTP	5			
10.0 mM	ddCTP		1		
10.0 mM	ddGTP			2	
10.0 mM	ddATP				5

As has already been mentioned, electrophoresis is performed using a standard procedure with the difference that connected to the instrument is an air-cooled argon laser, light collecting and imaging optics, a photomultiplier (detector), and a data acquisition system interfaced with a computer. Figure 6-34 shows schematically a system for automatic reading of DNA sequencing

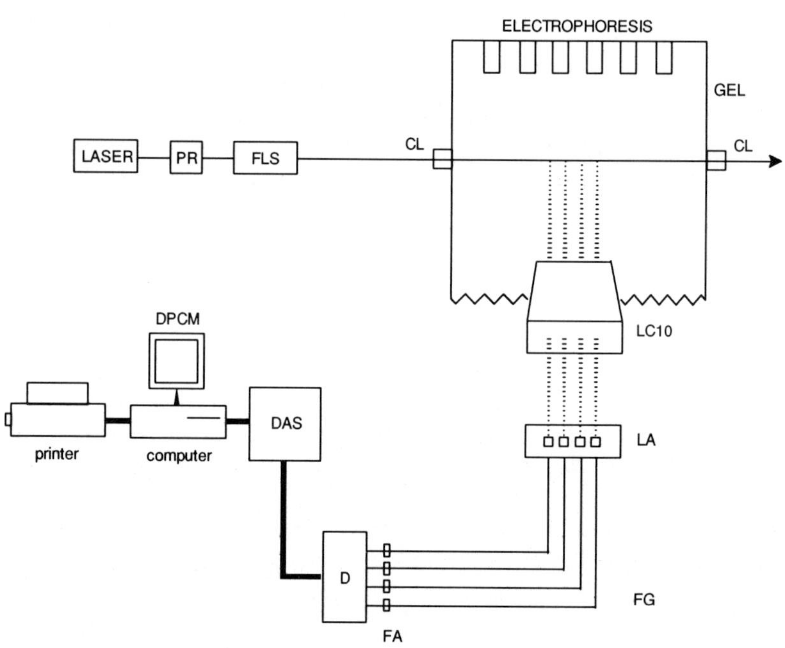

Fig. 6-34. Schematic diagram of the instrument for reading out electrophoresis data during DNA sequencing by the enzymatic method with a fluorescent primer (aminofluorescein). PR – polarization rotator; FLS – focusing lens system; CL – plates for coupling light into gel; LCIO – light collecting and imaging optics; LA – limiting aperture; LG – light guides; FA – filter assembly; D – detector (photomultipliers); DAS – data acquisition system; DPCM – data processing and control module (IBM PC).

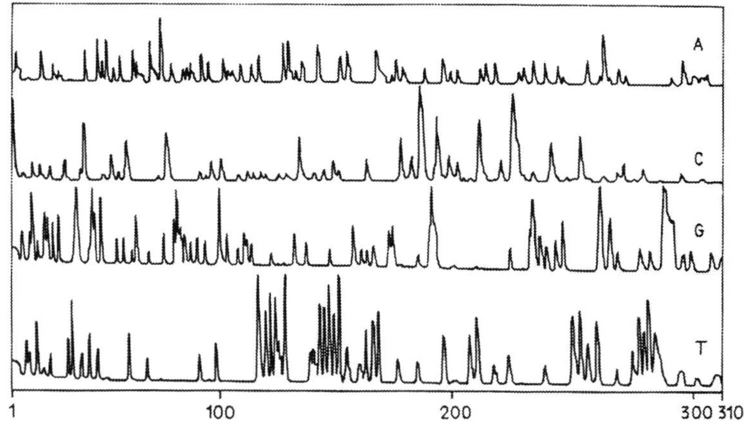

Fig. 6-35. Results of analysis of a sequencing gel (20 cm, 7 % polyacrylamide gel), obtained in enzymatic sequencing of recombinant M13 mp8 βSV DNA with an aminofluorescein-containing primer. The sequence was read from the gel using laser-induced fluorescence (for description of the instrument, see above).

data obtained during the electrophoresis (i.e. from sequencing gel). The plates for electrophoresis are made of non-fluorescent glass. The sensitivity for detecting macromolecules with a fluorescent label is $3 \cdot 10^{-18}$ moles/band. Within mere five hours one can read a sequence of 400 nucleotides (up to a maximum of 500 nucleotides) within a 20 to 30 cm long gel. The system may be used to handle anywhere from six to ten DNA samples per run (gel). The gel may be prepared from both polyacrylamide and agarose.

Figures 6-35 and 6-36 illustrate a partial nucleotide sequence of recombinant M13 mp8 βSV DNA containing a complex gene composed of rabbit β-globin genes and SV40 inserted into the restriction nuclease *Acc*I and *Eco*RI sites of vector M13 mp8.

Among the major advantages of this method, which was developed at the European Molecular Biology Laboratory (EMBL), are its simplicity, high speed and low cost. In spite of the advances in sequencing, associated with radioactive labels, serious difficulties are encountered when using the latter, including hazards in handling, storage and disposal of the waste (labeled products), high cost of radiolabeled compounds, as well as the extremely short lifetime of the ^{32}P-labeled compound usually employed in sequencing.

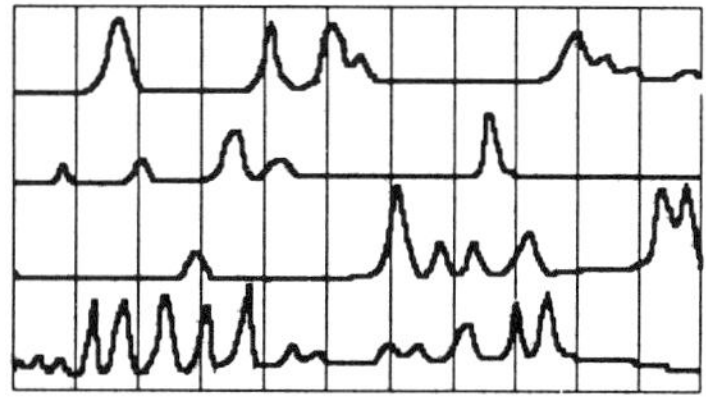

Fig. 6-36. Excerpt from Fig. 6-35. Shown here is a nucleotide sequence in a rabbit β-globin gene between poitions 189 and 153, which corresponds to bases 140–176 read from the gel.

Replacement of radioactive labels by fluorescent labels renders analysis of macromolecules simpler and less costly. The described instrument for continuous reading of sequencing gel data is rather simple and makes it possible to handle many DNA samples within a short period of time.

The company "Applied Biosystems" has started production of automatic DNA sequencing instruments. The procedure for analyzing oligo(poly)nucleotides is similar to the one just described, except that the detection of products of each of the four reactions during sequencing by the Sanger's method requires use of four structurally different fluorescent dyes (with four different fluorescence maxima) covalently bound to the primer oligonucleotide. Thus, distinction can be made between the sets of four reactions, which can be combined within a single cell, rather than being conducted in four as in the basic procedure described above. When four different fluorescent labels are selected, the following criteria must be met: (1) the fluorescence spectrum maxima must be clearly distinguishable: (2) all dyes must ensure almost the same high detection sensitivity (*ca.* 10^{-18} M); (3) their presence in the primer oligonucleotide should not affect the hybridization of the latter with DNA; and (4) the electrophoretic mobilities of the fluorescently marked DNA fragments should not differ widely because of the dissimilar structures of the fluorescent compounds. The following compounds were used by the authors as fluorescent labels more or less satisfying the above criteria:

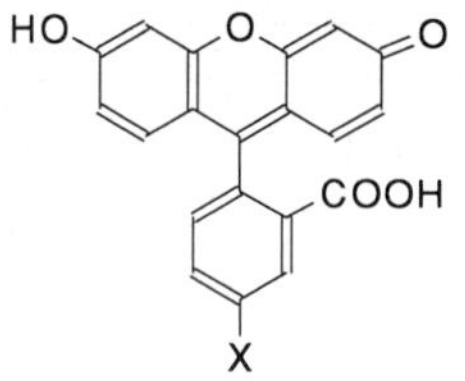

fluorescein (540 nm),
sensitivity 10^{-18} M

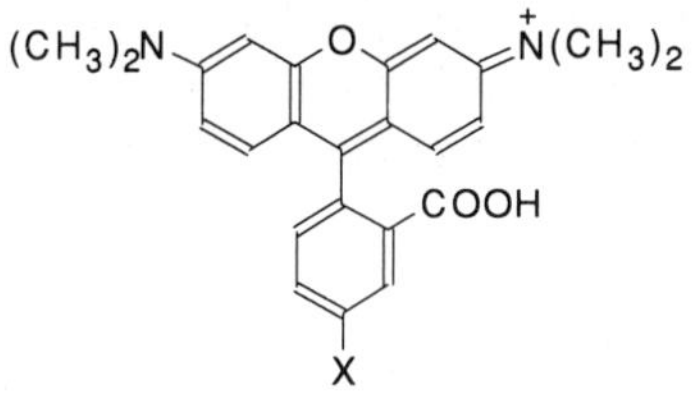

tetramethilrodamin (580 nm),
sensitivity $20 \cdot 10^{-18}$ M

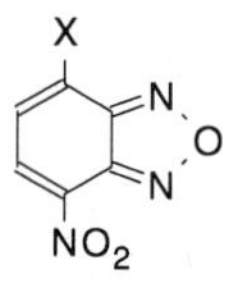

"NDB" (4- chloro-7-nitrobenz-2-oxa-1,3-diazol) (550 nm),
sensitivity $40 \cdot 10^{-18}$ M

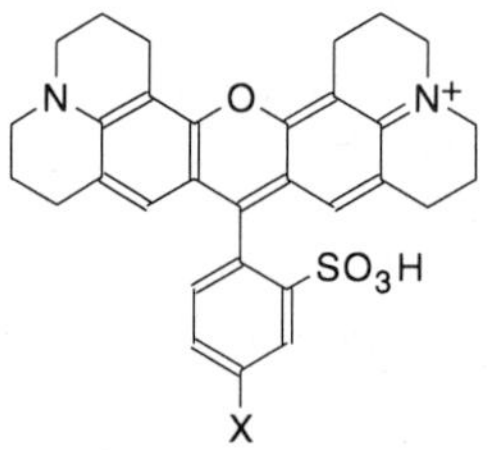

texas red (610 nm),
sensitivity $50 \cdot 10^{-18}$ M

The instrument for direct reading from the gel (electrophoresis is conducted in 50 cm long tubes with 8 % polyacrylamide gel is similar to the one described above. The laser beam passing through the gel at a distance of

40 cm from the top excites fluorescence which is then detected with the aid of four filters selected to match the spectra of each of the four dyes and changing periodically at a fast speed. More specifically, the fluorescence of the above-listed four dyes was detected at four wavelengths: 520, 550, 580 and 610 nm.

However, in spite of the successful implementation of such an approach for analysis of rather extended DNA fragments, which makes it possible to read simultaneously 16 gel tubes over a period of eight to ten hours, this sequencing procedure based on the automatic instrument of "Applied Biosystems" has some drawbacks. The fluorescence spectra of the dyes partially overlap so that the peaks corresponding to each dye appear in other channels as well. Moreover, the electrophoretic mobility of the labeled fragments varies depending on the label structure. For example, samples labeled with fluorescein and tetramethylrhodamine move in the gel so as if their length exceeds that of the NBD-labeled samples by one monomer, whereas samples labeled with Texas red appear to be one and a quarter monomers longer. In order to obviate such difficulties, special programs of a high degree of sophistication are employed using empirically selected correction factors. For all the attractiveness of the "a-label-for-each-type-of-nucleotide" approach, the method described at the beginning of this section – that is, the one involving a single dye species – is more likely to be universally used. It may be expected that in the near future scientists and companies working on such sequencers will concentrate their efforts on improvement of the existing designs, simplification of the chemical procedures for preparing samples for DNA and RNA analysis, complete automation of the entire sequencing process, and development of sequencers based on the Maxam–Gilbert method as well. The latter offers a number of advantages over Sanger's method, stemming, to be particular, from the use of less expensive reagents, more uniform intensity of the bands corresponding to all reaction products, and the fact that the analysis results are free of any influence exerted by the secondary structure of the DNA being sequenced. It is also possible that new, interesting results from the standpoint of full automation may be produced by combining the solid-phase technique with sequencing based on non-radiolabeled samples.

Doubtless, the automatic sequencing methods already in existence can already be used for determining the primary structure of DNA.

An alternative approach based on a fast, reliable and inexpensive DNA sequencing procedure has been proposed by Japanese scientists and companies belonging to the National Committee for Automation of DNA Sequencing. The general idea brought forth by the Committee boils down to setting up a facility performing assembly-line analyses of the primary DNA structure with the aid of automatic sequencers and robots. Individual components of the automated line for such analysis have already been developed. The process is based on Sanger's polymerase copying method which allows nucleotide sequences to be read at a rate of 0.28 second per base, in contrast to the Maxam–Gilbert method with its rate of 144 seconds per base (automatic sequencers operating on this principle are produced by "Seiko"). Technology has been elaborated for automatic production of ready-to-use polyacrylamide gel slabs (0.02 x 20 x 40 cm) wrapped in an acetate film which permits autoradiography to be performed immediately after electrophoresis. This enables one to get rid of the rather tedious and time-consuming manual operations of preparing gels and drying them after electrophoresis.

Electrophoresis can be sped up by designing units for simultaneous handling of hundreds of gels (each making it possible to analyze five DNA fragments about 250 bases long).

Autoradiograms are read using automatic scanning devices which ensure a rate of 1.4 seconds per base. To reduce the probability of a reading error (which is typically 1% per 250 bases), one can read two complementary strands at a time, thereby the error probability is brought down to 10^{-4}. There is every reason to expect faster and less costly analyses if the radioactive label is replaced by a fluorescent one.

The advantages offered by a well organized line for nucleic acid sequence analyses include high speed of the entire process, lower labor requirements, the possibility to maintain high quality of the analyses, ensure their reproducibility and, as a consequence, high reliability of the results. According to the Committee's estimates, at present a single skilled research worker can manually read a thousand base pair long sequence per day at a cost of a dollar per base. Through automation one can attain a rate of a million bases per day (per automatic unit). Such an increase in analysis rate must lower the cost by one order of magnitude.

6.11 Computers in Nucleic Acid Sequencing

The spectacular strides in nucleic acid sequencing have, in recent years, led to an exponential growth of knowledge about the primary structure of various nucleic acids. If sequences containing a total of at least a million nucleotides had been established by 1983, by the year 1985 this number had increased to four million, by September of 1986 – to nine million, and by the end of 1987 – to about ten million. As has already been mentioned, determination of the primary structure of the human genome (3 billion nucleotide pairs) is already being discussed today. In addition to automatic DNA sequencers, accomplishment of this task calls for powerful computers and software for data acquisition, storage and analysis. The extraordinary progress achieved over the past few years in microelectronics has led to the advent of mini- and microcomputers of the IBM PC type which, by virtue of their broad potential (memory size, speed, low cost and, as a result, ready availability to laboratories and even individual workers), have made computer analysis of the genome structure and functions routine for thousands of molecular biologists without any programming and computer background but capable of tackling any problem related to the functioning of biomolecules with the aid of computers. There is every reason to consider the late eighties as the dawn of the computer age in molecular biology. The possibility of producing "nucleotide texts" which encode genetic information provides new tools for its interpretation. What this means is new approaches that can be instrumental in finding not only loci of structural genes in any genetic information (since the protein code is known) but also regulation sites (promoters, operators, enhancers, splicing sites, etc.). All this is done using the information theory and statistical methods as well as such research tools as computers, nucleotide sequence data bases, and dedicated software packages.

Let us now return to the project aimed at establishing the primary structure of the human genome in its entirety. Remarkably, gene sequences account for only five to ten per cent of the genome. Knowledge of the primary structure of the entire genome (plus finding out what is encoded in the remaining sequences) opens up new possibilities in elucidating the genetic contribution to all aspects of normal and abnormal functioning of the living organism. Comparison of the findings with primary genome structures of other organisms is of greatest interest from the standpoint of evolution of the encoded genes and macromolecules. This may also be helpful in learning the language whereby the human gene expression is controlled. Since computers can store any amount of data on the primary genome structure and easily access the necessary nucleotide sequences, one can rapidly accomplish such tasks as determination of the primary structure of the protein encoded in the gene, comparison of the primary structures of genes (and, consequently, proteins), and identification of translation frames.

The most important application of computers has to do with the prediction of three-dimensional structures of ribonucleic acids and proteins. For instance, this approach enables one to predict three-dimensional structures of DNA introns acting as enzymes. It is widely used in gene engineering for proper selection of amino acid replacements.

The need to solve all these problems has led to the development of special centers for acquiring data on the primary nucleic acid structure as well as commercial dedicated software packages for complete analysis of such data.

The early stages of DNA and RNA sequencing were already marked by setting up of international systems for data storage. The most famous data banks of nucleotide sequences are located in the United States (Gene Bank established in 1982 on the basis of the Los Alamos data base) and in Europe (EBML bank created in Heidelberg in 1980). In the Soviet Union, the All-Union Bank of Nucleotide Sequences has been established under the auspices of the USSR Academy of Sciences. These and similar organizations specialize in collection of the published nucleotide sequences and releasing of the available information. The activities of such centers also include the following: data reduction and creation of directories; entering, editing and storage of primary structure data; and systematization of sequences, based on different parameters. The created data retrieval systems ensure rapid access to any sequence or accomplishment of any of the above tasks. The information is made available to users either on magnetic carriers or in the form of printouts or catalogues. The astounding rate of acquisition of DNA and RNA sequences is already challenging designers to find ways of speeding up their transmission to data banks and compressing the data for storage without adversely affecting their accessibility. Revolutionary changes are to be expected in the nearest future not only in experimental sequencing methods but also in the hardware, primarily development of new types of computers and their operating systems, wide-spread use of optical storage devices (CD-ROMs), and elaboration of electronic methods for data acquisition and distri-

```
LOCUS       HUMA1AT4      292 BP    DNA       UPDATED 03/12/84
DEFINITION  HUMAN ALPHA 1-ANTIYRYPSIN GENE: 3' TERMINUS.
ACCESSION   J00067
KEYWORDS    ALPHA-1-ANTITRYPSIN; ANTITRYPSIN; PROTEASE INHIBITOR.
SEGMENT     4 OF 4
SOURCE      HUMAN CDNA TO LIVER MRNA 1, 3; GENOMIC DNA 2.
ORGANISM    HOMO SAPIENS
            EUKARYOTA; METAZOA; CHORDATA; VERTEBRATA; TETRAPODA;
            MAMMALIA; EUTHERIA; PRIMATES.
REFERENCE   1 (BASES 1 TO 274)
AUTHORS     KURACHI, K., CHANDRA, T., FRIEZNER DEGEN, S.J., WHITE,
            T.T., MARCHIORO, T.L., WOO, S.L.C. AND DAVIE, E.W.
TITLE       CLONING AND SEQUENCE OF CDNA CODING FOR ALPHA 1-ANTI-
            TRYPSIN
JOURNAL     PROC NAT ACAD SCI USA 78, 6826-6830 (1981)
REFERENCE   2 (BASES 113 TO 292)
AUTHORS     LEICHT, M., LONG, G.L., CHANDRA,T., KURACHI, K.,
            KIDD, V.J., MAGE, M.JR., DAVIE, E.W. AND WOO, S.L.C.
TITLE       SEQUENCE HOMOLOGY AND STRUCTURAL COMPARSION BETWEEN
            THE CHROMOSOMAL HUMAN ALPHA 1-ANTITRYPSIN AND CHICKEN
            OVALBUMIN GENES
JOURNAL     NATURE 297, 655-659 (1982)
REFERENCE   3 (BASES 113 TO 287)
AUTHORS     ROGERS, J., KALSHEKER, N., WALLIS, S., SPEER, A.,
            COUTELLE, CH., WOODS, D. HUMPERIES, S.E.
TITLE       THE ISOLATION OF A CLONE FOR HUMAN ALPHA 1-ANTITRYPSIN
            AND THE DETECTION OF ALPHA 1-ANTITRYPSIN INMRNA FROM
            LIVER AND LEUKOCYTES
JOURNAL     BIOCHEM BIOPHYS RES COMMUN 116, 375-382 (1983)
COMMENT     ALPHA 1-ANTITRYPSIN IS AN IMPORTANT PROTEASE INHIBITOR
            PRESENT IN MAMMALIAN BLOOD. CORRESPONDING REGIONS OF
            HUMAN AND BABOON ALPHA 1-ANTITRYPSIN MRNAS AND THEIR
            AMINO ACID SEQUENCES ARE GREATER THAN 96% HOMOLOGOUS.
            SEE MNKA1AT AND AND OTHER HUM1AT LOCI. 3 CALCULATES
            THAT HUMAN LEUKOCYTES PRODUCE 0.15% AS MUCH ALPHA 1-AT
            MRNS AS HUMAN LIVER.
```

```
            THE MUTATION NOTED AT BASE 154 CHANGES A GLU CODON.
            TO AN ASP CODON. THE RESULTING CHANGE IS NEUTRAL.THIS
            IS ALSO A SLIM POSSIBILITY THAT THE BASE CHANGE IS
            MERELY DUE TO A CONFLICT.
FEATURES    FROM  TO/SPAN     DESCRIPTION
PEPT          1      211      ALPHA 1-ANTITRYPSIN (EXON 4, PARTIAL)
SITES
PEPT/PEPT     1        0      ALPHA 1-AT MATURE PEPT UNSEQUENCED/
                              SEQUENCED
REFNUMBER     2        3      NUMBERED CODON 326 IN 1
REFNUMBER   113        1      NUMBERED 1231 IN 2 ZERO NOT USED
REFNUMBER   113        3      NUMBERED CODON 363 IN 3
MUT         154        1      A IN 1, 2; C IN 3
PEPT -      211        1      ALPHA 1-AT MATURE PEPT (EXON 4) END
CONFLICT    258        1      C IN 1, 3;T IN 2
MRNA -      292        1      ALPHA 1-AT MRNA END (POLY-A SITE)
BASE COUNT   72A     94C      59G     67T
ORIGIN     ABOUT 3KB AFTER HUMA 1AT3.
     1  ACCCCTGAAG CTCTCCAAGG CCGTGCATAA GGCTGTGCTG ACCATCGACG
    51  AGAAAGGGAC TGAAGCTGCT GGGGCCATGT TTTTAGAGGC CATACCCATG
   101  TCTATCCCCC CCGAGGTCAA CTTCAACAAA CCCTTTGTCT TCTTAATGAT
   151  TGAACAAAAT ACCAAGTCTC CCCTCTTCAT GGGAAAAGTG GTGAATCCCA
   201  CCCAAAAATA ACTGCCTCTC GCTCCTCAAC CCCTCCCCTC CATCCCTGGC
   251  CCCCTCCCTG GATGACATTA AAGAAGGGTT GAGCTGGTCC CT
```

Fig. 6-37. Form sent to the Gene Bank (USA).

bution. Proposals are being made to send all newly decoded sequences directly to data banks without their publication in journals (to avoid errors arising from reprinting). It has been recommended to transmit data by telephone links. New forms of presenting nucleotide sequences have been developed, which lend themselves to automatic reading by computers; that is, not in the four-letter form but in binary code, in a graphic form (as four vectors differently oriented in two or three dimensions), and even as audio signals (of four different tones). At present, more or less universal rules govern the format in which sequences must be sent to data banks. Figure 6-37 illustrates a typical form filled out prior to sending to the Gene Bank (USA). Symbols and the corresponding codes have been developed for designating nucleotides, including minor ones. In addition to data acquisition and storage, computers perform another major function which is analysis of nucleic acid sequences. Today, a large number of software products (usually in the form of packages) are available for this purpose. They free research workers of many tedious routine operations, such as counting mono-, di- and trinucleotides, translation of a nucleotide sequence into that of amino acids, search for particular sites, comparison of sequences, and so on. Already in the early eighties, special issues of "Nuclear Acids Research" described software for storage and analysis of data on the primary structure of nucleic acids. All currently available software packages are tailored to particular computers, (users buy programs depending on the computer model at their disposal), and differ both in design and in the range of functions they can perform. Some of them include data from the Gene Bank or EBML. For example, four packages are commercially available at present for IBM PC or compatibles: DNASIS, DNASTAR, IBI, and MICROGENIE. The most popular software package in the Soviet Union was SEQBUS developed for PC "Iskra 226" at the Institute of Molecular Biology, USSR Academy of Sciences.

All programs can be conventionally divided into two categories: general-purpose and special or dedicated. The former are intended for the most common operations of data acquisition and analysis and perform the following functions: entering and editing of new sequences; direct reading of autoradiograms and gels with the aid of scanning devices; location of restriction endonuclease recognition sites and presentation of data in a convenient (tabular or graphic) form; location of sites with components of rotational and mirror symmetry (palindromes); translation of a nucleotide sequence into a protein one in all three reading frames; comparison of two sequences by the homology dot matrix method; comparison of a new sequence with all those in the Gene Bank; location of sites rich in particular nucleotides; calculation of a hypothetical DNA melting point; automatic assembly of sequenced fragments into a single structure (DNA molecule); translation of a protein sequence into a nucleotide one with due account for irregular usage of synonymous codons; determination of the molecular weight of nucleic acids and proteins; prediction of the secondary protein structure; calculation ot the free energy of hairpin formation; and many others.

Dedicated programs are developed to solve special, often more complicated problems and are of interest to a narrower group of specialists. They can perform e.g. a calculation of the length of DNA fragments on the basis of their electrophoretic mobility in gels; selection of hybridization probes; prediction of the secondary RNA structure; location of nucleotides in a gene, which can be altered (without changing the amino acid sequence) in order to introduce a restrictase recognition site; location of sites with a potentially possible Z-form of DNA structure; identification of functionally significant sites in an unkown newly decoded structure, based on a previous consensus (as a result of analysis of known structures performing the same function); location of protein-encoding sites; and so on. Figure 6-38 shows, by way of example, the menu of the MICROGENIE package to illustrate the general or special functions that can be chosen by the researcher when handling nucleotide sequences. The structure and functions of some programs will be covered at greater length in what follows.

The general-purpose program COMMON allows the user to enter a sequence into the computer and verify the entered data in an interactive

```
                                                    MICROGENIE


          Enter                 Analize               Learn
          Files                 Compare               Setup
          Merge                 Data Bank             Graphics
     Function:
```

Fig. 6-38. Menu of MICROGENIE software package.

mode. Other possibilities offered by the program include deletions and inserts as well as elimination of sequences to simulate hybrid and mutant DNA molecules. Sequences are usually entered manually using a keyboard, but in recent years instruments have been developed for automatic scanning of autoradiograms and sequencing gels, obtained with the use of fluorescent labels, and direct data transmission to the computer with subsequent analysis of the sequence using special software. The book "Sequencing Analysis of Nucleotide Acids and Proteins", published recently by IRL Press (Oxford), contains detailed descriptions of both scanning devices and software for reading autoradiograms. Programs have been written for determination of the primary structures of high-molecular weight DNAs based on the fragment sequencing data obtained during non-specific cleavage (e.g., by sonication). Whatever relationship exists between any two DNA fragments becomes known from coincidence of the nucleotide sequences in their structures, the coinciding sequences serving as overlapping sites, and the two fragments can be reunited in a more extended structure. The process of fragment selection and joining continues as long as it takes to determine the entire primary structure in the DNA under analysis. One such program is CONTIG (with a modification for IBM PC), created at F. Sanger's laboratory (Cambridge, England). Here are the basic operations that can be executed using CONTIG:

(1) Storage of sequences for each fragment

(2) Selection of adjacent fragments to assemble sequences

(3) Comparison of data obtained from reading of new autoradiograms with already established sequences

(4) Joining of two fragments with the aid of a third one which represents the site where the first two overlap

(5) Search for DNA sites complementary to the already established ones, which serves as a check ensuring that the full DNA structure has been assembled correctly.

After the above-described reconstruction of the DNA a check is carried out using the data on restriction sites in the starting DNA, obtained while preparing a restriction map. Coincidence of the restriction site location data (this conclusion is drawn from prediction of the DNA fragmentation picture, based on the established primary structure of the DNA) indicates that the primary structure of the DNA under investigation has been determined correctly. Testimony to the enormous potential of computers in solving such problems is provided by the fact, for example, that data from 12 gels, each containing 200 to 300 nucleotides, can be assembled automatically into a continuous sequence within just a few seconds.

Almost every software package includes a program for identifying sites recognized by restriction endonucleases. Each package usually has a full list of the known restriction endonucleases (at present more than 450), and some packages also include a special sublist of only commercially available enzymes (about 100). One of such programs, SITOR, allows the user to construct a restriction map and print out a table of fragments of a DNA restricted at a predetermined site. The table contains the length and molecular weight of each fragment. The corresponding autoradiogram can be synthesized and presented graphically. Figure 6-39 illustrates the search for one of the restriction sites with the aid of the SITOR program.

Data banks provide the foundation on which all research with nucleotide sequences, aimed at determining the biological importance of individual DNA fragments, is based. Such research comprises two phases. The first phase involves grouping and subsequent comparative analysis (using appro-

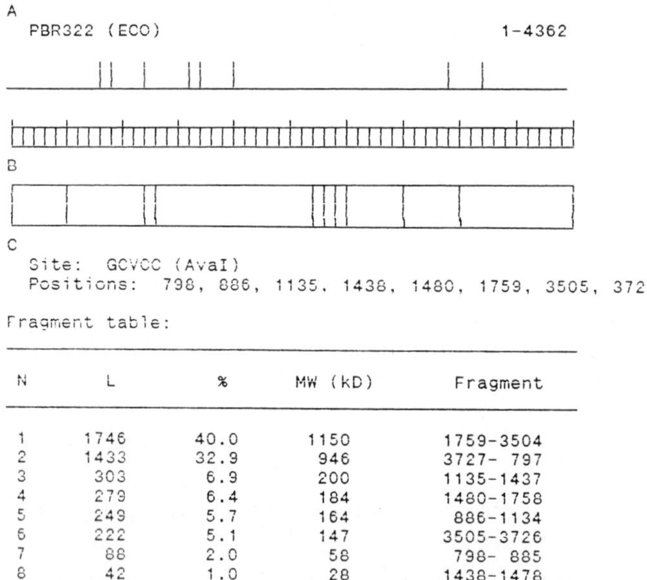

N	L	%	MW (kD)	Fragment
1	1746	40.0	1150	1759-3504
2	1433	32.9	946	3727- 797
3	303	6.9	200	1135-1437
4	279	6.4	184	1480-1758
5	249	5.7	164	886-1134
6	222	5.1	147	3505-3726
7	88	2.0	58	798- 885
8	42	1.0	28	1438-1478

Fig. 6-39. Search for pentanucleotide $GG^{A}_{T}CC$ recognized by restrictase *Ava*I in plasmid pBR322. A – restriction map; B – autoradiogram; C – fragment table.

priate criteria) of sequences performing the same biological function. More specifically, it includes identification of "consensuses" (generalized statistically average structures) at regulatory sites of DNA (promoters, operators, enhancers, splicing sites, and ribosome recognition sites) or determination of the frequency of occurrence of synonymous codons in genes encoding different groups of proteins. The second, more interesting phase includes attempts to identify, within an unknown (newly decoded) structure, sites performing a particular function; for instance, search, based on homology with previously established consensuses, for gene-controlling sequences or potential protein-encoding sites. Predictions of such DNA properties can be verified experimentally. The result may be a catalogue of properties of investigated DNAs.

The following programs may be mentioned as examples of software intended to determine the properties of particular sequences. One of the computer programs NAQ, for instance, has made it possible to establish the frequency of occurrence of codons in nucleotide sequences from yeast mitochondria, which encode proteins and do not do so, respectively. The resulting table for using codons later enabled investigators to identify polypeptide-encoding areas in unknown portions of the mitochondrial genome. Similar problems can be solved by means of the program PEPTIDOR. It permits the user to recognize opening reading frames on both DNA strands or, in other words, DNA portions encoding potential proteins from the arrangement of chain initiation and termination codons; to print out the amino acid sequence in the gene; to calculate the codon frequency; and to graphically represent the position of initiation and termination codons as well as the corresponding polypeptides. Figure 6-40 shows the distribution of encoded peptides in plasmid pBR322, the upward streaks standing for initiation codons, the downward streaks standing for termination codons, and the horizontal line standing for open reading frame.

It should be pointed out that to solve a more complex problem, such as identification of real protein-encoding sites among the potential ones, requires more sophisticated programs belonging to the category of dedicated software. They should also permit the user to perform more extensive analysis including search, near the boundaries of open frames, for so-called sequence signals homologous to regulatory sites (promoters, terminators, ribosome binding sites) typical of a particular organism, as well as analysis of the frequency of using synonymous codons within open reading frames and comparison with the data for known genes of the same organism (it has already been firmly established that the use of synonymous codons by all organisms is irregular, and this property can be used for identifying the encoding region for the organism in question). In the case of eukaryotic genes, the situation is complicated further due to the presence of introns. Yet the right approaches are already being worked out for them as well. A case in point is the program "ANALYSEQ" developed at Sanger's laboratory (it can be run on any graphic terminal) has such capabilities as search for sites encoding real proteins in both prokaryotes and eukaryotes.

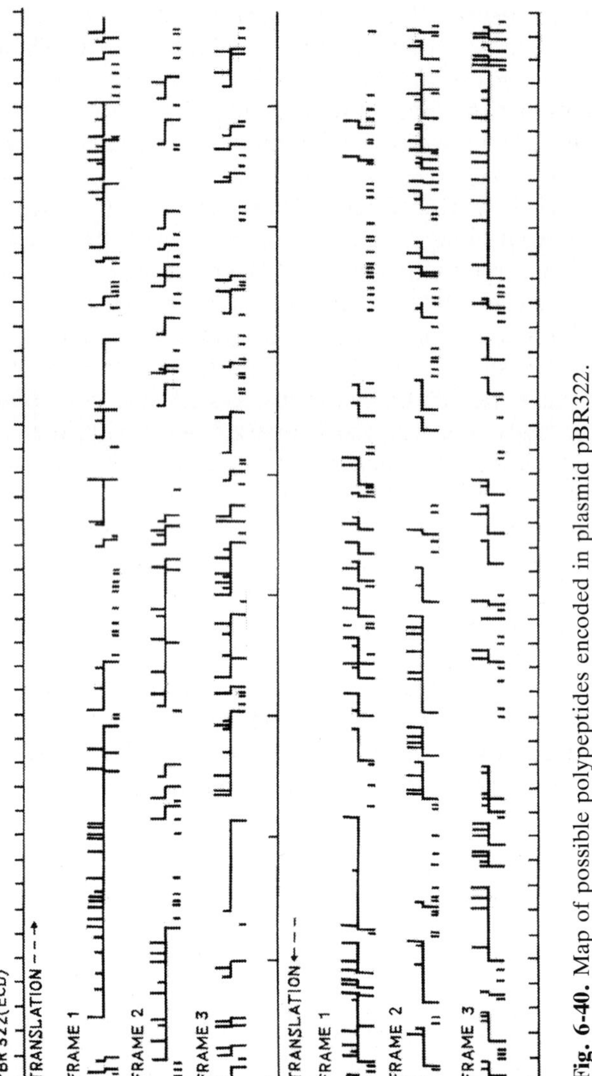

Fig. 6-40. Map of possible polypeptides encoded in plasmid pBR322.

An original approach to identification of functionally active sites is built in the program SASIP. It analyzes the frequency of octanucleotides in the Gene Bank sequences and generates the corresponding tables. According to postulates of the information theory, the most rarely occurring oligomers may perform the most important function. Comparison of the tabulated data with structures of fragments with known properties allows extremely important biological conclusions to be drawn as regards the structure and function of the genome.

Among the most important problems, (from the standpoint of understanding the molecular mechanism of biopolymer functions), which can be solved only with the aid of computers is construction of secondary RNA structures.

The currently existing programs, however smart they may be, are far from being perfect. But computer industry since it was established has been making so incredible progress that software producing companies have to do their best to stay in flow, and more powerful and sophisticated software packages are about to be released in the near future.

The development of fast DNA sequencing techniques gives a major impetus to molecular biology; it allows one, among other things, to elucidate the molecular mechanisms of the vital activity of organisms and their evolution.

Unfortunately, the rate of decoding sequences of nucleic acids lags far behind the current demand in analysis of their various fragments; each year, 10^6 bases are received by data banks. This is too little if it is remembered that the human genome alone contains $3 \cdot 10^9$ base pairs. Ways to speed up the reading of nucleic acid sequences should be sought in automation of the entire process, which will also free thousands of skilled operators from tedious work. Automation will be justified if the speed of analysis increases substantially (by several orders of magnitude) and its cost goes way down.

A prerequisite for solving the sequencing automation problem is availability of methods based on the simple principle of statistical degradation or biosynthesis. If in combination with such methods chemists and engineers pool their efforts, the problem of DNA and RNA sequencing may be solved in such a manner that before this century is over scientists will raise the veil of secrecy hiding not only the human genome but also that of other organisms.

References

1. A. M. Maxam and W. Gilbert, *Sequencing Endlabeled DNA with Base-specific Chemical Cleavages. Methods Enzymol.*, **65**, 499–560 (1980).
2. F. Sanger, *Determination of Nucleotide Sequences in DNA. Science,* **214**, 1205–1210 (1981).
3. F. Sanger, *Sequences, Sequences and Sequences. Ann. Rev. Biochem,* **57**, 1–28 (1988).
4. *Nucleic Acid and Protein Sequence Analysis: A Practical Approach.* W. J. Bishop and C. J. Rawlings (Eds.), IRL Press, Oxford (1989).
5. *Nucleic Acids Sequencing: A Practical Approach.* C. J. Howe and E. S. Ward (Eds.), IRL Press, Oxford (1989).
6. U. L. RajBhandary, *Recent Developments in Methods for RNA Sequencing Using in Vitro ^{32}P Labeling. Fed. Proc.,* **39**, 2815–2821 (1980).
7. J. Stanley and S. Vassilenko, *A Different Approach to RNA Sequencing. Nature,* **274**, 87–89 (1978).
8. D. A. Peattie, *Methods DNA, RNA Sequencing.* S. M. Weissman (Ed.), Praeger, New York, pp. 261–304 (1983).

9. A. Rosenthal, R. Jung and H.-D. Hunger, *Solid – phase Methods for Sequencing of Oligodeoxyribonucleotides and DNA. Methods Enzymol.,* **155**, 301 (1987).
10. W. Ansorge, B. S. Sproat, J. Stegemann and C. Schwager, *A Non-radioactive Method for DNA Sequence Determination. J. Biochem. Biophys. Methods,* 13, 315 (1986); NAR, **15**, No. 1, 4593 (1987).
11. J. Hindley and R. Staden, DNA Sequencing, in: *Laboratory Techniques in Biochemistry and Molecular Biology.* T. S. Work and R. H. Burdon (Eds.), Elsevier Biomedical Press, Amsterdam–N.Y.–Oxford (1983).

7 Conformation of Nucleic Acid Components. Macromolecular Structure of Polynucleotides

The current concepts concerning the structure of DNA and RNA provide the basis for research into the mechanisms of gene expression. In turn, elaboration of three-dimensional DNA and RNA structure models, their proof and further refinement are inexorably associated with advances in the structural chemistry of nucleic acids. Moreover, the fact that Watson and Crick had come up with the double helix model of DNA was testimony to the spectacular progress achieved in this branch of the chemistry of natural compounds in the early fifties.

It is absolutely clear that three-dimensional models of the structure of synthetic and natural polynucleotides can be constructed and correctly interpreted only with a profound knowledge of the conformation of monomer components of nucleic acids – nucleosides and nucleotides.

The conformation of nucleosides, nucleotides, oligo- and polynucleotides is studied using a wide range of physical and physico–chemical methods of which the most prominent are X-ray structural analysis and NMR spectroscopy. Also quite useful in these studies have been such simple spectral techniques as ultraviolet spectroscopy, optical rotatory dispersion (ORD) and circular dichroism (CD) as well as theoretical conformation analyses.

This chapter deals primarily with the final results produced by these approaches, without dwelling on the underlying theoretical concepts. Therefore, more detailed information on the latter can be obtained in the excellent textbook by C. Cantor and P. Schimmell (see references at the end of the chapter), which contains everything one needs to know about the theoretical fundamentals of these methods and their application to the structure of nucleic acids and their components.

7.1 Conformation of Nucleosides and Nucleotides

7.1.1 Heterocyclic Bases

All of the five major bases forming part of nucleic acids are more or less planar. As can be seen from the example illustrated in Figure 7-1, the atoms of heterocyclic rings are less than 0.01 nm (0.1 Å) away from the mean ring plane.

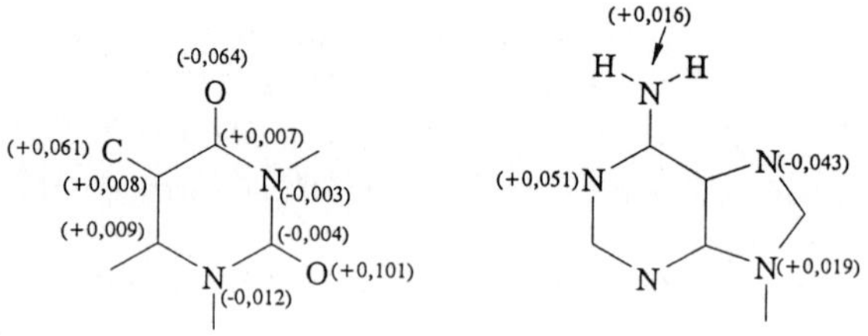

(-0,064) O (+0,007) (+0,061) C (+0,008) N(-0,003) (+0,009) (-0,004) N O(+0,101) (-0,012)

(+0,016) H N H (+0,051)N N(-0,043) N N(+0,019)

Fig. 7-1. Maximum displacement of atoms from a mean plane in thymines (deoxythymidine 5'-phosphate) and adenine (deoxyadenosine).

The planar conformations of heterocyclic bases are quite consistent with their chemical properties – all constituent bases of nucleic acids are quasiaromatic compounds.

A remarkable exception from this rule is dihydrouracil which lacks aromatic properties and has a half-chair conformation, the C^5 and C^6 atoms being spaced 0.03 nm (0.3 Å) apart (Fig. 7-2).

According to X-ray structural data, in most nucleosides and nucleotides the $C^{1'}$ atoms of the carbohydrate (ribose or deoxyribose) moiety, involved in glycosidic bonding, are 0.01 to 0.02 nm (0.1–0.2 Å) away from the mean plane of the heterocyclic ring.

The exocyclic amino groups of purine bases and cytosine are involed in the formation of resonance structures, and the C–NH bond has properties making it close to a double bond. Consequently, rotation about this bond is hindered, and the exocyclic amino groups together with the associated heterocycles are coplanar (lie in the same plane). At the same time, the exocyclic C=O group displays normal double-bond properties.

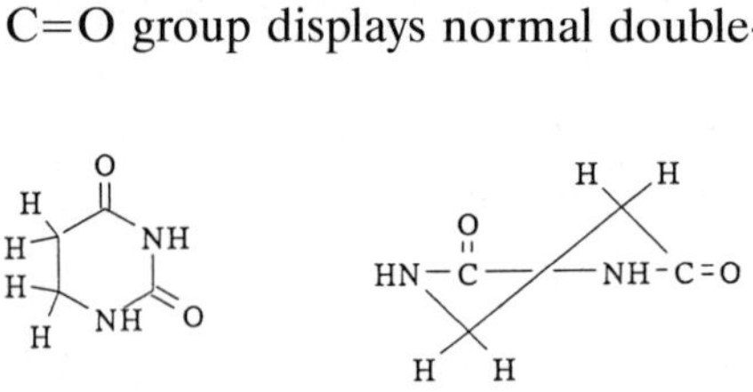

O H H NH H NH O H

H H O HN–C NH–C=O H H

Fig. 7-2. Puckered conformation of the dihydrouracil ring.

7.1.2 Ribose and Deoxyribose

As regards pentafuranoses, which form part of monomer units in nucleic acids, it has been established that four atoms of the five-membered ring are more or less coplanar in most cases, whereas the fifth atom, $C^{2'}$ or $C^{3'}$, is 0.05 to 0.06 nm (0.5–0.6 Å) away from the ring plane. In this case, if the $C^{2'}$ or $C^{3'}$ atoms lie on the same side of the ring plane as $C^{5'}$, we are dealing with so-called *C²'*- and *C³'-endo*-conformations, respectively; and if $C^{2'}$ or $C^{3'}$ are not on the same side as $C^{5'}$, we have *C²'*- and *C³'-exo*-conformations.

Such conformations in the ribose and deoxyribose moieties of nucleosides and nucleotides become clear if we address ourselves to such classical objects of stereochemistry as cyclopentane and tetrahydrofuran.

It is known that the planar form of cyclopentane is energetically unfavorable. Its molecule has conformations with one or two adjacent carbons away from the ring plane (Fig. 7-3).

In the case of tetrahydrofurane, whose derivatives are sugar moieties in nucleosides and nucleotides, similar non-planar conformations must occur as well. Calculations have shown that the most energetically favorable conformations for the pentafuranose ring are such in which the heterocyclic atom of oxygen and the neighboring $C^{1'}$ and $C^{4'}$ atoms are coplanar, while $C^{2'}$ and $C^{3'}$ do not lie in the ring plane. These conformations are shown in Figure 7-4.

An interesting exception is adenosine 3′,5′-cyclic phosphate which has a *C⁴'-exo*-conformation.

It should be pointed out that in the literature the *C²'-endo*-conformations of carbohydrate moieties in nucleic acids are often referred to as S-conformations, while the *C³'-endo*-conformations are referred to as N-conformations. The reason is that the values of the pseudorotational phase angle lie in the range of 0 to 30° for the *C³'-endo*-conformations and 150 to 180° for the *C²'-endo*-conformations or, in other words, in the northern (upper) and southern (lower) regions, respectively, of the pseudorotational cycle.

The majority of crystalline forms of nucleosides and nucleotides examined by X-ray structural analysis have either a *C²'-endo* or *C³'-endo*-conformation of their carbohydrate moiety. These conformations were identified in approximately equal numbers of cases, which suggested that the energy barrier

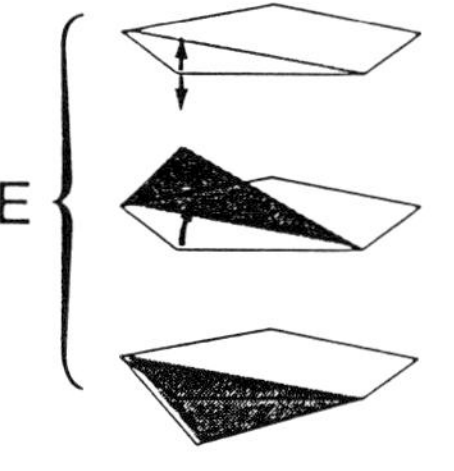

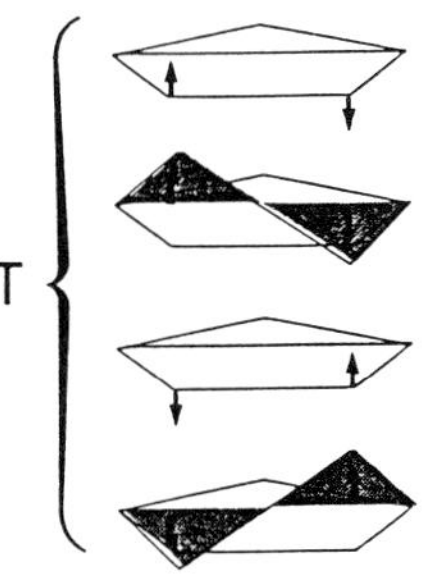

Fig. 7-3. Envelope (E) and twist (T) conformations of cyclopentane (adapted from W. Saenger, *Principles of Nucleic Acid Structure,* Springer Verlag, N.Y., 1984).

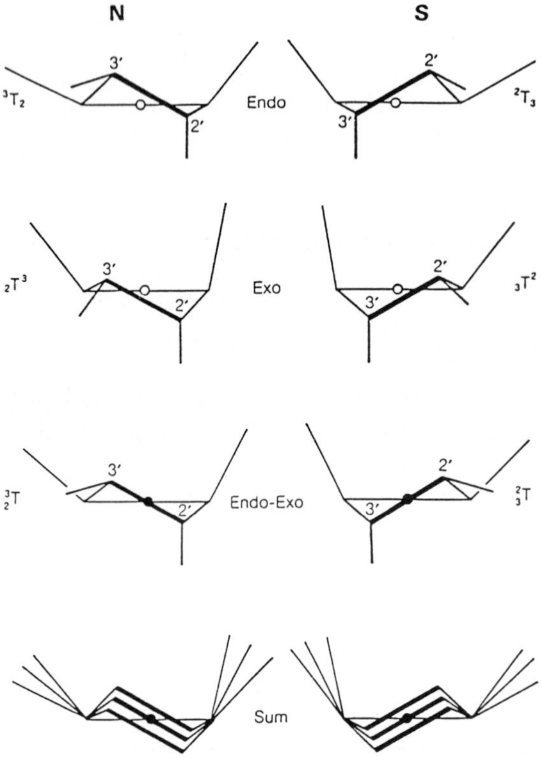

Fig. 7-4. Major puckered conformations of the pentafuranose ring of ribose and deoxyribose in nucleosides (adapted from W. Saenger, *Principles of Nucleic Acid Structure,* Springer Verlag, N.Y., 1984).

between the two is low. Theoretical calculations indicate that it is only 5–10 kJ/mol (1–2 kcal/mol) for deoxynucleosides. In the case of ribonucleosides, this barrier must be perceptibly higher: 25–30 kJ/mol (6–7 kcal/mol), to be precise. Experimental NMR data attest to a rapid equilibrium between the N and S conformers in solutions of nucleosides and nucleotides.

The readily observed ("primary") deformation (puckering) of the furanose ring usually goes hand in hand with another, slighter ("secondary") one. For instance, in cytidine 3′-phosphate whose $C^{2'}$ atom lies nearly 0.05 nm (0.5 Å) above the ring plane, $C^{3'}$ is 0.015 nm (0.15 Å) away from the latter in the opposite direction. Thus, the ribose moiety virtually has a conformation which is intermediate between the $C^{2'}$-*endo*- and $C^{2'}$-*endo*-$C^{3'}$-*exo*-conformations. An important consequence of deformation of the furanose ring is departure from the true *cis*-conformation of the 2′- and 3′-hydroxyl groups in ribonucleosides. The torsion angle between the bonds linking $C^{2'}$, $O^{2'}$ and $C^{3'}$, $O^{3'}$ is 43 to 54°, whereas for true *cis*-bonds it must be equal to zero (see also Fig. 7.6.A for a definition of torsion angles).

X-ray structural analysis reveals three types of conformation for the $C^{4'}$–$C^{5'}$ bond, which are illustrated in Figure 7-5.

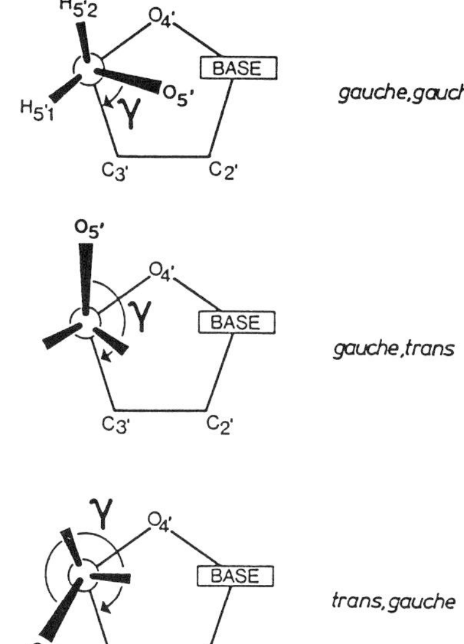

Fig. 7-5. Three possible conformations of ribose or deoxyribose due to different orientations of substituents about the C4′–C5′ exocyclic bond (adapted from W. Saenger, *Principles of Nucleic Acid Structure,* Springer Verlag, N.Y., 1984).

In ribo- as well as deoxyribonucleosides, all three conformations are found. At the same time, the presence of a phosphate group at $C^{5'}$ in nucleoside 5′-phosphates hinders rotation about the $C^{4'}$–$C^{5'}$ bond. As a result, all nucleoside 5′-phosphates are in a *gauche-gauche* conformation characterized by minimal interaction between the substituent at $C^{5'}$ and the furanose ring atoms.

As regards the conformation of the phosphate group itself, in nucleoside 5′-phosphates the phosphorus atom is usually in a *trans*-position with respect to $C^{4'}$.

7.1.3 Orientation of the Heterocyclic Bases Relative to the Sugar

An essential feature of the three-dimensional structure of nucleosides and nucleotides is the relative orientation of the base and sugar moiety, which is usually described by two parameters. The first parameter is the dihedral angle between the base and sugar planes (in this case, the mean square plane of the furanose ring is implied). In all nucleosides and nucleotides, it approaches, although never reaches, the right angle (ranging from 55° for adenosine 3′-phosphate to 81° for 5-iododeoxyuridine). The other major parameter is the torsion angle χ or the angle of rotation of the base about the glycosidic bond. In addition to the angle χ and puckered conformation of the sugar the three-dimensional structure of the nucleotide in the polynucleotide chain is also given by six torsion angles of the main chain (Fig. 7-6B).

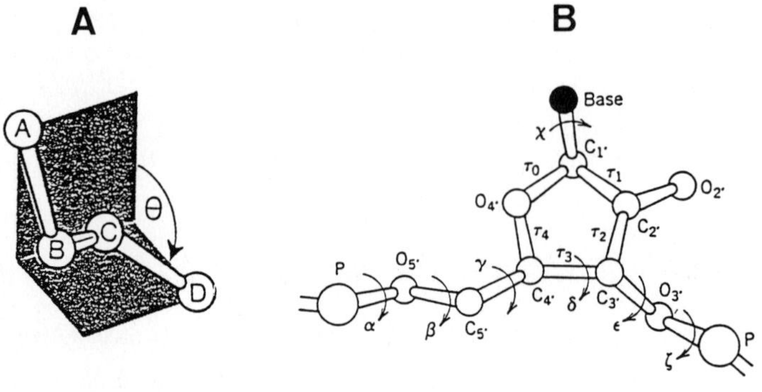

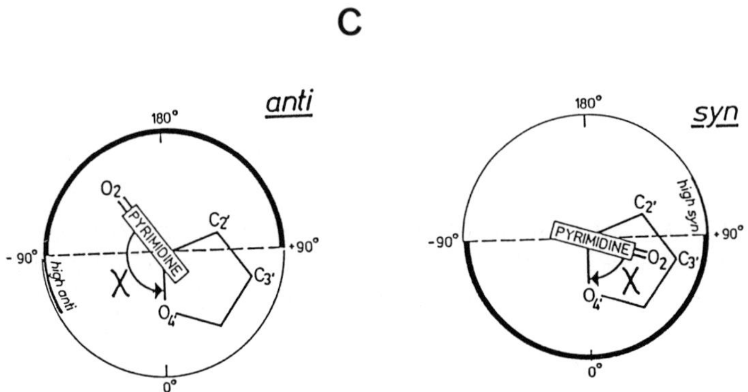

Fig. 7-6. System of torsion angles and their definition for a nucleotide in the polynucleotide chain. A – Definition of torsion angle θ (A–B–C–D) describing the orientation of bonds A–B and C–D relative to the central bond B–C. It can be seen that this angle is determined as that between planes A–B–C and B–C–D; B – Torsion angle χ and nomenclature of six torsion angles along the phosphate–sugar–phosphate chain (courtesy of John Wiley & Sons, Inc.); C – Determination of the magnitude and sign of angle χ for pyrimidine nucleotides in *syn*- and *anti*-conformations. The pyrimidine base is oriented toward the reader (consequently, atom $C^{1'}$ and bond N^1–C^2 are invisible) and rotated with respect to the immobile sugar (adapted from IUPAC–IUB Joint Commission on Biochemical Numenclature, Abbreviations and Symbols for the Description of Conformations of Polynucleotide Chains, *Eur. J. Biochem.*, **131**, 9–15, 1983).

Figure 7-6A shows how the torsion angles are determined. According to the rules adopted by the IUPAC–IUB in 1983, the torsion angle formed by the $O^{4\prime}$–$C^{1\prime}$–N^1–C^2 bonds in pyrimidines and $O^{4\prime}$–$C^{1\prime}$–N^9–C^4 bonds in purines equals 0° when the $O^{4\prime}$–C^1 bonds in the carbohydrate moiety lie in the same plane with the N^1–C^2 and N^9–C^4 bonds, respectively (Fig. 7-6C). It should be emphasized that one can find other (older) rules for determining the torsion angle in the literature.

The possible conformations of nucleosides, arising as a result of rotation of the base about the glycosidic bond, were first analyzed on molecular models using the so-called method of rigid spheres. This method resides in calculation of the distances between certain atoms in the base and carbohydrate moiety as functions of the angle χ.

As was established using this approach, the strongest hindraces to rotation about the glycosidic bond in derivatives of the pyrimidine series occur as a result of interaction between the oxygen at C^2 and hydrogen at C^6 with the substituents at $C^{2\prime}$ and $C^{3\prime}$, on the one hand, and the endocyclic oxygen of the sugar, on the other. In the purine series, hindrances are caused by interaction between N^3 and the same atoms of the furanose ring. As a general rule, however, the rotation of the base in purine nucleosides is hindered to a much lesser extent than in pyrimidine ones.

Careful conformation analysis of pyrimidine and purine nucleosides and nucleotides by this method has revealed two regions marked by the weakest intramolecular interactions. The first region corresponds to the angle equal to about −100° and the other, +100° (see also Fig 7-7C). The conformations associated with the first region are usually called *anti*-conformations and those associated with the second region are known as *syn*-conformations.

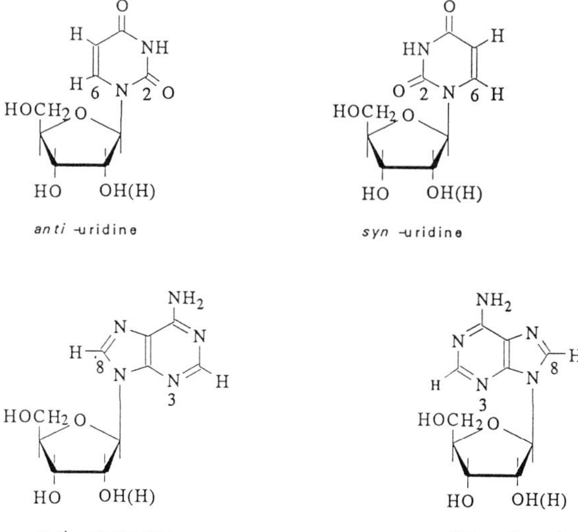

Fig. 7-7. *Syn*- and *anti*-conformations of nucleosides.

In other words, in the case of an *anti*-conformation in pyrimidine nucleosides it is the hydrogen at C^6 that is the closest to $C^{5'}$, whereas in the case of an *syn*-conformation the closest atom to $C^{5'}$ is the oxygen at C^2 (Fig. 7-7A).

In purine nucleosides, *anti*-conformation is marked by proximity of the hydrogen at C^8 to $C^{5'}$, and in the case of *syn*-conformation N^3 is above the plane of the carbohydrate moiety (Fig. 7-7B).

Later, theoretical conclusions were subjected to rigorous experimental verification by different methods. As indicated by X-ray structural analysis, crystalline pyrimidine nucleosides and nucleotides exist in *anti*-conformation in most cases. A classical exception from this rule is 4-thiouridine which crystallizes in the usual *syn*-conformation. It should be borne in mind, however, that when in solution, this nucleoside has an *anti*-conformation. *Anti*-conformation has also been found to occur in many purine nucleotides and nucleosides in the crystalline state. Guanyl ribo- and deoxyribonucleosides and -nucleotides present a special case because they are marked by *syn*-conformation both in the crystalline state and in solution. This, for example, is precisely what makes pdG and pG in oligonucleotides different from the rest of nucleotides. In this case, *syn*-conformation seems to be assured by the interaction between the 2-amino group of guanine and the 5′-phosphate group. Generally an *anti*-conformation can be converted in to a stable *syn*-conformation by inserting a large substituent at position 6 of the pyrimidine ring or position 8 of the purine ring.

The relative orientation of the base and sugar in nucleotides and nucleosides in solution was investigated in the early days primarily by optical activity measurements and, more recently, by NMR spectroscopy. In the former case, the object of the investigation was the optical rotatory dispersion or circular dichroism of nucleoside solutions in the UV region where these compounds are characterized by the Cotton effect. Since heterocyclic bases are optically inactive and ribose (or deoxyribose) does not absorb UV light in the region of 230 to 350 nm, it is believed that the sign and magnitude of the Cotton effect in nucleosides and nucleotides are determined chiefly by the relative orientation of the base and pentose. For example, pyrimidine nucleotides and nucleosides are characterized by positive Cotton effects near the absorption maximum of the pyrimidine base. A positive Cotton effect is a direct consequence of an *anti*-conformation in pyrimidine derivatives, since the cyclic derivative of uracil with a fixed *anti*-conformation also exhibits a positive Cotton effect.

Purine nucleotides display Cotton effects of smaller magnitude, as compared to pyrimidine ones. This finding confirms that the rotation about the glycosidic bond is easier in pyrimidine, rather than purine, derivatives.

More definite information on *syn*- and *anti*-conformations of monomer units in nucleic acids in solution can be obtained by NMR. This method makes it possible to easily identify NMR spectral peaks corresponding to protons associated with particular atoms of heterocyclic rings. As has already been mentioned, depending on whether a compound is in a *syn*- or *anti*-conformation, different protons of the heterocyclic base will be close to $C^{5'}$ (and the phosphate group in the case of nucleoside 5′-phosphates). Accordingly, the relationship between chemical shifts and pD (the NMR spectra were taken in heavy water) was studied for certain protons of the base. The results obtained for adenosine 5′-phosphate and its monomethyl ester are shown in Figure 7-8.

It can be clearly seen that at pD values exceeding 6.0, when secondary dissociation of the phosphate group begins, the value of the chemical shift changes only for the hydrogen atom linked with C^8 (just as expected, no changes take place in methyladenosine 5′-phosphate). A similar effect can be observed only if adenosine 5′-monophosphate is in an *anti*-conformation.

Similar measurements were carried out for other purine and pyrimidine nucleoside 5′-phosphates as well, and in each case these compounds were found to be in *anti*-conformation.

There is a certain correlation between the *syn*- and *anti*-conformations of nucleosides and nucleotides, on the one hand, and conformation of their sugars, on the other. For instance, purine nucleosides with the *C*$^{3'}$*-endo*-conformation of the sugar are usually in a *syn*-form. In the case of *C*$^{2'}$*-endo*-conformation of the sugar, *anti*- and *syn*-conformers occur with equal probability. As regards pyrimidine nucleosides and nucleotides which, as has already been pointed out, are usually in *anti*-conformation, the angle χ depends on the conformation of the sugar as follows:

$$-180^\circ \le \chi \le -138^\circ \text{ for } C^{3'}\text{-}endo$$
$$-144^\circ \le \chi \le -115^\circ \text{ for } C^{2'}\text{-}endo$$

Comparison of the conformations of monomer units in nucleic acids in the series: "nucleoside" → "nucleotide" → "nucleotide of the polynucleotide chain" indicates that their conformational rigidity increases. This has prompted Sundaralingam to formulate the "rigid nucleotide" concept according to which nucleotides in nucleic acids belong as a rule to one of only two classes of conformers:

C$^{2'}$*-endo, gauche-gauche, anti*

or

C$^{3'}$*-endo, gauche-gauche, anti*

shown in Figure 7-9.

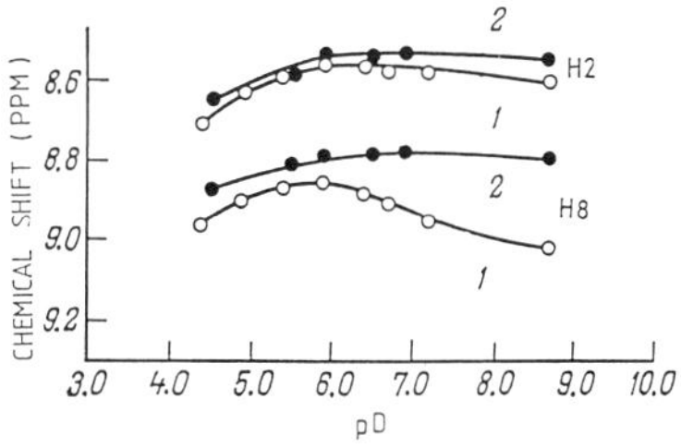

Fig. 7-8. Chemical shift δ of different protons as a function of pD for adenosine 5′-phosphate (1) and its monomethyl ester (2).

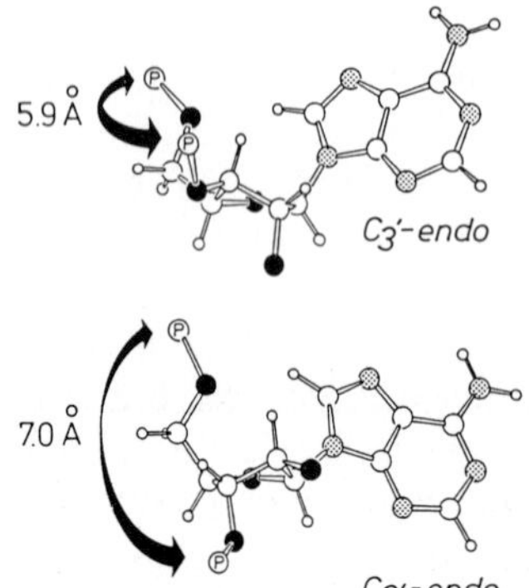

Fig. 7-9. Two conformations of a nucleotide in a polynucleotide chain. Note that in the case of the 2'-*endo*-conformation (bottom) the adjacent phosphates are spaced more widely apart, as compared to the 3'-*endo*-conformation (adapted from M. Sundaralingam, in: *Structure and Conformation of Nucleic Acids and Protein–Nucleic Acid Interactions,* M. Sundaralingam and S.T. Rao, eds., Univ. Park Press, Baltimore, pp. 487–524, 1974).

As will be seen from what follows, exceptions from this rule always stem from a nucleic acid having an unusual structure.

7.2 Intermolecular Interactions Between Heterocyclic Bases

The specific macromolecular structure of nucleic acids is determined by two types of interactions between heterocyclic bases: firstly, complementary interactions between bases lying in the same plane (complementary interactions), and, secondly, interplanar interactions between bases arranged one above another (vertical or stacking interactions).

As will be shown below, complementary interactions are based on formation of specific hydrogen bonds and the usual intermolecular forces determined by Van der Waals–London interactions. Stacking interactions are determined primarily by the latter.

It would be appropriate at this juncture to remind that the overall energy of Van der Waals–London interactions, E_M, is a sum of the energy of dipole interactions, $E_{\mu i \mu 2}$, the energy of a dipole-induced dipole type of interaction, $E_{\mu\alpha}$, and the energy of dispersion or London interactions, E_L, arising between a dipole due to fluctuation and a dipole induced by the former:

$$E_M = E_{\mu i \mu 2} + E_{\mu\alpha} + E_L,$$

Therewith,

$$E_{\mu\mu} \sim -1/R^3{}_{AB}\ \mu_A\ \mu_B,$$

where μ_A and μ_B are dipole moments, R_{AB} is the distance between the interacting dipoles.

$$E_{\mu\alpha} \sim -1 / R^6{}_{AB} (\alpha_A \mu_B + \alpha_B \mu_A),$$

where α_A and α_B are polarizabilities.

$$E_L = -3/2\ I_A I_B / I_A + I_B\ \alpha_A \alpha_B / R^6{}_{AB},$$

where $I_A$and I_B are ionization potientials.

It is extremely important that both complementary and stacking interactions can be studied on very simple systems, such as bases, nucleosides, and nucleotides. It is while studying such systems that the major factors responsible for a particular type of interaction have been identified. Let us now have a closer look at these interactions.

The magnitude of polarizability (usually determined from atomic refractions) is about 0.01–0.012 nm^3 (10–12 $Å^3$) for uracil, cytosine and thymine and about 0.014 nm^3 (14 $Å^3$) for adenine and guanine. The ionization potentials for all bases lie in range of 1.4–1.6 10^{-18} J (9–10 eV).

If two interacting bases are considered as two dipoles, the energy of their interaction can be estimated only very roughly. Therefore, the so-called monopole approximation was used to calculate the total energy and the contribution of individual types of interactions to it. The energy of interaction between two bases was regarded as a sum of interactions between *atoms* of one base and those of the other. Instead of constant dipole moments use was made of net atomic charges, and the total energy is expressed as follows:

$$E_M = E_{\varrho\varrho} + E_{\varrho\alpha} + E_L,$$

where $E_{\varrho\varrho}$ is the energy of the Coulombic interactions between net atomic charges of the two bases, $E_{\varrho\alpha}$ is the energy of the interaction between net charges of the atoms of one base and the dipoles induced thereby in the other base, the other symbols being the same as before.

The most difficult task is to calculate the total net charges – that is, to determine the σ- and π-electron densities at individual atoms of the bases. Used for this purpose are approximate quantum-mechanical calculations which are essentially modifications of molecular orbital theories. Figure 7-10 illustrates the calculated distribution of partial charges for nucleic acid bases; the presented values agree well with the known experimental data.

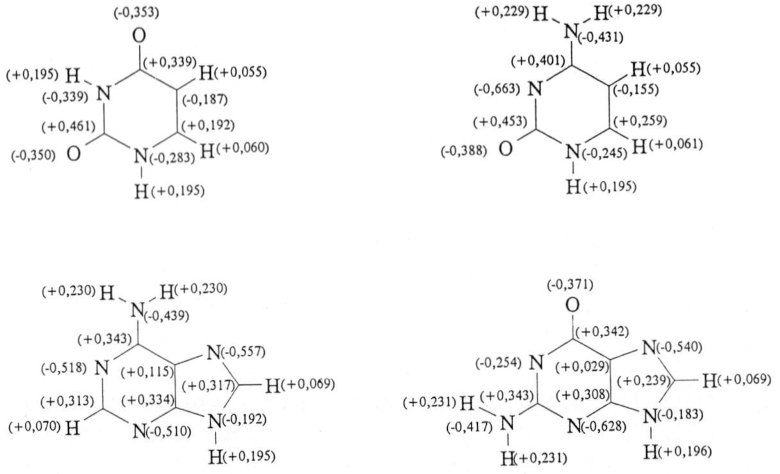

Fig. 7-10. Distribution of calculated total net charges in the nucleic acid bases. The net charges are expressed in fractions of one electronic charge (adapted from B. Pullmann and A. Saran, **Prog.** *Nucleic. Acid Res. Mol. Biol.*, **18**, 215–322, 1976).

7.2.1 Complementary Interactions

Such interactions were postulated for the first time by Watson and Crick. They had arrived at the conclusion that adenine in one DNA strand interacts specifically with the coplanar thymine in the other strand; guanine forms a specific pair with cytosine (Fig. 7-11).

This rule of formation of complementary pairs of bases is one of the fundamental principles of organization of the living matter. It is realized not only during formation of macromolecules of nucleic acids but also during their biosynthesis and protein synthesis in the cell. A better insight into the structure of complementary complexes of bases has become a possibility by virtue of the remarkable capacity of alkylated derivatives of purine and pyrimidine bases[1] to form specific pairs in the solid state. Single crystals of such complexes have been subjected to X-ray structural analysis.

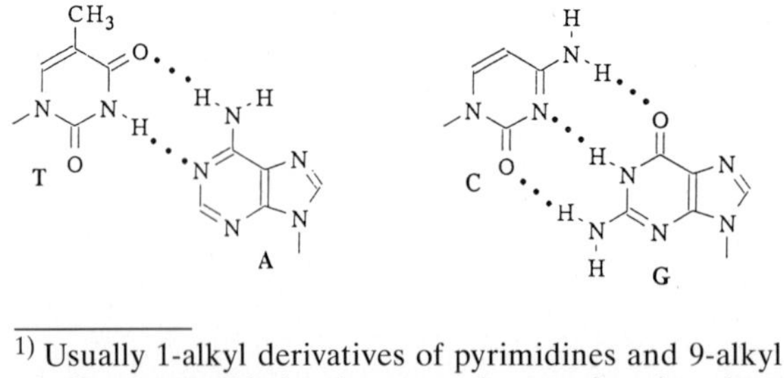

Fig. 7-11. Structure of Watson–Crick complementary base pairs.

[1] Usually 1-alkyl derivatives of pyrimidines and 9-alkyl derivatives of purines were used to rule out possible involvement of atoms forming glycosidic bonds in complexing.

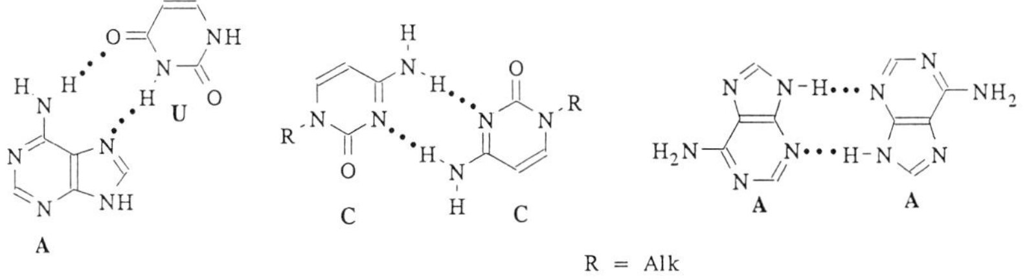

R = Alk

Fig. 7-12. Non–Watson–Crick mechanism of base pairing.

It has been established that adenine derivatives form pairs with those of thymine (or uracil), involving N^7 of the imidazole ring (so-called Hoogsteen pairs) (Fig. 7-12). The adenine-thymine (uracil) pairs of the Watson–Crick type occurring in the DNA molecule have never been observed in crystals of monomer units.

An important property of bases (and their derivatives) is their ability to form homoassociates, such as A·A and C·C pairs.

Formation of specific complementary base pairs is also observable in organic solvents (which is impossible in aqueous solvents because water forms strong hydrogen bonds with bases) by IR and NMR spectroscopy.

What is actually observed in the former case is the shift of the absorption bands corresponding to symmetrical and asymmetrical stretching vibrations of the NH and NH_2 groups of the bases at 3500–2800 cm^{-1}. Consider 1-cyclohexyluracil (I) and 9-ethyladenine (II) as examples.

In dilute chloroform solutions, the NH group in compound (I) is characterized by a band at 3395 cm^{-1}, while the NH_2 group in compound (II) is characterized by bands at 3527 and 3416 cm^{-1}. At such concentrations of compounds (I) and (II), no aggregates are formed. But if solutions of (I) and (II) are mixed, bands at 3490 and 3330 cm^{-1} appear in the IR spectrum of the mixture. The band shift is indicative of a complementary pair (I) (II) in solution as well as of involvement of the NH group of compound (I) and NH_2 group of compound (II) in hydrogen bonding. The composition of the resulting complex can be studied further by looking at the relationship between the intensity of the bands corresponding to the complex and the ratio of (I) to (II) in the solution (Fig. 7-13). It can be seen that the peak intensity of the band corresponds to a stoichiometric ratio of 1:1 between (I) and (II).

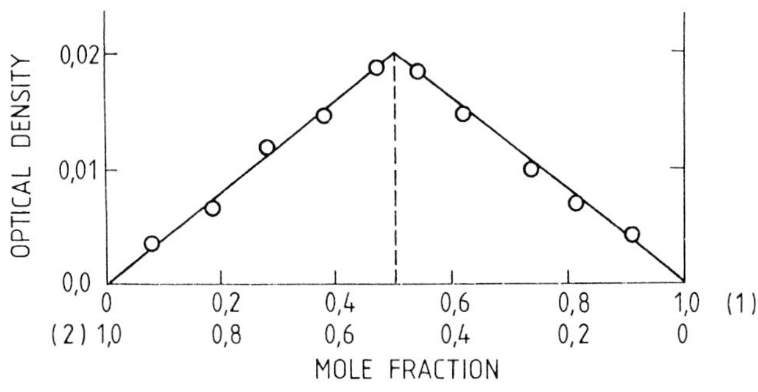

Fig. 7-13. Optical density of a 1-cyclohexyl-uracil (1) and 9-ethyladenine (2) mixture at 3260 cm^{-1} versus its composition (in $CDCl_3$).

Table 7-1. Total hydrogen-bonding ineraction energy in some base pairs (given in kcal/mole).

Base pair	E_{pp}	E_{pa}	E_L	E_M
A · T	–5.9	–0.2	–0.9	–7.0
A · A	–5.2	–0.1	–0.5	–5.8
T · T	–3.6	–0.4	–1.2	–5.2
G · C	–15.9	–2.0	–1.3	–19.2
G · G	–5.8	–0.7	–0.6	–7.2
C · C	–10.6	–1.1	–1.2	–13.0

The reason why NMR spectroscopy is used to study the formation of complementary base pairs is that the signals produced by the protons of the NH groups in guanine and uracil and those of the NH_2 groups in adenine and cytosine, if these groups participate in hydrogen bonding, shift toward low fields (Fig. 7-14).

The bulk of X-ray structural, IR and NMR spectroscopy data suggests that with heterocyclic bases occurring in any combination, the complementary pairing of A with T (U) and G with C is more advantageous because corresponding to this phenomenon is minimal energy of Van der Waals–London interaction, as compared to others. The results of calculating this total energy and its contributions (see above), based on net charges of the base atoms (see Fig. 7-10) are listed below in Table 7-1.

These results also indicate that the main stabilizing factors during complementary base pairing are forces of the *electrostatic* interactions between net charges of the atoms of both bases.

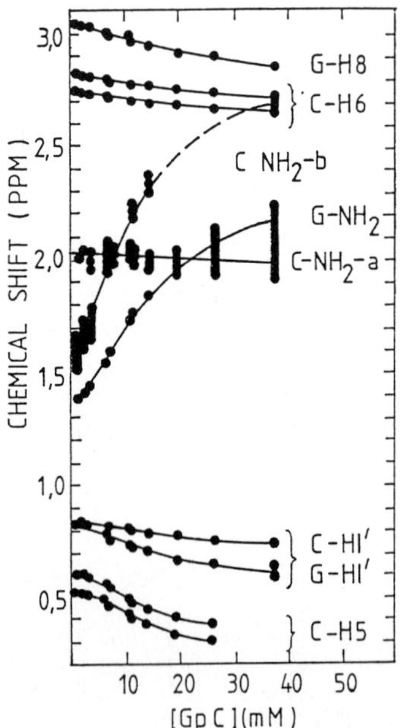

Fig. 7-14. The concentration dependence of the chemical shifts of the self-complementary ribonucleoside monophosphate GpC in aqueous solution (40 °C). Compare the behavior of an amino proton of G and one of the amino protons of C with that of protons at C^6 and C^8 of C and G, respectively (data from T.R. Krugh *et al.*, *Biochemistry,* **15**, 1224, 1976).

7.2.2 Stacking Interactions

In aqueous solutions, heterocyclic base, nucleoside and nucleotide derivatives form stacked complexes in which monomer units are organized into stack-type structures whose base planes are arranged one above another. The formation of associates is due to the fact that in the base (nucleoside)-water system there arise forces precluding contacts between the nonpolar (hydrophobic) base and water. The bases come closer together, their planes overlap, and the resulting conformation is stabilized by Van der Waals–London forces acting between the adjacent bases in a stack.

Thus, we are dealing here with a typical example of a structure resulting from hydrophobic interactions.

The advantage of formation of stack-like associates of nucleic acid bases in water from the energetic standpoint is determined by a number of stabilizing factors. One of the most important ones is believed to be a change in free energy, ΔF, due to reduction in size of the base surface interacting with water molecules; ΔF is first of all dependent on the amount of decrease in the surface tension of water as well as on changes in the degree of order displayed by the water molecules surrounding the base (i.e., on the entropy factor).

Associations of bases and nucleosides can be observed in their concentrated (almost saturated) aqueous solutions and assessed by measuring the osmotic coefficient Φ. It should be remembered that if the coefficient Φ of a substance decreases inversely with its concentration in the solution, this is a direct indication of aggregate formation. By measuring the relationship between Φ and concentration one can calculate the apparent equilibrium association constants K:

$$K = \frac{1-\Phi}{m\,\Phi^2},$$

where m is the molal concentration of the substance.

The values of Φ, K as well as polarizability α and dipole moments μ for some nucleosides are listed in Table 7-2.

Table 7-2. Parameters of stacking of nucleosides in water.

Nucleoside	K, molal^{-1}	Φ (at 25 °C)	α, A^3	μ, D
Uridine	0.61	0.943	10.2	3.9
Cytidine	0.86	0.935	11.0	7.2
Deoxythymidine	0.91	0.905	12.0	3.6
Inosine	–	0.888	13.0	5.2
2′-*O*-Methyladenosine	5.1	0.723	12.5	4.3
2′-Deoxyadenosine	4.5	0.688	13.9	3.0

Two extremely important conclusions follow from these data:

(1) The stacking of purine bases (and their derivatives) is more pronounced than that of pyrimidine ones (the values of K are higher and those of Φ are lower at the same molal concentration).

(2) Electrostatic dipole-dipole interactions are not essential to stabilize the bases in a stack (bases with higher values of μ are associated to a smaller degree); the stacking becomes more pronounced with greater polarizability α of the bases; that is, dipole (monopole)-induced dipole interactions and dispersion forces stabilize stack-like conformations substantially.

Attempts have been made to more accurately describe the mutual arrangement of bases in associates. For example, by studying the relationship between chemical shifts and concentration for various protons of heterocyclic bases and their derivatives, one can find out how closely the base atoms are brought together within a stack. The following model describing the interaction between two adenines within a dimer of adenosine 5′-phosphate (in D_2O) has been derived on the basis of the temperature dependence of the chemical shifts of the base protons in the NMR spectrum (this dependence has turned out to be significant only for the proton at C^2) (Fig. 7-15A).

The strong tendency of purine nucleotides toward formation of structures with marked overlapping of heterocyclic bases can be illustrated by yet another example.

In experiments with the hexanucleotide GmpApApYpApΨp, which is a fragment of $tRNA^{Phe}$ from yeast and contains the minor nucleoside Y, high-resolution NMR spectroscopy revealed that the plane of the base in the nucleoside Y overlaps significantly with those of adenines in the adjacent nucleotides. And if the base in Y is removed by mild acid hydrolysis, these the adenines fill the gap formed. As a result, the stacking conformation is maintained over the entire length of the oligonucleotide and it has a structure stable even at 20 °C.

The formation of stack-like structures is also observed in studying crystals of base, nucleoside, nucleotide and polynucleotide dervatives, however, the observations in this case include less complete overlapping of base planes, as compared to solution. Moreover, the crystal structures are marked by the ring

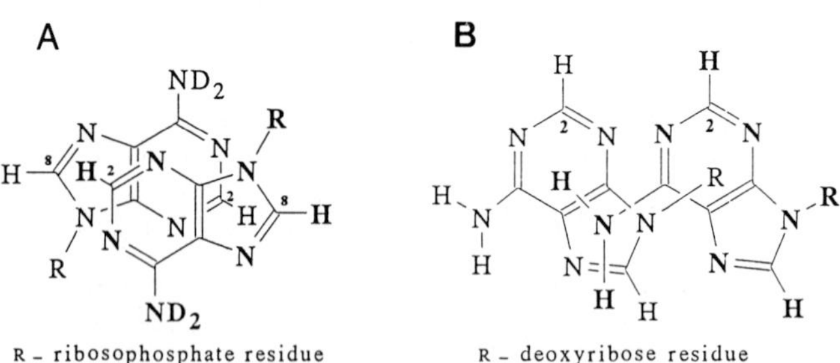

Fig. 7-15. Structure of adenosine associations in solution (A) and crystalline state (B).

plane of one base coming close to the polar groups of the other base, as is the case, for example, with crystals of deoxyadenosine monohydrate (Fig. 7-15B).

Thus, when crystalline associates are formed, the forces of a dipole (monopole)-induced dipole-like interaction gain significance.

7.3 Conformation of Single-Stranded Oligonucleotides and Polynucleotides

Stacking interactions between neighboring heterocyclic bases are the main factor determining the conformation of single-stranded oligo- and polynucleotides.

Since monomer units in oligo- and polynucleotides are linked by covalent phosphodiester bonds, the stacking interactions of bases are more pronounced than in nucleoside and nucleotide associates and, secondly, occurs at any concentrations. The latter is the reason why stacked conformations can be observed in dilute solutions where intermolecular interactions (i.e., association of individual oligo- and polynucleotides) are minimal.

In this case, use is made of two fundamental optical properties of these compounds, directly related to the existence of stacking conformations, namely, hypochromism and Cotton effect in the UV region of the spectrum. Let us now consider these phenomena at greater length.

Being essentially aromatic chromophores, purine and pyrimidine bases absorb light intensely at 180 to 300 nm. Their UV absorption spectra usually feature two bands with maxima near 200 and 260 nm (Fig. 7-16).

Each of these bands is due primarily to $\pi \rightarrow \pi^*$ electron transitions (i.e., excitation of π electrons) and, to a smaller extent, $n \rightarrow \pi^*$ transitions. As is usually the case, $\pi \rightarrow \pi^*$ electron transitions are polarized in the plane of heterocyclic bases.

The absorption spectra of heterocyclic bases are affected by all factors responsible for changes in their electron density distribution (introduction of substituents, protonation of bases, changeover from one tautomeric form to another, etc.).

It is important that the UV spectrum of a base changes as it comes closer together to another. In other words, the UV absorbance spectra of oligonucleotides (beginning with dinucleoside phosphates) differ from those of their constituent mononucleotides.

The absorption of oligo- and polynucleotides near 260 nm is less intense than that of all constituent monomer units taken together. This phenomenon has become known as the *hypochromic effect.* When the ordered conformation of oligo- and polynucleotides is upset or they are hydrolysed to nucleosi-

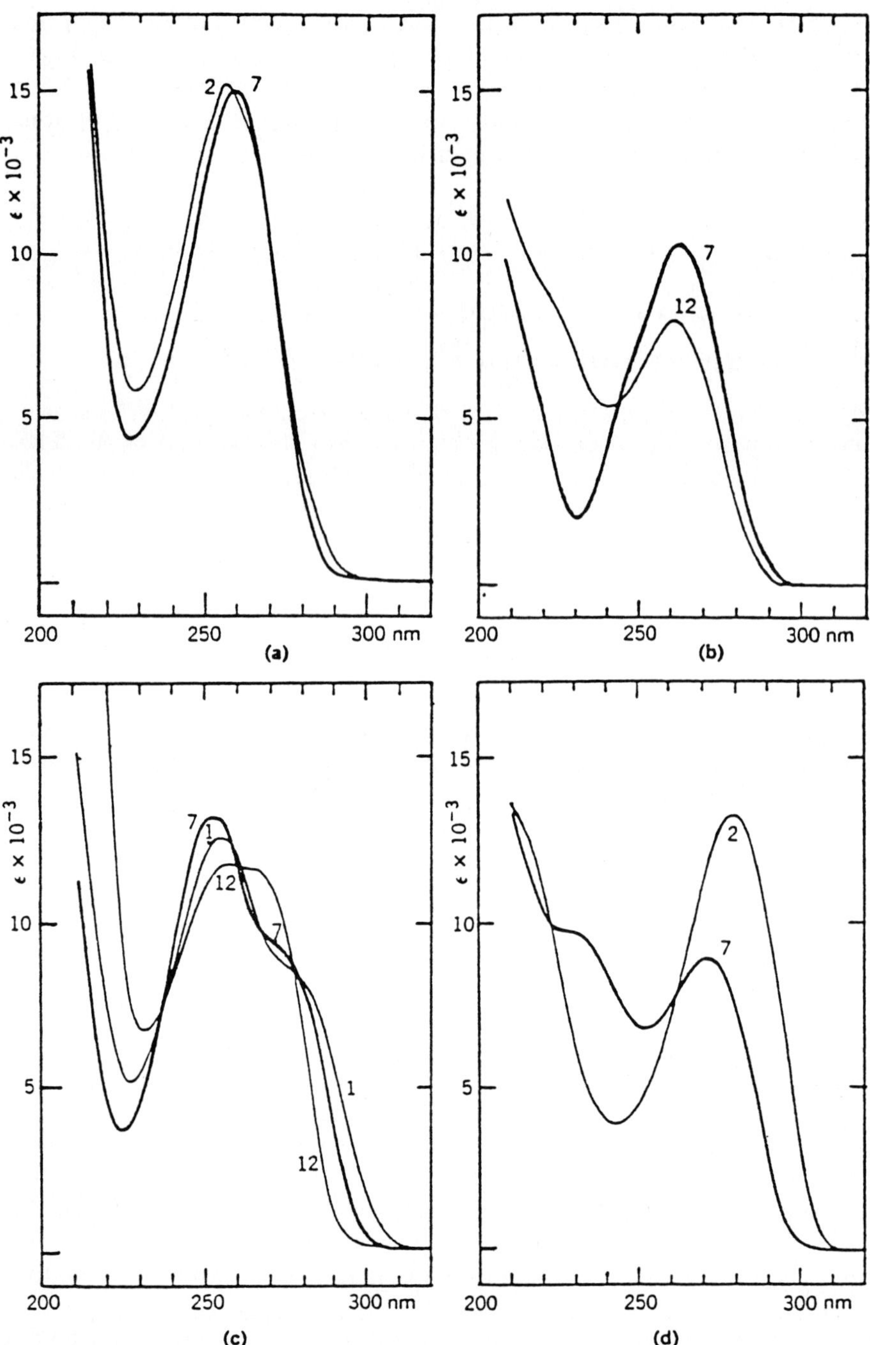

Fig. 7-16. UV spectra of AMP (a), GMP (b), CMP (c) and UMP (d) at various pH values.

des or nucleotides, increasing intensity of optical absorption known as the *hyperchromic effect* is observed.

The theory of hypochromism was elaborated in the early sixties by Tinoco. As detailed analysis of this theory goes beyond the scope of this chapter, we shall only dwell on the qualitative conclusions following from it and being of direct interest to us.

In its classical version the theory regards two closely spaced molecules interacting with light as two oscillators. The oscillators vibrate in the direction of transition moment vectors. The theory makes it possible to predict how the optical absorption of the system will vary as a function of the distance and angle between these vectors. In particular, if the oscillators are arranged one above the other in parallel planes, hypochromic effects will be observed in the long-wave part of the spectrum at certain values of the angle between their projections on the same plane (i.e., in the case of oligo- and polynucleotides in the region of the absorption band with a maximum near 260 nm).

The magnitude of the hypochromic effect is usually expressed as percentage hypochromism h calculated from the formula:

$$h = 100 \cdot \frac{1 - \varepsilon_p(\lambda)}{\varepsilon_m(\lambda)}$$

where $\varepsilon_p(\lambda)$ and $\varepsilon_m(\lambda)$ are the molar coefficients of extinction of the polymer and the constituent monomer units, respectively, at a given wavelenght λ (the measurements are usually made at the absorption maximum).

As was emphasized above in connection with association of nucleosides and nucleotides (7.2), their stacking interactions are strongly dependent on the heterocyclic base species. Hence, the magnitude of the hypochromic effect for oligo- and polynucleotides must be determined by their nucleotide composition. Table 7-3 shows the percentage hypochromism of some dinucleoside monophosphates (at 25 °C, ionic strength 0.1, pH 7).

Just as expected, the values of h for dinucleoside phosphates containing purine bases are usually higher.

As the polynucleotide chain increases in length, the values of h gradually go up and reach a maximum at a chain length of 7 to 10 nucleotides.

The ORD and CD spectra of oligo- and polynucleotides in the UV region (200–320 nm) also differ from those of their constituent monomer units, the

Table 7-3. Hypochromicity of selected dinucleoside monophosphates.

Compound	h, %	Compound	h %
ApA	9.4	CpG	3.2
ApG	5.8	CpU	6.3
ApC	7.6	UpC	2.4
ApU	5.0	UpU	1.7

differences in the magnitudes and positions of the maxima of the Cotton effects being usually greater, as compared to those in optical absorption of the same compounds (see Fig. 7-17). It should be pointed out that in the course of time the CD method almost completely displaced the ORD method by virtue of its higher sensitivity. However, we have included here some results obtained by the ORD method because it was rather important in studying the three-dimensional structure of nucleic acids and for quite some time was the principal tool for observing their conformational changes.

A remarkable property of the CD and ORD spectra is their sensitivity to nucleotide sequence. This important feature can be illustrated by CD spectra of two pairs of isomeric dinucleoside phosphates (see Fig. 7-18).

ORD and CD spectra have led to certain conclusions as regards the structure of oligo- and polynucleotides in aqueous solutions. Here are some of them.

(1) It has been confirmed that the stacking of purine oligonucleotides is more pronounced than that of pyrimidine ones; in this case, uracils and their derivatives interact with one another and with other heterocyclic bases to a much lesser degree than cytosine derivatives. This can be easily seen, for example, from comparison of ORD spectra of two isomeric trinucleotides – UpApGp and ApUpGp (Fig. 7-19). In the trinucleotide ApUpGp, the strongly interacting purine bases are separated by a uridine, and the Cotton effects of this compound are much less pronounced than in the case of UpApGp.

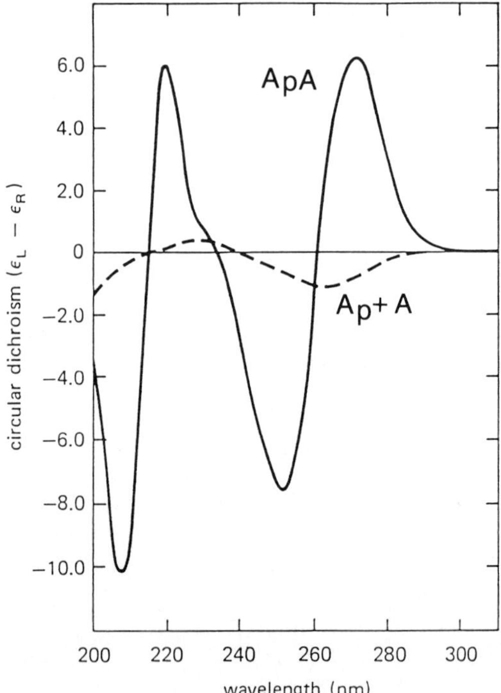

Fig. 7-17. CD spectra of dinucleoside phosphate and its constituent nucleotide and nucleoside (data from M. Warshaw and C. M. Canter, *Biopolymers* **9**, 1079, 1971).

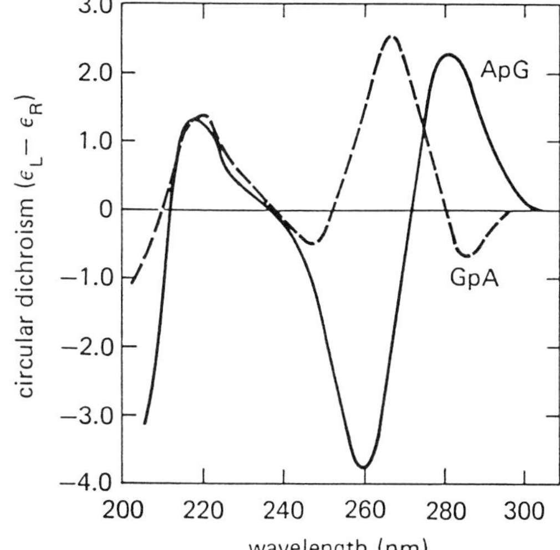

Fig. 7-18. CD spectra of isomeric dinucleoside phosphates (adapted from M. Warshaw and C. R. Cantor, *Biopolymers* **9**, 1079, 1971).

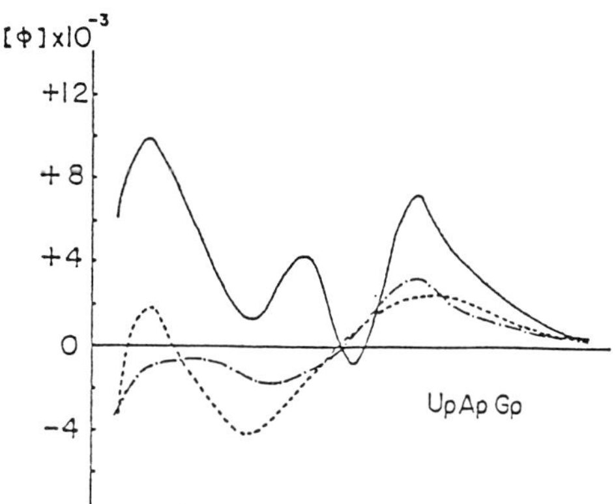

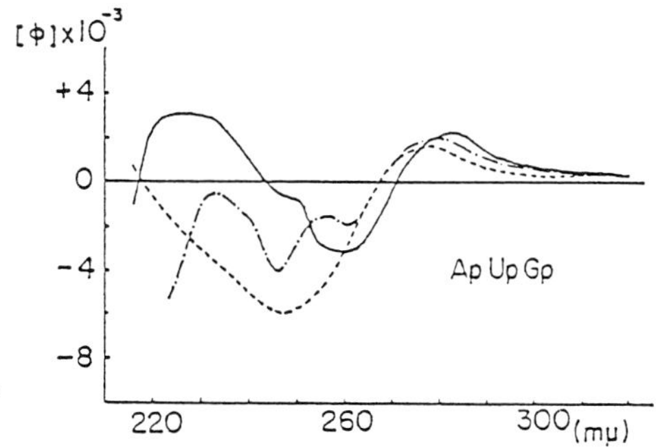

Fig. 7-19. ORD spectra of two isomeric trinucleotides (adapted from Y. Inoue *et al.*, *J. Amer. Chem. Soc.*, **89**, 5701, 1967).

(2) The ORD and CD spectra of oligoribo- and polyribonucleotides differ widely in amplitude and shape from those of oligodeoxyribo- and polydeoxyribonucleotides with the same nucleotide sequence, which is indicative of marked conformational dissimilarity. At the same time, the spectra of 2′-O-substituted oligoribonucleotides are closely similar to those of unsubstituted oligoribonucleotides. Thus, the most likely reason for the difference in conformation between single-stranded polyribo- and deoxyribonucleotides is the different conformation of the sugar in monomer units rather than involvement of the 2′–OH group of riboderivatives in intra-molecular hydrogen bonding.

(3) The ordered conformation of oligo- and polynucleotides, observable by ORD and CD methods, is determined chiefly by the interaction between adjacent bases in the polynucleotide chain. Long-distance interactions (even between monomers separated by a single nucleotide) are extremely weak. The lack of cooperativity during formation of an ordered structure of single-stranded oligo- and polynucleotides as well as the fact that the conformation of any two adjacent monomer units is the same both in di- and polynucleotides make it possible to calculate the CD and ORD spectra of the latter from those of the constituent dinucleoside phosphates.

(4) As the temperature increases, the ordered structure of single-stranded oligo- and polynucleotides is gradually destroyed (which is a direct indication of the non-cooperative nature of the intermolecular interactions in them). Since hydrophobic interactions are involved in stabilization of this structure, it is also disrupted when organic solvents are added to aqueous solutions of oligo- and polynucleotides.

The transition of oligo- and polynucleotides from ordered to disordered state can be easily followed by changes in their CD and ORD spectra (when the stacking conformations break down completely, the spectra of oligo- and

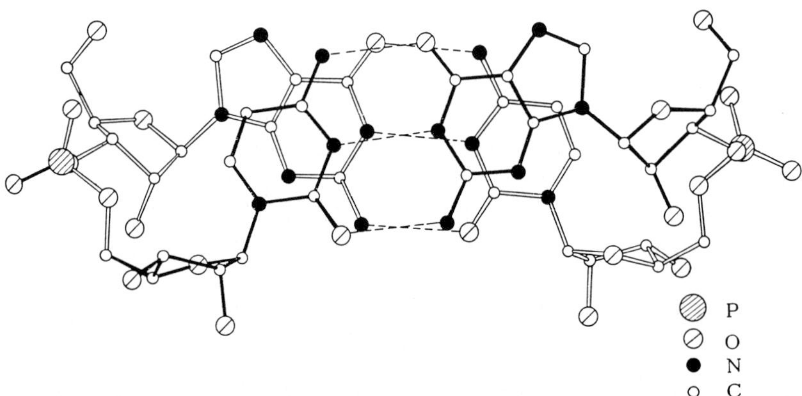

Fig. 7-20. Structure of self-complementary dinucleoside phosphate GpC, deduced from X-ray structure data (adapted from J. M. Rosenberg *et al.*, *J. Mol. Biol.*, **104**, 145–167, 1976).

polynucleotides become almost indistinguishable from the sum of spectra of the constituent monomer units). From changes in the spectra one can calculate the thermodynamic parameters of this transition, which are important characteristics of single-stranded oligo- and polynucleotides.

Detailed molecular models of di- and longer oligonucleotides can be proposed only if their fine structure is known well enough, which, in particular, becomes possible by resorting to NMR spectroscopy and X-ray structural analysis.

At present, X-ray diffraction patterns for many dinucleoside phosphates are available with a resolution of about 0.1 nm (1 Å). Examined in the first place were compounds capable of forming, during crystallization, intermolecular complexes with complementary base pairing. This has provided direct information on the precise geometry of Watson–Crick pairs. As regards stacking interactions, a rather interesting pattern has been revealed here as well. It has turned out that in dinucleoside phosphates, which contain both a purine base and a pyrimidine one, much more marked overlapping of the bases is observed in the case of isomers with a PupPy sequence. This can be illustrated using GpC as an example (Fig. 7-20).

As can be seen from the structure in Figure 7-20, if the positions of the bases in this dinucleoside phosphate are interchanged without affecting the sugar–phosphate backbone, the bases will overlap to much less extent.

7.4 Structure of Double- and Multiple-Stranded Polynucleotide Complexes

Single-stranded polynucleotides with complementary sequences may form helical complexes under appropriate conditions.

The simplest and most typical example of such structures is be provided by complexes formed by complementary homopolymers – that is, poly(A) and poly(U) or poly(G) and poly(C). The composition of a complex depends on the ionic strength of the solution in which it is formed. For instance, at a low ionic strength in the absence of magnesium ions, a double-stranded complex of poly(A) and poly(U) [i.e., a poly/A) · poly(U) complex] can be observed. In a 0.01 M magnesium chloride solution they form a complex made up of one poly(A) strand and two poly(U) ones [i.e., a poly(A) · 2poly(U) complex].

The complex formation process is always accompanied by a hypochromic effect. This phenomenon is put to practical use in determining the composition of complexes. By using this approach it was demonstrated that transitions of the types

$$\begin{gathered}2\text{poly(A)} + 2\text{poly(U)} \rightleftharpoons \text{poly(U)} \cdot \text{poly(A)} \cdot \text{poly(U)} + \text{poly(A)} \text{ and} \\ \text{poly(U)} \cdot \text{poly(A)} \cdot \text{poly(U)} + \text{poly(A)} \rightleftharpoons 2\text{poly(A)} \cdot \text{poly(U)}\end{gathered}$$

take place between different complexes.

Recently, a large number of double- and triple-stranded polynucleotide complexes have been investigated by X-ray structural analysis. The results of these investigations and some other examined properties to be discussed in what follows indicate that these compounds have the shape of regular double or triple helices. The stability of helical structures is determined by the stacking of adjacent bases and complementary interactions between bases facing each other in different strands.

In the case of double-stranded complexes, the second type of interaction is associated with formation of classical (or, as they are sometimes also referred to, canonical) Watson–Crick pairs (see 7.2). Polynucleotide strands in such complexes are antiparallel, which is to say that if we move along one of the strands in the direction of the phosphodiester $3' \rightarrow 5'$ bonds, in the opposite strand our movement will be in the direction of the phosphodiester $5' \rightarrow 3'$ bonds. In triple-stranded complexes, the pairing of bases involves the N^7 atom of the purine ring. An example of such a base-paired poly(A) · 2poly(U) complex is shown in Figure 7-21.

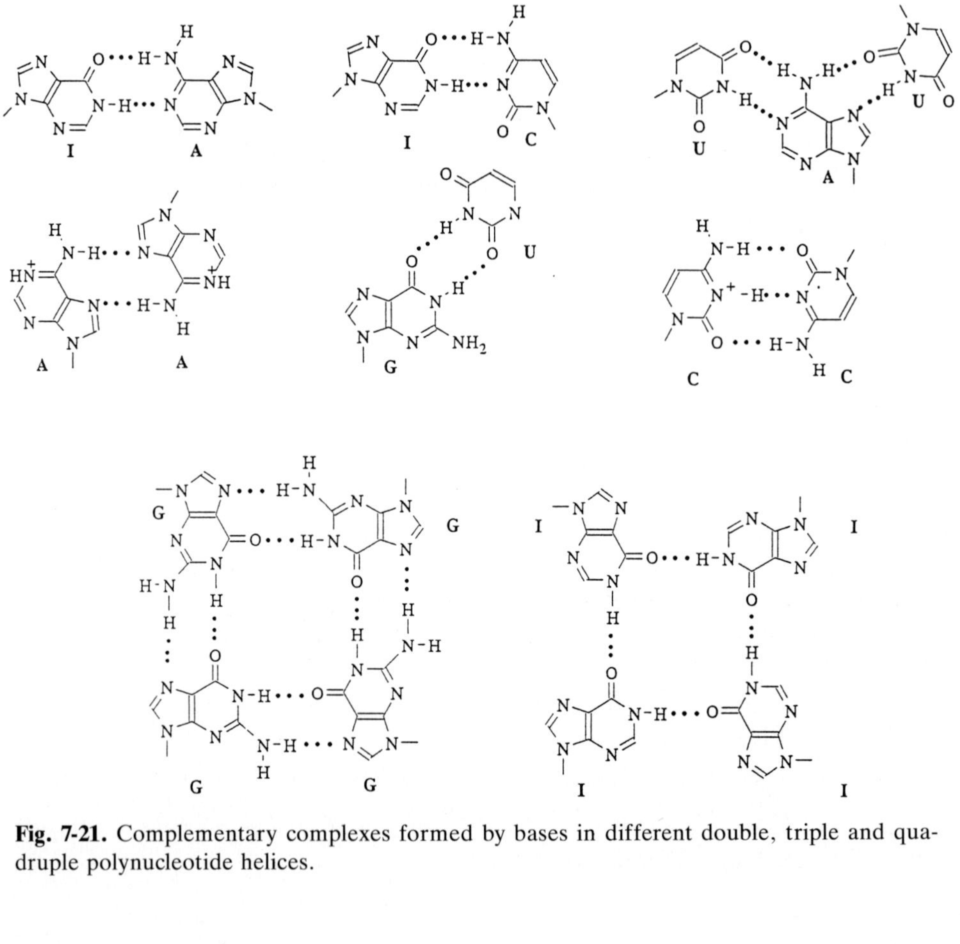

Fig. 7-21. Complementary complexes formed by bases in different double, triple and quadruple polynucleotide helices.

Interestingly, some homopolymers consisting of nucleotide units with bases that are non-complementary from the standpoint of the Watson–Crick concept are also capable of forming double-stranded complexes between themselves. A case in point is complexes of polyinosinic acid with poly(A) and poly(C) (Fig. 7-21).

What is more, at pH < 6.5 poly(A), poly(C) and poly(I) may form regular helical complexes through self-association. The complexes are additionally stabilized by the salt bonds between the phosphate groups of one strand and protonated bases of the other. In the case of poly(A) and poly(C), for example, the pairs shown in Figure 7-21 emerge. In contrast to the ordinary double-stranded complexes with Watson–Crick base pairs, the polynucleotide strands in complexes formed with the participation of protonated bases are parallel (i.e., the direction of their phosphodiester bonds is the same).

The most complex structure is observed in the highly stable complexes formed by poly(G). They are composed of four parallel strands. The ability of guanine derivatives to form quadruplex-type aggregates shown in Figure 7-21 manifests itself already at the mononucleotide level. In recent years, interest in polynucleotide complexes containing G-quartets has risen substantially because G-rich sequences have been revealed in terminal portions of linear chromosomes, known as telomers. Four-stranded DNA structures stabilized by cyclic hydrogen bonding of guanines are covered at greater length in Chapters 8 and 9.

Polyinosinic acid is also capable of forming four-strand complexes. The latter, however, are less stable than poly$(G)_4$ complexes because in this case the tetramer is stabilized only by four hydrogen bonds (Fig. 7-21).

Of particular interest are double-stranded complexes resulting from G pairing with U (or T). The structure of this pair is also illustrated in Figure 7-21. It occurs in double-stranded portions of RNA (Chapter 8) and forms during codon–anticodon interactions. The stability of such duplexes is low (much lower as compared, for example, to complexes formed by pairing of A with U), polyribonucleotides forming more stable complexes than their deoxy analogs.

At low temperatures (below 15 °C), poly(U) strands are capable of folding back upon themselves to form numerous imperfect double-stranded structures (of hairpin type) which are in equilibrium (Fig. 7-22). As the temperature rises, these helical structures transit into a single-stranded form in a cooperative manner (see below).

Evidence showing that helical complexes with two and more strands can emerge from various homopolynucleotides is summarized below ("+" indicates that a complex is formed, "–" indicates that no complex is formed, and "H" stands for a complex whose formation involves protonated forms of the base).

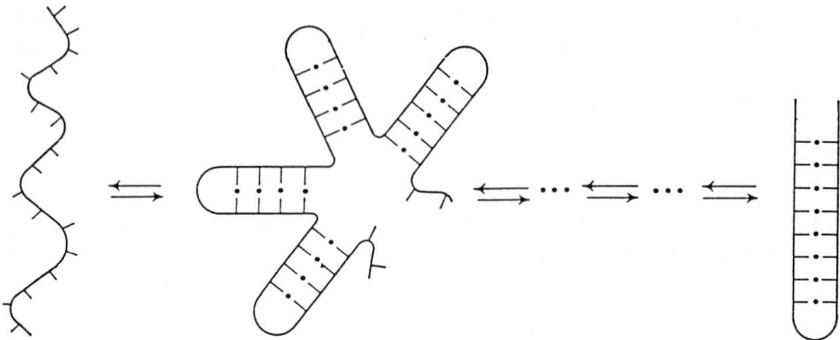

Fig. 7-22. Formation of helical structures in poly(U) at low temperature and high Mg^{2+} concentrations.

Polynucleotide base	A	C	U	G	I	X
X	+	−	+	−	+	+
I	+	+	−	−	+	
G	−	+	+	+		
U	+	−	+			
C	−	H				
A	H					

An important feature of the intramolecular forces stabilizing double- and triple-stranded helical structures is their cooperative nature. In the case of the simplest complexes formed by complementary homopolynucleotides, this means that under the effect of certain external factors denaturing the complex all base pairs in the helical structure break down simultaneously. Roughly, the term "phase transition" can be applied to this process.

In the case of double-stranded complexes with a more complicated nucleotide composition of complementary chains, the term "cooperativity" is used implying that the helical structure is also denatured within a rather narrow range of external changes.

Let us now consider the process of denaturation of double-stranded polynucleotide complexes in greater detail. Studying this process is extremely important because it provides a wealth of information on the nature of the factors determining the structure of these compounds. The process of denaturation of these complexes is often called helix–coil transition because the tertiary structure of single-stranded polynucleotides is close to a random coil.

The most commonly used method for breaking down double-stranded polynucleotide molecules is their thermal denaturation. Its course is usually monitored by changes in the optical properties of the complexes. Since the breakdown of an ordered helical structure is accompanied by a hyperchromic effect, it is more convenient and simpler to observe the changing optical absorption of these compounds in the UV region.

Since the denaturation of complexes occurs in a narrow temperature interval, the optical absorption-temperature curves are referred to as *melting curves.* Figure 7-23 represents a melting curve for a poly(A) · 2poly(U) complex.

The temperature at which the percentage of the helical and denatured portions are equal is known as *melting temperature* and is denoted T_m (on melting curves more complex than the one shown in Figure 7-23 T_m is determined as the temperature at which the increase in optical absorption is 50%

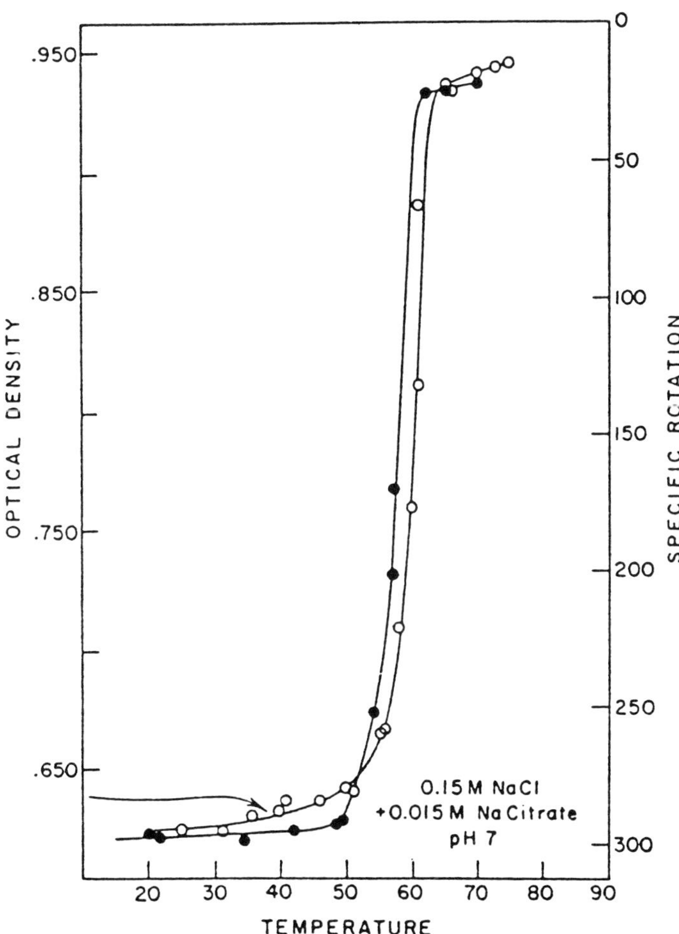

Fig. 7-23. Thermal denaturation profile of poly(A) · 2poly(U) complex at a high ionic strength, determined by UV absorption and ORD at 259 nm (adapted from J. R. Fresco, in: *Information Macromolecules,* H. J. Vogel, V. Bryson and J. O. Lampen, eds., Acad. Press, N. Y., pp. 121–142, 1963).

of the final level). Another characteristic of denaturation of a helical structure is the width of the melting range ΔT_m. In the case of the poly(A) · poly(U) complex, ΔT_m *is small* (about 2 °C), which is indicative of a *high degree of cooperativity* of melting of this complex.

Double-stranded polynucleotides with different nucleotide compositions have different T_m. As can be inferred from the above data (see, e.g., 7.2), G · C is a more stable pair than A · U. Besides, stacking interactions which involve cytosines are more pronounced than that involving uracils. Therefore, as the G · C pairs increase in number, T_m of the complex goes up.

It has been established that double-stranded complexes formed by complementary polyribonucleotides most commonly have higher T_m than those formed by the corresponding polydeoxyribonucleotides. We have already mentioned that the stacking of bases in oligoribonucleotides is more pronounced, as compared to oligodeoxyribonucleotides with the same nucleotide sequence.

The stability of double-stranded polynucleotides (as well as their T_m) increases with the polynucleotide chain length. For example, if we take a poly(A) · poly(U) system at room temperature, a double-stranded complex may exist if it is made up of hexa- and longer oligonucleotides. The differences in T_m diminish as the chain becomes longer and become negligible when the chain length reaches approximately 20 nucleotides.

T_m of double-stranded complexes is strongly dependent on the ionic strength of the solution. As a rule, it increases with salt concentration. The reason is that the electrostatic repulsion of negatively charged phosphate groups (located both in one strand and in opposite ones) exerts a strong destabilizing effect on the structure of double- and triple-stranded complexes. Moreover, these complexes may exist only if the phosphate groups are screened by cations. The screening effectiveness depends on the constants of dissociation of the corresponding metal phosphates. This is why most bivalent cations (Mg^{2+}, Ca^{2+}, Mn^{2+}, etc.) stabilize the structure of complexes much more efficiently than univalent ones. Usually, there is a linear relationship between the melting point of double-stranded polynucleotides and the logarithm of salt concentration in the solution. It should be pointed out that some bivalent cations (e.g., Cu^{2+}) lower the melting temperature of natural and synthetic double-stranded polynucleotides.

The stability of double-stranded helical polynucleotide complexes and their T_m are virtually independent on pH of the medium in the range from 5.5 to 9.0. However, at lower or higher pH values the complexes become less stable, which has to do with ionization of the bases.

T_m of double-stranded complexes is materially affected by the presence of organic solvents, urea, guanidine, and some anions (ClO_4^-, ClO_3^-, CCl_3COO^-) in high concentrations. All these substances upset the interplanar (stacking) and/or complementary interactions between bases.

While studying the process of thermal denaturation of complementary polynucleotide complexes, one can glean important quantitative characteristics of the intramolecular interactions in them. For instance, the following relation has been obtained for homopolynucleotide complexes:

$$12.4R \frac{T_m^2}{\Delta H} \sigma^{2/3},$$

where σ is the cooperativity factor, R is the universal gas constant, and ΔH is the enthalpy of the complex's transition from ordered to denatured state.

The cooperativity factor σ, in turn, is related quantitatively to the change in the free energy of stacking of two separate hydrogen-bonded base pairs, E, as follows (T is absolute temperature):

$$\sigma = e^{-E/RT}.$$

From the known value of ΔH and experimental melting curve one can calculate the value of E. For double-stranded homopolynucleotide complexes it is about 29.4 kJ/mol (7 kcal/mol). Knowing the basic factors responsible for denaturation of double-stranded polynucleotide complexes we now can formulate the optimal conditions for formation of such complexes: (1) the temperature must be substantially lower than T_m; (2) a solution with high ionic strength must be used; (3) the concentration of single-stranded polynucleotides must not be lower than 10^{-5} M.

7.5 Complexes of Polynucleotides with Mono- and Oligonucleotides

Single-stranded polynucleotides may form ordered helical complexes with complementary nucleosides, mono- and oligonucleotides. Their study is important for elucidating the mechanism of the interaction of nucleotide derivatives and short segments of nucleic acids with polynucleotide templates in the course of protein and nucleic acid synthesis. Moreover, as will be shown below, complexes of the oligomer–polymer type are widely used to investigate the secondary and tertiary structure of nucleic acids.

The formation of complementary monomer–polymer complexes can be observed only in the case of purine nucleosides and nucleotides. Pyrimidine nucleosides and nucleotides do not form such complexes. This suggests that stacking interactions (which are more pronounced in the case of purine derivatives) are essential for stabilization of the complex.

Analysis of the stoichiometry of monomer–polymer complexes indicates that under optimal conditions triplet complexes are usually formed, which are more stable than the corresponding double-stranded complexes. In other words, in the case of an adenosine–poly(U) system, for example, a complex is formed made up of two poly(U) strands and a sequence of adenosines (it can be regarded as poly(A) without phosphodiester bonds).

Naturally, in the case of mononucleotides (i.e., in the presence of charged phosphate groups), monomer–polymer complexes may form only in solutions with high ionic strength (when the phosphate groups are screened).

Oligonucleotides form much more stable complexes with the complementary polynucleotide, as compared to nucleosides and mononucleotides. Therefore, formation of oligomer–polymer complexes is observable not only when purine oligonucleotides but also pyrimidine ones are involved.

Oligomer–polymer complexes exhibit all the properties inherent in the polymer–polymer structures discussed in the previous section. The most interesting property is the dependence of the complex stability on the oligomer length. Here, the role of stacking interactions in complex stabilization becomes especially manifest. Indeed, as has been established for an oligo(A) · poly(U) mixture, all oligomer–polymer complexes beginning with the trinucleotide are characterized by hypochromism amounting to 35%. A similar hypochromic effect is observed when a triple-stranded poly(A) · 2poly(U) complex is formed. Thus, it may be assumed that in the oligo(A) · poly(U) complex oligomers form a continuous chain with the terminal links interconnected through stacking interactions.

On the other hand, in an oligo(U) [or oligo(dT)]-poly(A) system the oligomer–polymer complex displays the same percentage hypochromism as the poly(U) · poly(A) duplex, only beginning with the 14- to 16-unit oligonucleotides. It is believed that in the oligo(U) part of the complex two or three bases at each end are not involved in stacking, which precludes formation of a "continuous" helical complex.

Incorporation into the oligomer of bases noncomplementary with respect to those of the polymer disturb the ordered structure of a complex and destabilizes it.

Such complexes have been of particular interest because their formation underlies the site-specific mutagenesis technique. Experiments with short oligodeoxyribonucleotide complexes have shown that the behavior of an unpaired nucleotide is strongly dependent on the heterocyclic base species. For instance, in the duplex

```
d ( G A T G G - G C A G )
    · · · · ·   · · · ·
  ( C T A C C G C G T C ) d
```

the "extra" guanosine is intercalated into the double helix and participates in the stacking interactions with the adjacent GC-pairs. In this case, the helix is slightly bent at the site of the defect. Similar behavior is displayed by adenosines, for example, in the duplex

```
d ( C G C G – A A A T T T A C G C G )
    · · · ·   · · · · · ·   · · · ·
  ( G C G C A T T T A A A – G C G C ) d
```

when it is in solution. In the crystalline state, the duplex is marked by protrusion of the unpaired adenosines from the double helix. This is indicative of a certain conformational flexibility of double-helix polynucleotide complexes with unpaired bases.

An unpaired cytidine may also exist two states – outside the helix and incorporated into it, the transition from the former to the latter state occurring when the solution is heated. Unpaired uridines always tend to be outside the double helix. In contrast, thymidine, which is more likely to be involved in stacking interactions, is incorporated into the helix without causing any perceptible changes in its conformation.

It is known that complementary base pairs at the termini of duplexes are in a partially denatured state or, to be more precise, in equilibrium between the paired and unpaired states. Such termini are said to "breathe". If a helix terminates in several unpaired bases (as is the case, for example, with DNA fragments derived with the aid of restriction endonucleases), the stability of the double helix is adversely affected. Interestingly, however, single unpaired terminal nucleotides enhance the stability of the duplex. It has been speculated that this is due to additional stacking interactions between the unpaired and paired bases in the terminal complementary pair. For example, the melting point of the duplex

```
r ( G C G C )
    · · · ·
  ( C G C G ) r
```

in a buffer containing 1 M NaCl is 27.5 °C, whereas in duplexes

```
r ( U G C G C )
      · · ·
  ( C G C G U ) r
```

and

```
r ( G C G C G ) p
      · · ·
h ( G C G C G ) r
```

in the same buffer it is 31.5 and 51 °C, respectively. Consequently, the increase in melting point of the duplex is indeed in direct correlation with the ability of the unpaired base to participate in stacking interactions.

References

1. V. A. Bloomfield, D. M. Crothers, and I. Tinoco, Jr, *Physical Chemistry of Nucleic Acids.* Harper & Row, N.Y.–Evanston–San Francisco–London (1974).
2. C. R. Cantor & P. R. Schimmel, *Biophysical Chemistry,* parts I, II and III. W. H. Freeman and Co., San Francisco (1980).
3. W. Saenger, *Principles of Nucleic Acid Structure.* Springer Verlag, New York (1984).
4. W. Guschelbauer (1988), Polynucleotides, in: *Encyclopedia of Polymer Science and Engineering.* Vol. 12, John Wiley & Sons, Inc., New York, pp. 699–785 (1988).

8 Macromolecular Structure of DNA and RNA

A large body of literature is concerned with the structure of DNA and RNA. Some of the books and reviews dealing with the subject are listed at the end of this chapter. Here, we shall give a brief outline of the current concepts regarding the principles of spatial organization of DNA and RNA macromolecules and those of their features that offer an insight into their chemical properties and interactions with various ligands.

8.1 DNA

8.1.1 The Watson and Crick Model

In the spring of 1953, J. Watson and F. Crick published a short paper with a description of a new model of DNA structure. This report on one of the most important discoveries of our times had turned out to be seminal in many fields of natural sciences.

Proceeding from the available data on the structure of heterocyclic bases, conformation of nucleosides, the internucleotide linkage in DNAs and their nucleotide composition (Chargaff's rules), Watson and Crick decoded the X-ray fiber diffraction patterns representing the paracrystalline form of DNA emerging at a relative humidity in excess of 80 per cent and a high concentration of counter-ions (Li^+) in the sample. According to their hypothesis, the DNA molecule is represented by a symmetrical helix formed by two polydeoxyribonucleotide chains twisted one with respect to the other around a common axis (Fig. 8-1). The diameter of the helix is virtually constant over the entire length and equals 1.8 nm (18 Å). The pitch per helix turn, which corresponds to its identity period, is 3.37 nm (33.7 Å). There are ten bases per pitch in one strand. The base planes are thus spaced about 0.34 nm (3.4 Å) apart. They are perpendicular to the axis of the helix. The planes of the carbohydrate moieties are almost parallel to helix axis.

As can be seen from Figure 8-1, the sugar–phosphate backbone of the molecule points toward the outside of the helix, and two grooves differing in size can be discerned on its surface, the major groove being about 2.2 nm (22 Å) wide and the minor one, about 1.2 nm (12 Å) wide. The helix is right-handed. Its polydeoxyribonucleotide chains are *antiparallel:* this means that if we move along the axis of the helix from one end to the other, the phosphodiester bonds in one chain will be passed in the 3′ → 5′ direction and in the other, in the 5′ → 3′ direction. In other words, the linear DNA molecule terminates, in each direction, in the 5′ end of one chain and the 3′ end of the other.

The symmetry of the helix implies that for each purine base in one chain there is a matching pyrimidine base in the other chain. As has already been

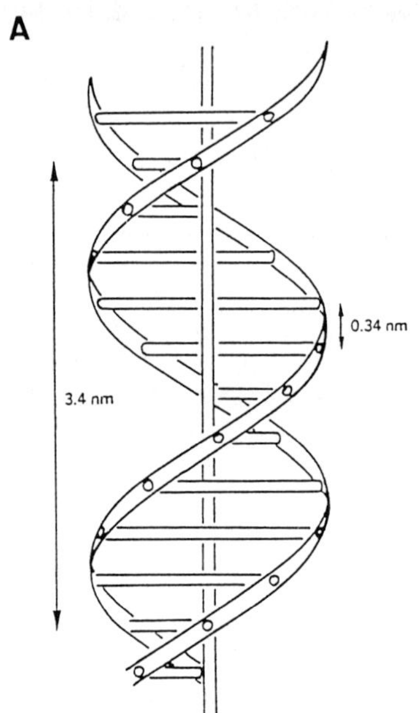

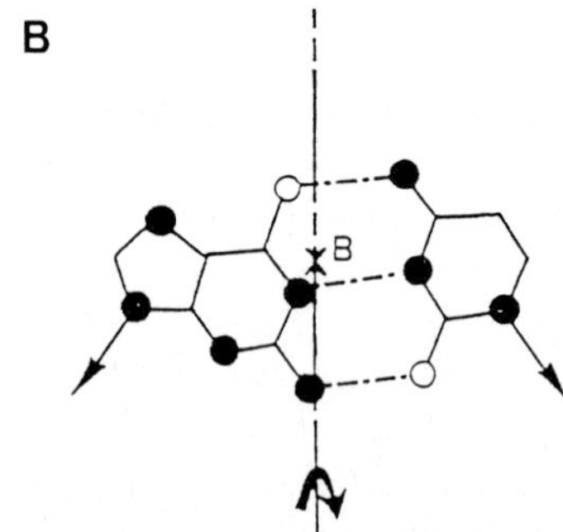

Fig. 8-1. (A) Watson and Crick model of B form of DNA and (B) location of helix axis (x) relative to a base-pair.

emphasized, this requirement is met by formation of complementary base pairs – that is, adenines and guanines of one chain are paired with thymines and cytosines of the other respectively (and vice versa).

Thus, the nucleotide sequence in one chain of the DNA molecule predetermines that in the other chain.

This principle is an important corollary of the Watson and Crick model as it provides a lucid explanation in chemical terms of the fundamental biological role of DNA – storage and faultless transmission of genetic information.

The only other thing to be added in conclusion of this brief description of the Watson and Crick model is the fact that the angle between the neighboring base pairs (the angle between the straight lines connecting the $C^{1'}$ atoms in the adjacent complementary pairs) is 36°. In subsequent writings, the double helix of DNA, whose model was proposed by Watson and Crick started being referred to as the B form of DNA or B-DNA. DNA exists predominantly in this form in the cell.

8.1.2 Polymorphism of the Double Helix of DNA

Although the structure of DNA helices is extremely rigid, it can be changed under certain conditions. At a relative humidity below 80 per cent or at a low concentration of counter-ions in the sample, DNA exists in a crystalline form also known as the A form. The latter is also a right-handed double helix with antiparallel chains. However, it differs markedly from the B form in many ways. If the helix axis in the B form passes through the center of complementary pairs, base pairs in the A form are nearly 0.5 nm (5 Å) off the axis and lie along the periphery of the macromolecule (which is clearly seen on the cross-sectional views of the helix in Fig. 8-2) with the result that DNA in the A form has the configuration of a spiral stair. Moreover, base pairs are not normal to the axis of the helix (the angle with the perpendicular is about 20°), and the bases themselves are not coplanar within a pair. The helix rise per base pair in A-DNA is much smaller and equals 0.23 to 0.26 nm (2.3–2.6 Å).

Such a significant difference between the A and B forms of DNA helices becomes better understood after analysis of the conformation of deoxyribose in the mononucleotide residues forming these helices. It has been found that the sugar in the A form of DNA has a C3'-*endo*-conformation, whereas in the B form it has a C2'-*endo*- (or the closely similar C3'-*exo*-) conformation. It is due to this difference in deoxyribose puckering that we have the two forms of DNA helices considered here. One should also remember that the distance between phosphates in nucleoside diphosphate depends largely on the sugar conformation (see Fig. 7-9). For the B and A forms, the difference in the distance between two adjacent phosphorus atoms on the DNA chain is about 0.11 nm. This difference greatly affects the shape of the helix.

The C2'-*endo*-conformation of deoxyribose has also been revealed in forms C (lithium salt of DNA at a humidity below 66%) and T (glucosylated DNA of bacteriophages) of DNA. At the same time, as will be shown below, ribose in double-stranded RNAs strongly reminiscent of A-DNA has a C3'-*endo*-conformation. This gives every reason to assume two families of helical nucleic acids – a family of A forms and that of B forms.

By virtue of the A and B forms of double-stranded DNAs having characteristic and widely different CD spectra, their interconversions can be observed in solution as well (Fig. 8-3A). Studies based on this approach have shown that the B–A transition (which occurs as the alcohol concentration in the solution increases) is essentially cooperative. At the same time, the change in the form of DNA helices within a family (i. e., without alteration of the sugar conformation) proceeds smoothly.

The capacity of the double helix of DNA for conformational changes becomes more manifest when it passes from the right- to left-handed form.

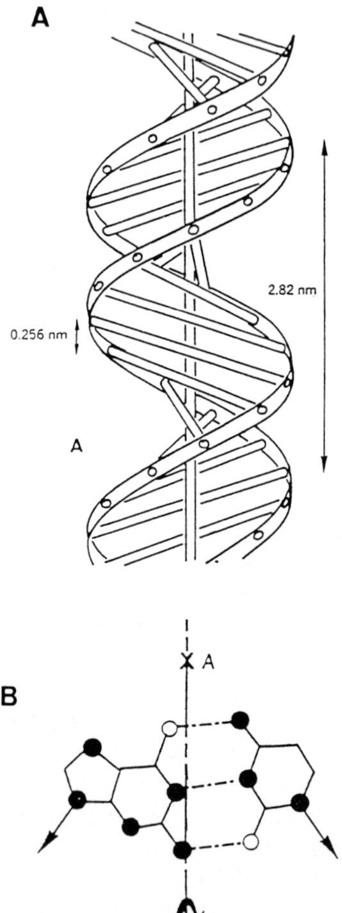

Fig. 8-2. (A) Model of DNA structure in A form and (B) location of helix axis (x) relative to a base-pair.

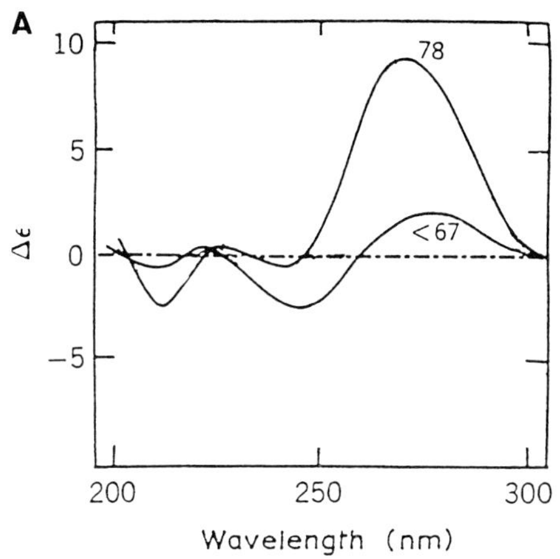

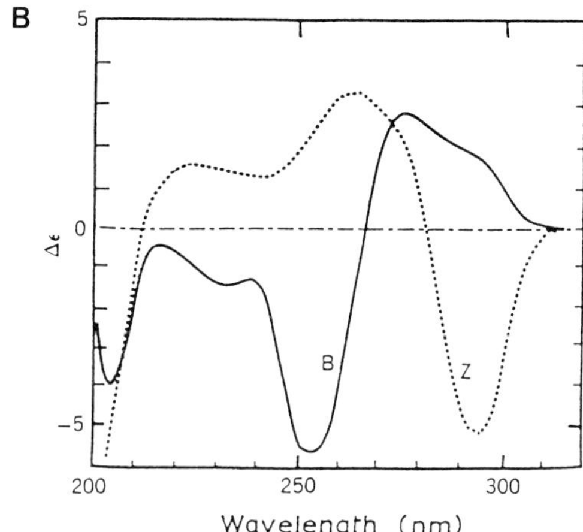

Fig. 8-3. CD spectra of DNA (A) – A and B forms; (B) – B and Z forms. Numbers at the curves indicate percentage of ethanol (adapted from W. Guschlbauer, *Polynucleotides,* in *Encyclopedia of Polymer Science and Engineering* **12**, 699–785 (1988)).

The earliest hypothesis of left-handed double helices was prompted by analyses of the CD spectra of a polynucleotide with an alternating d(GC)n sequence. Poly-d(GC) chains are self-complementary and form, in a solution of low ionic strength, duplexes producing the usual CD spectra typical of the B form. However, as the ionic strength goes up, their CD spectrum undergoes inversion (Fig. 8-3B). This may be indicative of the helix changing its handedness.

Indeed, when oligo- and polynucleotides with alternating G and C sequences were crystallized from solutions with high salt concentrations, X-ray structural analysis confirmed the existence of a left-handed helix which became known as Z-DNA (Fig. 8-4). It has a remarkable feature, namely, alternation of nucleotide conformations: if the sugar moieties in the dC units have a C2′-*endo*-conformation and the base, an *anti*-conformation, in the dG units deoxyriboses are in a C3′-*endo*-conformation and the base, in a *syn*-conformation.

Thus, the repeating unit in the Z form of DNA consists of two base pairs, and not one as in the B and A forms; as a result, the line connecting the phosphate groups takes a sharp turn and assumes a zigzag shape (hence the name Z form). As compared to the B form, the left-handed Z form is marked by a different pattern of base stacking: strong and weak interplanar interactions also alternate.

Analysis of the Z-DNA model illustrated in Figure 8-4 makes it easy to understand why the B → Z transition occurs only in solutions of extremely high ionic strength: in contrast to B-DNA, the distance between phosphate groups in the Z form across the groove is small – only 0.77 nm (7.7 Å) (in B-DNA it is 1.17 nm or 11.7 Å), and a left-handed helix may exist only provided the charges at the phosphates are screened.

The Z form of DNA is also stabilized by some chemical modifications of heterocyclic bases, such as bromination of deoxyguanosine at C^8 or methylation of deoxycytidine at C^5. In the former case, incorporation of a bulky sub-

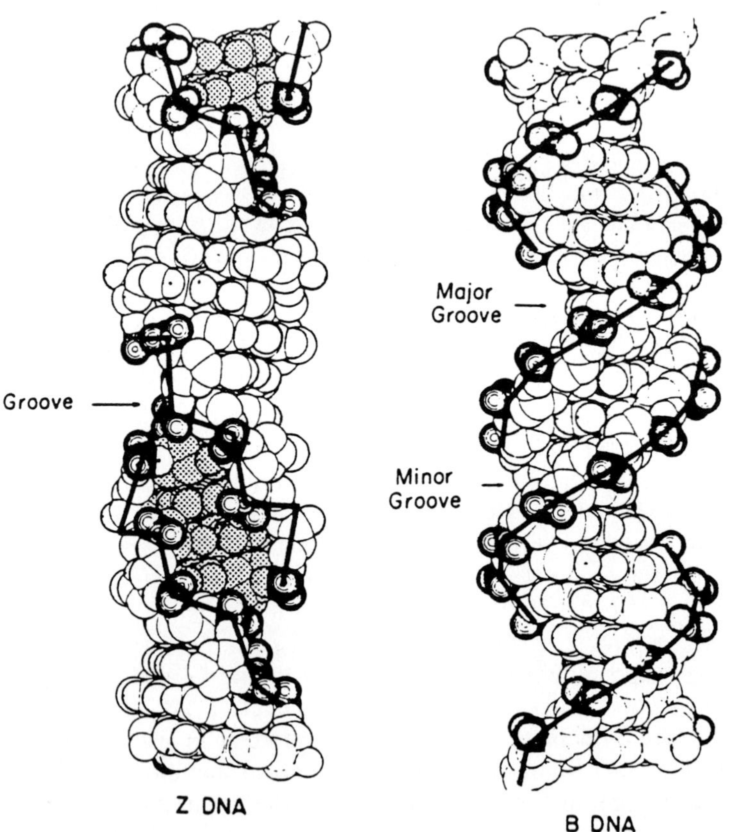

Fig. 8-4. Model of DNA structure in Z form. B form DNA is shown for comparison (adapted from A. Rich et al., *J. Biomol. Structures and Dynamics,* **1**, 1–20 (1983)).

stituent stabilizes the *syn*-conformation of G, typical of Z-DNA. In the latter case (which, incidentally, is nothing but a natural modification of C, regulating gene activity), the methyl groups come closer together in the groove of the left-handed helix and enter into hydrophobic interactions, thereby changing the hydration of the DNA helix. In both cases, the left-handed helix is stable at a moderate ionic strength, too.

The Z form is typical not only of dCdG units but also many other alternating sequences, for example, polyd(AC) and polyd(GT) units. Interestingly, a B-DNA helix may feature local segments in A or Z conformation. This has given rise to many hypotheses as regards the functional role of such segments. None the less, the question as to DNA having such conformations *in vivo* remains open.

8.1.3 Single-Crystal X-Ray Structures of DNA

For quite some time, X-ray diffraction analysis on fibres has been the primary method for studying helices formed by DNA and other complementary polynucleotides. Its resolution is not very high, which is why doubts arose from time to time about the Watson and Crick model. In the early eighties, single crystals of synthetic oligodeoxyribonucleotide duplexes were produced.

Investigation of single crystals by X-ray structural analysis with atomic resolution not only confirmed the basic characteristics of A- and B-DNA revealed by X-ray diffraction studies on fibres (Fig. 8-5 and Table 8-1), but also led to a discovery of a rather important and quite unexpected feature of the B form of DNA.

Unlike A- and Z-DNA, whose helices have parameters independent of the nucleotide composition and, therefore, are structurally homogeneous, in B-DNA marked differences between individual base pairs are observed. These differences manifest themselves primarily in the mutual arrangement of bases within a complementary pair and the mutual orientation of adjacent pairs. As a result, each particular dinucleotide sequence has its characteristic parameters in the B-helix.

For example, in the self-complementary dodecamer d(CGCGAATTCGCG) the mean turn angle between bases is 35.9° (Table 8-1). At the same time, differences in this parameter for individual base pairs range from 27.4° to 41.9°, which corresponds from 13.1 to 8.6 base pairs per turn of the helix. Single-crystal X-ray studies have also shown that base pairs in this duplex differ widely in the propeller twist or, in other words, the angle between the base planes within a complementary pair, arising as a consequence of departure of the bases from coplanarity within the pair. Such conformational mobility of the B form of DNA allows the neighboring bases of its chains to overlap more effectively.

There is no doubt that the sequence-specific microheterogeneity of DNA is extremely important for functional interactions with various ligands.

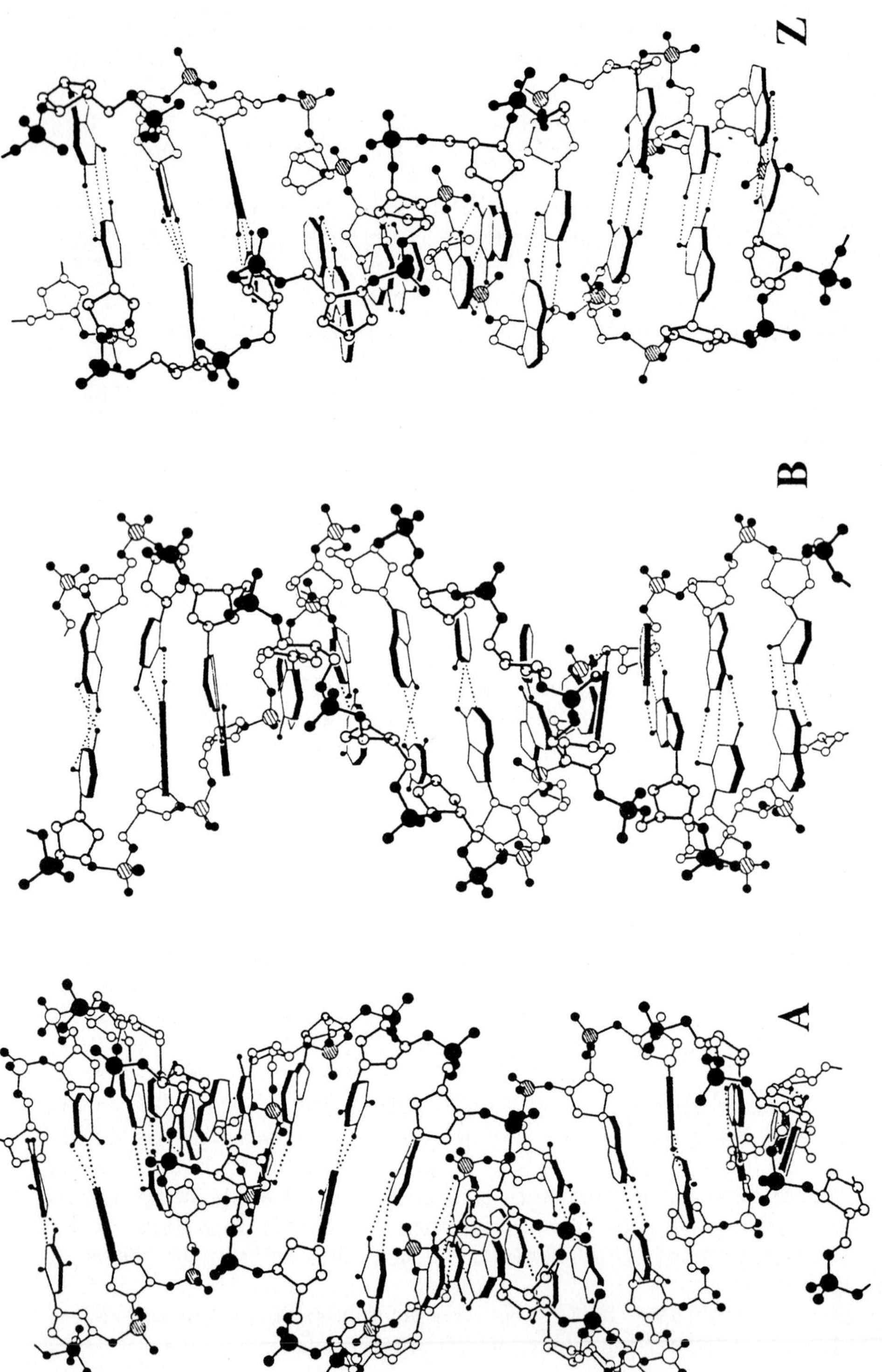

Fig. 8-5. Molecular models of B, A and Z forms of DNA from single-crystal studies.

Table 8-1. Structural Parameters of DNAs from Single-Crystal X-ray Structures*.

Property	Helix		
	A d(CCGG) d(CGCGTATACGCG)	B d(CGCGAATTCGCG)	Z d(CGCGCG)
Handedness of helix	right-handed	right-handed	left-handed
repeating unit	1 base pair	1 base pair	2 base pairs
rotation per base pair	33.6°	35.9° ± 4.2°	–60°/2
average base pairs per turn	10.7	10.0 ± 1.2	12
inclination of base normals to helix axis	+19°	–1.2° ± 4.1°	–9°
helix rise per base pair, nm	0.23	0.332 ± 0.01	0.38
pitch per helix turn, nm	2.46	3.32	4.56
glycosyl angle (χ)	(anti)	(anti)	dC (anti) dG (syn)
sugar pucker	18° (3′-endo)	90° (O′-endo) to 162° (2′-endo)	dC 162° (2′-endo) dG –18° (2′-exo)

* Data from R. E. Dickerson and H. R. Drew, *J. Mol. Biol.*, **152**, 723–736 (1981). R. E. Dickerson, *J. Mol. Biol.*, **166**, 419–441 (1983).

8.1.4 Denaturation and Renaturation of DNA

Since deoxyribonucleic acids belong to the category of double-stranded polynucleotides with complementary sequences, the stability of their molecules is determined by the same factors and the denaturation process is given by the same parameters as in the simple duplexes considered above.

However, as opposed to most synthetic double-stranded complexes made up of homopolynucleotides or polynucleotides with regularly alternating nucleotide sequences, DNA molecules are extremely heterogeneous in nucleotide composition. In other words, any natural DNA has segments more abundant either in G · C or A · T pairs. Therefore, the "helix → coil" transition in DNA is in marked contrast with processes of the "all" or "none" type (true phase transitions). It is believed that defects arise in the double helix in the course of denaturation of DNA (e.g., thermal) at temperatures well below the average melting temperature (T_m). The A · T-rich segments of the molecule are denatured first. As the temperature rises, defects grow in size and increase in number. The G · C-rich segments are the last to go. Thus, if the temperature dependence of the fraction of denatured DNA is carefully measured, the resulting curve will feature sharp peaks corresponding to the partially melted structures (Fig. 8-6).

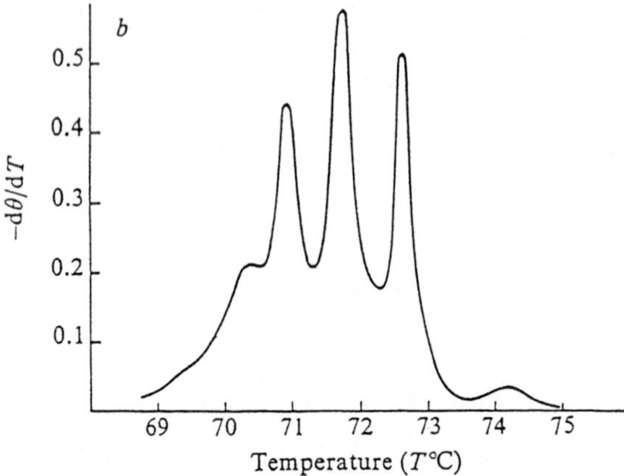

Fig. 8-6. Differential melting curve of the replicative form of ΦX174 DNA (courtesy of M. D. Frank-Kamenetskii).

The denaturation process ends in the strands coming fully apart. This can be demonstrated by direct methods, such as measurement of the molecular weight (in denatured DNA it is roughly half as high as in the native one), electron microscopy, separation of strands (if they differ in buoyant density) by ultracentrifugation in the cesium salt density gradient, and so on.

In the case of DNA, the above-mentioned linear relationship between T_m of helical double-stranded polynucleotides and the content of G · C pairs in their molecules has practical ramifications. Indeed, by experimentally determining T_m of DNA in a solution of a particular ionic strength one can find its overall nucleotide composition from a plot (e. g., that shown in Fig. 8-7).

Since T_m of helical double-stranded polynucleotides is also related linearly to the logarithm of salt concentration, the content of G · C pairs in DNA can be determined using a more general relation:

$$\text{content of G} \cdot \text{C pairs (\%)} = 2.44\,(T_m - 81.5 - 16.6 \log M),$$

where M is the molal concentration of a monovalent cation.

Alkaline denaturation is often used for full separation of DNA strands. To this end, the pH value of the DNA solution is brought to 12.5, and the latter is neutralized after a while.

Of particular interest is the process of DNA *renaturation,* whereby DNA molecules pass from the denatured into the original native state with more or less complete restoration of the secondary structure.

Renaturation of DNA calls for the same conditions as formation of helical complexes from synthetic complementary polynucleotides; that is, the temperature must be below T_m and the ionic strength of the solution must be sufficiently high. However, in this case, too, the renaturation of DNA proceeds in a much more complex manner by virtue of the heterogeneous nature of its polynucleotide chains.

If partially denatured DNA molecules (i. e., when polynucleotide strands have not yet been separated completely) undergo renaturation, if the strands are still held together after complete denaturation (i. e., by covalent cross-linking), or if we are dealing with covalently continuous cyclic DNA molecules, then the original secondary structure is restored quickly and completely.

But if renaturation takes place after complete separation of the strands, their recombination and restoration of the original structure are hampered by formation of intermediate structures through pairing of relatively short complementary segments in the same or different polynucleotide chains.

Therefore, the renaturation of DNA is conducted in such a way that the emerging spurious imperfect structures could fall apart again. A typical example is so-called renaturation by "annealing" when the DNA solution is very slowly cooled after denaturation.

It is clear that the greater the number of *repeating* complementary sequences in a DNA molecule and the longer they are, the lower the yield of rena-

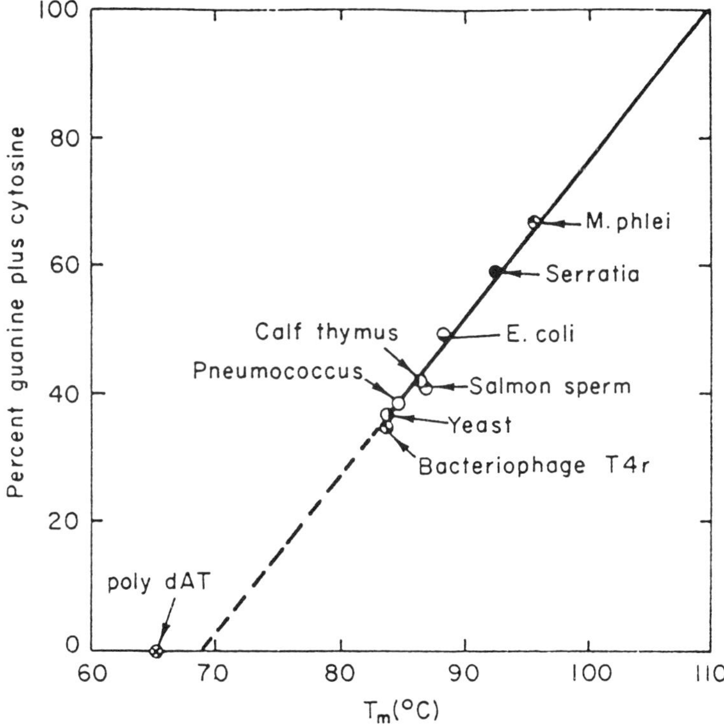

Fig. 8-7. Melting temperature of DNAs versus content of G · C pairs. The dots on the straight line correspond to different DNAs whose nucleotide composition has been determined by direct chemical methods. For a solution with ionic strength of 0.2, the straight line is given by the formula: content of G · C pairs = $(T_m - 69.3) \cdot 2.44$.

tured molecules will be. For instance, renaturation of animal DNAs in which such segments are numerous is slow and with a very low yield. On the contrary, in the case of bacteriophage DNAs practically free of long repeating sequences, renaturation proceeds with a high yield.

Renaturation of DNAs is used to solve quite a few biological problems. A case in point is experiments aimed at isolating so-called "individual" genes. Although such experiments are merely of historical interest, they provide a vivid illustration of the possibilities of the DNA denaturation-renaturation method. Very briefly, these experiments boil down to the following. Two totally different (usually viral) DNA molecules were selected. These DNAs differed in nucleotide sequence over the entire length with the exception of the segment encoding this common protein. Both types of DNA molecules were mixed, denatured, then placed under conditions optimal for their renaturation.

Among fully renatured molecules and those remaining in the denatured state there appeared molecules formed by strands of different DNAs, which had paired over a complementary segment (the latter being much longer than the duplex helical segments in imperfect intermediate structures).

The reaction mixture was then treated with DNase specifically hydrolysing single-stranded polynucleotides. Such treatment leaves fully renatured starting DNAs in the mixture along with double-stranded DNA fragments containing "individual" genes; both differ widely in molecular weight and can easily be separated from each other (Fig. 8-8).

It has become routine in molecular biology to study the kinetics of the process of renaturation (or reassociation) of DNAs as well as DNAs and complementary RNAs (the latter process is normally referred to as molecular hybridization). Such studies are usually conducted with a view to determining the degree of similarity or, in the case of RNA–DNA hybridization, complementarity of the nucleotide sequences of two nucleic acids. While studying the DNA reassociation kinetics one can also glean interesting information about the principles on which the genetic material is organized.

In addition to the usual parameters (ionic strength and temperature), the rate of reassociation of a denatured DNA is also dependent on its concentration and size. Therefore, DNA molecules are broken down in advance to fragments of a more or less the same size with a relatively small molecular weight. After denaturation, the DNA is placed under conditions optimal for renaturation (ionic strength of about 0.2; temperature 25 °C below the T_m of the native DNA in the same solvent). The degree of DNA reassociation at a given point in time is determined in different ways: by the hypochromic effect, by estimating the single-stranded DNA fraction hydrolysed with nucleases specific toward such DNA, and by chromatographic methods making it possible to separate the native (renatured) and denatured DNAs. The DNA–RNA hybridization process is normally monitored by separating the hybrids from the single-stranded RNA on membrane filters.

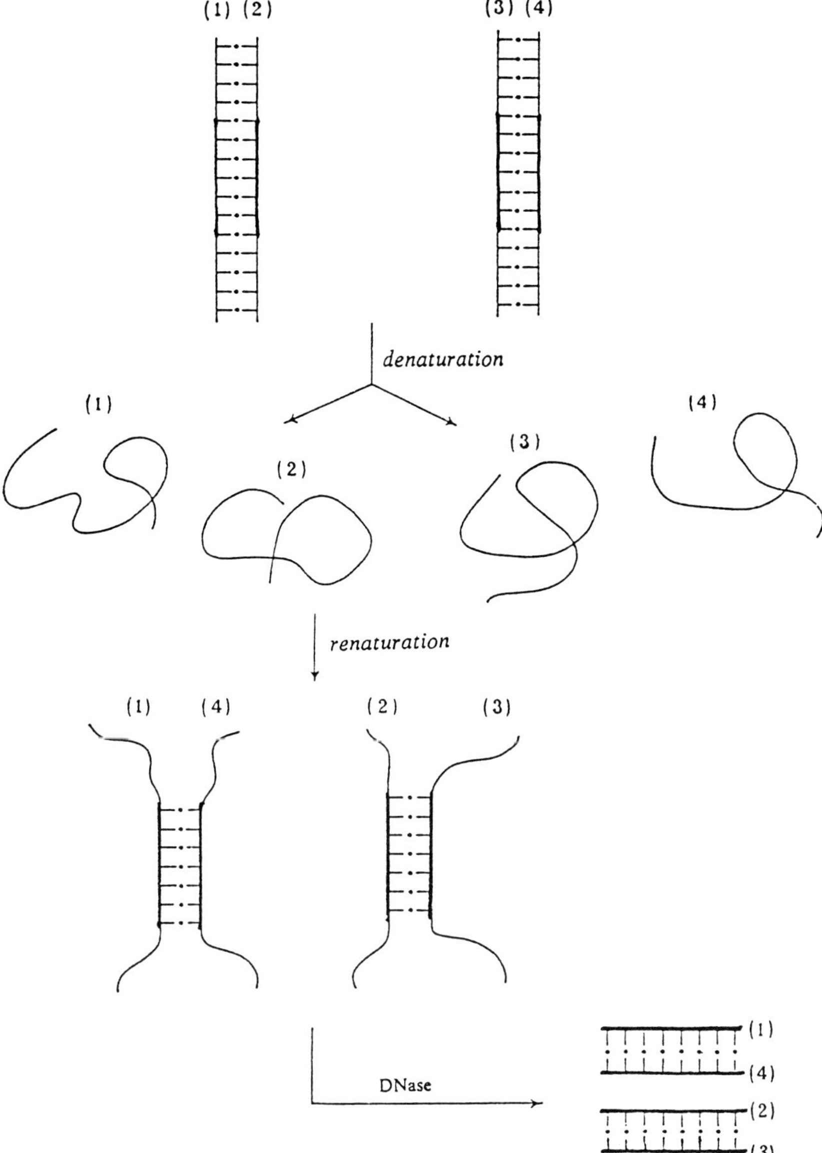

Fig. 8-8. Scheme illustrating how an individual DNA fragment is obtained by the denaturation-renaturation method.

Since the DNA reassociation process is a bimolecular reaction, the rate at which the amount of single-stranded DNA decreases is given by the equation

$$-\,\mathrm{d}C/\mathrm{d}t = kC^2, \tag{1}$$

where C is the single-stranded DNA concentration (mole of nucleotides per liter), t is time (s), and k is the second-order rate constant (l $\mathrm{mole}^{-1}\ \mathrm{s}^{-1}$).

Equation (1) can be easily converted into the following expression:

$$C/C_o = 1/(1 + kC_ot), \tag{2}$$

in which C_o is the total DNA concentration (mole of nucleotides per liter); at the initial moment of the reassociation, $C = C_o$.

As can be seen from this equation, the amount of denatured DNA not yet reassociated by instant t is a function of C_ot. Therefore, the process of DNA reassociation is usually given on a "degree of renaturation versus C_ot" plot (Fig. 8-9). The most important quantity on the latter is $C_ot_{1/2}$ at which the reassociation is 50% complete ($C/C_o = 1/2$) and which, as is shown by Eq. (2), equals the reciprocal of the reaction rate constant.

The shape of the reassociation curve depends on the number and size of the recurring nucleotide sequences in the nucleic acid molecule. In the case of viral and bacterial DNAs, the curve in its entirety fits the two logarithmic intervals C_ot (Fig. 8-10) because these DNAs are virtually devoid of long repeating sequences. The value of $C_ot_{1/2}$ for such nucleic acids is directly dependent on their size (and, vice versa, the size of such genomes may be determined from experimental values of $C_ot_{1/2}$), which is precisely the reason why the reassociation curves for different DNAs in Figure 8-10 are shifted

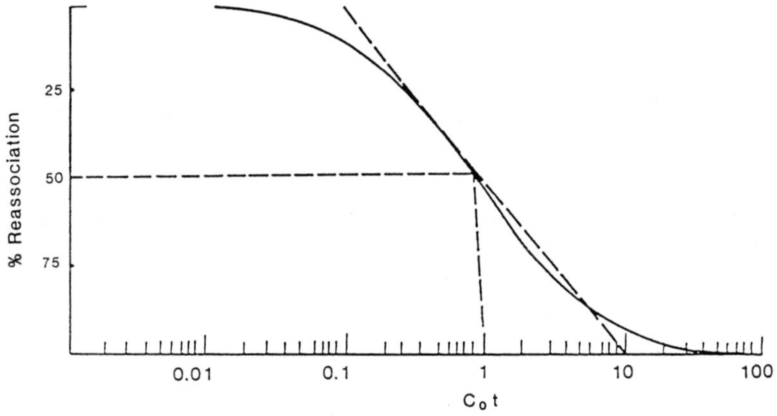

Fig. 8-9. Idealized reassociation curve for denatured DNA chains. Note that the reaction is 80% complete within the two logarithmic intervals C_ot.

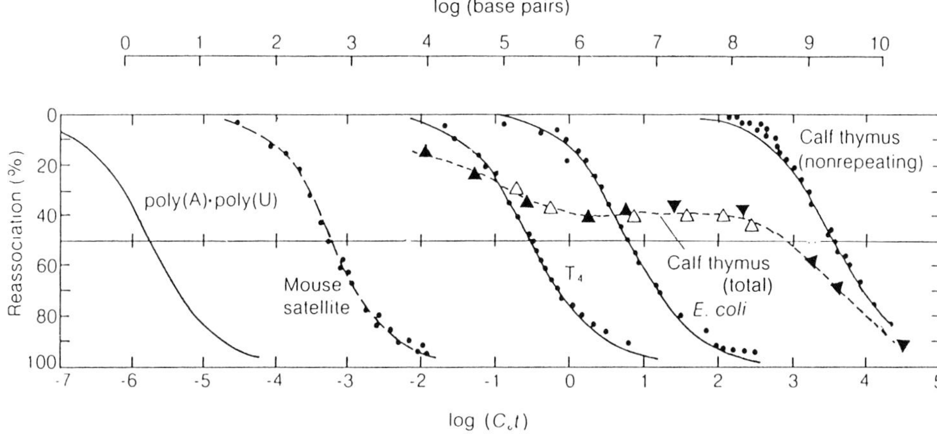

Fig. 8-10. Reassociation curves for denatured polynucleotides of different origin. Given on the upper abscissa is the number of complementary base pairs in the genome.

along the abscissa the way they are. This plot also shows an association curve for complementary polynucleotides that may be regarded as nucleic acids containing only one pair of nucleotides.

The reassociation curves for denatured DNAs of higher organisms indicate that their genome is composed of three main types of nucleotide sequences: those repeating frequently, those repeating at a moderate frequency, and those not repeating at all (unique sequences). Accordingly, one can discern three portions on the reassociation curves for such DNAs, each marked by its own value of $C_o t_{1/2}$.

Two reactions occur during DNA–RNA hybridization:

$$\text{DNA} + \text{DNA} \rightarrow \text{DNA} \cdot \text{DNA}$$
$$\text{DNA} + \text{RNA} \rightarrow \text{DNA} \cdot \text{RNA}$$

It can be easily demonstrated that when DNA is present in excess, the shape of the association curve for RNA with a denatured DNA is the same as that for the DNA–DNA association. Indeed, if the rate of the first reaction is determined from Eq. (1), the rate at which the content of single-stranded RNA decreases with concentration R is given by a similar equation:

$$-d\text{R}/dt = kRC \tag{3}$$

Substitution of the value of C derived from Eq. (2) into Eq. (3) gives

$$-\frac{dR}{dt} = kR\left(\frac{C_o}{1+kC_o t}\right) \tag{4}$$

or

$$\frac{dR}{R} = -k\left(\frac{C_o}{1+kC_o t}\right)\,\mathrm{dt} \tag{5}$$

At t = 0 and $R = R_o$, Eq. (5) integrates to the following expression:

$$\ln R/R_o = \ln\frac{1}{1+kC_o t} \tag{6}$$

or [cf. Eq. (2)]:

$$R/R_o = \frac{1}{1+kC_o t} \tag{7}$$

Therefore, one can determine, from the values of $C_o t_{1/2}$ for a given DNA–RNA hybridization process, the class of DNA sequences to which the RNA under analysis is complementary.

8.1.5 Some Aspects of DNA Behavior in Solution

The description of the hydrodynamic properties of DNA in solution constitutes a thorougly elaborated chapter of polymer chemistry. Here, we shall only dwell on some typical features of the tertiary structure of native DNAs determining their unusual (as compared to other biopolymers) behavior in aqueous solutions.

Native DNA molecules have a rigid secondary structure (DNA exists in the B form in acqueous solutions). Consequently, DNAs in solution cannot coil into dense structures of the Gaussian type. On the other hand, native DNA samples are characterized by an inordinate molecular weight. Depending on the source and isolation technique, it may vary from millions to billions. Molecules of that size cannot retain rigidity over the entire length (i. e., be in a "rigid rod" configuration). This is why double-stranded DNA molecules form extremely bulky (or, as they are often referred to, "rigid") coils in aqueous solutions. The "rigid coil" conformation occupies an intermediate position between those of the Gaussian coil and rigid rod.

Indeed, if we consider such classical relations between the basic hydrodynamic parameters (sedimentation constant S, intrinsic viscosity $[\eta]$, and radius R_g of gyration) and molecular weight M of the polymer as

$$S = kM^{\alpha_S}$$
$$[\eta] = kM^{\alpha_\eta}$$
$$R_g = kM^{\alpha_R}$$

then, in the case of Gaussian coils, α_S, α_η and α_R are almost equal to 0.5. For molecules in a rigid rod conformation, these quantities are 0.2, 1.8 and 1.0, respectively. For native DNAs (with a molecular weight ranging from 3 to $130 \cdot 10^6$), α_S, α_η and α_R were found equal to 0.4, 0.7 and 0.67.

The denaturation of DNA and breakdown of the rigid ordered structure are accompanied by drastic changes in its hydrodynamic properties because the separated polynucleotide chains have a compact tertiary structure in solution, approaching that of a Gaussian coil.

The rigidity of the secondary structure and high molecular weight account for another important property of DNA. In an aqueous solution, DNA molecules "break" very easily even when the slightest hydrodynamic action is exerted. Therefore, isolation of intact native DNA molecules with a molecular weight in excess of 20 million is a challenging experimental task and calls for special precautions.

8.1.6 Supercoiling of DNA

In the early sixties, J. Vinograd and coworkers found that the DNA of some bacteriophages and mitochondria may exist in the form of cyclic molecules. As was established later, most viral and many cell DNAs have a circular form. When both strands in a circular molecule formed by a duplex DNA are covalently closed (a similar situation arises when the ends of loops formed by a duplex DNA are linked by proteins), they are topologically constrained and can no longer be separated.

The most important parameter of each DNA molecule covalently closed into a circular structure is the linking number. It is denoted Lk^o (or α) and equals, to a first approximation, the number of intersections of the two strands in a double-stranded ring (Fig. 8-11). Thus, Lk^o has integer values. The most important point is that Lk^o is invariant for a given covalently closed circular DNA.

If such a DNA is in a *relaxed state* – that is, it can lie in one plane without any constrain – then $Lk^o = Tw$, where the winding number Tw (or β^o) equals the number of turns in the double helix of the DNA.

However, if the number of turns in the double helix is increased or decreased without affecting the covalent continuity of both strands, then, by virtue of the constant nature of Lk^o, such a change will be compensated by forma-

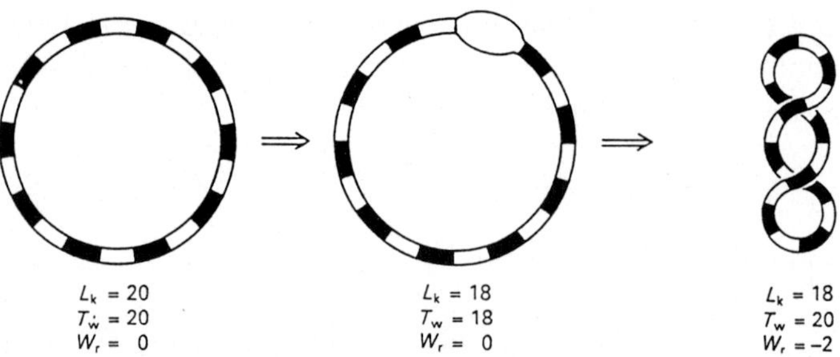

Fig. 8-11. Introduction of two supercoils into DNA with twenty duplex turns (adapted from C. R. Cantor and P. R. Schimmel, *Biophysical Chemistry*, part III, W. H. Freeman and Co., San Francisco, 1980).

tion of *supercoils* (Fig. 8-11). The number of supercoils in a superhelical DNA is denoted Wr (from "writhe") (or τ) and can be written as

$$\mathrm{Wr} = \mathrm{Lk}^{\mathrm{o}} - \mathrm{Tw}.$$

Wr can be a positive or a negative quantity. Covalently closed circular DNAs isolated from cells are negatively supercoiled as a rule. When Wr is negative, the linking number of the duplex DNA is lower than the number of double helix turns in a linear DNA of the same length or, in other words, the two strands in a negatively supercoiled DNA are "underwound" with respect to each other. The degree of supercoiling of a particular DNA can be best characterized by the density of supercoils (also called "specific linkage"), σ, which equals Wr/Tw. For most naturally occurring supercoiled DNAs σ is in the neighborhood of -0.05. It should be borne in mind, however, that in reality we always deal with populations of DNA molecules with different Lk^{o}. Therefore, σ is the mean density of supercoils for a given DNA preparation.

Supercoiled DNA can be observed by electron microscopy (Fig. 8-12). The number of supercoils in DNA is usually determined by gel electrophoresis. The separation is based on the fact that the greater the value of Wr, the more compact the DNA macromolecule and, consequently, the greater the electrophoretic mobility in the gel (see, for example, Fig. 8-14 in section 8.1.7).

From the functional standpoint it is important that supercoiled DNA is higher in free energy than the relaxed form. Hence, local unwinding of a double helix of DNA with negative supercoils will lead to a release of the stress due to supercoiling and, therefore, it is energetically advantageous. Its clear manifestation can be seen in the fact that negative supercoiling strongly stimulates the transition from the right-handed B form of DNA into left-handed Z-DNA. Indeed, already under normal physiological conditions segments with sequences $d(CG)_n \cdot d(CG)_n$ and $d(AC)_n \cdot d(GT)_n$ incorporated into DNA with $\sigma = -0.06$ are transformed into a left-handed Z form.

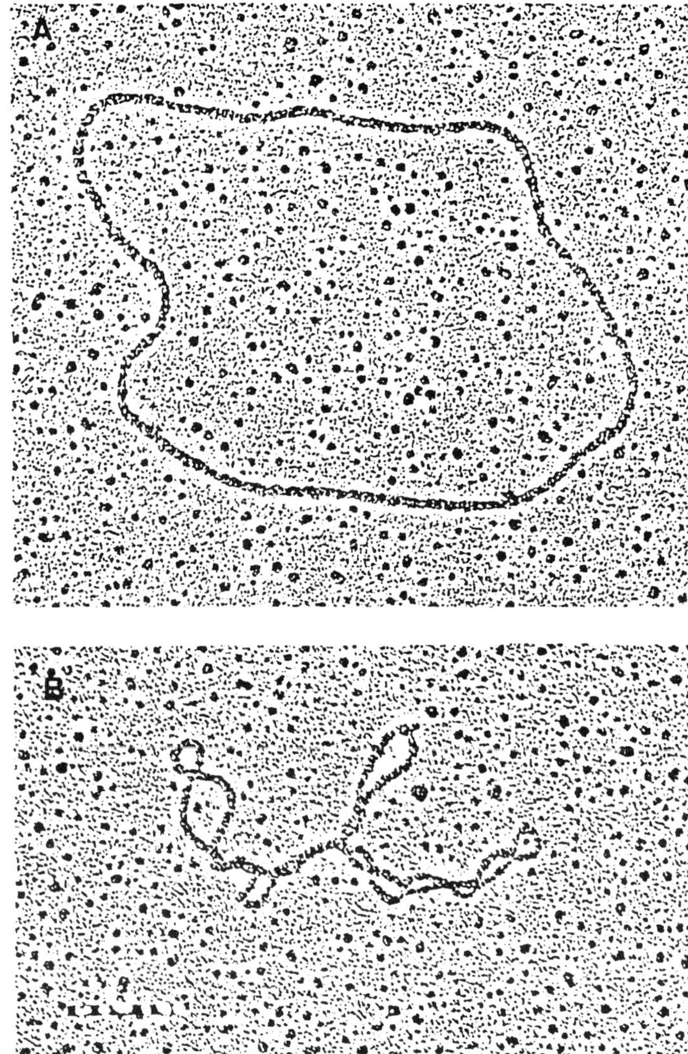

Fig. 8-12. Electron micrographs of (A) relaxed and (B) supercoiled circular DNAs (courtesy of G. G. Gause).

In the cell, the negative supercoiling of DNA is provided by DNA gyrases which belong to the category of topoisomerases. Topoisomerases are enzymes capable of altering the number of supercoils in covalently closed circular DNAs. They are subdivided into two groups: topoisomerases I and II. The substrate for topoisomerase I is a supercoiled DNA in which they reduce the number of supercoils and eventually turn it into relaxed circular DNA. Topoisomerases II (DNA gyrases), on the contrary, can create DNA supercoils.

When this occurs, every single step of DNA gyrase action results in two negative supercoils and its activity is associated with hydrolysis of ATP. Topoisomerases are present in every living cell and play a vital role in the processes of DNA replication and recombination.

8.1.7 Unusual DNA Structures

Another example of the influence exerted by supercoiling on the structural changes in a double helix of DNA is provided by formation of *cruciform* structures. Virtually every DNA contains inverted, or palindromic, repeating sequences varying in length from several to many thousands of base pairs.

5' AAAGTCCTAGCAATCCAAATGGGATTGCTAGGACCAA 3'
3' TTTCAGGATCGTTAGGTTTACCCTAACGATCCTGGTT 5'

A

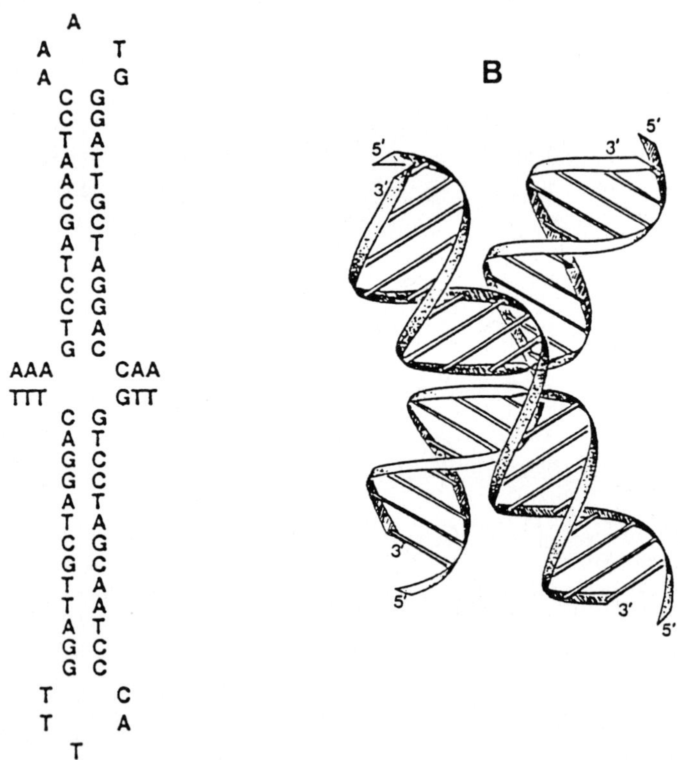

Fig. 8-13. (A) Linear and cruciform structures of the inverted repeat sequence of Col E1 DNA. (B) 3D model of the cruciform structure (adapted from A. I. H. Murchie and D. M. J. Lilley, *Methods in Enzymol.*, **211**, 158–180 (1992)).

Theoretically, one can picture transformation of a linear duplex palindrome to a cross (Fig. 8-13). In the case of a relaxed DNA, the probability of such a transformation is negligible. Since it is energetically advantageous in DNA with negative supercoils, cruciforms *in vitro* have been observed in all examined supercoiled DNAs with a normal density of supercoils. (Experimentally,

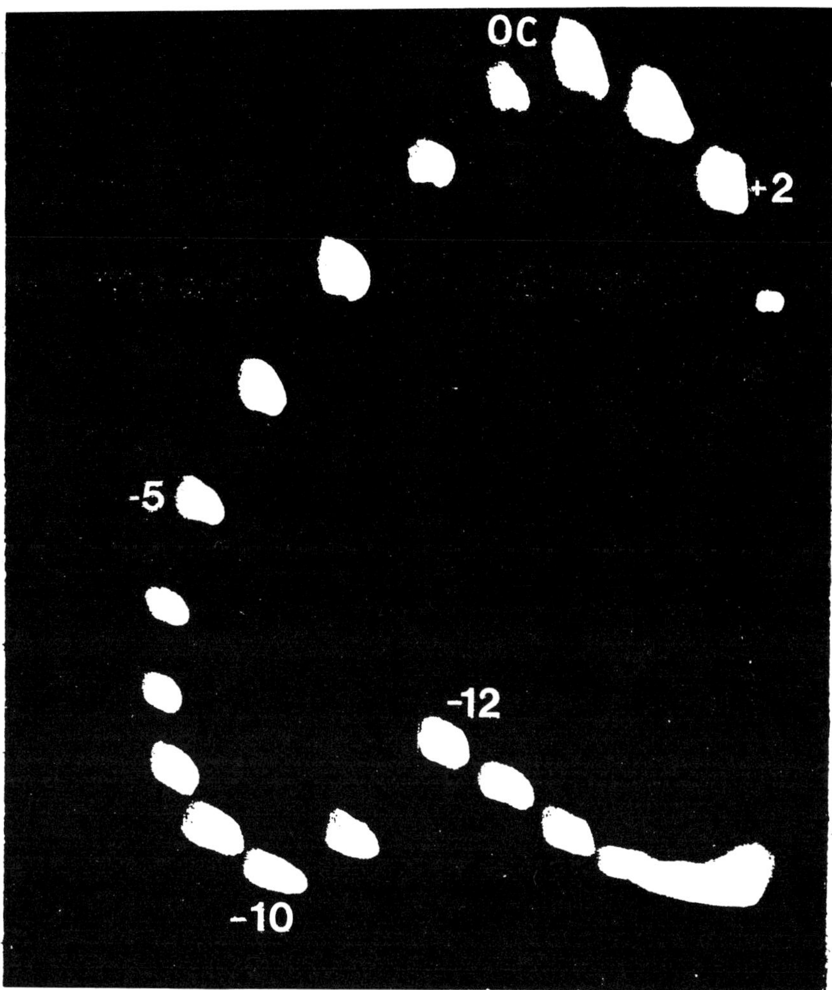

Fig. 8-14. Separation of DNA topoisomers by two-dimensional gel electrophoresis. After separation in the first direction (from top to bottom along the left edge), the gel was saturated with an intercalator (see 8.1.8) which reduced the number of supercoils in the DNA. Then, electrophoresis was conducted in the second direction (from left to right). The upper spot (oc) corresponds to relaxed DNA. Positively supercoiled topoisomers move clockwise; negatively supercoiled topoisomers move counterclockwise. The difference in mobility between topoisomers –10 and –12 means that topoisomers –12, –13, and so on have a cruciform structure (courtesy of M. D. Frank-Kamenetskii).

cruciform structures are identified as having single-stranded loops in the apices of "hairpins" which are cleaved by nucleases specific with respect to single-stranded DNA.) The formation of cruciforms can also be demonstrated by gel electrophoresis (Fig. 8-14). The question as to occurrence of cruciform DNA structures *in vivo* remains open. The rate of their formation is extremely low, which is possibly why nobody has been able to observe it in the cell.

If DNA contains homopyrimidine-homopurine sequences, negative supercoiling may turn it into a form illustrated in Figure 8-15. Since in the case of sequences of the $d(AG)_n \cdot d(CT)_n$ type such a transition occurred at low pH values (and at pH 4.3 it is observed even at $\sigma = 0$), it was termed the H form. The H form has been given as the reason for existence of segments in natural DNAs overly sensitive with respect to nucleases specific toward single-stranded polynucleotides.

Unusual structures of DNA which do not require supercoiling to arise by virtue of the ability of guanine bases to form tetrades. They are responsible for the structural organization of telomers – special segments created at the ends of linear chromosomes by the telomerase enzymes. Telomers consist of repeating sequences which contain blocks of three or four guanines. G-rich sequences associate with one another to form four-stranded structures known

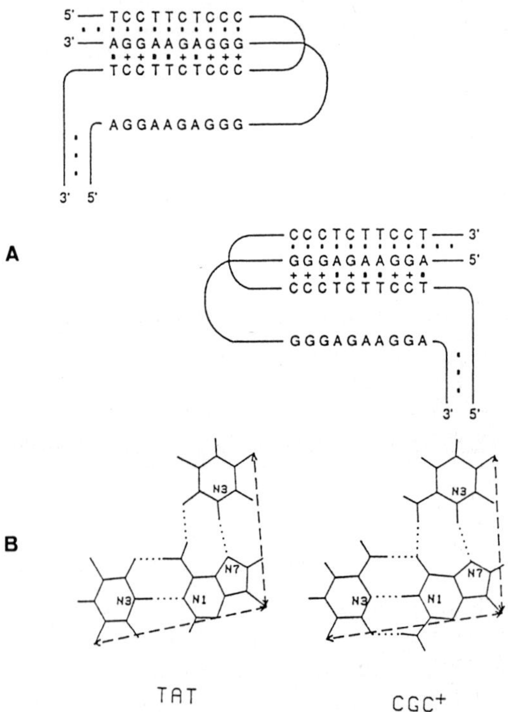

Fig. 8-15. (A) Two isoforms of H-DNA. (B) Scheme of base triads (adapted from M. D. Frank-Kamenetskii, *Methods in Enzymol.*, **211**, 180–191 (1992)).

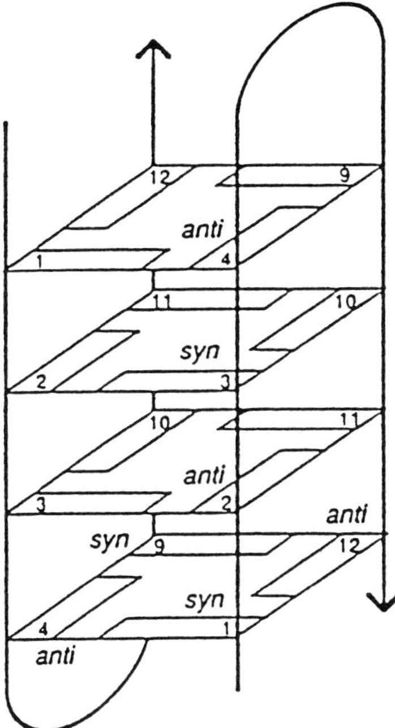

Fig. 8-16. Structure of G-DNA formed with $(dG)_n$ stretches within single-stranded DNA segments (adapted from J. R. Williamson, *Curr. Opinion in Stuct. Biol.*, **3**, 357–362 (1993)).

as G-DNA. Among the many possible G-DNA structures we selected as an example the antiparallel dimers shown in Figure 8-16, which contain guanosines in both *syn-* and *anti*-conformation. It can be seen that at first single-stranded G-rich segments fold back and then the two hairpin loops associate into G-DNA.

Unusual DNA structures were studied using a great variety of techniques. Chapter 9 will present some evidence of formation of such DNAs through chemical modification.

8.1.8 Interaction of Ligands with Double Helices of DNA

DNA forms complexes with proteins both *in vivo* and *in vitro* which affect its function. Moreover, numerous low-molecular weight substances, primarily antibiotics and carcinogens, interact with DNA and materially affect its functional properties. Many of such substances are mutagens. With few exceptions, proteins as well as low-molecular weight ligands are bound to DNA in a noncovalent fashion.

The ligands interacting with DNA can be divided into two groups according to the manner in which they are bound to the double helix – *groove binders* and *intercalators.* The first group includes a great variety of proteins, primarily those regulating transcription processes. Some low-molecular weight substances containing positively charged groups are also bound to grooves of the

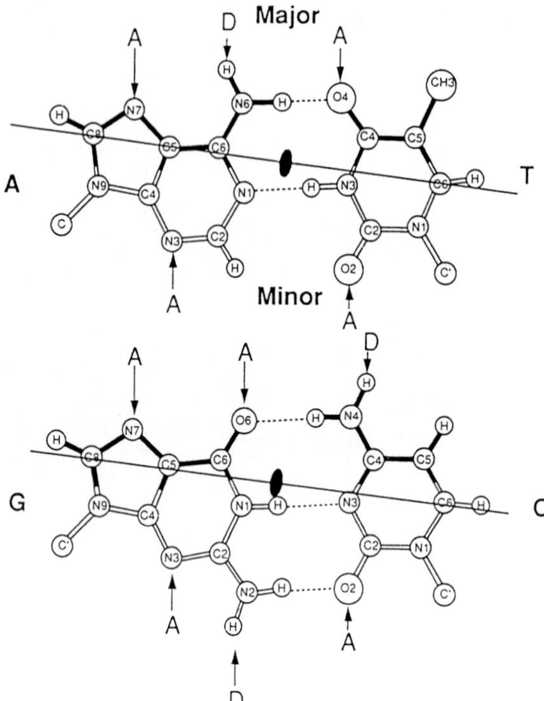

Fig. 8-17. Donors (D) and acceptors (A) of hydrogen bonds in grooves of the double helix of DNA (adapted from T. A. Steitz, *Quart. Rev. Biophys,* **23**, 205–280, (1990)).

double helix. A highly distinctive feature of the second group of ligands is their having polycyclic aromatic groups with various side chains.

In view of the great functional importance of DNA–ligand interactions it would be appropriate here to discuss their mechanism in greater detail.

Groove Binders. A remarkable feature of this group of ligands is their ability to recognize nucleotide sequences in DNA without unwinding its double helix. This recognition is possible due to formation of specific hydrogen bonds between the functional groups of the ligands and those of the heterocyclic bases "looking" into grooves of the double helix.

Analysis of Figure 8-17 which illustrates Watson–Crick pairs easily brings to light a marked difference between the patterns of proton-donor and -acceptor sites in the minor and major grooves of the double helix of B-DNA. It can be seen that the protein in the major groove (or any other groove binder) is capable of readily recognizing any of the four base pairs (A·T, T·A, G·C and C·G). On the other hand, the ligand in the minor groove can discriminate base pairs only by composition (i. e., G- and C-containing ones from those containing A and T). Therefore, the recognition of the nucleotide sequence in DNA by regulatory proteins takes place in the major groove of the double helix.

Figure 8-18 offers an example of specific contacts formed by the amino acids of a regulatory protein with the heterocyclic bases in the major groove

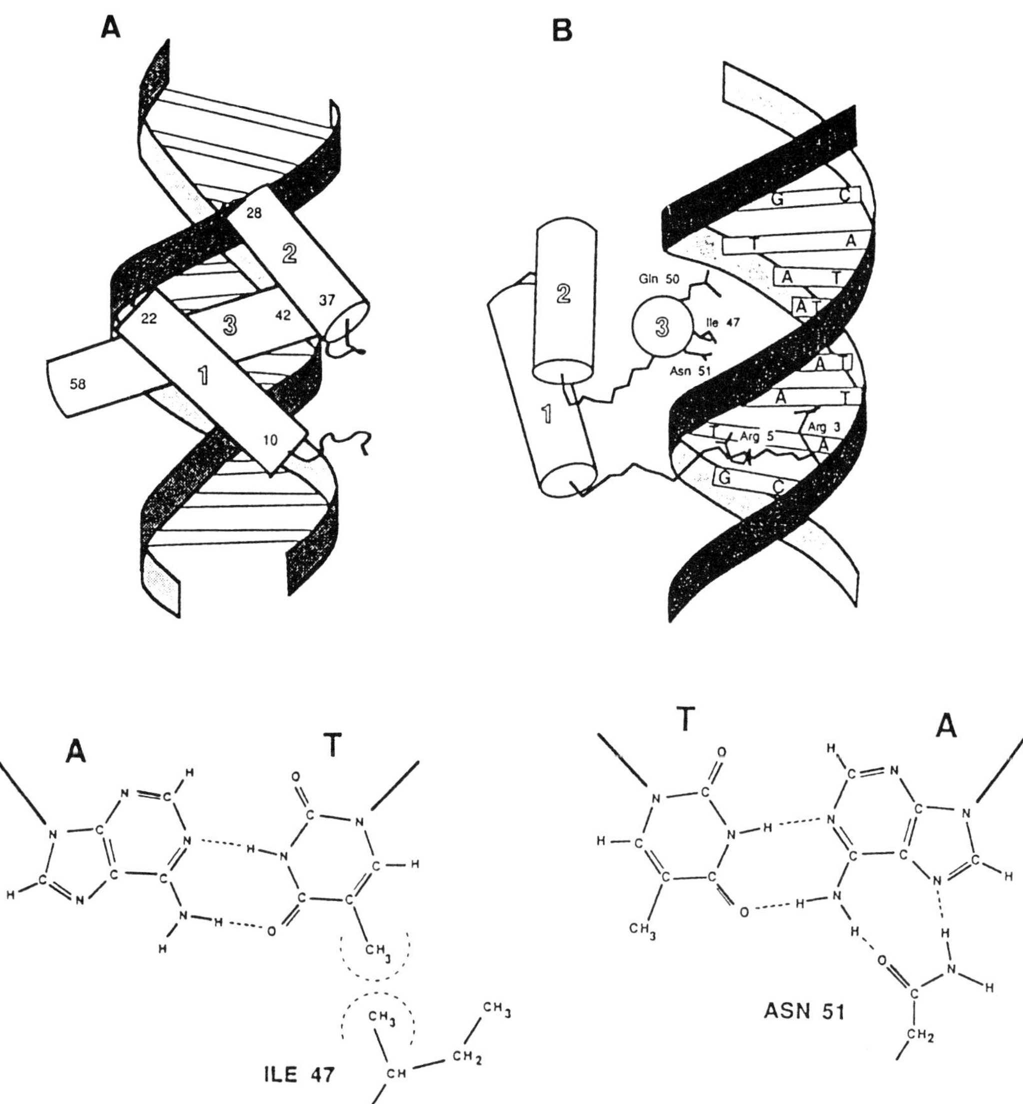

Fig. 8-18. DNA-protein complexes with a helix-turn-helix motif of so-called homeodomain in a *Drosophila* protein that regulates development. (A) and (B) – two projections of the complex showing the arrangement of helices (shown as cylinders) and the *N*-terminal arm with respect to the double-helical DNA; C – formation of hydrogen bonds between T·A pair and Asn-51 at hydrophobic contact between A·T pair and Ile 51 respectively (adapted from C. K. Kissinger et al., *Cell,* **63**, 579–590, (1990)).

of DNA. The protein shown here regulates the transcription process and features what is known as helix-turn-helix motif. One of the α helices of this feature fits into the major groove of DNA. Note that the proteins also interact with the minor groove, which renders the complexes highly stable.

Another example of binding between a ligand and the minor groove of the double helix is provided by the complex of DNA with the antibiotic netropsin (Fig. 8-19). The peptide groups of the antibiotic form hydrogen bonds with the 2-keto group of thymine and N^3 of adenine. Therefore, the antibiotic becomes specifically bound to A–T-rich regions of DNA. The netropsin–DNA complex is also stabilized by electrostatic interactions between the positively charged guanidinium groups and negatively charged phosphate groups of DNA.

It would be wrong to assume that groove binders do not affect at all the initial conformation of the double helix of DNA. As the double helix of DNA interacts with regulatory proteins, it exhibits a rather important ability to bend (bendability) and even form kinks, which enhances its contacts with the protein molecule. Moreover, groove binders bring about local changes in the mutual arrangement of the bases in complementary pairs, and that results in, among other things, local changes in groove size. Since, as we have already pointed out, B-DNA displays microheterogeneity dependent on the nucleotide sequence, deformations of the double helix are sequence-dependent, too.

Intercalators. Intercalation implies incorporation of a polycyclic aromatic compound between two base pairs into DNA without affecting not only the covalent linkages in their polynucleotide chains but also the Watson–Crick pairing. As far back as 1961, L. Lerman proposed a model according to which intercalation of a planar polycyclic dye into the double helix of DNA

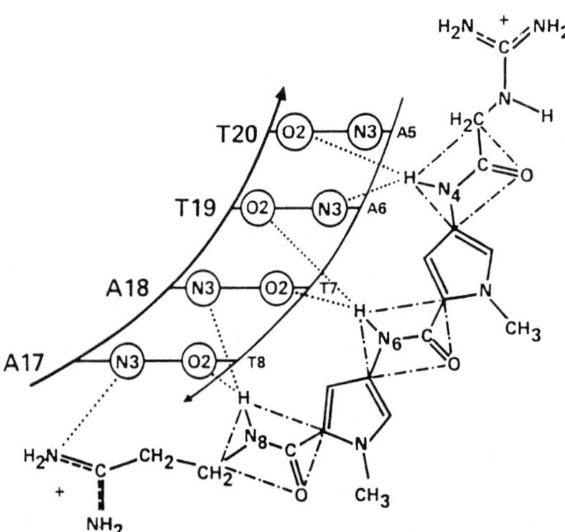

Fig. 8-19. Scheme of specific interactions between netropsin and A·T pairs in the minor groove of the double helix of DNA (adapted from M. L. Kopka et al., J. Mol. Biol., **183**, 553, (1985)).

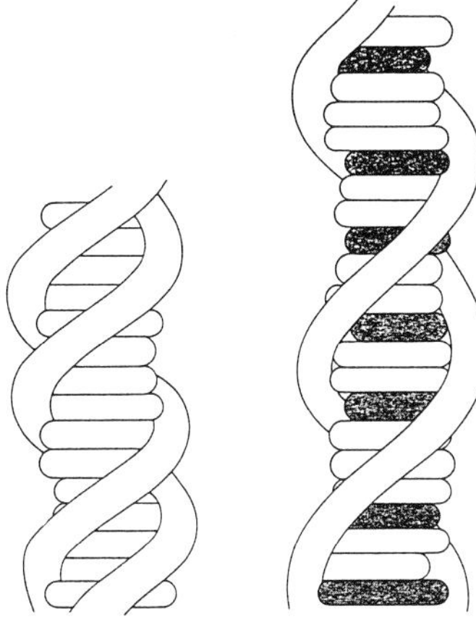

Fig. 8-20. Lerman's intercalation model: (A) native DNA; (B) complex of DNA with a polycyclic aromatic dye (black disks).

increases the distance between neighboring base pairs from 3.4 to 6.8 Å (Fig. 8-20). The model suggested that intercalation must be accompanied by a change in the angle of rotation of base pairs about the helix axis, which is to say that the helix is unwound. Besides, intercalation is, of course, accompanied by elongation of the helix. Later, the model was fully borne out by direct X-ray structural analysis of single crystals of complexes formed by oligonucleotide duplexes with various intercalators. One such complex is illustrated in Figure 8-21.

In the case of covalently linked cyclic DNA molecules, intercalation alters the number of superhelices in their supercoiled forms. For instance, intercalation of the dye ethidium bromide, widely used for this purpose, into the double helix of each DNA molecule unwinds it by 26°. As can be inferred from the relation between Wr, Lk^o and Tw (8.1.6), binding of ethidium bromide to a natural supercoiled DNA (for which, as is known, $Wr < 0$) will first lead to disappearance of the superhelices and emergence of a relaxed form, followed by coiling of the double helix in the opposite direction. Indeed, according to the results of sedimentation analysis (Fig. 8-22), as the intercalator concentration increases, the supercoiled DNA molecule becomes less and then more compact.

Some intercalators exhibit rather high selectivity of binding to DNA. For example, the antibiotic actinomycin D widely used as a transcription inhibitor binds exclusively to C·G sites of B-DNA. If an intercalator contains side substituents, they occupy the grooves of the double helix to form, as a rule, specific contacts with heterocyclic bases and, thereby, make the interaction between the intercalator and DNA even more specific. A similar complex is formed by DNA with the antibiotic daunomycin which contains a sugar moiety.

Finally, the ability to be incorporated into the double helix of DNA may be displayed by an intercalator together with the capacity to form covalent bonds

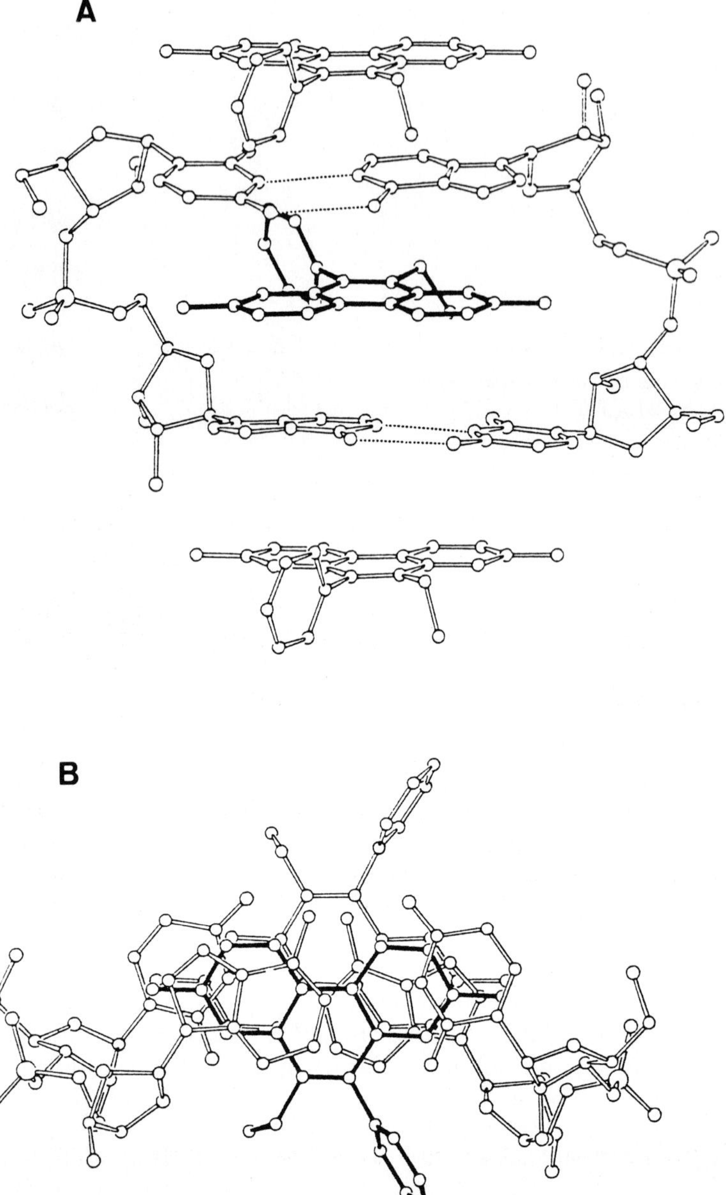

Fig. 8-21. Two projections of a ethidium-dinucleotide complex, drawn on the basis of X-ray analysis; (A) looking along the mean planes of the base pairs; (B) looking down on these planes (adapted from S. C. Jain and H. M. Sobell, *J. Biomol. Structure and Dynamics,* **1**, 1161–1194 (1984)).

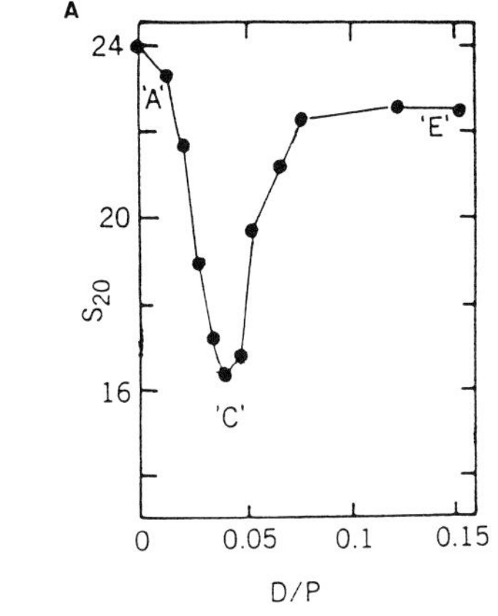

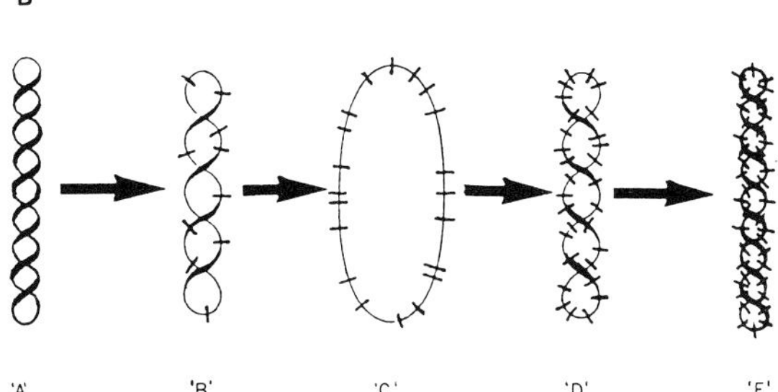

Fig. 8-22. (A) Titration of supercoiled DNA with ethidium bromide; (B) Interpretation of the results presented in (A). D/p = dye bound/nucleotide. 'A'–supercoiled DNA; 'B'–partially unwound DNA; 'C'–relaxed DNA; D'–partially overwound DNA; 'E'–overwound DNA (adapted from W. Guschelbauer, in: *Encyclopedia of Polymer Science and Engineering,* **12**, 699–785 (1988)).

with DNA. If an intercalator is a bifunctional cross-linking agent, it forms cross-links between the DNA strands. The most widely used for this purpose are both natural and synthetic psoralens – linear bifunctional tricyclic furocoumarines which photoreact with DNA to form cyclobutane adducts with pyrimidine bases.

8.2 RNA

8.2.1 Secondary Structure of RNA. Fresco–Alberts–Doty Model

As had been established by the late fifties, in contrast to DNA, macromolecules of the overwhelming majority of cellular and viral RNA are single-stranded. None the less, in solutions of sufficiently high ionic strength RNA exhibited pronounced Cotton effects as well as hyperchromicity in the UV region. This was unambiguous evidence that RNA has an ordered secondary structure.

At the same time, the melting curves of RNA differed markedly from those of DNA and other complementary double-stranded polynucleotides. The transition of RNA from ordered to denatured state occurred in a rather broad temperature range (Fig. 8-23). What is more, it was found that RNAs in an ordered state (at "physiological" temperatures and salt concentrations) have a compact tertiary structure approaching that of the Gaussian coil. This was construed as evidence of presence of flexible single-stranded segments in RNA molecules. Notably, denaturation of ordered components of the RNA macromolecule is also followed by breakdown of this compact structure.

In 1960 such experimental evidence lead J. Fresco, B. Alberts and P. Doty to propose a model of secondary structure of single-stranded RNAs (Fig. 8-24). According to the model, the secondary structure of a single-stranded RNA is a combination of double-helical segments (widely differing in length and composition) and interlinking single-stranded ones. Thus, it was postulated that double-helical RNA segments result from base pairing in complementary nucleotide sequences belonging to the same strand. Fresco and coworkers hypothesized that this pairing obeys the Watson and Crick rule or, in other words, canonical G·C and A·U pairs are formed. The polynucleotide chain segments forming double helices are antiparallel. At one end or, more precisely, where the polynucleotide chain of RNA folds back on itself, the paired regions are connected by single-stranded hairpin loops of different length. Fresco and coworkers assumed that the double-stranded segments of RNA are not perfect double helices but contain internal loops and all kinds of bulges.

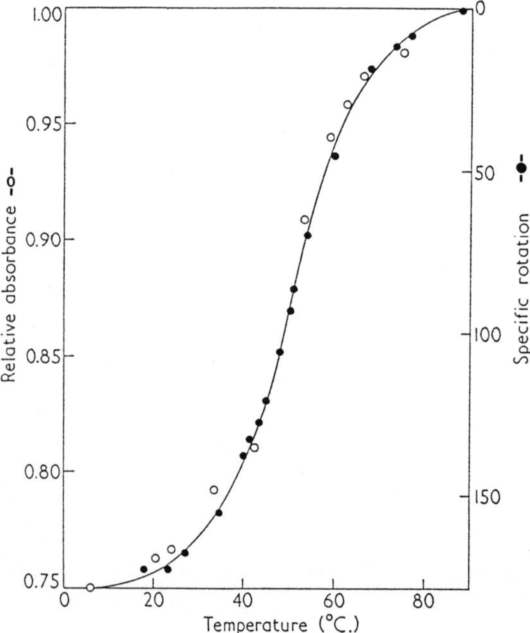

Fig. 8-23. Typical melting curve of single-stranded RNA (from tobacco mosaic virus) in 0.1 M phosphate buffer, pH 7.0 (adapted from P. Doty et al., *Proc. Natl. Acad. Sci. USA*, **45**, 482–499 (1959)).

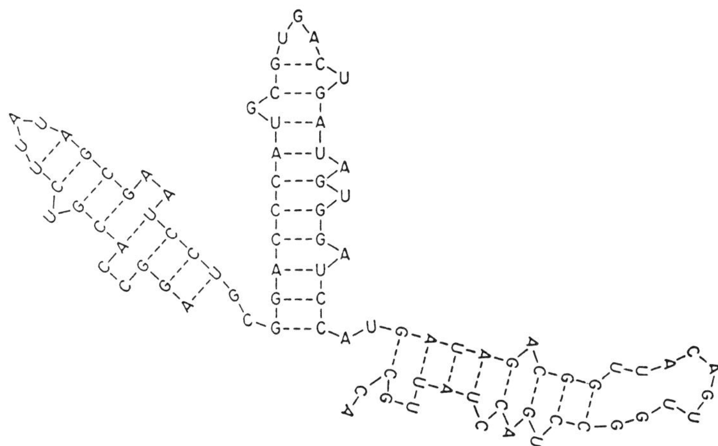

Fig. 8-24. Fresco-Alberts-Doty model of the secondary stucture of a single-stranded RNA (reproduced with permission from J. Fresco, B. Alberts and P. Doty, *Nature,* **188**, 98–101 (1960)).

All subsequent studies into the macromolecular structure of RNA supported the Fresco–Alberts–Doty model, and it may be considered as a universal description of the secondary structure of single-stranded RNAs. At the same time, the model was expanded by addition of significant details. It has turned out that in addition to classical Watson–Crick pairs double-stranded segments of RNA often contain G·U pairs whose structure was discussed above. Formation of other noncanonical pairs has also been assumed. Moreover, as was demonstrated for many RNAs, double-stranded segments may result from pairing of not only neighboring complementary segments but also sequences spaced rather far apart in the polynucleotide chain. Consequently, double-stranded segments of RNA form branches of different kinds. All these features of the secondary structure can be observed in fragment 23S of ribosomal RNA, illustrated in Figure 8-25.

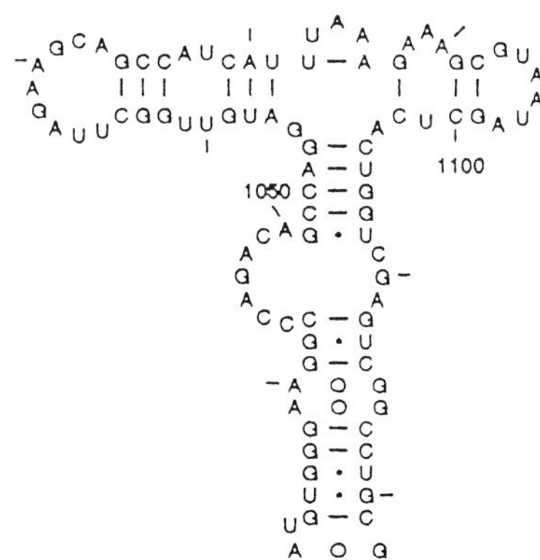

Fig. 8-25. Secondary structure of the segment of the *E. coli* 23S ribosomal RNA (adapted from R. R. Gutell et al., *Nucl. Acid Res.*, **20**, *Supplement,* 2095–2109, (1992)).

8.2.2 Elements of Secondary Structure

Let us now consider at greater length the properties of the structural elements constituting RNA macromolecules. Most of information on the subject is provided by the three-dimensional structure of transfer RNAs (tRNAs) as well as the results of studies into the conformation and thermodynamic stability of synthetic oligoribonucleotides which serve as models of a particular secondary structure motif of RNA.

Transfer RNAs were the first nucleic acids whose nucleotide sequences had been established. Moreover, several representatives of tRNA had been crystallized and investigated by high-resolution X-ray structural analysis. The results will be discussed in greater detail in 8.2.3. Here it should simply be pointed out that the secondary structure of any tRNA regardless of its nucleotide sequence is consistent with the "clover-leaf" model (Fig. 8-26). It can be seen that in essence the clover-leaf structure is quite compatible with the overall Fresco–Alberts–Doty model of RNA secondary structure.

Base-paired Regions. X-ray structural analysis of double helices of RNA has shown that they belong to the A family (i. e., are similar to A-DNA in many respects). Indeed, complementary base pairs in double-stranded segments of RNA lie on the periphery of the helix (Fig. 8-27) and the base planes are significantly (almost by 15°) deflected from the axis of the helix. Two forms of RNA helices are known: form A with 11 base pairs per turn and form A′ with 12 base pairs per turn. Form A may be converted into form A′ if the salt (NaCl or KCl) concentration in the sample goes up.

As has already been mentioned, a characteristic structural feature of the A family of double helices of nucleic acids is the C3′-*endo*-conformation of the sugar. Insofar as helical RNA is concerned, such a conformation of the ribose

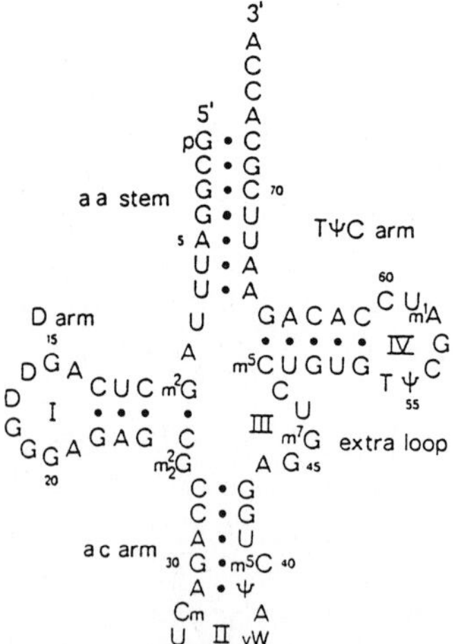

Fig. 8-26. Clover-leaf model of the secondary structure of yeast phenylalanine tRNA.

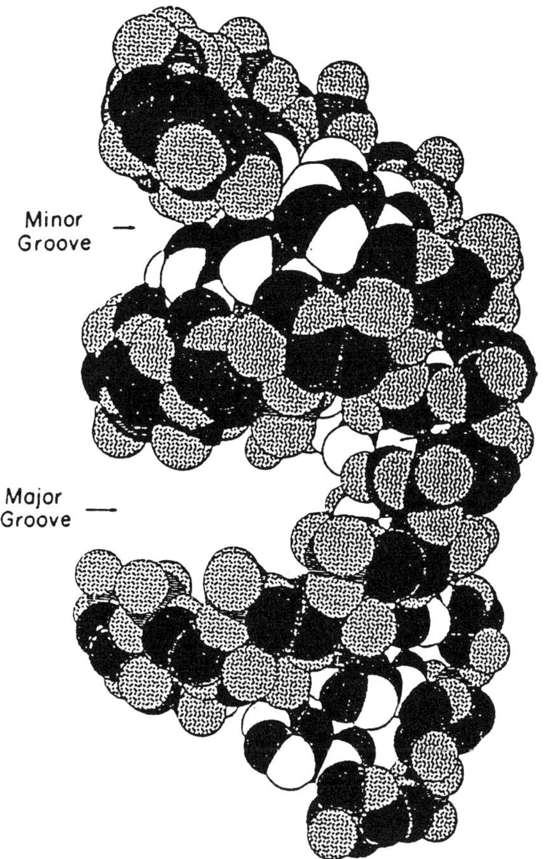

Fig. 8-27. Model of the RNA double helix structure (A-form) (adapted from M. Chastain and I. Tinoco, Jr. *Progr. Nucl. Acid Res. Mol. Biol.*, **41**, 131–177 (1991)).

seems to be the only possible one and, consequently, transition A → B for this class of nucleic acids is forbidden. The reason is that the ribose cannot change from C3'-*endo*- into C2'-*endo*-conformation because of the contacts between the hydroxyl group at $C^{2\prime}$ and other atoms (primarily oxygen in the phosphodiester bond) being too close.

The distance between phosphate groups in base-paired regions of RNA is 0.59 nm (5.9 Å). The grooves of the double helix in RNA differ considerably from those of B-DNA: its major groove is deep and narrow, while the minor groove is shallow and wide. As a result, in contrast to DNA, the main site of interaction with ligands in double-stranded RNA is its minor groove.

If the A → B transition for double helices of RNA is forbidden, the A → Z like transition is possible; it has been demonstrated that polyribonucleotide duplexes with a repeating GC sequence are in the form of a left-handed helix at a high concentration of salts. Just as in B-DNA and in contrast to A-DNA, a certain degree of microheterogeneity can be discerned in base-paired regions of RNA, which is dependent on the sequence of nucleotides. It manifests itself, for example, in the fact that the anticodon stem of tRNA (Fig. 8-26 and Section 8.2.3) is in the classical A form (11 base pairs per turn), whereas the stem of the D arm contains only about 10 base pairs per turn.

Hairpin Loops. Hairpin loops are elements invariably present in the secondary structure of any single-stranded RNA. In the RNAs whose three-dimensional structure has been studied in greater detail, their length varies from three to 15–17 base pairs. There is every reason to assert that the stability of hairpin loops is determining by stacking interactions and hydrogen bonding within the loops. However, there are very few cases where the secondary structure of hairpin loops has been definitively established. At the same time, it is extremely important because, on the one hand, many hairpin loops are involved in tertiary interactions and, on the other, they are the principal components of functional sites in most single-stranded RNAs.

Analysis of the thermodynamic stability of many synthetic hairpins has shown that the most stable structures are those with loops of four or five nucleotides. If a loop consists of only three nucleotides, their ribose moieties

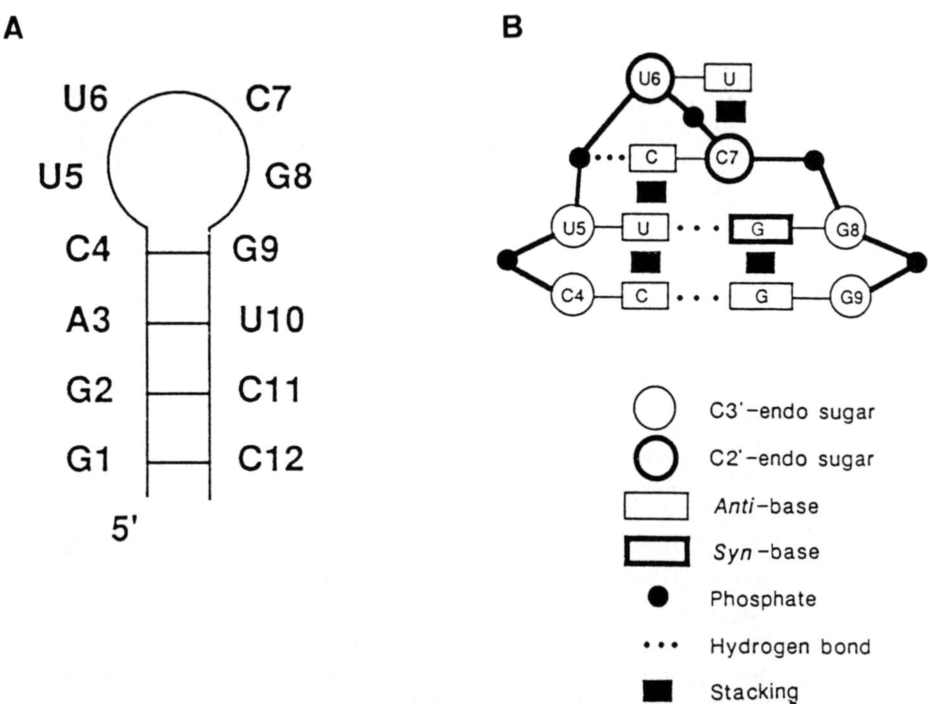

Fig. 8-28. Unusually stable RNA hairpin; (A) secondary structure of the hairpin; (B) schematic diagram of base-pair interactions in its loop (reproduced with permission from C. Cheong, G, Varani, and I. Tinoco, Jr., *Nature*, **346**, 680–682 (1990)).

must acquire the C2′-*endo*-conformation, which allows individual nucleotides to be spaced as widely apart as possible. Interestingly, the stability of a hairpin as a whole is almost independent of the nucleotide composition of its loop if the conformation of the latter is ensured only by ordinary stacking. However, if more complex intramolecular interactions occur in the loop, its nucleotide sequence may affect structural stability as a whole in a most dramatic fashion.

A good case in point is the hairpin shown in Figure 8-28. It occurs in quite a few RNAs and is marked by high stability. At an ionic strength of about 0.1, its melting point (T_m) is 80.3 °C. If a C is replaced by a U in the loop, T_m will drop by 5 °C at once. Investigation of the three-dimensional structure of this hairpin by high-resolution NMR has shown that it has a rather unusual secondary structure which is illustrated in Figure 8-28. Since G8 acquires *syn*-conformation, it becomes capable of forming a G·U pair with U5 and being heavily involved in stacking with G9. Nucleotides U6 and C7 are in the C2′-*endo*-conformation with the result that the loop becomes more elongated and C7 can enter into stacking with U5 and form a hydrogen bond with the phosphate group. The latter possibility deserves special attention because we have not yet discussed here any example of formation of specific hydrogen bonds between bases and phosphates, and this, in combination with stacking, may be of great significance for compact packing of loops.

Particular attention should also be given to the anticodon loop of tRNA. It consists of seven nucleotides, and its most distinctive feature is the sharp turn the chain takes between the second and third nucleotides from the 5′ end. The other five nucleotides form strong stacking contacts with one another, and the overall conformation of this portion of the loop seems to continue the helical structure of the anticodon stem (Fig. 8-29). The overall conformation of the loop, including the bend, is also stabilized by the intra-loop hydrogen bonds whose formation involves the 2′-OH groups of ribose.

Bulges. Similarly to loops, the bulges forming over double-stranded segments of RNA have been attracting attention as potentially functional sites. This is true, primarily, for single-base bulges containing adenosines whose participation in the RNA-protein interaction was proved on several occasions and which are present in the secondary structure of many RNAs. As was already pointed out in Chapter 7, the extra single nucleotide either protrudes from the double helix or intercalate in it, being involved in stacking with the paired bases. This leads to a certain degree of destabilization of the double helix, which is dependent on the nature of the bulge nucleotide and the composition of the neighboring base pairs. It has been demonstrated that the destabilizing action in the case of bulge purine is more significant than in the case of bulge pyrimidine.

Very little is known about the conformation of larger bulges. There is reason to believe that they destabilize double-stranded segments of RNA to a greater degree than single-base bulges. As regards much larger bulges, their

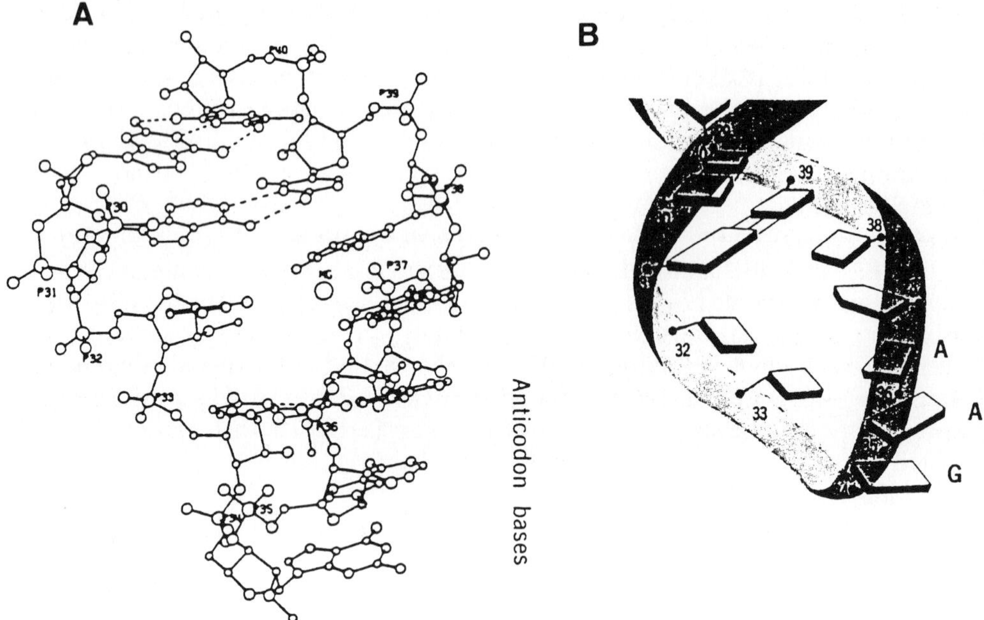

Fig. 8-29. (A) 3D structure and (B) schematic diagram of three-dimensional structure of the anticodon loop of yeast phenyl-alanine tRNA (adapted from S.-H. Kim and J. L. Sussman, *Nature*, **260**, 645–646 (1976)).

conformation is likely to be stabilized by intra-bulge interactions typical of hairpin loops.

Internal Loops. These motifs arise whenever bases that are not complementary in the Watson–Crick sense find themselves opposite each other in a double-stranded segment of RNA. Therefore, the stability of an internal loop will depend on whether the bases in the opposite strands become paired as well as on the degree to which such pairing disturbs the geometry of the helix. The least disturbance in the double helix is caused by formation of a G·U pair. It is thermodynamically more advantageous than formation of any other noncanonical pair. What is more, X-ray structural analysis of short oligonucleotide duplexes with G·U pairs has shown that they do not cause any perceptible disturbances in the geometry and stacking of other pairs in the duplex. This is precisely why, if G and U are in complementary strands of RNA opposite each other, they are represented in a base-paired form. It should be remembered, however, that incorporation of a G·U pair always destabilizes the double helix of RNA and its replacement by an A·U pair raises the T_m of this RNA region by several degrees.

It has been speculated that other bases may also be paired in internal loops, for example, A with G or another A, a protonated A with C, or U with another U. The formation of such pairs is usually postulated proceeding from the assumption that their bases do not lend themselves to modification with chemical agents. However, the screening of bases may also stem from the fact that a particular internal loop is involved in tertiary interactions.

Junctions. If we take portions of the RNA secondary structure where branches are linked together, the most interesting and not yet discussed type of interaction is coaxial stacking of double-helical segments. It occurs by virtue of the fact that the base pairs at the ends of the double-stranded segments are in contact with each other, become involved in stacking, and one helix thus becomes a continuation of the other. The most vivid and yet rigorously proved example of such junction is provided by paired coaxial interactions between helical portions of tRNA. Similar junctions are believed to exist in many other RNAs as well.

8.2.3 Macromolecular Structure of Transfer RNAs

On several occasions we already mentioned the "clover-leaf" model and its constituent secondary structure elements. It should be noted, however, that after the nucleotide sequence of the first tRNAs had been established, this model ceased to be as obvious as it was purported to be and serious efforts had to be made to corroborate it. Indeed, analysis of the nucleotide sequences of tRNA available at that time had led to three hypothetical models of secondary structure: (a) a long hairpin with defects; (b) several (two or three) separate short hairpins; and (c) several (three or four) short hairpins forming a "clover leaf" (Fig. 8-30).

These three hypothetical models were subjected to thorough verification by various chemical and physical methods. It has been found that the "clover-leaf" model is most consistent with the experimental results.

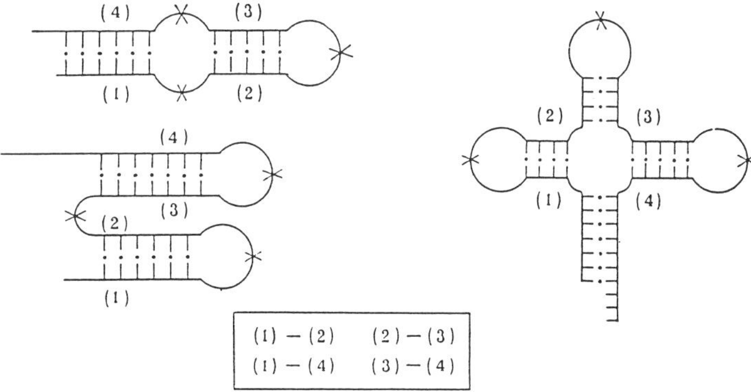

Fig. 8-30. Three possible ways of organization of the secondary structure of tRNA and proof of validity of the clover-leaf model.

Firstly, the "clover-leaf" model fits well the physico-chemical data on the degree of coiling (number of paired bases) and the length of helical segments in different total and individual tRNAs.

Secondly, as indicated by theoretical calculations, the "clover-leaf" model for most tRNAs with a known primary structure is a rather advantageous type of secondary structure in thermodynamic terms (i. e., corresponds to the minimum of free energy).

Thirdly, strong arguments in favor of this model were supplied by analysis of oligo- and polynucleotide fragments resulting from hydrolysis of tRNAs with specific RNases. It is known that most RNases hydrolyse primarily the phosphodiester bonds in single-stranded segments of RNA. Thus, partial enzymatic digestion of tRNAs makes it possible to obtain large fragments retaining their capacity for reassociation and restoration of the double-stranded segments that existed in the starting tRNA molecule. Figure 8-30 shows schematically that the pattern in which quarters of tRNA can be reassociated is consistent with only one of the three possible models of secondary tRNA structure, namely, the "clover-leaf" model.

It should be remembered that in accordance with the functions performed by particular tRNA regions or their exact nucleotides distinction is made, in the tRNA molecule, between single-stranded anticodon loop, dihydrouridyl loop (D-loop), extra loop, TψC-loop, and acceptor end (see Figure 8.26).

Numerous experiments have shown that most bases in single-stranded portions of tRNA, inferred from a "clover-leaf"-type model, are not amenable to chemical modification. A case in point is bases in the TψC-loop.

The different accessibility of nucleotide sequences in single-stranded segments of tRNA can also be demonstrated by the method of complementary

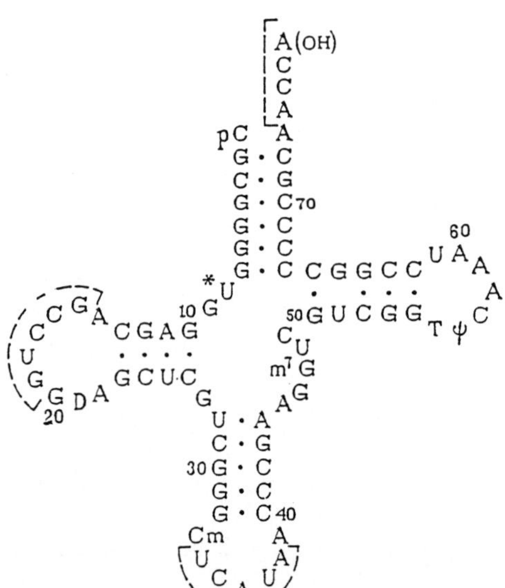

Fig. 8-31. Probing of three-dimensional structure of *E. coli* $tRNA^{Met}$ by association with complementary oligonucleotides. Sites available for binding of oligonucleotides are shown by broken lines.

oligonucleotide binding, which is quite popular in studying the macromolecular structure of nucleic acids. The method involves the following procedure.

Having a set of short oligonucleotides (most commonly, penta- and hexanucleotides) complementary with respect to the single-stranded segments predicted by the secondary structure of a particular RNA, one can measure the association constant (K_{ass}) of these nucleotides with the RNA. The measurements are conducted under conditions optimal for formation of double-stranded complexes. Surface portions of the macromolecule will be characterized by maximum value of K_{ass} for oligonucleotides. In this case, the nucleotide composition of the probed segments is taken into account, of course, in view of the fact that the presence of G·C pairs in the emerging oligonucleotide–RNA complex may bring K_{ass} up by several orders of magnitude. The results produced by thus method agree well with those of chemical modification of tRNA. Moreover, they provide information on accessibility not only of individual nucleotides but also some RNA segments (Fig. 8-31).

Studies of tRNA macromolecules by such methods have led the investigators to a firm conclusion that the separate loops of the "clover-leaf" form a specific tertiary structure common to all tRNAs.

In 1973–75, the tertiary structure of tRNA was solved in brilliant experiments conducted by A. Rich's and A. Klug's groups who investigated, by X-ray structural analysis [with a resolution of up to 0.25 nm (2.5 Å)], two crystalline forms of yeast phenylalanine tRNA ($tRNA^{Phe}$) and arrived at virtually identical tertiary structure models for this tRNA (Figs. 8-32 und 8-33).

According to the Rich–Klug model, the macromolecule of tRNA has a simplified L shape, its major functional sites, anticodon loop and acceptor end, terminating the L at both limbs. The distance between the two is 7.6 nm (76 Å). The double-stranded segments adjacent to the acceptor end and TψC-loop form a single double helix. This helix forms almost a right angle with the double-stranded segment of the D-loop and the following double-stranded part of the anti-codon "hairpin". The two main linear portions of the macromolecule are held together by complex intertwining of the D- and TψC-loops. All double-stranded segments of the molecule belong to the A form (i. e., their riboses are in 3′-*endo*-conformation). At the same time, some nucleotides have been found to contain a ribose moiety in 2′-*endo*-conformation in single-stranded segments.

By far the most important result of X-ray structural analysis of tRNAs seems to be evaluation of the specific internucleotide interactions responsible for the tertiary structure of their macromolecules. The main types of intermolecular interactions have turned out to be the same as in formation of the classical secondary structure of nucleic acids – that is, stacking and complementary interaction of heterocyclic bases. However, base pairing and the associated hydrogen bonding in this case involve donor and acceptor groups other than those participating in Watson–Crick pairing. Consider a few examples.

Figures 8-32 and 8-33 illustrate nucleotides occupying different parts of the "clover-leaf" but coming close together and interacting during formation of

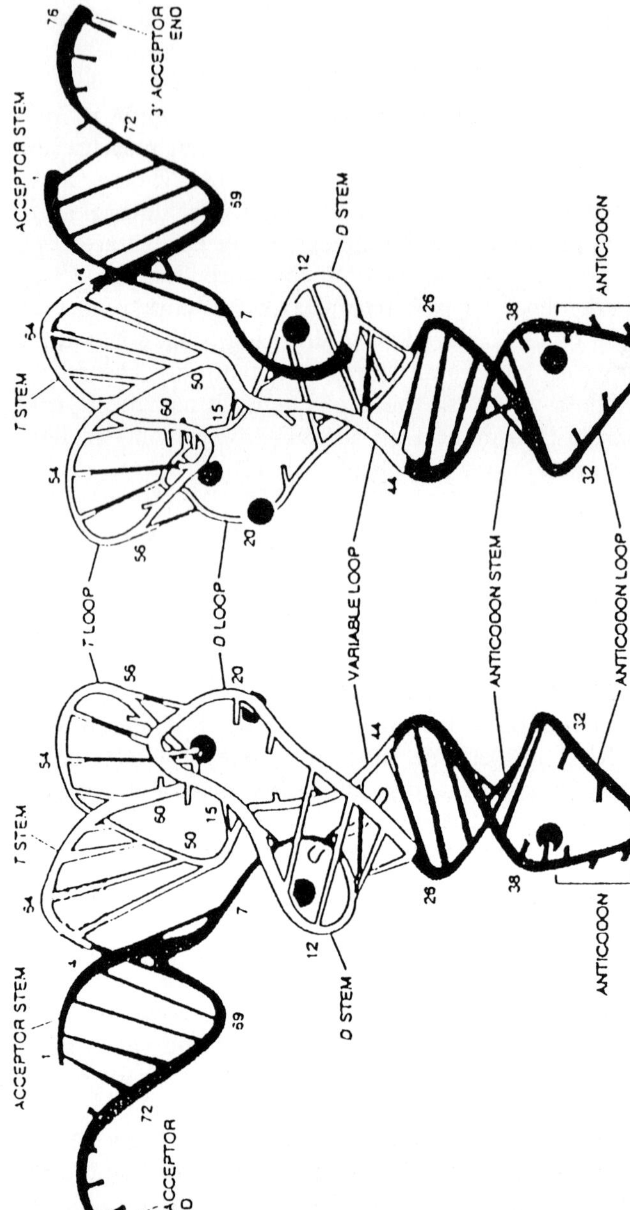

Fig. 8-32. Tertiary structure model proposed by Rich and coworkers for $tRNA^{Phe}$. The sugar-phosphate backbone of the molecule takes the form of a ribbon. The macromolecule is shown from two opposite sides. Hydrogen bonds are shown as junctions between the strands. The molecular fragments and bonds involved in the tertiary structure are shown in black.

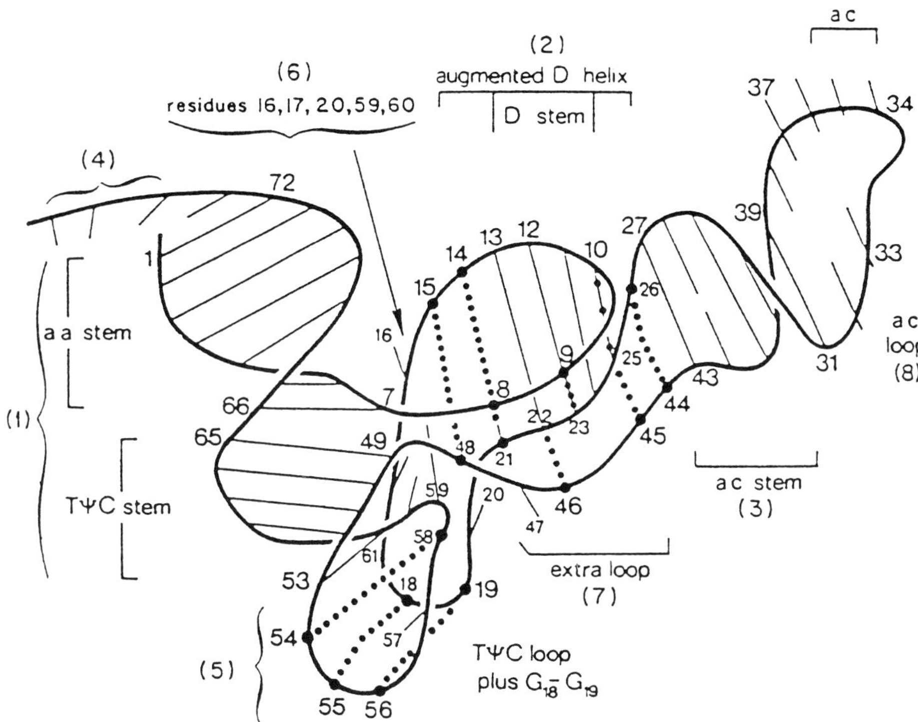

Fig. 8-33. Tertiary structure model proposed by Klug and coworkers for $tRNA^{Phe}$. The sugar-phosphate backbone is shown as a thick line; the long thin lines stand for hydrogen bonds; the dotted lines stand for the hydrogen bonds involved in the tertiary structure; the short thin lines indicate the position of bases not linked by hydrogen bonds.

the tertiary structure of $tRNA^{Phe}$. For instance, pseudouridylic acid $\psi^{(55)}$ (from the TψC-loop) forms an inverted Hoogsteen pair with $G^{(18)}$ from the D-loop, whereas the 1-methyladenine $m^1A^{(58)}$ is paired with the thymine $T^{(54)}$ (Fig. 8-34).

$G^{(57)}$ finds itself between the $G^{(18)}–\psi^{(55)}$ and Watson–Crick pairs; $G^{(19)}$ (from the D-loop) and $C^{(56)}$ (from the TψC-loop) become involved in pronounced stacking with two neighboring guanines $G^{(18)}$ and $G^{(19)}$ (Fig. 8-35).

Another interesting contact arises at the site where the double-helical segment adjacent to the D-loop approaches the extra loop. Here, the minor base $m^7G^{(46)}$ forms a pair, across the broad groove of the helix, with $G^{(22)}$ which, in turn, is paired with $G^{(13)}$ (Fig. 8-34). In this case, the positive charge at $m^7G^{(46)}$ is neutralized by the phosphate group at $A^{(9)}$.

A number of unusual base pairs emerge in that part of the tRNA molecule where its chain forms a deep fold (between the 7th and 10th b.p.). This part features, for example, an inverted Hoogsteen pair $U^{(8)}–A^{(14)}$ and an unusual pair between $C^{(48)}$ and $G^{(15)}$ (Fig. 8-34).

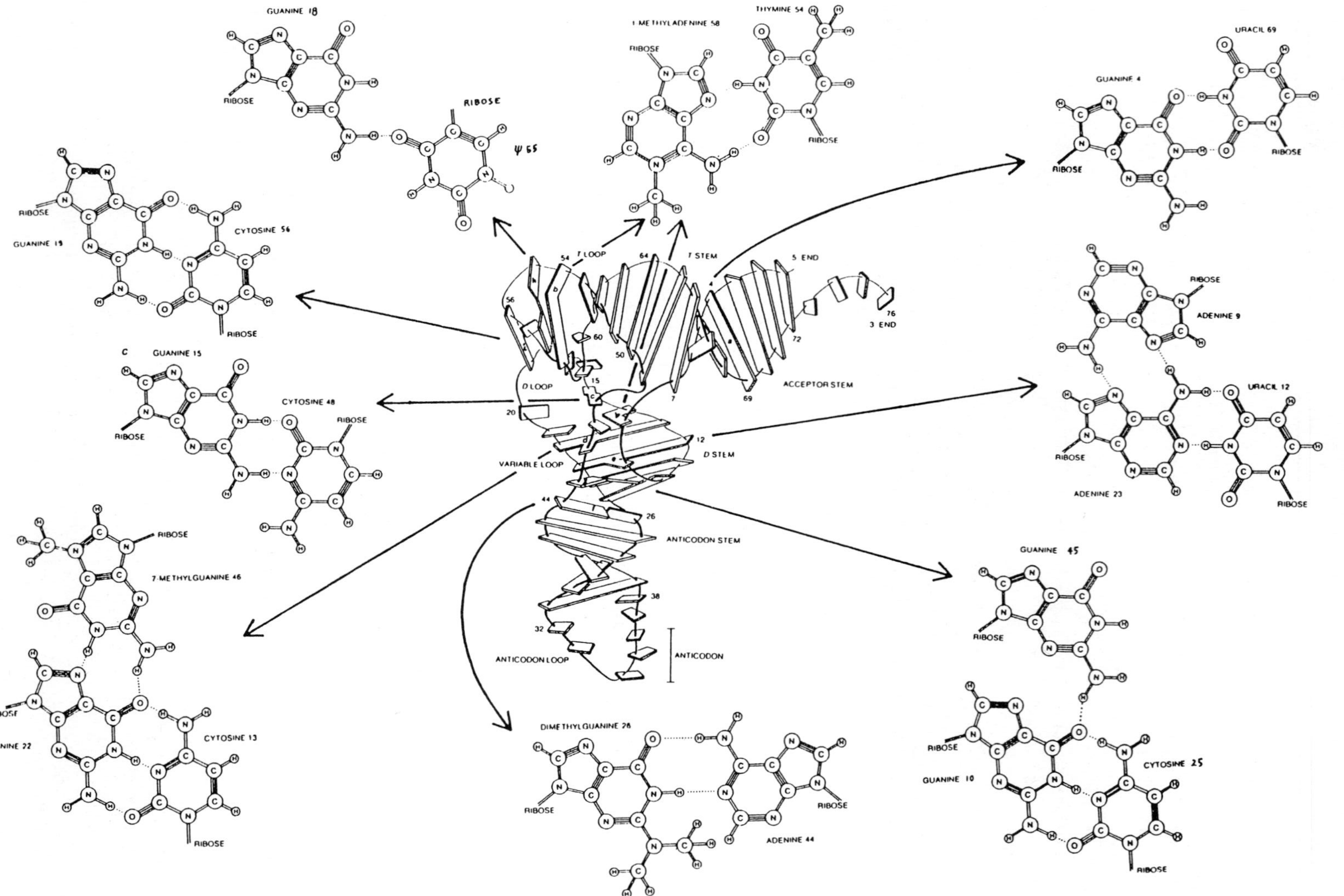

Fig. 8-34. Tertiary base pairs in the crystal structure of yeast tRNA^{Phe} (reproduced with permission from S.-H. Kim in: S. Neidle, ed., *Topics in Nucleic Acid Structure*, part I, pp. 83–112, MacMillan 1981)

In addition, certain parts of the molecule become sites of hydrogen bonding between neighboring riboses as well as binding of neighboring phosphate groups by ions of bivalent metals.

As already mentioned, significant progress has been made over the past few years in studying tRNA by NMR spectroscopy. This technique makes it possible to resolve signals from a large number of protons, including those involved in hydrogen bonding and belonging to NH groups practically in all base pairs emerging during formation of the secondary and tertiary tRNA structures. NMR spectra indicate that tRNA molecules in solution and in the crystalline form have closely similar, if not identical, structures. Thus, they allow tRNA samples that cannot be crystallized to be investigated with a nearly atomic resolution. Moreover, analysis of NMR spectra permits one to

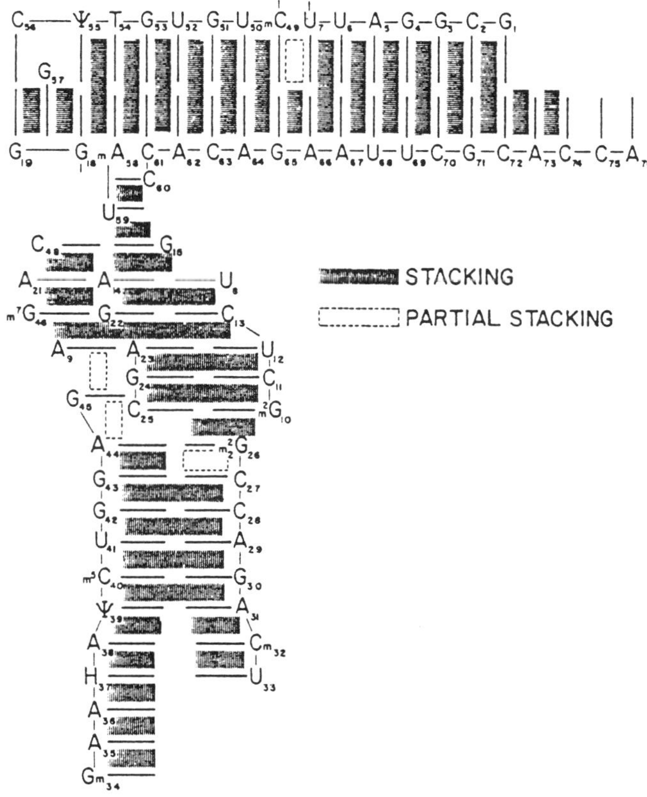

Fig. 8-35. Stacking interactions in $tRNA^{Phe}$. The TψC-loop occupies the upper left corner. The thin lines interconnect adjacent bases in the polynucleotide chain. The thick lines indicate paired bases; bases not involved in stacking are not shown. It can be seen how helical portions associated with the acceptor end and TψC-loop are interlinked as well as how the D- and TψC-loops are intertwined (reproduced with permission from S.-H. Kim, *Progr. Nucl. Acid Res. Mol. Biol.*, **17**, 181–216 (1976)).

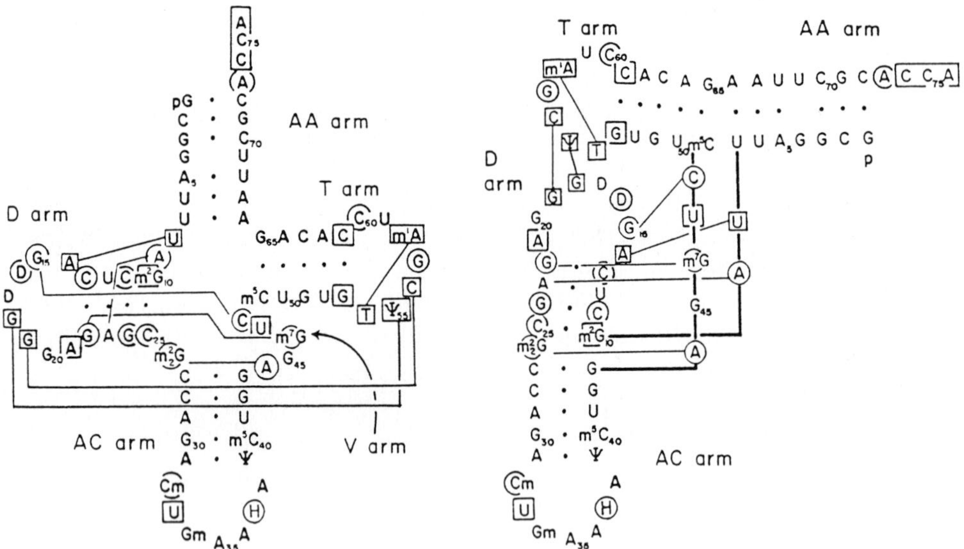

Fig. 8-36. Scheme showing the conservation of tertiary interactions in tRNA. The solid lines interconnect bases paired in the tertiary structure of tRNA. The boxed bases occupy permanent sites in all tRNAs, while the encircled ones occupy those sites in the polynucleotide chain where all tRNAs contain only purines or only pyrimidines.

separately observe almost every base pair in a tRNA molecule, which is extremely important in studying the mechanism of its functioning.

Importantly, many bases involved in the formation of the tertiary structure of $tRNA^{Phe}$ are universal in the sense that they occur at the same positions of the polynucleotide chain in all tRNAs that have been studied (Fig. 8-36). Hence, all tRNAs are characterized by a macromolecular structure consistent with the Rich–Klug model. This has been convincingly supported by the results of X-ray structural analysis of other crystalline tRNA structures as well as those in solution. Minor dissimilarities have been revealed between different tRNAs, primarily near the anticodon loop and acceptor end of the molecule. It has also been found that the angle made by two axes of the limbs of the L-shaped $tRNA^{Asp}$ structure is apparently 10° greater than that of $tRNA^{Phe}$ (Fig. 8-37).

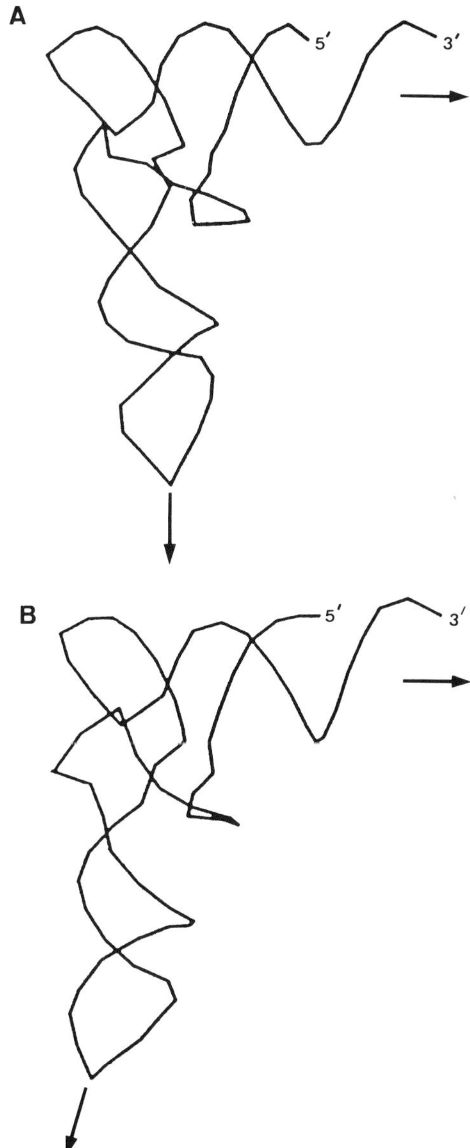

Fig. 8-37. Comparison of backbone conformation of (A) $tRNA^{Phe}$ and (B) $tRNA^{Asp}$ (adapted from D. Moras, *Curr. Opinion in Struc. Biol.*, **1**, 410–415 (1991)).

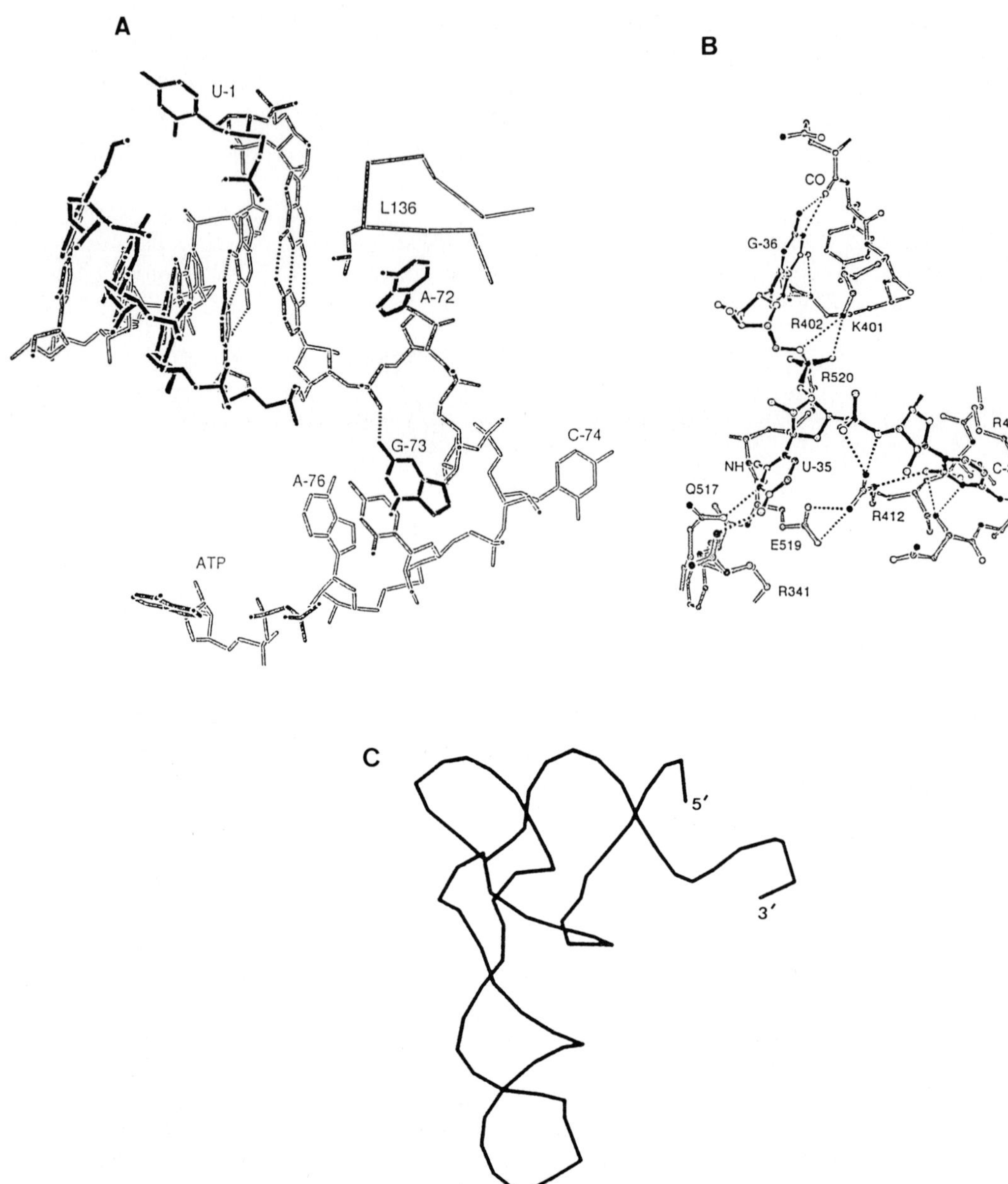

Fig. 8-38. (A) Structure of the acceptor strand of the *E. coli* $tRNA^{Gln}$ in the complex with glutaminyl-tRNA synthetase (adapted from M. A. Rould, *Science*, **246**, 1135–1142, 1989); (B) structure of an anticodon loop of *E. coli* $tRNA^{Gln}$ in the complex with glutaminyl-tRNA synthetase (adapted from M. A. Rould, *Nature*, **352**, 213–218, 1991); (C) schematic representation of the $tRNA^{Gln}$ backbone in the complex with glutaminyl-tRNA synthetase (adapted from D. Moras, *Curr. Opinion in Struc. Biol.*, **1**, 410–415, 1991).

It has often been suggested that the tertiary structure of a functioning tRNA undergoes drastic changes. Indeed, when a complex of $tRNA^{Gln}$ with glutaminyl-tRNA synthetase is formed, the last base pair in its acceptor stem breaks down and the acceptor end forms a loop directed toward the anticodon (Fig. 8-38). The bases in the anticodon loop, forming part of the anticodon, pass into a more open conformation, which ensures a more intimate contact with the protein. What is not known, however, is how common such conformational changes are. For example, in a complex with specific aminoacyl-tRNA-synthetase, the tertiary structure of $tRNA^{Asp}$ is virtually indistinguishable from that of free tRNA.

8.2.4 Three-Dimensional Structure of High-Molecular Weight RNAs

Transfer RNAs are the only representatives of this class of nucleic acids that have been successfully crystallized and studied by X-ray structural analysis with atomic resolution. In the case of RNAs comparable in size with tRNA, such as 5S ribosomal RNAs or small nuclear RNAs, high-resolution NMR was applied only to study the conformation of rather short fragments (the separate elements) of their molecules.

However, as we go to larger RNAs, rather serious difficulties arise and in order to establish their three-dimensional structure we have to use many indirect mutually complementary methods.

As a rule, the investigator begins with establishing the nucleotide sequence of the RNA. The next step is a computer search for possible models of secondary structure of the same RNA. Such search involves well proven algorithms for constructing various possible elements of the RNA secondary structure, considered in 8.2.2, and evaluating their thermodynamic stability. They are based on a wealth of experimental data obtained on synthetic oligoribonucleotides and make it possible to estimate the length of double-stranded segments, the sequence of G·C, A·U and G·U pairs in the latter, the size of the bulges, and so on.

Unfortunately, calculations of this kind usually result in numerous secondary structure models with closely similar values of free energy for a given RNA (the number of such models being extremely high for high-molecular weight RNAs). However, the theoretical search for the model may be continued if the nucleotide sequences for at least one more (better several) RNA of the same type but isolated from a different source are known. In such a case, one can successfully resort to the so-called “phylogenetic” approach which is based on the assumption that the RNAs performing the same function in the cell must have identical secondary and tertiary structures. Then, one must select models of the same type among those calculated for two (or more) RNAs.

The applicability of the "phylogenetic" approach was for the first time demonstrated while studying the structure of tRNA. It was found that the secondary structure of any tRNAs, no matter how their nucleotide sequences vary, is consistent with the universal "clover-leaf" model. Even more impressive are the results of using this approach to ribosomal RNAs, when a single model had been selected for RNA of each type among hundreds of theoretically possible ones. The exact way in which such selection is carried out at the level of individual elements of the RNA secondary structure is shown in Figure 8-39, while Figure 8-40 illustrates the similarity of secondary structures of the RNA component of ribonuclease P.

The theoretical model of the RNA secondary structure must be experimentally verified. Indeed, one can select a chemical reagent for each heterocyclic base, which interacts selectively only with the atoms or groups involved in hydrogen bonding during complementary pairing of nucleotides. For instance, ketoxal reacts selectively with N^1 and the $2\text{-}NH_2$ group of guanine, dimethyl sulfate does so (under certain conditions) with N^1 of adenine and N^3 of cytosine, and carbodiimide reacts selectively with N^3 of uracil. Consequently, if a particular nucleotide in RNA is modified by one of these reagents, it belongs to the single-stranded segment of RNA. Instances where no modification takes place are harder to interpret – the nucleotide may form part of the double helix but be screened by tertiary interactions as well.

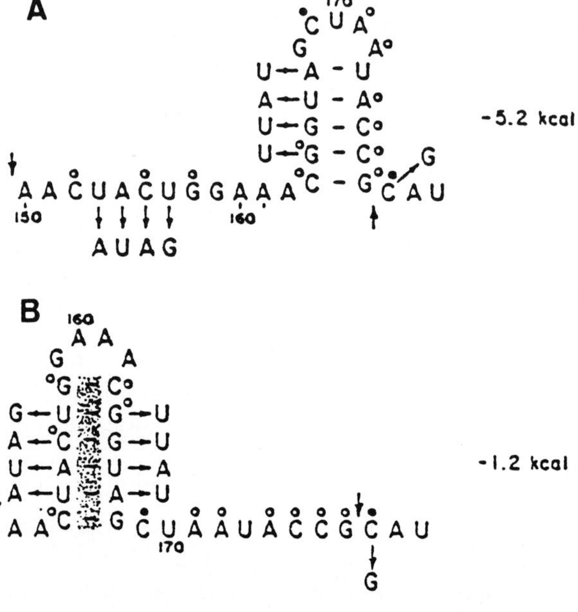

Fig. 8-39. Comparative (phylogenetic) analysis used to select the most probable secondary structure of RNA. Two possible secondary structures of a portion of 16S ribosomal RNA of *E. coli* (numbering from the 5′ end of RNA) are shown. In spite of the fact that the energy of formation of structure (A) is lower, preference should be given to structure (B). Indeed, substitutions in the nucleotide sequence of a similar portion of the molecule of 16S ribosomal RNA of *Bac.brevis* (shown by arrows) rule out the existence of structure (A). At the same time, in the case of structure (B), the double-stranded portion remains in both RNAs in spite of the nucleotide substitutions. Such substitutions are referred to as compensatory (adapted from R. R. Gutell et al., *Progr. Nucl. Acid Res. Mol. Biol.*, **32**, 156–217 (1985)).

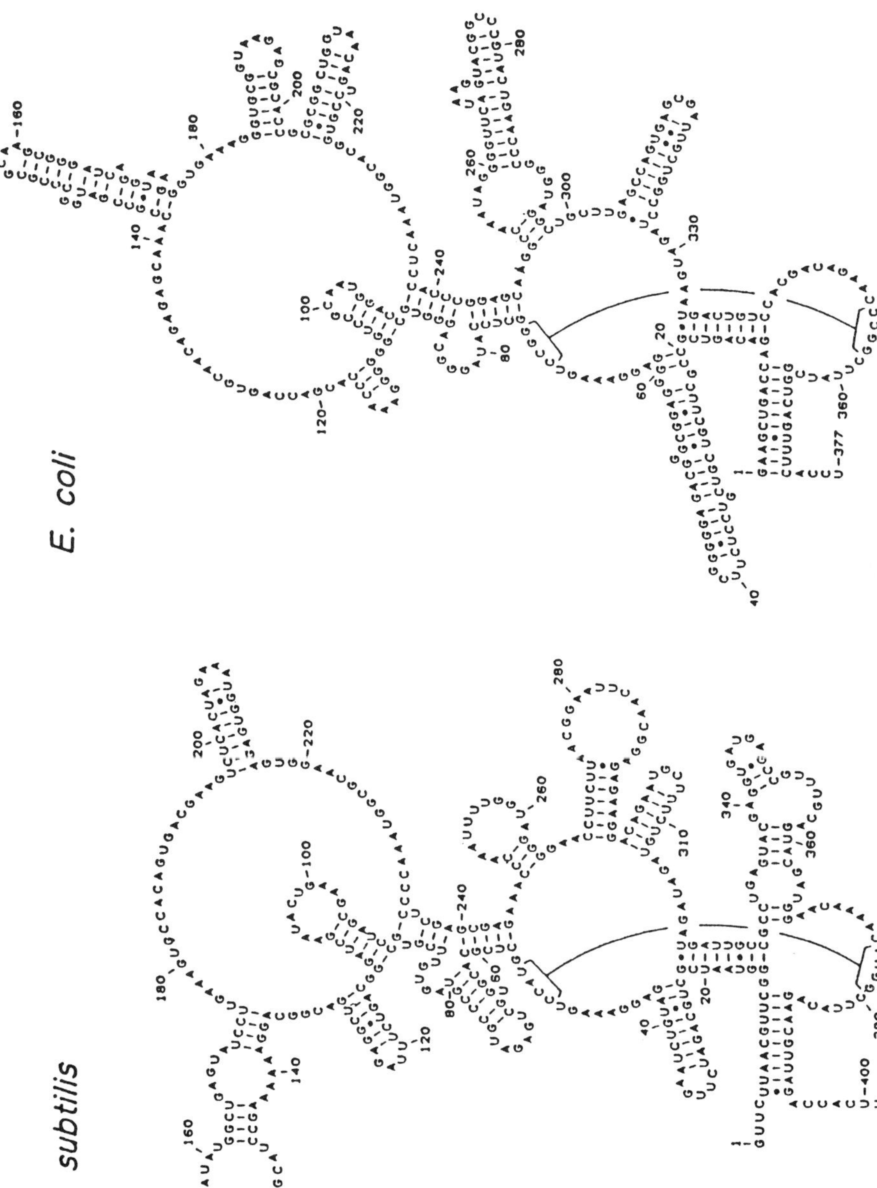

Fig. 8-40. Secondary structures of RNA components of *E. coli* and *B. megaterium* RNase, proposed on the basis of phylogenetic comparison (reproduced with permission from B. D. James et al., *Cell*, **52**, 19–26 (1988)).

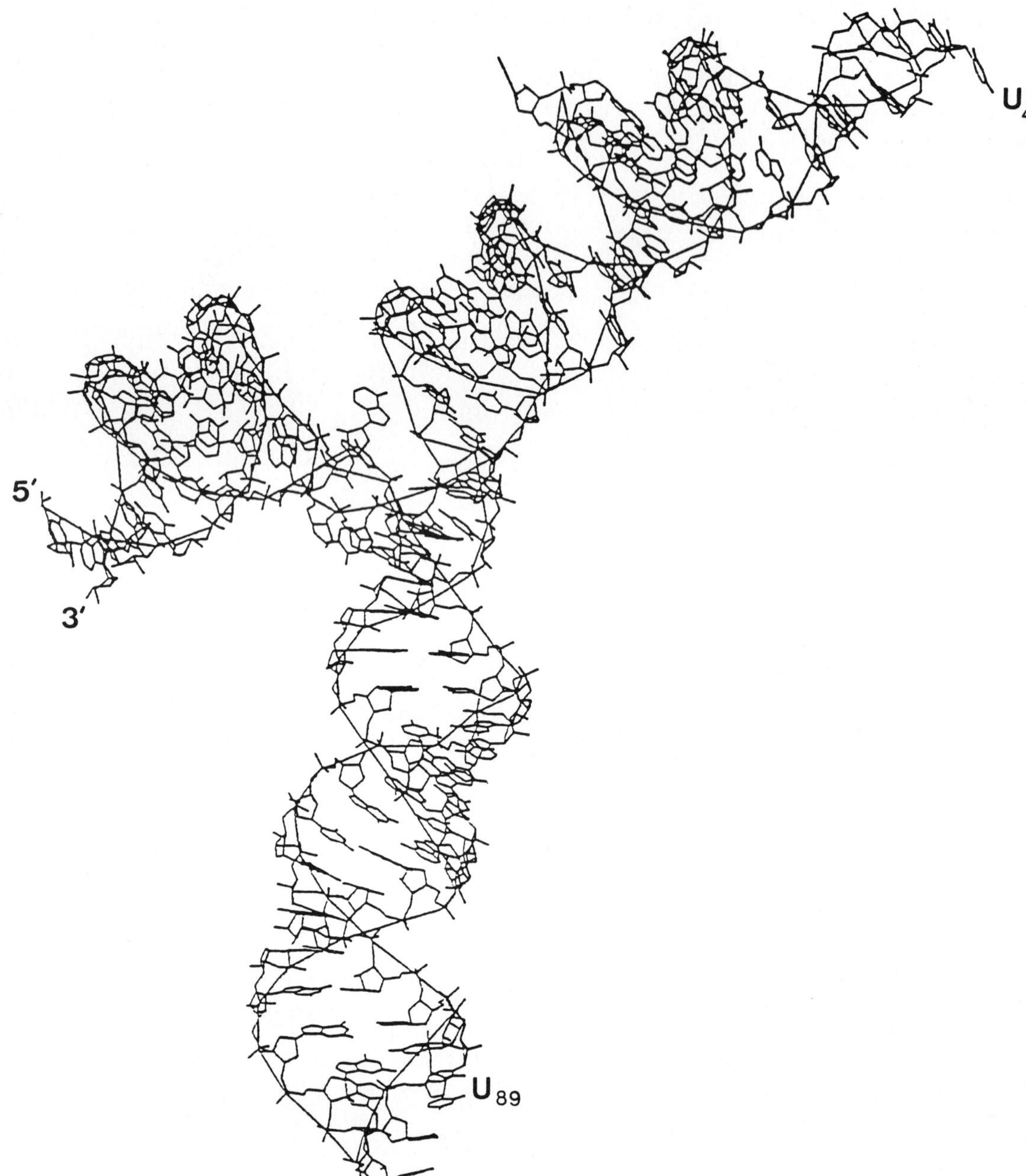

Fig. 8-41. Three-dimensional structure of 5S ribosomal RNA, based on chemical and enzymatic probing (courtesy of E. Westhof).

Similar principles underlie the enzymatic methods of RNA secondary structure analysis. A large number of RNases are known which provide directed hydrolysis of the phosphodiester bonds in single-stranded segments of RNA. Several nucleases have also been discovered which exhibit pronounced affinity toward double-stranded RNAs. Analysis of RNA hydrolysis by these enzymes will reveal the type of secondary structure to which particular nucleotides belong. The results of chemical and enzymatic methods complement each other rather well. Fig. 8-41 represents a 3D structural model of the E. coli 5S ribosomal RNA, derived by combination of the methods described above.

These approaches have been used to derive secondary structure models for many viral and cell RNAs. In spite of the fact that the length of such RNAs equals hundreds of nucleotides, reliable models of their secondary structure have been produced in full agreement with experimental data. The new property brought to light by secondary structure analysis of such large RNAs is the subdivision of their macromolecules into structural domains or, in other words, large molecular fragments having a more or less independent three-dimensional structure. Figure 8-42 shows the universal structure of an RNA from a small ribosomal subunit.

Trying to establish the tertiary structure of high-molecular weight RNAs is an even more challenging task. For want of a better approach, one must proceed from the principles on which the tertiary structure of tRNAs is based. In other words, the following assumptions are to be made as regards any single-stranded RNA:

1. Elements of the RNA secondary structure are mutually arranged in such a fashion so as to ensure a maximum degree of base stacking in the macromolecule as a whole.
2. Contacts between individual secondary structure elements are based on several types of so-called "tertiary" intramolecular interactions: (a) those involving formation of additional, often non-Watson–Crick base pairs between nucleotides in single-stranded segments spaced widely apart (in the primary and secondary structures) and base triplets between nucleotides in single- and double-stranded segments; in the latter case, participating in hydrogen bonding are the N^7 atoms of purine bases as well as the functional groups at their C^6 atoms, already taking part in formation of a single hydrogen bond in the Watson–Crick pair; (b) those involving additional ("tertiary") stacking interactions after intercalation of bases from one segment between two neighboring bases of another single-stranded segment; and (c) those involving formation of additional hydrogen bonds between the 2′-OH groups of ribose and bases as well as other groups of the sugar–phosphate backbone.
3. The tertiary structure of RNA is stabilized by bivalent metal ions which become bound not only with phosphate groups but also with bases.

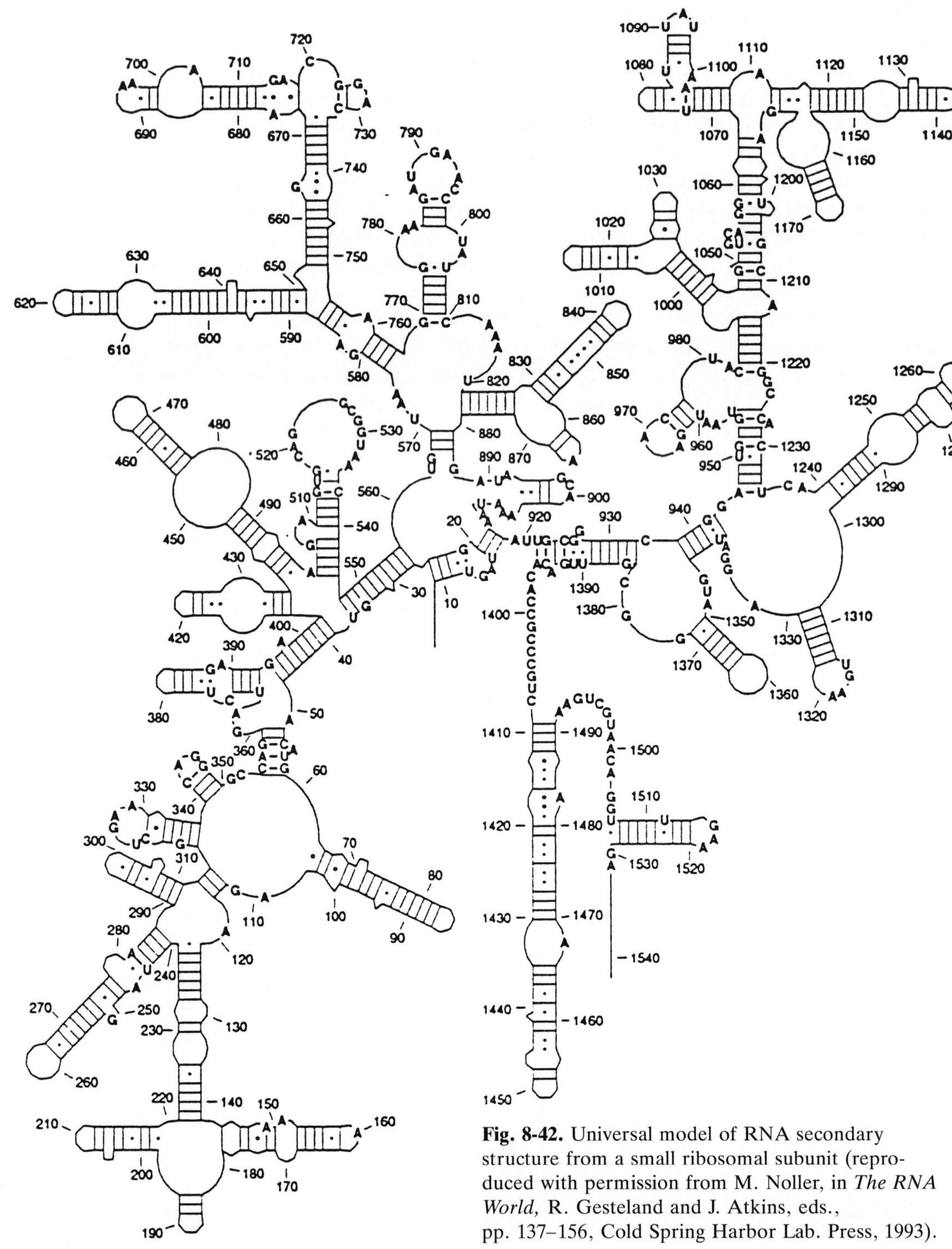

Fig. 8-42. Universal model of RNA secondary structure from a small ribosomal subunit (reproduced with permission from M. Noller, in *The RNA World,* R. Gesteland and J. Atkins, eds., pp. 137–156, Cold Spring Harbor Lab. Press, 1993).

These principles are not sufficient, of course, to deduce the three-dimensional structure of an RNA from its primary and secondary structures alone. As will be demonstrated in the following chapter, chemical modification data are instrumental, too. Valuable information is also provided when bifunctional alkylating agents, such as bis(2-chloroethyl)methylamine or "nitrogen mustard", are used. They cross-link bases spatially arranged next to each other but belonging to different segments of the RNA chain. Then the RNA is fragmented and the cross-linked oligonucleotides are isolated and sequenced.

There is no doubt that the "phylogenetic" principle must also hold for the tertiary RNA structure. What is more, it can be assumed that two or more RNA molecules differing in secondary structure but performing similar functions in the cell will have similar tertiary structures. A case in point is RNAs of plant viruses. Some of them, such as turnip yellow mosaic virus (TYMV) RNA, are aminoacylated at the 3' end by aminoacyl-tRNA-synthetases, similarly to tRNA. The secondary structures of the 3′-terminal segment of TYMV RNA and tRNA are similar and yet display significant differences. As can be seen from Figure 8-43, these differences disappear almost completely when this segment of the viral RNA is integrated into the tertiary structure. Here we encounter yet another essential element of macromolecular RNA structure, known as the pseudoknot. Pseudoknots result from complementary pairing of hairpin-loop bases with those in the single-stranded segment outside the hairpin structure (Fig. 8-44). The possibility of such pairing has been demonstrated on model oligoribonucleotides. Pseudoknots represent a widely occurring and highly important element of three-dimensional structure of many RNAs.

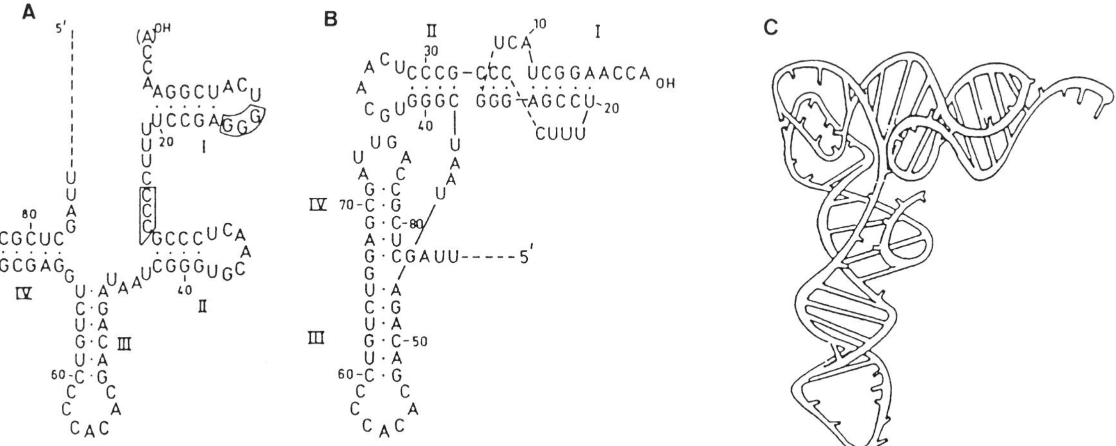

Fig. 8-43. Secondary and tertiary structure of the 3′-terminal segment of turnip yellow mosaic virus (TYMV) RNA. A – Clofer-leaf structure; the regions capable of forming pseudoknots are boxed; B – L-shaped structure; C – three-dimensional model (adapted from C. Pleij, *TIBS*, **15**, 143–147, 1990).

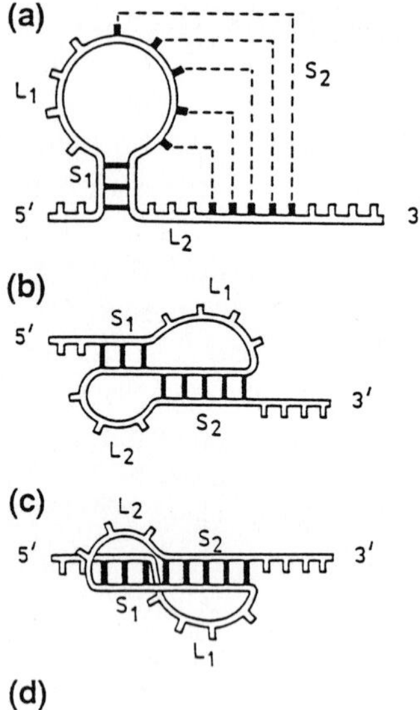

Fig. 8-44. Formation of a pseudoknot (a) – starting structure; the broken line indicates complementary bases in the potential pseudoknot; (b) and (c) – formation of the S2 stem. Note the coaxial stacking of S1 and S2 in (c); (d) – three-dimensional structure of the pseudoknot (reproduced with permission from C. Pleij, *TIBS,* **15**, 143–147 (1990)).

References

1. J. D. Watson and F. H. C. Crick, *A Structure of Deoxyribose Nucleic Acid. Nature,* **171**, 737–738 (1953).
2. J. R. Fresco, B. M. Alberts, and P. Doty, *Some Molecular Details of the Secondary Structure of Ribonucleic Acid. Nature,* **188**, 98–101 (1960).
3. C. Cantor and P. Schimmel, *Biophysical Chemistry,* Parts I (Chapters 3 and 6) and III (Chapters 19, 22, 23 and 24). W. H. Freeman and Co., San Francisco (1980).
4. S.-H. Kim, *Transfer RNA: Crystal Structures,* in: *Topics in Nucleic Acid Structure.* S. Neidle (Ed.), Part I, MacMillan Publishers Ltd, pp. 83–122 (1981).
5. R. E. Dickerson *et al., The Anatomy of A-, B- and Z-DNA. Science,* **216**, 475–485 (1982).
6. W. Saenger, *Principles of Nucleic Acid Structure.* Springer Verlag, New York (1984).
7. W. Guschelbauer, Polynucleotides, in: *Encyclopedia of Polymer Science and Engineering.* vol. 12, John Wiley & Sons, Inc. (1988).
8. M. Chastain and I. Tinoco, Jr., *Structural Elements in RNA, Progr. Nucl. Acid Res. Molec. Biol.,* **41**, 131–177 (1991).
9. *Nucleic Acids in Chemistry and Biology.* G. M. Blackburn and M. J. Gait (Eds.), Chapters 2, 7 and 8, IRL Press, Oxford–N. Y.–Tokyo (1990).
10. T. A. Steitz, *Structural Studies of Protein-Nucleic Acid Interaction: the Sources of Sequence-specific Binding. Quart. Rev. Biophys.,* **23**, 205–280 (1990).

9 Chemical Properties of Polynucleotides. Modification of Nucleic Acids

9.1 Introduction

Ever since nucleic acids were discovered, the efforts of all those involved in investigation of their chemical properties had been focused, primarily at methods for breaking these compounds down as well as separation and identification of the resulting fragments. Intensive studies into the reactivity of nucleic acids began only after their primary structures had been definitively established. This field of organic chemistry received even greater attention since methods for isolating individual molecules of various types of RNA and DNA had been elaborated. Naturally, the first objects of directed chemical modification were tRNAs which are the simplest of all known nucleic acids. The functional importance of tRNAs was established soon after their discovery, and that immediately gave impetus to further research in this area.

As regards nucleic acids, our knowledge has grown tremendously about the major aspects of the reactivity of their constituent heterocyclic bases, *N*-glycosidic bonds, carbohydrate moieties, as well as internucleotide and terminal phosphate groups.

The macromolecules of nucleic acids enter into many reactions usually involving their monomer units. The mechanism of reactions of heterocyclic bases, *N*-glycosidic bonds, carbohydrate moieties and terminal phosphate (phosphomonoester) groups is generally the same as in the case of monomers. The internucleotide phosphate groups, which introduce a new element into the nucleic acid structure, impart new properties to nucleic acids. Other factors that render the reactivity of polynucleotides somewhat different also include their polyfunctionality and polymer structure. First of all, the presence of a great number of identical and structurally similar reactive sites in a nucleic acid molecule calls for judicious selection of the right conditions for conducting reactions in the desired direction. A major influence on the reactivity of heterocyclic bases is exerted by the steric factors associated with the presence of secondary and tertiary structures in nucleic acids. When reactions are conducted in aqueous solutions under conditions when the nucleic acid retains its ordered structure, some of the groups may be accessible to reagents, while others may be blocked. Thus, when reactions usually involving monomer units of nucleic acids are considered in the context of the nucleic acids themselves, one must first of all take into account the stereochemical factors governing their course.

It is known, for example, that a heterocyclic base may be attacked by a reagent at a right angle to the plane of the base, or the attack may proceed in the same plane. Certain reagents will readily interact with the base of a monomer, but in the broader environment of the entire nucleic acid such a base may be blocked in a particular way and thus inaccessible to one of the reagents (normally of the first type or even both). Such evidence accumulated while studying the chemical properties of nucleic acids has been extremely useful in investigating their structure. As has already been mentioned, it formed the basis of some methods for determining the nucleotide sequence of both RNA and DNA. Equally important are findings as regards the reactivity of heterocyclic bases for establishing the secondary and tertiary structure of nucleic acids. This approach is also widely used as a means to determine the structure of mixed biopolymers, such as nucleoproteins (in trying to elucidate the nature of the contacts between functionally significant nucleic acid sites of particular elements of the secondary and tertiary structure of proteins).

Binding heavy metal atoms to certain constituent bases of nucleic acids is instrumental in determining their structure from the results of X-ray analysis and electron microscopy. Controlled incorporation of paramagnetic and fluorescent groups into nucleic acids facilitates their investigation by spectroscopic techniques.

Knowledge about the reactivity of polynucleotides and nucleic acids has turned out to be beneficial not only for elucidating their structure, but also in many other ways. A case in point is activation of terminal groups in nucleic acids with the aid of special reagents. Nucleic acids modified in this manner may then be ligated (immobilized) to water-insoluble polymers for subsequent use in the separation of individual compounds from complex mixtures, the preparation of sorbents for affinity chromatography, and so on. In recent years, nucleic acids with groups capable of entering into covalent reactions have found wide application in biospecific marking of biopolymers. Such groups are usually coupled to terminal parts of nucleic acids.

Success of chemical modification of nucleic acids depends primarily on the specificity of the reagents used.

This chapter deals with the reactivity of individual constituents of nucleic acids, then some applications of the chemical nucleic acid modification method are described. Emphasis is placed on the reactivity of polynucleotides in aqueous solutions. There are two reasons for this: firstly, nucleic acids are readily soluble in water by virtue of their polyelectrolytic behavior and almost insoluble in organic solvents; secondly, it is only in aqueous solutions that the chemical modification method can be regarded as a tool for studying the three-dimensional structure and functioning of native nucleic acids (in organic solvents the native structure of nucleic acids is disturbed).

9.2 Reactions of Heterocyclic Bases of Polynucleotides

9.2.1 General Concepts

Nature of Reactive Sites. As can be inferred from previous chapters, the properties of the pyrimidine and purine bases of monomer components are first of all determined by the electronic structure of the atoms and groups of atoms constituting the heterocycle. Before discussing their chemical properties in the context of a polynucleotide (RNA, DNA, or a synthetic polymer), in addition to the concepts outlined in Chapter 2 one must also consider the stereochemistry of the reactions involving the respective bases and, in this connection, the spatial distribution of electron density at each reactive atom of the base. Since the tautomeric equilibrium in heterocyclic bases is normally shifted toward one of the forms, which seems to be responsible for the reactivity of a given base under particular conditions, the electron density distribution will be discussed only in the context of stable forms. Naturally, when we are speaking about the reactivity of a given heterocyclic base at pH values markedly different from 7, it should be remembered that the solution contains other forms that may react in an entirely different fashion.

The following scheme is based on the results of studies into the chemical properties of pyrimidine and purine bases of nucleic acids in aqueous media where the native structure of the latter usually remains intact. In all instances, the reactive sites of similar nature are marked by the same symbol (square, circle, triangle, diamond).

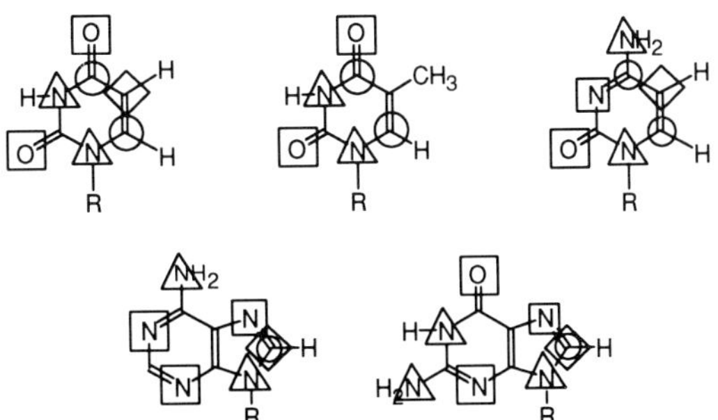

R=polynucleotide chain

In the past, the influence of steric factors on the reactivity of pyrimidine or purine bases was virtually ignored because in monomer units this influence is minimal and, as a rule, does not determine the course of the reaction. If a base is not hindered sterically, it does not make any difference from which direction its atoms are attacked by the reagent – from the plane of the base or at a right angle thereto. The situation is entirely different when we are speaking about the reactivity of bases within the polymeric nucleic acid molecule. In this case, individual portions of the base may be substantially screened as a result of their involvement in hydrogen bonding or interplanar interactions. This is why the reagent cannot attack a particular atom of the heterocycle from any direction. In this connection we must figure out which direction of attack on each of the reactive sites marked on the generalized scheme will produce any results, and also which of the possible routes is open or blocked in the secondary and tertiary structures of the nucleic acid. Answers to these questions will tell us, firstly, which atoms of a particular pyrimidine or purine base may be modified irrespective of the three-dimensional structure of the corresponding nucleic acid fragment and, secondly, which of these atoms can react only as parts of single-stranded fragments of the polymer chain.

Let us now examine each of the reactive sites separately. The above scheme shows that they may coincide with nitrogen atoms (two species marked by square and triangle), the carbonyl oxygen (marked by square), and carbons (three species marked by circle, diamond, and also by circle and diamond at the same time).

The pyridine nitrogens and carbonyl oxygens are highlighted by the square symbol. The pyridine nitrogens form two σ-bonds with two carbons (the orbitals lie in the nucleus plane between the corresponding atoms) and one π-bond (the corresponding orbitals occupy a plane perpendicular to that of the nucleus). The orbital in the plane of the aromatic ring and at an angle of 120° to the N–C bonds of the aromatic system accommodates two lone-pair electrons.

The carbonyl oxygen has a similar electronic structure. This atom forms a σ-bond with carbons. The other bond is formed by π-electrons whose orbitals lie at a right angle to the plane of the aromatic system. Each of the two orbitals lying in the plane of the aromatic ring and forming an angle of 120° with each other and with the O–C bond contains two lone-pair electrons.

The above-described nitrogen and oxygen atoms apparently may be subject predominantly to electrophilic attack which must be aimed at their lone-pair electrons. Since, as has already been pointed out, the orbital, or orbitals, accommodating these lone-pair electrons lies in the plane of the heterocyclic (or aromatic) nucleus of the base, the reagent's attack must be directed toward this plane to have any results.

The second type of reactive sites is represented by pyrrole nitrogens and those of the amino group (marked by triangle). The pyrrole nitrogens form three σ-bonds at 120° angles with two carbons and a hydrogen (the orbitals lie in the plane of the aromatic ring). There are two electrons on the *p* orbital of the nitrogen overlapping with that of one of the carbons. This orbital forms a right angle with the plane of the heterocyclic nucleus.

A similar electronic structure is also observed in the nitrogen of the exocyclic amino group: the *p* orbital with a free pair of electrons, conjugated with the aromatic system, is perpendicular to the nuclear plane as well. Nitrogen atoms of this type must also be vulnerable to electrophilic attack, however, depending on the orientation of the orbital accommodating the lone-pair electrons of nitrogen, the attack will be productive only if the reagent approaches from the direction perpendicular to the plane of the heterocyclic base.

Thus, the reactive sites in triangles will be affected by electrophilic reagents when the attack is directed at a right angle to the plane of the heterocyclic base.

Insofar as the reactivity of carbons in pyrimidine and purine bases of nucleic acids is concerned, it should first of all be pointed out that they are in the sp^2-hybrid state because each of them forms part of a $\gtrdot$C=Y group (where Y = O, N and C). Accordingly, reagents will always attack the carbons (C^2, C^4, C^5, C^6, C^8) of these bases at a right angle to the plane occupied by the group and, consequently, to the plane of the corresponding base.

As is known, the reaction of electrophilic substitution occurs at C^5 in pyrimidine nucleosides and at C^8 in purine nucleosides. These bases react similarly within nucleic acids. The corresponding atoms in the generalized scheme are marked by diamonds and circles.

The response of carbons in the $\gtrdot$C=Y group to different reagents depends both on the position of this group in the molecule of the base and on the species of atom Y.

In purine bases, for example, the carbons shared by the pyrimidine and imidazole rings are not reactive. In pyrimidine bases, the reactivity of the carbonyl C^2 atoms bound to two nitrogens is low. These atoms are left unmarked in the scheme.

Just as expected, the species of atom Y markedly influences the reactivity of the associated carbon.

In cases where Y is oxygen, we are dealing with a carbonyl group ($\gtrdot$C=O) in which the carbon is known to carry a partial positive charge and to be easily attacked by nucleophilic reagents. Uracil has two such carbonyl carbons, C^2 and C^4, whereas cytosine has only one, C^2. It has already been mentioned that the reactivity of the C^2 atoms is low because of the electron-donor influence of the two adjacent nitrogens. Being not reactive per se, C^4 in uracil none the less renders C^6 sensitive to nucleophiles, as a result of its electron-acceptor effect being transmitted via the vinyl fragment, just as in the case of α,β-unsaturated carbonyl compounds. Nothing is known about the reactivity of the carbonyl C^6 carbon in guanines of nuclei acids.

When Y is nitrogen, we have a $\gtrdot$C=N– group in which the nitrogen exerts the same influence on the associated carbon as the carbonyl oxygen does, albeit not so markedly. Consequently, it may sometimes be expected that the corresponding carbons in pyrimidine and purine bases will be sensitive to nucleophilic attack. C^4 is such a carbon in the case of cytosine; its reactivity seems to be due to more pronounced electronacceptor behavior of N^3 under the influence of the neighboring carbonyl group. In the case of adenine and guanine, nucleophilic attack is aimed at C^8.

Thus, only carbon atoms are prone to nucleophilic attack in the pyrimidine and purine bases of nucleic acids. They are shown encircled in the scheme.

Effect of Ionization on Reactivity. In the above discussion of reactivity of pyrimidine and purine bases in nucleic acids it was assumed that the bases are in a neutral form. Yet under the conditions prevailing in reactions of nucleic acids (aqueous solutions, varying pH values), each of the bases may accept or donate a proton, by virtue of its acid–base properties, and thus become charged. This, of course, must affect the course of reactions in which bases are involved to some or other extent.

As has already been mentioned, the proton is attached at the pyridine nitrogens or, in other words, the attachment proceeds as an electrophilic attack in the plane of the heterocyclic base. According to their ability to accept a proton, or be protonated, heterocyclic bases (for their pK_a values, see Table 4.1) can be arranged in the following order even in nucleic acids: cytosine > adenine > guanine. Only uracils, thymines and guanines can give up a proton (see Table 4-1).

A protonated base becomes more immune toward electrophilic agents and more vulnerable toward nucleophiles. When bases give up a proton, their reactivity varies inversely.

An example illustrating the effect of protonation on the chemical properties of a base is the inertness of the pyridine nitrogen N^3 in a protonated form of cytosine with respect to electrophilic reagents and simultaneous mitigation of nucleophilic attack at carbon C^4 next to the protonated nitrogen.

Thus, the acid–base properties of a heterocyclic base can be used to predict its reactivity, depending on the pH value of the solution in which the reaction is conducted. Since, as has already been mentioned, reactions with nucleic acids are usually carried out in aqueous solutions, the effect of pH on the reactivity of heterocyclic bases must always be taken into account.

Here are some examples of reactions involving constituent bases of nucleic acids in neutral and charged forms.

Example 1. Only the protonated form of the base enters into the reaction with a nucleophile – that is, the reaction rate is almost zero at a pH value when the base is in a free state and maximum at a pH value when practically all bases are in a protonated form. Take the simplest case where the nucleophile Nu is not capable of attaching a proton. Such a reaction may be written as

$$^{+}BH + Nu \rightarrow \text{reaction product}$$

At pH equal to pK_a of the base, the reaction mixture will contain equal amounts of the protonated and uncharged forms. Apparently, at this point on the reaction rate (v) versus pH curve v will equal 0.5 of v_{max}.

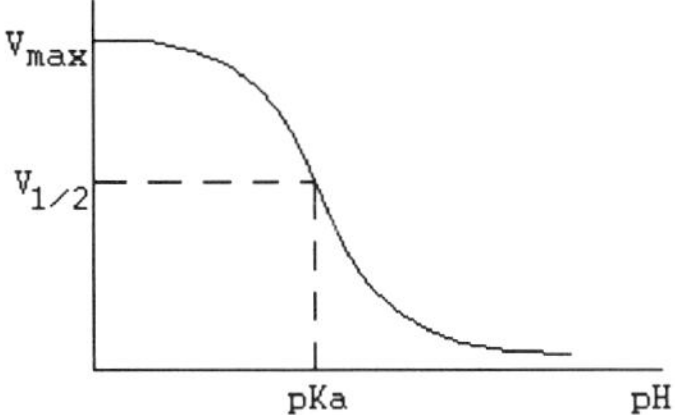

In this case, the reaction rate changes sharply with varying pH in the region of pK_a of the base. Such reactions, for example, for guanine ($pK_a \sim 3.0$), must proceed at a satisfactory rate at pH < 3 and very slowly already at pH 4–5.

Example 2. The protonated form of the base reacts with the nucleophile, but in this case the nucleophile itself can attach a proton: pK_a of Nu > pK_a of the base.

In a strongly acidic medium, the reagent and base are protonated, therefore the reaction rate is virtually nil. The reaction proceeds at a maximum rate at a pH value when the reagent is already deprotonated, while the base still contains a proton, but there comes a moment at which the base and reagent lose their protons, and the reaction rate drops once more.

Evidently, the reaction rate will be maximum, v_{max}, at a pH value equal to half the sum of pK_a values of the reactants (pH_a):

$$\frac{pK_a^{base} + pK_a^{reagent}}{2}$$

The relationship between the reaction rate and pH will be represented by a bell-shaped curve.

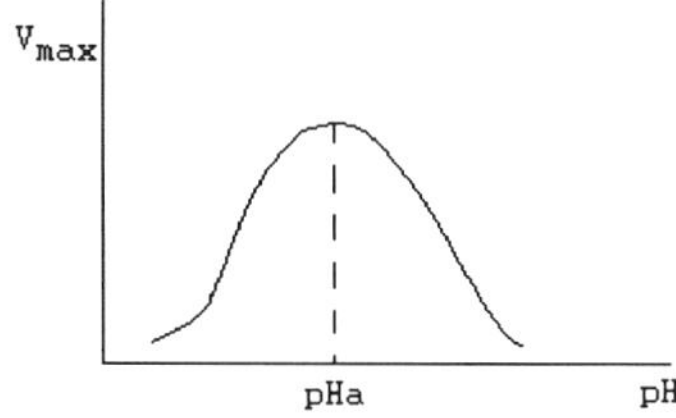

Example 3. The base reacts with an electrophile only in a deprotonated form. The reaction may take the following general form:

$$B^- + E \rightarrow \text{reaction product}$$

In this case, the reaction rate is zero in the pH range where the base is protonated. It reaches its peak at a pH value when the base loses its proton.

Here, too, the dependence of the reaction rate on pH of the reaction mixture is maximum when pH approaches pK_a of the reacting base.

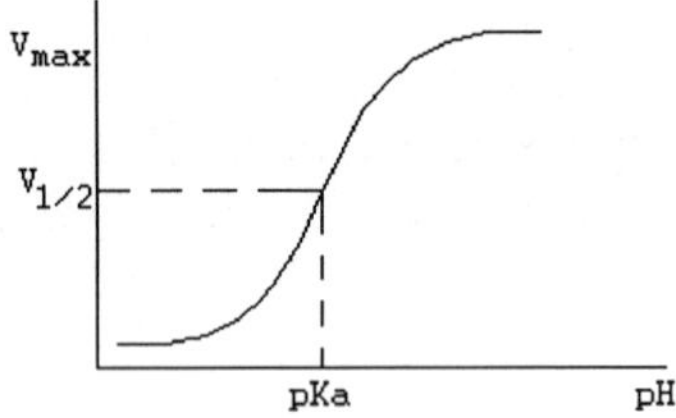

Many more examples could be given showing the relationship between the reactivity of nucleic acid bases and their form – protonated or deprotonated. The relationship between the reaction rate and pH is usually more complex than in the above examples. This depends on the reaction mechanism, ionic strength of the protonating agent, temperature, and many other factors. There is every reason to believe that the rate of reactions between heterocyclic bases of nucleic acids and various reagents is to some extent influenced by negatively charged phosphate groups of the carbohydrate–phosphate backbone.

Effect of the Macromolecular Structure of Nucleic Acids on Reactivity. Different aspects of the ability of heterocyclic bases in nucleic acids to undergo chemical modifications become important when they are regarded as keepers and carriers of genetic information. Since the coding capacity of nucleic acids is determined by the sequence of heterocyclic bases in the polynucleotide chain, the vulnerability of the bases to chemical attack is a measure of stability of the genetic material. Knowledge of the mechanisms underlying the action of mutagens (substances capable of altering the structure of heterocycles) on nucleic acids not only may give an insight into the alterations occurring when the genes become active in the cell, but also provide a tool for their manipulation in an intrusive and directed manner.

It is quite obvious that the reactivity of pyrimidine and purine bases in nucleic acids must be strongly influenced by the three-dimensional structure of the polymer, including its secondary, tertiary and quaternary components. Since the main features of this structure are already a matter of common knowledge for both DNA and RNA, one can predict the effect of conformation of various polynucleotide chain fragments on the reactivity of the heterocyclic bases present in these fragments of located in the neighborhood.

Let us now consider the three basic types of three-dimensional structure of nucleic acids: (a) ideal single-stranded polynucleotide without double-stranded fragments; (b) ideal double-stranded polynucleotide; and (c) polynucleotide with both single- and double-stranded fragments (RNA).

Single-Stranded Polynucleotide. It was already shown in Chapter 7 (Section 3) that the main type of intermolecular interaction responsible for the secondary structure of single-stranded polynucleotides is base stacking.

As can be inferred from the above-described behavior of reactive sites in heterocyclic bases, stacking will hinder a reagent's attack if it is directed at a right angle to the plane of the base. The hindrances will be especially pronounced in reactions where the most fully overlapping reactive sites must be involved. Those sites of bases which are practically not blocked as a result of stacking may react with the corresponding reagents, just as in momomer units but at a slower rate. The reagents attacking a base along the plane of the ring must interact with it at the same rate, irrespective of whether the base forms part of the single-stranded polynucleotide or a monomer unit – a particular nucleotide (with due account, of course, for the polyanionic nature of the polymer). For instance, the rate of reactions at the pyridine nitrogen or carbonyl oxygen, whose orbitals with free electron pairs lie in the plane of the ring, is not affected by stacking.

Ideal Double-Stranded Polynucleotide. Most natural DNAs are double-stranded. Only some viruses have single-stranded DNAs.

In this case, base pairs are stacked to form a helix. The stacking interactions in double-stranded nucleic acids are much stronger than in single-stranded ones. Yet the double-stranded helix is also a dynamic structure. At a point in time, the chains may separate (as a result of breaking hydrogen bonds), and the bases may be turned inside out. How reactive, then, are the bases in double-stranded nucleic acids? Obviously, the reactivity of stacked bases is minimal. Practically no reactions with nucleophilic and electrophilic agents attacking the heterocyclic nucleus at a right angle to the plane of the base are initiated. Only the pyridine nitrogen and carbonyl oxygen atoms are reactive, just as in single-stranded polynucleotides. This applies only to those atoms that are not involved in hydrogen bonding – that is, N^7 in A and G as well as $C^2{=}O$ in T. Reactions on the side of the major groove proceed more easily, apparently because of the facilitated approach of the reagent.

However, in spite of their stable secondary structure, high-molecular weight DNAs are partially vulnerable also to reagents attacking bases from a plane perpendicular to that of the heterocyclic nucleus. The primary attack may be aimed at any base which has become vulnerable as a result of fluctuation. The structure of such a segment becomes loose (hydrogen bonding and interplanar interactions are hindered by a new substituent incorporated into the nucleus of the heterocycle). Naturally, looseness of structure renders the neighboring bases vulnerable to attack as well.

Ribonucleic Acids. As has already been mentioned in a previous chapter, the polynucleotide chain of a single-stranded RNA has alternating single- and double-stranded segments. A classical example is tRNA shaped as a cloverleaf folded into an L-shaped tertiary structure. The tRNA portions prone to reagent attack are not involved in its three-dimensional structure.

Consequently, knowing the reactivity of a polynucleotide, one can distinguish those segments of its polymer chain which take part in formation of the secondary or tertiary structures, as well as segments and individual nucleotides of the RNA molecule not involved in intramolecular interactions. This principle underlies studies into the three-dimensional structure of nucleic acids by the chemical modification method.

9.2.2 Modification of Heterocyclic Bases

As is known, the order in which heterocyclic bases alternate in nucleic acids as well as their nature exert a decisive influence on the functioning of nucleic acids, particularly, on the way they pass on genetic information. So when we examine the ways in which the chemical nature of pyrimidine and purine bases in nucleic acids, changes, that is, their chemical modification, we find ourselves face to face with the most exciting opportunity in the organic chemistry of nucleic acids – the possibility of tinkering with the structure of the genetic material.

The reactivity of pyrimidine and purine bases in nucleosides and nucleotides is not affected in any significant way by three-dimensional structure, and from the standpoint of chemical behavior they are similar to the corresponding free pyrimidines and purines (under conditions when the carbohydrate moiety or phosphate group is not involved). At the same time, as follows from the foregoing, the steric factors of the structure of nucleic acids significantly influence the reactivity of their constituent bases. Knowing these factors and also the nature of the reactive sites of the bases and the mechanisms of their reactions with various substances, one can select the right conditions for the modification to proceed in the easiest way along a definite route. This means that the reaction should not affect internucleotide linkages plus carbohydrate moieties and phosphate groups. Another implication is that the reagent's attack must be aimed at particular reactive sites of heterocyclic bases. This section deals successively with transformations involving each of the above discussed reactive sites. Hopefully, by the end of the chapter the reader will have a better understanding of how the structure of bases in nucleic acids can be modified in a controlled manner.

9.2.2.1 Reactions at Carbon Atoms

Interaction with Nucleophilic Reagents. The most reactive fragment in pyrimidine bases of nucleic acids, under conditions when the latter retain their three-dimensional structure (aqueous solutions, pH values close to neutral, moderate temperature), is the double bond $C^5{=}C^6$ deficient in electrons because of the effect exerted by the carbonyl groups.

The uracil and cytosine systems are in certain ways reminiscent of compounds in the ethylene series, containing an electron-acceptor substituent at the α-carbon. Just as such compounds are attacked by nucleophilic reagents at the β-position, the nuclei of uracil and cytosine react with nucleophiles at C^6. Since, as has already been pointed out, the nucleophile attacks the base at a right angle to the plane of the heterocyclic ring, such reactions are possible only if the base occupies a single-stranded segment of the nucleic acid. Stacking makes them impossible over double-stranded segments of RNA and DNA.

It should be borne in mind that the interaction of nucleic acids with various reagents is much slower, as compared to nucleosides and nucleotides. Consequently, in order to ensure sufficiently fast reaction rates, nucleic acids are modified using reagents in excess amounts. As a result, the concentration of ions in the reaction mixture may reach a rather high level. In some cases, this disturbs the three-dimensional structure of the nucleic acid involved. This possibility must always be taken into consideration when nucleic acids are modified with a view to elucidating their spatial organization.

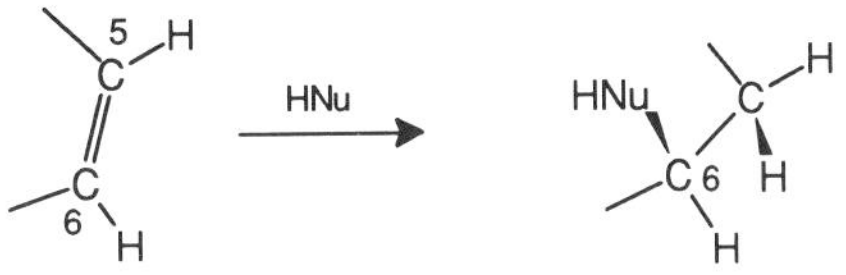

As a consequence of addition, the pyrimidine nucleus is not planar any more (it acquires the half-chair conformation similar to that in cyclohexene) and loses its aromaticity.

Introduction of alkyl substituents at C^5 and C^6 in the pyrimidine ring renders the latter significantly more stable toward nucleophiles. This is why thymines usually enter into such reactions with much greater difficulty. Purine bases are not readily involved in them either (the double C=C bond in the purine nucleus, corresponding to the $C^5{=}C^6$ bond in pyrimidines, is integrated into the aromatic system of imidazole).

Under the effect of nucleophilic agents, cytosines and adenines may undergo substitution of the exocyclic amino group. In the case of adenine, however, this reaction is extremely slow.

Thus, no acceptable methods for modifying purine bases in nucleic acids by nucleophiles have been devised so far.

For modification of uracils and cytosines use is made of *O*-methylhydroxylamine, sodium bisulfite and, less commonly hydrazine or alkylhydrazines. Depending on the reaction conditions and the reagents used, the initial addition at the double bond $C^5{=}C^6$ may be followed by both amino group substitution (for cytosine) and cleavage of the heterocyclic ring.

Let us, first, consider the addition and substitution reactions.

In the case of uracils, the addition of nucleophiles, such as hydroxylamine and bisulfite, proceeds at the fastest rate in a weakly alkaline medium (pH ~ 8) when the heterocyclic base is not ionized. *O*-Methylhydroxylamine does not enter into the reaction.

X=X'=NHOH; X= $-O-SO_2^-$; X'= $-SO_3^-$

R - polynucleotide chain

The resulting dihydrouracil derivatives are rather unstable and readily enter into a back reaction. The rate of hydroxylamine detachment is virtually independent of the pH of the medium, at pH values ranging from O to 6, but increases sharply with temperature (in an alkaline medium at pH ≥ 10 the uracil nucleus breaks down). The bisulfite anion is easily detached in an alkaline medium at pH 9–10; the elimination reaction slows down at pH ≥ 11. This indicates that dianion (1) does not participate in the reaction. There is evidence that the process takes the following course:

Consequently, the above reactions involving uracil derivatives cannot be used to obtain and subsequently investigate nucleic acids containing bases in a modified form; however, they must be taken into account in the calculations necessary for modifying nucleic acids by reagents.

The addition of such nucleophiles as hydroxylamine and its *O*-methyl ester as well as bisulfite at the double C^5–C^6 bond in cytosines proceeds easily at pH 5–7; that is, under conditions when the latter are partially protonated at N^3. It is the protonated form that is the most active in this case.

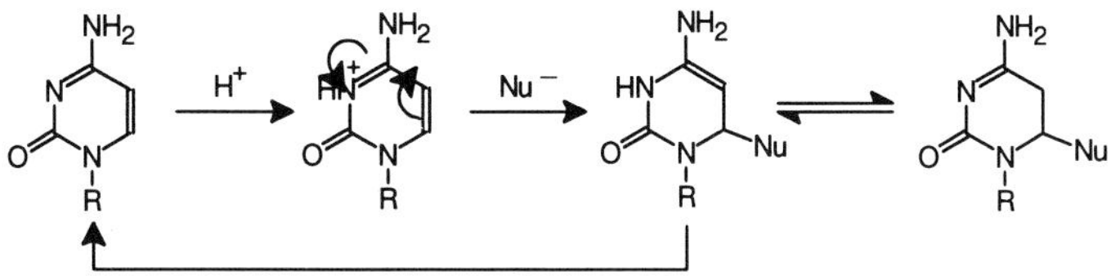

R=polynucleotide chain

What sets the modification of cytosines apart is the fact that immediately after addition of the nucleophile at the C^5–C^6 bond (this reaction is reversible just as in the case of uracils) the amino group readily undergoes nucleophilic substitution. The following scheme summarizes all possible transformations affecting cytosines in the presence of nucleophilic reagents (HX and HY, see Table 9-1).

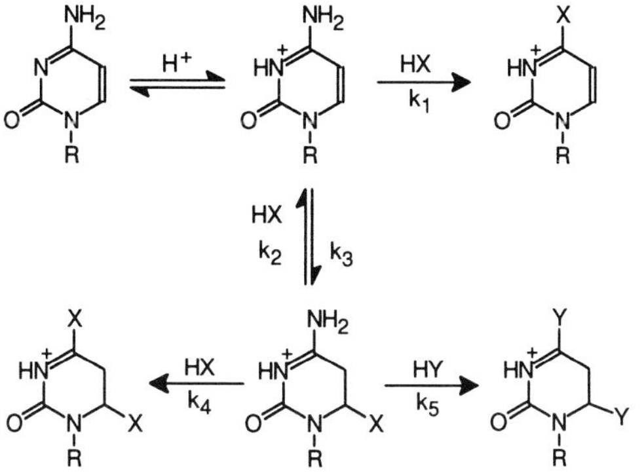

R=polynucleotide chain

Addition at the double C^5–C^6 bond in the reaction with bisulfite anion proceeds at the fastest rate. If the reaction medium contains another nucleophilic reagent (HY), the amino group may be substituted by both nucleophilic groups X and Y. The nucleophilic substitution in the aromatic ring of cytosine is rather slow ($k_1 \ll k_2$). Hydroxylamine, *O*-methylhydroxylamine, bisulfite ion, and some combinations of these reagents are usually the nucleophiles of choice for this type of nucleic acid modification (see Table 9-1). Of particular interest as a modification agent is the bisulfite ion. Its use makes it rather easy to attain an equilibrium which is markedly shifted toward adduct formation if the conditions are right (by more than 70% in a 1 M solution of bisulfite at 20 °C and pH 5). Under the same conditions, substitution of a hydroxyl for the amino group takes place at a rate slower than that of adduct formation (water in the above scheme acts as HY; $k_1 = k_4 = 0$, $k_5 < k_2$ and k_3). The resulting deamination product is similar to the adduct of uracil and bisulfite and easily loses the bisulfite anion at pH 8–9 to turn into a respective uracil derivative. Knowing the optimal pH value for each step of this process, one can convert nucleic acid cytosines into uracils under sufficiently mild conditions. This method of nucleic acid modification can be used, for example, in secondary structure studies.

It should be borne in mind that if the reaction with bisulfite is conducted at pH 7, virtually no cytosines are converted into uracils.

A rather important application has been found for a reaction between cytosines and bisulfite in the presence of nucleophiles more active than water. Such nucleophiles have usually been various amino compounds of the NH_2R.

HSO_3^- → $R'NH_2$ → pH ≥ 8 →

R=polynucleotide chain

R'= H, CH_3, C_6H_5, OH, OCH_3, CH_2COOH, $CH_2CONHCH_2COOH$

The optimal pH value for such a modification of cytosines depends, naturally, on p*K* of the amino compound used (the latter acts as a nucleophile when the amino group is not protonated).

Thus, in the case of cytosines the above reactions can, according to the end results, be classified as nucleophilic substitution at C^4.

Table 9-1 summarizes reactions between cytosines of nucleic acids and nucleophiles, proceeding at C^4 (k_1, k_4, k_5) and C^6 (k_2).

As has already been mentioned, addition of nucleophiles at the double $C^5{=}C^6$ bond may in certain instances be followed by rupture of the heterocyclic rings in pyrimidines. Such transformations involving uracils, thymines

Table 9-1. Reaction of the Cytosine Fragment in Nucleic Acids with Nucleophilic Reagents "+" – the reaction proceeds; "–" – the reaction does not proceed.

Reagents		Direction of the reaction				
HX	HY	(k_1)	(k_2)	(k_3)	(k_4)	(k_5)
NH_2OH	HOH	+	+	+	+	–
NH_2OCH_3	HOH	+	+	+	+	–
HSO_3^-	HOH	–	+	+	–	+
	NH_2OH	–	+	+	–	+
	NH_2R	–	+	+	–	+

and cytosines have been observed while treating nucleic acids with hydrazine at pH > 8.

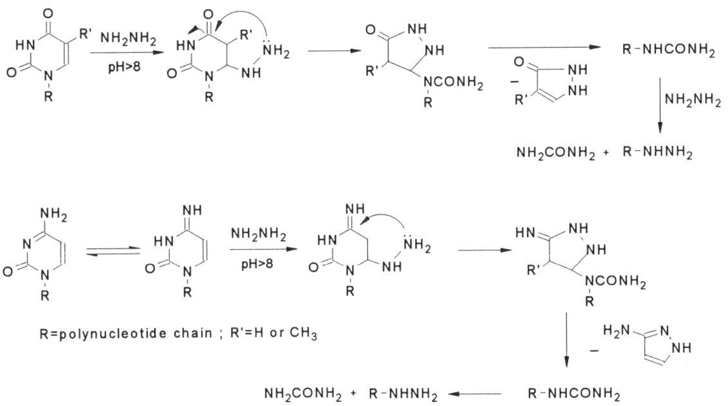

R=polynucleotide chain ; R'=H or CH_3

For removal of pyrimidine bases nucleic acids are normally treated with anhydrous hydrazine (60 °C, 20 h) to avoid cleavage of phosphodiester bonds in an alkaline medium. A similar result can be achieved when aqueous solutions of hydrazine are used at controlled pH values (pH 9.5, 0 °C). Hydrazine treatment of DNA leads to complete elimination of cytosines and thymines. As regards RNA with its more reactive uracils (in comparison with thymines), complete elimination of pyrimidines is much faster. When, for example, tRNA is treated with anhydrous hydrazine (37 °C, 15 h), all uracils and cytosines are eliminated. A side process of note is cleavage of internucleotide linkages or, in other words, depolymerization.

The residual hydrazine can be eliminated from nucleic acids depyrimidinated in this fashion with the aid of aldehydes (ketones) so that the site previously occupied by the pyrimidine unit in the nucleic acid now accommodates heterocycle-free ribose whose phosphodiester bonds are highly labile and can break under mild conditions. This procedure is widely used to break nucleic acids down in determining their primary structure.

Hydroxylamine acts on nucleic acids in an alkaline medium similarly to hydrazine but in a more specific manner. At pH ≥ 10 it breaks down only uracils. This reaction resembles that of cytosines and uracils with hydrazine under identical conditions.

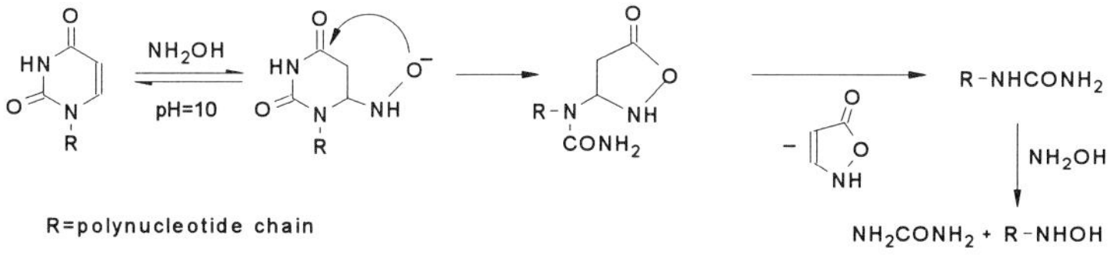

R=polynucleotide chain

The specificity of this reaction makes it possible to selectively modify uracil nuclei in RNA. The optimal conditions for the reaction are as follows: pH 10, 10 M solution of hydroxylamine, 10 °C, 150 h. The last step boils down to substitution of hydroxylamine for urea in the corresponding segments of the polynucleotide chain. Under the above optimal conditions, the nonspecific modification of the polynucleotide chain is minimum.

Just as hydrazinolysis, the reaction of DNA and RNA with hydroxylamine is employed for controlled cleavage at base-free units. The reaction with hydroxylamine is sensitive to the secondary nucleic acid structure (under the reaction conditions the secondary structure seems to persist) so that only uracils outside double-stranded segments are cleaved. In the case of tRNA, for instance, only those uridine units undergo modification under normal conditions which are located in loops, whereas at 37 °C or at a low temperature but in the presence of urea which destroys the secondary structure uracils are cleaved completely.

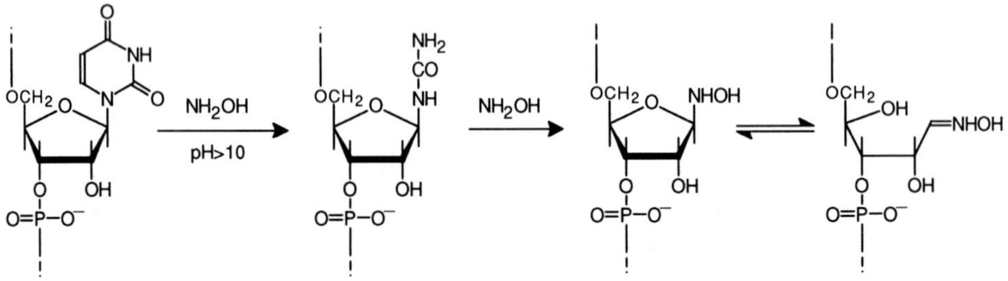

NH_2OH, pH>10 → ; NH_2OH → ; ⇌

Interaction with Electrophilic Reagents. As is known, electrophilic substitution in pyrimidine bases forming part of nucleotides occurs at C^5, while in purine bases it occures at C^8. These bases behave similarly in nucleic acids as well. This section deals only with reactions taking place in water under mild conditions favorable for nucleic acids to retain their three-dimensional structure.

These conditions are ideally provided by halogenation and mercuration. The halogenation reaction involves uracil, cytosine, thymine, and guanine. Just as in the cases described above, the halogenating agent attacks the reactive site at an angle normal to the plane of the heterocyclic ring, which is why the secondary structure and other structural features of nucleic acids impose the same limitations on the reactivity of the constituent bases as the use of nucleophilic reagents.

The bromination of uracil and cytosine in nucleic acids can be achieved by adding bromine to water at pH values close to neutral.

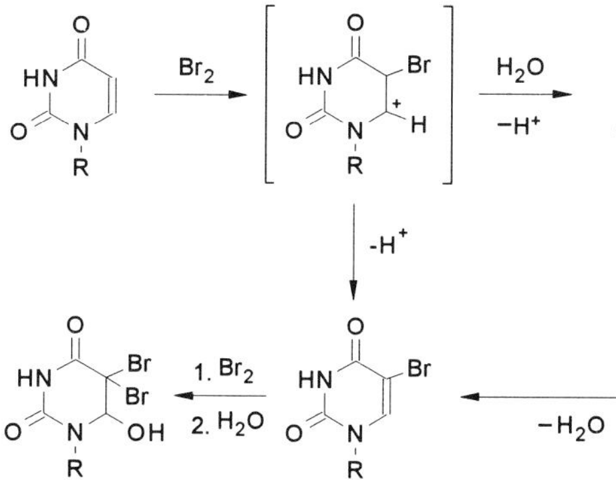

R=polynucleotide chain

In nucleic acid cytosines whose modification also proceeds according to the above scheme, the bromohydrin which is formed first easily undergoes substitution of a hydroxyl for the amino group (just as the adduct of cytosine with bisulfite) with subsequent release of water to give 5-bromouracil:

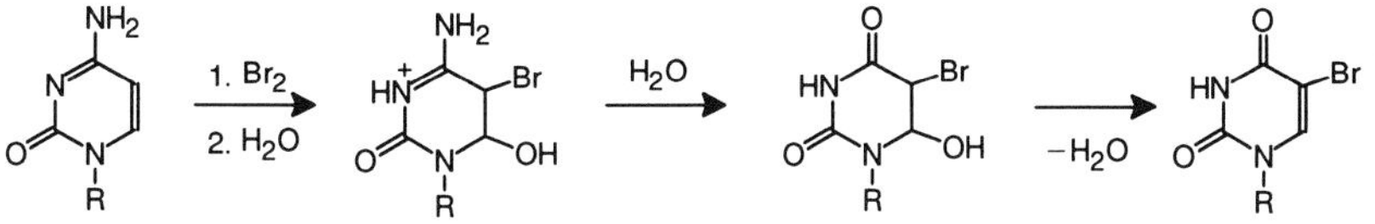

R=polynucleotide chain

The modification of nucleic acid thymines comes to a halt at the step of formation of the respective bromohydrin.

Treatment of nucleic acids with bromosuccinimide modifies cytosines and guanines, the modification of the former occurring predominantly at pH 9 and that of the latter at pH 7.

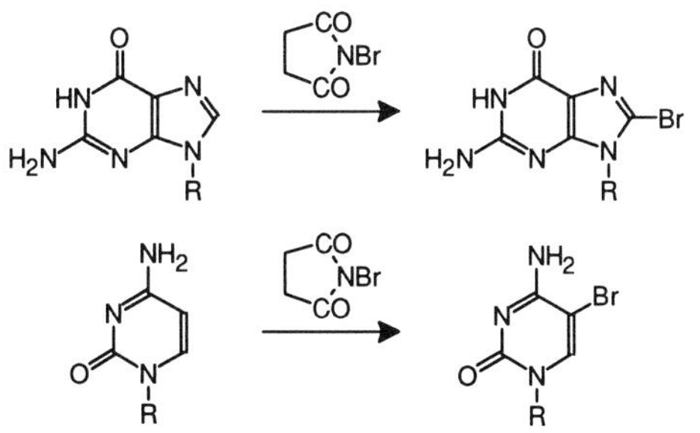

R=polynucleotide chain

To incorporate iodine into polynucleotides ICl or I^- is normally used in the presence of $TlCl_3$ (mild oxidizing agent) because free iodine is not reactive enough. Iodine itself, however, is a specific agent with respect to some tRNAs because some constituent minor bases of the latter do enter into a reaction with it. For example, the reaction with 6-isopentenyladenosine yields a new cyclic system:

R=residue of ribose

When iodine chloride is employed as the iodination agent, the halogenation proceeds in the same fashion as bromination. The wide use of iodination to modify nucleic acids stems from the fact that this technique allows relatively stable isotopes (the half-life of ^{125}I, for example, is about 25 days) to be incorporated into them under mild conditions, which enables one to use radiometric methods and handle nucleic acids in ultramicroamounts. Since incorporation of an iodine isotope does not alter the specific interactions in nucleic acids (three-dimensional structure), this approach is highly promising for structural and functional studies.

Another reaction of electrophilic substitution – mercuration – is also used for modification of nucleic acids apart from halogenation. The mercurating agent, which is usually mercuric acetate, reacts with pyrimidine bases at position 5 without affecting purines and thymines.

R=polynucleotide chain

Uracils and cytosines, which react in a similar manner, are converted into the corresponding 5-mercury substituents as part of both mononucleotides and nucleic acids. The reaction proceeds in buffer solutions at pH 6–7. The covalent (or, to be more precise, coordination covalent) C–Hg bond is rather strong in such compounds under conditions usually maintained while conducting biochemical reactions. Treatment of nucleic acids with excess mercuric acetate leads to substitution for the hydrogens at C^5 of all pyrimidines. It should be remembered that the course taken by mercuration of nucleic acids is independent of their secondary structure, although this reaction must proceed, similarly to halogenation, as an attack perpendicular to the base plane. This may stem from the fact that the ion Hg^{2+} with its pronounced capacity for complexing is retained, at first, by the nitrogens of the heterocyclic bases through noncovalent interactions and then gets sandwiched (intercalates) between base planes which are involved in stacking. When a limited number of bases are to be modified, the reagent is used in a limited amount. Any RNAs and DNAs (including synthetic polynucleotides) may enter into the reaction.

Mercurated polynucleotides (even completely mercurated DNAs and RNAs) retain their capacity for complementary interactions (the HgX moiety lies in the major groove) and may serve as templates and primers in certain nucleic acid biosynthesis reactions involving polymerases.

Mercurated double-stranded DNAs are characterized virtually by the same melting points and display the same kind of hypochromism as the starting DNA molecules.

Mercurated nucleotides and polynucleotides readily react with mercaptans, which becomes useful in immobilization of polynucleotides on sulfhydryl Sepharose®:

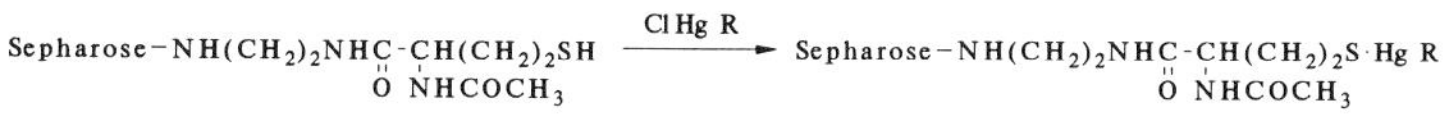

R=polynucleotide chain

This reaction, yielding a rather stable C^5–Hg–S group, shows a great deal of specificity and proceeds at a high rate: the immobilization is rather easy even for preparations with only one mercury atom per 200 nucleotides. Elution with a buffer containing mercaptan (mercaptoethanol, etc.) removes the polynucleotide from Sepharose (as a result of exchange). This technique for immobilizing polynucleotides on and removing them from the carrier under very mild conditions is employed to separate and isolate mercurated polynucleotides from complex mixtures where the latter are used as templates or primers in biosynthesis of DNAs and RNAs.

Mercurated nucleotides rather easily exchange a mercury atom for a halogen or tritium if the corresponding compound is treated with such agents as iodine, *N*–bromosuccinimide, sodium borohydride, and so on. For instance:

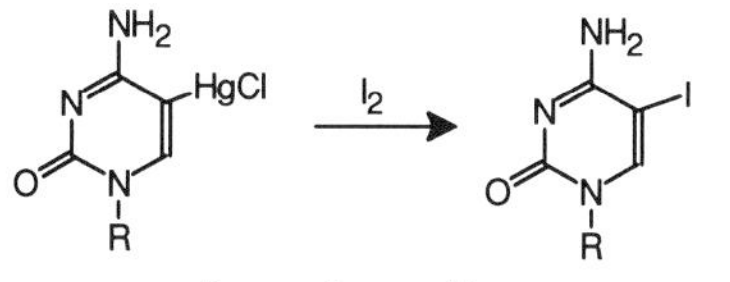

R = pentose residue

Polynucleotides can easily exchange a mercury atom as well, which is used when isotopes of iodine (reagent $^{125}I_2$) or tritium (reagent $NaBT_4$) are to be incorporated into DNA and RNA molecules. Such a labeling method is of particular importance for RNAs because the corresponding forward reactions involving the latter are accompanied either by formation of side products (e.g., dihydrouracil derivatives) or cleavage of internucleotide linkages (when the exchange takes place in an alkaline solution of T_2O).

The mercuration of RNAs and DNAs as well as synthetic oligo- and polynucleotides may be used in studying the nucleic acid structure by electron microscopy and other techniques.

Isotopic Exchange of Hydrogen. Polynucleotides are typically involved in isotopic exchange of hydrogen atoms in aqueous media containing deuterium and tritium. Distinction is made between "fast" and "slow" isotopic exchange. During fast isotopic exchange, the state of equilibrium is attained within seconds or a few minutes even at 0 °C. Groups participating in fast isotopic exchange include the phosphates and hydroxyls of the carbohydrate–phosphate backbone as well as the amino and imide groups of heterocyclic bases. The high rates of exchange observed in all cases are due to the high rates of protonation and deprotonation of oxygen and nitrogen atoms. Thereby, any alteration in the electronic state or accessibility of oxygen or nitrogen (e.g., when exocyclic amino and other groups as well as heterocycles are involved in complementary pairing) is followed by a drastic change in the isotopic exchange rate.

Analysis of the kinetic data pertinent to isotopic equilibrium makes it possible to identify rather clearly defined groups with characteristic isotopic change rates. Such analysis permits one, for example, to estimate the involvement of particular nucleic acid bases in complementary interactions, to determine the relative number of single- and double-stranded segments in RNAs and DNAs, and to monitor changes in the intramolecular organization of nucleic acids and nucleoproteins.

By slow exchange in nucleic acids is meant the reaction of isotopic exchange of the hydrogens bound to carbons of heterocyclic bases. Such an exchange is observed when a nucleic acid is heated for many hours in deuteriated or tritiated water at elevated (about 100 °C) temperatures. It has been found that only the hydrogen at C^8 in purines and C^5 in pyrimidines is capable of entering into the isotopic exchange reaction. The hydrogen at C^6 in pyrimidines also tends to enter into such a reaction but at an extremely slow rate and requires rather rigorous conditions. The rest of the hydrogens bound to carbons do not enter into isotopic exchange reactions in aqueous solutions[1)].

The slow isotopic exchange in purines is essentially a first-order reaction and proceeds at a rate 30 to 100 times faster than in pyrimidines. The so-called ilide mechanism has been proposed to account for the isotopic exchange in purine derivatives. According to this mechanism, the following sequence of reactions leading to the isotopic exchange can be written for guanines in a polynucleotide chain:

[1)] When dry nucleic acid preparations are irradiated with tritium or deuterium atoms – that is, under conditions of heterogeneous exchange – all hydrogens in nucleic acids may enter into isotopic exchange reactions. The fastest exchange involves the methyl hydrogens in thymines.

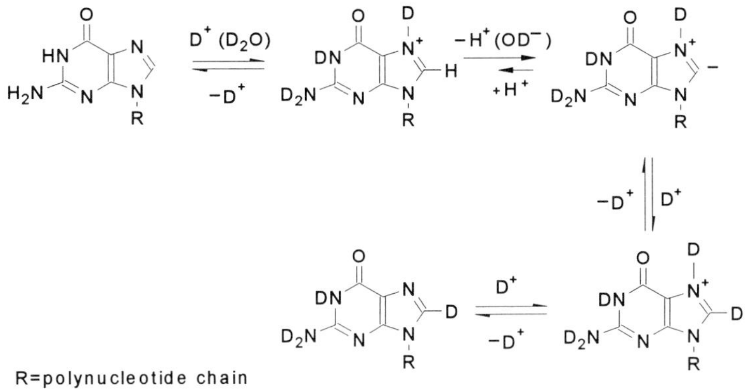

R=polynucleotide chain

The first step of the process is protonation of the imidazole ring at N^7. This is one of the two steps limiting the entire process (the protonation itself is rather fast but the equilibrium concentration of the protonated form which continues to react is small). The second, slow step is proton detachment from C^8, mediated by a hydroxyl ion, to form an ilide. The resulting carbanion abstracts a proton (deuterium or tritium) from water, which eventually leads to isotopic exchange at C^8.

A similar mechanism of isotopic exchange of hydrogen at a respective carbon has been demonstrated for imidazole and related five-membered heterocyclic systems.

Interestingly, favorable conditions for the first limiting step are created in an acid medium, while those for the second step are created in an alkaline one. This results in a rather complex relationship between the isotopic exchange rate and pH. In the case of guanines, for instance, an increase in pH from 4 to 8 affects only insignificantly the rate of isotopic exchange but at pH ranging from 8 to 11 the reaction rate increases sharply, as much as almost ten-fold. A further increase in the concentration of hydroxyl ions up to 1 N does not bring about any significant changes in the isotopic exchange rate.

As regards adenines, the isotopic exchange in neutral and weakly alkaline solutions is based on the same mechanism but proceeds at a perceptibly slower rate than in the case of guanines, which may be explained by the diminishing positive charge at N^7 due to intramolecular protonation of the exocyclic amino group.

The rate of exchange in adenines varies slightly in the pH range of 4 to 12, but in concentrated alkalies (> 0.1 N) it increases drastically in proportion to the hydroxyl ion concentration. The mechanism of this phenomenon is yet to be elucidated.

Purine derivatives carrying a positive charge at N^7 (e.g., N^7-methylguanine, N^1, N^7-dimethylguanine, N^7-methylinosine) have been found to display rates of the isotopic exchange at C^8 comparable to those of fast isotopic exchange. This principle underlies an original method for determining the degree of methylation of guanines in DNAs. Taken for the test is a ^{3}H-DNA sample with tritium atoms at C^8 of guanines, which is then treated with dimethyl sulfate, and the percentage of the tritium capable of rapid exchange with water is determined as a measure of modification (methylation) of N^7 of guanine in the DNA.

+ H⁺ −H⁺
isotope exchange
R=polynucleotide chain

As has already been mentioned, the rate of the slow isotopic exchange at C^5 in pyrimidines is much lower than in purines. This exchange, however, is accelerated significantly in the presence of compounds capable of undergoing reversible addition at the double $C^5{=}C^6$ bond. Such compounds may be citrates, bisulfite ion, 2-mercaptoethylamine, ethylamine, ethanolamine, cysteine, and so on. In this case, the isotopic exchange rate becomes proportional to the concentration of the corresponding homogeneous catalyst and may be tens and hundreds of times higher. The mechanism of this homogeneous catalysis resides in reversible addition of respective compounds at the $C^5{=}C^6$ bond in pyrimidines. The exchange in a cytosine with bisulfite ion serving as the catalyst can be written as follows:

R=polynucleotide chain

It can be seen that the addition proceeds in two steps: 1,4-addition to a conjugated system of double bonds (N^3, C^6), followed by intra- or intermolecular transfer of the isotope from N^3 to C^5. The last step is not sterochemically directed so that the deuterium added to C^5 may find itself either in *syn* or *anti* position with respect to the remaining hydrogen atom (remember that the emerging hydrogenated pyrimidine ring is not planar). The subsequent β-elimination seems to proceed in a single step and in a stereochemically directed manner with removal of the bisulfite anion and hydrogen that are either in *anti–anti* or *syn–syn* positions (*trans*-detachment). Thus, during regeneration of the starting cytosine some hydrogens become substituted by deuterium. When the addition-detachment cycle is repeated over and over, isotopic equilibrium is achieved, which is to say that we are dealing with an isotopic exchange reaction.

A similar mechanism underlies exchange in uracils as well, with the difference that the exocyclic oxygen at C^4 is involved in the 1,4-addition.

Reversible addition of some nucleophiles at the double $C^5{=}C^6$ bond has also been described for thymines; however, it must not (and does not) lead to isotopic exchange of hydrogen at carbon atoms.

Interestingly, the best catalysts of isotopic exchange at C^5 in pyrimidines are bifunctional reagents of the cysteine or 2-mercaptoethylamine type. The catalysis of isotopic exchange in uracil by deuteriated 2-mercaptoethylamine can be represented as follows:

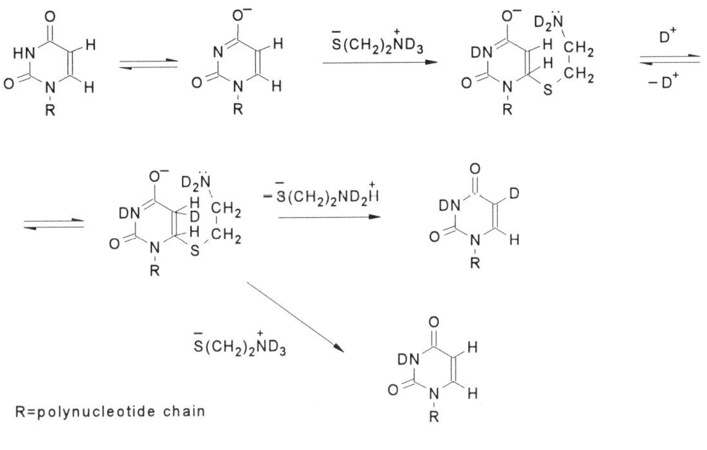

R=polynucleotide chain

It is evident that in the case of bifunctional agents of the 2-mercaptoethylamine type, where the mercapto group acts as a nucleophile and the amino group acts as a base, intramolecular catalysis of detachment of the added nucleophile is possible, which increases the rate of addition-detachment or, in other words, isotopic exchange.

The different optimal pH values for these reactions, depending on whether they involve cytosines (pH 5–7) or uracils (pH 7–9), make it possible to conduct isotopic exchange selectively for each of the two bases in the heteropolynucleotide. Moreover, it should be pointed out that the reaction of isotopic exchange at C^5 in pyrimidine derivatives is contingent on possible attack of C^6 by a nucleophile at a right angle to the ring plane. The reaction is virtually at a standstill in native double-stranded DNAs and double-stranded segments of single-stranded DNAs and RNAs where such an attack is hindered. This property is often put to practical use when it is necessary to determine the involvement of a particular pyrimidine in the formation of secondary RNA or DNA structure.

The slow isotopic exchange of hydrogen at C^6 in pyrimidines, which may occur without any specific catalyst as well, is based on the same addition-detachment mechanism with water acting as a nucleophile and proton donor.

The slow isotopic exchange reactions described in this section are widely used in preparing tritium-labeled DNA and RNA. To this end, a nucleic acid preparation is kept at high temperature in tritiated water till isotopic equilibrium is reached and the rapidly exchangeable tritium is eliminated. The nucleic acid isolated after such treatment contains only slowly exchangeable tritium (tritium atoms at restrictive positions). Being easy to prepare, nucleic acids labeled with tritium in this fashion do not undergo any marked degradation and modification, whereas the loss of tritium in aqueous solutions (pH 3–10) is negligible due to slow reverse isotopic exchange.

9.2.2.2 Reactions at Pyridine Nitrogens

As a consequence of the presence of lone-pair electrons not participating in the formation of an aromatic sextet and the lack of bonds with hydrogen atoms, the pyridine nitrogens acting as reactive sites in nucleic acid bases react primarily with electrophiles. The most commonly used procedures for modification at these atoms are alkylation, production of N-oxides, and the reaction with diethylpyrocarbonate.

Alkylation. In assessing the possibilities of alkylating heterocyclic bases in nucleic acids, one must first of all compare the nucleophilic properties of the reactive sites in their molecules. It has already been mentioned that such properties must be most pronounced at pyridine nitrogens. In order to determine which of them will be alkylated most readily (exhibit nucleophilic properties by giving up a pair of electrons for binding with a carbon) and in what bases, it is advisable to bear their capacity for protonation in mind (i.e., capacity to give up an electron pair for binding with a hydrogen, thereby displaying basic properties), since it is konwn that the nucleophilic and basic behavior of nitrogen atoms in organic compounds varies concomitantly. As has been pointed out, cytosines, adenines and guanines in nucleosides and nucleotides are protonated, respectively, at N^3, N^1 and N^7 (the reasons why the exocyclic amino groups in cytosines and adenines do not lend themselves to protonation have already been discussed), whereas uracils and thymines do not exhibit any basic properties at all. However, the differences in the behavior of adenines and guanines during protonation are yet to be explained. The only assumption that can be made is that the lower basicity of N^7 in adenines, as compared to guanines, has to do with screening by the exocyclic amino group.

According to the ease with which the above bases enter into the reaction, they can be arranged in the following order: guanine (N^7) > adenine (N^1) >

cytosine (N^3). The involvement of other pyridine nitrogens is much more insignificant.

Just as with other types of nucleic acid modification, alkylation is usually conducted in aqueous solutions at neutral or slightly acid pH values. As the medium becomes more acidic (pH < 6), when the bases are predominantly in a protonated form, alkylation naturally comes to a halt.

At alkaline pH values, $-NH-\underset{|}{C}=O$ groups in uracils thymines and guanines lose their proton to pass into the anionic form $-N=\underset{|}{C}-O^-$ which, as has already been mentioned, is alkylated at the nitrogen atom. For instance, the alkylation of uracils and thymines at alkaline pH values can be represented as follows:

1. ^-OH; 2. $-H_2O$ → ; $R''X$, $-X^-$ →

R=polynucleotide chain
R'=H or CH_3 ; X=Hal , OTs
R"=Alk

Similarly, at pH > 8, guanines are alkylated at N^1 (along with formation of an N^7-alkyl derivative).

Under normal alkylation conditions (pH 6–7, aqueous solutions, 20–37 °C), the alkylating agent attacks the heterocyclic pyridine nitrogen in the base plane. This is why stacking does not interfere with the reaction and the modification may involve bases forming part of double-stranded segments of nucleic acids. In this case, however, attack of these nitrogens by the alkylating agent is impossible (unless the double helix is damaged) because N^1 of adenine and N^3 of cytosine participate in hydrogen bonding. Therefore, when native DNAs are alkylated, the modification must occur primarily at N^7 of the guanine located in the wide groove and, to a lesser extent, at the N^3 of adenine. The alkylation of RNAs involves guanines, adenines and cytosines in single-stranded segments and only guanines in double-stranded ones. Since this method allows the investigator to find differences in the three-dimensional structure of nucleic acids, it is considered to hold a certain degree of promise for structural studies. More particularly, such an approach can be used in studying the structure of nucleoproteins where failure of guanines to be alkylated may suggest that a given portion of the nucleic acid is blocked by a protein. Another area of application of alkylation reactions is selective rupture of internucleotide linkages in nucleic acids. Since the glycosidic bond in an alkylated guanosine (which, as has been pointed out, is alkylated at a maximum rate) can be cleaved rather easily, one can produce polymers without guanines at selected points.

pH=4-5

R=Alk

β-elimination (cleavage of polymer)

The same applies to segments with adenines alkylated at N^3. Such selective cleavage of the polynucleotide chain forms the basis of a new chemical method for determining the nucleotide sequence in DNAs.

The most frequently employed alkylating agents include alkyl halides, dimethyl sulfate, methyl methanesulfonate, and diazomethane. Analogues of nitrogen mustards, sůch as *N*-aryl-*β*-chloroethylamines, are also used for the purpose. The alkylation with alkyl halides as well as sulfates and sulfonates is based on the usual mechanisms of nucleophilic substitution.

Which bases are methylated in the presence of diazomethane depends, as has already been indicated, on the reaction conditions: in the absence of proton donors it is pyrrole nitrogens that are methylated, whereas with protonation pyridine nitrogens are involved. It should be noted that the methylation of polynucleotides with diazomethane is always accompanied by cleavage of phosphodiester bonds (i.e., degradation of the polymer chain) and methylation at the 2′-hydroxyl groups of carbohydrate moieties (if such are present). At the polymer level this reaction is used when it becomes necessary to methylate uracils or thymines; it should be remembered, though, that the reaction is not selective – it also involves guanines as well as, to a lesser extent, cytosines and adenines.

Alkylation with *N*-arylchloroethylamines initially yields an imonium cation which then reacts wiht the base acting as the nucleophile Nu.

R = H or Alk;
Nu = heterocyclic base in polynucleotide chain

In this case, steric factors must have a more tangible effect on the course of alkylation, as compared to methylation, by virtue of the larger size of the electrophilic particle (imonium cation).

Production of N-Oxides. Pyridine nitrogens in some bases (of nucleic acids) are easily oxidized with peracids to yield N-oxides. The mechanism of this reaction is similar to that of alkylation because added to the atomic oxygen generated by the peracid is a free electron pair of the nitrogen being attacked (i.e., the attack proceeds in the plane of the heterocyclic nucleus).

Nucleic acids are usually modified by means of monoperphthalic acid which is one of the most stable peracids. The reaction is conducted in aqueous solutions at neutral pH values and at temperatures ranging from 0 to 20 °C. In the case of oligonucleotides, whose reactivity is not affected by steric factors in any significant manner, adenines react with particular ease under such conditions to yield an N^1-oxide. Cytosines react at a much slower rate and yield an N^3-oxide. It should be borne in mind that the reactivity of adenines and cytosines in a nucleic acid may vary substantially. Guanines and uracils (as well as thymines) do not react under the above conditions.

Since, as has already been mentioned, the nitrogen atom is attacked during oxidation in the ring plane, this reaction affects only single-stranded segments of nucleic acids. As regards double-stranded segments, the oxidation of N^1 in adenines and N^3 in cytosines with oxygen is inhibited practically a hundred per cent because these atoms are involved in hydrogen bonding.

Reaction with Diethylpyrocarbonate. Diethylpyrocarbonate is a rather important reagent used for modifying heterocyclic bases in nucleic acids. As it reacts with nucleophiles, the compound yields carbethoxylation products, the ethoxycarbonate anion acting as the departing group. This anion easily breaks down in an aqueous solution, whereas the resulting carbethoxy derivative is rather stable toward hydrolysis. *N*-Carbethoxyimidazole, for example, is a hundred times more stable than *N*-acetylimidazole.

$$\mathrm{Nu{:}} + C_2H_5O{-}C(=O){-}O{-}C(=O){-}OC_2H_5 \longrightarrow \underset{\text{product of carbethoxylation}}{Nu{-}C(=O){-}OC_2H_5} + {}^{-}O{-}C(=O){-}OC_2H_5 \xrightarrow{\text{fast decay}} CO_2 + {}^{-}OC_2H_5 \xrightarrow[-OH^-]{H_2O} C_2H_5OH$$

Thus, if the pyridine nitrogen of a purine or pyrimidine base acts as a nucleophile, diethylpyrocarbonate attacks it in the base plane to yield an amide-type derivative in which the positively charged nitrogen forms a rather strong linkage with the carbethoxy group; at the same time, the rapid degradation of the departing group will lead to a situation in which the hydroxyl anion becomes a counterion with respect to the resulting ammonium cation. For instance:

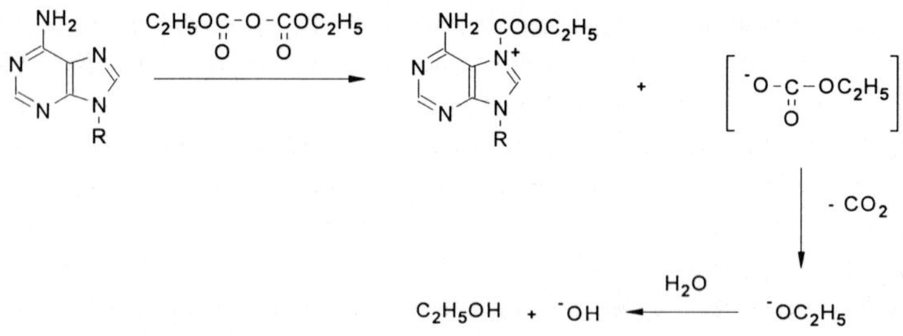

Being a strong nucleophile, the hydroxyl anion immediately attacks the neighboring C^8 atom. In this case, the attack is perpendicular to the heterocyclic base plane. The subsequent transformations result in opening of the imidazole ring (R – polynucleotide chain):

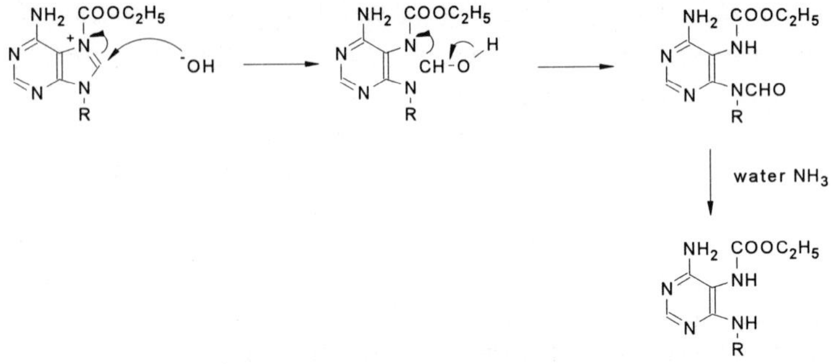

When the reagent is present in excess and the reaction takes a sufficiently long period of time, the exocyclic amino group is acylated as well.

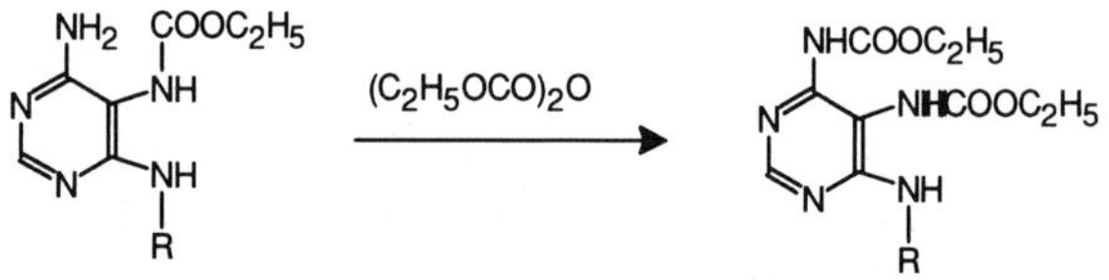

Guanines react similarly, but at a slower rate, with the difference that the intermediate formyl derivative is so unstable that it is immediately hydrolysed by water:

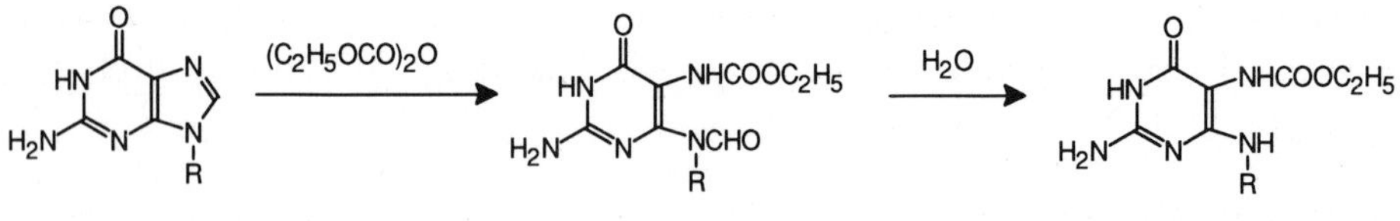

Cytosines are acylated with diethylpyrocarbonate at the exocyclic amino group and only if the reagent is used in an amply excess amount. Uracils and thymines do not get modified by diethylpyrocarbonate.

Since diethylpyrocarbonate is moderately soluble in water and easily hydrolysable, it is normally used as an emulsion and in a markedly excess amount. The reaction is carried out in neutral or almost neutral media and at room temperature.

Since, as has already been pointed out, the base is attacked during the reaction not only in the heterocycle plane (the first step of modification of adenines and guanines is acylation of N^7 atoms), but also at a right angle thereto (the second step of modification of adenines and guanines is nucleophilic addition of the hydroxyl anion at C^8 plus acylation of the exocyclic amino groups in adenines and cytosines), diethylpyrocarbonate can be used to modify only single-stranded segments of nucleic acids. Consequently, the above reactions may involve oligonucleotides, denatured DNAs and RNAs. As a rule, more complex molecules enter into a reaction with diethylpyrocarbonate more sluggishly. For example, if the modification of a dinucleotide with adenines is a matter of just a few minutes, that of an RNA takes more than three hours to produce the same results. Since the reaction of adenines and guanines with diethylpyrocarbonate is accompanied by opening of the purine ring, it can be easily monitored by decreasing intensity of the UV absorption maximum typical of purine derivatives (259 nm in the case of adenine derivatives).

Studies into the behavior of homopolyribonucleotides in this reaction have shown that poly(U) does not enter into the latter, while poly(A), poly(G) and poly(C) are modified. Under conditions when the secondary structure cannot be upset, each base in poly(A) is involved in the reaction, whereas in poly(G) and poly(C) this is true only for one out of three and one out of five bases, respectively. Such a difference in behavior is due not only to the lower reactivity of guanines and cytosines, as compared to adenines, but also to the fact in poly(G) and poly(C) some bases participate in hydrogen bonding. The latter assumption is borne out by evidence showing that the poly(U) · poly(A) complex does not react with diethylpyrocarbonate at all.

9.2.2.3 Reactions at Pyrrole Nitrogens and Exocyclic Amino Groups

When the types of reaction sites in the heterocyclic bases of nucleic acids were discussed, it was pointed out that pyrrole and amine nitrogens must be attacked by reagents in a direction normal to the plane of these bases. Consequently, such nitrogens may be reactive only when their base occupies single-stranded segments of the nucleic acid.

Addition at Active Double Bonds. The most frequently used compounds of this type are water-soluble acrylonitrile and carbodiimide salts. It has already been mentioned that the reaction involves bases containing a –CO–NH–

group in the heterocycling ring, capable of being ionized at alkaline pH values. Thus, the reaction proceeds as nucleophilic addition of the anion emerging from the base to an active multiple bond. In this case, the base reacts as an ambidentate ion at a nitrogen marked by maximum nucleophilic activity, which is N^3 in uracils and thymines and N^1 in guanines.

The rate of the reaction is maximum at a pH value of about 9. As the medium becomes more basic (at pH > 10), the resulting adduct undergoes almost complete degradation. This allows the investigator to come back to the initial structure, after having studied the modified nucleic acid, and see how well it has been retained.

Guanines react with water-soluble carbodiimides at a much slower rate; some minor bases in tRNAs are modified as easily as uracils and thymines.

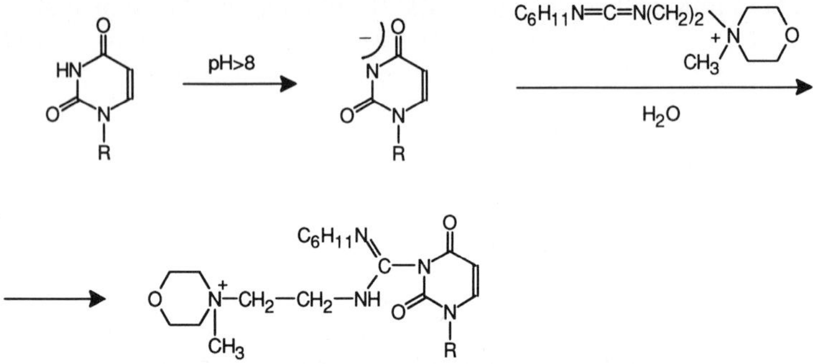

pH>8

$C_6H_{11}N{=}C{=}N(CH_2)_2$ N^+ CH_3 (morpholinium)

H_2O

$C_6H_{11}N$ C—N N^+—CH_2—CH_2—NH CH_3 R

Pseudouridylic and inosinic acids provide good examples:

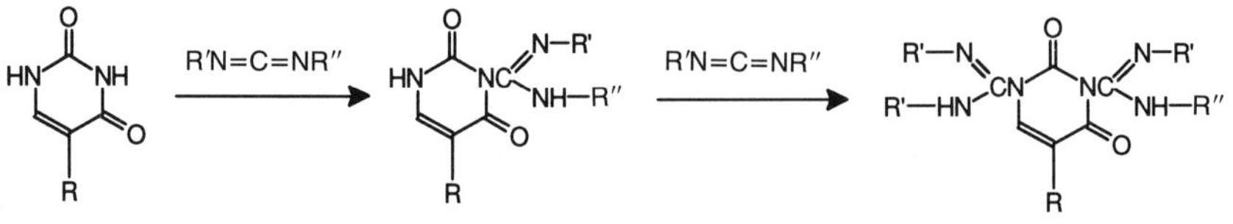

R′N=C=NR″ → R′N=C=NR″ →

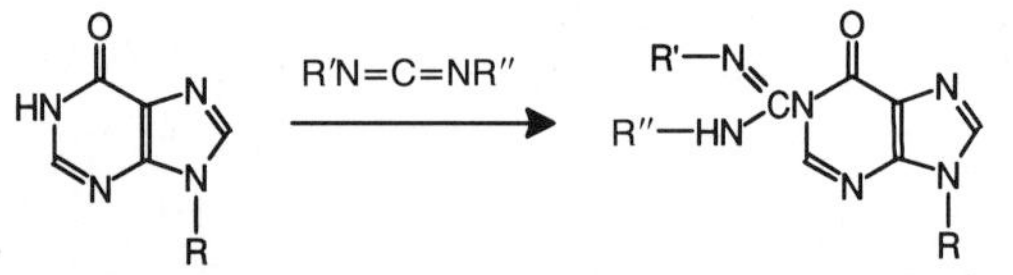

R′N=C=NR″ →

Significantly, internucleotide linkages formed by modified uracils, are not broken by pancreatic RNase.

Acrylonitrile reacts with bases, just as carbodiimides. The reaction proceeds at the highest rate at pH 8–9, the most easily modifiable minor components of tRNA being pseudouridines, 4-thiouridines and inosines. Pseudouridine, for example, reacts at a rate 20 times faster than uridine; similarly to the reaction with carbodiimides, pseudouridine may have two molecules of the reagent added thereto. The following adducts are formed:

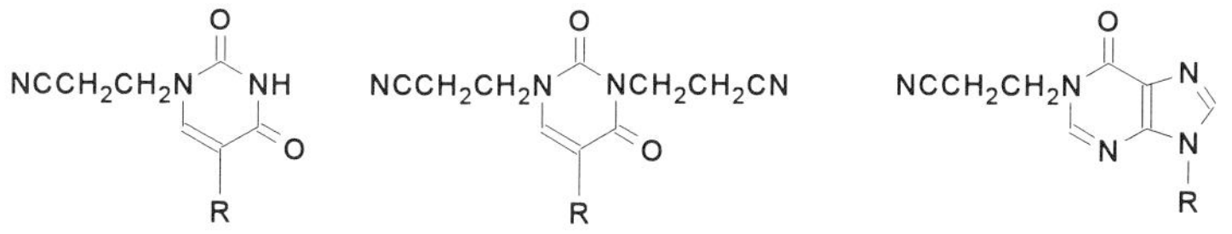

R-polynucleotide chain

Just as expected, the 4-thiouracil ionized in a weakly alkaline medium reacts at a site corresponding to maximum nucleophilic activity – that is, at the sulfur atom:

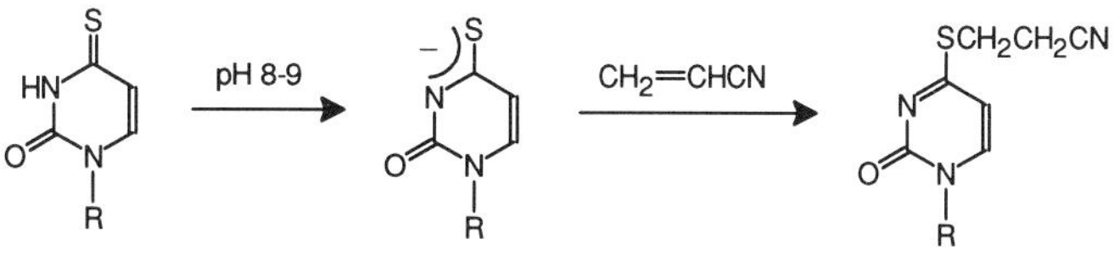

R-polynucleotide chain

2-Thiouridines do not react with acrylonitrile under the same conditions.

As regards uracils, they are modified very slowly in a weakly basic medium, whereas cytosines, adenines and guanines are not modified at all.

Reactions with Aldehydes and Ketones. The interaction of nucleotides and nucleic acids with the simplest carbonyl compound, formaldehyde, has been studied in depth. It has been established that uridines, thymidines, pseudouridines and inosines in oligonucleotides and nucleic acids yield the corresponding methylol derivatives as a result of substitution at the imine nitrogen adjacent to one or two carbonyl groups. The reaction proceeds at a rather fast rate in a moderately alkaline medium, slows down in a moderately acidic medium, and is completely inhibited when the acidity of the medium is high. This suggests that in this case, just as in a reaction with compounds having active C=C and C=N bonds, the base reacts as an anion with addition at the C=O bond, for example:

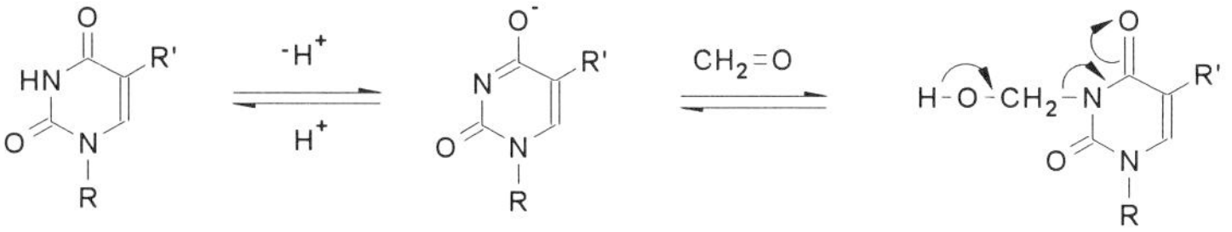

R-polynucleotide chain
R′-H or CH_3

In the case of pseudouridines, mono- and dimethylol derivatives are formed:

HN NCH₂OH   HOCH₂N NCH₂OH

R-polynucleotide chain

For methylol groups to be eliminated from the above heterocyclic bases, it is quite sufficient to gradually remove formaldehyde (e.g., by dialysis) from the reaction mixture. This is another proof of reversibility of the reaction under consideration.

While interacting with the formaldehyde of heterocyclic bases containing exocyclic amino groups, the methylol group becomes incorporated into the latter:

$$R{-}NH_2 + H_2CO \rightleftharpoons R{-}NHCH_2OH$$

R-adenine or guanine residue

The most thoroughly studied compounds include adenine derivatives. It is the above example that served as demonstration of cross-linking of polynucleotide chains via purine bases in an alkaline medium under the effect of formaldehyde. The methylol derivative formed during the first step of the reaction may have a proton detached, in the presence of an alkali, from the oxygen of the methylol group (in which case it breaks down into the starting substances) and also, to a lesser extent, from the nitrogen with which the group is linked. In the latter case, a rather active Schiff's base is formed to react immediately at the exocyclic amino group with the second purine base:

$$R{-}NHCH_2{-}O{-}H$$

$$\xrightleftharpoons[-H^+]{H^+} [R{-}NHCH_2{-}O^-] \xrightleftharpoons[H^+]{-H^+} R{-}NH_2 + CH_2O$$

$$\xrightarrow{-H^+} R{-}\bar{N}{-}CH_2{-}OH \xrightarrow{^-OH} [R{-}N{=}CH_2] \xrightarrow{RNH_2} R{-}NH{-}CH_2{-}NH{-}R$$

R - adenine residue

It has been found that the cross-linking may involve not only two adenines but also an adenine and a guanine:

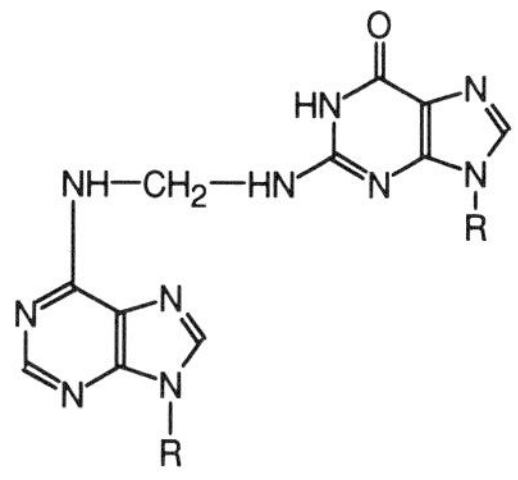

R-polynucleotide chain

Such cross-links are quite stable in alkaline media, which has been demonstrated in experiments with certain RNAs. The rate of the reaction between formaldehyde and polynucleotides is strongly dependent on the structure of the latter. Double-stranded nucleic acids react with formaldehyde at a very slow rate (the reaction seems to proceed at fluctuation points). Thereofore, formaldehyde treatment is widely used in studies of the secondary structure of nucleic acids, the size and stability of double-stranded segments in RNAs, as well as the structure of denatured and native DNAs (to determine the number of defects in the helical structure). Modification of polynucleotides with formaldehyde has also been used to elucidate the functioning of tRNAs and to determine the template activity of polyadenylic acid.

Fairly recently an interesting observation was made regarding modification of nucleic acids with formaldehyde. It was found that during treatment of denatured nucleic acids with a mixture of formaldehyde and glycine adenines are eliminated from the polynucleotide chain already at neutral pH values, with subsequent cleavage of the chain at depurination sites. This reaction may be instrumental, for example, in controlled cleavage of polynucleotide chains at adenines during investigations of the primary structure.

A highly promising procedure is using aldehydes with additional functional groups as modification agents. The most frequently used reagents include glyoxal, α-ketoaldehydes and similar compounds forming stable, usually cyclic derivatives with heterocyclic bases. Among the most thoroughly studied is the reaction with glyoxal, pyrotartaraldehyde, and 3-ethoxybutan-2-on-1-al (ketoxal) which react at pH 7–7.5 only with guanines to yield a tricyclic system.

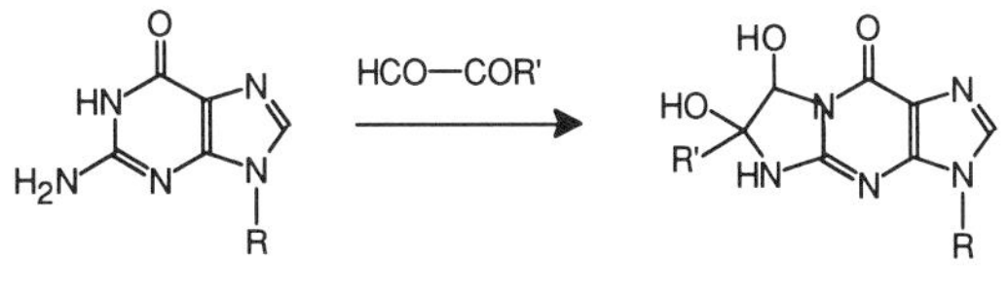

R' = H, CH_3 and $CH_3CH\ (OC_2H_5)$

R-polynucleotide chain

The fact that glyoxal reacts with 2-*N*-methylguanine but not with 2-*N*, *N*-dimethylguanine and N^1-methylguanine suggests that the reaction involves two nitrogen atoms: the heterocyclic N^1 atom and amino nitrogen – that is, the reaction calls for a –NH–C(NHR)=N– group.

The structure of products of glyoxal, pyrotartaraldehyde and ketoxal addition to guanines has been established by oxidation with periodate, followed by reduction with lithium aluminum hydride to 2-*N*-alkylguanine derivatives.

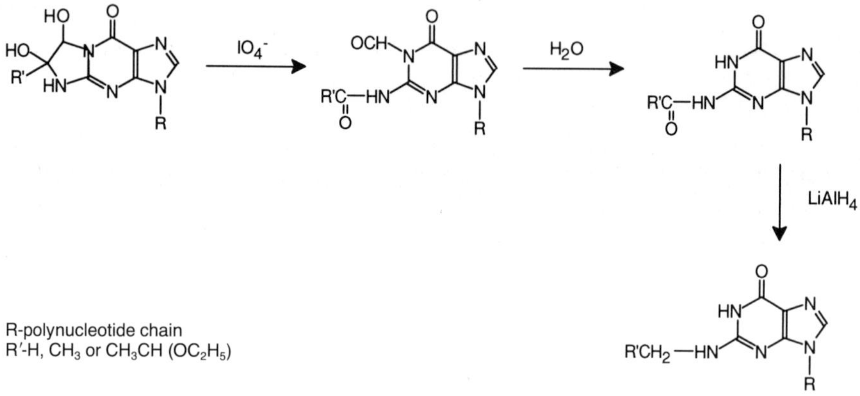

R-polynucleotide chain
R′-H, CH_3 or $CH_3CH\ (OC_2H_5)$

These transformations showing the position of the alkyl radical in the reaction product suggest that the modification begins with reaction at the most active aldehyde group [at R′ = CH_3 and $CH(CH_3)(OC_2H_5)$] which is added at the heterocyclic nitrogen, and then the unstable alkylol derivative is converted into a stable cyclic diol. For instance:

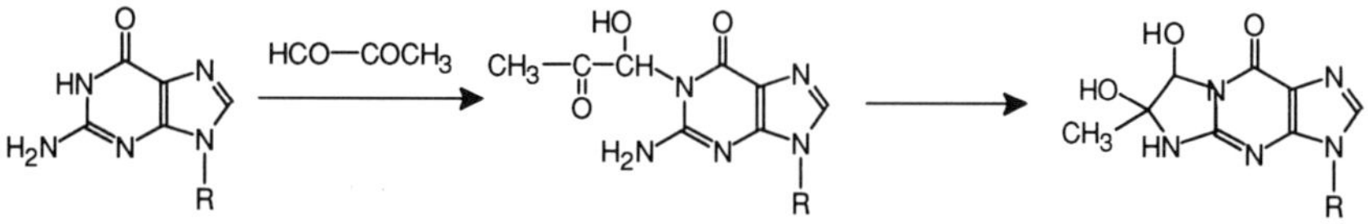

R-polynucleotide chain

Hence, the initial attack of glyoxal, too, proceeds at N^1 (pyrrole nitrogen) and, consequently, perpendicularly to the heterocyclic base plane. PMR spectra are indicative of *trans*-position of hydroxyl groups in the cyclic diol which is the glyoxal addition product.

Guanines modified with glyoxal or ketoxal are stable compounds, do not degrade during dialysis, but when the solution gets alkaline they are quickly regenerated, which is used to eliminate the modification agent (the half-life at pH 10 and 20 °C is 24 h) if needed. This property of guanines modified with glyoxal and ketoxal makes modification of this kind rather convenient for certain experiments. Since, as has already been mentioned, this modification is carried out as an attack perpendicular to the heterocyclic base plane, glyoxal and ketoxal will modifiy only those guanines that occupy single-stranded segments in tRNAs, for example. If a tRNA is modified with glyoxal or ketoxal at 70 °C (under conditions upsetting the secondary structure), all guanines undergo modification. This property of glyoxal and ketoxal – that is, inability to react with guanines involved in complementary interactions, is put to practical use in studies into the three-dimensional structure of nucleic acids. The modification with glyoxal (ketoxal) is instrumental in studying the primary structure of nucleic acids because guanyl RNase (RNase T_1) does not cleave RNAs at modified units and, therefore, the RNA can be divided into larger fragments (the enzyme exhibits greater specificity). All this renders modification of nucleic acids with glyoxal a rather useful tool for studying the structure and functions of nucleic acids.

This group of modification agents also includes chloroacetaldehyde and similar compounds. It has been found that chloroacetaldehyde with its two reactive groups reacts selectively with adenines and cytosines at pH 4–5, also yielding polycyclic derivatives:

NH2 N N N N R → $ClCH_2CHO$ → N N N N N R

NH2 N O N R → $ClCH_2CHO$ → N N O N R

R-ribose residue

In single-stranded polynucleotides (e.g., polyadenylic and polycytidylic acids), the bases are modified quantitatively under rather mild conditions (the optimal pH of the reaction medium being 3.5 for modifying cytosines and 4.5 for adenines).

The condensed system emerging during treatment of adenines has become known as the ethenoadenine system (its symbol being εA for ethenoadenosine and pεA for ethenoadenosine 5′-phosphate). What sets ethenoadenine derivatives distinctly apart is their capacity for vigorous fluorescence. This permits their use as fluorescent markers in structural studies of nucleic acids (especially nucleoproteins) and also in ivestigations of the mechanism underlying the action of enzymes of nucleic acid metabolism, where the respective ethene-containing monomer units and oligopolymers may be used as substrates.

Interaction with Nitrous Acid. It was one of the first reactions used to modify nucleic acids. Its result is deamination of the corresponding heterocyclic bases. As is known, treatment of viral RNAs with nitrous acid leads to their inactivation. The main reason for the mutagenic action of nitrous acid seems to be deamination of cytosines (substitution of uracils for cytosines). Moreover, at acid pH values (usually pH 4.2–4.5) partial depurination takes place in DNAs, especially that involving guanines. The rate at which bases in nucleic acids are deaminated drops as follows: G > A > C. Guanines are usually deaminated at a rate two to six times faster than cytosines (in tRNA, viral RNA). The deamination rate increases rapidly with decreasing pH; for example, the deamination of guanines with pH decreasing from 4.5 to 3.75 becomes five times faster. This may be due to the fact that the presence of a positive charge in the heterocycle promotes the hydrolysis of the diazonium salt, which proceeds as a nucleophilic attack of water at C^6.

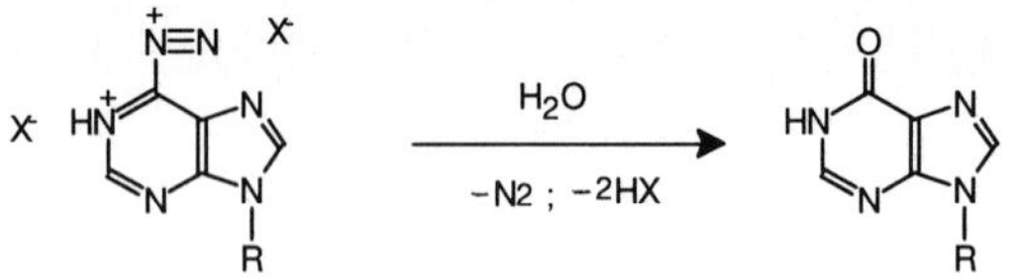

H_2O; $-N_2$; $-2HX$

R-polynucleotide chain
X-phosphate

Appropriate conditions have been selected for complete deamination of DNAs, but there are still some side processes taking place, namely, cleavage of *N*-glycosidic bonds and formation of covalent cross-links between the strands. Most likely, a reaction occurs similar to formation of cyclic nucleosides with the diazonium salt acting as an alkylating agent.

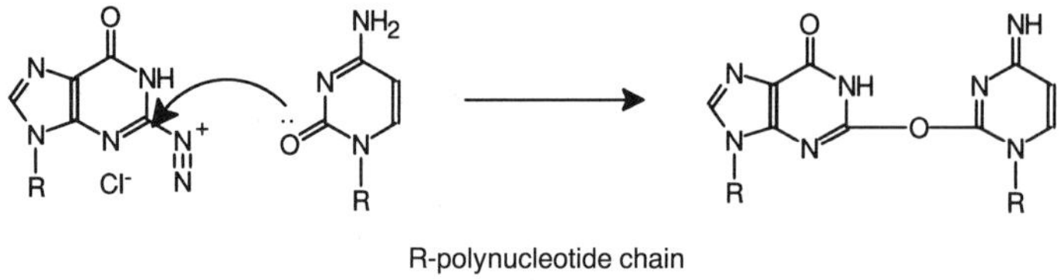

R-polynucleotide chain

Complete deamination of RNAs can be achieved by using sodium nitrite in glacial acetic acid, however, the polynucleotide chain undergoes marked degradation. Under milder conditions, with side processes with reduced to a minimum, the deamination of both DNAs and RNAs is only partial.

Because of a number of steric reasons (the amino group must be attacked by nitrous acid only at a right angle to the heterocycle plane), the secondary structure (stacking and involvement in hydrogen bonding) must inhibit the reaction with nitrous acid. Yet the reaction does take place in the case of double-stranded nucleic acids. For instance, native calf thymus DNA is deaminated exactly as a denatured one but at different relative rates for guanines, adenines and cytosines. By virtue of the small size of the nitrosonium cation the reaction is possible in double-stranded complexes as well.

The possibility of partial denaturation of double-stranded segments of nucleic acids cannot be ruled out altogether in view of the fact that the reaction proceeds at acid pH values. Table 9-2 lists some reagents commonly used for modifying bases in nucleic acids.

Table 9-2. Some Reagents Used for Modification of Heterocyclic Bases in Nucleic Acids.

Modifying agent	Modification conditions		Specificity of action	
	optimal pH	temperature, °C	modifying base	site of attack
	Nucleophilic agents			
Hydroxylamine or *O*-alkylhydroxylamines	5–7	20	C	C^6 and C^4 (NH_2OH)
Ditto	5–7	20	U	C^4 (NH_2OAlk)
Hydroxylamine	10	10	U	Cleavage
Bisulfite ion	5–7	20	C	C^6
Ditto	8	20	U	C^6
Bisulfite ion in a mixture with *O*-methylhydroxylamine (and other amines)	7	20	C	C^6 and C^4
Hydrazine and alkylhydrazines	6	20	C	C^4
Hydrazine	9.5	0	C, U, T	Cleavage
	Electrophilic agents			
Halogens	–	0	U, C G	C^5 C^8
Bromosuccinimide	7	20	C	C^5
Ditto	9	20	G	C^8
Mercuric acetate	6–7	50	C, U	C^5
Tritiated water	7	37	G, A	C^8
Ditto	7	37	C, U	C^5
Alkylating agents	6–7	20–37	G A C	N^7 (slowly N^1 and N^3) N^1 (slowly N^3) N^3
Peracids (usually monoperphthalic acid)	~7	0–20	A C	N^1 N^3
Diethylpyrocarbonate	7	20	A G	N^7 and C^8 N^7 and C^8
Compounds with active double bonds (watersoluble carbodiimides, acrylonitrile)	8–9	20	U, T G	N^3 N^1

Table 9-2. (Continued).

Modifying agent	Modification conditions		Specificity of action	
	optimal pH	temperature, °C	modifying base	site of attack
	Aldehydes and ketones			
Formaldehyde	8	10–50	U (Ψ)	N^3 (N^1 and N^3)
			A	NH_2 group
Glyoxal, pyruvaldehyde, etc.	7–7.5	20–37	G	N^1 and NH_2 group
Chloraldehyde	3.5	37–50	C	N^3 and NH_2 group
	4.5	37–50	A	N^1 and NH_2 group
Nitrous acid	Weakly acidic values	0–50	G	Deamination
			A	Deamination
			C	Deamination

9.3 Hydrolysis of *N*-Glycosidic Bonds

The *N*-glycosidic bond, especially that formed by purine bases, is one of the most reactive covalent bonds in nucleic acids.

Just as in monomer units, *N*-glycosidic bonds in ribonucleosides of nucleic acids are more stable than in deoxyribonucleosides; within each class of nucleic acids, purine *N*-glycosidic bonds are more easily cleaved than pyrimidine ones. At present, cleavage of *N*-glycosidic bonds is broadly employed in analysis of the primary structure of nucleic acids. There are two methods for cleaving these bonds in the latter: direct, when the *N*-glycosidic bonds is hydrolysed by the action of an acid, and indirect, when hydrolysis is preceded by modification of particular bases so that the linkage between them and the carbohydrate moiety is significantly labilized.

9.3.1 Direct Hydrolysis Methods

The hydrolysis of *N*-glycosidic bonds in polymeric molecules is accompanied by cleavage of internucleotide linkages. These processes may be independent in RNAs, whereas in DNAs hydrolysis of *N*-glycosidic bonds is followed by depolymerization. Most likely, the mechanism of such hydrolysis in RNAs and DNAs is similar to that described in the context of nucleotides. In determining the nucleotide composition of DNA, *N*-glycosidic bonds are hydrolysed by 72% chloric acid (110°C, 1 h) or 98–100% formic acid (175°C, 2 h); in this case, all *N*-glycosidic bonds get cleaved. The use of dilute acids to depolymerize DNA results in polypyrimidine blocks and purine bases. For instance, partial depurination of DNA is observed when aqueous solutions are heated at pH 6–7.5 (65–100 °C) and complete depurination occurs at pH 2 (37 °C, 24 h, dilute hydrochloric acid). *N*-Glycosidic bonds in RNA are more difficult to break (just as in monomer units). When RNA is treated with 1 N hydrochloric acid (100 °C, 1 h), pyrimidine nucleoside 3′(2′)-phosphates and purine bases are formed. Such acid hydrolysis is used to determine the nucleotide composition of RNAs.

9.3.2 Indirect Hydrolysis Methods

Since the alkylation of purines at N^3 and N^7 leads to a marked labilization of the *N*-glycosidic bond, nucleic acids can be alkylated to be depurinated. The hydrolysis of *N*-glycosidic bonds in DNA at methylated units occurs already at pH 7. When a methylated DNA is maintained at 37 °C (pH 7.2), the half-period of glycosidic bond hydrolysis is 25 h for 3-methyladenine and 140 h for 7-methylguanine. The hydrolysis rate is to some extent dependent on the species of the introduced alkyl group. For instance, adenines with an $HOCH_2CH_2SCH_2CH_2$ group at N^3 are detached from the alkylated DNA with a half-life of 8 h under the same conditions.

These findings suggest that the *N*-glycosidic bond is labilized to a much greater degree after N^3-alkylation of adenine, as compared to N^7-alkylation of guanine. Comparison of the ease with which N^7-alkylated adenines and guanines detach themselves from the nucleic acid indicates that the *N*-glycosidic bond formed by the former is more reactive in this case as well. Interestingly, the N^1-alkylation of adenine virtually does not affect the strength of the glycosidic bond it forms. However, because of the much faster rate of DNA alkylation at N^7 in guanines, the subsequent hydrolysis of *N*-glycosidic bonds in DNA proceeds primarily at guanines.

Worthy of mention among other reactions in which glycosidic bonds are hydrolysed is the one with preliminary opening of the heterocyclic rings. A rather common procedure is treatment of RNAs and DNAs with hydrazine or hydroxylamine to open pyrimidine rings. For example, the hydroxylamine linked to the ribose phosphate chain in uracil-free RNAs resulting from hydroxylamine treatment can be eliminated by hydrolysis in a weakly acidic (pH 5) medium in the presence of cyclohexanone. Uridines of the starting RNA are replaced in the resulting polymer by ribosyls with free glycosidic centers.

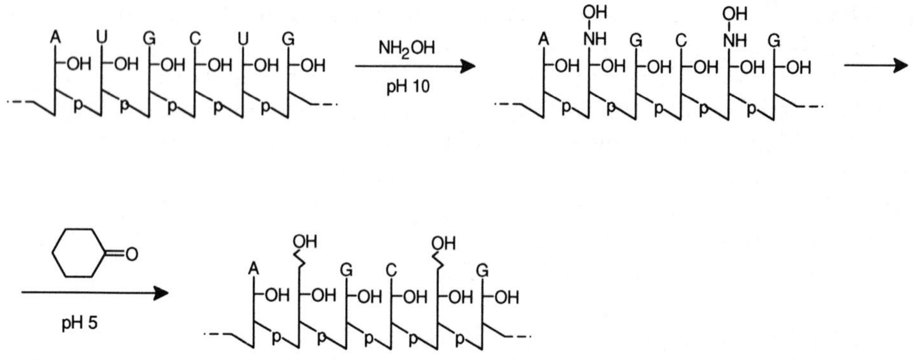

DNA does not contain bases reactive enough with respect to hydroxylamine at pH 10; however, the reaction can be conducted after deamination of DNA in the presence of nitrous acid, with cytosines being converted into uracils. In this way DNA can be cleaved specifically at cytosines.

9.4 Reactions of Carbohydrate Moieties

If we take the reactions involving the ribose and deoxyribose of polynucleotides, two of them have been studied more or less in depth – substitution for hydrogens in hydroxyl groups and oxidation of the 2′,3′-*cis*-glycol group in terminal nucleotides, leading to rupture of the $C^{2'}$–$C^{3'}$ bonds. This statement should be qualified by adding that the reactions proceed in aqueous solutions. The reactions of oligo- and polynucleotides in organic solvents were studied, as a rule, in the context of development of synthetic methods. This section deals only with those reactions in organic solvents which have been studied on natural nucleic acids in search for ways to modify carbohydrate moieties in a controlled fashion.

9.4.1 Substitution for Hydrogen Atoms in Hydroxyl Groups

Acylation. Reactions of this type have been studied for RNAs having a free hydroxyl at $C^{2'}$ of ribose in each monomer unit in the middle of the polymer chain and an unsubstituted 2′,3′-*cis*-glycol group at the 3′ end of the molecule. DNA molecules contain only one hydroxyl at the 3′ end of the polymer. Just as in RNAs, the terminal 5′-hydroxyl in DNAs usually contains a phosphate. Since the hydroxyls of carbohydrate moieties are not the only nucleophilic groups in the polynucleotide molecule, acylation may involve not only the hydroxyl groups of the sugar but also the amino groups of the heterocycle and phosphates. Therefore, selective acylation of carbohydrate moieties requires a carefully chosen set of conditions.

The optimal conditions for acetylation of the internal 2′-hydroxyls in RNAs include reaction with acetic anhydride in an aqueous solution at pH 7 (5–20 °C). To ensure homogeneity of the reaction medium and better reproducibility of the results, the modification should be carried out with acetic anhydride in a 5 % aqueous solution of dimethylformamide. This procedure allows up to 70 per cent of the 2′-hydroxyls in tRNA to be easily modified without any perceptible degradation of the polymer chain.

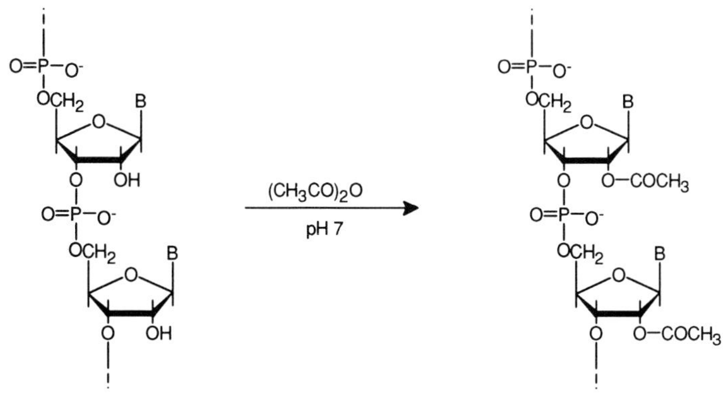

$(CH_3CO)_2O$, pH 7

The integrity of the polynucleotide chain in tRNA during acetylation under the above conditions together with the fact that the internucleotide linkage does not undergo any isomerization indicate that the internucleotide phosphate group is not acylated. Otherwise, the following reactions would take place:

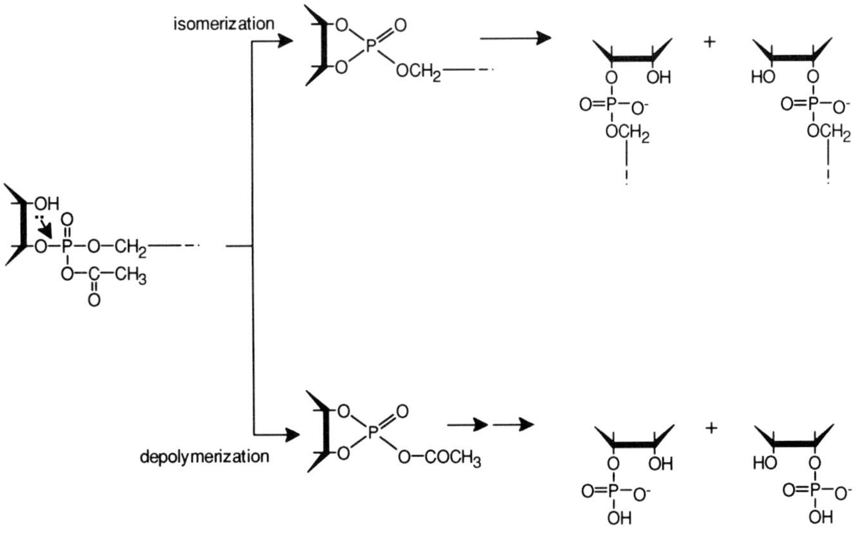

isomerization

depolymerization

Aminoacylation. Amino acid esters of RNA and oligonucleotides have attracted a great deal of interest because tRNAs acylated with an amino acid at one of the hydroxyls of the 3′-terminal *cis*-glycol groups are intermediate compounds in the activation of amino acids during protein biosynthesis in ribosomes. The following scheme illustrates biosynthesis of a peptite bond in ribosomes accomodating a constantly growing polypeptide chain, linked via an ester bond to the tRNA specific for the peptide in question, as well as constantly "arriving" 3′-*O*-aminoacyl derivatives. As a result of growth of the peptide chain, the 3′ end of the tRNA serves as the site for formation of a specific protein which has the information necessary for its snythesis stored in the mRNA; the small straight arrows show the polarity of the respective nucleic acids.

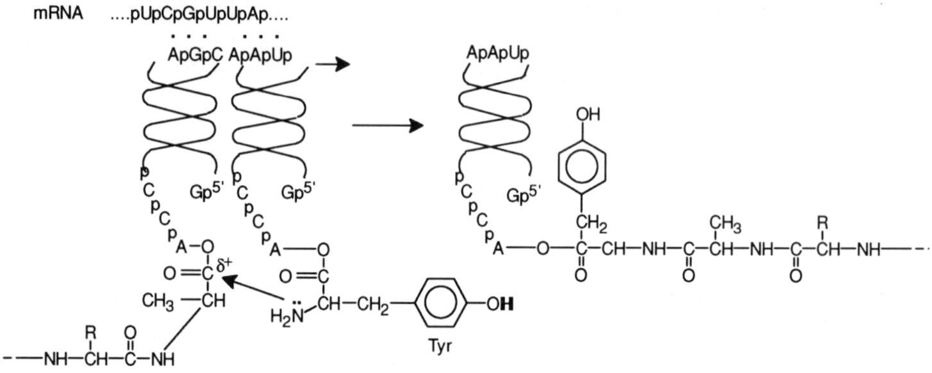

Thus, the polypeptide chain is assembled stepwise from the N-end. Studies into the properties of 3′-aminoacyl(peptidyl)-tRNA and similarly structured model compounds have shown that their ester bond is activated by the pentose moiety (such amino acid derivatives belong to the category of activated esters) and the energy it contains is quite sufficient for the peptide synthesis to take place.

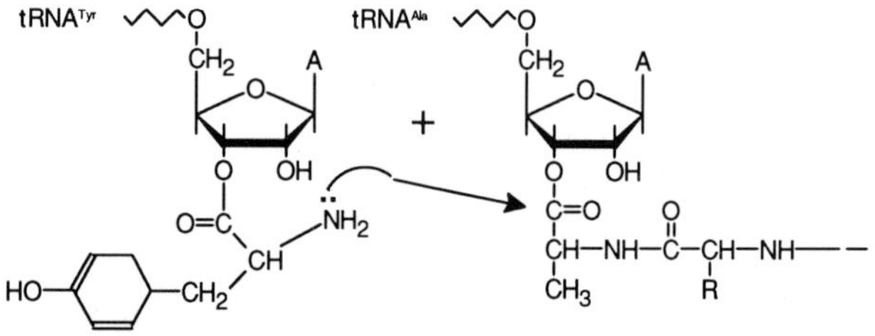

At present, this process can be replicated *in vitro,* both with the ribosome functioning continuously on a natural or synthetic RNA template, when a polypeptide chain made up of at least several peptide bonds is formed, and in steps, with each ribosome functioning event being adequately controlled.

Along with natural participants of polypeptide synthesis on the ribosome, amino acid esters of nucleosides, nucleotides and oligonucleotides (analogues of 3′-aminoacyl-tRNA) have been successfully used as peptide acceptors in discrete events of peptide bonding, while acyloaminoacyl-oligonucleotides (analogues of peptidyl-tRNA) have been used as peptide donors. The use of such model compounds has been instrumental in elucidating the topography of the ribosomal portion (so-called peptidyl-transferase site) where the peptide bond is synthesized, and also in gleaning information on the substrate specificity of this site. In view of this and the fact that the ribosomal synthesis of peptides is widely applied to studies into the functioning of the ribosome, it would be most appropriate to develop methods for chemical synthesis of aminoacyl- and peptidyl-tRNAs and their low-molecular analogues. This is why such modification of tRNAs and oligonucleotides (as tRNA models) has received particular attention.

Selective aminoacylation of the hydroxyls in ribose becomes increasingly difficult as one goes from nucleosides to nucleotides and, to a greater degree, nucleic acids. For instance, aminoacylation of RNA may give derivatives containing an aminoacyl in the carbohydrate moiety (ester bond formed by internal 2′- or terminal 2′- or 3′-hydroxyl groups of the carbohydratephosphate backbone in RNA), in the heterocyclic base (amide bond involving the heterocyclic amino groups), and in phosphate groups (anhydride bond associated with the aminoacylation of internucleotide and terminal phosphate groups).

The aminoacylation of tRNA in organic solvents by amino acids activated at the carboxyl group yields compounds such as mixed anhydrides (reaction at phosphate groups) and amides (aminoacylation of heterocyclic bases).

In aqueous solutions, however, the aminoacylation of tRNA, just as acylation, proceeds at the hydroxyls of the carbohydrate moieties. Imidazolides of amino acids have turned out to be the best aminoacylation agents in this case. For example, condensation of *N*-butoxycarbonylic (Boc) aminoacylimidazolides with total tRNA from baker's yeast in a strongly aqueous medium at 20 °C for several hours gives compounds with an ester (rather than anhydride or amide) type of bond. Stable and high yields were attained when the reaction was conducted in a 1 M imidazole buffer:

tRNA 3′-terminal adenosine (HO, OH) + Boc—NH—CH(R)—C(=O)—imidazolide → 3′-O-(Boc-aminoacyl)-tRNA + 2′-isomer $\xrightarrow{F_3C\text{—}COOH}$ 3′-O-aminoacyl-tRNA ($O{=}C\text{—}CH(R)\text{—}NH_2$) + 2′-isomer

R=H or $C_6H_5CH_2$

However, as the Boc group was being eliminated, the tRNA became inactivated, probably due to the rigorous conditions of the reaction. It was therefore necessary to drop this step of the synthesis. This was successfully accomplished using imidazolides of *N*-protonated amino acids reacting with tRNA to yield aminoacyl-tRNA in just one step:

O—tRNA
O=P—OCH2 A
O-
HO OH
+
N—C(=O)—CH(R)—NH3+ · F3C—COO-

→ O—tRNA
O=P—OCH2 A
O-
O OH
O=C—CH—NH3+
R
+ 2'- and other isomers

The reaction takes several hours under mild conditions (1 M aqueous buffer solution of imidazole, pH 7, 4 °C). As has been demostrated using radiometric methods, 3–10 % of the amino acid is incorporated into the terminal 3′-*cis*-glycol group. Moreover, the internal 2′-hydroxyl groups of ribose are aminoacylated, too.

Since special experiments with nucleotides have shown that no aminoacylation of heterocyclic bases takes place under such conditions, it may be assumed that no amide bonds are formed in the case of tRNA either. What makes this method so important is the fact that it allows the experimenter to have "hybrid" molecules with any amino acid and tRNA combination (an amino acid which is not specific with respect to a given tRNA can be added thereto).

Aminoacylation or acylation of 3′-*O*-aminoacyl-tRNAs (including those resulting from enzymatic phosphorylation) may yield the corresponding peptidyl(acylaminoacyl)-tRNAs. This process can be represented by the following general scheme:

O—tRNA
O=P—OCH2 A
O-
O OH
O=C—CH—NH2
R
R'—C(=O)—OY →
O—tRNA
O=P—OCH2 A
O-
O OH
O=C—CH—NH—C(=O)—R'
R

For selective acylation of amino acids use is normally made of *para*-nitrophenyl (Y = $C_6H_4NO_2$-*p*) or *N*-hydroxysuccinimide (Y = $\overline{NCO(CH_2)_2CO}$) esters because, as is known, such acid derivatives do not acylate hydroxyl and amino groups in nucleic acids.

The synthesis of peptide derivatives of tRNA usually involves *N*-hydroxysuccinimide esters of *N*-methoxytritylamino acids:

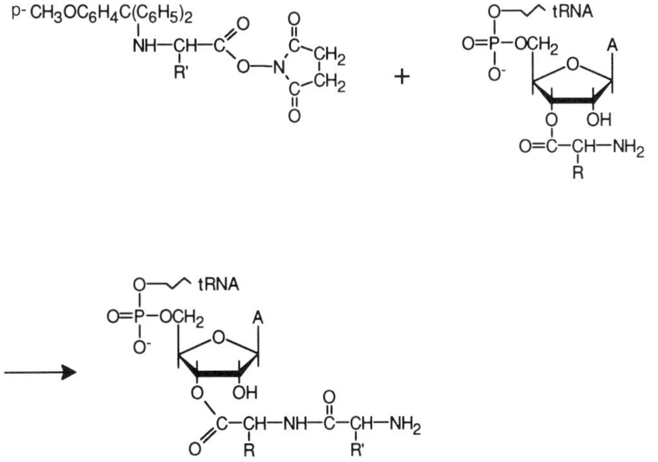

Selecting the right protective group for the amino acid is not as simple as it may seem because it should be eliminated only under very mild conditions (in order not to affect the many labile bonds in tRNA). The monomethoxytrityl group meets this criterion since it can be eliminated in the presence of 5% dichloroacetic acid at 4°C (5 min). The resulting 3′-*O*-peptidyl-tRNAs remain active in the ribosomal system.

The peptide chain can be built up unit by unit with dipeptide fragments being added to aminoacyl-tRNA. The yields of peptidyl-tRNA range from 80 to 90 per cent in this case.

9.4.2 Oxidation of 3′-Terminal *cis*-Glycol Group in RNA

The 3′-terminal *cis*-glycol group in RNA is readily oxidized with periodic acid. As a result, dialdehydes are formed, just as when monomer units are oxidized:

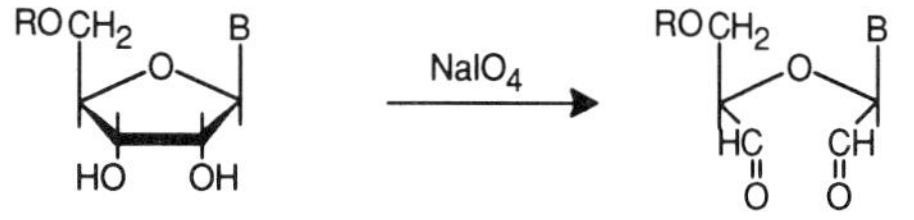

R-polyribonucleotide chain

The oxidation is conducted in dilute aqueous solutions at 0–20 °C. At a low temperature, the reaction proceeds quantitatively within an hour. Reduction of dialdehydes with sodium borohydride in a neutral medium yields the corresponding diols:

$NaBH_4$

R-polyribonucleotide chain

Thus, a tritium label is incorporated into RNAs or oligoribonucleotides with the aid of sodium borotritide $NaBT_4$.

Dialdehydes easily enter into other reactions typical of carbonyl compounds. A case in point is their interaction with hydrazines and amines:

$2\ C_6H_5NHNH_2$

NH_2CH_3

R-polyribonucleotide chain

A characteristic property of periodate oxidation products is their capacity for easy degradation in an alkaline medium with β-elimination of the phosphate group.

Oxidation of the 3′-terminal glycol group with subsequent immobilization on hydrazide columns is used to isolate individual tRNAs.

9.5 Reactions of Nucleotide Phosphate Groups

The reactions involving internucleotide phosphate groups may either be accompanied by degradation of the polynucleotide chain (as a result of cleavage of internucleotide linkages) or leave the latter intact.

9.5.1 Cleavage of Internucleotide Linkages

Obviously, the stability of a polymeric nucleic acid molecule is determined by the reactivity of the internucleotide node in aqueous solutions. It is most important for an experimenter working with nucleic acids to know the limits of stability of internucleotide linkages in both DNA and RNA. To a certain degree, this has to do with the fact that the result of biological research often hinge on the integrity of the native primary structure of the biologically active nucleic acid molecule: cleavage of just one of a thousand internucleotide linkages may lead to a complete loss of activity. Thus, reactions leading to cleavage of phosphodiester (internucleotide) bonds occupy a special position among those involving nucleic acids.

Reactions in which the polynucleotide chain undergoes degradation may be based on two different mechanisms. Reactions of the first type, with cleavage of the P–O bond, are essentially nucleophilic substitution at the internucleotide phosphorus atom which, in this case, exhibits electrophilic properties.

$$RO{-}P(=O)(O^-){-}OR + :X \longrightarrow RO{-}P(=O)(O^-){-}X + {}^-OR$$

R-polynucleotide chain

The nucleophilic attack of the phosphorus atom may be inter- or intramolecular.

Reactions of the second type, with cleavage of the C–O bond, occur when a carbonyl group emerges in the dinucleoside phosphate fragment of the polynucleotide at the β-position with respect to the phosphodiester group. They take the form of β-elimination catalyzed by bases:

$$RO{-}P(=O)(O^-){-}O{-}C{-}C(H){-}C(=O){-} \quad (:B) \xrightarrow{(-BH^+)} RO{-}P(=O)(O^-){-}O^- + C{=}C{-}C(=O){-}$$

Just as with cleavage of the P–O bond, the motive force behind the β-elimination is the electrophilic behavior of the phosphorus, the capacity for cleaving the C–O bond being in this case determined by the stability of the departing anion. Consequently, this type of cleavage of internucleotide linkages is indirect, dependent on some kind of primary transformations giving rise to a carbonyl at the β-position with respect to the internucleotide phosphodiester group. This is the mechanism governing acid hydrolysis of DNAs, hydrolysis of depyrimidinated RNAs by primary amines, as well as detachment of the 3′-terminal nucleosides in RNAs after periodate oxidation.

Reactions of Nucleophilic Substitution at the Internucleotide Phosphorus Atom (cleavage of the P–O bond). The stability of internucleotide linkages in RNAs and DNAs has already been discussed at length. It was pointed out that the sharp difference between the two in this respect stems from the presence or absence of a hydroxyl at the $C^{2'}$ atom next to the internucleotide node.

Internucleotide linkages in DNAs are extremely stable in both acid and alkaline media. In an acid medium, their hydrolysis takes place at $pH < 3$. As regards alkaline media, the hydroxyl ion attack at the phosphorus atom is hindered, just as in the case of simple dinucleoside phosphates, by the negative charge at the phosphate group (pK_a^1 of the phosphate group does not exceed unity at $pH > 3$, and the internucleotide phosphate group is ionized almost completely). Internucleotide linkages in RNAs are much less stable. If you remember, this is due to the presence of a hydroxyl at position 2′ in ribose, which acts as an internal nucleophile in this case. The *cis*-configuration of the glycol group in the furanose ring of ribose is the reason for the ease of nucleophilic substitution even at pH values that are not too alkaline. Already at pH 10, internucleotide linkages in RNAs start breaking, which is indicative of the fact that this process has to do with dissociation of the 2′-hydroxyl group ($pK \approx 12$).

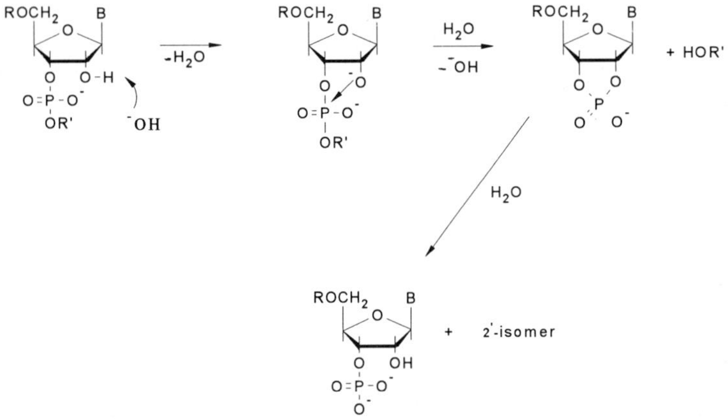

The rate of this process is strongly influenced by the nature of the nucleotides forming the dinucleoside phosphate fragment under investigation. This influence seems to be due to stacking interactions and the presence of charged groups in the heterocycles. For instance, studies into the rate of alkaline hydrolysis of internucleotide linkages in dinucleoside phosphates have shown that they are more stable in purine derivatives and less stable in pyrimidine ones where stacking is not pronounced. Stacking interactions are likely to result in such a conformation of the internucleotide node when the phosphorus atom is remote from the OH group at $C^{2'}$ and the rotation about the $C^{3'}$–O bond is slowed down. This inhibits the formation of the pentacoor-

Table 9-3. Partial Hydrolysis of RNA in an Alkaline Medium Under Different Conditions.

pH	Temperature, °C	Ionic strength of buffer solution	Phosphodiester bond cleavage time, min		
			1 %	10 %	50 %
7.6	90	0.2	420	260,000	–
9.0	70	0.2	30	300	–
9.0	100	0.01	–	17	100
11.0	70	0.02	–	20	120
13.0 (0.1 N KOH)	60	–	–	0.8	5
14.0 (1 N KOH)	27	–	0.7	7	40

dinate phosphorus in the transient state, emerging during intramolecular nucleophilic attack of the internucleotide phosphate group by the 2′-hydroxyl. The rate of hydrolysis of interribonucleotide linkages suggests that stacking reduces the rate of their cleavage by one order of magnitude.

The difference in the rates of interribonucleotide linkages is maximum at pH 11, when the ionization of uracils and guanines upsets the course of stacking interactions. Under such conditions, polyguanyl fragments are hydrolysed faster than polyadenyl ones and polyuridyl fragments are hydrolysed faster than polycytidyl ones.

The alkaline hydrolysis of RNA to mononucleotides is usually carried out using a 0.3 N KOH solution at 37 °C for 20 h. Under milder alkaline conditions, RNA yields oligonucleotides. Table 9-3 summarizes the effect of such factors as pH of the medium, temperature and ionic strength on the rate of hydrolysis of internucleotide linkages in RNA.

Ions of many heavy metals exert catalytic action on the hydrolysis of internucleotide linkages in RNA, catalysis being the most effective at neutral and weakly alkaline pH – that is, under conditions conducive to formation of metal salts with phosphate groups. Such formation of salts (complexes) seems to promote the nucleophilic attack of the phosphorus atom by the hydroxyl at $C^{2'}$ which is possibly involved in the complex formation as well, whereby its nucleophilic activity is enhanced.

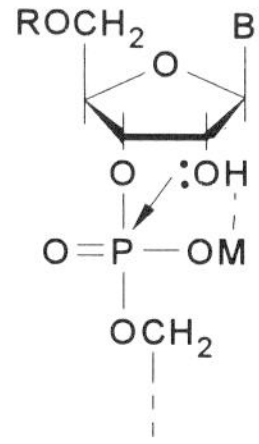

$ROCH_2$ B
O
O :OH
O=P—OM
OCH_2

The participation of the neighboring 2′-hydroxyl in this process is inferred from the fact that DNA is not hydrolysed under similar conditions; RNA methylated at 2′-OH groups is not hydrolysed either.

Comparison of the catalytic activity of ions of different metals suggests that compounds of lead and lanthanum are the most active catalysts. Ions of bivalent zinc and cadmium are less active. The rate of RNA hydrolysis in the presence of ions of heavy metals is also materially affected by the nucleotide composition of RNA. Purine polynucleotides are hydrolysed more slowly than pyrmidine ones. Hydrolysis of RNAs with a more stable secondary structure proceeds at al slower rate. The effect of ions of heavy metals on the strength of internucleotide linkages in RNAs should be taken into account in experiments with biologically active molecules even in cases where the experimental conditions are rather mild (neutral pH values, room temperature). For example, ions of bivalent lead, nickel and cobalt (10^{-3}–10^{-4} M, pH 7, 20 °C) catalyze the polymerization of RNA from tobacco mosaic virus, yeast RNA, and other polyribonucleotides.

Intermolecular attack of the internucleotide phosphorus by a nucleophile may also lead to degradation of the polymer. For instance, treatment of RNA with a 1 N solution of sodium methylate in a mixture of dimethylformamide with methanol yields ribonucleoside 3′(2′)-methyl phosphates.

O OH
O=P–O⁻
OCH2
$\xrightarrow{CH_3ONa}$
O OH
$O{=}P{-}OCH_3$
O^-
+ 2′-isomer

The mechanism of such hydrolysis may include not only intermolecular attack of the phosphorus by the methylate ion but also that of the nucleoside 2′,3′-cyclic phosphate forming in advance.

In an acid medium (pH $\leq$ 3), RNA is also hydrolysed to a mixture of ribonucleoside 3′(2′)-phosphates, the mechanism of such hydrolysis being similar to that of alkaline rupture of RNA. A distinctive feature of acid hydrolysis of RNA is the simultaneous isomerization of the internucleotide (3′–5′ $\rightleftharpoons$ 2′–5′) linkages. The factors responsible for such isomerization have already been discussed in the context of alkyl phosphate migration in ribonucleoside 3′(2′)-alkyl phosphates. The rates of acid hydrolysis and isomerization of internucleotide linkages in RNA are strongly influenced by the nucleotide composition, this influence being similar to that in the case of alkaline hydrolysis already described.

Hydrolysis of RNA in 1 N hydrochlorid acid (100 °C, 1 h) is used for analytical purposes. During such treatment, cleavage of all internucleotide linkages goes hand in hand with hydrolysis of the *N*-glycosidic bonds in purine nucleoside 3′(2′)-phosphates so that acid hydrolysis of RNA yields purines and pyrimidine ribonucleoside phosphates. Quantitative analysis of these compounds

Table 9-4. Partial Hydrolysis of RNA in an Acid Medium Under Different Conditions.

pH	Temperature, °C	Phosphodiester bond cleavage time, min		
		1 %	10 %	50 %
1	20	10	100	600
2	100	–	4	50
3	100	4	40	–

forms the basis of the method for determining the nucleotide composition of RNA.

Acid hydrolysis of RNA may be interrupted at the step of formation of oligonucleotides which, as can be seen from Table 9-4, determine the dependence of the time it takes for internucleotide linkages in RNA to be cleaved on pH and temperature.

Isomerization of internucleotide linkages is an undesirable side process during acid hydrolysis of RNA to oligonucleotides.

β-Elimination Reaction (cleavage of the C–O bond). It has already been pointed out that the cleavage of the C–O bond in an internucleotide node and, consequently, rupture of the polynucleotide chain occurs when there is a carbonyl group in the carbohydrate moiety at the β-position with respect to the phosphodiester group. The reaction proceeds as β-elimination of the phosphate group under very mild conditions and almost quantitatively. The carbonyl is introduced at the first step usually in two ways: elimination of the base, which is normally used for degradation of both DNAs and RNAs, or oxidation of the 3′-terminal *cis*-glycol group in RNA. Let us first consider the first procedure. Hydrolysis of the *N*-glycosidic bond, accompanied by elimination of the base, yields an aldehyde group in the carbohydrate moiety. This may lead to β-elimination of the phosphomonoester group (polynucleotide chain with a 5′-terminal phosphate group) or, in other words, cleavage of the polynucleotide chain at the C–O bond. The reaction is catalyzed by acids, alkalies and also amines.

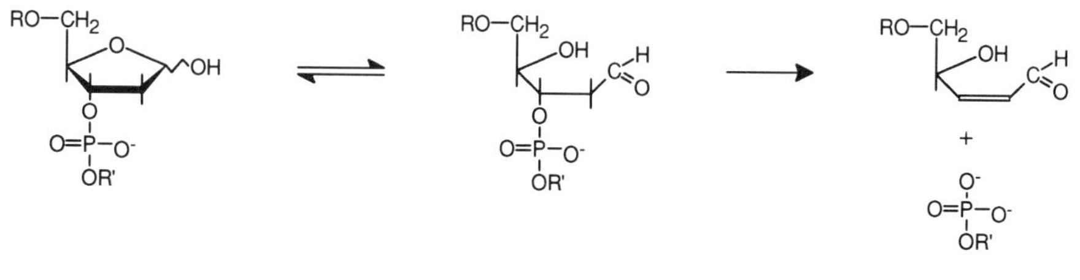

In the case of DNA, purine bases are eliminated under rather mild conditions. The resulting depurinated nucleic acids readily undergoing β-elimination under the same conditions may be isolated and subjected to exhaustive

β-elimination in the presence of aromatic amines in an aqueous solution of formic acid. The above-described method for hydrolysis of DNAs to polypyrimidine units, known as the Burton method, resides in incubation of DNA with a 2% solution of diphenylamine in 66% formic acid at 30 °C for 17 h. Under the same conditions, hydrolysis of *N*-glycosidic bonds formed by purine bases takes place along with β-elimination of a poly(oligo)-pyrimidine unit:

The detachment of α,β-unsaturated aldehydes from the poly(oligo)nucleotide chain proceeds quantitatively as a result of the second elimination event. The latter involves the same mechanism as β-elimination since, in this case, the electron-acceptor effect of the aldehyde group, enhancing the protonic mobility of the corresponding hydrogen atom, is to a considerable degree transmitted through the vinyl group (vinylogy).

R-poly(oligo)nucleotide chain

Diphenylamine, without which practically no cleavage of phosphodiester bonds is possible, seems to form a salt together with the carbonyl of the acyclic form of deoxyribose, which facilitates the β-elimination:

Subsequently, poly(oligo)pyrimidine fragments are, as a rule, divided along the chain and then according to composition (resulting in so-called isoplates). Hydrolysates of φX174 phage DNA have yielded pyrimidine units up to 11 nucleotides long, those of f1 phage DNA, up to 19 nucleotides, and those of fd phage DNA, up to 20 nucleotides.

Alkaline treatment of depurinated nucleic acids also breaks internucleotide linkages as a result of β-elimination of the phosphate group. However, because of the facilitated course of side reactions into which aldehydes are likely to enter in an alkaline medium, some phosphate groups remain linked to the carbohydrate moiety after β-elimination of the 3′-phosphate, for example:

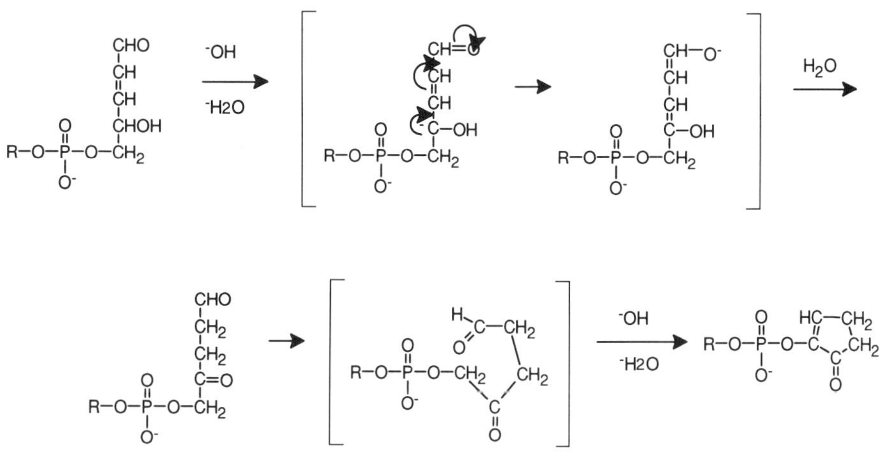

The second procedure, which permits partial elimination of pyrimidine bases from DNA, boils down to its treatment with hydrazine. The resulting depyrimidinated DNAs cannot be subjected to hydrolysis under conditions of acid catalysis because of the extremely high lability of the glycosidic bonds in purine nucleotides. This difficulty, however, has been successfully obviated by resorting to pretreatment of depyrimidinated nucleic acids with benzaldehyde (to eliminate the residual hydrazine) with subsequent degradation of the polymer under the effect of weak bases. This is how polypurine isoplates are obtained, albeit not quantitatively, under very mild conditions.

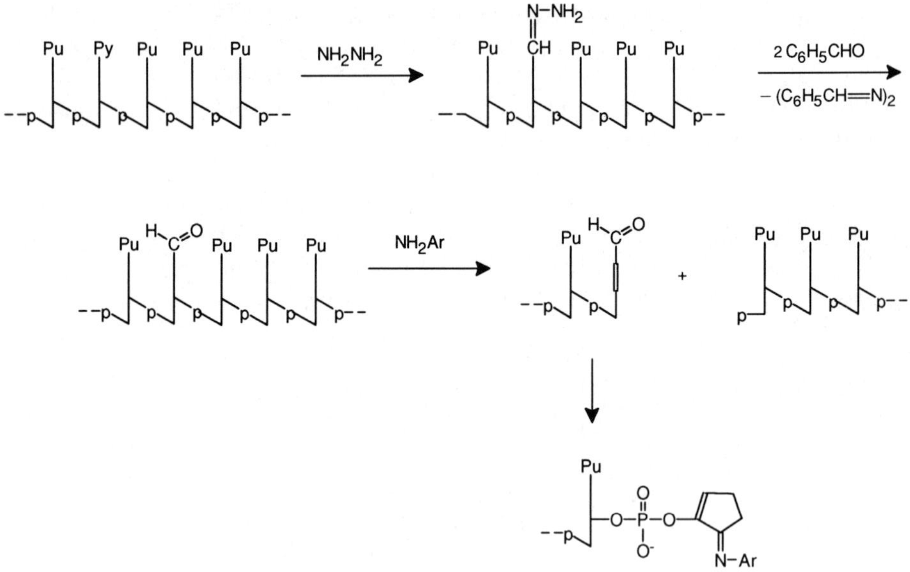

Good results were obtained when aromatic amines (aniline, *para*-anisidine) were used as the weak bases for β-elimination. After separating isoplates one can derive oligopurine units which, as can be seen on the above scheme, have a blocked phosphate group at the 3′ end. Similar approaches for degrading RNAs are of little practical use because internucleotide linkages in the latter are much more labile, while *N*-glycosidic bonds are much more stable. However, after elimination of uracils from an RNA by means of hydroxylamine at pH 10, uracil-free nucleic acids may undergo, after acid hydrolysis under mild conditions (to eliminate the hydroxyamine residue), degradation catalyzed by aromatic amines (e.g., *para*-anisidine at pH 5). The transformations occurring in the RNA unit being hydrolysed can be represented as follows:

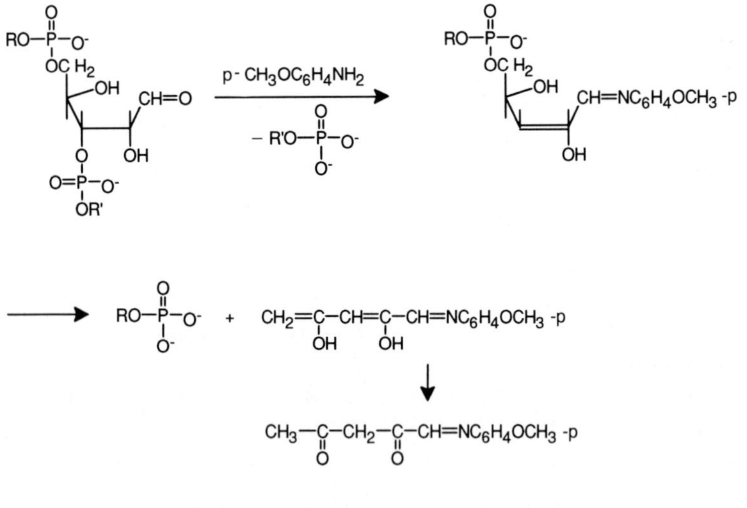

Degradation of $tRNA^{Phe}$ can serve as an example of controlled cleavage of the polyribonucleotide chain at a nucleotide with an eliminated base. The RNA, which is 76 base pairs long, contains base Y at position 37 (see the following scheme). This base is rather easily eliminated from $tRNA^{Phe}$ during hydrolysis with 0.1 N hydrochloric acid at room temperature, without any other bonds in the tRNA being affected.

5' pGp—OCH2 ... pCpCpA 3' —(⁻YH)→ ... → 5' pGp—OCH2 ... + ⁻O—P(O)(O⁻)—OCH2 ... pCpCpA 3' + ribose degradation product

A similar degradation can also be carried out in the case of RNAs containing N^7-methylated guanines and N^3-methylated adenines. A common approach recommended for controlled cleavage of the polynucleotide chain at a particular nucleotide includes preliminary modification of a base in the latter with the result that the *N*-glycosidic bond is considerably labilized and it becomes possible to cleave both the glycosidic bond and the internucleotide linkages at this nucleotide (by the β-elimination mechanism).

The rupture of internucleotide linkages as a result of β-elimination can be put to practical use for step-by-step degradation of RNAs from the 3′ end of the polyribonucleotide chain. The first step of such a reaction is periodate oxidation of the 3′-terminal *cis*-glycol group. The resulting 2′,3′-dialdehyde contains a 3′-aldehyde group at the β-position with respect to the 5′-phosphodiester bond. The polynucleotide unit is eliminated under the effect of various primary amines (cleavage of the $C^{5'}$–O bond). As a consequence, a polyribonucleotide shortened at the 3′ end is formed.

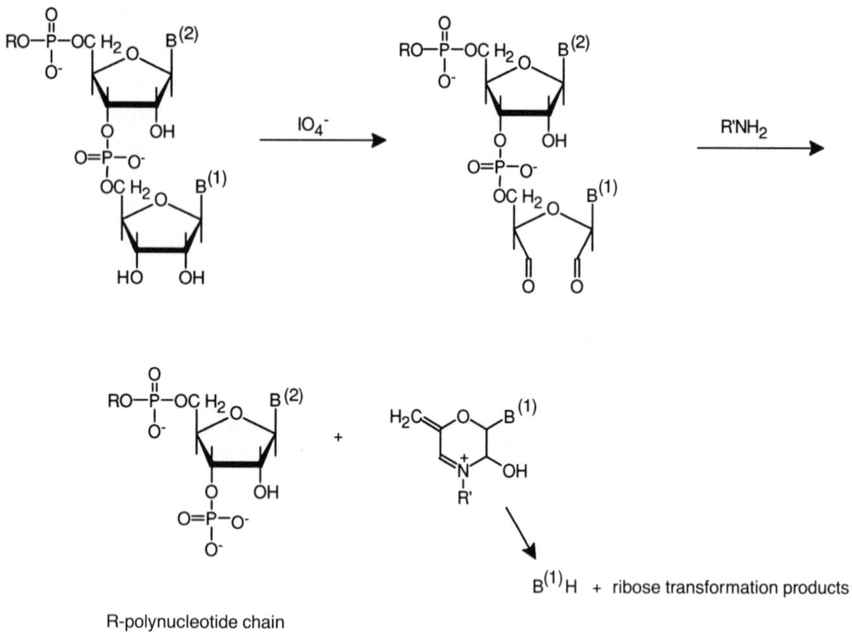

IO_4^- → $R'NH_2$ →

$B^{(1)}H$ + ribose transformation products

R-polynucleotide chain

The entire cycle of transformations can be repeated. To this end, the phosphate group is enzymatically removed (with the aid of PME) from the 3′-terminal nucleotide. Acid ($pH \leq 3$) or alkaline ($pH > 11$) hydrolysis of the former 3′-terminal nucleotide yields a pyrimidine or purine base that can be identified by conventional analytical methods. As has already been mentioned, this procedure is used to determine the primary structure of oligoribonucleotides. Quantitative periodate oxidation and β-elimination are achieved by conducting the reaction at pH values close to neutral. The primary amine seems to perform two functions: it participates in the formation of active intermediates and catalyzes the β-elimination process. The results of studies into the effect of pH, amine structure and other factors on the course of degradation of the polynucleotide chain, based on the β-elimination mechanism, together with the data on the structure of the intermediate compounds, obtained in experiments with simpler model systems, suggest that apart from direct β-elimination this process also includes formation of some cyclic intermediates whose subsequent transformations lead to shortening of the polyribonucleotide chain by one nucleotide.

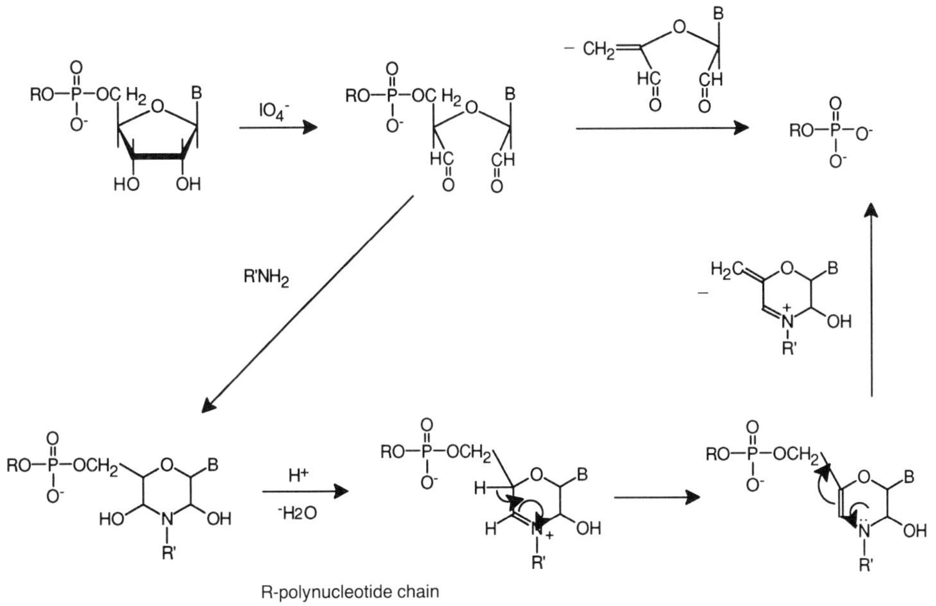

R-polynucleotide chain

9.5.2 Cleavage of the Sugar–Phosphate Backbone with the Aid of "Chemical Nucleases"

Bonds in the sugar–phosphate backbone can be cleaved in the presence of O_2 or H_2O_2, reducing agents, and chelated ions of transition metals. The oxidative degradation of nucleic acids was studied most thoroughly in experiments where used as chelating agents were some antibiotics, such as bleomycin (Blm) and its analogues, as well as ethylenediaminetetraacetic acid (EDTA), *ortho*-phenanthroline (Phen), and porphyrins.

Molecular oxygen is a four-electron oxidizing agent and its stepwise reduction to water can be written as follows:

$$O_2 \xrightarrow{1e^-, 1H^+} HO_2 \xrightarrow{1e^-, 1H^+} H_2O_2 \xrightarrow{1e^-, 1H^+} OH + H_2O \xrightarrow{1e^-, 1H^+} 2H_2O$$

Molecular oxygen in an aqueous solution is a potent oxidizing agent, during reduction to both H_2O and H_2O_2, the values of $E^{\circ\prime}$ being 0.815 and 0.3 V, respectively. However, direct oxidation with oxygen, whether a two- or four-electron one, fails to materialize. This is first of all due to spin forbiddenness, because the ground state of O_2 is triplet. At the same time, one-electron reduction, which is the first step in the above scheme, is endothermic (E° = –0.16 V). The role of the metal ion or complex boils down first of all to enabling this step. The resulting superoxide radical turns into H_2O_2 either by way of disproportionation or a reaction with a supplementary reducing agent

H_2A (ascorbic acid, thiols, etc.). Hydrogen peroxide is not capable of oxidizing most organic compounds at a measurable rate under mild conditions either, and its one-electron reduction also calls for a transition metal ion. Formed in this case are hydroxyl radicals that act as most potent agents ($E° = 1.9$ V) capable of oxidizing any organic compounds. This process, just as the first step of oxygen reduction, also requires a conjugated easily oxidizable substrate H_2A for reduction of the metal ion to the lowest state of oxidation. The active forms of oxygen emerging in the course of the process are in a free state as well as coordinated with the metal ion. By way of an example, here is the most probable scheme of oxidation of organic substrate H_2X in the presence of metal ion or complex M^{+n} and a conjugated reducing agent H_2A (the oxygen particles are not coordinated):

$$M^{+(n-1)} + O_2 \rightarrow M^{+n} + O_2^-$$

$$O_2^- + H_2A \rightarrow HO_2 + HA^-$$

$$M^{+(n-1)} + H_2O_2 \rightarrow M^{+n} + OH^{\cdot} + OH^-$$

$$M^{+n} + H_2A\ (\text{or } HA^{\cdot}) \rightarrow M^{+(n-1)} + HA^{\cdot}\ (\text{or } A) + H^+$$

$$OH^{\cdot} + H_2X \rightarrow OH^- + HX^{\cdot} + H^+$$

$$HX^{\cdot} \rightarrow \text{oxidation products}$$

It can be seen that the metal complex acts as a reducing agent in reactions with O_2 and H_2O_2 and, after subsequent regeneration, as an oxidizing agent. Consequently, the metal complex must have an intermediate value of the redox potential, which is to say that its behavior as a reducing or oxidizing agent must be moderate. Moreover, the ligands must readily lend themselves to substitution, because some of the electrontransfer reactions shown in the above scheme are intraspheric. These conditions are most fully met by copper complexes whose standard redox potentials are dependent on the ligand species but insignificantly and range from 0.1 to 0.2 V and which are highly labile with respect to substitution, as well as iron complexes, such as those with EDTA or porphyrins, whose redox potentials range from 0.15 to 0.30 V. At the same time, the iron complex with *ortho*-phenanthroline cannot act as a catalyst because of the very high redox potential ($E° = 1.06$ V). In principle, complexes of other transition metals, such as cobalt, nickel, and manganese, may also catalyze redox reactions involving nucleic acids.

Two pathways of degradation have been identified, different in composition of the end products. The first pathway leads to direct degradation of the chain with formation of 3′-phosphoglycolates at the 5′ end and a phosphate group at the 3′ end, accompanied by liberation of propenal in the corresponding heterocycle. The second pathway leads to liberation of a heterocycle with the deoxyribophosphate backbone remaining intact but being prone to hydrolysis in a weakly alkaline medium. Such hydrolysis is typical of deoxyribophosphate fragments with a liberated heterocycle.

The mechanism of the process seems to involve detachment of the hydrogen at $C^{4'}$ during the limiting step. Direct degradation occurs when the emerging radical turns into a hydroperoxide one and then forms a hydroperoxide group. The latter undergoes rearrangement with the oxygen of the hydroperoxide group being inserted between $C^{3'}$ and $C^{4'}$. As a result, the above-mentioned products of direct degradation are formed.

The main features of the mechanism of DNA cleavage with the aid of Fe(III)-EDTA derivatives were studied in an experiment with an ethidiumpropyl-EDTA-Fe(III) derivative. In this case, the 5′-terminal unit of the chain turned into an oligonucleotide with a 3′-terminal phosphate or phosphoglycolate, whereas the 3′-terminal unit carried a 5′-phosphate at the end. The process was accompanied by liberation of a heterocycle. The emerging three-carbon fragment was not identified. The degradation products were the same as during gamma radiolysis of DNA. This is indicative of direct involvement of free OH radicals at the first step of degradation.

Analysis of the products of DNA hydrolysis in the presence of a Cu(II)-$(Phen)_2$ complex has shown that the attack of the active oxygen-containing particle begins at $C^{1'}$ of deoxyribose. The hydrolysis products are fragments with 5′- and 3′-phosphate groups, a free heterocycle, and 5-methylenefuranone. The differences in products of degradation by $Cu(Phen)_2$ and Fe(III)-EDTA complexes indicate that the active oxygen-containing particles belong to different species.

This process results in cleavage of DNA with formation of 3′- and 5′-phosphorylated fragments at the site where the DNA chain is attacked. An alternative pathway comprises an attack of C^4 in the nucleoside with liberation of a base and formation of a fragment with a 5′-phosphate group, while the second (5′-terminal) fragment has a substituted 3′-phosphate group.

Given below are the formulas of the chelate complexes of various metals most frequently used for degradation of DNAs under physiological conditions.

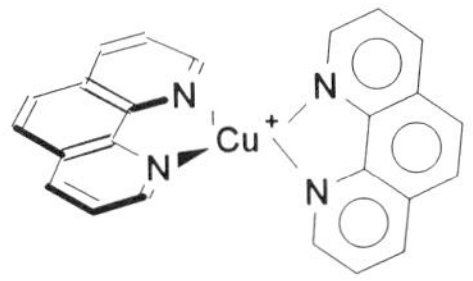

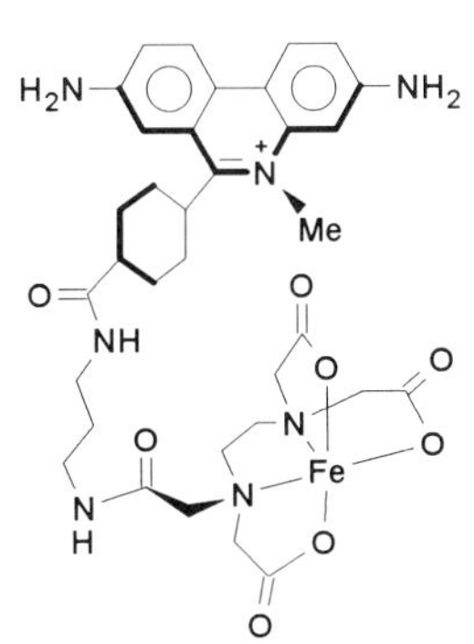

c)

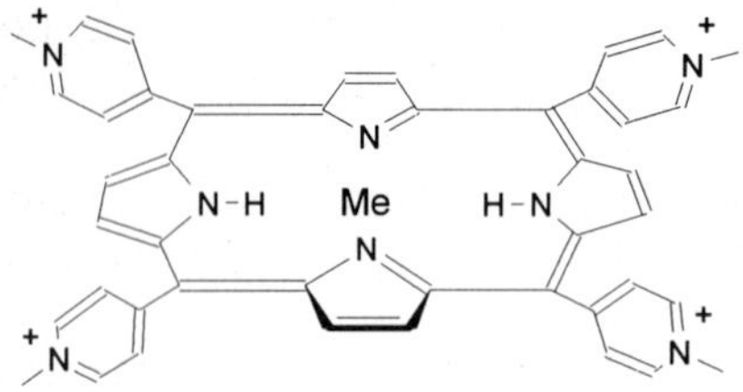

d)

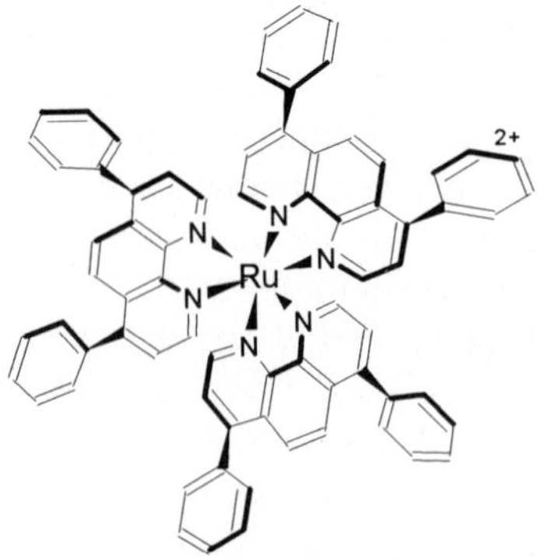

These and other similar compounds have become known as "chemical nucleases". They include, among others, the natural antibiotic bleomycin.

9.5.3 Reactions with Internucleotide Linkages Remaining Intact

Reactions of this type, involving internucleotide phosphate groups, have not been studied as thoroughly as those accompanied by cleavage of the polynucleotide chain. Only the substitution reaction has been described in this category.

Internucleotide phosphate groups in nucleic acids under normal conditions (aqueous medium, 20 °C, pH $\sim$ 7) exist in anionic form (pK_a $\simeq$ 1) and, therefore, display nucleophilic behavior. This property is used in alkylation, whereby an internucleotide phosphate group is converted into a triester one. Along with alkylation of heterocyclic bases, conversion of this type involves attack of nucleic acids by such alkylating agents as nitrosoalkylurea. The latter is known as one of the most potent mutagenic compounds and as an antitumor drug. Studies of reactions between nitrosoethylurea and nucleic acids have demonstrated that this reagent is capable of alkylating internucleotide linkages (phosphodiester bonds) more effectively as compared to a reaction at heterocycles.

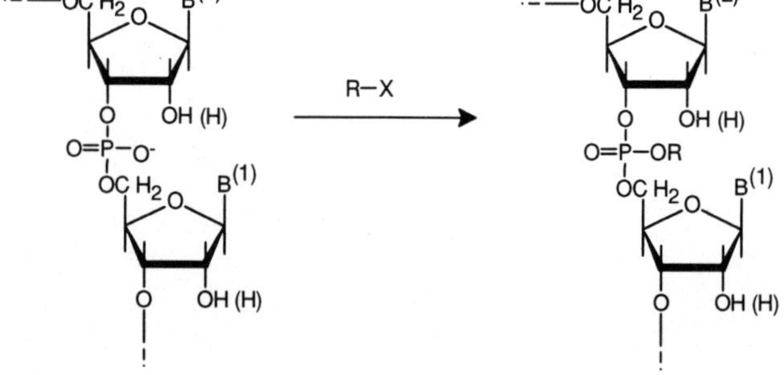

For instance, when viral RNA is modified with nitrosoethylurea by 70 %, the alkylation takes place at internucleotide phosphate groups. Carbamoylation occurs as a side reaction.

The phosphotriester internucleotide groups yielded by the reaction with nucleic acids are labile, especially in the case of RNAs. Such an alkylated RNA may be easily cleaved at alkylated internucleotide groups in a weakly alkaline medium. Intramolecular nucleophilic attack by the 2′-hydroxyl group ruptures internucleotide linkages.

9.6 Reactions of Terminal Phosphate Groups

Differences in the reactivity of terminal and internucleotide phosphate groups in nucleic acids have not been determined in direct experiments because of the extremely low relative amount of the former. However, there is every reason to believe that the differences here will be similar to those observed in the case of mono- and diphosphates; the former are more active as nucleophiles because they carry practically two negative charges at neutral pH.

Nonetheless, it is virtually impossible to conduct such a reaction as alkylation of a nucleic acid selectively at the terminal phosphate group. For the reaction to take place, the terminal phosphate group is activated with the nucleic acid being treated with carbodiimide; in this case, internucleotide phosphate groups are practically not affected. The resulting derivative is not separated. If the reaction is carried out in an alcohol, the terminal phosphate is immediately alkylated. This is how, for example, 5′-terminal phosphate groups in oligoribo- and oligodeoxyribonucleotides were selectively methylated:

$$\xrightarrow[C_6H_{11}N{=}C{=}NC_6H_{11}]{CH_3OH}$$

X=H or OH

The reaction proceeds quantitatively in methanol at room temperature (24–48 h); however, heterocyclic bases also undergo partial methylation. The same method was used to introduce substituents into 5′-phosphate groups in tRNA. For instance, treatment of a trialkylammonium salt of tRNA with diisopropylcarbodiimide in the presence of aniline in a mixture of water, dimethylformamide and *term*-butyl alcohol at pH 8 yielded an anilide of tRNA. At room temperature, the reaction was 60–80 % complete within 24–48 h.

Activation of the terminal phosphate in oligo(poly)nucleotides is a widely used procedure for their immobilization on cellulose or Sepharose. Particularly good results are obtained by activation with water-soluble carbodiimides. In this case, the immobilization on hexamethylene diamino-Sepharose, for example, can be performed in aqueous solutions:

$$\text{—NH—(CH}_2\text{)}_6\text{—NH}_2 + \text{pNpNpNp...} \xrightarrow{\text{R—N=C=N—R'}} \text{—NH—(CH}_2\text{)}_6\ \text{pNpNpNp...}$$

$R = C_6H_{11}$; $R' = (CH_2)_2$–$\overset{+}{N}$(morpholino)–CH_3

This is how oligo(dT), poly(U), poly(A), and other polynucleotides bound to Sepharose (or cellulose) are prepared.

Another method for selective activation of the terminal phosphate in oligo- and polydeoxyribonucleotides was developed recently. A careful study of the reaction between oligodeoxyribonucleotides and mesitylenecarboxyl chloride has shown that only the terminal phosphate is acylated under certain conditions. The reaction is conducted in absolute pyridine for 10 min (20 °C). Under such conditions, heterocyclic bases and pentose are not acylated. Internucleotide phosphate groups are acylated, but subsequent treatment with water (when the reaction products are formed) leads to almost instantaneous hydrolysis of the resulting acyl phosphates with departure of the carboxyl anion.

Analysis of the properties of anhydrides of oligo(poly)nucleotides with the sterically hindered carboxylic acid has revealed a number of specific features. First of all, such anhydrides are good phosphorylating agents for sterically unhindered nucleophiles; the reaction proceeds only at the phosphate group. No hydroxamic acids are formed even with hydroxylamine which reacts with acyl phosphates at the phosphate group only. Hence, the steric hindrances for the carbonyl group in mesitylenecarboxylic acid remain and become even more pronounced in the mixed anhydride whose formation involves the terminal (3′ or 5′) nucleotide phosphate group.

This conclusion confirms the second characteristic property of such mixed anhydrides, which is stability in aqueous solutions. Unlike mesitylenecarboxyl chloride which is readily hydrolysed with water, the corresponding mixed anhydrides of oligodeoxyribonucleotides are rather stable in aqueous solutions (the half-period of hydrolysis at pH 7 varies from 3 to 4 days). Such mixed anhydrides readily react with amines in aqueous solutions and with good yields:

Reactions with weaker nucleophiles (alkyl mercaptans, alcohols, phosphates) proceed more easily in organic solvents:

R'SH R'OH

R-oligonucleotide chain
R′-Alk

In this way oligo(poly)deoxyribonucleotides are easily immobilized on hexamethylene diamino-Sepharose:

$-NH-(CH_2)_n-NH_2$ + $1,3,5\text{-}(CH_3)_3C_6H_2CO\text{-pdNpdNpdNp...}$

$\longrightarrow$ $-NH-(CH_2)_n-NHpdNpdNpdNp...$

As opposed to the above-described method for activating the terminal phosphate in polynucleotides by carbodiimides, that based on mixed anhydrides with mesitylenecarboxylic acid is rather simple and much more effective. There is every reason to consider it as one of the most promising ways of controlled modification of nucleic acids at terminal phosphate groups, including introduction of various "active" non-nucleotide groups to create affine reagents[2].

An alternative method for selective modification of the terminal phosphate group in oligo(poly)nucleotides is the reaction with morpholides of alkyl phosphates.

$$R'O-P(=O)(O^-)-N\langle\text{morpholine}\rangle + {}^-O-P(=O)(OH)-OR \longrightarrow R'O-P(=O)(O^-)-O-P(=O)(OH)-OR + HN\langle\text{morpholine}\rangle$$

Condensation of methylphosphomorpholide with tri-*n*-hexylammonium salt of tRNA in pyridine (20 °C, 6–7 days) makes it possible to convert the 5′-terminal phosphate in tRNA into methylpyrophosphate:

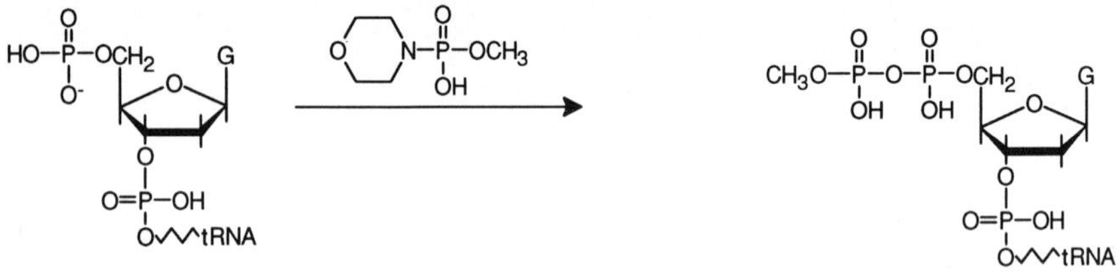

Similarly, an alkylphosphomorpholide containing a dansyl (5-dimethylamino-naphthalene-1-sulfamido-methyl) group was used to obtain *E. coli* $tRNA^{fMet}$ with a fluorescent marker at the 5′-terminal phosphate group (the dansyl group is broadly employed as a fluorescent marker in studying the three-dimensional structure of biopolymers and for other purposes).

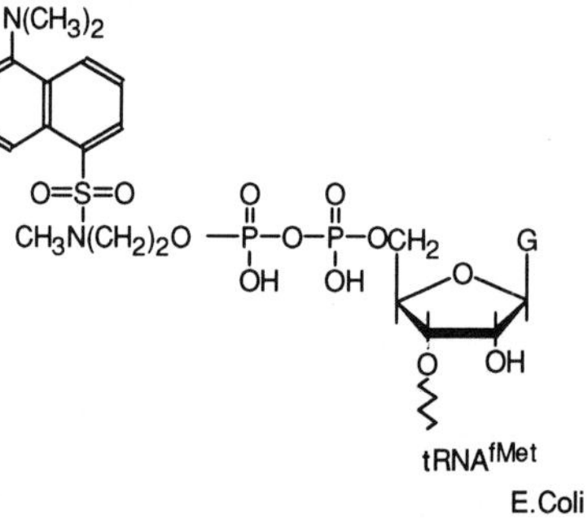

[2] Affinity reagents are compounds consisting of two fragments. One of them is "recognized" by the biopolymer or, in other words, forms a stable complex with it through noncovalent interactions (affinity interactions). The other is an ordinary active group sufficiently stable in aqueous solutions but capable of forming covalent bonds with functional groups of the biopolymer in the complex by virtue of close proximity thereto. Oligonucleotides with their inherent capacity for complementary interactions provide ideal building blocks for reagents recognized both by nucleic acids (their respective complementary portions) and by proteins exhibiting affinity for oligonucleotides (e. g., restrictases, other nucleases, etc.). This technique, known as biospecific marking, is widely used in molecular biology.

Other non-nucleotide groups are selectively introduced at the terminal phosphate in nucleic acids by this method as well.

In recent years, a universal method for modifying the terminal phosphate of oligo(poly)nucleotides in aqueous solutions has been developed.

Two alternative approaches have been devised for selective activation of the phosphomonoester group: the direct derivatization under the effect of water-soluble 1-ethyl-3(3′-dimethylaminopropyl)-carbodiimide (EDC) and the substitution reaction of *N*-hydroxybenzotriazole phosphodiesters (both types of activation are obtained in an aqueous solution on unprotected oligonucleotides).

EDC-induced derivatization is very simple in experimental terms and proceeds according to the pathways:

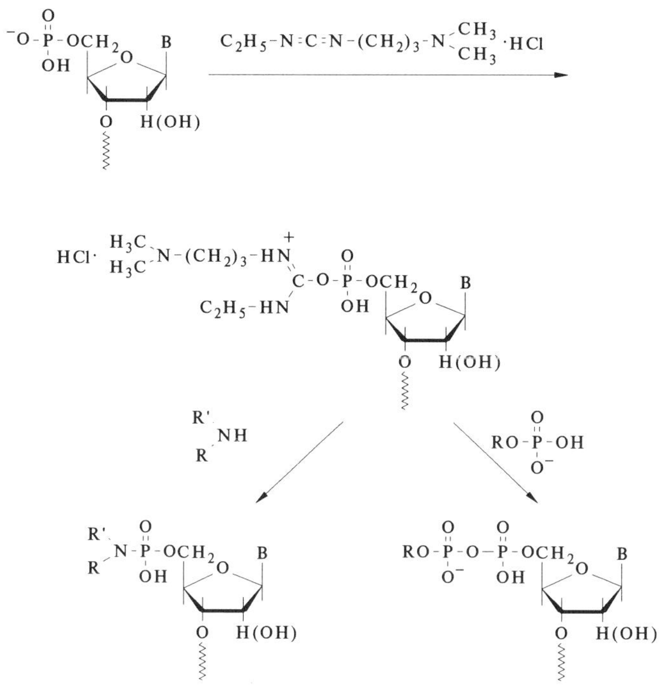

$$C_2H_5{-}N{=}C{=}N{-}(CH_2)_3{-}N(CH_3)_2 \cdot HCl$$

It was demonstrated that EDC is the only reagent capable of synthesizing phosphoramidate as well as phosphodiester derivatives in an aqueous solution.

Reactions performed in the optimal pH range and in the presence of a high (2–3 M) concentration of a nucleophile give high yields of products. This method can be recommended as an optimal one for inserting such spacer groups as residues of aliphatic diamines etc. Non-substituted amino groups of the latter may be used for subsequent derivatization.

The oligonucleotide *N*-hydroxybenzotriazole phosphodiesters are also very promising intermediate compounds for designing a broad series of oligo(poly)nucleotides substituted at the terminal phosphate. These compounds are hydrolytically stable at pH ≤ 8 and at the same time are readily subjected to aminolysis and alcoholysis in an aqueous solution:

$C_2H_5-N=C=N-(CH_2)_3-N(CH_3)_2 \cdot HCl$

aqueous NH_3

aqueous C_2H_5OH

In contrast to the EDC-induced condensation, when all the groups that can be activated by a condensation agent should be protected, in this particular case, one may use amino acids and peptides unprotected at the carboxyl. The nucleophile concentration can be brought down to 0.2–0.3 M.

Both methods of derivatization are simple, effective and mutually complementary.

Thus, the difference in reactivity between the terminal phosphate in polynucleotides and other nucleophilic groups (amino groups, hydroxy groups in carbohydrate moieties, phosphodiester groups), established for some reactions conducted under rigorously defined conditions (solvent, temperature, reaction time), allows one not only to selectively modify the terminal phosphate group but also to activate, also selectively, a given polynucleotide and convert it into a compound capable of forming covalent bonds. This may form the basis of using appropriately modified nucleic acids to solve many problems in molecular biology, such as defining the pathways of nucleic acid functioning, elucidating the topography of cell components (ribosomes, etc.), and so on.

9.7 Chemical Probing of Nucleic Acid Structure

We shall now discuss the use of reactions aimed at chemical modification of heterocyclic bases and the sugar–phosphate backbone of nucleic acids for studying their macromolecular structure. As was already mentioned in the beginning of this chapter, the reactivity of a particular nucleotide in DNA or RNA is to a certain degree dependent on the secondary or tertiary structure element of the nucleic acid accommodating it. Consequently, by studying the degree of modification of particular functional groups of bases as well as the carbohydrate moieties and phosphate group of each nucleotide in a nucleic acid molecule one can obtain valuable information about its three-dimensional structure. This approach has become known as chemical probing. In combination with other physical methods, it allows the macromolecular structure of DNA and RNA to be described with high resolution. Moreover, by comparing the reactivity of nucleotides in DNA or RNA in a free state and in a complex with specific ligands, one can identify, also with a high resolution, nucleic acid sites responsible for specific binding and recognition of a particular ligand.

Chemical probing is especially valuable in cases where X-ray structural analysis or NMR spectroscopy are not applicable, for example, when analyzing large RNAs or multicomponent complexes of nucleic acids with proteins. Furthermore, in contrast to X-ray structural analysis, chemical probing makes it possible to study the behavior of nucleic acids in solution as well as the dynamics of their interaction with proteins over the course of time. Another important consideration is the fact that the sample amount for this technique can be hundreds or even thousand of times smaller than for many physical measurements.

9.7.1 Location of Modified Nucleotides

The improvement of methods for locating modified nucleotides in DNA and RNA has gone hand in hand with advances in nucleic acid sequencing because the objectives are similar or identical in both cases.

Whereas sequencing of nucleic acids had been based on identification of the primary structure of each fragment, in order to identify a modified base one also had to find the host oligonucleotide, then isolate the modified base by further cleavage and fractionation. Therefore, the search for a particular modified nucleotide had always been an extremely difficult task. This inevitably led to disturbance of its macromolecular structure and, ultimately, to wrong conclusions. A case in point is attempts to determine the tertiary structure of tRNA in the years preceding elucidation of its crystal structure. The chemical modification data had formed the basis of many models of the tRNA tertiary structure proposed at that time, and every model turned out to be incorrect.

The situation changed drastically with the development of rapid methods of gel sequencing of DNA and RNA (see Chapter 6). Today, the following two strategies of modification and location of modified nucleotides are used (see Fig. 9-1).

In the first case, just as during nucleic acid sequencing by the Maxam–Gilbert method, subjected to modification is a homogeneous nucleic acid, or its fragment, labeled at one end (usually with radioactive phosphate). The reaction is conducted in such a manner that not more than one nucleotide, on the average, becomes modified in each nucleic acid molecule, which is achieved by lowering both the concentration of the reagent and the reaction temperature. The direction of the reaction is thus determined only by the starting three-dimensional structure of the nucleic acid. Then the polynucleotide chain is cleaved at the modidfied base (if the cleavage does not take place during the modification itself, for example, when "chemical nucleases" are used), and the length of the resulting fragment, measured during gel electrophoresis, is used to locate the base to within a single unit.

As can be seen, the first strategy is restricted to use of reagents modifying DNA or RNA so that it can be specifically cleaved and to homogeneous polynucleotides with a length not exceeding 300 to 400 nucleotides. These restrictions are removed in the second strategy of locating modified nucleotides, which has turned out to be especially useful in studying large RNAs.

In this case, an oligodeoxyribonucleotide complementary to a nucleic acid segment located not far from and downstream of the target sequence is synthesized. After the modification reaction is over (it is also conducted so as to modify not more than one base), the oligonucleotide is used as a primer for reverse transcriptase. If the chemical modification of nucleotides is carried out in a way that inhibits Watson–Crick hydrogen bonding, the enzyme will not be able to read modified bases and will stop at a modification site (Fig. 9-1). Of course, it will also stop if the modification leads to rupture of the polynucleotide chain. Then, the length of the synthesized cDNA, determined by means of dideoxy sequencing ladders (see Chapter 6), is used to locate the modified nucleotide. And if the modification involves a single-stranded DNA, one of the DNA polymerases is used instead of reverse transcriptase.

Fig. 9-1. Two strategies of localization of modified nucleotide residues in a nucleic acid molecule. ▶

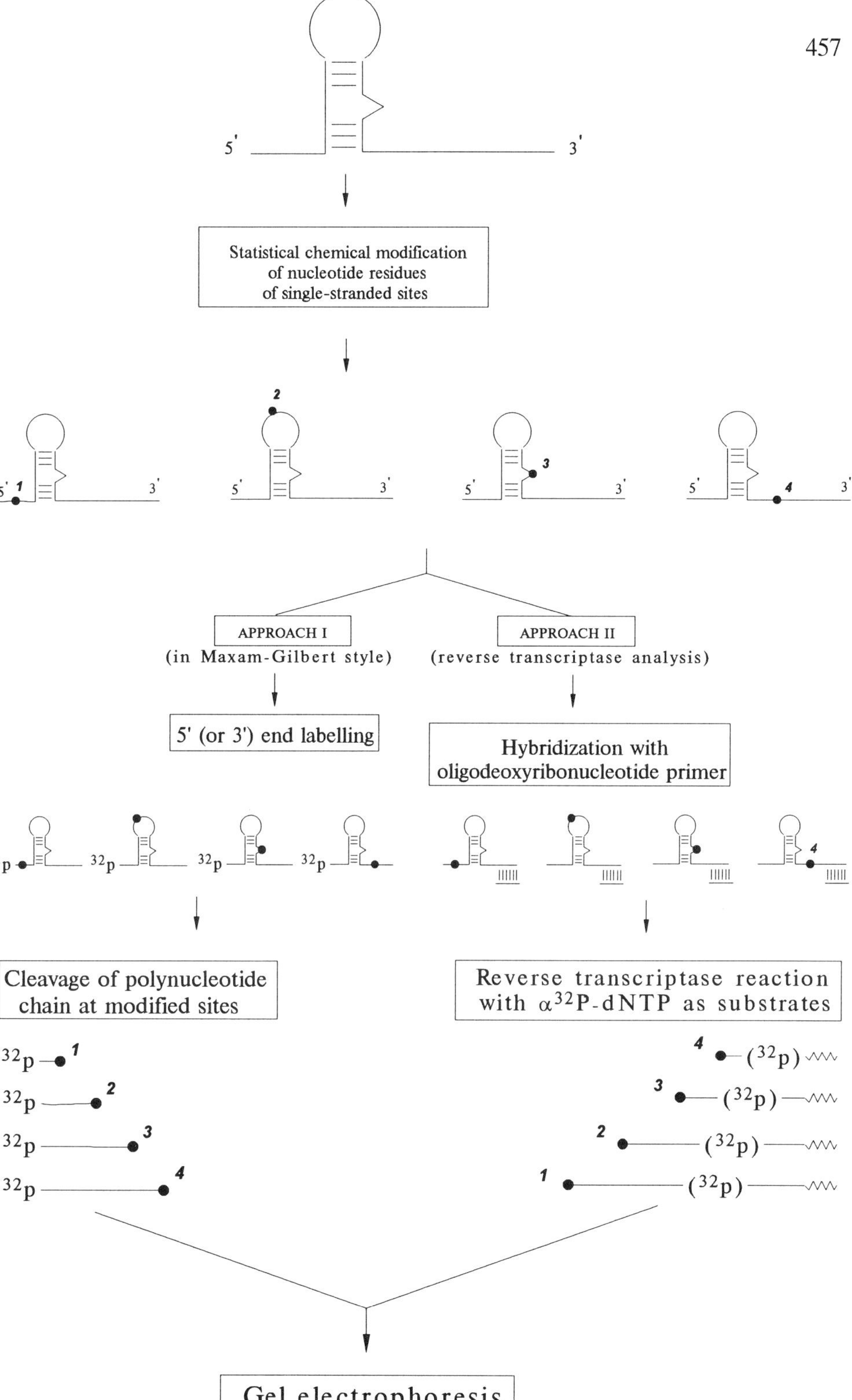

5'
3'
Statistical chemical modification of nucleotide residues of single-stranded sites
1
2
3
4
APPROACH I
(in Maxam-Gilbert style)
APPROACH II
(reverse transcriptase analysis)
5' (or 3') end labelling
Hybridization with oligodeoxyribonucleotide primer
32p
Cleavage of polynucleotide chain at modified sites
Reverse transcriptase reaction with α32P-dNTP as substrates
(32p)
Gel electrophoresis

Optimally, both strategies enable one to monitor every base over the nucleic acid segment under investigation. Some problems may arise in interpretation of the results. In particular, it is not always clear why a particular nucleotide is not reactive or, to be more precise, whether it is involved in secondary or tertiary interactions. Investigating RNAs under partially denaturing conditions, when the tertiary structure is disturbed to some degree whereas the secondary is yet intact, often helps in interpreting the results of chemical probing.

The same strategies are used in studying nuclein–protein complexes. To this end, the chemical modification of DNA or RNA in a free state is in most cases compared with that in a complex with a protein. In this way it becomes possible to detect the trace left by the protein on the nucleic acid molecule, which is why this technique has become known as footprinting.

9.7.2 Systematic Chemical Probing of RNA Secondary and Tertiary Structure

9.7.2.1 Transfer RNA

Chemical probing in combination with gel sequencing was used for the first time by Peattie and Gilbert in 1980 to monitor the $tRNA^{Phe}$ conformation in solution. They employed two reagents – dimethylsulfate (DMS) to probe position N^7 of guanine and N^3 of cytosine as well as diethylpyrocarbonate (DEPC) to probe position N^7 of adenine (for mechanisms of these reactions, see 9.2.2.2). Since the use of these reagents allowed the investigators to subsequently cleave phosphodiester bonds at modification sites, tRNA was labeled with ^{32}P at one end and the first strategy shown in Fig. 9-1 was used ot locate a modified base (see Fig. 9-2). The result was information about 51 out of 76 bases in tRNA. It had been established that most bases behave as could be expected from the crystallographic three-dimensional structure of tRNA. However, at least three bases (A73, C72 and G19) did not fit the pattern, and eight guanines were assumed to have been involved in hitherto unknown tertiary interactions.

Fig. 9-2. Secondary and tertiary structure of the T arm part of $tRNA^{Phe}$ according to the results of modification with DMS and DEPC.
A. The cloverleaf structure of yeast $tRNA^{Phe}$; lines connect the bases involved in tertiary contacts.
B. The portions of gel autoradiograms illustrating chemical probing of the native, semi-denaturated and denaturated forms of the T arm.
C. Tertiary structure of the T arm part of yeast $tRNA^{Phe}$ (shaded area); tertiary base-base hydrogen bonds are marked by black rods (adapted from D. A. Peatie and W. Gilbert, *Proc. Nat. Acad. Sci. USA*, **77**, 4679–4682 (1980)). ▶

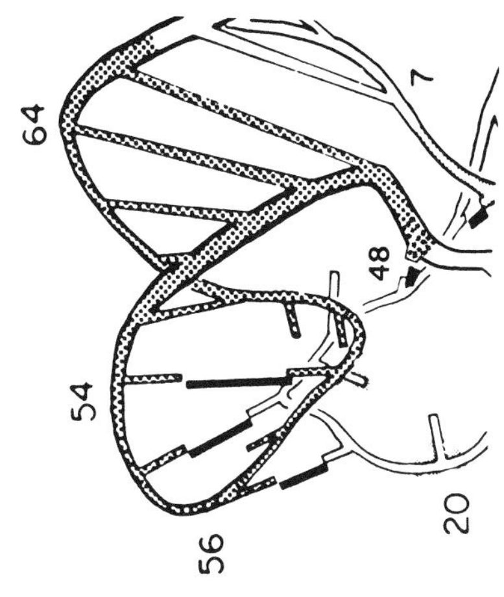
B
NATIVE
SEMIDENATURED
DENATURED
A G C
C48
G51
G53
C56
G57
U59
G65
U50
U52
xc

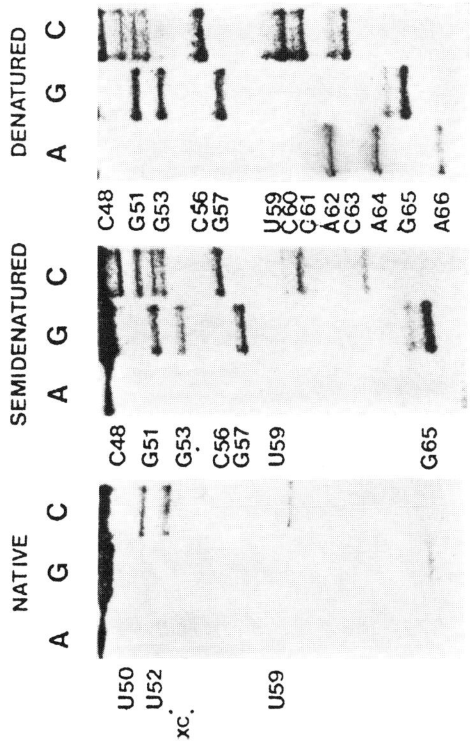
C
64
54
56
48
20
7

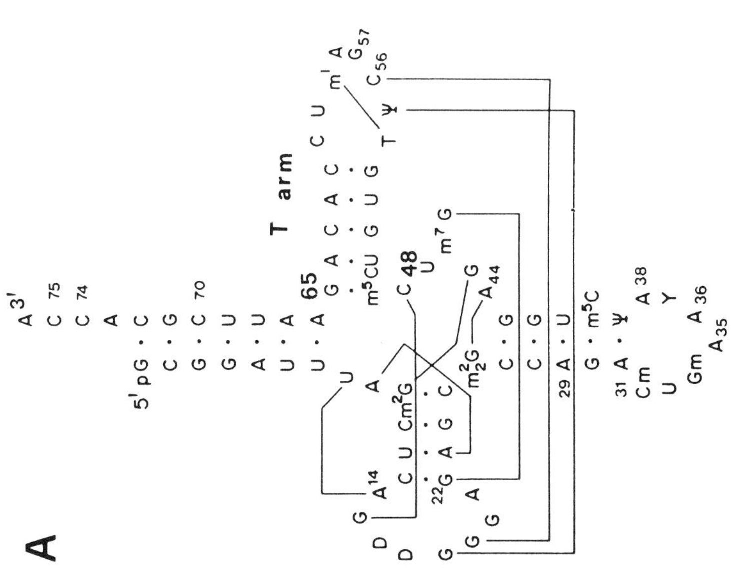
A
T arm
65
48

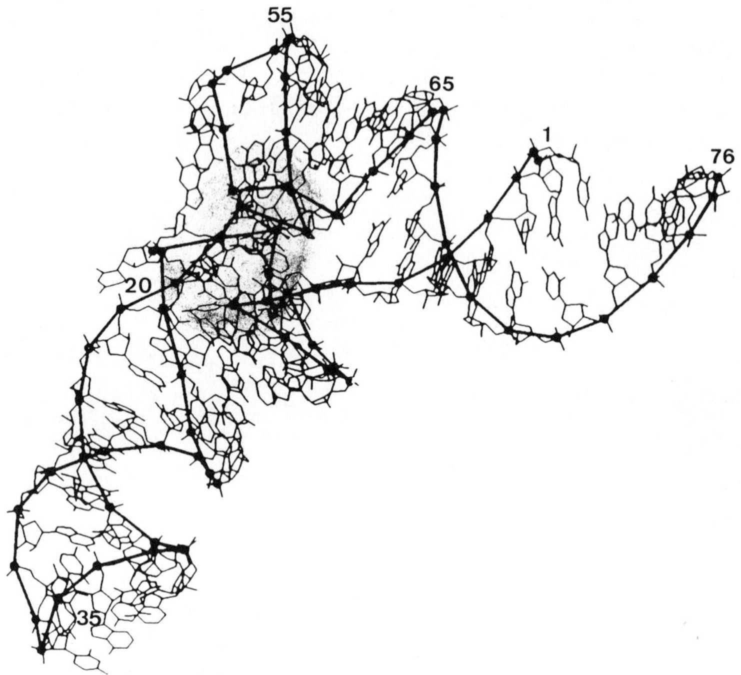

Fig. 9-3. Three-dimensional structure model of initiator E. coli $tRNA^{Met}$ based on the results of the detailed chemical probing; the shaded area is the region where phosphate groups are protected from the attack with ethylnitrosourea (adapted from H. Wakao et al., *J. Biol. Chem.*, **264**, 20363–20371 (1989)).

Later, many tRNA were subjected to chemical probing. Figure 9-3 illustrates the results of monitoring of *E. coli* initiator tRNA. In this particular case, added to the above-mentioned DMS and DEPC was 1-cyclohexyl-3-(2-morpholinoethyl)-carbodiimide metho-*para*-toluene sulfonate (CMCT) which modifies Watson–Crick positions N^1 in guanine and N^3 in uridine. Moreover, adenines were monitored by DMS at N^1. A radically new approach here is monitoring of phosphates by ethylnitrosourea (ENU). Modifications at phosphate groups as well as at C(N^3), G(N^7) and A(N^7) were identified after tRNA cleavage. Modifications at A(N^1), G(N^1) and U(N^3) could only be identified by primer extension. Combination of all these reactions has given a good insight into the tertiary structure of initiator tRNA, which has turned out ot be instrumental in studying the interaction between this tRNA and proteins involved in protein biosynthesis.

Such a detailed picture provided by chemical monitoring of tRNA can only be enhanced by the results of $tRNA^{Phe}$ cleavage by Fe(II)-EDTA (see also 9.5.2). Fe^{2+} ions in a complex with EDTA in the presence of O_2 and a reducing source (e.g., dithiothreitol) generate hydroxyl radicals attacking ribose in

RNA and deoxyribose in DNA. The attack occurs near the site of binding of the redox-active metal complex with the nucleic acid. As can be inferred from Figure 9-4, several short tRNA segments are protected against "chemical nuclease". Just as expected, the tertiary interactions observed in this experiments are stabilized by magnesium ions.

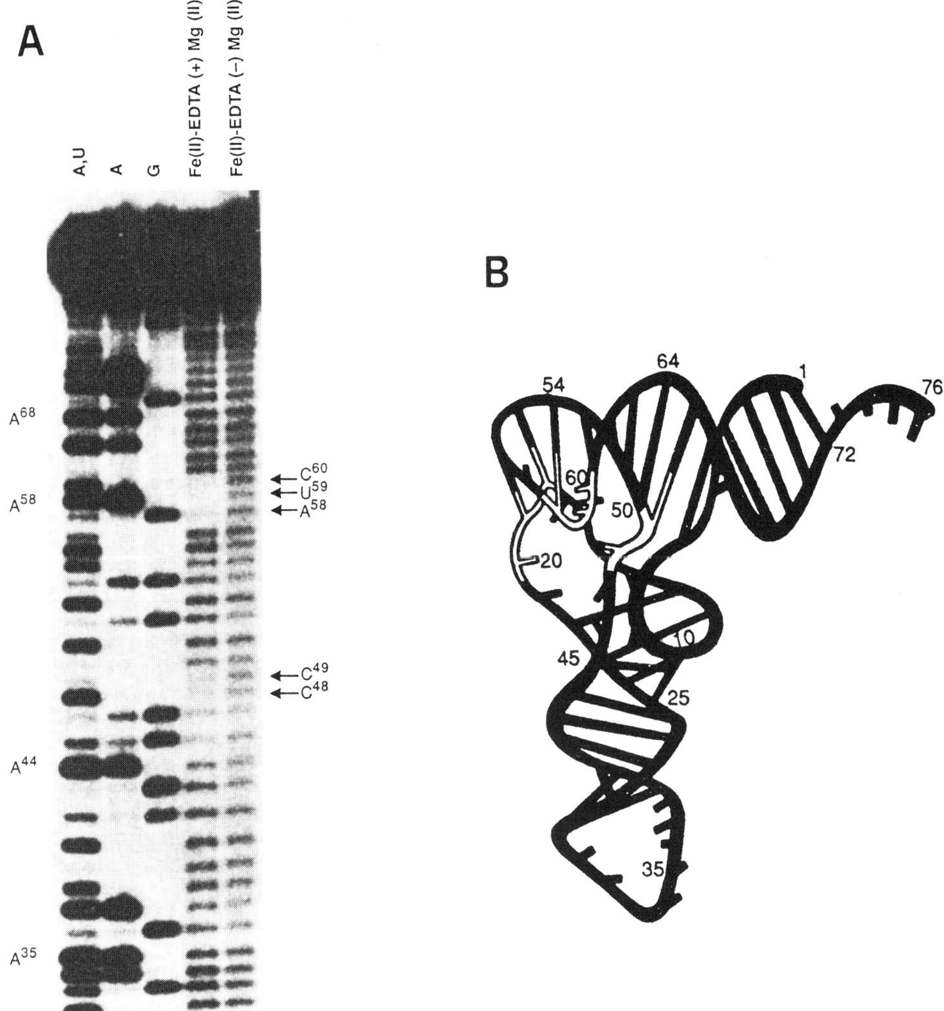

Fig. 9-4. Cleavage of $tRNA^{Phe}$ by Fe(II)-EDTA.
A. Autoradiogram of the polyacrylamide gel; A+U, A and G are sequencing lines; oxidative degradation of the tRNA chain is clearly seen in the lane corresponding to the native form of $tRNA^{Phe}$ (marked with arrows).
B. Three-dimensional structure of $tRNA^{Phe}$; white backbone indicates protection from Fe(II)-EDTA. (adapted from J. A. Latham and T. R. Chech, *Science,* **245**, 276–282 (1989)).

9.7.2.2 Ribosomal RNA

We have already mentioned that the secondary structure models for all types of rRNA are based primarily on comparative (phylogenetic) analysis of the primary structures of rRNAs in many organisms (see 8.2.3). These models were experimentally tested and refined by chemical probing.

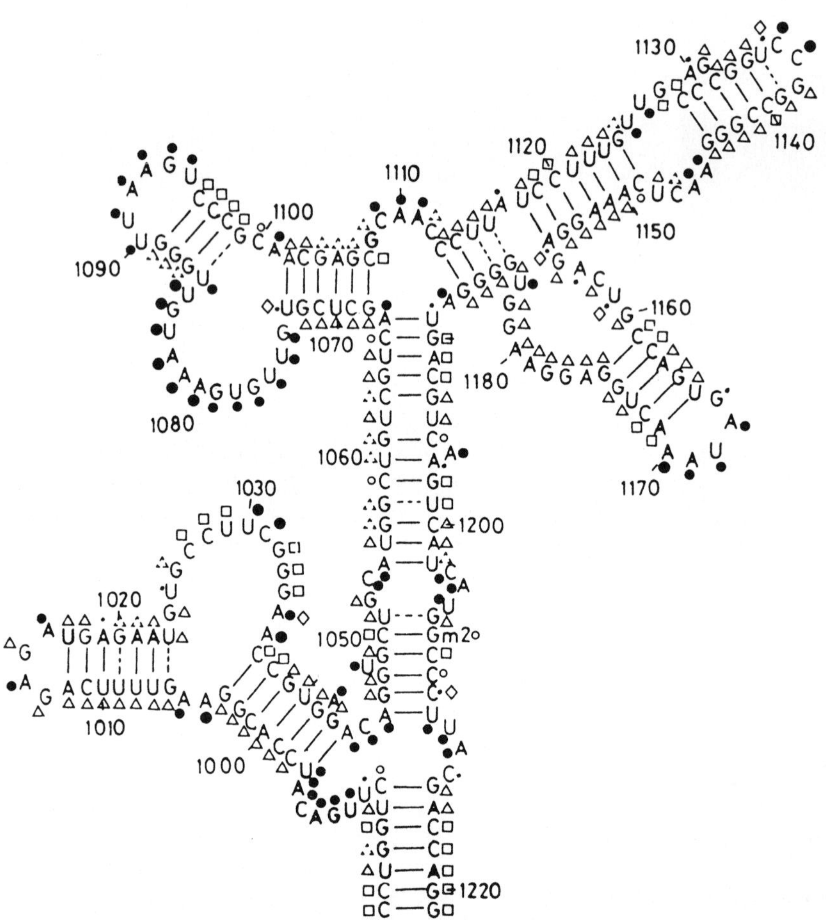

Fig. 9-5. Scheme illustrating the exhaustive chemical modification of the segment (nucleotides 984–1221) of the 16S RNA from E. coli ribosomes by DMS and CMCT. Reactive under native conditions: (●) strong hit, (•) moderate hit, (·) marginal hit; unreactive under native conditions but reactive under semi-denaturing conditions: (△) moderate hit, (∴) marginal hit; unreactive under both native and semi-denaturing conditions: (□); (✧) denotes increased reactivity under semi-denaturing conditions, as compared to native conditions. (∘) denotes not determined positions (reproduced with permission from C. Ehresmann et al., *Nucl. Acids Res.*, **15**, 9109–9128 (1987)).

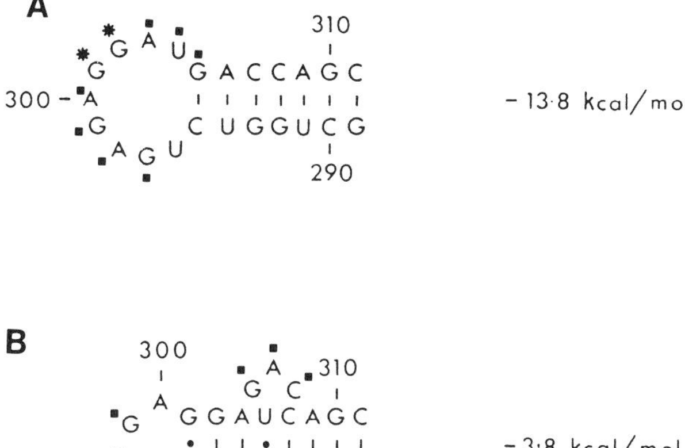

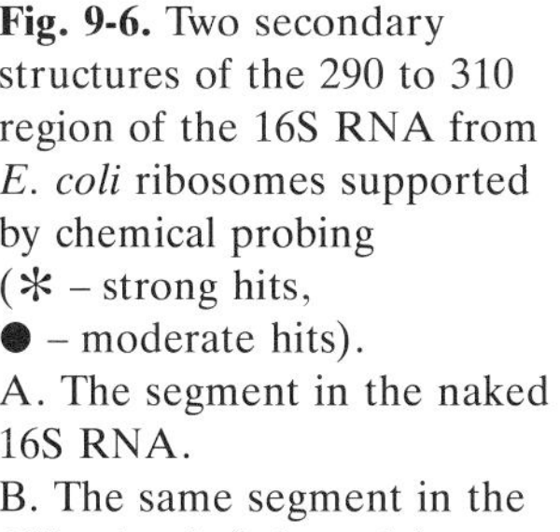
Fig. 9-6. Two secondary structures of the 290 to 310 region of the 16S RNA from *E. coli* ribosomes supported by chemical probing (✻ – strong hits, ● – moderate hits).
A. The segment in the naked 16S RNA.
B. The same segment in the 30S subunit (adapted from D. Moazed, S. Stern and H. Noller, *J. Mol. Biol.*, **187**, 399–416 (1986)).

Chemical monitoring of the rRNA structure was most widely used in Noller's and Ebel's laboratories. In addition to the reagents mentioned in the previous section use was also made of kethoxal (KE) which modifies guanine bases at N^1 and N^2. The modified bases were identified by the primer extension approach. Used for this purpose was a series of oligodeoxyribonucleotides that prime reverse transcription at 150–200 nucleotide intervals along the RNA chain. As a result, it became possible to monitor the overwhelming majority of bases in rRNA and to demonstrate convincingly that the theoretical models quite adequately describe the macromolecular structure of rRNA both in a free state and as part of ribosomal subunits (see Fig. 9-5).

It should be mentioned that reverse transcriptase makes quite a few spontaneous stops on unmodified rRNA. Although they make it more difficult to interpret the results of chemical probing, every such stop carries additional information about the rRNA structure, insofar as it is indicative of incorporation of the nucleotide in question into a stable element of the secondary or tertiary structure of rRNA.

Another example, shown in Figure 9-6, illustrates how chemical probing can be used to corroborate a local conformational change occurring in the short hairpin loop of *E. coli* 16S rRNA forming part of the 30S ribosomal subunit. Interestingly, this rRNA secondary structure element acquires a conformation which is energetically less advantageous.

The method just described has found broad application in monitoring the structure of large RNAs of other types, including viral and messenger RNAs.

9.7.2.3 Searching for Unusual RNA Structures

Quite often, chemical probing reveals unusual features in RNA structure, such as non-Watson–Crick base pairing.

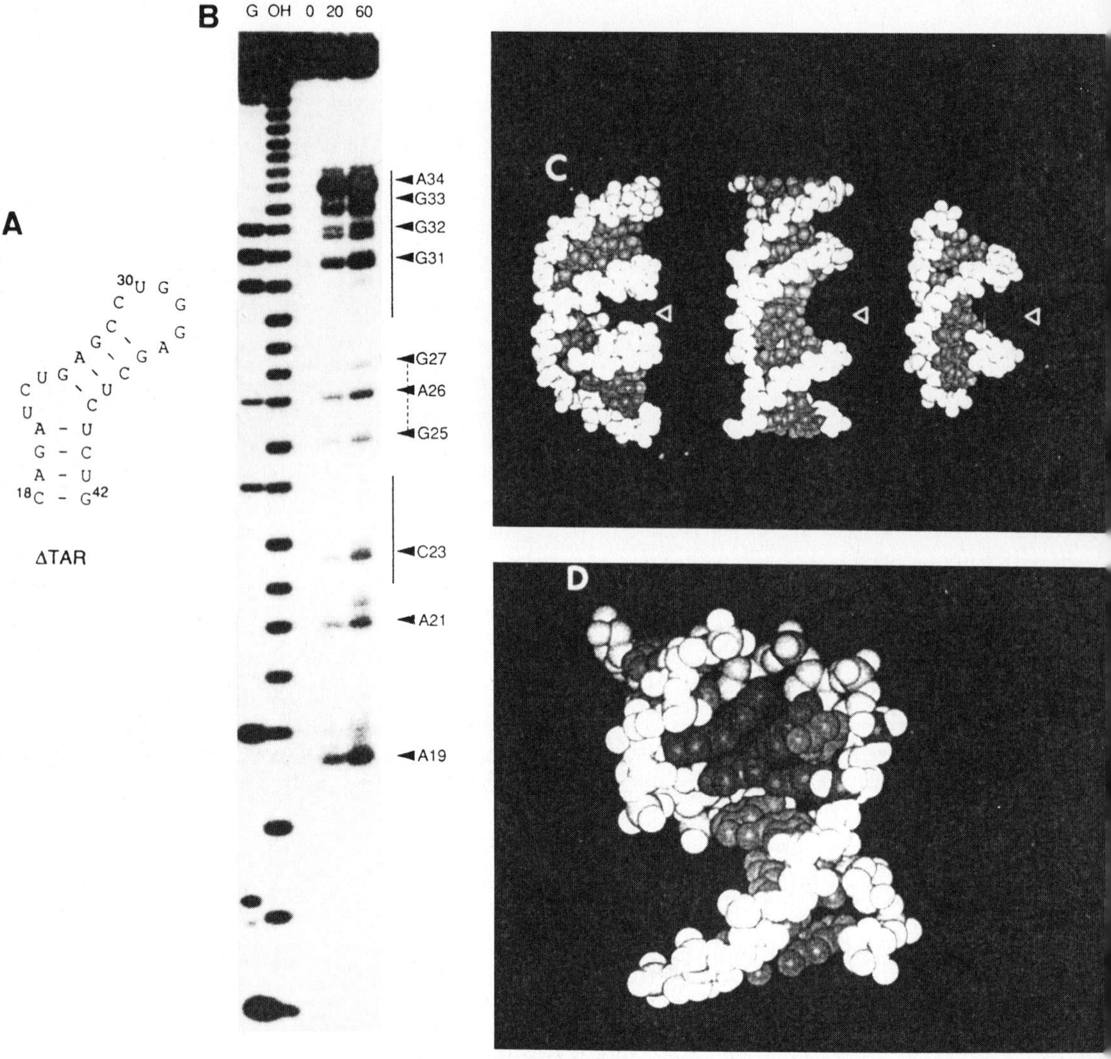

Fig. 9-7. A. Secondary structure of △TAR.
B. Autoradiograms demonstrating accessibility of bases in △TAR to DEPC.
C. Three projections of RNA models in A form, DNA in B form and △TAR.
D. View into the major groove of the △TAR model (adapted (A, B) and reproduced with permission from R. M. Weeks and D. M. Crothers, *Cell,* **66**, 577–588 (1991)).

Let us now consider a case where the chemical modification method was instrumental in bringing to light an unusual conformation in a short double helix forming part of TAR – transactivation response RNA sequence of the human immunodeficiency virus type 1 (HIV-1). TAR is located at the 5′ end of the untranslated leader region of all viral messenger RNAs. It comprises nearly 60 nucleotides and has a stable stem-loop structure. The TAR segment consisting of 27 nucleotides and marked ΔTAR (Fig. 9-7A) forms a specific complex with HIV-1 Tat protein, and this binding is essential for virus replication.

The conformation of ΔTAR was studied by subjecting it to chemical modification by DEPC. It has already been mentioned that DEPC is used for monitoring adenines at N^7. DEPC also reacts, albeit less effectively, with N^7 in guanines. If adenine or guanine take part in Watson–Crick base pairing, their N^7 are unreactive because they are buried deep in the narrow major groove of the RNA helix. Besides, DEPC reacts to some extent with pyrimidine bases as well, although the mechanism of this reaction is yet to be elucidated.

It can be seen from Figure 9-7A that ΔTAR contains a bulge consisting of three nucleotides. In experiments with modification by DEPC two mutant structures were used in addition to ΔTAR. One of them had no trinucleotide bulge at all, whereas in the other bases 23 and 24 were removed from the bulge. The chemical modification results indicate (Fig. 9-7B) that the presence of a bulge radically alters the conformation of the adjacent double-helix segments. If in the bulge-free mutant ΔTAR base-paired purines were unaffected by DEPC treatment, just as expected, in the wild type variant of ΔTAR several purine bases were modified at the atoms that were supposed to occupy the major groove of the double helix. To explain this inconsistency it was speculated that the A form of the RNA double helix in ΔTAR acquires a conformation which looks more like the B form (Fig. 9-7C,D). In this case, the major groove widens and positions N^7 in purines become accessible to DEPC.

9.7.3 Studying Conformational Changes in DNA

As already mentioned in Chapter 8, the double helix of DNA is in B form under normal conditions. However, if DNA in a superhelical state contains certain specific nucleotide sequences, its macromolecules may give rise to unusual structures (Z form, cruciform, H, form [see 8.1.7]).

The chemical modification method has turned out to be a powerful tool for identifying such forms.

It should be remembered that in the early experiments in which cruciforms were revealed in a palindromic DNA use was made of nuclease S1 which cleaves single-stranded DNA segments inevitably produced during formation of cruciforms (Fig. 8-13). It might be assumed, however, that the protein interacting with the DNA somehow stabilizes the cruciform. Therefore, taken as direct proof of emergence of such a form could be only modification of the corresponding DNA loop segments by reagents specific toward single-stranded polynucleotides.

For instance, a cruciform was found in a plasmid DNA, whose single-stranded segment was expected to contain cytosines, as a result of their conversion into uracils by sodium bisulfite treatment (i.e., as a result of oxidative deamination). This plasmid was used to transform *E. coli* cells deficient in uracil-*N*-glycosidase. After replication of the plasmid, U · G pairs were substituted by T · A ones. These mutations were detected further by sequencing the corresponding DNA segments. The same experiments also demonstrated that the cytosines at the junction between the four helices of the cruciform were not modified by bisulfite, the implication being that the DNA helices are joined together without unwinding.

A simpler way to find single-stranded loops in a DNA cruciform is through carboxyethylation of positions $N^7$of their purine bases with DEPC, followed by cleavage of the chain at the modification sites. Modification of purine bases by DEPC does not reveal any single-stranded segments at the junction between the four helices either. Moreover, it has led to a discovery that the optimal size of a loop in the DNA cruciform is four to six nucleotides.

The left-handed Z form of the double helix in DNA can be detected by several procedures based on chemical modification. Firstly, the DNA segments lying at the boundary between B and Z forms are single-stranded and can be revealed with the aid of DEPC (purines), hydroxylamine (cytosines) or OsO_4 (thymidines). Secondly, as has already been pointed out (see 8.1.2), pyrimidine bases in DNA segments that are in Z form are in an *anti*-conformation, whereas purine bases in the same segments are in a *syn*-conformation. As a consequence, the imidazole ring of purines is closer to the double helix surface and becomes vulnerable to DEPC attack. For example, when the $d(A–C)_{32}$-$d(G–T)_{32}$ sequence is incorporated into a cyclic DNA and turns into a superhelical molecule with a supercoiling density necessary for transition B $\rightarrow$ Z, A and G in this sequence start being modified. Interestingly, at the same time purines in the segments lying next to the incorporated one are modified as well. This indicates that the Z form spreads over regions lying beyond the alternating sequence. Finally, there is a reagent that can be used for directed probing of the left-handed DNA helix, namely, Λ-$Co(DiP)_3^{3+}$ [tris(4,7-diphenyl-1,10-phenanthroline)cobalt (III)]; it is specifically bound with DNA in Z

Fig. 9-8. Chemical probing of H form of DNA.
A. H form of the DNA segment incorported into a superhelical plasmid.
B. Autoradiogram showing the results of modification of the DNA segment with DEPC (line 3).
C. Autoradiogram showing the results of modification of the DNA segment with OsO_4. Modification of the T residue in the loop is clearly seen (line 1) (courtesy of O. N. Voloshin). ▶

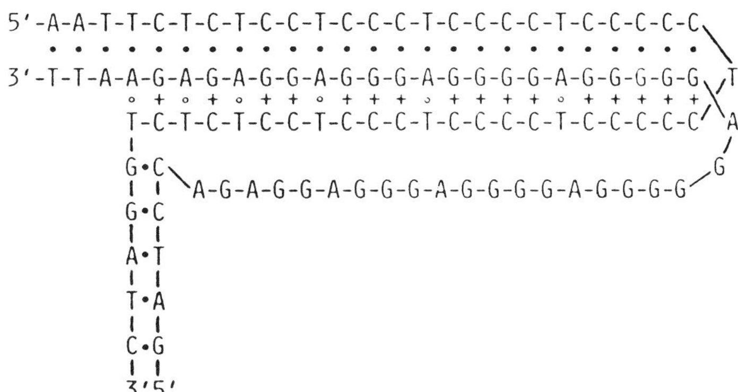
A
5'-A-A-T-T-C-T-C-T-C-T-C-C-T-C-C-C-T-C-C-C-C-T-C-C-C-C-C
3'-T-T-A-A-G-A-G-A-G-A-G-G-A-G-G-G-A-G-G-G-G-A-G-G-G-G-G
T-C-T-C-T-C-C-T-C-C-C-T-C-C-C-C-T-C-C-C-C-C
T
A
G
A-G-A-G-G-A-G-G-G-A-G-G-G-G-A-G-G-G-G
G•C
G•C
A•T
T•A
C•G
3'5'

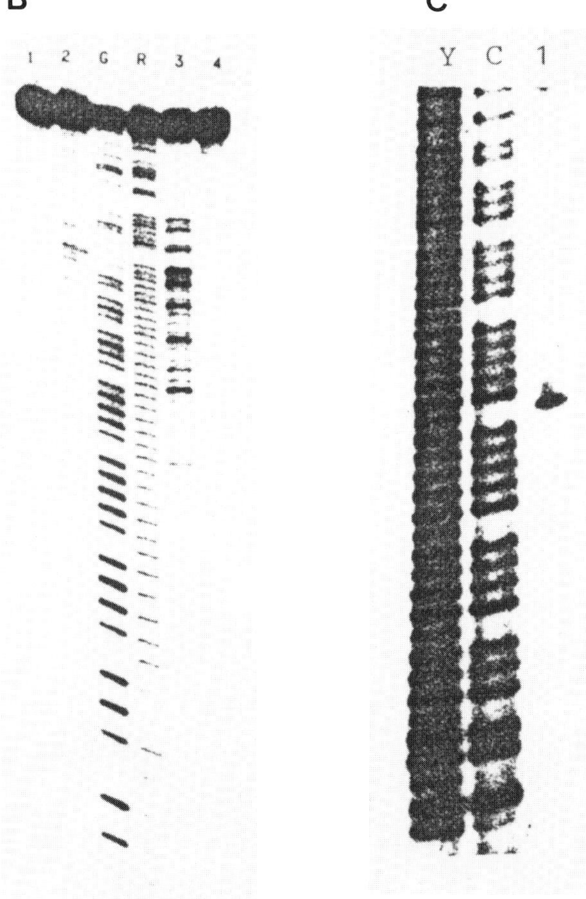
B
1 2 G R 3 4
C
Y C 1

form and cleaves it after photoactivation. It is experiments with this particular reagent that have suggested possible presence of the Z form in functionally important DNA segments.

Since transformation from the B into the H form of DNA is accompanied by formation of a triple helix so that one of the polypurine strands is in a single-stranded state (Fig. 8-15), it is absolutely clear that this unusual form can also be detected with the aid of reagents attacking adenine and guanine only in single-stranded segments of nucleic acids. Figure 9-8 demonstrates the results of such an experiment.

9.7.4 Evaluation of Four-Stranded G4-DNA

Guanine-rich sequences, especially common in telomeric portions of chromosomes, are also found in certain recurring DNA segments, in promoters of some genes and immunoglobulin-switch regions. On several occasions we pointed out the capacity of guanine bases for forming tetrads, which is the reason for formation of four-stranded DNA regions known as G4-DNA.

As can be seen from Figure 7-27, in a G-tetrad all N^7 atoms of the imidazole ring in guanines are involved in hydrogen bonding. It is therefore to be expected that they will be fully protected against modification by DMS. This simple test is used today as proof of formation of G4.

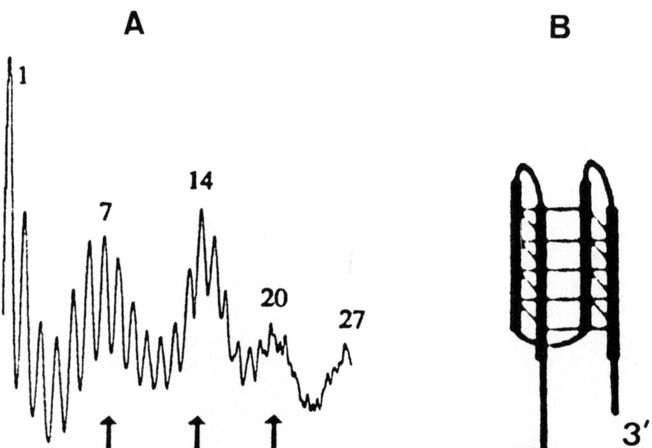

Fig. 9-9. Evaluation of G-DNA by chemical probing.
A. Distribution of guanine methylation within the $(dG)_{27}$ insert into a plasmid.
B. Model of the G-DNA structure. Three loops in the structure correspond to the three peaks (marked by arrows) in A. (adapted from I. G. Panyutineal et al., *Proc. Nat. Acad. Sci. USA*, **87**, 867–870 (1990)).

Figure 9-9 shows the results of methylation of the poly (G) insert into a circular DNA, clearly indicating that guanine-rich sequences acquire a four-stranded form.

References

1. N.K. Kochetkov and E.I. Budovskii, *Organic Chemistry of Nucleic Acids.* Plenum Press, London–N.Y. (1972).
2. D.M. Brown, *Chemical Reactions of Polynucleotides and Nucleic Acids,* in: *Basic Principles in Nucleic Acid Chemistry.* Vol. II, P.O.P. Ts'o (Ed.), Academic Press, London, pp. 2–90 (1974).
3. H.G. Zachau and H. Feldman, in: *Progress in Nucleic Acid Res. and Mol. Biol.* Vol. 4, J.N. Davidson and W.E. Cohn (Eds.), Academic Press, N.Y.–London, pp. 217–230 (1965).
4. R. Shapiro, in: *Progress in Nucleic Acid Res. and Mol. Biol.* Vol. 8, J.N. Davidson and W.E. Cohn (Eds.), Academic Press, N.Y.–London, pp. 73–112 (1968).
5. N.K. Kochetkov and E.I. Budowsky, in: *Progress in Nucleic Acid Res. and Mol. Biol.* Vol. 9, J.N. Davidson and W.E. Cohn (Eds.), Academic Press, N.Y.–London, pp. 403–438 (1969).
6. H. Hayatsu, in: *Progress in Nucleic Acid Res. and Mol. Biol.* Vol. 16, W.E. Cohn (Ed.), Academic Press, N.Y.–San Francisco–London, pp. 75–124 (1976).
7. E.I. Budowsky, *ibid.,* pp. 125–188.
8. L. Ehrenberg, I. Fedorcsak, and F. Solymosy, *ipid.,* pp. 189–262.
9. A.M. Maxam and W. Gilbert, *Sequencing Endlabeled DNA with Base-specific Chemical Cleavages. Methods Enzymol.,* **65**, 499–560 (1980).
10. *Nucleic Acids in Chemistry and Biology.* G.M. Blackburn and M.J. Gait (Eds.), IRL Press, Oxford–N.Y.–Tokyo, pp. 261–288 (1990).
11. Z.A. Shabarova, *Chemical Development in the Design of Oligonucleotide Probes for Binding to DNA and RNA. Biochemistry,* **70**, 1323–1334 (1988).
12. C. Ehresmann, F. Baudin, M. Mougel, P. Romby, J.-P. Ebel, and B. Ehresmann, *Probing the Structure of RNAs in Solution. Nucleic Acids Res.,* **15**, 9109–9128 (1987).

10 Catalytic Activity of Nucleic Acids

The totally unexpected discovery of the catalytic activity of RNAs was one of the highlights of molecular biology in the eighties, has turned out to be a rather common phenomenon. The RNA enzymes isolated from different sources have become known as ribozymes.

Ribozymes catalyze reactions of specific cleavage or exchange of phosphodiester bonds in polyribonucleotides and in some instances also in polydeoxyribonucleotides. They catalyze transfer of the phosphate group and nucleotides; in the latter case, the processes may include both specific cleavage of the polynucleotide chain and its extension. As of today, the substrates of ribozymes in the overwhelming majority of cases are nucleic acids themselves. In other words, the organization of active centers of ribozymes and binding of substrates by them are due to specific RNA–RNA or RNA–DNA interactions. Ribozymes satisfy all the criteria met by enzyme proteins: the reactions they catalyze are highly specific, and the rate of these reactions is increased by them as much as 10^{11}-fold with their initial structure remaining virtually intact.

10.1 RNA subunit of Ribonuclease P

The first and only catalytic RNA known to act as an enzyme *in vivo* was discovered in 1983 by S. Altman, N. Pace and coworkers. It had turned out to be an RNA subunit of ribonuclease P–the key enzyme in the processing of tRNAs, instrumental in the maturation of their 5′ ends (Fig. 10-1A). At first it was found that RNase P is a ribonucleoprotein containing, along with a protein (ca. 14 kD), a small RNA subunit (ca. 400 nucleotides). It is only the latter that displayed catalytic activity. In contrast to the complete enzyme, this activity manifested itself only at high concentrations of magnesium ions which, on the one hand, were conducive to compact folding of the RNA structure and, on the other, screened the negative charges of the phosphate groups in the enzyme and substrate and participated *directly* in the chemical

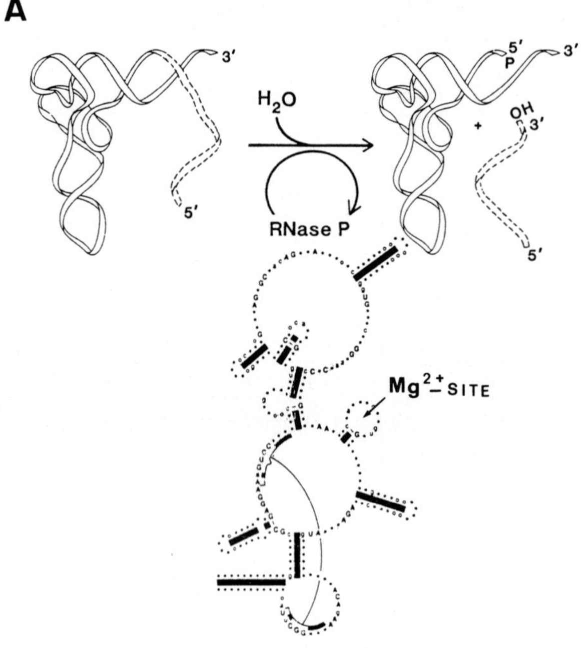
A
3'
5'
H2O
RNase P
P
OH
Mg2+ SITE

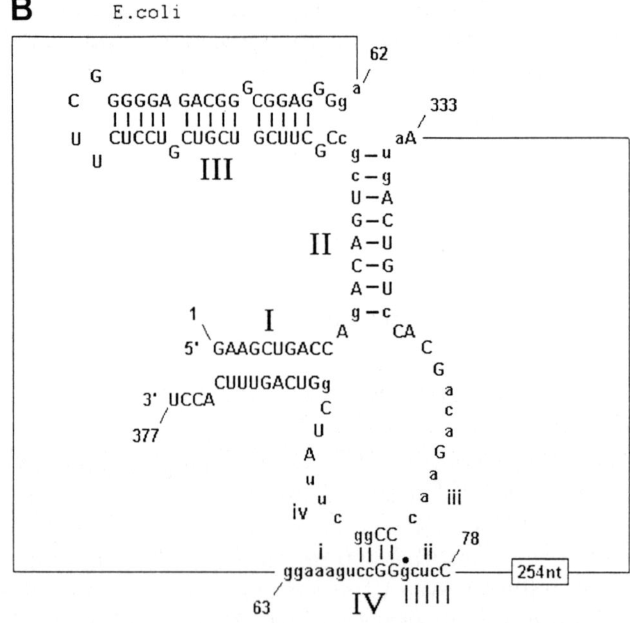
B
E.coli
62
333
GGGGA GACGG CGGAG Gg
CUCCU CUGCU GCUUC Cc
III
II
I
IV
1
5' GAAGCUGACC
CUUUGACUGg
3' UCCA
377
63
78
ggaaaguccGGgcucC
254nt
i
ii
iii
iv

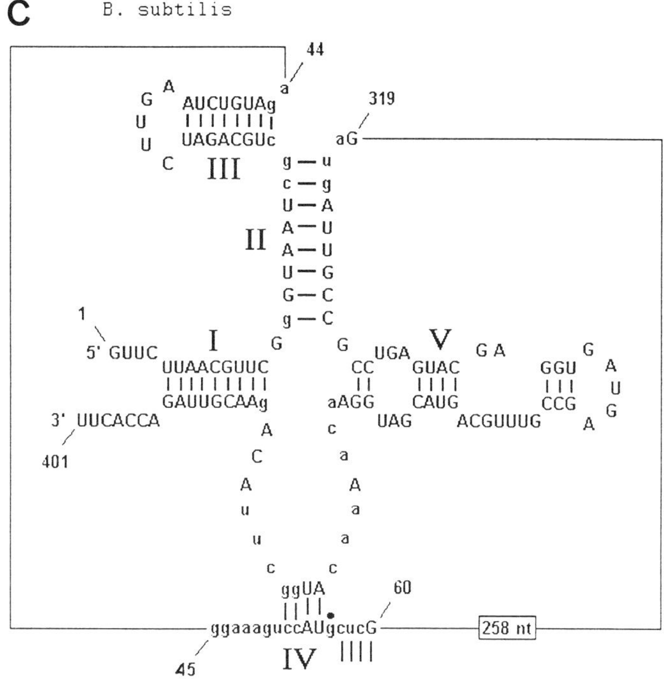

Fig. 10-1. Ribonuclease P. A. Scheme illustrating participation of the catalytic RNA subunit of RNase P in tRNA maturation. The consensus secondary structure of eubacterial RNA is shown. Filled rectangles represent phylogenetically proven helices. Invariant nucleotides are indicated by upper letters. The shaded areas include RNA sequences which are essential for catalysis (adapted from J. W. Brown and N. Pace, *Nucleic Acids Res.,* **20**, 1451–1456 (1992). B and C. Hypothetical cage structures for RNA subunits of RNase P of *E. coli* (B) and *B. subtilis* (C). Nucleotides depicted in lowercase letter are concerved. Nucleotides 74–78 in B and 57–60 in C form base pairs with regions in the central portions of the RNA (adapted from A. C. Forster and S. Altman, *Cell,* **62**, 407–409 (1990)).

step of the cleavage reaction. The ribozyme–substrate interactions were characterized by a value of K_m close to that of the holoenzyme, the ribozyme cleaving the pre-tRNA at a rate only half as high as RNase P itself. Just as the complete enzyme, ribozyme cleaves the pre-tRNA in such a way that the phosphate group remains at the 5′ end of the tRNA.

The secondary structure of the RNA subunit of RNase P was established by phylogenetic analysis. Figures 10-1 B and C illustrate the arrangement of polynucleotide chains in those portions of the RNA subunits of the enzymes isolated from *E. coli* and *B. megaterium* which, judging from the results of phylogenetic analysis, are especially important for the enzymatic activity of ribozymes. Their structures were given in their entirety in Chapter 8 as an illustration of comparative (phylogenetic) analysis in selecting the secondary

RNA structure model (Fig. 8-40). Interestingly, when applied to RNase, such analysis also revealed that the helical elements present only in *E. coli* RNA and only in *B. megaterium* RNA are not essential insofar as the enzymatic activity of ribozymes is concerned. This finding has made it possible to derive, from a gene of the RNA subunit of *E. coli* RNase P, a so-called mini-ribozyme 263 nucleotides long, which did not differ much from the initial ribozyme in terms of catalytic properties. Nevertheless, attempts to identify the real catalytic center of the RNA subunit of RNase P are yet to be successful. The only thing that has become clear is that these ribozymes recognize in their substrates a certain motif in the three-dimensional structure of pre-tRNAs, rather than their primary structure, because the sites of pre-tRNA cleavage by the ribozyme differ widely in nucleotide sequence.

10.2 *Tetrahymena* Ribozyme

A lot more is known about the mechanism of catalytic action of the ribozyme isolated by T. Cech and coworkers from the intervening sequence RNA (IVS RNA) released during self-splicing of the 26S precursor of ribosomal RNA (pre-rRNA) from *Tetrahymena.* As shown in Figure 10-2, IVS RNA results

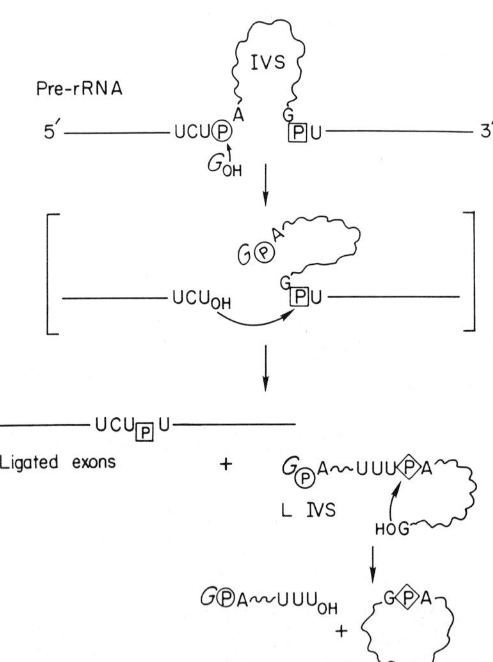

Fig. 10-2. Mechanism of formation of the IVS RNA ribozyme as a result of self-splicing of the *Tetrahymena* pre-rRNA. Straight lines – exons (mature rRNA sequences); wavy lines – IVS; circle – 5′ splice site phosphate; square – 3′ splice site phosphate; diamond – cyclization site phosphate (reproduced with permission from T. R. Cech, *Science,* **236**, 1532–1539 (1987)).

from a sequence of transesterification reactions which begin with the phosphodiester bond between the 3′ end of exon sequence and 5′ end of the intron of the pre-rRNA being attacked by the 3′ hydroxyl of guanosine or guanosine 5′ phosphates. All these reactions proceed *in vitro* with intramolecular catalysis and without any proteins being present whatsoever. The end product of the reaction, IVS RNA, is an intron of pre-rRNA shortened from the 5′ end by 19 nucleotides. It is this particular product that is the true enzyme involved in numerous experiments carried out to elucidate the mechanism of action of ribozymes of this class. At present, an intron shortened from the 5′ end by 21 nucleotides in a feat of gene engineering is available. The structure of this region of pre-rRNA from *Tetrahymena* is shown in Figure 10-3.

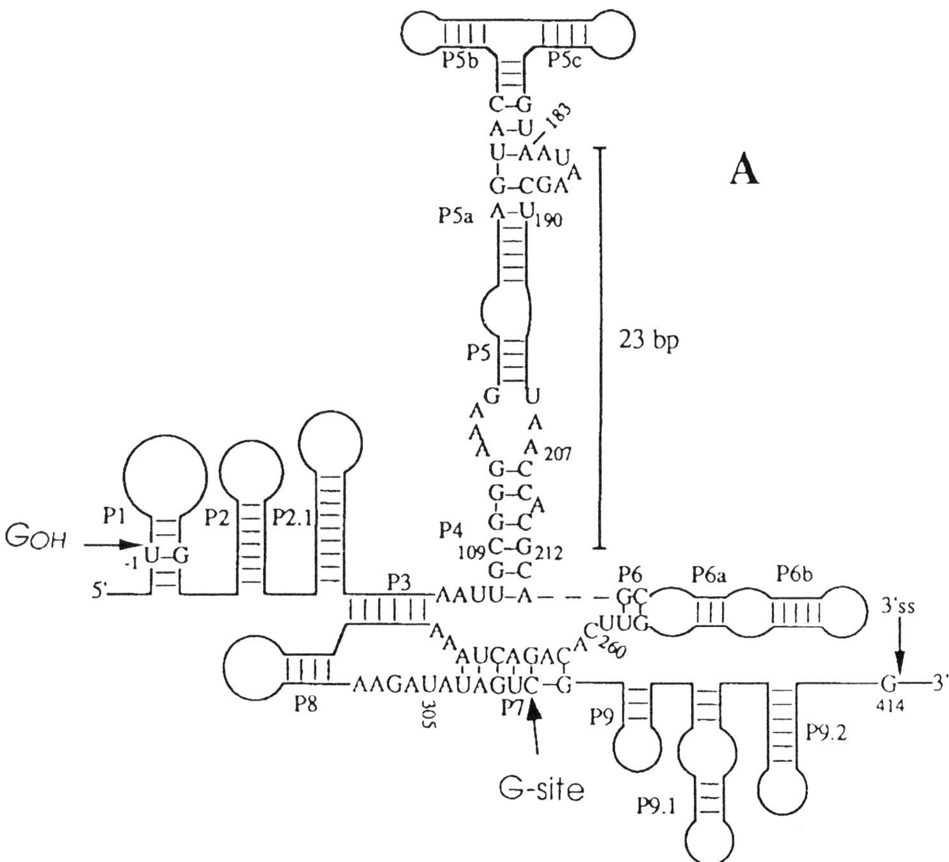

Fig. 10-3. Model of secondary structure of IVS RNA. Letters stand for nucleotides highly conserved among group I introns. Also shown are the guanosine-binding site and the site at which pre-rRNA is attacked by guanosine during self-splicing (adapted from P. T. Flor et al., *EMBO J.*, **8**, 3391–3399 (1989)).

The IVS RNA contains ten double-helical regions on which P9 can be deleted without materially affecting the activity of the ribozyme. The short helical region P7 accomodates the site of specific binding of the guanosine which is involved in the first transesterification reaction during self-splicing (Fig. 10-2) and some other reactions catalyzed by the IVS RNA. Most likely, a guanine base in the guanosine forms two hydrogen bonds with another G in the G264–C311 pair (Fig. 10-4).

There is convincing experimental proof of this binding pattern. Firstly, 1-methylguanosine and N^2-methylguanosine are poor substrates for selfsplicing, whereas the bulky substituents at C^8 and N^9 of the guanine ring do not alter the substrate properties of guanosine. Secondly, the amino acid arginine is an effective competitive inhibitor of guanosine in these reactions. Thirdly, directed substitution of the A264–U311 base pair for G264–C311 results in the guanosine losing its substrate properties, but it can be effectively replaced by 2-aminopurine with citrulline, rather than arginine, acting as a competitive inhibitor of the latter (Fig. 10-5). The binding of the guanosine with the IVS

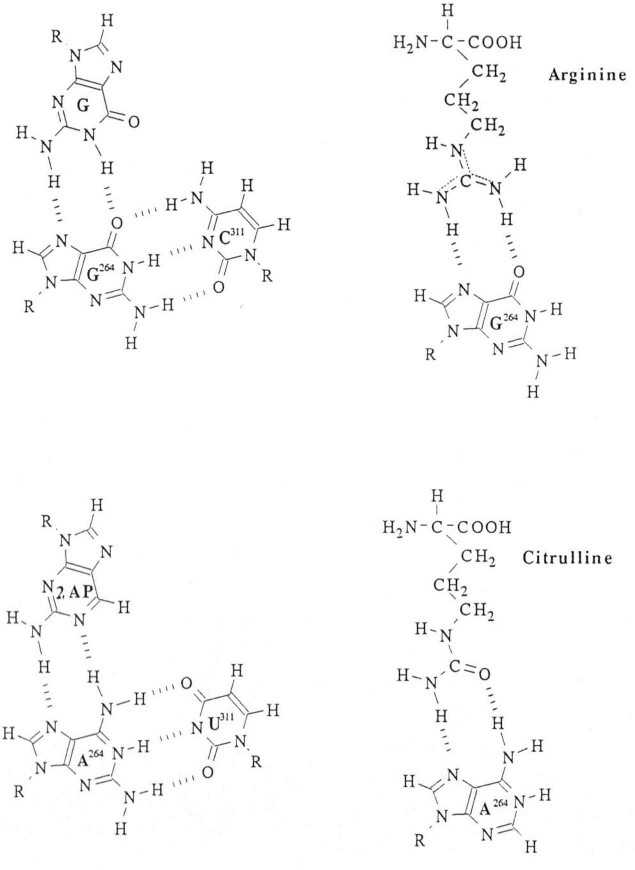

Fig. 10-4. Possible mechanism of binding of guanosine with the G264-C311 pair in IVS RNA. Also shown is pairing of 2-aminopurine with mutant IVS RNA. The scheme explains why the binding of the nucleotide is inhibited by arginine in the first case and by citrulline in the second (adapted from F. Michel et al., *Nature,* **342**, 391–395 (1989)).

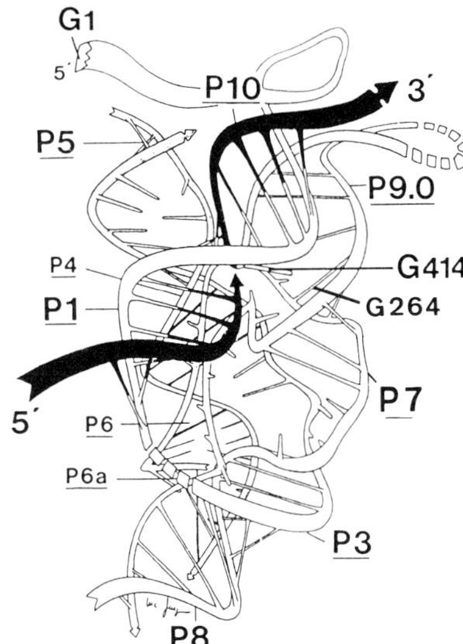

Fig. 10-5. Michel-Westhof model of the three-dimensional structure of the *Tetrahymena* group I intron. Black wriggled arrows indicate exons (reproduced with permission from F. Michel and E. Westhof, *J. Mol. Biol.*, **216**, 585–610 (1990)).

RNA also seems to involve the 2′-OH group of ribose, in view of the fact that deoxyguanosine derivatives may compete with it only in high concentrations.

Then, for the first transesterification reaction to take place, the guanosine linked to G264 must be appropriately oriented with respect to the phosphorus in the phosphate group between the 3′ end of exon and 5′ end of intron. This is attained by fitting the IVS RNA into a specific tertiary structure whose hypothetical version is represented in Figure 10-5.

There is a good reason (the inversion of the configuration at the phosphorus) to believe that the nucleophile, 3′ hydroxyl of guanosine, attacks the phosphorus atom by an in-line, S_N2 (P) mechanism. It was shown by Cech and co-workers that the ribozyme-bound Mg^{2+} in the transition state is in direct contact with the pro-Sp oxygen of the phosphate group. This metal ion is coordinated to the 3′ OH group of uridine to facilitate leaving of its 3′-oxyanion and to stabilize the trigonal bipyramidal transition state (Fig. 10-6). The stereochemistry of Mg^{2+} binding was evaluated by comparison of the activity of ribozymes after substitution of the phosphoryl group at the cleavage site to phosphorothioate group. Substitution of the pro-Sp oxygen atom with sulfur reduces the rate of the cleavage step more than 1,000-fold, whereas thio-substitution of the pro-Rp oxygen gives a small effect. The similar approach was used to prove the direct contact of the ribozyme-bound Mg^{2+} with 3′ oxygen atom of the uridine residue at cleavage site: thio-substitution of this atom also dramatically reduces the ribozyme activity. It is important that this thio-effect was partially relieved when Mg^{2+} was replaced by Mn^{2+} which coordinates sulphur much stronger than Mg^{2+}does.

T. Steiz and J. Steiz suggested that a second ribozyme-bound Mg^{2+} ion can activate 3′ hydroxyl of the guanosine for nucleophilic atack of the phosphoryl

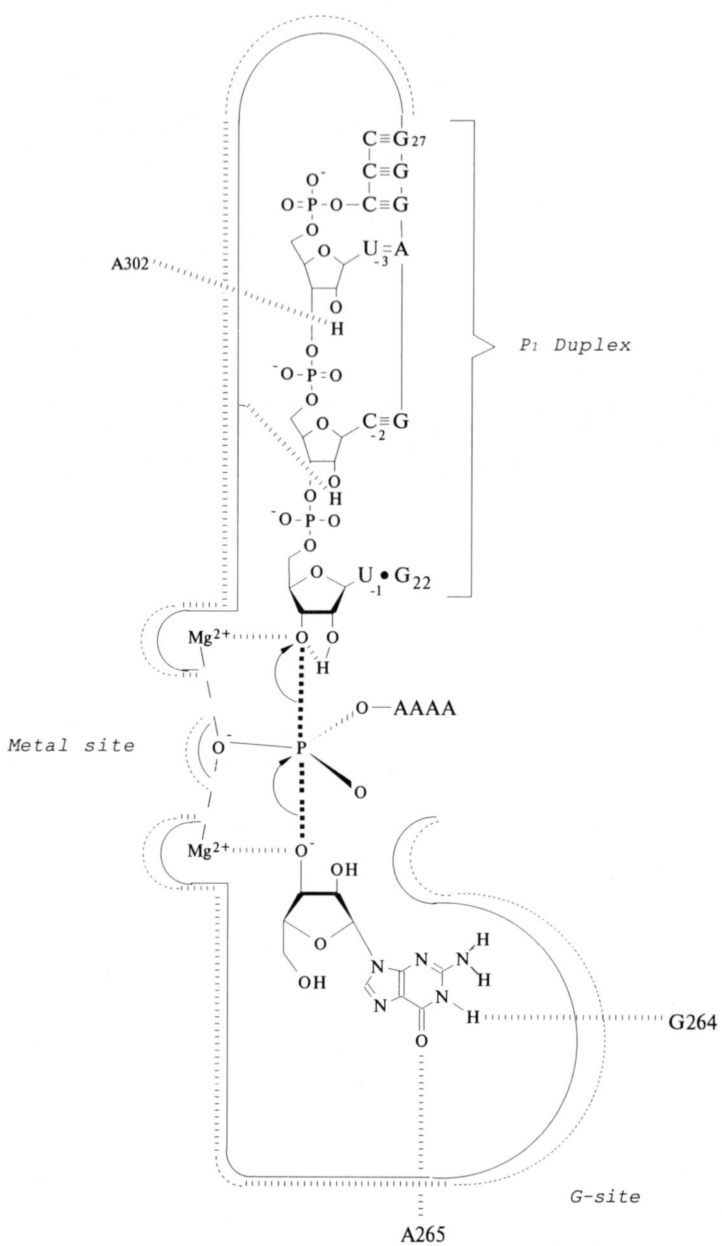

Fig. 10-6. Cech-Steiz and Steiz mechanism of transition state stabilization by *Tetrahymena* ribozyme catalytic center. Hydrogen bonds and Mg^{2+}-oxygen coordination are shown as dashed lines. Dotted P-O bonds indicate bonds partially formed or partially broken in the transition state (adapted from J. Piccirilli et al., *Nature*, **361**, 85–88 (1993); T. Cech, in "The RNA World" (R. Gesteland and J. F. Atkins, eds), pp. 239–269, CSH, 1993; T. A. Steiz and J. A. Steiz, *Proc. Natl. Acad. Sci. USA*, **90**, 6498–6502 (1993)).

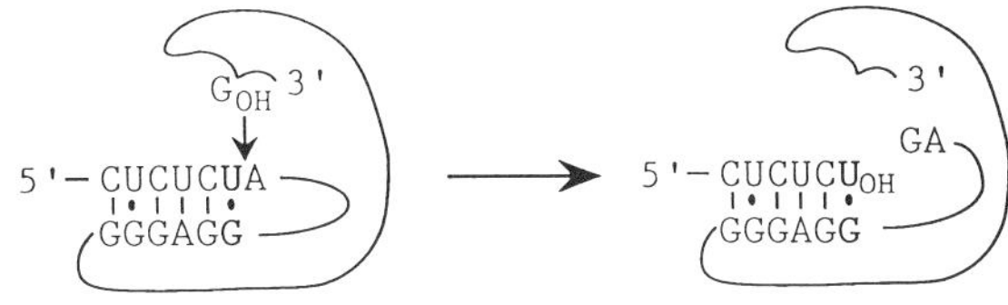

Fig. 10-7. Interaction of the internal guide sequence (IGS) of IVS RNA with the 5′ splice site. It can be seen that base-pairing of IGS with the 5′ splice site guides the guanosine attack of the phosphodiester bond between U and A (adapted form F. L. Murphy and T. R. Cech, *Proc. Natl. Acad. Sci. USA,* **86**, 9218–9222 (1989)).

group. The two-metal-ion-mechanism they proposed (Fig. 10-6) could be common for other ribozymes and certain protein enzymes involved in phosphoryl transfer reactions.

Notably, both 5′ and 3′ splice site phosphodiester bonds of the *Tetrahymena* 26S pre-rRNA are hydrolysed at a much faster rate than average phosphodiester bonds. There is reason to believe that their hydrolysis as well as the specific hydrolysis of the cyclic IVS shown in Figure 10-2 are based on the same general mechanism as the transesterification involving the guanosine, since the phosphate group also remains at the 5′ end of the polynucleotide chain. A similar mechanism of hydrolysis must be involved in the cleavage of pre-tRNA by ribonuclease P.

Another important functional site in the IVS RNA of *Tetrahymena* is the GGGAGG sequence termed "internal guide sequence" (IGS), which occupies the 3′ end of P1. During self-splicing it participates in the complementary pairing with the 5′ splice site (Fig. 10-7). The formation of a non-canonical U–G base pair immediately preceding the phosphate group attacked by the guanosine is highly important for the reaction. It may be substituted only by another non-Watson–Crick pair.

In spite of the fact that the IGS is substantially screened by the ribozyme structure, it can be paired with an exogenous RNA containing a region with a nucleotide sequence complementary to the IGS. In the presence of guanosine and magnesium ions this RNA undergoes specific cleavage (Fig. 10-8). By substituting nucleotides in the IGS one can alter the specificity of the ribo-

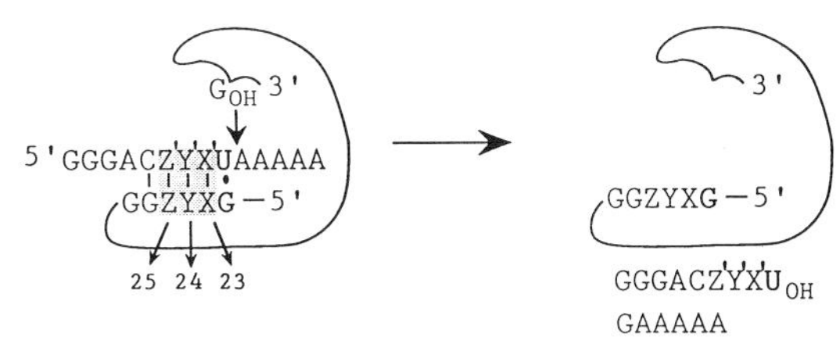

Fig. 10-8. Construction of RNA restriction endonucleases on the bases of IVS RNA. Note that the 5′ splice site is replaced by an exogenous RNA. The XYZ sequence varies in such a manner as to retain the complementarity of the Z′Y′X′ sequence (adapted from F. L. Murphy and T. R. Cech, *Proc. Natl. Acad. Sci. USA,* **86**, 9218–9222 (1989)).

zyme. For example, by varying the trinucleotide sequence immediately adjacent to the G–U pair in the IGS one can create 64 ribozymes specifically cleaving respective regions of the RNA. They only differ in the magnesium ion concentration optimal for the transesterification reaction, due to differences in the stability of ribozyme–substrate duplexes. By analogy with DNA restriction enzymes these ribozymes have been termed RNA restriction endonucleases.

The binding of an exogenous RNA with the IGS in the ribozyme is a rather complex process. In any event, both the substrate and the reaction product are bound with the ribozyme, the binding being about 10^4 times stronger than might have been expected had the process confined itself to formation of an RNA–RNA duplex only. Apprarently, this binding also involves tertiary interactions.

Interestingly, the pre-rRNA self-splicing process considered here has turned out to be reversible. If, after the splicing is over, excess RNA with a sequence complementary to the IGS is added to the IVS RNA with an additional G at its 5′ end (L-19 IVS RNA), the ribozyme will be incorporated into the substrate RNA as shown in Figure 10-9. And if a ribozyme with IGS at the 5′ end is used, linked to its 3′ end will be only the 3′-terminal portion of the cleaved substrate. This reaction provided the basis for an interesting method for selective amplification of ribozymes capable of cleaving the nucleotide sequences of interest. Its principle is illustrated in Figure 10-10. The key point here is the fact that the amplification is triggered by a cleavage reaction. Consequently, if a wild ribozyme is replaced by a set of its mutants, it is possible to choose one that will cleave the desired substrate. This is precisely how a ribozyme cleaving single-stranded polydeoxyribonucleotides rather effectively has been selected.

Along with specific cleavage of RNA, ribozymes produced from the IVS RNA may also catalyze a number of important biochemical reactions. It has been found that the substrates for it may include not only phosphodiesters but also -monoesters. For example, the ribozyme may transfer a phosphate

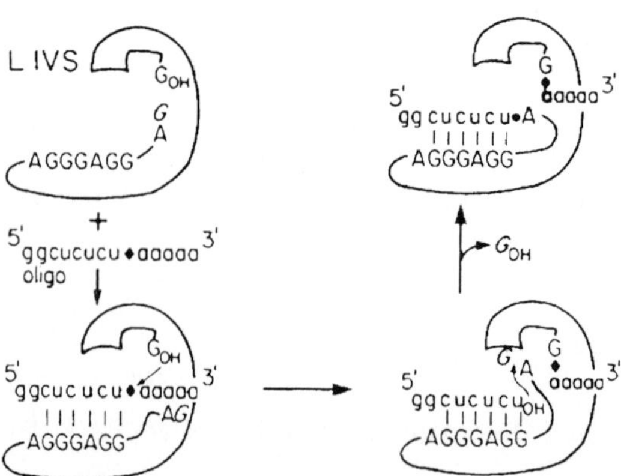

Fig. 10-9. Reverse splicing of the *Tetrahymena* group I intron (compare with the direct splicing illustrated in Fig. 10-2). The diamond denotes the 3′ splice phosphate that becomes the ligation junction phosphate. The releasing G, which initially was at the 5′ end of IVS RNA, is shown in italics (adapted from S. A. Woodson and T. R. Cech, *Cell*, **57**, 335–345 (1989)).

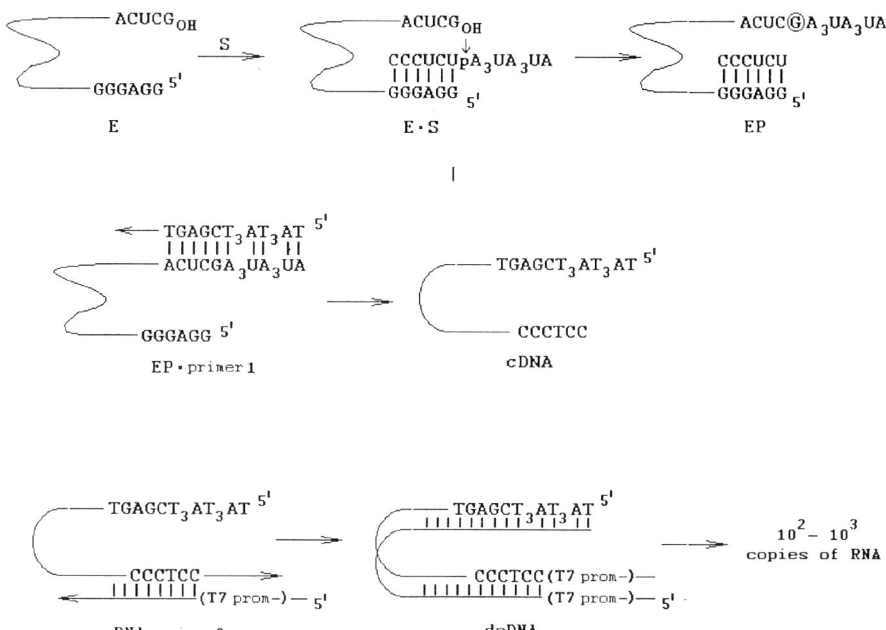

Fig. 10-10. Selective amplification of L-21 IVS RNA. Note that the process is based, firstly, on the ribozyme capacity for reversible self-splicing (compare with the first step of the reaction shown in Fig. 10-9) and, secondly, on the fact that only ribozymes capable for transesterification (i.e. active ribozymes) can be amplified. An important step of this process is also the formation of a double-helical DNA containing a T7 RNA polymerase promoter (adapted from D. L. Robertson and G. F. Joyce, *Nature,* **334**, 467–468 (1990)).

(i.e., act as phosphotransferase) reversibly from the 3′ end of the oligo-Cp substrate to that of its own chain. The optimal pH value for the reaction is 5– that is, the ribozyme exhibits higher activity with respect to the monoanionic form of the 3′-phosphate, as compared to the dianion. At pH 4–5, the phosphoribozyme is hydrolysed with liberation of an inorganic phosphate or, in other words, the ribozymes display acid phosphatase activity.

The guanyl residue, at the 3′ end of the ribozyme may also participate in a nucleotidyl transferase reaction which, in the final analysis, results in extension of the oligonucleotide chain (Fig. 10-11).

The IVS RNA ribozyme may also catalyze transesterification reactions between two dinucleotides:

$$\text{pCpU} + \text{GpC} \xrightarrow{\text{ribozyme}} \text{pCpUpC} + \text{G},$$

in which case its substrate may only be a dinucleotide with a natural 3′, 5′-phosphodiester bond.

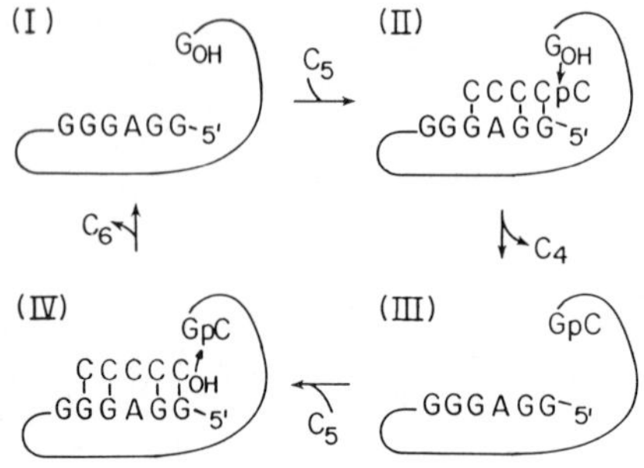

Fig. 10-11. Mechanism of the nucleotidyltransferase reaction catalyzed by L-19 IVS RNA (adapted from A. L. Zaug and T. R. Cech, *Science,* **231**, 470–475 (1986)).

These reactions can be regarded as simplest instances of enzymatic polymerization of ribonucleotides, although the length of the product is limited to about 15 nucleotides.

However, the IVS RNA can be converted into a ribozyme with an activity similar to that of RNA replicase, which is capable of catalyzing template-directed RNA polymerization on a template not linked to the ribozyme. To this end, the ribozyme is synthesized *in vitro* in the form of two parts one of which is a P1-like stem-loop element and the other comprises the rest of the ribozyme slightly shortened from the 3′ end. Thus, the IGS becomes separated from the main part of the ribozyme.

The truncated ribozyme may catalyze (albeit with a much higher k_{cat} than in the case of normal self-splicing) specific cleavage of P1 at the usual site in the presence of guanosine (Fig. 10-12). What is more important, it has been found to be capable of ligating two oligonucleotides in P1-like structures shown in Figure 10-12.

In the presence of spermidine, the polyamine stabilizing the RNA secondary structure, the ribozyme catalyzed ligation of oligonucleotides in structures with any Watson–Crick or wobble base pair at the ligation junction (right substrate). It is also shown that the presence of a loop in the P1 element is of no importance for the transesterification reaction. This allowed the ligation of

```
                       5'G         A G A
5'GGGAGAGGCUOH           pGGAG  U        A  U
  ||||||                  ||||              A
 3'CUCCGG - - - - CCUC C              U
                              C  A  U  U

                                       A G A
                                    U        A
RIBOZYME    5'GGGAGAGGCUpGGAG                  U
---------->     |||||| ||||                    A
  - GOH       3'CUCCGG-CCUC C                  U
                              C  A  U  U
```

Fig. 10-12. Ligation reaction catalyzed by a truncated IVS ribozyme (deprived of P1, P9.1 and P9.2 elements) (adapted from J. A. Doudna and J. W. Szostak, *Nature,* **339**, 519–522 (1989)).

```
5'G                  5'G                     5'G
   pGGAGUAGCAU_OH      [pGGAGUAGCAU]            pGGAGUAGCAU
3'CCUCAUCGUG--         [CCUCAUCGUG]_2 --CCUCAUCGUA 5'

RIBOZYME        5'GGAGUAGCAUp[GGAGUAGCAU]  3'
--------->
- 3G_OH         3'CCUCAUCGUG[CCUCAUCGUA]   5'
                                        3
```

Fig. 10-13. Multiple oligonucleotide ligation catalyzed by the truncated ribozyme (adapted from J. A. Doudna and J. W. Szostak, *Nature,* **339**, 519–522 (1989)).

oligonucleotides catalyzed by the ribozyme to be conducted on linear templates (Fig. 10-13).

The *Tetrahymena* IVS has turned out to be the first representative of a large group of introns (usually referred to as group I and Ia introns) similar to it in terms of secondary and tertiary structure. Their guanosine-binding centers are also similar. They have been discovered in a great variety of organisms, including fungal mitochondria and chloroplasts as well as such evolutionarily distant representatives of procaryotes as bacteriophages. Most of them have been found to exhibit enzymatic activity *in vitro.*

10.3 Ribozymes Based on Self-Cleaving RNAs

The capacity for specific self-cleavage is a rather common phenomenon among viral RNAs. As a rule, this reaction completes the replication of a viral genome, and its biological significance resides in fragmentation of an oligomeric RNA chain formed by the so-called rolling circle mechanism.

The first self-cleaving RNAs were discovered among precursors of viroids (relatively small RNAs self-replicating in plant cells) and so-called viroid satellite RNAs (emerging in plant cells only in the presence of a helper virus). The self-cleavage of multimeric forms of these RNAs occurs at a unique site and results in 2′, 3′-cyclic phosphate and 5′-hydroxyl termini. Magnesium ions act as the cofactor in this highly specific hydrolytic reaction.

Comparison of the nucleotide sequences of several self-cleaving RNAs near the unique site has revealed a consensus domain accommodating the active center of the ribozyme. It consists of 50 to 60 nucleotides, and its secondary structure is represented by three helical regions joined by single-stranded segments, which is why it has been termed "hammerhead" (Fig. 10-14). The domain was found to contain 13 to 15 conserved nucleotides whose position with respect to the cleaved phosphodiester bond is strictly predetermined. The model of hammerhead secondary folding shown in Figure 10-14, constructed originally on the basis of comparative philogenetic analysis, was subsequently supported by high-resolution NMR spectroscopy.

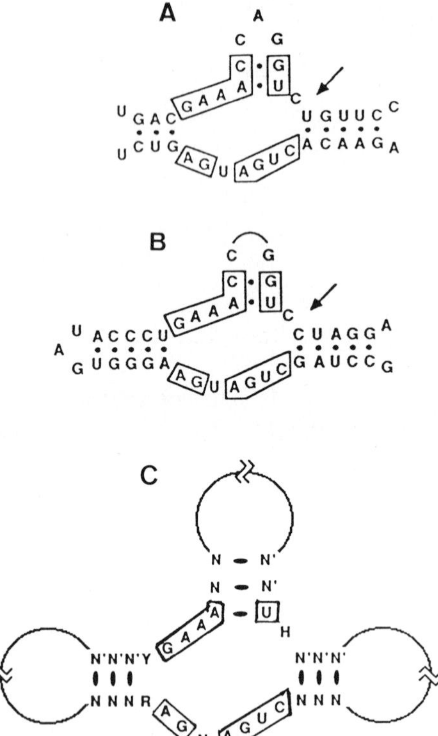

Fig. 10-14. Proposed secondary structure of hammerhead ribozymes. The arrows indicate the site of cleavage.
A. Self-cleavage site of avocado sunblotch viroid (adapted from A. C. Forster and R. H. Symons, *Cell*, **49**, 211–220 (1987).
B. Self-cleavage site of the new satellite 2 RNA (adapted from L. M. Epstein and J. G. Gall, *Cell*, **48**, 535–543 (1987).
C. A hammerhead consensus structure. The boxed positions are believed to be essential due to their conservation in many hammerhead domains. Y, R, N and H designate either pyrimidine or purine and any nucleotide or any nucleotide except G, respectively (adapted from Ruffner et al., *Biochemistry,* **29**, 1065–1072 (1990)).

It was found that the self-cleavage of such RNAs can be modeled in a rather simple system consisting of two oligoribonucleotides partially complementary with respect to each other, whose pairing reproduces the hammerhead (Fig. 10-15). These oligonucleotides had been synthesized with the aid of T7 RNA-polymerase. The cleavage of a longer oligonucleotide took place precisely where expected, and the shorter oligonucleotide 19 nucleotides long, remaining intact after the reaction was over, acted as an enzyme (ribozyme). The substrate cleavage rate reached a maximum at 55 °C and remained almost the same as pH increased from 7.5 to 9.

The discovery of a simple synthetic system modeling a self-cleaving hammerhead had several important consequences. Firstly, it allowed the construction of a sufficiently simple ribozyme (Fig. 10-16) instrumental in cleavage of the internucleotide bond after C in the GCU sequence of virtually any RNA. The central portion of this ribozyme comprises an active center and, therefore, is identical in all ribozymes of this type. The specific binding of the ribozyme to a particular RNA is determined by its 3′ and 5′-terminal regions. Naturally, the number and species of the base pairs formed by these regions with the RNA determines both K_m and k_{cat} of the reaction. It was then demonstrated that substitution of GUA or GUU (but not GUG) for GUC does

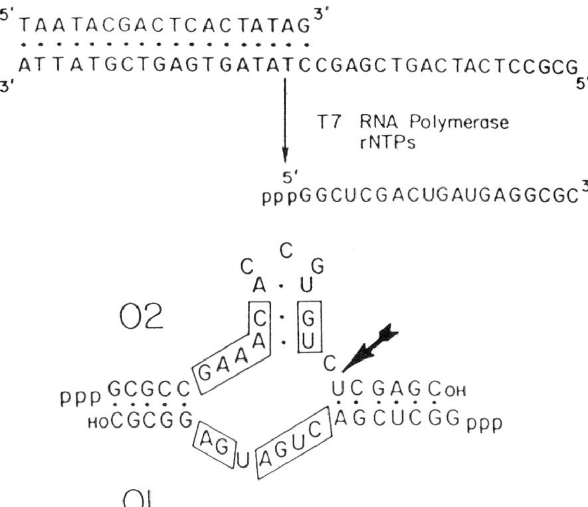

Fig. 10-15. Synthesis of a 19-nucleotide ribozyme and the structure of its complex with the substrate. Oligonucleotide O1 can be referred to as enzyme and oligonucleotide O2 as substrate only conventionally because both are involved in the hammerhead structure formation. The designations are the same as in Figure 10-14 (adapted from O. C. Uhlenbeck, *Nature,* **328**, 596–600 (1987)).

not affect the activity of this ribozyme in any perceptible manner. Moreover, the ribozyme cleaved the RNA in respective complexes after CUC, AUC and UCU, albeit to a lesser degree. Therefore, such ribozymes have the potential for creating agents capable of cleaving, in a controlled manner, the genomes of RNA-containing viruses *in vivo*. If the problem of delivering such a ribo-

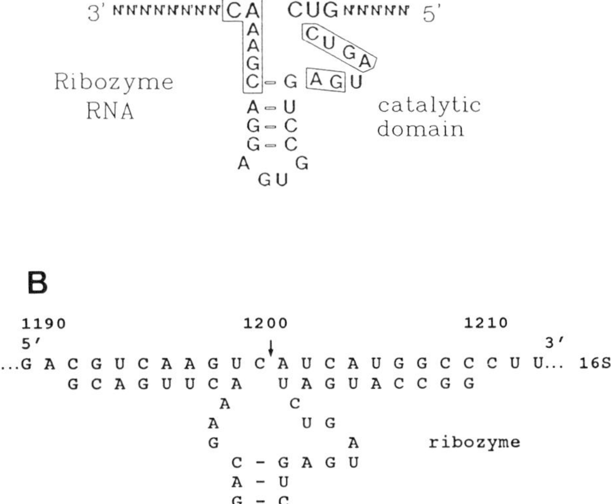

Fig. 10-16. Cleavage of any exogenous RNA containing a GUC sequence by a hammerhead-like ribozyme.
A. General structure of the ribozyme-substrate complex. (adapted from J. Haseloff and W. L. Gerlach, *Nature,* **334**, 585–591 (1988)).
B. Cleavage of *E. coli* 16S ribosomal RNA (1542 nucleotide long) at single site by a ribozyme specially designed for this RNA (courtesy of S. L. Bogdanova).

zyme into target cells is put aside, it becomes evident that the limiting factor in this case will be the secondary and tertiary structures of the viral RNA region with which the complementary pairing of the ribozyme must take place.

Secondly, experiments with a simplified version of the ribozyme have revealed nucleotides whose replacement or modification materially affects its activity, which was a serious step toward understanding the mechanism of self-cleavage. It was found, for example, that the hammerhead may be represented by a more rigorous consensus than expected earlier (Fig. 10-14C). It contains only 11 nucleotides whose replacement is not permissible, and all nine conserved nucleotides in the single-stranded regions of the ribozyme are necessary for self-cleavage. Furthermore, replacement of some nucleotides in the ribozyme by their deoxy analogues has shown that its activity is strongly dependent not only on the 2′-OH group next to the cleaved phosphodiester bond but also the 2′-hydroxyls of some "internal" nucleotides. At the same time, many other ribozyme residues can be substituted by the corresponding deoxynucleotides without any tangible effect on its activity. It allowed R. Cedergren and co-workers to propose the mechanism of hammerhead ribozyme action which suggests that several specific 2′-hydroxyls from both the substrate and the ribozyme coordinate a magnesium ion:

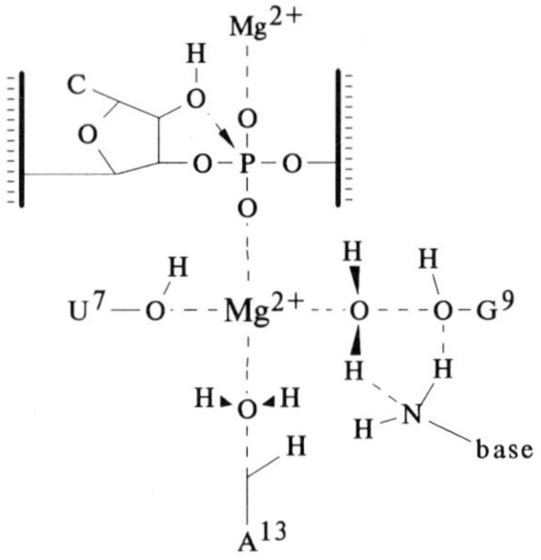

An interesting approach to identifying functionally essential ribozyme fragments is based on substitution of a phosphothioate for the usual phosphate group in its phosphodiester bonds. Among the essential features of nucleoside phosphothioates are the chirality of the phosphorus atom and the fact that the negative charge of the phosphothioate group is located primarily at the oxygen (the sulfur atom being practically neutral). T7 RNA-polymerase normally employed in the synthesis of simple ribozymes uses only the Sp isomer of NTPaS as the substrate. However, when this isomer is incorporated into the polynucleotide chain, its configuration is inverted to give the Rp isomer of the phosphothioate linkage. If the emergence of such bonds in the ribozyme alters its enzymatic activity, it may be assumed that the active center accomodates an appropriate phosphate group. Phosphothioate linkages are easily identifiable through alkylation and subsequent hydrolysis of the emerging triester group. Therefore, one of the four nucleotides in the ribozyme is statistically replaced by its phosphothio analog (under conditions when one nucleotide on the average is substituted by a molecule), then the slowly cleaving molecules are isolated for identification of the nucleotides participating in enzymatic catalysis.

All this suggests that conserved nucleotides in single-stranded regions of the hammerhead form a tertiary structure organized on principles yet to be understood. It is assumed that there appears an ensemble of functional groups retaining the bivalent metal ion near the phosphodiester bond being cleaved. As has already been mentioned, the self-cleavage of RNAs gives termini containing 2′, 3′-cyclic phosphate and 5 ′-phosphodiester groups. Thus, the role of the metal (usually magnesium ions) seems to boil down either to enhancing the nucleophilic activity of the neighboring 2′-OH group or to making the phosphorus more electrophilic. A good analogy is provided by the specific cleavage of one of the internucleotide linkages in tRNA, catalyzed by lead (Pb II) ions. As can be seen from Figure 10-17, being linked to nucleosides 59 and 60 in the TΨC loop, the Pb (II) ion is a source of hydroxyl ions linked to the metal. The special system was developed on the basis of these data to isolate very simple polyribonucleotides that undergo autocatalytic cleavage with Pb^{2+} in the presence of Mg^{2+} (Fig. 10-17 B). It has also been speculated that the metal ion in the self-cleaving domain may be involved in stabilization of the transient state giving rise to 2′, 3′-cyclic phosphate.

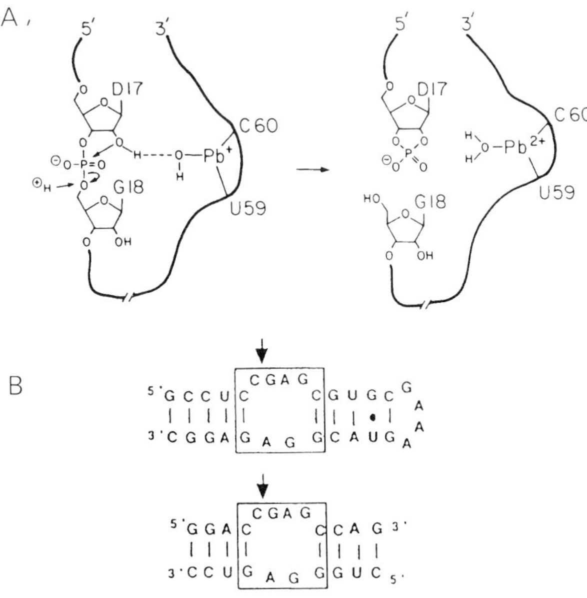

Fig. 10-17. A. Hypothetical mechanism of site-specific hydrolysis catalyzed by lead ions (reproduced from T. R. Cech and B. L. Bass, *Ann. Rev. Biochem.*, **55**, 599–629 (1986)).
B. Short RNA constructs that undergo Pb^{2+} cleavage at sites indicated by the arrows (adapted from T. Pan and O. Uhlenbeck, *Nature,* **358**, 560–563 (1992)).

Some other instances of self-cleavage of RNA have been reported apart from RNAs having a consensus structure of the hammerhead type. Examples are illustrated in Figure 10-18. Judging from most parameters of these reactions, they are governed basically by the same mechanism, although in each case the active center of the ribozyme has a different individual structure.

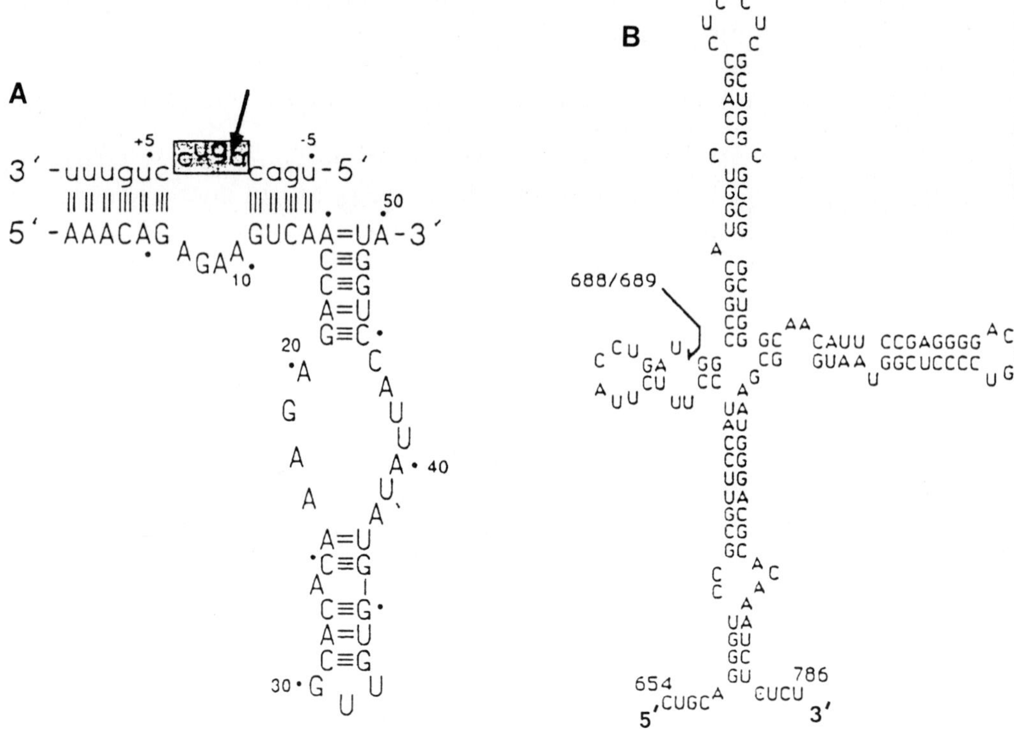

Fig. 10-18. Self-cleaving RNA domains having a secondary structure different from the hammerhead one.
A. the hairpin ribozyme constructed on the basis of self-cleaving domain of the negative strand of the satellite RNA of tobacco ringspot virus (adapted from B. M. Chowrira et al., *Nature*, **354**, 320–322 (1991)).
B. The smallest self-cleaving fragment of human hepatitis δ virus RNA (adapted from H.-N. Wu et al., *Proc. Natl. Acad. Sci. USA,* **86**, 1831–1835 (1989)). The arrows mark cleavage sites.

References

1. C. Guerrier-Takada, F. Gardnier, T. Marsh, N. Pace and S. Altmann, *The RNA Moiety of RNase P As the Catalytic Subunit of the Enzyme. Cell*, **35**, 849–857 (1983).
2. S. Altman, L. Kirsebom and S. Talbot, *Recent Studies of Ribonuclease P. FASEB J.* **7**, 7–14 (1993).
3. K. Kruger, P. J. Grabowski, A. J. Zaug, J. Sands, D. E. Gottschling and T. R. Cech, *Autoexcision and Autocyclization of the Ribosomal RNA Intervening Sequence of Tetrahymena. Cell* **31**, 147–157 (1982).
4. T. R. Cech and B. L. Bass, *Biological Catalysis by RNA. Ann. Rev. Biochem.* **55**, 559–629 (1986).
5. T. R. Cech, *The Chemistry of Self-splicing RNA and RNA Enzymes. Science* **236**, 1532–1539 (1987).
6. J. A. Latham and T. R. Cech, *Defining the Inside and Outside of a Catalytic RNA Molecule. Science* **245**, 276–282 (1989).
7. R. H. Symons, *Self-cleavage of RNA in the Replication of Small Pathogens of Plants and Animals. TIBS* **14**, 445–450 (1989).
8. C. Wilson and J. W. Szostak, *Ribozyme Catalysis, Curr. Op. Struct. Biol.* **2**, 749–756 (1992).

11 Synthesis of Nucleic Acids

11.1 Introduction

Advances in DNA synthesis, which have made it possible to widely use man-made DNAs in molecular biology, genetics, virology, and other fields, represent the most important contribution to the chemistry of nucleic acids in recent years. Along with achievements in molecular cloning and nucleic acid sequencing, DNA synthesis opens up broad possibilities for future progress in biology. Synthetic DNAs and their fragments are already extensively employed in the synthesis and isolation of genes, in protein engineering, as probes in the diagnosis of infectious and genetic diseases, and, last but not least, as gene-targeted agents.

The availability of synthetic DNA fragments with a predetermined monomer sequence, which has made the above advances in molecular biology, genetic engineering and other related fields possible, was ensured by automation of their chemical synthesis and developments in purification of the preparations. The entire process of synthesizing gene fragments, assembling them into genes, cloning, selection of clones and structural analysis is now a matter of just a few weeks. Armed with powerful tools, the organic chemist is entering a new era. It has become possible to study the structure–activity relationship in proteins as well as to shape their three-dimensional structures in the course of DNA-dependent synthesis *in vitro.* According to predictions, new enzymes with predetermined functions will be designed and synthesized by the turn of the century.

Emphasis in this chapter will be placed on the chemical principles underlying synthesis of oligodeoxyribonucleotides.

Currently, synthesis of oligonucleotides in automatic synthesizers is becoming routine even in laboratories without specialists in synthetic organic chemistry. This places special importance on reviews and textbooks summarizing the latest knowledge about DNA synthesis not only for the benefit of students but also scientists involved in synthetic oligo(poly)nucleotide research.

Chemical synthesis of oligo- and polynucleotides represents the most exciting and challenging field of synthetic chemistry of nucleic acids. In establishing an internucleotide bond (chemical linkage of monomer units), the problem boils down to using the phosphate of one monomer unit (nucleotide component) to acylate the hydroxyl group of another unit (nucleoside component):

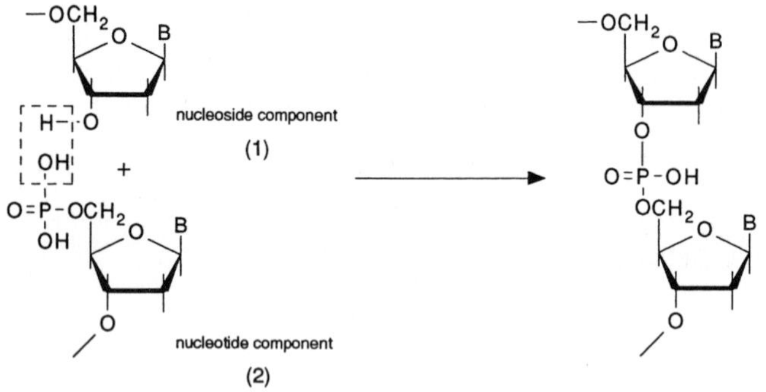

nucleoside component (1)

nucleotide component (2)

Enzymatic linking of monomer units involves a "directed" reaction or, to be more precise, activation of the phosphate and formation of a natural 3′-5′-internucleotide bond. The enzyme "takes care" of directed activation and linkage of groups capable of intermonomer bonding. To achieve the same end by chemical synthesis requires accomplishment of at least two tasks:

(1) protection of all functional groups of the monomer, not subject to phosphorylation;
(2) activation of the phosphate group in the nucleotide or, in other words, transforming the nucleotide to a phosphorylating agent.

If these tasks are compared with the similar ones arising during peptide bond synthesis, it becomes evident that the synthesis of oligonucleotides gives rise to many more difficulties. There are basically two reasons for that:

(a) a great number of functional groups in the monomer units of nucleic acids and presence of sufficiently labile bonds in the latter; and
(b) much lower activity of the reacting groups (the hydroxyls of ribose make a much weaker nucleophile in comparison with the amino group of amino acids, whereas the phosphate group in nucleotides is a weaker electrophile than the carbonyl of amino acids).

As a result, the preparation of monomers – nucleoside (1) and nucleotide (2) components for directed synthesis of the 3′-5′-internucleotide bond – is extremely difficult. The highly selective blocking and deblocking of the hydroxy groups in pentose, the amino groups of the heterocycles, and the phosphate groups in the nucleoside component (which must remain intact) should be carried out under very mild conditions; at the same time, the phosphate group must be active enough to acylate the desired hydroxyl of the sugar. Another difficulty in oligonucleotide synthesis stems from the presence of two hydroxyls in the nucleotide's phosphate group.

Once an oligonucleotide has been synthesized, the internucleotide phosphate does not remain neutral. It may react, for example, with the activated phosphate present in the reaction mixture, which significantly affects the oligonucleotide yields. To circumvent this difficulty, methods have been developed to ensure internucleotide linkage not of the phosphodiester type (3) (phosphodiester method of synthesis) but of the phosphotriester type (4) (phosphotriester method).

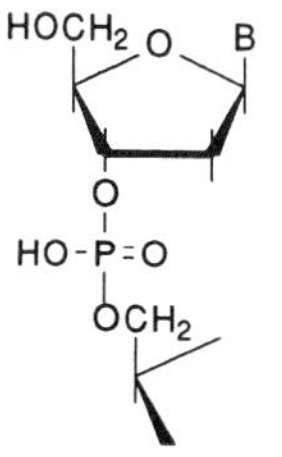

(3)

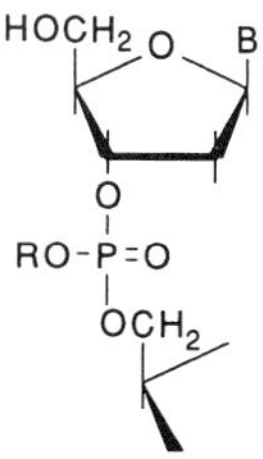

(4)

To create natural 3′-5′-internucleotide bonds one can, in principle, resort to phosphorylation of the corresponding nucleoside components with both 5′-phosphates (path I) and 3′-phosphates (path II):

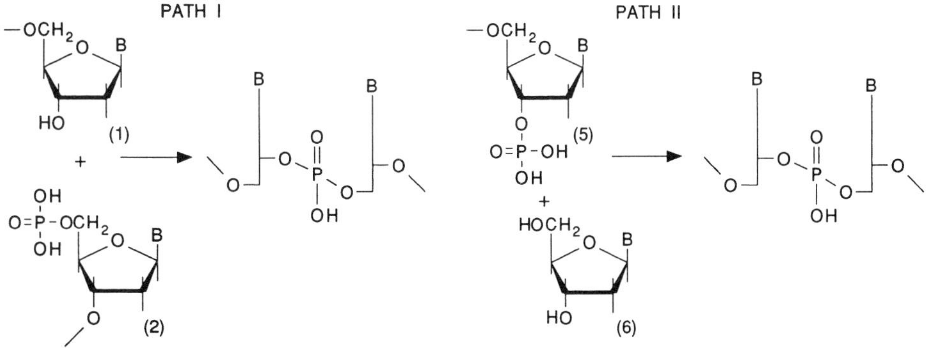

PATH I

PATH II

(1) (2) (5) (6)

11.2 Historical Background

The history of oligonucleotide synthesis goes back over thirty years. The first report on synthesis of dinucleoside phosphate was published by Todd and Michelson in 1955. This synthesis performed by the triester method, however, was ineffective (the yield was 1–2%) and its historical significance lies in that it is precisely this method (albeit in a new modification) that is the most widely used at present.

From the sixties onward, the most important contribution to DNA synthesis procedures must be credited to Khorana and coworkers who have elaborated the phosphodiester method. From the very outset, the method made it possible to use relatively simple condensation agents (carbodiimide and arylsulfonyl chlorides) and not only produce 10- to 12-unit oligodeoxyribonucleotides but also "assemble" them into genes. In 1970, for example, Khorana synthesized the gene of alanine tRNA. This second gene was incorporated into a bacteriophage, and its activity was demonstrated by gene engineering techniques. This work laid the foundation of DNA synthesis, and Khorana's writings are now used as standard guides both in preparation of the starting material for synthesis and in assembly of double-stranded DNAs from synthetic oligonucleotides. The historical importance of Khorana's work also lies in its having demonstrated the possibility of creating synthetic genes even before gene engineering became routine in molecular biology. In the late seventies it had already become evident that gene engineering might be instrumental in obtaining proteins with synthesized genes. This realization provided a major impetus to the development of synthetic methods. Ever since, organic chemists have been concentrating their efforts in two areas: improvement of techniques for creating internucleotide linkages in solution and solid-phase synthesis. By the end of the seventies, Saran Narang and Reese had come up with a fast and economic phosphotriester method and thereby laid the groundwork for the up-to-date procedures which, in combination with the solid-phase approach, have made it possible to automate synthesis. We should also mention the work by R. L. Letsinger (USA) who developed a solid-phase technique similar to Merrifield's method in peptide chemistry, as well as the phosphite-triester method (M. H. Caruthers, USA) to create internucleotide linkages. In the late seventies, the solid-phase method underwent further improvements proposed by M. J. Gait (UK), Z. A. Shabarova and V. K. Potapov (USSR). The early eighties saw a merger of both approaches to synthesis.

This marked a new period which ended, by the mid eighties, with automation of oligonucleotide synthesis, development of several models of automatic gene fragment synthesizers (USSR, USA, UK), and a highly effective modification of the phosphoramidite-triester method of building up oligonucleotide chains. Among the later entries was the hydrophosphoryl method which allows one to automatically synthesize not only natural oligonucleotides but also those with internucleotide linkages not existing in nature.

The two principal methods built into the modern synthesizers are the phosphoramidite-triester and the hydrophosphoryl methods. However, the earlier phosphodiester and -triester methods continue to play a major role in synthetic research. For instance, when Khorana developed the phosphodiester method, techniques were elaborated for introduction of protective groups into and their removal from nucleosides and nucleotides. These techniques are still in wide use. The importance of the triester method of synthesis in solution stems from the fact that it is instrumental in producing preparative amounts (up to hundreds of mg) of oligonucleotides.

Therefore, this chapter will cover all of the above-mentioned methods. In conclusion, we shall offer a list of currently produced synthesizers and a list of genes synthesized by the late eighties.

The chapter begins with discussion of the strategies used in assembling genes from chemically synthesized fragments. The complementarity principle underlying such assembly is also applicable in all other cases where oligonucleotides serve as primers in nucleic acid biosynthesis, probes in RNA isolation, tools for diagnosis of certain sequences in DNA and RNA, gene-targeted agents, genetic engineering tools, and so on.

11.3 General Principles of Chemoenzymatic Synthesis of Duplex DNAs

As already mentioned, these principles have been formulated and implemented in the early gene syntheses by Khorana.

It became clear from the very outset that the likelihood of working out a procedure for chemical synthesis of sufficiently long DNA chains is rather small. Therefore, synthesis of extended polynucleotides had to combine chemical methods with some other approach. According to Khorana himself, this approach, which finally turned out to be so obvious and used by nature in every biological aspect of nucleic acids, of course rested on the principle of complementarity of bases.

11.3.1 Implementation of the Principle of Complementarity of Heterocyclic Bases in Nucleic Acids

Practically at the same time (in the late sixties) with development of the first effective methods for synthesizing short (approximately 10 b.p.) oligonucleotides and emergence of various complementary oligonucleotide complexes, there appeared reports announcing the discovery of the enzyme DNA ligase (polynucleotide ligase) which restores the ruptured internucleotide linkages in native DNAs. The enzyme catalyses the formation of internucleotide linkage with participation of the 5′-phosphate group and 3′-hydroxyl brought closer together by complementary interactions. This reaction can be illustrated by an example with a hypothetical DNA fragment:

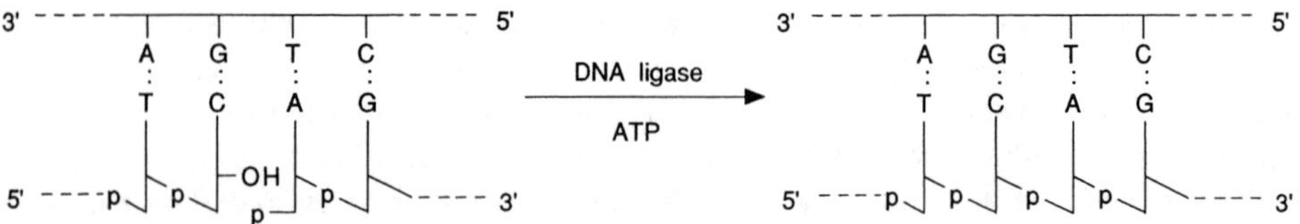

For obvious reasons DNA ligase has proved to be a real godsend for synthesists as it enabled them to combine chemically synthesized oligonucleotides within complementary complexes.

The very first attempts to use DNA ligase were aimed at finding out the minimal length of complementary oligonucleotides that can be joined together. It turned out that the latter are limited in length to 8–10 b.p., for example:

```
   20 19 18 17 16 15 14 13 12 11 10 9  8  7  6  5  4  3  2  1
                          32p
                           |
3'  C-T-A-A(8)G-G-C-C  T-G-A-G-C-A-G(9)G-T-G-G-T
               | | | |  | | | | | |
5'             C-C-G-G-A-C-T-C-G-T
                  (7)
                        |  1. DNA ligase
                        |  2. isolation of (10)
                        v
    C-T-A-A-G-G-C-C*-T-G-A-G-C-A-G-G-T-G-G-T
                   (10)
                        |  phosphorylated
                        |  oligonucleotides (7) and (11)
                        v
    C-T-A-A-G-G-C-C*-T-G-A-G-C-A-G-G-T-G-G-T
            | | | |  | | | | | |  | | | | | |
            C-C-G-G-A-C-T-C-G-T  C-C-A-C-C-A
            |     (7)            |   (11)
           32p                  32p
                        |  DNA ligase
                        v
                   (10)
3'  C-T-A-A-G-G-C-C*-T-G-A-G-C-A-G-G-T-G-G-T   5'
            | | | |  | | | | | |  | | | | | |
5'          C-C-G-G-A-C-T-C-G-T-C-C-A-C-C-A    3'
            |                  *
           32p          (12)
```

Ligation takes place when two adjacent portions of one oligonucleotide (7) are complementary to the termini of two other nucleotides (8) and (9) which, by virtue of complementary interactions, overlie the first as if it were a template. In this case, the termini are fixed in a position most favorable for covalent phosphodiester bonding.

After the ligase joins the two ends, only two polynucleotides, (7) and (10), representing a double-stranded DNA-like structure are left in the complex.

The course of the reaction is controlled by monitoring the ability of the substrate (8) carrying the 5′-terminal phosphate group (the 5′-phosphate is usually incorporated as ^{32}P with the aid of polynucleotide kinase) to be broken down by phosphomonoesterase to give inorganic phosphate whose quantity is measured in the reaction medium. Compounds (10) and (7) are then isolated individually to be used in subsequent buildup of the polynucleotide chain. Similarly, one can join oligonucleotide (11) to (7), in which case synthesized icosanucleotide (10) serves as the template. The ligation site and, consequently, the structure of the polynucleotides constructed in this fashion are determined (after their separation from the reaction mixture and purification) by the "closest neighbor" method.

As has already been mentioned, the starting substances for such polycondensation are oligonucleotide blocks with phosphate groups at the 5′ end. Assaying by this method involves incorporation of ^{32}P into the 5′-terminal groups [blocks (8) and (11) on the above scheme], followed by joining with the help of DNA ligase. The separated and purified polynucleotides [such as (10) and (12)] are subjected to complete hydrolysis in the presence of spleen PDE, and the nucleotide with ^{32}P is found. The ^{32}P-labeled phosphate group is at the 3′-terminal monomer of the oligonucleotide, to which the 5′-terminal phosphate of the second oligonucleotide has been joined with the aid of DNA ligase, and thus the ligation site can be identified. If, for example, ^{32}P becomes bound with thymidine as a result of such treatment of polynucleotides (10) and (12), the above-shown linkage of the oligonucleotide blocks (formation of the –Cp̆T– bond) can be considered to be correct beyond any doubt.

The ligase-mediated joining does not always proceed quantitatively (sometimes its effectiveness is low) and seems to be governed by the structure of the emerging complementary complex. This is why in selecting the right conditions for enzymatic ligation of oligodeoxynucleotide blocks due regard should be given to such factors affecting the complexing process as relative concentration of the interacting components, temperature, and dilution. The effect of temperature seems to involve possible formation of a secondary structure for each substrate, for example, (7), (8) and (9), which hinders emergence of the linear DNA-like complex recognized by the enzyme.

DNA ligase may join preassembled duplexes (A, B and C) if they have sticky ends:

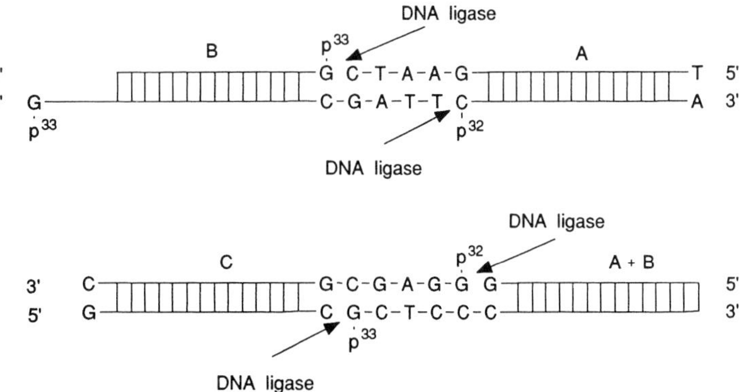

Another type of DNA ligase activity has also been discovered; namely, in the presence of an enzyme isolated from *E. coli* cells infected with phage T4, two duplexes lacking sticky ends are ligated. Such joining of blunt-ended DNA fragments becomes more effective if RNA ligase[1] is added to the reaction mixture.

11.3.2 Ways to Produce Duplex DNAs

It has already been pointed out that for polynucleotides to be linked together in complementary complexes (duplexes) the length of their chain must be sufficiently small (8–10 b. p.) and lie within the possibilities of chemical synthesis which existed in the sixties. Proceeding from these assumptions, Khorana formulated and then, together with his coworkers, developed a three-step procedure for duplex DNA synthesis schematically shown below:

T A C A T G C C A G
HO p p p p p p p p p OH

^{32}pppA
polynucleotide kinase

T A C A T G C C A G
^{32}p p p p p p p p p p OH

oligonucleotides
DNA ligase

5' T–G–T–T–A–A–G–C ^{32}p–T–A–C–A–T–G–C–C–A–G 3'
3' A–T–T–C–G————A–T–G–T–A 5'

The first step is the complete chemical synthesis of oligonucleotides corresponding to both target strands. The oligonucleotides must contain free 3′- and 5′-terminal hydroxyl groups; their length usually varies from 10 to 12 nucleotides. Taken together they must represent a double-stranded DNA fragment. The overlap between fragments belonging to the opposite strands must be 5–6 b. p.

The second step involves phosphorylation of the 5′-hydroxyl groups of oligonucleotides with the aid of (γ-^{32}P) ATP in the presence of polynucleotide kinase. Thus, all synthetic oligonucleotides become 5′-labeled compounds, which renders analysis during the third step much easier.

[1] This enzyme joining single-stranded oligo(poly)ribonucleotides was discovered much later.

The third step is ligation of three or more oligonucleotides in the presence of DNA ligase. To this end, segments of (7–9) and (11) types (see above scheme) with overlapping complementary sequences are mixed at an appropriate temperature and ionic strength to obtain a double-stranded complex. To ensure covalent binding of the adjacent segments resulting in a continuous duplex, the mixture is treated with DNA ligase.

In addition to the "closest neighbor" method (see above), other techniques can be used to characterize the newly formed chains by determining their length, such as polyacrylamide gel electrophoresis (PAGE). The complete nucleotide sequence of chains constructed by ligase-assisted linkage is currently determined by the Maxam–Gilbert method.

The reliability of the above-described chemoenzymatic method for producing double-stranded DNA fragments was demonstrated in Khorana's laboratory, in the synthesis of the complete 207 b. p. gene of suppressor thyrosine RNA, as well as in other laboratories, in the synthesis of protein genes and other functionally important DNA fragments.

The scheme that follows shows the structure of the suppressor thyrosine tRNA gene synthesized in Khorana's laboratory, which contains, apart from the structural gene, a promoter and a terminator.

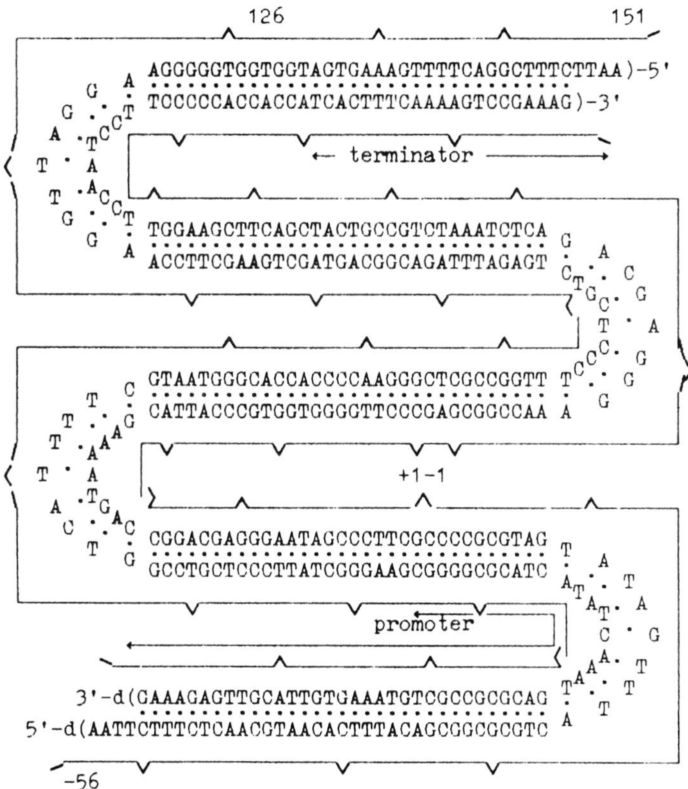

11.3.3 Enzymatic Methods for Construction of Duplex DNAs

So far, a large number of genes have been assembled from synthetic fragments to be subsequently used in microbiological synthesis of peptides and proteins by genetic engineering techniques. Table 11-3 lists such genes described in 1989.

Today, when extended (20–100 b. p) synthetic polynucleotides are available, genes are not always constructed by the method described above, often referred to as step-by-step assembly. One of the recent procedures is so-called "one-step ligation". In this case, all synthetic oligonucleotides constituting a double-stranded DNA and prepared for enzymatic ligation are placed in a vessel, T4 DNA ligase is added after heating (denaturation) and cooling, and the mixture is subjected to cloning. This is how the gene of α-neoendorphin was produced in 1982, to be followed later by other genes. Sometimes, a similar approach is used, with the difference that the ligation is carried out directly in the cell in the presence of a suitable vector. Then, DNA ligase of the cell is used (Narang, 1986).

Some of the reports published in recent years deal with step-by-step solid-phase construction of duplex DNAs. The basic principle behind this approach is that a terminal oligonucleotide is immobilized on an insoluble polymer with two more added to form a complementary complex, and the break in the latter is eliminated using DNA ligase. After filtration and washing, another oligonucleotide is added to the polymer and treated with DNA ligase again. Thus, both chains are built up step by step (as shown schematically below). This assembly technique is yet to be refined, but it is hard to deny the promise it holds. Once removed from the polymer (cross-hatched circle in the picture), the DNA duplex with its protruding ends is incorporated into an appropriate vector by usual means. None of the defective duplexes find their way into the vector because of lack of the right terminators. In spite of the low effectiveness of such assembly, it is quite applicable in genetic engineering. Table 11-3 shows examples of enzymatic assembly of genes from synthetic oligonucleotides using the above methods.

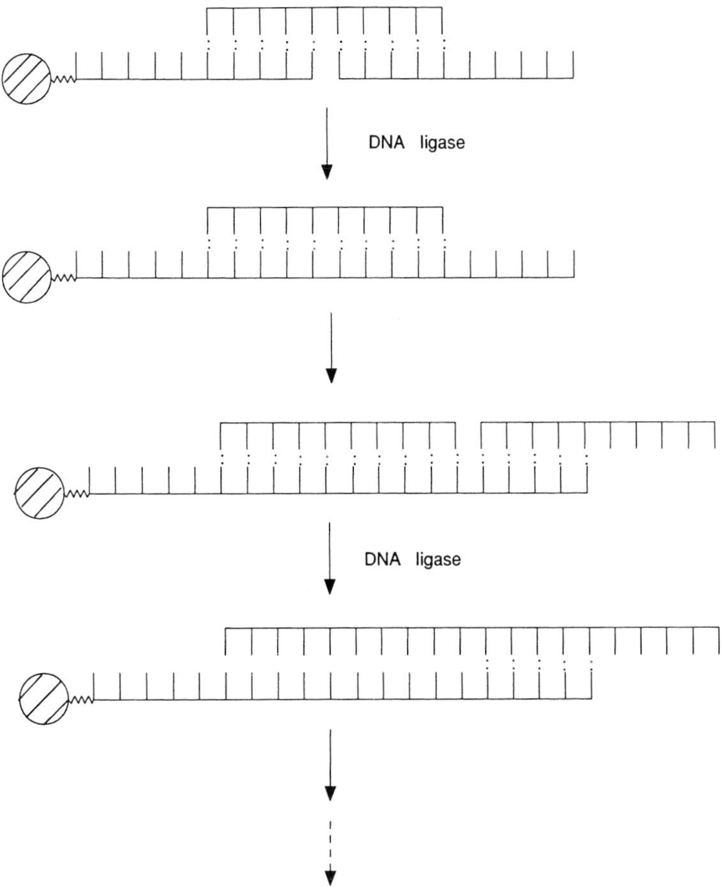

11.3.4 Chemical Methods for Assembling Duplex DNAs

Instead of an enzymatic reaction (DNA ligase) for assembling double-stranded DNAs from synthetic single-stranded oligo(poly)deoxyribonucleotides, chemical reactions resulting in internucleotide linkage can be used. This method named "chemical ligation" has been developed by Shabarova and coworkers (Moscow State University). The linkage of single-stranded fragments within a DNA duplex is achieved through activation of the phosphate groups at cleavage sites in each strand under conditions making existence of a complementary complex possible (aqueous solution, a certain ionic strength, pH, temperature). It should be emphasized here that complementary complexes are highly ordered systems in which the terminal phosphate and hydroxyl groups at cleavage sites are brought close enough to be separated by the covalent bond distance and spatially oriented in a definite way by virtue of forces stabilizing the double helix. As a consequence, one can attain chemical reaction selectivity and rates too high to be achievable for similar transformations outside the complementary complex.

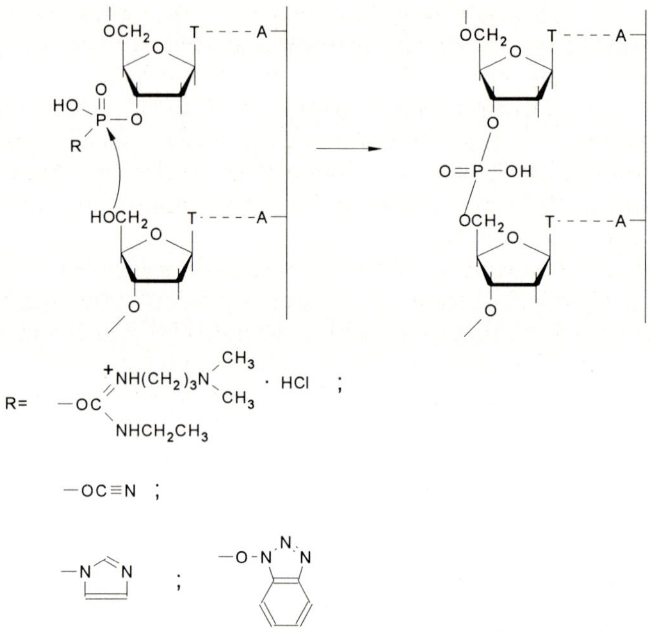

The reaction rate is increased not only because of a locally higher concentration of the reacting groups, but also as a result of their particular orientation which facilitates the transition state. As can be seen from the above scheme, two types of terminal phosphate activation are used: with the aid of condensation agents or by way of pre-activation (by obtaining potentially active oligonucleotides in advance). The first type involves reactions in which activation of the phosphorus at the junction between oligonucleotides occurs after formation of the duplex, as a result of introduction of water-soluble condensation agents, such as carbodiimides or cyanogen bromide. They easily react with the terminal monophosphate of the oligonucleotide to yield an active derivative which reacts only with the closest hydroxyl in the duplex, as is the case with enzymatic ligation. It has been established that if in the presence of carbodiimides this process takes two to three days, in the presence of cyanogen bromide its duration is as short as a minute.

By "pre-activation" (second type) is here meant use of groups covalently bound with the terminal phosphate and activating it, namely, oligonucleotide derivatives stable in aqueous solutions but exhibiting phosphorylating activity within the duplex. They include imidazolides and *N*-hydroxybenzotriazole phosphodiesters of oligonucleotides. Condensation in the presence of such derivatives proceeds under optimal conditions (i. e., when a stable duplex is formed and the steric and chemical factors are favorable) with good yields (up to 80–90 %). Simple methods have been elaborated for synthesis of imidazolides and other azolides of oligonucleotides as well as the more active *N*-hydroxybenzotriazole phosphodiesters.

Condensation agents for activating the phosphate during chemical ligation were used in studying the following chemical and steric factors of this process: position of the phosphate group in the oligonucleotide (5′ and 3′), nucleophilicity of the reacting groups, position of the hydroxyl in the *endo*- and *exo*-conformation as well as its *cis*- and *trans*-positions in the oxide ring of the carbohydrate moiety. It has been found that the optimal conditions for a quantitative course of chemical ligation in DNA duplexes are provided by using 3′-phosphates of natural oligonucleotides.

The above factors have been taken into consideration to produce extended DNA duplexes by chemical ligation. In the early eighties, when synthetic oligonucleotides were not as readily available, a rather successful model was proposed for DNA duplex construction, which turned out to be extremely useful not only in studying the reaction of template ligation but also in performing complete chemical synthesis of long DNA duplexes from short (9–15 b. p.) synthetic oligonucleotides. The system represents a DNA-like

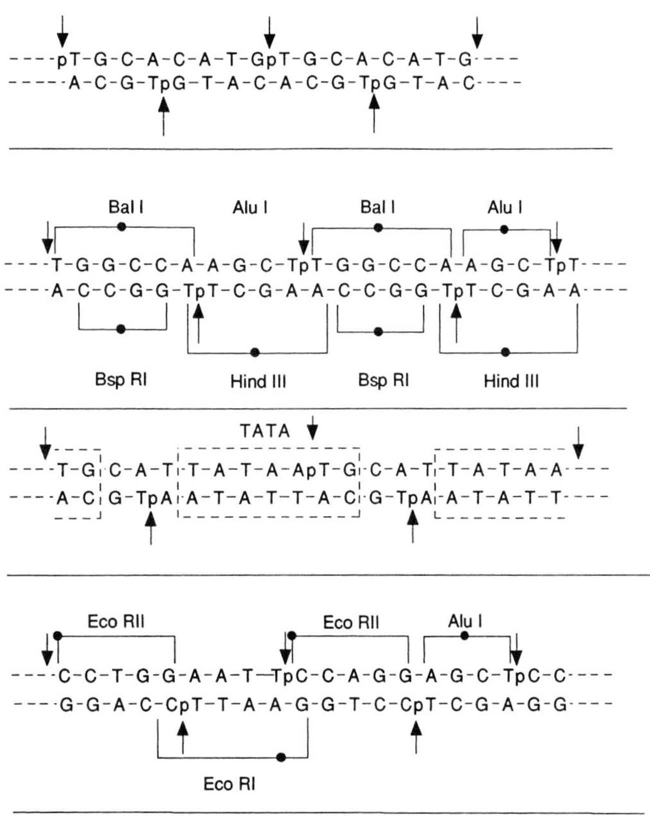

(concatemeric) complex with a recurring sequence of only one or two oligonucleotides with alternating and non-coinciding breaks in each strand. The scheme that follows illustrates such DNA duplexes. The examined complexes were different in composition and, consequently, had different theoretical stabilities at the same length of the oligomer. The physical and chemical properties of such double-stranded structures are closely similar to those of the corresponding DNAs in the B form. The above-described chemical ligation procedures were employed to assemble extended (more than 300 b. p.) duplex DNAs containing sites for recognition of various restriction endonucleases and gene promoters.

A good example of chemical ligation in double-stranded DNAs consisting of heterogeneous oligodeoxyribonucleotides (see below) is the assembly, with the aid of cyanogen bromide, of a 200 b. p. gene containing various regulatory sites and intended for modular mutagenesis while studying the functioning of different gene portions.

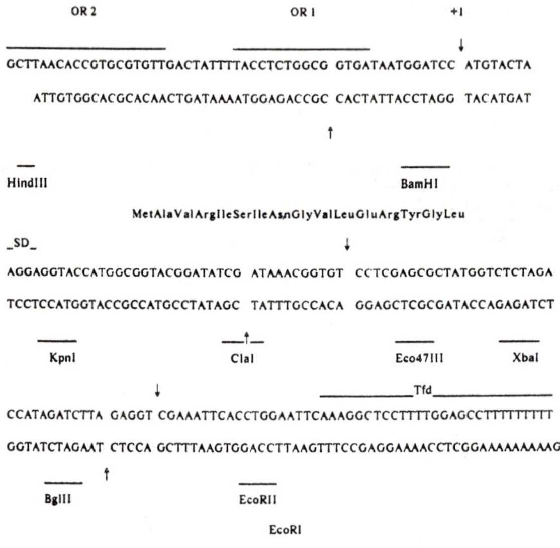

Arrows in the above scheme indicate the cleavage sites at which the chemical ligation takes place.

Thus, it was demonstrated that natural DNA duplexes with a predetermined primary structure can be constructed not only enzymatically using DNA ligase but also with the aid of chemical activation procedures.

An important advantage of the chemical assembly of oligonucleotides into duplexes is the possibility of introducing modified bonds (not occurring naturally) into them. As can be inferred from the references cited at the end of this chapter, when oligonucleotides modified at the 3′ or 5′ ends are used for ligation, phosphoamide, pyrophosphate and polyphosphate bonds formed by anomalous nucleosides (containing xylose, arabinose, lyxose) can easily and effectively be incorporated into predetermined portions of DNA, to say nothing of non-nucleoside inserts. Modified interoligomer bonds can be introduced only chemically. What makes this method especially valuable is that it is a simple and in most cases the only way to create substrates and inhibitors of nucleic exchange proteins. Now that synthetic oligonucleotides are becoming more readily available, the chemical method for constructing modified DNAs opens up wide possibilities for studying the mechanism of action of the proteins and enzymes recognizing nucleic acids. The strategy and tactics of such studies are described in the review by Ye. S. Gromova and Z. A. Shabarova published in *Progress in Nucleic Acids Res. and Mol. Biol.* (see references).

11.4 Methods for Chemical Synthesis of Oligodeoxyribonucleotides

As already mentioned in the introduction, each internucleotide (intermonomer) linkage event during chemical synthesis boils down to acylation of the hydroxyl group of one monomer (nucleoside component) with the phosphate of another (nucleotide component):

nucleoside component

nucleotide component

Here, at least two things must be accomplished: (1) protection of all functional groups of the nucleoside and nucleotide components, not subject to phosphorylation; and (2) activation of the phosphate in the nucleotide component.

We shall now consider the most widely used methods for introducing protective groups and preparing nucleoside and nucleotide components. Covered as a separate topic will be methods for activating the phosphate group and the synthesis procedures involved in the diester, triester and phosphite-triester methods for producing oligodeoxyribonucleotides.

11.4.1 Nucleoside and Nucleotide Components for Synthesis by the Phosphodiester and Phosphotriester Methods

As can be seen from the above schemes illustrating the formation of natural internucleotide linkages, the nucleotide chain grows in length as a result of the interaction between the groups in the carbohydrate moiety: namely, 3′- or 5′-hydroxyl group of the nucleoside component and, respectively, 5′- or 3′-phosphate group of the nucleotide component.

To make sure that the reaction proceeds in a predicted fashion, the other reactive sites in these components (hydroxyl and phosphate groups in the carbohydrate moieties, which must remain intact, as well as amino groups of the heterocyclic bases) must be protected or, in other words, blocked. This can be accomplished in two ways: (1) by blocking all reactive groups in the nucleoside and nucleotide components with subsequent selective deblocking of the hydroxyl and phosphate groups, respectively, to create an internucleotide linkage during the following step of the synthesis; and (2) by selective protection of those groups in both components which should not participate in the reaction. Examples illustrating these two approaches will be given in what follows.

The protective groups introduced into the nucleotide and nucleoside components during the steps preceding the internucleotide linkage formation must meet the following basic requirements: (1) the introduction and removal conditions must be mild to avoid undesired side processes; (2) no elimination must take place during formation of internucleotide linkages and subsequent isolation of the reaction products; and (3) ways must be devised for selective elimination and, sometimes, introduction of a given protective group.

Distinction between protective groups is usually made according to the ways of their elimination. The two most common types include those eliminated by acid and alkaline hydrolysis. The third type includes protective groups eliminated by other means (hydrogenolysis, with the aid of anions, in the presence of a zinc–copper couple in dimethylformamide, etc.).

11.4.1.1 Blocking of the Hydroxyl Groups of Pentose and Amino Groups in Heterocyclic Bases

It has already been pointed out that the two types of groups to be blocked in the nucleoside and nucleotide components prior to formation of an internucleotide linkage are, firstly, the hydroxyls that must remain intact and, secondly, the amino groups of heterocyclic bases, which may interact with the activated phosphate group of the nucleotide component.

Notably, acylation of the amino groups in heterocyclic bases enhances the solubility of components in absolute pyridine or another organic solvent in which the oligonucleotide synthesis is carried out. The acylation of nucleotides has already been covered at length with examples of complete selective blocking of the above groups and principles of selection of blocking groups for different heterocyclic bases. Reactions with acid halides or anhydrides proceed quantitatively for all practical purposes, both at the amino groups of heterocycles and hydroxy groups of pentose. Selective removal of protective groups from carbohydrate moieties is achieved by virtue of different rates of cleavage of ester and amide bonds. Protective groups meeting all requirements have been identified with due attention paid to the stability of the amide bonds in the nucleoside and nucleotide components acylated at their heterocyclic bases under conditions of oligonucleotide synthesis as well as deblocking (usually involving treatment with concentrated ammonia) for each constituent heterocyclic base of the nucleic acid. These are isobutyric, 2-methylbutyric and, less commonly, benzoyl group in the case of guanine, benzoyl group in case of adenine, and anisoyl or benzoyl group in the case of cytosine.

Table 11-1 gives some information (introduction and removal conditions) about the acyls most frequently used as protective groups in oligonucleotide synthesis.

Acetylation is a common procedure for blocking the secondary hydroxyl of pentose in the nucleotide and nucleoside components. The primary (5′) hydroxyl group can be blocked selectively in a forward reaction with trityl chloride and its substituents. Methoxytrityl groups are used most often because they are easier to remove from oligonucleotides than an unsubstituted trityl group.

Table 11-1. Major Protective Groups Used in Oligonucleotide Synthesis.

Group	Formula	Abbreviation	Elimination conditions
Blocking of Hydroxyls in Deoxyribose			
Formyl	$HC(=O)-$	–	Dilute aqueous alkali, aqueous pyridine
Acetyl	$CH_3C(=O)-$	Ac	Aqueous alkali, aqueous ammonia
Phenoxyacetyl	$C_6H_5OCH_2C(=O)-$	–	Aqueous alkali, aqueous ammonia
Benzoyl	$C_6H_5C(=O)-$	Bz	Aqueous alkali, aqueous ammonia
Levulinyl	$CH_3COCH_2CH_2CO-$	Lev	NH_2NH_2

Table 11-1. (Continued).

Group	Formula	Abbreviation	Elimination conditions
Trimethylsilyl	$(CH_3)_3Si–$	TMS	Dilute aqueous alkali or weak acid
Trityl	$(C_6H_5)_3C–$	Tr	
Monomethoxytrityl	*p*–$CH_3OC_6H_4C(C_6H_5)_2$	MeOTr, MTr	80 % acetic acid, trichloroacetic acid, trifluoroacetic acid, benzenesulfonic acid,
Dimethoxytrityl	(*p*–$CH_3OC_6H_4)_2CC_6H_5$	$(MeO)_2Tr$, DMTr	$ZnBr_2$
Blocking of Amino Groups in Heterocyclic Bases			
Acetyl	$CH_3C(=O)–$	Ac	9N NH_4OH (50 °C, 3–18 h)
Isobutyryl	$(CH_3)_2CHC(=O)–$	Ibu	dito
Benzoyl	$C_6H_5C(=O)–$	Bz	dito
Anisoyl	*p*–$CH_3OC_6H_4C(=O)–$	An	dito
Blocking of Phosphate Groups			
β-Cyanoethyl	$NC(CH_2)_2–$	CNEt	NH_3, NaOH, triethylamine
Phenyl	$C_6H_5–$	–	0.1 N NaOH (20 °C, half-life in triesters: 3 or 4 min)
p-Chlorophenyl	*p*–$ClC_6H_4–$	ClPh	0.1 N NaOH (20 °C, half-life in triesters: 2–4 min) *syn*-2-nitrobenzaldehyde or pyridine-2-carbaldehyde
Trichloroethyl	$Cl_3CCH_2–$	TCEt	Zn/Cu $HCON(CH_3)_2$ (50 °C, 1 h) or 0.1 N NaOH (20 °C)
Methyl	$CH_3–$	Me	$NEt_3H^+C_6H_5S^-$ in dioxane, 1h, 20 °C

11.4.1.2 Protection of Phosphate Groups

The incorporation of substituents into the phosphate groups of nucleosides is usually achieved through reactions with alcohols or amines in the presence of condensation agents. The most widely used protective group is the β-cyanoethyl one which is removable by mild alkaline treatment (see Table 11-1). Another method for obtaining nucleotides protected at the phosphate group is phosphorylation of nucleosides with active phosphate derivatives, such as halides of *p*-chlorophenylphosphate. Selective removal of the *p*-chlorophenyl group can be effected during alkaline hydrolysis without any cleavage of the phosphate–nucleoside linkage (see Table 11-1). The table also shows some other groups used for protection of phosphate groups in nucleotides.

11.4.1.3 Preparation of the Nucleoside and Nucleotide Components

As has already been mentioned, the nucleotide chain usually can have its length increased during oligonucleotide synthesis as a result of interaction between the 3′- or 5′-hydroxyl of the nucleoside component and, respectively, the 5′- or 3′-phosphate group of the nucleotide component.

The function of both components can be performed by monomeric nucleosides and nucleotides as well as oligomers, or oligonucleotides, insofar as any one of these can be entered into phosphorylation reactions of the above-mentioned type after appropriate preparation.

Described below are the most widely used approaches for the preparation of the nucleoside and nucleotide components.

Nucleoside Component. By definition, a nucleoside component is a compound undergoing phosphorylation in the course of internucleotide linkage formation by a nucleotide component. Accordingly, nucleoside components are prepared in such a manner that all of their reactive groups are blocked, except for one of the hydroxyls which must be involved at a later stage in internucleotide linkage formation. Here are some examples (for symbols, see Table 11-1) illustrating the protection of the heterocyclic amino groups.

dA $\xrightarrow{BzCl}$ [structure: N,N-dibenzoyl-3′,5′-di-O-benzoyl deoxyadenosine; labels: N–C(=O)–, N–C(=O)C$_6$H$_5$, CO, CO] $\xrightarrow{^{-}OH}$ A^{bz} (dbzA), HO, OH

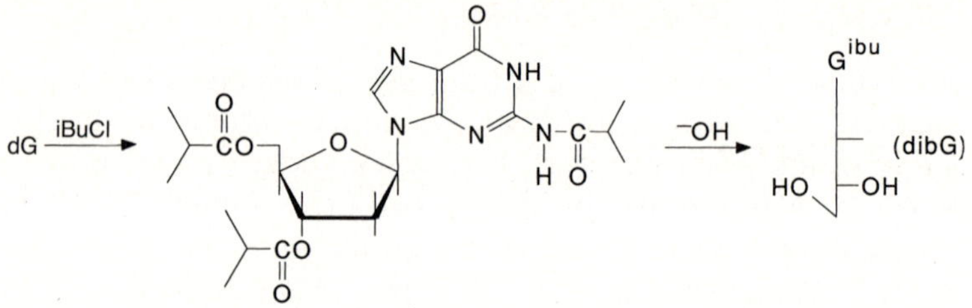

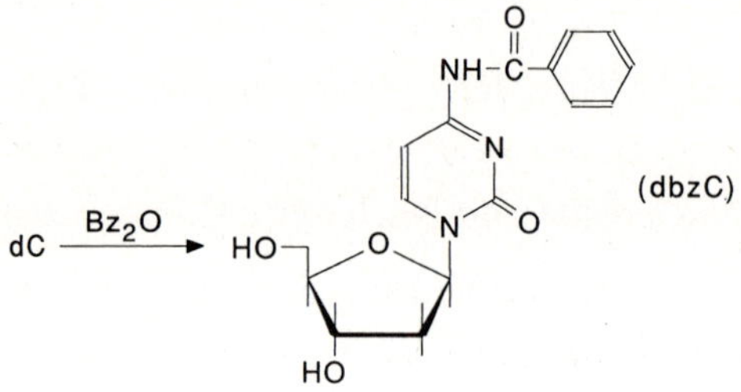

Today, the most effective yet simple way to produce *N*-benzoyl derivatives of nucleosides is a one-step procedure with intermediate silylation (without separation of the reaction products) of deoxyribose hydroxyls. The synthesis of *N*-benzoyldeoxyadenosine by this method is as follows:

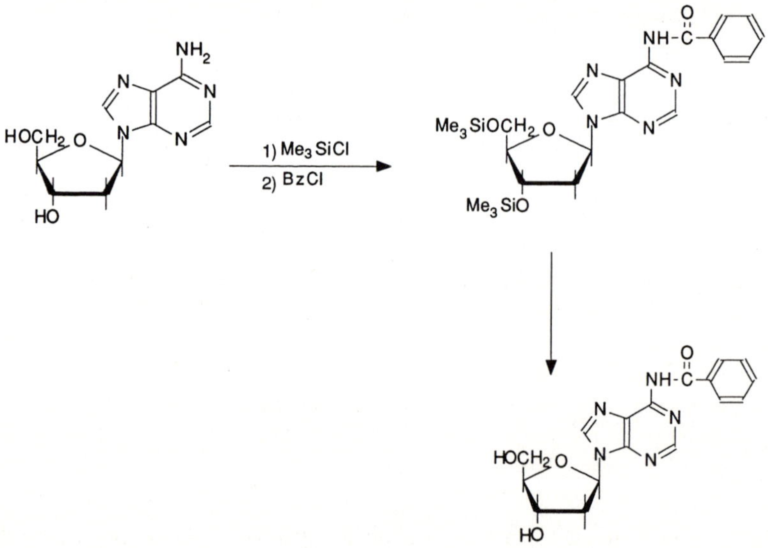

Nucleosides containing a single deprotected hydroxyl (secondary or primary) group, at which phosphorylation during oligonucleotide synthesis may occur, are produced according to scheme (a) or (b):

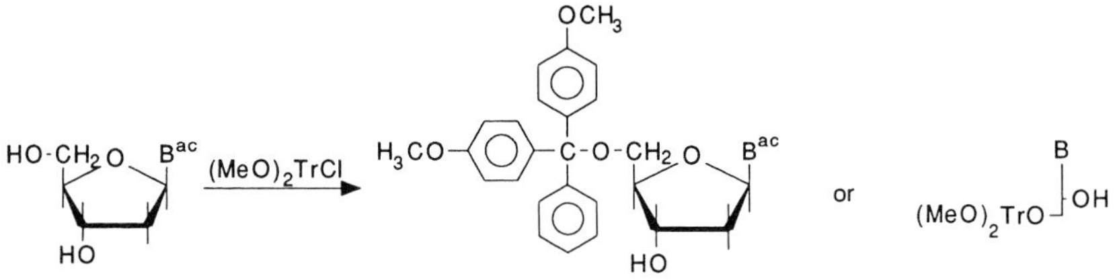

a

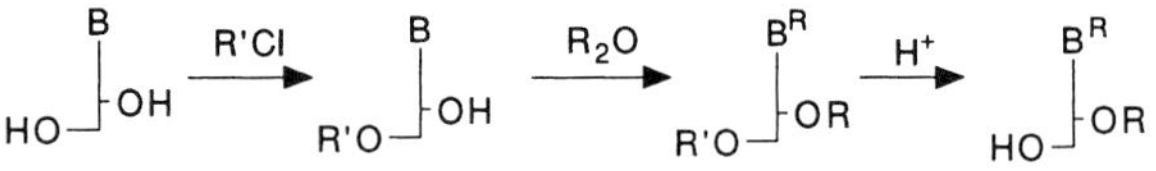

R = CH_3CO, $(CH_3)_2CHCO$, C_6H_5CO, n-$CH_3OC_6H_4CO$; R' = Tr, MeOTr, $(MeO)_2Tr$

b

If a nucleotide is used as the starting nucleoside component, as is often the case with oligodeoxyribonucleotide synthesis, its phosphate group must be blocked. To this end, the reaction with β-cyanoethanol in the presence of dicyclohexylcarbodiimide (DCC) is usually conducted.

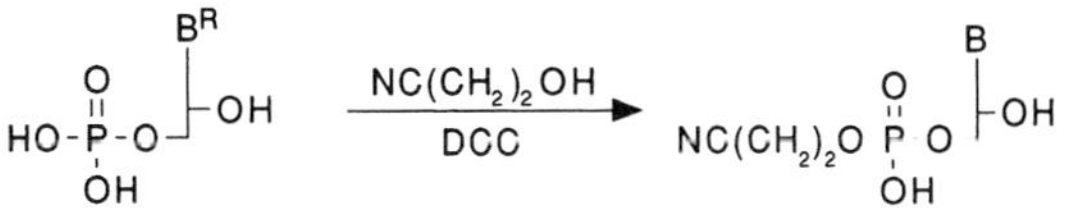

R = Acyl

Nucleotide Component. Usually, the respective 3′- or 5′-phosphates called, as we already know, nucleotide components, act as phosphorylating agents during synthesis by the phosphodiester and -triester methods. The procedures for preparing these phosphates depend on the exact way in which the nucleotide chain is to be elongated.

The most readily available source of the nucleotide component is 5-deoxythymidylic acid:

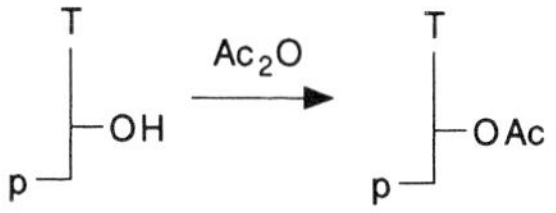

In other deoxyribonucleotides, the amino group of the heterocyclic base residue is blocked first, then the 3′-*O*-acyl group is replaced by a 3′-*O*-acetyl one, because the latter can be easily eliminated in a selective manner during alkaline treatment of the synthesized oligonucleotide, which is required for subsequent elongation of the oligonucleotide chain.

$$\mathrm{pC\text{-}OH} \xrightarrow{n\text{-}CH_3OC_6H_4COCl} \mathrm{pC^{an}\text{-}OAn} \xrightarrow{^-OH} \mathrm{pC^{an}\text{-}OH} \xrightarrow{Ac_2O} \mathrm{pC^{an}\text{-}OAc}$$

or pdanC(Ac)

An = $COC_6H_4OCH_3$-p

Similarly, sequential acylation, deacylation and acetylation give d[pibG(Ac)] and d[pbzA(Ac)] (for symbols designating protective groups, see Table 11-1).

The above procedures for obtaining nucleotide components by introducing protective groups are by far the most common approaches to ester synthesis. There is a great many of blocking techniques differing by the type of protective groups as well as ways to introduce and remove them. These techniques, however, are not as frequently used in oligonucleotide synthesis and are preferred only by a few groups of specialists.

11.4.1.4 Mechanisms of the Internucleotide Linkage Formation

We already know that the chemical synthesis of oligonucleotides involves condensation of specially prepared nucleoside and nucleotide components. For the latter to react effectively enough, the phosphate group of the nucleotide component must be activated. The activating agents in the phosphodiester method are arylsulfonyl chlorides and, primarily, 2,4,6-tri-isopropylphenylsulfonyl and mesitylenesulfonyl chlorides. At the early stages of the method's evolution *N,N*-dicyclohexylcarbodiimide was used.

The mechanisms of the reactions involved in activation of the phosphate group by arylsulfonyl chlorides have been described elsewhere, but nothing was said about the paths of internucleotide linkage formation and the role of absolute pyridine in which the reaction is conducted. The latter compound stabilizes the intermediately forming "active metaphosphate". It has been speculated that the stable form emerging in this fashion is essentially an internal salt (betaine) in which the pyridine residue is rather rapidly exchanged. This is borne out by the fact that the stable form of metaphosphate can be obtained only if the reaction is conducted in absolute pyridine or if it is present in the reaction mixture in sufficient amounts. This exchange probably involves a transition state having a trigonal bipyramid structure, and two pyridine molecules are most likely to participate in the stabilization.

RO-P(=O)(O⁻)-N⁺(pyridinium) or RO-P(=O)(O⁻) with two pyridine Nδ+ ligands

Phosphorylation of the nucleoside component occurs at the moment when the trigonal bipyramid structure receives the hydroxyl of the nucleoside component instead of a pyridine molecule (a similar pattern is observed in the interaction between metaphosphate and water (yielding an appropriate nucleotide). In what follows, the pyridine-stabilized metaphosphate will be designated as $ROPO_2 \cdot C_5H_5N$.

Condensation of the nucleotide and nucleoside components in the presence of triisopropylphenylsulfonyl (or mesitylensulfonyl) chloride leads to formation of a dinucleoside phosphate in absolute pyridine.

$$RO{-}\overset{O}{\overset{\|}{P}}{-}OH \ (OH) \xrightarrow[C_5H_5N]{ArSO_2Cl} ROPO_2^- \cdot C_5H_5N \xrightarrow{R'OH} RO{-}\overset{O}{\overset{\|}{P}}{-}OR' \ (OH) \longrightarrow RO{-}\overset{O}{\overset{\|}{P}}{-}O{-}\overset{O}{\overset{\|}{P}}{-}OR \ (OR',\ O^-)$$

$$RO{-}\overset{O}{\overset{\|}{P}}{-}O{-}\overset{O}{\overset{\|}{P}}{-}OR \xrightarrow[-\,RO{-}\overset{O}{\overset{\|}{P}}{-}OR' \ (O^-)]{R'OH} RO{-}\overset{O}{\overset{\|}{P}}{-}OR' \ (OH)$$

However, the phosphodiester group in the latter is a stronger nucleophile than the hydroxyl of the nucleoside and immediately reacts with the next molecule to yield a trisubstituted pyrophosphate. This compound is a less active phosphorylating agent but reacts (albeit slowly and non-quantitatively) with the nucleoside to give a dinucleoside phosphate; that is, the phosphorus of the ionized phosphate group is attacked, and the departing group is the represented by anion of the stronger acid.

As can be inferred from the findings by Knorre and coworkers concerning the mechanism of internucleotide linkage formation, this reaction is accompanied by several side processes, primarily hydrolysis of betaine and formation of a triphosphate.

$$RO{-}\overset{O}{\overset{\|}{P}}{-}OR' \ (O^-) \xrightarrow{ArSO_2Cl} RO{-}\overset{O}{\overset{\|}{P}}{-}O{-}\overset{O}{\overset{\|}{P}}{-}OR \ (OR',\ OR') \xrightarrow[-\,RO{-}\overset{O}{\overset{\|}{P}}{-}OR' \ (O^-)]{R'OH} RO{-}\overset{O}{\overset{\|}{P}}{-}OR' \ (OR')$$

The latter is rather unstable and, for example, can be easily hydrolysed with formation of a dinucleoside phosphate having a 3′–3′ internucleotide linkage not existing in nature; thus, this side process is not only accompanied by the cleavage of the internucleotide linkage formed during synthesis, but also "contaminates" the end product by giving a compound with a closely similar structure, which is difficult to eliminate.

No satisfactory condensation agents for phosphotriester synthesis existed by the late seventies when Narang proposed to use tetrazolide of triisopropylsulfonic acid (TPSTe). Later, other azolides of arylsulfonic acids were proposed, such as

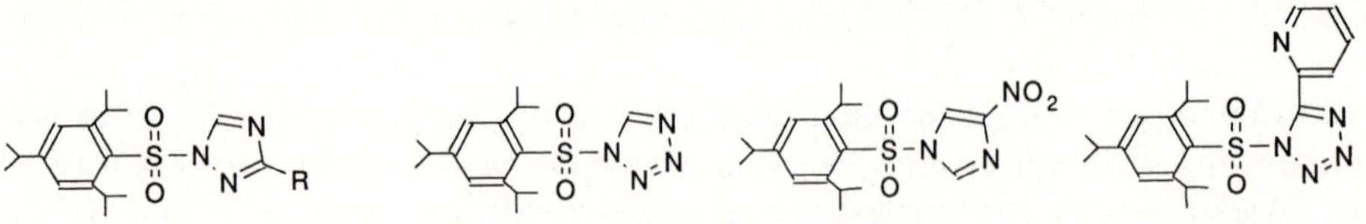

(MST ,R=H)
(MSNT , R=NO_2)

(TPSTe)

(TPSNI)

(TPSPy)

The mechanism of phosphodiester activation by arylsulfonyl azolides seems to include formation of a mixed phosphodiester anhydride with subsequent nucleophilic substitution at the phosphorus atom to give a diphosphate azolide. The latter reacts with the hydroxyl of the nucleoside component to form internucleotide linkage.

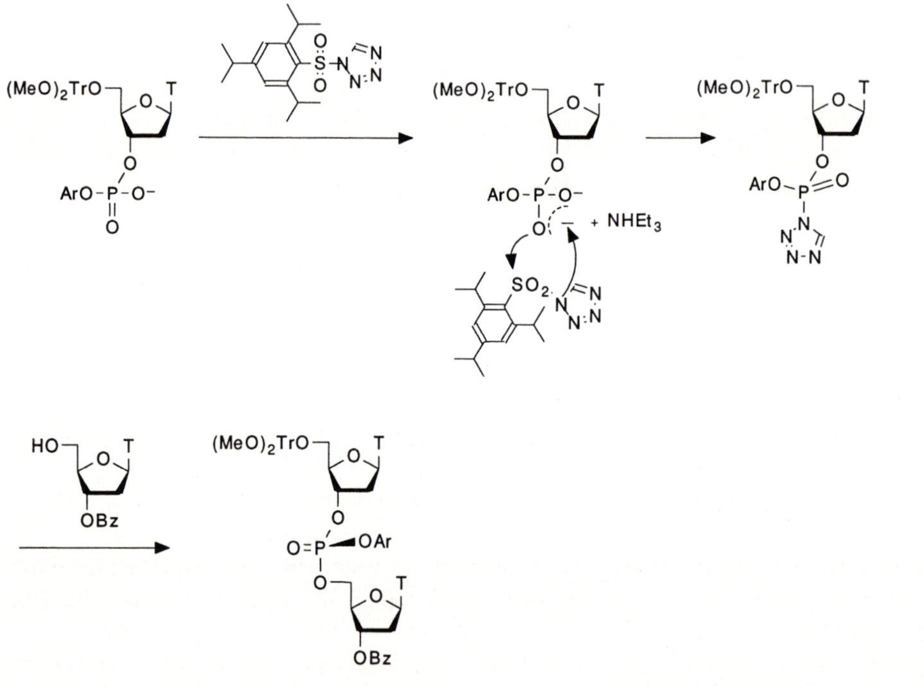

11.4.2 Chemical Synthesis of Oligodeoxyribonucleotides in Solution

11.4.2.1 Phosphodiester Method

This method was the first to be used for synthesis of oligo- and polydeoxynucleotides and, therefore, has been studied most thoroughly. Most of the credit for its development, including preparation of the nucleotide and nucleoside components, activation of the phosphate group with the aid of DCC and Ar_2SO_2Cl (see above), as well as elaboration of procedures for elongating the nucleotide chain, must be given to Khorana and coworkers.

Synthesis of oligodeoxyribonucleotides with a chain 8 to 12 monomer units long by Khorana's method includes, at the early stages, condensation of the corresponding (5′-terminal) deoxyribonucleoside (13), in which the 5′-hydroxyl is blocked by a *para*-methoxytrityl group, together with the nucleotide (14) in which the 3′-hydroxyl is protected by an acetyl group. The condensation is performed in the presence of 1,3,5-triisopropylphenylsulfonyl or mesitylenesulfonyl chlorides:

MeOTrOCH2 O B^ac; HO (13) + OH-P(=O)(OH)-OCH2 O B^ac; AcO (14) —ArSO2Cl, C5H5N→ MeOTrOCH2 O B^ac; O-P(=O)(OH)-OCH2 O B^ac; AcO (15) —1. NH3, 2. H+→ HO-[B]p[B]-OH; —⁻OH→ MeOTrO-[B]p[B]-OH (16)

After removal of the 3′-acetyl group, the resulting protected dinucleoside phosphate (15) is subjected to alkaline treatment to be converted into the nucleoside component for subsequent elongation of the oligonucleotide chain. Treatment of the dinucleoside phosphate (15) with ammonia (to eliminate the acetyl and protective groups from the heterocyclic bases) and dilute acid (to eliminate the monomethoxytrityl group) gives a deprotected dinucleoside phosphate.

The condensation is carried out in absolute pyridine which, as has already been pointed out, is indispensable for activating the nucleotide and also capable of dissolving even high-molecular weight oligonucleotides (in the form of the corresponding derivatives). Another advantage of this solvent is its relatively high volatility, which makes its removal from the reaction mixture possible under very mild conditions.

The resulting partially protected oligonucleotide (16) is purified by extraction with organic solvents, which is quite possible because it contains a lipophilic trityl group, or by ion-exchange chromatography on DEAE cellulose.

Subsequent elongation of the oligonucleotide chain is possible by stepwise addition of monomers of type (14). This is what is known as the step-by-step method. It is generally used to obtain three- to four-unit oligonucleotides. Step-by-step synthesis of longer oligonucleotides creates difficulties in separation. The reason is that every new unit added to the nucleotide chain in such synthesis leads to smaller charge and molecular weight increments. This is precisely why the block method had to be adopted. It resides in attachment to the nucleoside component (16) of a nucleotide one which is essentially a di- or trinucleotide (at later stages this component may comprise even a greater number of monomer units). Synthesis of the latter may proceed as follows, for example:

$NC(CH_2)_2O$–P(O)(OH)–OCH_2–[sugar, B^{ac}, 3'-HO] (17) + (HO)$_2$P(O)–OCH_2–[sugar, B^{ac}, 3'-AcO] (18) $\xrightarrow{ArSO_2Cl}$ $NC(CH_2)_2O$–P(O)(OH)–OCH_2–[sugar, B^{ac}]–O–P(O)(OH)–OCH_2–[sugar, B^{ac}, 3'-AcO] (19) $\xrightarrow{OH^-}$ p–[B^{ac}]–p–[B^{ac}]–OH (20) $\xrightarrow{Ac_2O}$ p–[B^{ac}]–p–[B^{ac}]–OAc (21)

The condensation is conducted as described above; it should be noted here that the nucleoside component (17) is also a nucleotide. Its difference from the nucleotide component (18) is that its phosphate group is blocked by cyanoethyl. After condensation, the protected dinucleotide (19) is treated with an alkali (both the cyanoethyl and acetyl or acyl groups are removed) then, after chromatography on DEAE cellulose, with acetic anhydride which selectively acetylates the 3′-hydroxyl group of the dinucleotide (20). The emerging partially protected dinucleotide (21) is used for the elongation of the oligonucleotide chain block by block. This approach was implemented in Khorana's laboratory for preparing large quantities of 8- to 12-unit oligodeoxyribonucleotides from which the complete $tRNA^{Tyr}$ gene was then constructed. The following scheme illustrates the synthesis of undecadeoxyribonucleotide, which is one of the fragments if this gene.

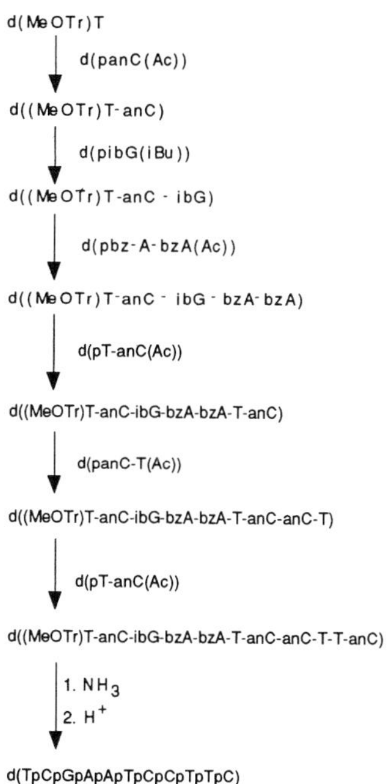

What sets this synthesis path apart is the fact that the 5′ end of the chain always retains an alkali-stable *para*-methoxytrityl group, while the 3′ end contains, after each step of synthesis, an acetyl group which is selectively eliminated by alkaline treatment prior to the next step (see Table 11-1). Thus, the chain is built up from the 5′ to the 3′ end. When the synthesis is over, sequential treatment with ammonia and a dilute acid removes all protective groups, and the oligonucleotide is carefully parified and analyzed by ion-exchange chromatography. This scheme has been successfully used in the synthesis of many biologically active 8- to 12-unit oligodeoxyribonucleotides.

If polydeoxyribonucleotides containing more than 12 monomer units are to be synthesized, the effectiveness of the block method also goes down for two reasons. Firstly, the yields at every subsequent step of condensation drop drastically. This decrease in yields stems from side processes, primarily those due to cleavage of internucleotide linkages via intermediates of triester type (7) and also, in all likelihood, steric hindrances to the condensation.

When oligonucleotides containing more than 10 to 12 units are joined, the yields of condensation products become extremely low and cannot be raised even if the nucleotide component is used in markedly excess amounts (up to 200-fold). The steric hindrances to condensation during synthesis of polynucleotides longer than 10 units are most likely due to formation of intermediate compounds of type (6) (products of addition of an active nucleotide component to every internucleotide phosphate group). The result may be a globule in which the 3′-hydroxyl end of the polynucleotide may be drawn into the molecule by virtue of non-covalent interactions (e.g., hydrogen bonding). The other reason for low effectiveness of block-by-block elongation of the chain has to do with difficulties in purification and separation of the desired polynucleotide by ion-exchange chromatography.

The above method has been used for one-step synthesis of oligo- and polydeoxyrubonucleotides by way of polycondensation of appropriately protected mono- and oligonucleotides. In the simplest case, if only the heterocyclic base is blocked in the nucleotide, homogeneous oligo- and polynucleotides are formed in the presence of a condensation agent (the apostrophe indicates the presence of a protective group in the base residues):

$$\text{pdN}' \rightarrow \text{d[pN}'(\text{pN}')_{\text{n}}\text{pN}']$$

A similar course is taken by the polycondensation of oligonucleotide segments in which only heterocyclic bases are protected:

$$\text{d(pN}^{(1)}\text{pN}^{(2)}) \rightarrow \text{d[(pN}^{(1)}\text{pN}^{(2)})_n]$$
$$\text{d(pN}^{(1)}\text{pN}^{(2)}\text{pN}^{(3)}) \rightarrow \text{d[(pN}^{(1)}\text{pN}^{(2)}\text{pN}^{(3)})_n]$$
$$\text{d(pN}^{(1)}\text{pN}^{(2)}\text{pN}^{(3)}\text{pN}^{(4)}) \rightarrow \text{d[(pN}^{(1)}\text{pN}^{(2)}\text{pN}^{(3)}\text{pN}^{(4)})_n]$$

This is how homogeneous oligonucleotides or oligo- and polynucleotides with alternating di-, tri- and tetranucleotide sequences can be obtained at a sufficiently fast rate. A drawback of the method is the possibility of many side processes. In addition to those already described, we may have formation of cyclic oligonucleotides[2)] and various pyrophosphates as a result of intramolecular phosphorylation. Serious difficulties arise during separation of oligo- and polynucleotides as well as their identification. None the less, such synthetic oligo- and polynucleotides with alternating di-, tri- and tetranucleotide segments became available in the mid sixties and played a key role in establishing the structure of the amino acid code. Homo- and heterogeneous polycondensation products continue to be extremely important research tools in molecular biology.

11.4.2.2 Phosphotriester Method

The above-mentioned drawbacks of the phosphodiester method have been obviated to a large extent by blocking the phosphate group prior to synthesis of the internucleotide linkage; as a result, the emerging intermonomer linkage has a phosphotriester structure:

[2)] This reaction can be partially averted by adding no more than 20% of mono- or polynucleotide with a protected 3′-hydroxyl (terminating component of polycondensation) to the reaction mixture.

$$RO{-}\overset{O}{\overset{\|}{P}}{-}OH + HO{-}R' \longrightarrow RO{-}\overset{O}{\overset{\|}{P}}{-}OR'$$
(with OX on P in both)

It took quite some time for the method to gain popularity, although, historically, this is precisely how the first oligonucleotide was obtained by Michelson and Todd. There are many reasons for this lack of popularity till 1973. First of all, it was the fact that no condensation agent capable of activating the diphosphate as effectively as a monophosphate was available.

It was also important to find suitable protective groups to block the internucleotide phosphate. In 1965, Letsinger and coworkers came up with a proposal to use 2-cyanoethyl, which can be easily eliminated by mild alkaline treatment (β-elimination), as the protective group and TPS as a condensation agent. In 1967, cyanoethyl was replaced by trichloroethyl which can be quantitatively eliminated by treatment with zinc powder in 80% acetic acid, and the following year it was superseded by an alkali-labile phenyl group. None of the above groups, however, could meet all the requirements of oligonucleotide synthesis (ease of incorporation, stability during synthesis, ease of removal). In some cases, by-products with internucleotide (3′–3′ and 5′–5′) bonds not existing in nature were formed, in others, the internucleotide bonds were ruptured in a most pronounced manner. Nevertheless, it was at that stage that the basic principles of the general strategy for building up oligonucleotide chains by the triester method were formulated. It became obvious that deoxyribonucleosides were the starting compounds for synthesis and that the best way to prepare the nucleotide component is through phosphorylation of the 5′-protected nucleoside. However, the continuing application of the proven approach (diester method) to elongate oligonucleotide chains and use of arylsulfonyl chlorides slowed down implementation of the triester method. A major breakthrough for the latter took place in 1973–74 when a new type of arylsulfonic acid derivatives emerged as condensation agents. This is when the main advantage of imidazolides, namely, their much lower aggressivity (the reaction mixture becomes strongly resinous in the presence of TPS) was pointed out, although the internucleotide linkage formation proceeds at a slower rate. In 1974, Narang and coworkers described the highly successful use of triazolides of arylsulfonic acids in the triester synthesis. From that moment on, the triester method, which is much more effective than the diester one, started making deep inroads into oligonucleotide synthesis.

Diphosphate activation techniques have been rapidly improving. Tetrazolides of arylsulfonic acids have turned out to be the best condensation agents.

In the meantime, modifications of the method appeared. For example, in 1982 V. A. Yefimov and coworkers proposed the use of $ArSO_2Cl$ in the presence of *N*-methylimidazole in neutral organic solvents (acetonitrile, nitromethane, etc.). In 1985, the same team introduced more powerful nucleophilic catalysts into the phosphotriester method, namely, N-oxides of 4-substituted pyridines in the combination with $ArSO_2Cl$, which enhances considerably the rate of phosphodiester bonding (1 min) and brings it closer to the rates typically observed in phosphite methods.

In Krahmer's laboratory there was proposed an integrated approach to synthesizing oligonucleotides by the triester method, which, in its general form after undergoing some changes, underlies the current strategy of synthesis by this method.

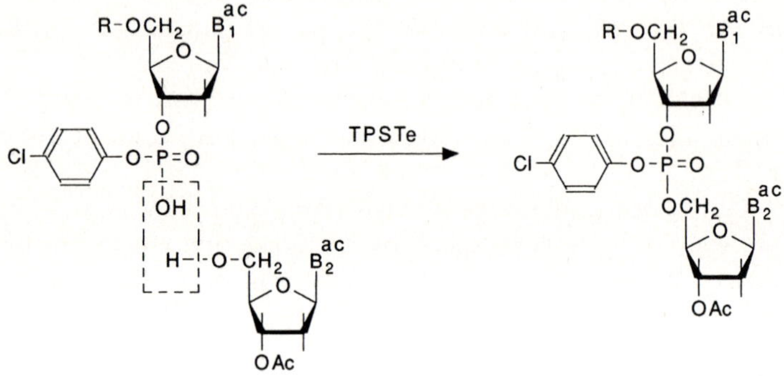

As has already been mentioned, a major contribution to this oligonucleotide synthesis method was made by Narang who further elaborated it and proposed a new group (substituted aryl), in addition to the new condensation agent, for blocking the internucleotide phosphate:

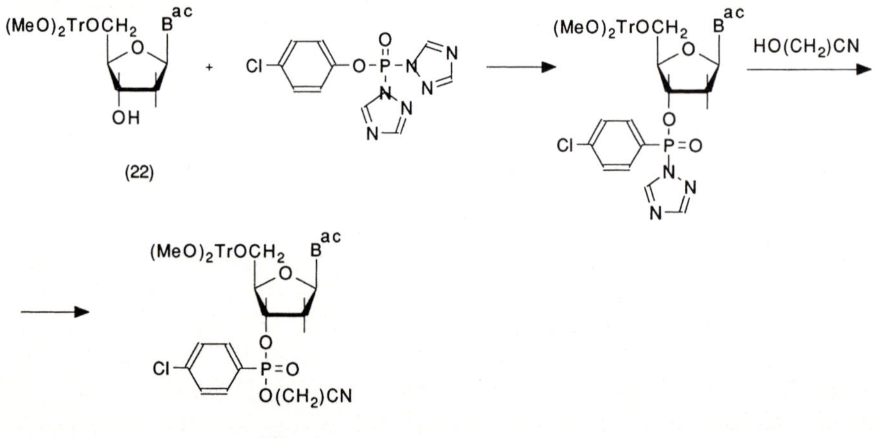

The key compounds in this scheme are deoxynucleosides with protected 5′- and 3′-hydroxyl groups. Deoxyribonucleoside (22) with a di(*para*-methoxy)trityl group at position 5′ is phosphorylated with bis-triazolide of *para*-chlorophenylphosphoric acid, and fully protected nucleotide (23) is obtained after treatment with ethylene cyanohydrin. The phosphorylating agent is derived from *para*-chlorophenyl dichlorophosphate by treatment with two aliquots of 1,2,4-triazole in the presence of triethylamine. Because of its extraordinary lability it is not isolated but entered immediately into the reaction. The protected nucleotide (23) is a key compound and can serve as a source, after selective elimination of protective groups, of both nucleoside (24) and nucleotide (25) components:

(23) —H^+→ HO–[B^{ac}]–O–$P(=O)(OC_6H_4Cl\text{-}p)$–$OCH_2CH_2CN$ (24)

(23) —^{-}OH→ $(MeO)_2TrO$–[B^{ac}]–O–$P(=O)(OH)$–$OC_6H_4Cl\text{-}p$ (25)

Compounds (24) and (25) are then entered into a reaction with a condensation agent, usually tetrazolide of arylsulfonic acid. The reaction is conducted in absolute pyridine for one or two days as follows:

(24) + (25) —1. $ArSO_2Te$; 2. Silica gel chromatography; 3. ^{-}OH→ $(MeO)_2TrOCH_2$ B^{ac} … Cl–C_6H_4–$P{=}O$ … O–CH_2 B^{ac} … Cl–C_6H_4–$P{=}O$, O^- or $(MeO)_2TrO$–[B^{ac}]–O–$P(=O)(OC_6H_4Cl)$–[B^{ac}]–O–$P(=O)(O^-)$–OC_6H_4Cl (26)

In this case, in contrast to the phosphodiester method, the condensation proceeds at a slower rate. This seems to be due to the steric hindrance to the reagents (condensation agent and nucleotide component) and (which may be the main reason) a different mechanism of substitution at the nucleotide phosphate. The active form of the phosphorylating agent here is tetrazolide (27) of the nucleotide component, which is formed immediately after elimination of tetrazole according to the scheme:

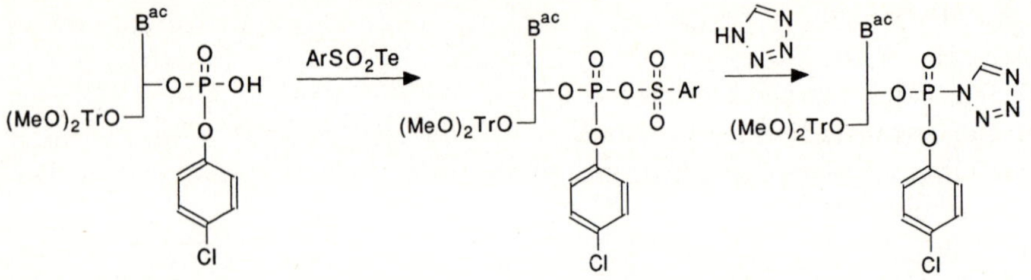

Compound (27) must be much less reactive than the pyridine-stabilized metaphosphate used as the condensation agent in the diester version of synthesis.

Oligonucleotide (26) may be further entered into the condensation reaction either with a nucleotide of type (24) to give oligonucleotides containing a phosphate group at the 3′ end or with the protected nucleoside (29) to yield trinucleoside diphosphate (30) according to the following scheme:

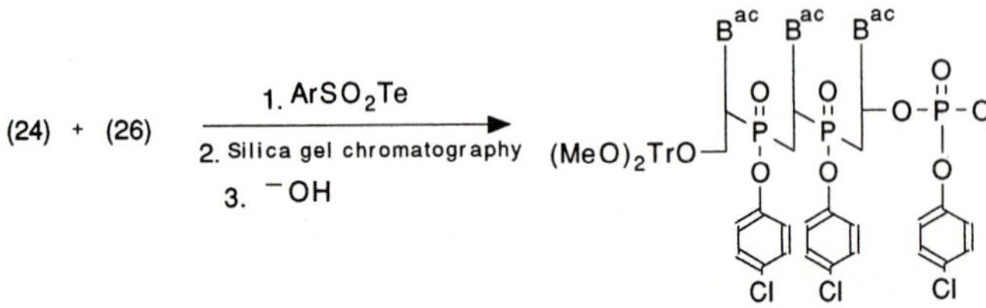

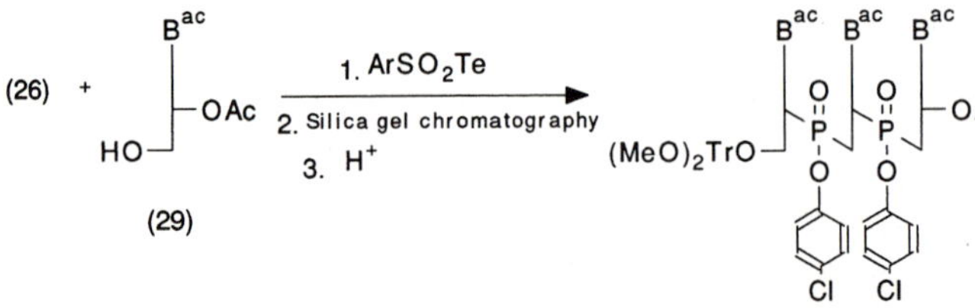

Condensation of (28) and (30) type compounds results in oligonucleotides with six monomer units which can be used in a similar fashion for further elongation of the oligonucleotide chain. Here is a standard scheme of the synthesis of 10- to 20-unit oligonucleotides.

(31)

(32)

IV)-VII) → T-G-A-G-T-G-A-G-C-T-A-A-C-T-C-A

The protective group for the internucleotide phosphate is selected in such a way as to ensure its easy removal at the end of the synthesis. For selective removal of the acid-labile $(MeO)_2Tr$ group an oligonucleotide is treated with benzenesulfonic acid in chloroform at 0 °C. Under such conditions, the depurination of a protected oligonucleotide is virtually nil.

A selective removal of the 2-cyanoethyl group in compounds of type (24) is achieved by treatment with 0.2 N sodium hydroxide in dioxane, when oligonucleotides are short, and with anhydrous triethylamine in pyridine, when they are long. All condensations are carried out using equivalent amounts of the nucleoside and nucleotide components in 2,3,5-triisopropylbenzenesulfotetrazolide (TPSTe) as the condensation agent. After each condensation, the oligonucleotide is separated by silica gel chromatography.

A comparison of the triester and diester methods reveals the most important advantages of the former: fewer side reactions, stability of yields in each step of the synthesis even when long oligonucleotides are formed, and ease of their purification by adsorption chromatography on silica gel.

Within a short period of time, however, the accumulated experience with the above-described triester synthesis procedure brought to light some of its drawbacks. The main problem, lowering the synthesis rate, has to do with the separation of fully protected oligonucleotide segments by silica gel chromatography, and the longer the chain, the worse the separation (this depends also on the nucleotide sequence). Itakura and coworkers proposed to eliminate the intermediate purification of oligonucleotide segments and, accordingly, altered slightly the condensation conditions. In order to bring the condensation to a quantitative level, the nucleotide component was entered into the reaction in a doubly excess amount. Notably, the entire synthesis was conducted by combining trinucleotide segments as shown below:

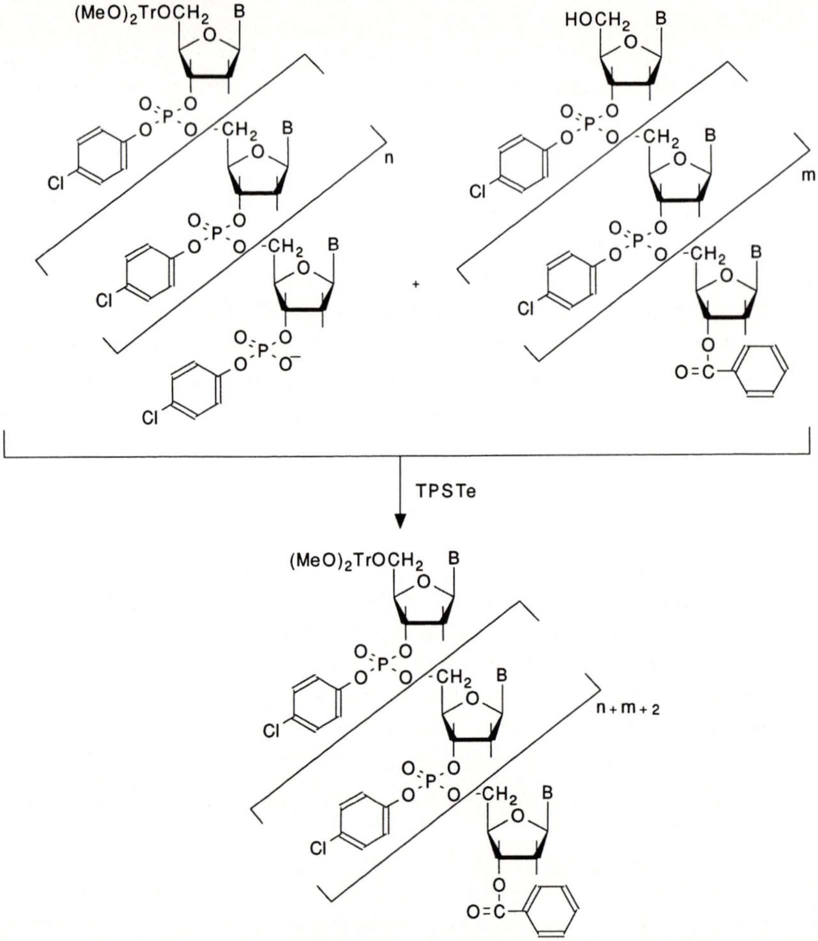

TPSTe

where B = T, A^{bz}, C^{bz}, G^{ib}
n = 1 or 4
m = 4.7 or 10

The excess of the nucleotide component was removed by treatment with an ion-exchange resin (DOWEX 1x8 or DEAE cellulose) in aqueous organic solvents, and a fully protected oligonucleotide was separated after chromatography on very wide (7 cm in diameter) and short (3 cm) columns. When this oligonucleotide was treated with a 2% solution of benzosulfonic acid in a methanol–chloroform (3:7) mixture to remove the $(MeO)_2Tr$ group, the resulting one (nucleoside component) was used to further build up the chain without additional purification. Assembling nucleotides from trinucleotide segments makes it possible to easily separate the end product from the mixture of intermediate oligonucleotides at the last stage, after elimination of all protective groups, by liquid chromatography on Permaphase AAX. According to the authors, all these improvements allow one, when trinucleotide segments are available, to speed up several times the synthesis of oligodeoxyribonucleotides of a desired structure.

As was pointed out in the introduction to this section, new nucleophilic catalysts (*N*-methylimidazole, N-oxides of substituted pyridines) have enabled the phosphodiester method to be considerably improved and expand the spectrum of modern techniques based on trivalent phosphorus compounds.

When *N*-methylimidazole is used (in an amount twice that of arylsulfonyl chloride) in synthesis according to the scheme just described, the reaction takes 5 to 10 minutes in the case of short oligomers and 20 to 40 minutes in the case of 16- to 20-unit ones. The level of modification of the heterocyclic bases has gone way down, which makes purification of the target oligonucleotides easier.

Even more advanced is the procedure proposed by V. A. Yefimov and coworkers, whereby condensations are carried out in the presence of *O*-nucleophilic calalysts of the pyridine N-oxide type. The reaction rate was maximum when N-oxides of 4-dimethylamino- and 4-alkoxypyridines were used. The internucleotide condensation takes only a minute in the presence of arylsulfonyl chloride and such a catalyst – that is, it proceeds at a rate several times faster, as compared to *N*-methylimidazole. In spite of their low basicity (pK_a 2–4), N-oxides of pyridines exert not only nucleophilic but also basic catalytic effect and act as acceptors of the acids evolving in the course of the reaction. This is why the latter can be conducted not only in pyridine but also in other organic solvents. Here is a possible course that the reaction may take:

RO-P(O)(OR')-OH
RO-P(O)(OR')-O-N⁺(pyridine)-X
R''OH
RO-P(O)(OR')-OR''
Ar-SO₂-Cl
+
O←N(pyridine)
Ar-SO₂-O-N⁺(pyridine)-X
⁻Cl

A further improvement of this technique was using a protective phosphate group as a catalyst. Such a group speeds up substantially the internucleotide condensation in the presence of arylsulfonyl chlorides. The increase in the phosphodiester bonding rate is most likely due to formation of an active cyclic intermediate:

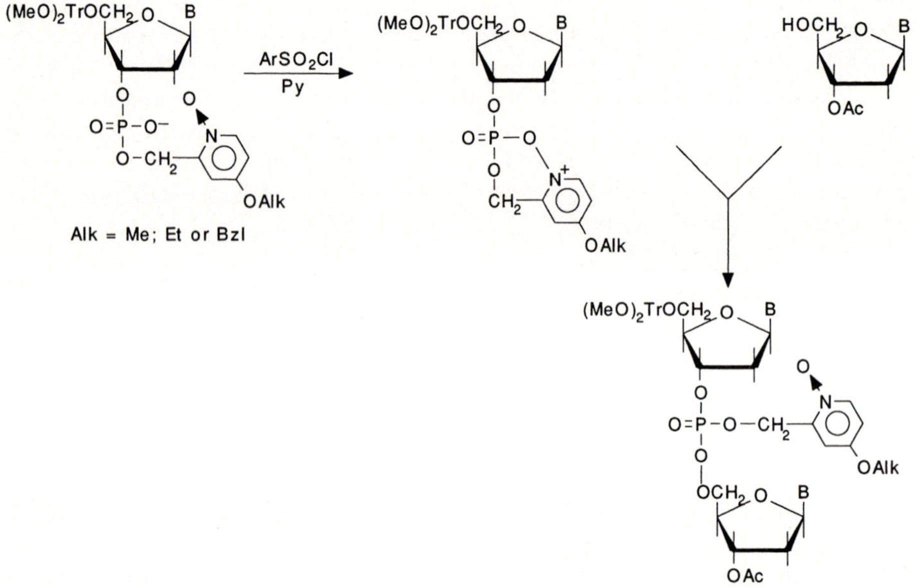

Condensation involving such intramolecular catalysts and $ArSO_2Cl$ is over within less than a minute. Studies into the stability of *O*-nucleophilic protective groups have shown that they are stable toward acid reagents used to remove the 5′-trityl groups and also during internucleotide condensations. They are eliminated, upon completion of oligonucleotide synthesis, in the presence of triethylammonium thiophenolate or piperidine. Advantages of this modification of the phosphotriester method include not only a very high speed of synthesis but also a very low level of side processes (5′-sulfonylation, heterocycle modification).

Monomers carrying catalytic phosphate-protective groups are prepared from the corresponding fully blocked nucleoside phosphotriesters by selective removal of the noncatalytic aryl or 2-cyanoethyl phosphate-protective group. Synthesis of the starting triester may take one of the following routes:

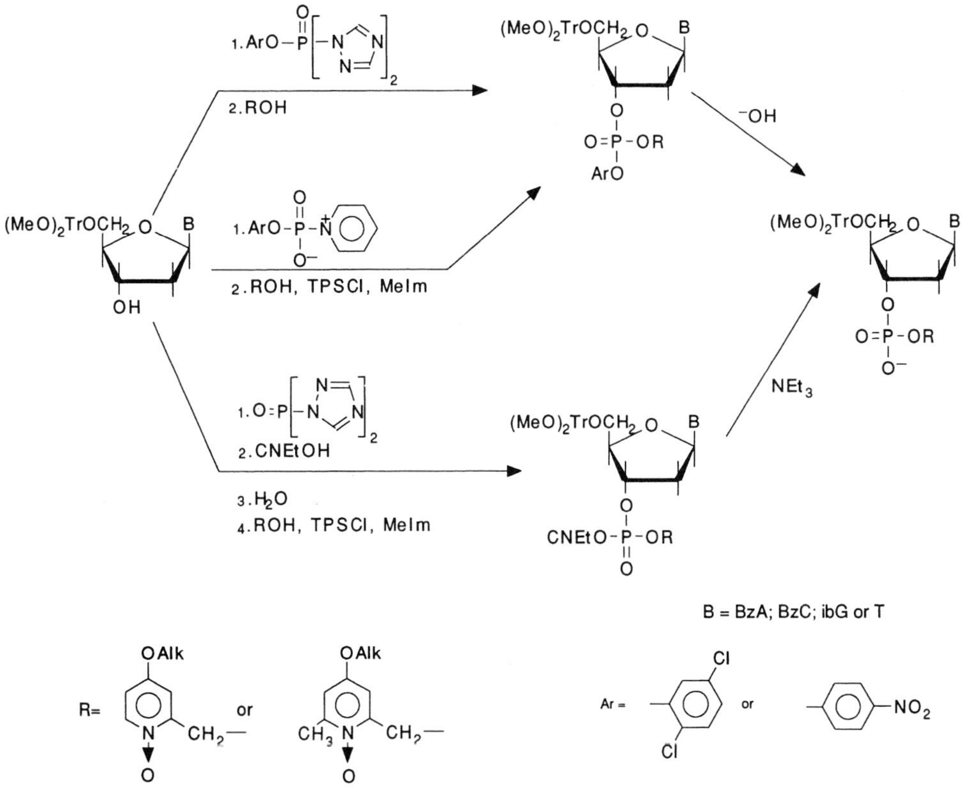

This particular modification of the phosphotriester method has been successfully used for automated solid-phase synthesis as well.

Now that simpler and more effective methods for creating internucleotide linkages with the aid of trivalent phosphorus derivatives are available, interest in the phosphotriester method has faded perceptibly. When the latter is used for synthesis in solution, however, its importance remains unabated for preparation of moderately long (15 to 20 b. p.) oligonucleotides, which is essential in developing new generations of oligonucleotide-based drugs.

In this section we have not touched upon the highly important aspects of individual steps of synthesis, such as selection of protective groups, their selective and complete removal, and the side processes accompanying each of these steps. Detailed and competent discussion of these matters can be found in Reese's review (see References).

11.4.2.3 Phosphite triester Method

In 1975, Letsinger proposed a new method for creating internucleotide linkages. Its main difference from the previously described methods was the use of trivalent phosphorus derivatives which are highly effective phosphorylating agents. What makes this method especially attractive is the fact that phosphorylation of the hydroxyl group proceeds at a rather fast rate (1–2 min) and a very low temperature (–78 °C):

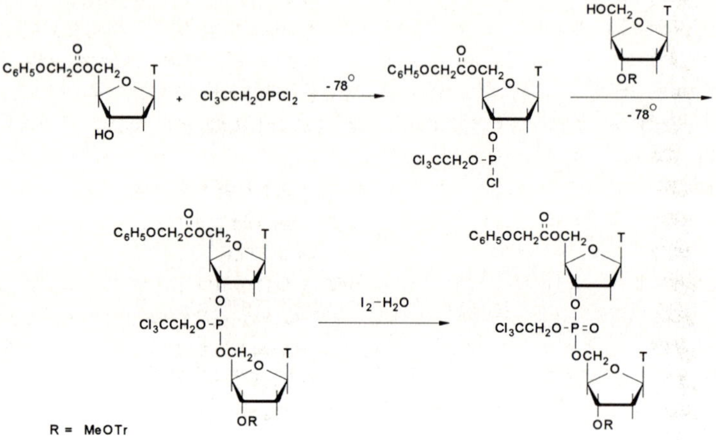

That first attempt had laid the foundation for the most fruitful approach in oligonucleotide synthesis, which became the most widely used method to synthesize oligo- and polynucleotides already in the early eighties.

As can be seen from the above scheme, the method includes the phosphorylation of a 5′-protected nucleoside with alkyl dichlorophosphate and another nucleoside with the resulting active derivative of the 5′-hydroxyl, followed by oxidation of the internucleotide phosphite group to the corresponding phosphate. Various alkyls and aryls were used at the early stages as protective groups in phosphites. The methyl group has turned out to be the most convenient by virtue of its easy removability after synthesis through treatment with trialkylammonium thiophenolate. Moreover, methylphosphite diazolides marked by greater stability started being used:

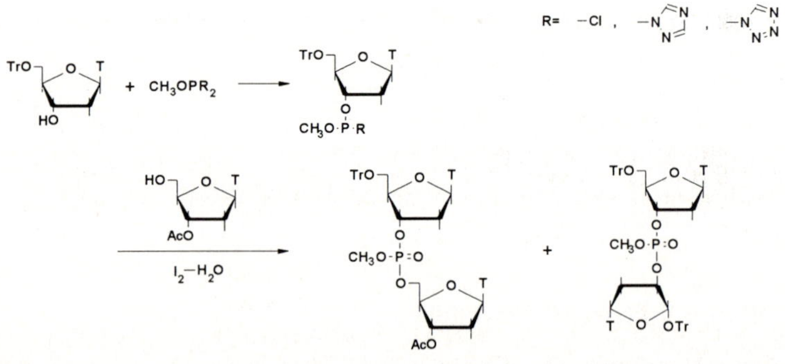

The most serious drawbacks of the processes illustrated above is the need to conduct the reactions at an extremely low temperature, as well as formation of by-products with an internucleotide 3′–3′ bond. Besides, phosphite derivatives were found to be rather unstable and sensitive to even traces of moisture.

A major contribution to the evolution of the phosphite method for creating internucleotide linkages has been made by Caruthers. The di(isopropyl)amide or morpholide derivatives of phosphites proposed by him in the early eighties turned out to be much more convenient in use. They are characterized by moderate reactivity (reactions involving these compounds proceed effectively at room temperature), long shelf life even at room temperature, and lack of hygroscopicity.

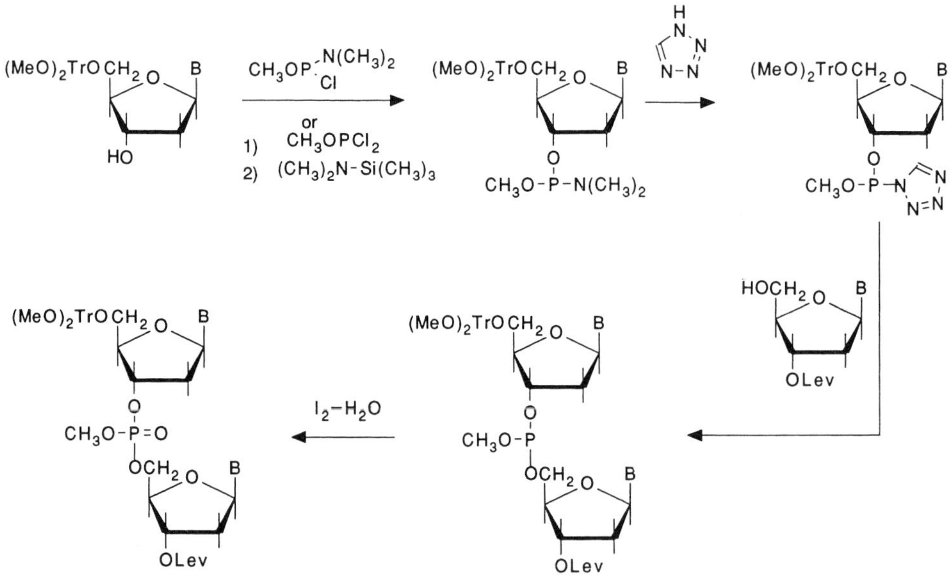

Lev – levulinic acid residue

These and similar phosphorylating agents are employed most extensively in solid-phase synthesis of oligonucleotides.

11.5 Solid-Phase Synthesis of Oligodeoxyribonucleotides and Its Automation

Beginning from the mid sixties, the classical organic synthesis of oligonucleotides in solution evolved hand in hand with synthesis in a biphase system or, to be more precise, on insoluble polymer supports. This radical approach was for the first time proposed by Merryfield in 1962 for polypeptide synthesis.

Not long after the first results of its implementation as a means for building up a peptide chain had been obtained it became clear that it suits ideally oligonucleotide synthesis as well.

To begin with, no racemization can take place at the intermonomer node during formation of an internucleotide linkage (at least after removal of the protective group from the internucleotide phosphate in its synthesis by the triester method), and, consequently, there is no need to separate isomers after each monomer addition step. This drawback of solid-phase peptide synthesis at the early stages did not allow optically active polypeptides to be produced with good yields. Secondly, early in the evolution of oligonucleotide, as opposed to peptide, synthesis, long polymers were not needed during DNA synthesis. It was quite sufficient to synthesize 8- to 12-unit fragments to be used as building blocks for extended DNA duplexes. This is precisely why work on solid-phase synthesis of oligonucleotides began already when the phosphodiester method provided the only possibility to form internucleotide linkages – a rather ineffective procedure which did not allow each reaction of monomer addition to the emerging oligomer chain to be conducted with a good yield (exceeding 90%). In spite of the rather modest results, three university laboratories embarked on search for solutions to the many problems involved in oligonucleotide synthesis on polymer supports: in Evanstone (Letsinger), in Moscow (Shabarova) and in Heidelberg (Krahmer). The efforts of their research workers (Caruthers, Potapov, Köster, etc.) turned out to be consequential in the subsequent development of the most effective up-to-date method for automatic synthesis of oligo(poly)nucleotides. Later (in the mid seventies), this group of scientists was joined by representatives of the Cambridge school in England (Gait), where advances in the phosphotriester method were implemented within a short period of time. As far back as 1977, Gait published a seminal paper in NAR, describing the synthesis of 7- to 9-unit oligonucleotides some of which contained all the four DNA monomers, while synthesis of only 3- to 7-unit fragments, containing primarily the thymidine nucleotide, was reported by that time. The authors of the second paper on the subject, which made its appearance in NAR in 1979, were from Moscow University. They not only reported on synthesis of 10- to 15-unit nucleotides, including heterogeneous ones, but also presented the first map showing the operations involved in solid-phase synthesis and a schematic illustration of a semiautomatic synthesizer. The latter was a prototype of the synthesizer "Victoria" designed soon after in Novosibirsk. The next important event after these publications was the international symposium in Hamburg (Germany) held at Köster's initiative. It brought together all leading authorities on oligonucleotide synthesis. Many scientists with previous contributions to synthesis in solution spoke about synthesis on polymer supports. The culmination of the symposium was recognition of the solid-phase method as the most promising because it lends itself ideally to automation. Caruthers presented an important paper on highly effective application of the phosphite triester procedure to solid-phase synthesis. The years that followed were marked by extraordinary strides in solid-phase oligonucleotide synthesis and its automation.

11.5.1 Basic Principle of Solid-Phase Synthesis

The basic principle of the method is that synthesis is conducted in a two-phase system. One of the components (usually the nucleotide one) is covalently bound on a specially modified insoluble polymer and, consequently, is on the solid phase. The rest of the components are in solution. After condensation it is the oligonucleotide that is on the solid phase.

P ~ R-O + p OAc → P ~ R-O p OAc —⁻OH→

→ P ~ R-O p OH → P ~ R-dN¹pdN²pdN³p -- Nⁿ —H+→

→ P ~ R + oligodeoxyribonucleotide

P — polymer support R - anchor group

By simple filtration and washing it is separated from the reaction mixture and subjected to subsequent manipulations. Once the 3′-blocking group is removed, such a polymer-oligonucleotide is suitable for further condensation. As the chain continues to be built up, the target oligonucleotide is always on the solid phase from which it is separated only at the end of the synthesis. The complex oligonucleotide molecule is "assembled" in this manner, too.

What makes solid-phase synthesis of oligonucleotides particularly attractive is the drastic decrease in the time it takes, which is made possible by the extremely simple procedure of treating the reaction mixture and, what is most important, the fact that the reaction product does not have to be isolated after condensation (this is what takes most time). The simplicity of the procedure involving a polymer support makes it possible to make use of any excess reagents and also to repeat the treatment of immobilized molecules. Evidently, apart from taking much less time on the whole, the solid-phase method is the only one in which all operations can be standardized and the entire process automated.

The method, however, suffers from some shortcomings.

The steric hindrances inherent in the synthesis slow down somewhat the rates of all chemical reactions. The nonquantitative course of the reactions results in products with shorter chains accumulating on the polymer in addition to the target oligonucleotide. It is thus a rather serious problem to separate oligonucleotides after their removal from the polymer.

As suggested by the described principle of the solid-phase synthesis (we are considering it only in the context of oligonucleotides but it is equally applicable to any organic compounds), this process can be easily automated because it is essentially an iteration of three basic operations:
(1) covalent bonding;
(2) washing of the polymer;
(3) preparation of the immobilized molecule for the next step.

Elaboration of an effective procedure of solid-phase oligonucleotide synthesis, which would lend itself readily to automation, boils down to solution of at least two key problems:
(1) optimization of the support's macromolecular structure;
(2) selection of the right method to create the internucleotide linkage.

Solving the second problem was especially important because the entire synthesis hinged on it. If any of the synthesis steps is ineffective, the probability of obtaining an extended oligonucleotide is extremely low. For example, the yield in each step being 70%, the overall yield of a hexanucleotide will be 12%. At an 80% yield, one can obtain a decanucleotide with the same yield. And if the yield during the formation of each internucleotide bond is 90%, a 20-unit polynucleotide can be obtained with the same yield. In other words, solid-phase synthesis of oligonucleotides calls for condensation procedures with yields of at least 90%.

11.5.2 Polymer Supports and Immobilization of the First Monomer Thereon

For successful outcome of reactions involving a polymer support, the latter must meet some general requirements. It must not influence the course of the synthesis. This requirement is practically hard to meet because condensations occur within the polymer which acts as some kind of a solvent and whose interactions with the immobilized nucleotide component and molecules in solution materially affect their properties. The support must exhibit minimal adsorptivity to ensure rapid and quantitative washing of the polymer to get rid of excess reagents and solvents; it must also be large enough to permit free diffusion of the reagents. At the same time, it must be mechanically and chemically stable as well as "compatible" with the solvents in which the synthesis takes place (solvatable) and also with the growing oligonucleotide chain. Another important consideration is how quantitative all reactions involving immobilized molecules (condensation, removal of the protective groups) are.

Practical introduction of polymer supports into oligonucleotide synthesis must be credited to Letsinger who did it in the mid sixties. Before the seventies, polystyrene-type supports were employed in all experiments. These are readily available polymers carrying aromatic groups and can be very easily modified through introduction, usually rather simple, of an anchor group (to which the first nucleotide unit is attached). Such polymers have a long shelf life and undergo no changes in the course of oligonucleotide synthesis.

In the early seventies, V. K. Potapov proposed a new type of support for oligonucleotide synthesis, in which the best properties of previously studied polymer matrices were combined. The support was produced by chemical modification of polystyrene grafted onto the surface of polytetrafluoroethylene under radiation. The amount of grafted polystyrene can be easily controlled within 10 to 50 % of the teflon weight.

TEFLON $-(CF_2)_n-$ $-CH_2-CH-CH_2-CH-$ anchor group MeOTr OCH_3 $C-NpNpNp\cdots N$
chain of oligodeoxynucleotide
$CH_2CHC_6H_5$
[wwww] — polystyrene chain

Transferring the reaction sites to the bead surface minimizes the role of diffusion processes, whereas the high conformational mobility of polystyrene chains not joined by divinylbenzene cross-linking mitigates steric hindrances both during modification of the polymer and in the course of oligonucleotide synthesis. The support displays low adsorption activity and swelling because the teflon matrix is totally inert with respect to all kinds of reagents and solvents. An outstanding feature of grafted supports, setting them apart from any cross-linked polymers, is the possibility of a high degree of modification (30–50 %) by introducing anchor groups into such a support. Already in 1975–78, heterogeneous 8- to 10-unit oligonucleotides were obtained on this polymer support. Another support with a similar performance in oligonucleotide synthesis was polydimethylacrylamide proposed by Gait, which is embedded in macroporous kieselguhr. Just as the two supports mentioned earlier, this polymer may be quite useful in preparative synthesis of oligonucleotides. Yet another support, cellulose, is employed, similarly to polystyrene grafted to teflon and polyacrylamide/kieselguhr, only in the laboratories of origin. However, cellulose is used in the method of simultaneous synthesis on disks (see below). In the early eighties there appeared new types of supports having similar properties but more readily available, which accounts for the wide popularity they have gained in oligonucleotide synthesis. The main two types of supports are:

Controlled Pore Glass. This support is used more often than anything else for producing microamounts of oligonucleotides in automatic synthesizers. The size of pores in such a glass may be 240, 500, 1400 or 3000 Å. Glass with 500 Å pores is most frequently used. Aminopropyl is employed as the "anchor" group (a group mediating between the polymer and the first nucleotide molecule from which the oligonucleotide chain starts to grow).

Silica Gel. This is another widely used support in automatic synthesis with aminopropyl performing the function of the anchor group.

The almost universally adopted procedure of building up an oligonucleotide chain in solid-phase synthesis is based on a scheme with the starting point at the 3′ end of the oligonucleotide; in other words, the synthesis proceeds in the 3′–5′ direction. The main reason for this choice is the experimentally established fact that the rate of the reaction between the 3′-phosphate of the nucleotide component and 5′-hydroxyl of the nucleoside component is four times that between the 5′-phosphate and 3′-hydroxyl. This is why the first nucleotide unit is immobilized on the polymer support via its 3′-hydroxyl. To this end, one must first of all prepare a succinate of 5-dimethoxytrityl nucleoside, in which the second carboxyl is activated (in the form of an activated ester). The following scheme illustrates the sequence of steps in the preparation of modified silica gel and immobilization of the first nucleoside on its surface.

As a result of acylation of the amino group in the modified silica gel, the polymer and the first nucleotide unit are linked via an amide bond known for its high stability.

11.5.3 Phosphoramidite-Triester Method

Solid-phase synthesis of oligonucleotides has become much more effective since the early eighties, after radically new methods for forming internucleotide linkages were introduced, based on trivalent phosphorus compounds. The first to be introduced (see scheme below) was the so-called phosphoramidite method in which the nucleoside component is deoxynucleoside 3′-phosphoramidite.

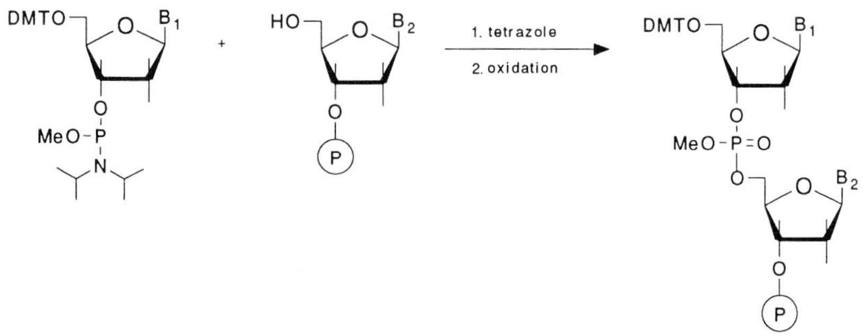

The phosphorylating function in the nucleotide component is performed by the trivalent phosphorus atom. Being intrinsically stable, phosphoramidites may enter into nucleophilic substitution reactions only after protonation. Caruthers, who elaborated this method for practical applications, proposed tetrazole (a very mild reagent and a proton donor) to be used for acid catalysis. The condensation proceeds almost quantitatively within 30 to 40 seconds. The next step is oxidation of the internucleotide phosphite group or, in other words, conversion of the trivalent phosphorus into a pentavalent one, which is also very fast (1–2 min) and quantitative in the presence of iodine and water. Significantly, the end product of such a two-step process (condensation, oxidation) is phosphotriester – that is, a compound structurally similar to the internucleotide phosphate forming during synthesis by the triester method. The protective group at the phosphate is a methyl easily and quantitatively removable by thiophenol. Notably, by virtue of the fast reaction rates this scheme can be used with any success only in solid-phase synthesis of oligonucleotides.

The following scheme with abridged formulas illustrates a single reaction cycle whereby a monomer becomes coupled to the polymer chain. Repetition of such cycles leads to extension of the oligonucleotide chain.

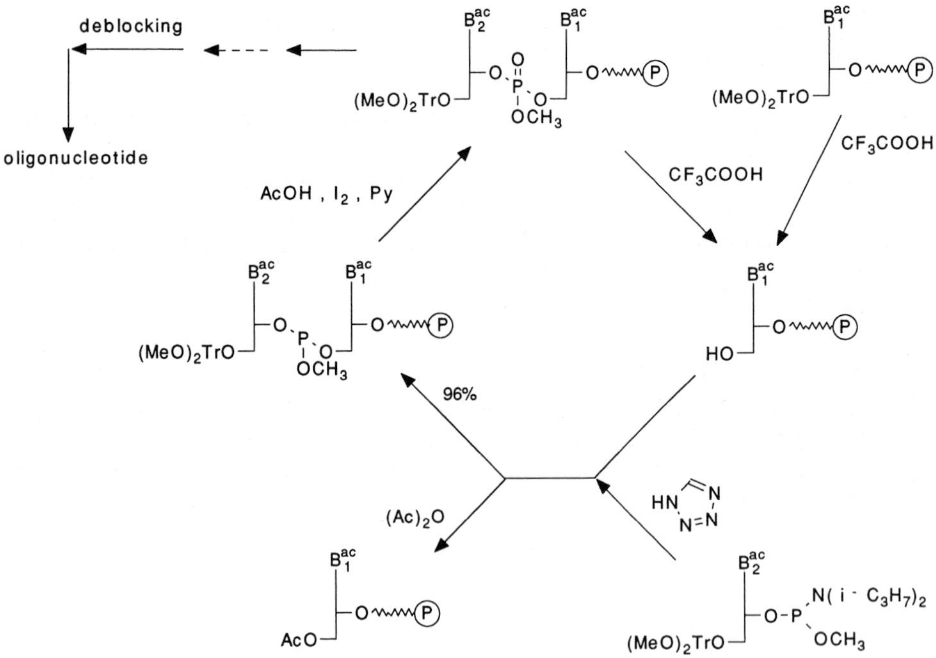

The first step of the cycle is removal of the $(MeO_2)Tr$ group from the 5′-hydroxyl (here, from the nucleoside coupled to the polymer support). To accomplish this, the polymer is treated with trichloroacetic acid (very mild conditions, not exceeding 1 min). After washing, protected nucleoside phosphoramidite and tetrazole are coupled to the support. The protonated amino group departs easily, and the phosphorus atom becomes sensitive to nucleophilic attack. As a result, a second monomer in phosphite form is coupled to the 5′-hydroxyl of the nucleoside associated with the polymer support. At this stage, the internucleotide phosphorus is still trivalent.

It may happen that the coupling reaction will not be quantitative (96–98 %). Some molecules of the nucleoside component coupled to the support may stay out of a reaction with phosphite. If they are left unreacted, their 5′-hydroxyls may enter into the reaction during the next step of coupling. This will give rise to an oligomer shortened by a unit (and so in each cycle), which will make it difficult to separate the mixture after synthesis. In order to rule out such impurities, the unreacted nucleoside component is blocked, or "capped". This is done by acetylation with acetic anhydride in the presence of *N*-methylamidazole and 2,6-lutidine. The 5′-acylated nucleoside component coupled to the support is now eliminated from the synthesis and remains uninvolved all the way until the oligo(poly)nucleotide chain is fully extended. After dimethoxytritylnucleoside phosphoramidite has been coupled, the phosphorus is trivalent. Therefore, the last step is oxidation, for which purpose iodine and water are used in the presence of a base (lutidine).

These five steps constitute a chain extension cycle which keeps being repeated.

Thus, the method resides in sequential extension of an oligonucleotide chain by using a respective activated monomer. The complete cycle takes three to seven minutes.

Once synthesis is over, the protective methyl groups are removed from the internucleotide phosphates (thiophenol), the oligonucleotide is removed from silica gel (aqueous ammonia, 20 °C, 20 min), and, finally, the heterocycles are deprotected (concentrated ammonia, 60 °C, 24 h). The dimethoxytrityl group blocking the 5′-hydroxyl serves several purposes:

(1) blocking of the 5′-hydroxyl throughout the reaction cycle when a nucleotide is coupled;
(2) determination of the yield of each step after decoupling of the $(MeO_2)Tr$ cation whose characteristic absorption occurs at 490 nm;
(3) as a lipophilic group during chromatographic separation of the reaction products (the target product is the only one having a $(MeO_2)Tr$ group at the end of the synthesis). This is especially important in the synthesis of extended oligonucleotides (with 20 and more units).

The advantages of the phosphoramidite method include the following:

(1) high yields in every step (96–98 %), which allow, firstly, the oligonucleotide chain to be extended by adding monomers, and that means easier preparation of the starting materials (there is no need to synthesize dimers or trimers in advance), and, secondly, the synthesis to be conducted at the micromole scale and purified 30- to 40-unit oligonucleotides to be obtained in an amount of 0.1 to 0.2 mg;
(2) lack of hygroscopicity of diisopropylamides of the protected nucleoside phosphites (there is no need to dry them prior to synthesis) and their stability in an anhydrous solution for two and weeks more, which allows the nucleotide component to be used in an absolute organic solvent over a long period of time;
(3) it takes only five to ten minutes for an internucleotide linkage formation cycle to be completed in the synthesizer.

The above scheme of extending an oligo(poly)nucleotide chain facilitates purification of the desired product. The post-condensation capping of the unreacted 5′-hydroxyl of the nucleoside component precludes growth of defective chains. It is only the target product that has a dimethoxytrityl group at the end of its chain, which makes its separation by reversed-phase high-performance liquid chromatography extremely easy. Detritylation (80 % acetic acid, 20 °C, 2 min) yields a rather pure oligonucleotide. This procedure for building up oligonucleotide chains has become routine in laboratories equipped with automatic synthesizers.

11.5.4 Hydrophosphoryl Method

The other method based on trivalent phosphorus compounds is the hydrophosphoryl one shown schematically below:

(MeO)$_2$TrOCH$_2$ O B

O

O=P-O$^-$

H +

HOCH$_2$ O B

O

P

(CH$_3$)$_3$C-C(=O)Cl

(MeO)$_2$TrOCH$_2$ O B

O

O=P-H

OCH$_2$ O B

O

P

oxidation

(MeO)$_2$TrOCH$_2$ O B

O

$^-$O-P=O

OCH$_2$ O B

O

P

The phosphate in the nucleotide component is activated with the aid of such condensation agents as triisopropylbenzenesulfonyl chloride or tetrazole and *N*-methylimizadole. Experience has shown, however, that the best agent is pivaloyl chloride (PvCl) which brings the condensation yield up to 97–100%. The oxidation in the next step is performed with iodine in the presence of water. It has been found that dialkylphosphonates are oxidized with halogens during base catalysis. The general equation of the reaction involving iodine takes the form:

$$(RO)_2PHO + I_2 + 3^-OH \rightarrow (RO)_2PO_2^- + 2I^- + 2H_2O$$

It has been demonstrated that the oxidation of the internucleotide phosphate group with iodine proceeds faster in the presence of *N*-methylimidazole, *N*-methylmorpholine, or triethylamine. The most universally accepted procedure involves a 2% iodine solution in a pyridine–water (98:2) system. Here is a scheme with abridged formulas illustrating a reaction cycle in which a monomer is coupled to a nucleoside immobilized on a polymer support:

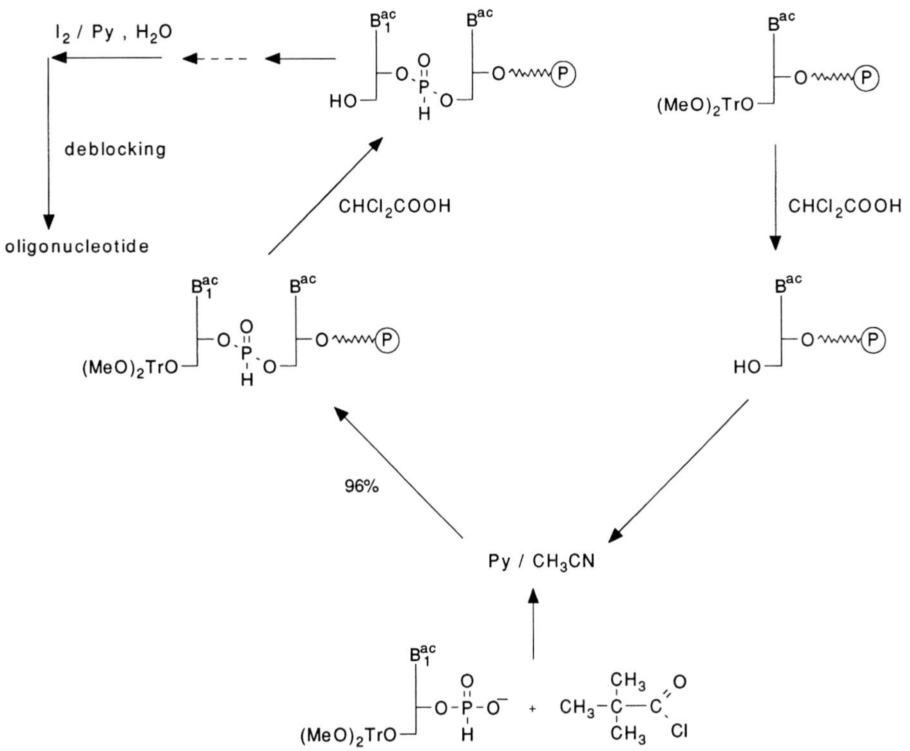

In the first step of this cycle, five aliquots of the phosphonate component and an aliquot of the 5′-hydroxyl nucleoside component condense in the presence of 50 mM PvCl in an anhydrous pyridine-acetonitrile (1:1) mixture for five minutes. After preliminary detritylation of the nucleoside component in the presence of 2.5% dichloroacetic acid in methylene chloride, the next condensation is carried out for two minutes. Then, after repeating the required number of cycles, the next step is oxidation with the following mixture: (a) 0.1 M I_2 in pyridine/*N*-methylimidazole/H_2O/tetrahydrofurane (5/1/5/90) or (b) 0.1 M I_2 in triethylamine/H_2O/tetrahydrofurane (5/5/90). In either case, the oxidation is over within two and a half minutes. Removal of the oligonucleotide from the polymer and protective groups from the heterocycle is achieved by treatment with concentrated ammonia (5 h, 55 °C). The main difference between the hydrophosphoryl method of solid-phase synthesis of oligonucleotides and the above-described phosphoramidite one is the absence of a protective group at the internucleotide phosphate in the product of condensation and subsequent oxidation. Moreover, the reactions of the cycle proceed with 98 to 100% yields. This allows the cycle to be shortened by several steps, and the absence of the step in which the protective group is removed from the internucleotide phosphate lowers the probability of side processes and the presence of lingering methyl groups in the sugar-phosphate backbone. All this has made the hydrophosphoryl method a procedure of choice for automatic synthesis of oligodeoxyribonucleotides.

The starting hydrophosphoryl derivatives of protected deoxynucleosides are synthesized by treatment of the corresponding *N*-acylated 5′-dimethoxytrityl nucleosides with a mixture of phosphorus trichloride, 1,2,4-triazole and *N*-methylmorpholine in anhydrous methylene chloride. The resulting hydrophosphoryl derivatives are rather stable and can be stored for a long period of time at room temperature.

11.5.5 Phosphotriester Method

The above-described phosphotriester method of oligonucleotide synthesis, based on intramolecular catalysis, has also proved successful with polymer supports:

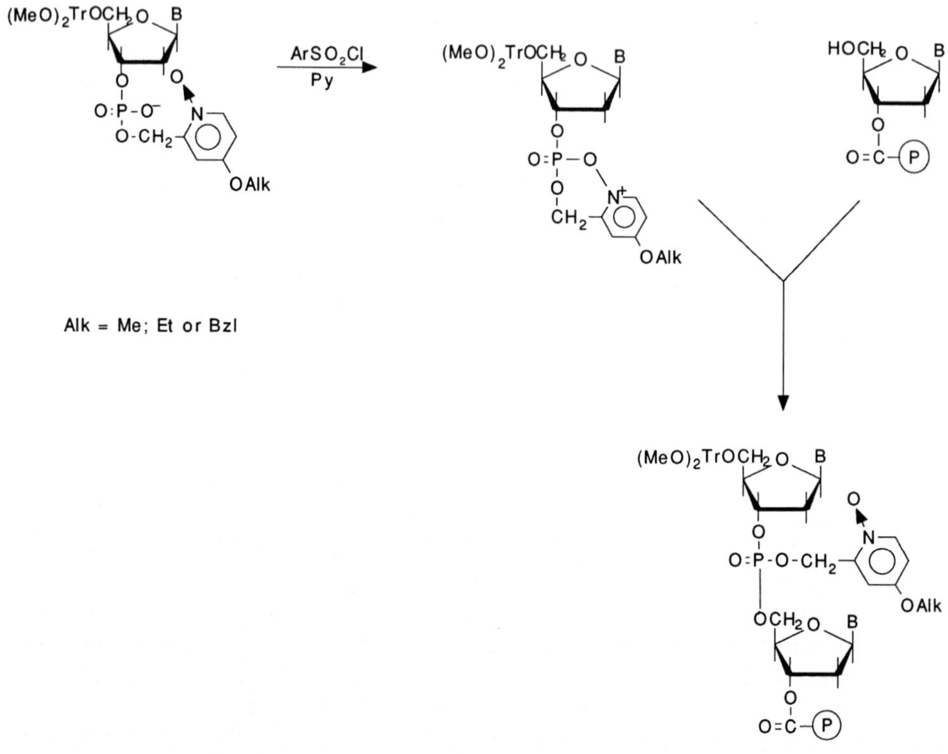

Alk = Me; Et or Bzl

In this case, the fastest rate (30–60 s) of phosphotriester bond formation is ensured by using 1-oxide-4-alkoxy-2-picolyl derivatives in the presence of TPS. The yield of each condensation step is 98 to 99%. The duration of one cycle is six minutes – that is, it is the same as in the phosphoramidite method.

The phosphate-protecting groups are removed from the synthesized oligonucleotides by treatment with piperidine. Notably, the synthesis can be performed virtually on any type of modern synthesizers, and the requirements of solvent drying are not as stringent as in the phosphoramidite case.

11.5.6 Automation of Synthesis

The standard nature of operations in the solid-phase method has led to automation of chemical oligonucleotide synthesis. The following diagram illustrates a general approach used in automatic synthesizer design.

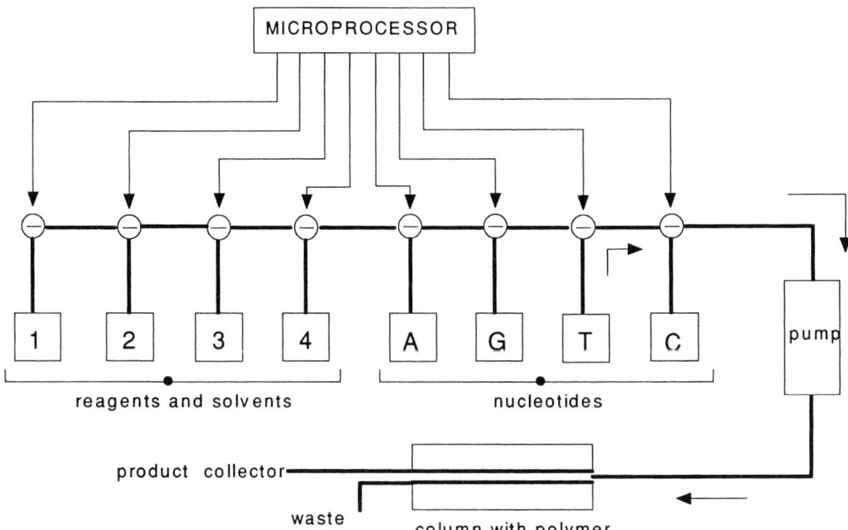

The operating principle of such a synthesizer is as follows. A microprocessor controls the pumping of protected nucleotide components (A, G, C and T), reagents, and solvents (1–4). All reagents are successively fed to a column containing a polymer support with the first nucleoside immobilized thereon. The synthesis and separation of the fully protected oligonucleotide from the polymer support are followed by deprotection, purification and analysis of the synthesized DNA fragments. Table 11-2 presents, by way of example, the sequence of operations to be performed in a synthesizer by the hydrophosphoryl method. The times listed in the table can be shortened. For some synthesizers a cycle takes three to five minutes.

The first Soviet synthesizer “Victoria” was developed already in 1980. At present, synthesizers “Victoria-5M” are produced by a special electronics and analytical instrumentation bureau under the USSR Academy of Sciences. In 1981–85, what are considered to be the best synthesizers in the West were

Table 11-2. Solid-phase Synthesis of Oligodeoxyribonucleotides by the Hydrophosphoryl Method.

Step	Reagents and solvents	Time
	Chain Extension Cycle	
Detritylation	2.5 % $CHCl_2COOH$ in CH_2Cl_2	2 min
Washing	CH_3CN	1 min
Washing	CH_3CN–Py (1:1)	1 min
Condensation	monomer (40 mM solution) and PvCl (200 mM solution) in CH_3CN–Py (1:1) (9 times 13 s each)	54 s
Washing	CH_3CN–Py (1:1)	30 s
Washing	CH_3CN	1.5 min
	End of Cycle	
Oxidation	2 % I_2 in Py–H_2O (98:2)	10 min
Washing	CH_3CN	2 min
Detritylation	2.5 % $CHCl_2COOH$ in CH_2Cl_2	2 min
Washing	CH_3CN	3 min

purchased from such companies as "Applied Biosystems", "Biosearch" and, later, "Pharmacia". A number of other companies also supply synthesizers.

Some models permit simultaneous synthesis of three or four oligonucleotides. Automation allows 30- to 40-unit oligodeoxyribonucleotides to be synthesized within just a few hours.

11.5.7 Multiple Simultaneous Synthesis on Polymer Segments

An alternative promising approach to solid-phase oligonucleotide synthesis involves polymer segments or, more specifically, cellulose (paper) disks. This procedure, sometimes also referred to as segment method, makes it possible to obtain a great number of oligonucleotides at a time. In this case, an oligonucleotide with a predetermined primary structure is synthesized not throughout the entire volume of insoluble polymer beads but on a small piece (segment) of cellulose or another material. Such a segment is used to immobilize the first nucleoside by the same chemical procedures as in conventional solid-phase synthesis. Elongation of the oligonucleotide chain does not necessarily have to involve just one segment. It becomes possible to attach the next monomer (one out of four) to several immolized nucleosides. One can use several segments with coinciding first and second nucleotides from the 3′ end of the chain (the elongation proceeds in the 3′–5′ direction, just as in conventional solid-phase synthesis). To perform the first two elongation steps, the disks are placed in columns isolated from atmospheric moisture, and the nucleotide component is passed through them under the usual conditions. Simultaneous synthesis of many oligonucleotides is conducted in four columns accommodating all segments, in which the chain starts to grow from the same monomer (i.e., there is coincidence with the structure of the second nucleotide from the 3′ end). Each subsequent elongation step involves the same cycle. For this purpose, all segments intended for synthesis are marked with the numbers of the oligonucleotides to be synthesized, then four types of segments are prepared by immobilizing four nucleosides (the synthesis gives four types of oligonucleosides differing in the 3′-terminal nucleoside structure). After that, the segments are sorted out and placed in columns A, G, C and T, depending on the structure of the next nucleotide unit. Similarly, after the second step, the segments are sorted out and placed together in the same columns depending on the structure of the next unit. These operations are repeated as many times as necessary. This principle of simultaneous elongation of chains on six segments is illustrated below:

1 – T G A A A A . . .
2 – T C T T T T . . .
3 – T G C A A A . . .
4 – T C T T T G . . .
5 – T G A C A A . . .

	Steps					
	1	2	3	4	5	6
Column dA			1,5	1,3	1,3,5	1,3,5
Column dG		1,3,5				4
Column dC		2,4	3	5		
Column dT	1,2,3,4,5		2,4	2,4	2,4	2,4

Synthesis is conducted simultaneously in this case – that is, all disks are involved in attachment of the next monomer (2nd, 3rd, 4th, etc.) at a time. The asynchronous elongation method has been proposed for synthesis of a great number of rather large (up to 30–32 b.p.) oligonucleotides. In this case, one can use a single column with the four nucleotide components being passed through it consecutively one at a time. To this end, an automatic one-column synthesizer is employed. The following example illustrates a plan for synthesizing ten fragments of two genes, g_1 and g_3, of melanocyte-stimulating hormones (g_1-MSH and g_3-MSH).

A

```
 (1) TGG GCG GCC TGG
 (2) TAG GGT TGC CTT CAC TGG GTA TTG CAT GCC
 (3) GAT AAT TGG CTT TGC
 (4) A CGT CCA GGC CGC CCA GGC ATG CAA TAC CCA
 (5) GTG AAG GCA ACC CTA GCA AAG CCA ATT ATC TTA
 (6) ACG ACG ACG ACG ACA ATG CTG CTG GCT TTG C
 (7) GAT AAT GAC TGG GCG TGG CGA CGA TGG
 (8) GTG AAG GCA ACC CTA GCA AAG CCA GCA GCA
 (9) TTG TCG TCG TCG TCG CCA TCG TCG CCA CGC
(10) CCA GTC ATT ATC TTA A
```

B

```
 010203040506070809101112131415161718192021222324252627282930313233

T    GG       G   C   G     GCCT  G  G

T A  GG        GT      T    GCCT          TCA
G AT      A  A  T      T    G       GC    T
AC   G TC  CAG  GC  CG    C  CC  AG  G      CA
G   TG    A  AG  GCA     AC   CCTAGC  AA     A
C   G     AC  G     ACG  ACG      A  C  AAT
G AT      A  A  TG  AC   T    G       G  G    C
G   TG    A  AG  GCA     AC   CCTAGC  AA     A
T   TG  TC     GT  C   GT  CG    T   CG      C
CCA  G  TCA      T       TA      T   C       T

 34353637383940414243444546474849505152535455565758596061

T
T  CTG       G  GTA  T  TGCATGC  C
G   T       TGC
A   TGCAAT          AC  C    CA
GGC   CAAT       TA  TCT      T   A   A
CGCTGC    TG  G    CT  T      TGC
GG  TG       GCG  AC      G  ATG     G
T  C     A  T  CGT  C      GC     CACG  C
C   T    AA
```

Given in column A is the primary structure of ten 12 to 34 b. p. oligonucleotides constituting both genes. Column B is a printout representing an optimal procedure for the simultaneous asynchronous elongation of each of the ten chains of the oligonucleotides listed in column A. It can be seen that a cytosine nucleoside is coupled to segments 4 and 10 during the first step (01), an adenine nucleoside is coupled to segments 2, 3, 7 and 10 during the second step (02), and so on. In other words, if dimers are formed at segments 2, 3, 4 and 7 during these two steps, we have already a trimer at segment 10. What makes the asynchronous elongation so attractive is the possibility to use a single-column synthesizer with all the advantages involved (automatic feed of the reagents, washing of the fluid, isolation from atmospheric moisture).

In the early work, the segments were of Whatman 3M or similar paper as well as mechanically stronger cellulose fabrics. The condensations were carried out by the usual triester method. Because of the relatively low yields (70–80%), the condensation by this method yielded only 10 to 12 b. p. sequences. In recent years, Seliger's laboratory (Germany) introduced synthesis on Fractosil 500 support (the segments are prepared in a special way by the phosphoramidite-triester method). In this case, the yields at the elongation step average 95% with the result that chains with up to 30–35 units can be produced.

There is every possibility that the triester method with a phosphate-protective group capable of *O*-nucleophilic catalysis will also be rather effective.

11.5.8 Separation and Purification of Synthetic Oligonucleotides

Liquid chromatography is the basic method for separating and purifying oligonucleotides. In the phosphodiester case, oligonucleotides undergo ion-exchange chromatography. If synthesis is done by the triester method, partition chromatography on silica gel is used. Separation of oligonucleotides with protected internucleotide phosphates is often accomplished by reversed-phase chromatography. This approach is especially useful if the oligonucleotide contains a 5′-dimethoxytrityl group. In this case, it is easily separated from less lipophilic oligonucleotides lacking such a group on silica gel C-18. High-performance liquid chromatography (HPLC) on silica gel C-18 is also recommended for the same purpose. This chromatographic technique is instrumental in analysis and fractionation of deprotected oligonucleotides as well as separation and purification of compounds produced by solid-phase synthesis. Oligonucleotides separated by reversed-phase chromatography must be assayed by ion-exchange chromatography or gel electrophoresis for homogeneity (separation by charge or chain length). Another method for identification and purification of synthetic oligonucleotides is electrophoresis in polyacrylamide gel. This technique provides for separation by charge and allows rather extended (up to 200 b. p.) sequences to be obtained. Among its drawbacks is that it is impossible to separate preparative amounts of oligonucleotides. Moreover, quantitative separation from the gel is also impossible. PAGE is especially good for separation and purification of oligonucleotides synthesized on polymer segments. It is conducted under denaturating conditions (to avoid complexing) in 7 M urea or in the presence of formamide.

11.6 Synthetic Genes and Their Cloning

11.6.1 Methods of Duplex DNA Construction

Section 11.3 outlined the basic principle of gene formation with the aid of the enzyme T4 DNA ligase, and an example was given illustrating assembly of a thyrosine suppressor tRNA gene. The latter was assembled from synthetic oligonucleotides step by step. At first, short double-stranded fragments of types A, B and C, having sticky ends, were produced before assembling the complete gene.

An alternative approach involves direct ligation of all oligonucleotide fragments constituting a gene (one-pot ligation). To this end, a mixture of fragments is heated, cooled and then subjected to ligation. Some examples of such assembly are given in Table 11-3. Direct ligation is sometimes performed in the course of cloning; that is, the mixture of oligonucleotides with a linear vector is incorporated into competent cells already containing DNA ligase.

Recently, subcloning has become a widely used procedure. What it boils down to is preliminary cloning of intermediate DNA duplexes (with 60 to 100 b. p.) to be assembled into a gene. The intermediate double-stranded fragments must have sticky ends corresponding to a particular restriction endonuclease. This procedure allows rather pure fragments to be obtained. Their structure is determined by sequencing.

11.6.2 Synthesis of Genes and Their Functionally Important Segments

It has already been mentioned that the first chemoenzymatic synthesis of a complete thyrosine suppressor tRNA gene was performed in Khorana's laboratory. In addition to the structural gene of pre-tRNA, the synthesized 207 b. p.-long duplex also included a promoter (51 nucleotides) and a segment with a signal for processing the primary transcript of the gene (25 nucleotides) into functional tRNA. The synthesized double-stranded polynucleotide had sticky ends resulting from digestion of the corresponding sequence by restriction endonuclease *Eco*RI (AATT).

The gene was assembled, according to the scheme presented earlier, from duplexes I–V, P_{1-3} and P_{4-10}.

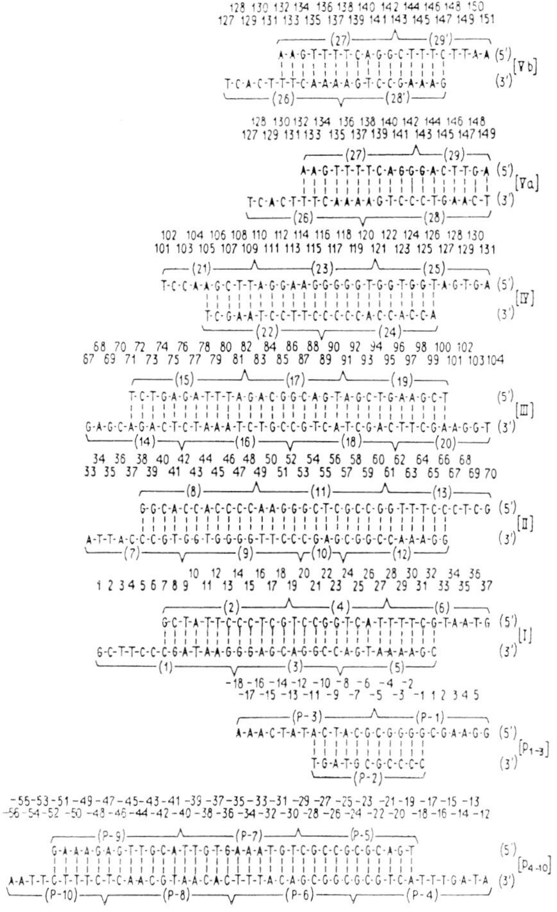

The scheme below illustrates the structure of the gene: symbols (⌢) are used to indicate the sites at which the oligonucleotides are linked; also shown is the structure of the RNA-transcript obtained by transcription with a primer, namely, tetraribonucleotide C–C–C–G complementary to the gene fragment (–3)–(+1) (marked by symbol –).

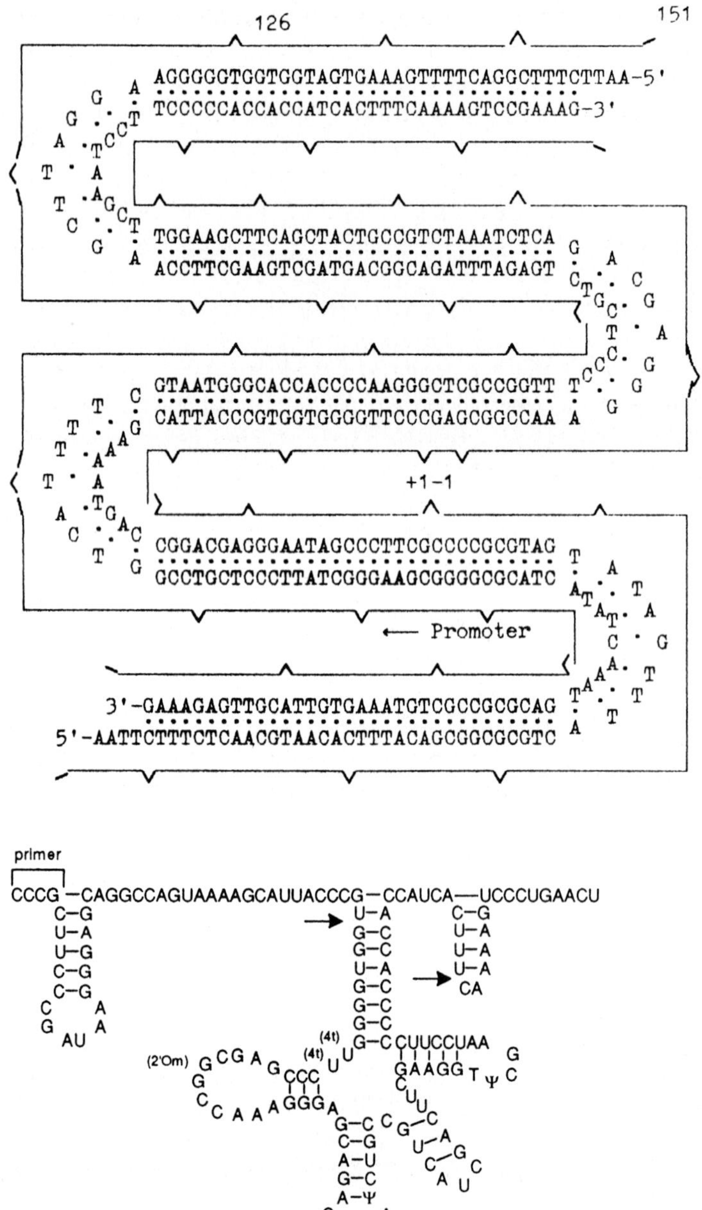

126
151
AGGGGGTGGTGGTAGTGAAAGTTTTCAGGCTTTCTTAA-5'
TCCCCCACCACCATCACTTTCAAAAGTCCGAAAG-3'
TGGAAGCTTCAGCTACTGCCGTCTAAATCTCA
ACCTTCGAAGTCGATGACGGCAGATTTAGAGT
GTAATGGGCACCACCCCAAGGGCTCGCCGGTT
CATTACCCGTGGTGGGGTTCCCGAGCGGCCAA
+1-1
CGGACGAGGGAATAGCCCTTCGCCCCGCGTAG
GCCTGCTCCCTTATCGGGAAGCGGGGCGCATC
Promoter
3'-GAAAGAGTTGCATTGTGAAATGTCGCCGCGCAG
5'-AATTCTTTCTCAACGTAACACTTTACAGCGGCGCGTC
primer
CCCG-CAGGCCAGUAAAAGCAUUACCCG-CCAUCA-UCCCUGAACU
(4t)
(2'Om)
(ms²i⁶)

Analysis of the primary transcript has demonstrated that the nucleotide sequence of a gene whose chemical synthesis required 400 condensations was correct. Addition of an extract from *E. coli* S-100 to the primary transcript leads to its processing and modification. The primary transcript is digested with RNAse P to yield a 5′-terminal sequence (left arrow). During 3′-terminal processing (right arrow), the endonuclease breaks the linkage between the seventh and eighth nucleotides from CCA, then it sequentially detaches the seven "excess" nucleotides. Modification of the primary transcript is not consistent. The rTΨCG segment shows complete conversion of uridine into pseudouridine, but only 20% of uridine are converted into ribosylthymine. In the anticodon loop, the rate of conversion of uridine into pseudouridine is 30%. Formation of 2′-*O*-methylguanosine and 2-methylthio-6-isopentenyladenosine has not been observed. Incomplete modification of the tRNA obtained *in vitro* must be the reason why such tRNA is not aminoacylated. Transcription of the promoter-containing synthetic gene is controlled by a promoter 51 b. p. long. The produced transcript begins with the 5′-terminal pppQ, and the entire 5′-terminal sequence corresponds to what was expected.

The thyrosine suppressor tRNA gene synthesized by the chemoenzymatic procedure was built into two vectors: plasmid (ColE1) and a specially constructed vector Charon 3A based on bacteriophage λ. Both vectors had amber mutations. The incorporation into the vector followed its digestion with restriction endonuclease *Eco*RI.

After the transformation of *E. coli* with recombinant DNAs carrying the synthetic suppressor tRNA gene, amber mutations were suppressed, both bacterial and in the case of bacteriophage λ; this indicates that the processing of pre-tRNA *in vivo* is complete and yields mature thyrosine suppressor tRNA.

After cloning, the synthetic gene was separated from the vector by treatment with restriction endonuclease *Eco*RI and underwent analysis, including determination of its capacity to suppress bacterial mutations after incorporation into the chromosome of *E. coli* as part of bacteriophage λ.

The transcription of the cloned gene *in vitro* yielded a product which was converted, after treatment with a crude *E. coli* extract, into a precursor of suppressor $tRNA^{Tyr}$. The structure of the latter was confirmed by *in vitro* digestion by RNAse P to a 41-unit 5′-terminal precursor fragment. It turned out that the cloned gene had the structure of the natural one, as was intended by the synthesis, its expression was controlled by the promoter, and it was functionally active.

This experiment demonstrated already in the late seventies that high-molecular weight polynucleotides identical to the corresponding portions of natural DNAs can be synthesized.

In 1975, Köster and coworkers synthesized the first protein gene in a similar fashion. What they actually produced was a "minigene" or, to be precise, the gene of the peptide hormone angiotensin II. Unlike Khorana, who had synthesized his gene proceeding from the known primary structure of tRNA and established sequence of the promoter and terminal segments, Köster was guided by the amino acid sequence of angiotensin II. The genetic code was used to establish the nucleotide sequence of the structural gene. The "minigene" also included a translation initiating segment (ATG) and two stop codons.

```
    stop  start  ASPARGVALTYRILEHISPROPHEstop
    UAA   AUG    GAUCGCGUUUAUAUUCAUCCCUUUUAA
5'  ATT   TAC    CTAGCGCAAATATAAGTAGGGAAAATT   3'
3'  TAA   ATG    GATCGCGTTTATATTCATCCCTTTTAA   5'
```

As was intended by the experimenters, the presence of a stop codon before the structural gene had to ensure the onset of translation.

In 1977, Itakura and coworkers synthesized a second minigene – that of somatostatin (see scheme below):

```
   EcoRI    Ala Gly Cys Lys Asn Phe Phe Trp Lys Thr Phe Thr Ser Cys stop   BamHI
 ←——(A)——→ ←——(B)——→ ←——(C)——→ ←———(D)———→
AATTCATGGCTGGTTGTAAGAACTTCTTTTGGAAGACTTTCACTTCGTGTTGATAG
        GTACCGACCAACATTCTTGAAGAAAACCTTCTGAAAGTGAAGCACAACTATCCTAG 5'
        ←——(E)——→ ←——(F)——→ ←——(G)——→ ←———(H)———→
```

Shown on top of the scheme is the primary structure of somatostatin, while the bottom structure corresponds to the structural gene of the peptide hormone. The parenthesized letters from A to H stand for each of the 9 oligonucleotide segments synthesized by the triester method. The codons corresponding to each amino acid have been selected proceeding from the primary structure of the peptide hormone. The selection was not random but based on the frequency at which the codons occurred in the genome of phage M2. The double-stranded polynucleotide contained a methionine codon at the 5′ end of the coding sequence (to ensure directed cleavage of the polypeptide at this point after translation) and had sticky ends corresponding to restrictases *Eco*RI and *Bam*HI. The somatostatin gene was inserted, within a specially constructed plasmid vector, into the β-galactose operon of the β-galactosidase gene. Thus, the regulatory apparatus of the β-galactosidase gene was used for regulating the synthetic gene. Amplification of the selected plasmid carrying the peptide gene in *E. coli* brought about generation of a chimeric protein (β-galactosidase bound via methionine to somatostatin) from which the peptide was separated using cyanogen bromide. The somatostatin produced in this manner was identical to its natural counterpart in terms of antigenic properties.

Itakura's work can be regarded as seminal. To begin with, for the first time a bacterial cell was made to produce animal protein. Secondly, the expression of the synthesized somatostatin gene testified to correct planning of the synthetic procedure and the experiment as a whole. Thus, Itakura's efforts have

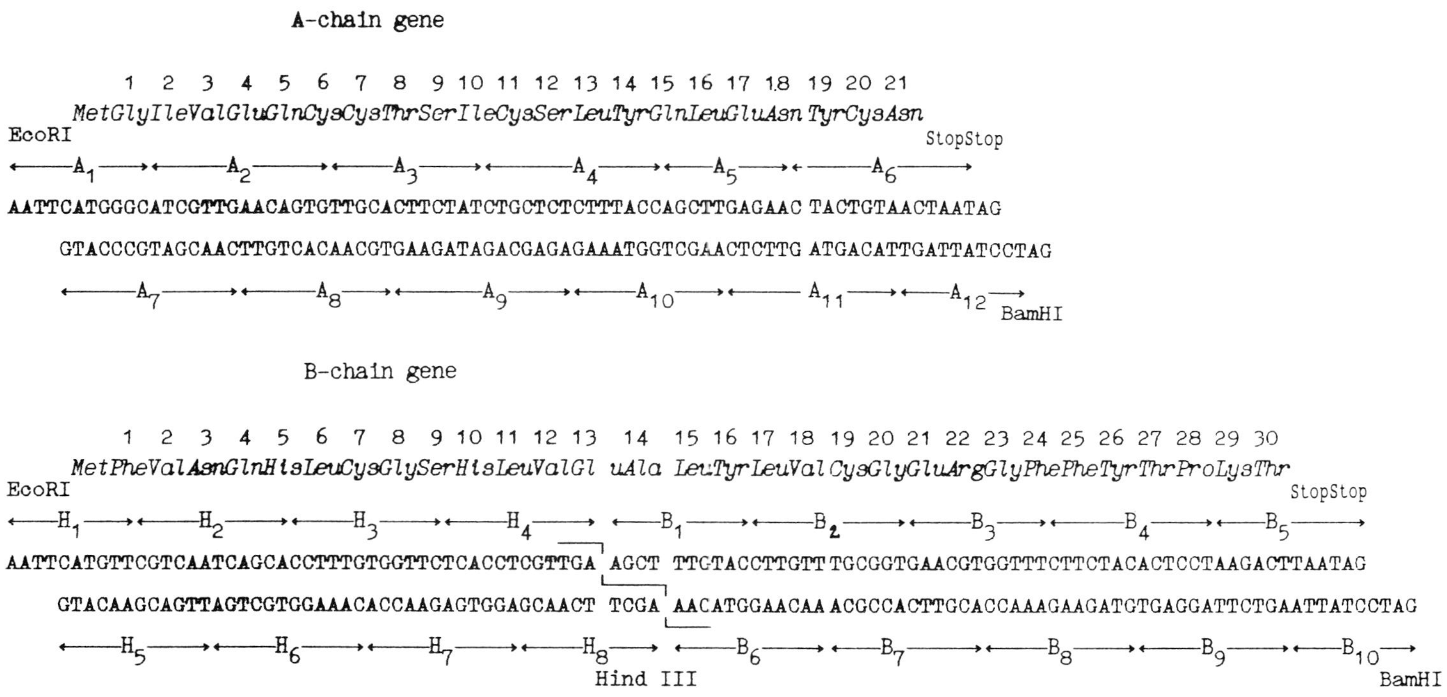

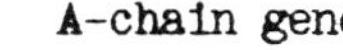

A-chain gene

```
    1  2  3  4  5  6  7  8  9 10 11 12 13 14 15 16 17 18 19 20 21
    MetGlyIleValGluGlnCysCysThrSerIleCysSerLeuTyrGlnLeuGluAsn TyrCysAsn
EcoRI                                                               StopStop
<--A1--><--A2--><--A3--><--A4--><--A5--><--A6-->
AATTCATGGGCATCGTTGAACAGTGTTGCACTTCTATCTGCTCTCTTTACCAGCTTGAGAAC TACTGTAACTAATAG
    GTACCCGTAGCAACTTGTCACAACGTGAAGATAGACGAGAGAAATGGTCGAACTCTTG ATGACATTGATTATCCTAG
    <--A7--><--A8--><--A9--><--A10--><--A11--><--A12-->
                                                          BamHI
```

B-chain gene

```
    1  2  3  4  5  6  7  8  9 10 11 12 13  14  15 16 17 18 19 20 21 22 23 24 25 26 27 28 29 30
    MetPheValAsnGlnHisLeuCysGlySerHisLeuValGl uAla LeuTyrLeuVal CysGlyGluArgGlyPhePheTyrThrProLysThr
EcoRI                                                                                      StopStop
<--H1--><--H2--><--H3--><--H4-->  <--B1--><--B2--><--B3--><--B4--><--B5-->
AATTCATGTTCGTCAATCAGCACCTTTGTGGTTCTCACCTCGTTGA AGCT TTGTACCTTGTT TGCGGTGAACGTGGTTTCTTCTACACTCCTAAGACTTAATAG
    GTACAAGCAGTTAGTCGTGGAAACACCAAGAGTGGAGCAACT TCGA AACATGGAACAA ACGCCACTTGCACCAAAGAAGATGTGAGGATTCTGAATTATCCTAG
    <--H5--><--H6--><--H7--><--H8-->  <--B6--><--B7--><--B8--><--B9--><--B10-->
                                Hind III                                         BamHI
```

opened up new possibilities in the synthesis of physiologically active peptide genes and making of peptide and protein producers.

A logical extension of the work done by Itakura and coworkers was synthesis of the human insulin gene, accomplished by the same group in 1978. As is known, the insulin molecule consists of two polypeptide chains A and B. In separate syntheses, genes of both chains were produced, their primary structure being predetermined, just as in the case of somatostatin, from the amino acid sequence of the corresponding chains of insulin (see scheme on page 551). It can be seen that in addition to the structural portion both genes comprise codons of methionine, stop codons, and sticky ends corresponding to restriction endonucleases *Eco*RI and *Bam*HI. In the middle of the chain B gene there is a site recognized by restriction endonuclease *Hind*III (AAGCTT).

The synthesis of both genes had been completed within a record period of time: it took a mere six months for four workers to synthesize 77 and 104 b. p. duplexes. The synthetic work was distinguished by use of trinucleotides as "building blocks". A library of 45 trinucleotides was created within three months.

Both genes were inserted within different plasmid vectors into the β-galactosidase segment and, after amplification within *E. coli,* chains A and B of insulin were produced separately in purified form to synthesize the protein itself.

The next achievement by the same group was the synthesis in 1979 of an even more complex gene of the human growth hormone somatotropin. This protein consists of 191 amino acids, and in spite of the spectacular advances in the triester block method synthesis of its gene would take much more time than that of the insulin gene. Therefore, the synthesis was performed only partially, the major portion of the gene being produced through reverse transcription of mRNA from somatotropin extracted from the pituitary. However, such a cDNA could not be used in its entirety to produce the hormone because it had a leader sequence, and the corresponding polypeptide is detached from the protein only in human cells (naturally, a bacterial cell lacks the enzymes necessary for processing of the human gene). This is why the cDNA was treated with restriction endonuclease, and part of the required sequence was removed (69 b. p.) then restored using a synthesized polynucleotide having an initiation codon and a methionine one. The rest of the experiment was conducted as described above with the difference that the gene was inserted directly into the promoter. The hormone yield turned out to be quite high – 3 mg per liter of the culture fluid, although its presence was proved only by radioimmunoassay.

As automation proliferated more widely in the eighties, the number of synthesized genes increased from one year to another. 1000 b. p. genes became routine. Whatever complications were involved had to do with their composition and assembly techniques. Here is how the bovine rhodopsin gene was assembled by Khorana and coworkers in 1986.

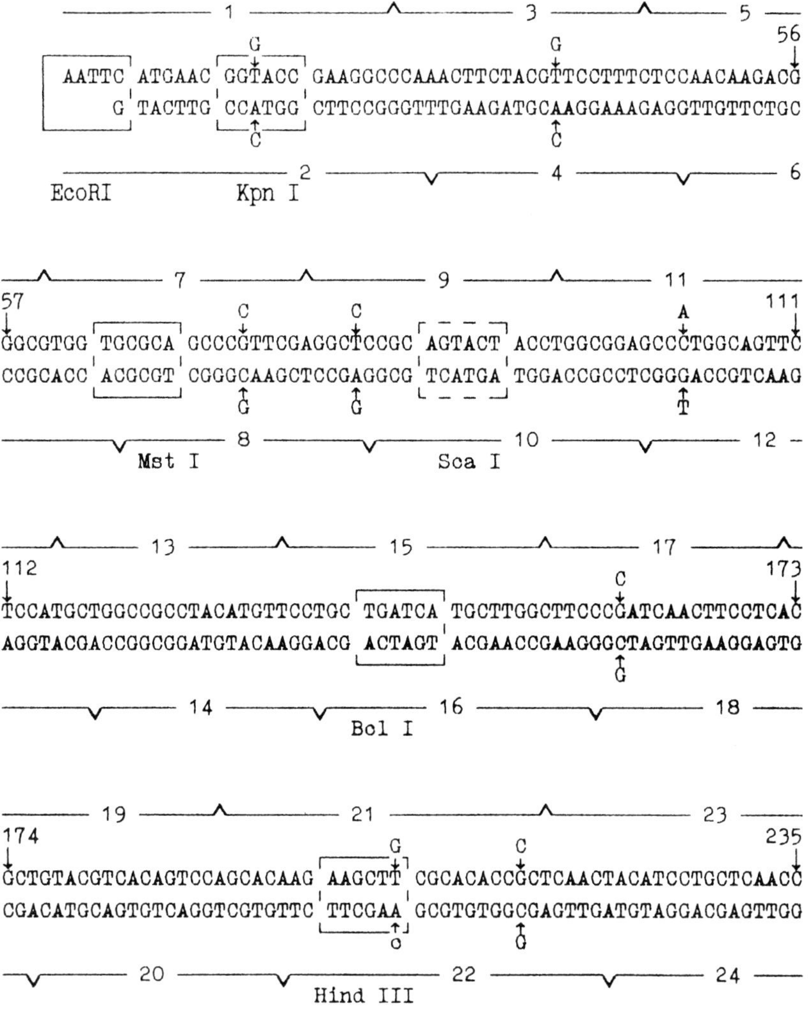

1
3
5
56
AATTC ATGAAC GGTACC GAAGGCCCAAACTTCTACGTTCCTTTCTCCAACAAGACG
G TACTTG CCATGG CTTCCGGGTTTGAAGATGCAAGGAAAGAGGTTGTTCTGC
2
4
6
EcoRI
Kpn I
7
9
11
57
111
GGCGTGG TGCGCA GCCCGTTCGAGGCTCCGC AGTACT ACCTGGCGGAGCCCTGGCAGTTC
CCGCACC ACGCGT CGGGCAAGCTCCGAGGCG TCATGA TGGACCGCCTCGGGACCGTCAAG
8
10
12
Mst I
Sca I
13
15
17
112
173
TCCATGCTGGCCGCCTACATGTTCCTGC TGATCA TGCTTGGCTTCCCGATCAACTTCCTCAC
AGGTACGACCGGCGGATGTACAAGGACG ACTAGT ACGAACCGAAGGGCTAGTTGAAGGAGTG
14
16
18
Bcl I
19
21
23
174
235
GCTGTACGTCACAGTCCAGCACAAG AAGCTT CGCACACCGCTCAACTACATCCTGCTCAACC
CGACATGCAGTGTCAGGTCGTGTTC TTCGAA GCGTGTGGCGAGTTGATGTAGGACGAGTTGG
20
22
24
Hind III

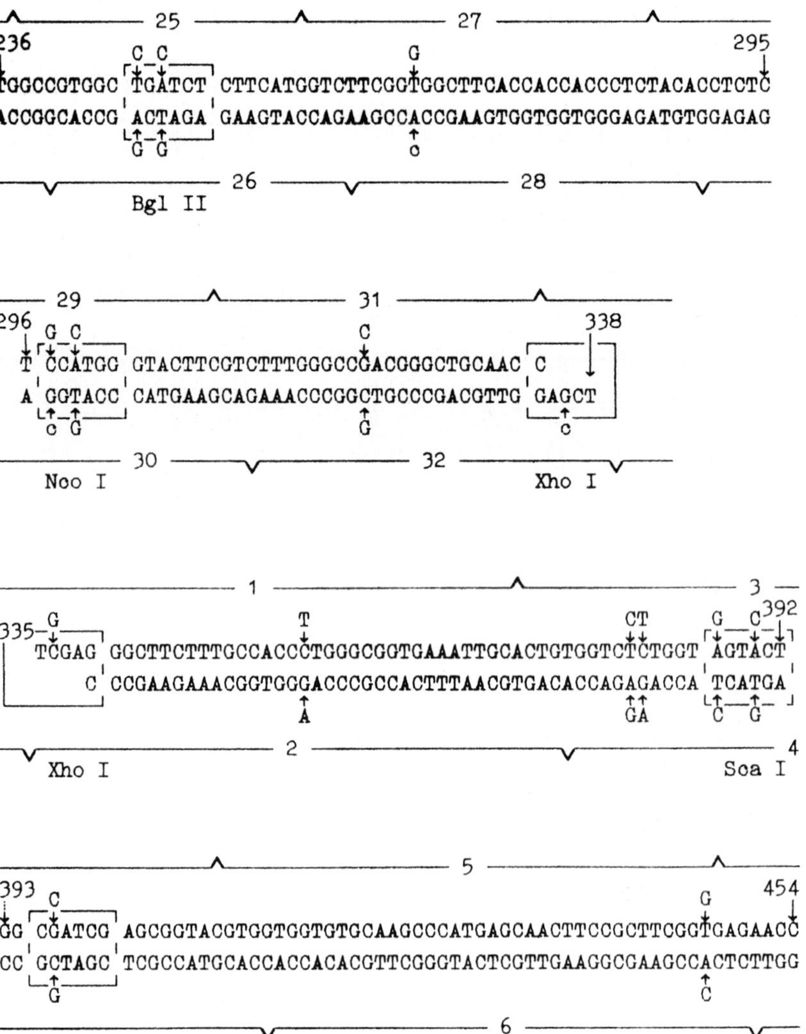

25
27
236
295
C C
G
TGGCCGTGGC TGATCT CTTCATGGTCTTCGGTGGCTTCACCACCACCCTCTACACCTCTC
ACCGGCACCG ACTAGA GAAGTACCAGAAGCCACCGAAGTGGTGGTGGGAGATGTGGAGAG
G G
C
26
28
Bgl II
29
31
296
338
G C
C
T CCATGG GTACTTCGTCTTTGGGCCGACGGGCTGCAAC C
A GGTACC CATGAAGCAGAAACCCGGCTGCCCGACGTTG GAGCT
C G
G
C
30
32
Nco I
Xho I
1
3
335
G
T
CT
G C
392
TCGAG GGCTTCTTTGCCACCCTGGGCGGTGAAATTGCACTGTGGTCTCTGGT AGTACT
C CCGAAGAAACGGTGGGACCCGCCACTTTAACGTGACACCAGAGACCA TCATGA
A
GA
C G
2
4
Xho I
Sca I
5
393
454
C
G
GG CGATCG AGCGGTACGTGGTGGTGTGCAAGCCCATGAGCAACTTCCGCTTCGGTGAGAACC
CC GCTAGC TCGCCATGCACCACCACACGTTCGGGTACTCGTTGAAGGCGAAGCCACTCTTGG
G
C
6
Pvu I

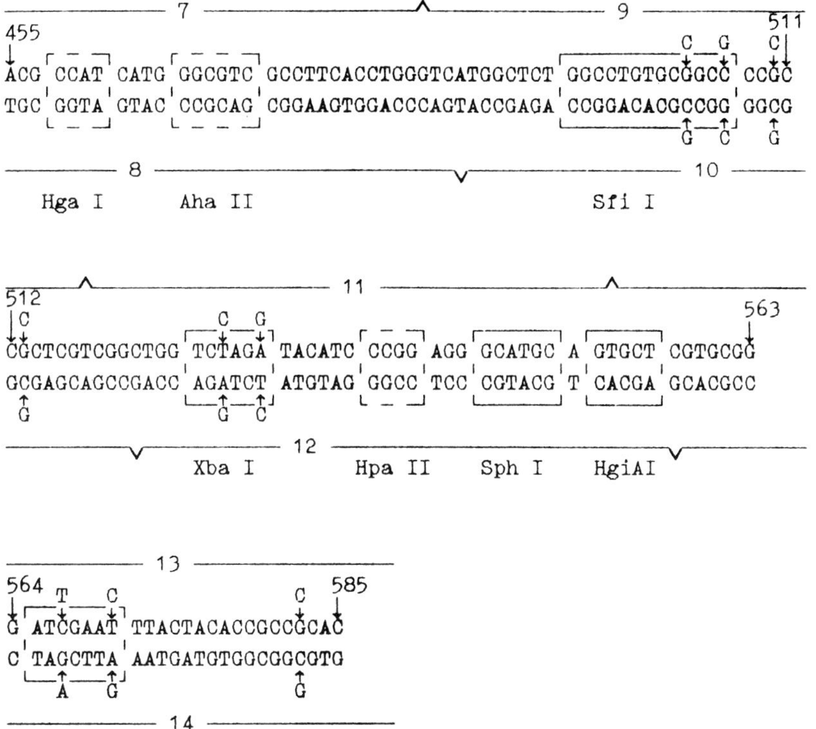
7
9
455
511
ACG CCAT CATG GGCGTC GCCTTCACCTGGGTCATGGCTCT GGCCTGTGCGGCC CCGC
TGC GGTA GTAC CCGCAG CGGAAGTGGACCCAGTACCGAGA CCGGACACGCCGG GGCG
8
10
Hga I
Aha II
Sfi I
11
512
563
CGCTCGTCGGCTGG TCTAGA TACATC CCGG AGG GCATGC A GTGCT CGTGCGG
GCGAGCAGCCGACC AGATCT ATGTAG GGCC TCC CGTACG T CACGA GCACGCC
12
Xba I
Hpa II
Sph I
HgiAI
13
564
585
G ATCGAAT TTACTACACCGCCGCAC
C TAGCTTA AATGATGTGGCGGCGTG
14
Cla I

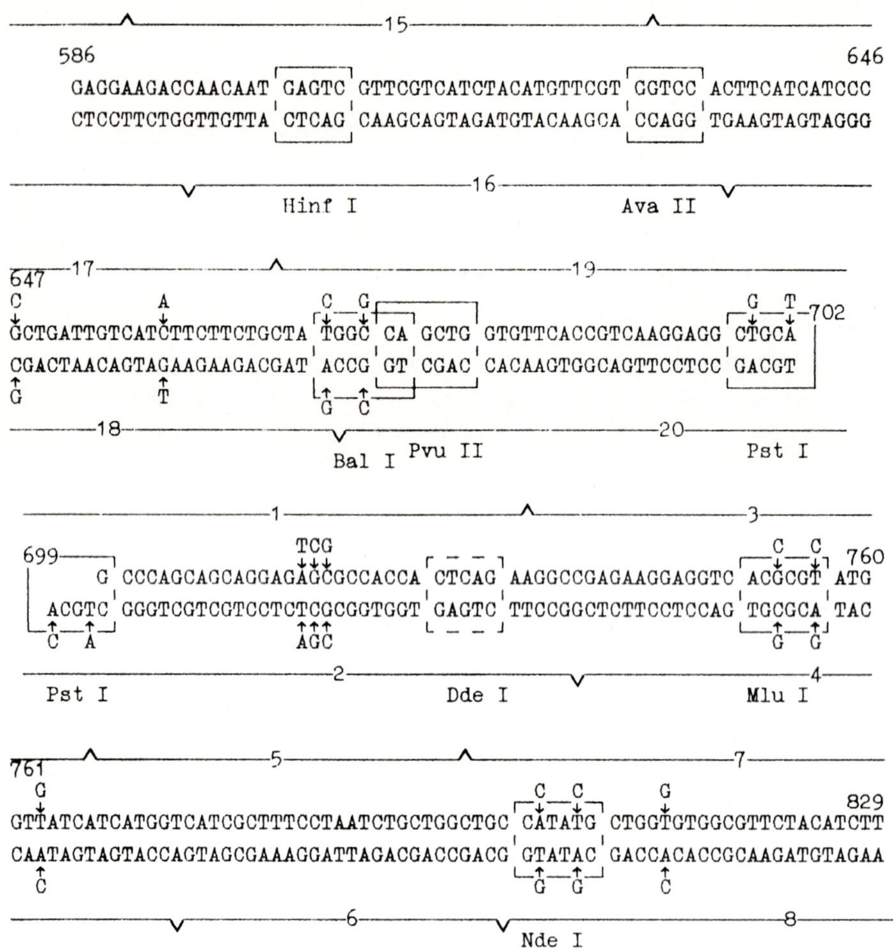
15
586
646
GAGGAAGACCAACAAT GAGTC GTTCGTCATCTACATGTTCGT GGTCC ACTTCATCATCCC
CTCCTTCTGGTTGTTA CTCAG CAAGCAGTAGATGTACAAGCA CCAGG TGAAGTAGTAGGG
16
Hinf I
Ava II
647
17
19
702
GCTGATTGTCATCTTCTTCTGCTA TGGC CA GCTG GTGTTCACCGTCAAGGAGG CTGCA
CGACTAACAGTAGAAGAAGACGAT ACCG GT CGAC CACAAGTGGCAGTTCCTCC GACGT
18
20
Bal I
Pvu II
Pst I
1
3
699
760
G CCCAGCAGCAGGAGAGCGCCACCA CTCAG AAGGCCGAGAAGGAGGTC ACGCGT ATG
ACGTC GGGTCGTCGTCCTCTCGCGGTGGT GAGTC TTCCGGCTCTTCCTCCAG TGCGCA TAC
2
4
Pst I
Dde I
Mlu I
761
5
7
829
GTTATCATCATGGTCATCGCTTTCCTAATCTGCTGGCTGC CATATG CTGGTGTGGCGTTCTACATCTT
CAATAGTAGTACCAGTAGCGAAAGGATTAGACGACCGACG GTATAC GACCACACCGCAAGATGTAGAA
6
8
Nde I

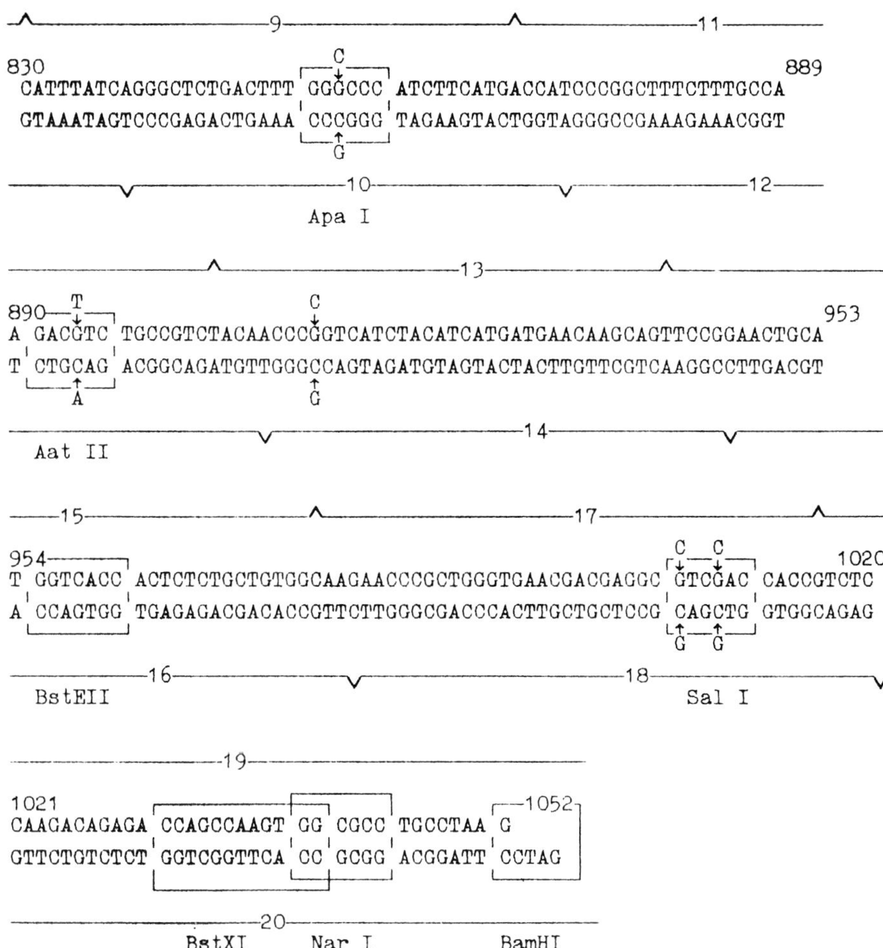

The gene is divided into three fragments. Fragment EX contains nucleotides 5 to 338, fragment XP, 335–702, and fragment PB, 699–1052. The synthesized oligonucleotides of each fragment are numbered (the numbers are indicated above and below duplexes). The vertical arrows mark nucleotides that were replaced in comparison with the natural gene. The unique restriction sites in the synthetic gene are in solid squares. The restriction sites unique in the fragments but not in the gene itself are in dashed squares. Finally, the restriction sites present in the natural gene and removed from the synthetic one are in cross-hatched squares. Site Dde1 in fragment PB has Latin letter designations because there is yet another site Dde1 which overlaps with the unique site BamHI. Therefore, the second site Dde1 is not recognized if the fragment containing it is treated with the enzyme *Bam*HI.

Table 11-3 illustrates genes of various length, obtained before 1988 from synthetic oligonucleotides.

Table 11-3. Synthetic Genes.

Gene	Lenght of separately cloned synthetic fragment (b.p.)	Design of gene sequence (selection of codons; design of synthetic fragment ends; restriction sites (RS)	Length of synthetic oligonucleotides necessary to design the gene; method of their chemical synthesis and purification
A. Genes from a Single Cloned Synthetic Fragment.			
Somatostatin	56	Initiation and stop codons; sticky ends for cloning at different RS	8 oligonucleotides 10–15 b.p. long; synthesis by phosphotriester method in solution
α_1-Thymosine	98	Codons often encountered in highly expressed *E. coli* genes; sticky ends for cloning at different RS	16 oligonucleotides 10–15 b.p. long; synthesis by phosphotriester method in solution; separation by anion-exchange HPLC and PAGE in 7 M urea
Human insulin, chains A and B	I 77 II 104	Initiation and double stop codons; sticky ends for cloning at different RS	12 and 18 oligonucleotides 10–15 b.p. long; synthesis by phosphotriester method in solution; separation by anion-exchange HPLC
Salmon calcitonin-^{33}Gly	113	Codons often encountered in highly expressed *E. coli* genes; sticky ends for cloning at different RS; unique RS in the middle of gene for subsequent cassette mutagenesis	2 oligonucleotides 109 and 117 b.p. long; synthesis by phosphoramidite method on automatic synthesizer; separation by PAGE in 7 M urea
Salmon calcitonin-^{33}Gly	113		One oligonucleotide 140 b.p. long; synthesis by phosphoramidite method on automatic synthesizer; separation by PAGE in 7 M urea
β-Urogastron	170	Codons of higly expressed yeast genes; initiation and double stop codons; sticky ends for cloning at different RS	16 oligonucleotides 12–33 b.p. long; synthesis by phosphoramidite method; separation by PAGE in 7 M urea

Method of obtaining DNA fragment from synthetic oligonucleotides for cloning	Cloning vector	Expression vector; regulatory portion of the gene and yield of expression product	Reference
Two-step ligation: I. Assembly of 2 DNA blocks from 4 oligonucleotides; II. Assembly of target DNA duplex from 2 blocks	pBR322;	pBR322; *lac*-promoter; yield of chimeric protein with long β-galactosidase fragment – 0.03 % of total *E. coli* cell protein	Itakura K., Hirose T. *et al. Science,* 1977, **192**, pp. 1056–1063
Two-step ligation: I. Assembly of 4 DNA blocks from 4 oligonucleotides; II. Assembly of target DNA duplex from 4 blocks	pBR322	pBR322; *lacUVS*-promoter; yield of chimeric protein with long β-galactosidase fragment – 294 mg per 24 g of *E. coli;* 4.5 mg of α_1-thymosine after BrCN hydrolysis	Wetzel R., Heyneker H. L. *Biochemistry,* 1980, **19**, pp. 6096–6104
Two-step ligation: I. Assembly of 5 DNA blocks from 5–8 oligonucleotides; II. Assembly of 2 target DNA duplexes from 2–3 blocks	pBR322	pBR322; *lac*-promoter; yield of chimeric proteins with long β-galactosidase fragment – 10 mg per 24 g of wet *E. coli* cells	Goeddel O. V., Kleid O. G. *et al. Proc. Natl. Acad. Sci. USA,* 1979, **79**, pp. 106–110
Hybridization of 2 extralong oligonucleotides by formation of DNA duplex with sticky ends	M13mp18am		Hein F., Jansen H. W., Uhlmann E., *Nucleotides,* 1988, **7**, pp. 497–510
3′-terminal part of oligonucleotide is self-complementary, addition of polymerase to complementary chain and treatment with restriction endonuclease to form sticky ends	M13mp18am		Uhlmann E., Hein F., *Nucl. Acids Symp. Ser.*, 1987, **18**, pp. 237–240
Two-step ligation: I. Assembly of 3 DNA blocks from 3–7 oligonucleotides; II. Assembly of target DNA duplex from 3 blocks	pBR328	pJDB219; *GAPDH*-promoter; β-urogastrone yield – 30 μg per liter of yeast culture at a culture density of 2 O. U.$_{650}$/ml	Urdea M. S., Merryweather J. P. *et al., Proc. Natl. Acad. Sci. USA,* 1983, **80**, pp. 7461–7465

Table 11-3. (Continued).

Gene	Lenght of separately cloned synthetic fragment (b.p.)	Design of gene sequence (selection of codons; design of synthetic fragment ends; restriction sites (RS)	Length of synthetic oligonucleotides necessary to design the gene; method of their chemical synthesis and purification
Pancreatic secretory inhibitor of human trypsin	184	Cloning with 3-sticky and 5-blunt ends	25 oligonucleotides 7–16 b.p. long; synthesis by phosphotriester method on segments; separation by PAGE in 7 M urea or TLC on cellulose
Tyr-tRNA^{Sup}	200	Natural sequence; sticky ends for cloning at one RS	39 oligonucleotides 4–13 b.p. long; synthesis by phosphodiester method in solution; separation by anion-exchange chromatography
Somatomedin-C–L$^{0-51}$ Thr	226	Codons often encountered in highly expressed *E. coli* genes; Thr is substituted for ^{51}Met to separate somatomedin from BrCN leader sequence	18 oligonucleotides 16–32 b.p. long; synthesis by solid-phase phosphoramidite method; separation by PAGE in 7 M urea
Somatomedin-C	234	Codons often encountered in highly expressed *E. coli* genes; sticky ends for cloning at different RS	23 oligonucleotides 10–23 b.p. long; synthesis by solid-phase phosphoramidite method; separation by anion-exchange HPLC
Cow colostrum trypsin inhibitor	239	*Hind*III CP at 5′ end of gene for solid-phase ligation of synthetic gene to vector; polylinker at 3′ end for removal of gene ligated to vector from support	19 oligonucleotides 28–36 b.p. long; synthesis by phosphoramidite method on automatic synthesizer; separation by PAGE in 7 M urea
Ubiquitin	240	16 nucleotide substitutions in natural sequence to create RS dividing gene into 8 modules for subsequent cassette mutagenesis; sticky ends for cloning at different RS	8 oligonucleotides 50–64 b.p. long; synthesis by phosphoramidite method on automatic synthesizer; separation by PAGE in 7 M urea

Method of obtaining DNA fragment from synthetic oligonucleotides for cloning	Cloning vector	Expression vector; regulatory portion of the gene and yield of expression product	Reference
Two-step ligation: I. Assembly of 2 DNA blocks from 11 and 13 oligonucleotides; II. Assembly of target DNA duplex from 2 blocks	pGV451, pUC8	pIN-III-ompA; *IpplacZpo*-promoter; leader sequence of protein from outer membrane of *E. coli,* yield of properly processed peptide – 2.5–10 μg/ml of *E. coli* supernatant	Maywald F., Böldicke T. *et al. Gene,* 1988, **68**, pp. 357–369
Multiple-step ligation: I. Assembly of 7 DNA blocks from 3–7 oligonucleotides; II. Step-by-step assembly of target duplex from DNA blocks	Φ–80	Bacteriophage; synthetic gene suppressed amber mutation	Khorana H. G. *Science,* 1979, **203**, pp. 614–624
One-step ligation; assembly of target fragment from 9 DNA duplexes with sticky ends (from 9 oligonucleotide pairs)	M13mp9	pCFM414; *trp*-promoter; leader sequence (8 aa)–N-end of β-galactosidase	Peters M. A., Lau E. P. *et al Gene,* 1985, **35**, pp. 83–89
One-step ligation; assembly of target duplex from 23 oligonucleotides	M13mp8		Sproat B. S., Gait M. J. *Nucl. Acids Res.,* 1985, **15**, pp. 2959–2977
Solid-phase assembly; multiple step-by-step extension of DNA duplex by ligation of hybridized oligonucleotides (3–4) to an oligonucleonucleotide immobilized on polymer support. At the last step, a vector linearized by appropriate restrictases is ligated to the resulting DNA duplex	M13mp8		Hostomsky Z., Smrt J. *Nucl. Acids Res.,* 1987, **15**, pp. 4849–4856
One-step ligation; assembly of target DNA duplex from 4 oligonucleotide pairs	pUC	pM627N-S; p^{L}-promoter; yield – 10–15 % of total *E. coli* protein	Ecker D. J., Butt T. R. *et al. J. Biol. Chem.,* 1987, **262**, pp. 3524–3527, 14213–14221

Table 11-3. (Continued).

Gene	Lenght of separately cloned synthetic fragment (b.p.)	Design of gene sequence (selection of codons; design of synthetic fragment ends; restriction sites (RS)	Length of synthetic oligonucleotides necessary to design the gene; method of their chemical synthesis and purification
Neutrophileactivating factor (NAF)	244	Natural sequence, sticky ends for cloning at different RS; RS for subsequent cassette mutagenesis of 5′ end	6 oligonucleotides 7–83 b.p. long; synthesis by phosphoramidite method on automatic synthesizer; separation by PAGE in 7 M urea
Transactivating protein of human immunodeficiency virus	287	Preferential codons either in *E. coli* or in yeast; sticky ends for cloning at different RS; some RS are included for cassette mutagenesis; inverted repeats and other secondary structure elements in gene are excluded	2 oligonucleotides 139 and 144 b.p. long and 3 oligonucleotides (11, 14 and 16 b.p.); synthesis by phosphoramidite method on automatic synthesizer; separation by PAGE in 7 M urea
Human lysozyme	418	Codons often encountered in highly expressed yeast genes; Shine–Dalgarno sequence	56 oligonucleotides 11–15 b.p. long; synthesis by solid-phase phosphotriester block method; separation by reversed phase HPLC
Interferon	453	Codons for Arg–CGT and Leu–CTG; sticky ends for cloning at different RS	66 oligonucleotides 12–16 b.p. long; synthesis by solid-phase chlorophosphite method; separation by anion-exchange and reversed phase HPLC
Bovine pancreatic phospholipase A_2 (PLA_2)	456	Codons of highly expressed yeast genes; sticky ends for cloning at different RS	22 oligonucleotides 33–48 b.p. long; synthesis by phosphoramidite method on automatic synthesizer; separation by reversed phase HPLC and PAGE in 7 M urea
Interferon-α_2	511	Codons often encountered in highly expressed *E. coli* genes; 20 % nucleotide substitutions, as compared to natural gene; sticky ends for cloning at different RS	68 oligonucleotides 10–16 b.p. long; synthesis by solid-phase phosphotriester method; separation by anion-exchange HPLC

Method of obtaining DNA fragment from synthetic oligonucleotides for cloning	Cloning vector	Expression vector; regulatory portion of the gene and yield of expression product	Reference
One-step ligation; assembly of target DNA duplex from 3 oligonucleotide pairs	pBSM13	pIL402; *trp*-promoter; isolation of NAF from 516 g of *E. coli;* quantity not specified	Lindley I., Aschauer H. *et al. Proc. Natl. Acad. Sci. USA,* 1989, **85**, pp. 9199–9203
In vivo repair of *gap*-duplex resulting from ligation of 2 long oligonucleotides to vector using 3 additional template oligonucleotides	pUC18	pKV461; *hCMW*-promoter; splicing and polyadenylation signals from SV40; gene biologically active in HeLa cells	Adams S. E., Johnson, I. O. *et al Nucl. Acids Res.,* 1988, **16**, pp. 4287–4298
Three-step ligation; I. Assembly of 9 DNA blocks from 4–9 oligonucleotides; II. Assembly of 3 gene parts from DNA blocks; III. Assembly of target DNA duplex	pBR322	pMY12-6; p^L-promoter; yield of lysozyme with *N*-terminal Met – a few per cent of total *E. coli* cell protein	Muraki M., Jigami Y. *et al. Agric. Biol. Chem.,* 1986, **50**, pp. 713–723
Two-step ligation; I. Assembly of 6 DNA blocks from 9–14 oligonucleotides; II. Assembly of target DNA duplex from 6 DNA blocks	pBR322	pBR322; *bla*- and *T5*--promoter; yield – 16.3 % of total *E. coli* cell protein	Jay E., Rommens J. *et al. Proc. Natl. Acad. Sci. USA,* 1984, **81**, pp. 2290–2294
Two-step ligation; I. Assembly of 3 DNA blocks from 3–4 oligonucleotide pairs; II. Assembly of target DNA duplex from 3 DNA blocks	pUC119	pAT405; *PHO5*-promoter; yield – 2.8 μg/ml of yeast supernatant of active form of PLA_2	Tanaka T., Kimuro S., Ota Y. *Gene,* 1989, **64**, pp. 257–264
Three-step ligation; I. Assembly of 13 DNA blocks from 5–8 oligonucleotides; II. Assembly of 3 gene parts from 4–5 DNA blocks; III. Assembly of target duplex from 3 parts	pPM60	pAT153; *trp*-promoter; yield – 5×10^7 units of IFN-α_2 per liter of *E. coli* culture at a culture density of 1 O.U./ml	Edge M. D., Greene A. R. *et al. Nucl. Acids Res.,* 1983, **11**, pp. 6419–6435

Table 11-3. (Continued).

Gene	Total gene length (b.p.)	Lenght of separately cloned synthetic fragments (b.p.)	Design of gene sequence (selection of codons; design of synthetic fragment ends; restriction sites (RS)	Length of synthetic oligonucleotides necessary to design the gene; method of their chemical synthesis and purification
B. Genes from Several Subcloned Synthetic Fragments.				
β-Urogastrone	174	A. 87 B. 87	Codons often encountered in highly expressed *E. coli* genes	23 oligonucleotides 11–16 b.p. long; synthesis by phosphotriester method in sulution
Artificial protein	190	A.83 B.93 C.64	Codons often encountered in highly expressed *E. coli* genes; RS flanked at synthetic fragment ends by additional G/C pairs, permitting subcloned fragments to be joined after treatment with appropriate restrictases and cloning at different RS; verification of hybridization of oligonucleotides before polymerase reaction	6 oligonucleotides 38–53 b.p. long; synthesis by phosphoramidite method on automatic synthesizer; separation by PAGE in 7 M urea
Chirudin	226	A.99 B.70 C.84	Shine–Dalgarno sequence, initiation and stop codons; unique RS to cut out and join subcloned fragments	3 oligonucleotides 49, 60 and 84 b.p. long, 11 oligonucleotides 12–31 b.p. long; synthesis by phosphoramidite method on automatic synthesizer (long nucleotides) and manually (short ones); separation by PAGE in 7 M urea and reversed phase HPLC
Eglin C	232	A.123 B.106	Codons often encountered in highly expressed *E. coli* genes; initiation and stop codons; RS flanked by additional b.p. at 5′ end of fragment A and 3′ end of fragment B, RS to cut out and join fragments A and B	6 oligonucleotides 34–61 b.p. long; synthesis by solid-phase phosphotriester block method manually; separation by reversed phase HPLC and PAGE in 7 M urea

Method of obtaining DNA fragment from synthetic oligonucleotides for cloning	Cloning vector	Expression vector; regulatory portion of the gene and yield of expression product	Reference
A. Four-step ligation and treatment of resulting dimer of fragment A with restrictase B. Three-step ligation and treatment of resulting dimer of fragment B with restrictase	pAT153 pUB110	*trp-promoter;* fragment trpE (7 aa of N end); chimeric product yield – 2–3 mg/l of *E. coli* culture 4467–4482	Smith, J., Cook, E. *et al Nucl. Acids Res.*, 1982, **10**, pp. 4467–4482
A,B,C. Addition of PolkI to duplex made up of 2 partially complementary oligonucleotides, treatment with restrictases to produce necessary sticky ends for ligation into vector	pKK233-2	pGF211N; *tac*- and *bla*-promoters; pB1052; *tac*-promoter, atpE gene translation initiation region	Biernat J., Köster H. Protein eng., 1987, **1**, pp. 345–351, 353–358
A. Addtion of AMV revertase to duplex made up of two partially complementary oligonucleotides, cloning with the aid of linkers B,C. One-step ligation	pUC18 pVH27	pWH701, pMR3; p^L-promoter; yield – 28 ng per 1 O.U.$_{578}$ of *E. coli* cells	Bergmann C., Dodt J. *et al. Biol. Chem.* Hoppe-Seyler, 1986 **367**, pp. 731–740
A. Addition of PolkI to 2 duplexes made up of 2 pairs of partially complementary oligonucleotides, ligation of resulting duplexes at blunt ends. B. Addition of PolkI to duplex made up of two partially complementary oligonucleotides	pBR322	pBR322; *trp*-promoter; yield – 3×10^5 molecules per *E. coli* cell	Rink H., Liersch M. *et al. Nucl. Acids Res.*, 1984, **12**, pp. 6369–6387

Tab.11-3. (Continued).

Gene	Total gene length (b.p.)	Lenght of separately cloned synthetic fragments (b.p.)	Design of gene sequence (selection of codons; design of synthetic fragment ends; restriction sites (RS)	Length of synthetic oligonucleotides necessary to design the gene; method of their chemical synthesis and purification
Third factor of human anaphylatoxin complement system *C3a*	263	A.89 B.88 C.96	Codons often encountered in highly expressed *E. coli* genes, as many unique RS as possible; *lac*-gene ribosome binding site, initiation and stop codons	3 oligonucleotides 122–125 b.p. long; synthesis by phosphoramidite method on automatic synthesizer; separation by PAGE in 7 M urea
Transforming rat growth factor (*TGF-α*)	267	A.140 B.60 C.120	Preferable codons for higher eucaryotes, gene sequence for pre*TGF-α*	16 oligonucleotides 22–40 b.p. long; synthesis by phosphoramidite method on automatic synthesizer; separation by PAGE in 7 M urea
Human proinsulin	277	A.140 B.63	Rat proinsulin codons, initiation and stop codons, fragment A; sticky ends for cloning at different RS, fragment B; cloning with the aid of MboII adapter	42 oligonucleotides 11–17 b.p. long; synthesis by phosphotriester method in solution separation by reversed phase HPLC
Cytochrome B_5	330	A.96 B.143 C.90	Synthetic Shine–Dalgarno sequence, initiation and double stop codons; RS dividing the sequence into fragments A, B and C; fragments synthesized with corresponding sticky ends	16 oligonucleotides 31–46 b.p. long; synthesis by phosphoramidite method on automatic snythesizer; separation by PAGE in 7 M urea

Method of obtaining DNA fragment from synthetic oligonucleotides for cloning	Cloning vector	Expression vector; regulatory portion of the gene and yield of expression product	Reference
Three-step bridge mutagenesis: in each step, vector is linearized by BsmI restrictase and hybridized with a synthetic oligonucleotide whose ends are complementary to those of vector (18 b.p.) and transform *E. coli* without any manipulations *in vitro*	pUC9		Hayden M. A., Mandecki W. *DNA*, 1988, **7**, pp. 571–577
A. Addition of PolkI to 2 duplexes made up of 2 pairs of partially complementary oligonucleotides, ligation of resulting duplexes at blunt ends B. 4 inserts each 14–16 b.p. long, by site-specific mutagenesis, into cloned sequence of fragment A C. One-step ligation of 4 pairs of complementary oligonucleotides	M13mp18	pSW272 (retrovirus vector); yield – 0.3 ng/1 ml of medium	Watanabe S., Lazar, E., Sporn M. B. *Proc. Natl. Acad. Sci. USA*, 1987, **84**, pp. 1258–1262
A. Three-step ligation B. Addition of AMV revertase to duplex made up of 2 partially complementary oligonucleotides, cloning with the aid of adapters	pBR322	pBR322; *lac*-promoter; 590 a.a. fragment of β-galactosidase; yield of chimeric protein – 41.7 mg per g of total *E. coli* protein	Brousseau R., Scarpulla R. *et al Gene,* 1982, **17**, pp. 279–289
A, B, C. One-step ligation of 2, 4 and 2 pairs of complementary oligonucleotides, respectively	pUC13	pUC13; *lac*-promoter; yield – 8 % of total *E. coli* cell protein (cells stained red)	Von Bodman S. B., Schuller M. A. *et al. Proc. Natl. Acad. Sci., USA,* **83**, pp. 9443–9447

Tab.11-3. (Continued).

Gene	Total gene length (b.p.)	Lenght of separately cloned synthetic fragments (b.p.)	Design of gene sequence (selection of codons; design of synthetic fragment ends; restriction sites (RS)	Length of synthetic oligonucleotides necessary to design the gene; method of their chemical synthesis and purification
Bovine hepatitis B virus protein	560	A.349	Based on natural sequence with 56 b.p. substitutions to create 34 RS; double stop codon, synthetic transcription termination signal	22 oligonucleotides 36–64 b.p. long; synthesis by phosphoramidite method on automatic synthesizer; separation by PAGE in 7 M urea
Human growth hormone	584	A.415 B.176	Codons of *E. coli* elongation factor Tu gene	78 oligonucleotides 10–25 b.p. long; synthesis by solid-phase phosphotriester block method; separation by anion-exchange HPLC
Transducin (guanosine-binding protein)	1076	A.637 B.451	38 unique RS, consensus sequence CCACC before initiation codon	44 oligonucleotides 39–59 b.p. long; synthesis by phosphoramidite method on automatic synthesizer; separation by PAGE in 7 M urea
Bovine prochimosin	12a 1146 12b 1146	A.87 B.89 C.307 D.267 E.150 F.246 A.468 B.348 C.318	12a: Codons of highly expressed yeast genes, RS for subcloning, a total of 25 % of nucleotide substitutions as compared to natural sequence;	12a: 28 oligonucleotides 59–102 b.p. long; 12b: 26 oligonucleotides 71–102 b.p. long; synthesis by phosphoramidite method on automatic synthesizer; separation by PAGE in 7 M urea

Method of obtaining DNA fragment from synthetic oligonucleotides for cloning	Cloning vector	Expression vector; regulatory portion of the gene and yield of expression product	Reference
A. Two-step ligation B. One-step ligation C. 2 complementary synthetic oligonucleotides	pLT1	pLT1; p^{L}-promoter; Shine–Dalgarno sequence; yield – 1.2–4 % of total soluble *E. coli* protein	Nassal M. *Gene*, 1988, **66**, pp. 279–294
A, B. Two-step ligation A. I. Assembly of 5 DNA blocks from 9–14 oligonucleotides; II. Assembly of fragment A from 5 blocks B. I. Assembly of 3 DNA blocks from 5–6 oligonucleotide pairs; II. Assembly of fragment B from 3 blocks	pBR322	pBR322; *trp*-promoter and Shine–Dalgarno sequence; yield – 169 μg/ml of *E. coli* culture	Ikehara M., Ohtsuka, E. *et al. Proc. Natl. Acad. Sci. USA,* 1984, **81**, pp. 5956–5960
A, B. Two-step ligation I. Assembly of 2 DNA blocks from 5–6 oligonucleotide pairs; II. Ligation of 2 DNA blocks with vector	pT2	pMT-2, main late adenovirus promoter; level of expression in COS-1 cells – 0.3 % of total cell protein	Sakmar, T. P., Khorana, H. G. *Nucl. Acids Res.*, 1988, **16**, pp. 6361–6372
12a: A, B, D, E, F. Ligation of 1–3 pairs of oligonucleotides with linearized vector; 12a: C. Assembly of a fragment from 4 pairs of oligonucleotides, ligation with a vector after separation by electrophoresis in native PAGE; 12b: A, B. Assembly of fragments from 5 and 4 pairs of oligonucleotides, respectively; separation by electrophoresis and ligation with linearized vector in low-melting agarose gel; 12b: C. Two-step ligation, separation of two intermediate DNA blocks and target fragment by electrophoresis in low-melting agarose gel	M13mp18 M13mp19	pMV-1; promoter and signal sequences of melibiose	Wosnick M. A., Barnett R. W. *et al. Gene,* 1987, **60**, pp. 115–127

Tab.11-3. (Continued).

Gene	Total gene length (b.p.)	Lenght of separately cloned synthetic fragments (b.p.)	Design of gene sequence (selection of codons; design of synthetic fragment ends; restriction sites (RS)	Length of synthetic oligonucleotides necessary to design the gene; method of their chemical synthesis and purification
Tissue type plasminogen activator *t-PA*	1730	A.710 B.650 C.290 D.134	Codons of highly expressed yeast genes, insertion of RS at equal intervals, gene encodes pre-pro-*t-PA*	101 oligonucleotides 17–37 b.p. long; synthesis of 6 oligonucleotides at a time by phosphoramidite method on specially designed automatic synthesizer; separation by PAGE in 7 M urea

Method of obtaining DNA fragment from synthetic oligonucleotides for cloning	Cloning vector	Expression vector; regulatory portion of the gene and yield of expression product	Reference
A, B, C. Two-step ligation; I. Assembly of DNA blocks from 5 oligonucleotides; II. Assembly of fragments from 4–8 blocks	pAT153	Vector with origin pBR322 and SV40; long terminal repeats RSV; polyadenylation and splicing sites SV40; expression in cells COS-7; yield – 20 ng *t-PA* in 1 ml of culture supernatant	Bell L. O., Smith J. C. *et al. Gene,* 1988, **63**, pp. 155–163

The table lists not only the methods used for chemical synthesis of oligonucleotides but also some details of how the gene sequences were divided into fragments as well as procedures for their assembly. Besides, some details of cloning are also included, such as types of vectors used, structure of promoters and, wherever relevant, fusion proteins.

References to sources are given for those interested in more detailed information about synthesis of a particular gene. Seliger's laboratory (Ulm, Germany) possesses information on all genes synthesized before 1988. It is summarized in his paper in *AP*, **16**, 7763–7769 (1988).

Conclusion

The importance of organic synthesis of DNA fragments has been realized for quite some time. As early as the first brilliant work by Khorana, in which decoding was done by direct experiments based on synthetic oligonucleotides, it became absolutely clear that directed synthesis of DNA fragments (practically the only way to obtain nucleotides with a strictly predetermined structure) would be a mainstay of molecular biology for years to come.

Several reviews offer detailed discussion of work in which relatively short synthetic oligonucleotides with a predetermined primary structure are employed to solve certain problems in molecular biology and genetic engineering.

In the eighties, a large number of rather extended protein genes, regulatory portions of genes, and similar genetic structures had been synthesized. Evidently, further advances in the synthesis of genes and their fragments in combination with proven techniques of genetic engineering will make it possible not only to obtain producers of biologically active peptides and proteins as well as easily reproduce the newly created genetic information, but also make broad use of the latter to study the functioning mechanisms of genes. It is becoming feasible to plan experiments aimed at elucidating such fundamental aspects of molecular biology as functioning of genes, recognition of different gene segments by respective regulatory proteins, polymerases, processing enzymes, hydrolases, and the like. Using natural DNA fragments for this purpose is most unlikely in the foreseeable future. Even if recourse is had to this approach, it will not provide experimenters with a set of genetic structures differing in the chemical makeup of individual fragments. Preplanned creation of genetic structures is possible only through organic synthesis, controlled replacement of natural monomers or their fragments by various analogues, inversion, insertion, fission, and other procedures.

All this stimulates organic chemists to constantly improve chemical synthesis methods. Before the eighties, organic synthesis of DNA fragments could be accomplished only in very few laboratories. With the advent of automation, organic synthesis is becoming routine – it takes much less time and methods are more easily reproducible.

Currently, genes comprising up to 2000 base pairs are being synthesized. Gene synthesis along with site-directed mutagenesis have become essential prerequisites for progress in genetic engineering as means for producing innumerable genes with single and directed substitutions and, consequently, sets of proteins with different amino acids in particular positions. These techniques will also make it possible to synthesize proteins yet to be isolated, perhaps even nonexistent, in nature. New vistas are being opened up in the development of proteinaceous catalysts for new reactions as well as all kinds of proteins to meet daily needs of man.

References

1. H. G. Khorana, *Some Recent Developments in the Chemistry of Phosphate Esters of Biological Interest.* John Wiley, New York (1961).
2. D. G. Knorre and V. F. Zarytova, *Reactive Phosphosylating Intermediates in the Nucleic Acid Chemistry,* in: *Phosphorus Chemistry Directed Towards Biology.* Ed. W. J. Stec, Pergamon Press, Oxford and New York, pp. 13–31 (1980).
3. *Nucleic Acids in Chemistry and Biology.* Ed. G. M. Blackborn and M. J. Gait, JLR Press, Oxford – New York – Tokyo (1991).
4. *Oligonucleotide Synthesis. A Practical Approach.* Ed. M. J. Gait, JRL Press, Oxford and Washington, DC (1985).
5. *Synthesis and Applications of DNA and RNA.* Ed. S. A. Narang, Academic Press Inc., Orlando, San Diego, New York, Austin, Boston, London, Sydney, Tokyo and Toronto (1987).
6. M. H. Caruthers, *Gene Synthesis Machines: DNA Chemistry and Its Uses.* Science, **230**, 281–285 (1985).
7. Z. A. Shabarova, *Chemical Development in the Design of Oligonucleotide Probes for Binding to DNA and RNA. Biochemistry,* **70**, 1323–1334 (1988).
8. Z. A. Shabarova *et. al., Chemical Ligation of DNA: the First Non-Enzymatic Assembly of a Biologically Active Gene. Nucleic Acids Res.* **19**, 4247–4251 (1991).

Index